GROUP THEORY FROM A GEOMETRICAL VIEWPOINT

INTERNATIONAL CENTRE FOR THEORETICAL PHYSICS

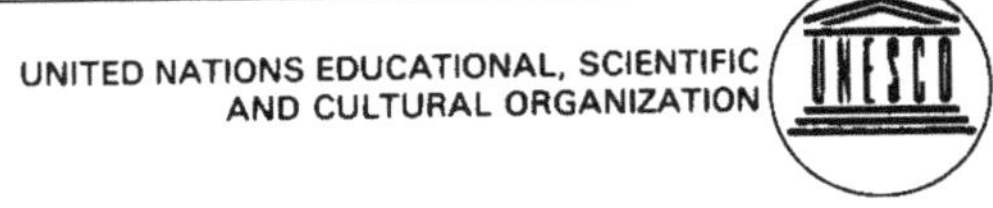

Group Theory from a Geometrical Viewpoint

26 March — 6 April 1990 ICTP, Trieste, Italy

Editors

E. Ghys
Ecole Normale Supérieure de Lyon
France

A. Haefliger
Université de Genève
Switzerland

A. Verjovsky
ICTP, Trieste
Italy

World Scientific
Singapore • New Jersey • London • Hong Kong

Published by

World Scientific Publishing Co. Pte. Ltd.
5 Toh Tuck Link, Singapore 596224
USA office: 27 Warren Street, Suite 401-402, Hackensack, NJ 07601
UK office: 57 Shelton Street, Covent Garden, London WC2H 9HE

British Library Cataloguing-in-Publication Data
A catalogue record for this book is available from the British Library.

GROUP THEORY FROM A GEOMETRICAL VIEWPOINT

ISBN-13 978-981-02-0442-6
ISBN-10 981-02-0442-6
ISBN-13 978-981-02-1430-2 (pbk)
ISBN-10 981-02-1430-8 (pbk)

Printed in Singapore

PREFACE

The workshop on Group Theory from a Geometrical Viewpoint, held at the International Centre for Theoretical Physics in Trieste, Italy, took place from 26 March to 6 April 1990. Altogether 44 lectures were held. There were 105 participants, the great majority of whom were from developing countries. Group theory has always been central in mathematics. In this Workshop, material at the frontier of research was presented.

One of the purposes of the Workshop was to introduce the participants to recent developments in combinatorial group theory with emphasis on Gromov's theory of hyperbolic groups. The lectures were delivered by leading specialists in this new and very active field. The lectures were very well attended and there was a clear demand for the publication of the proceedings, including — besides survey and expository articles — research papers. We hope that this volume fulfils this demand.

Acknowledgements.

First, we would like to thank all the lecturers of the Workshop for offering so much of their time and energy in delivering their lectures. Thanks are especially due to all the contributors of this volume for their efforts in preparing their papers, under the pressures of publication deadlines.

We are most grateful to Professor Abdus Salam, Director of the ICTP, for making possible the realization of the Workshop.

It is a pleasure to thank Alessandra Bergamo for her unstinting efforts and invaluable help before, during and after the Workshop, and Bonaventure Loo whose expertise in TeX helped transform manuscripts sent by electronic mail and computer diskettes into beautifully typeset, camera-ready documents, and for his general assistance in the assembly of this volume.

André Haefliger
Université de Genève

Etienne Ghys
Ecole Normale Supérieure de Lyon

Alberto Verjovsky
International Centre for Theoretical Physics

CONTENTS

I. HYPERBOLIC GROUPS

NOTES ON WORD HYPERBOLIC GROUPS

BY

J. M. ALONSO, T. BRADY,
D. COOPER, V. FERLINI,
M. LUSTIG, M. MIHALIK, M. SHAPIRO, H. SHORT.
EDITED BY H. SHORT

September 1990

This is a record of a series of seminars held at the Mathematical Sciences Research Institute during the spring of 1989, as part of the program on Combinatorial Group Theory and Geometry. This series followed on from, and interacted with, the previous series of seminars on J.W. Cannon, D.B.A. Epstein, D.R. Holt, M.S. Paterson and W.P. Thurston's work on Automatic groups [CEHPT], under the direction of M. Shapiro. In those seminars, M. Gromov's hyperbolic groups were frequently cited as examples (see [BGSS]). Also S. M. Gersten's work on isoperimetric inequalities in groups at M.S.R.I stimulated interest in Gromov's work. These notes were subsequently revised after the meeting in Trieste in March 1990 for inclusion in the proceedings of that meeting.

The object of the seminars was to gain some understanding of the class of groups studied by Gromov in his important (and difficult) paper 'Hyperbolic groups' published in the volume "Essays on Group Theory" [G]. The class of groups studied is defined in geometric terms, usually making reference to the Cayley graph of a finitely generated group. The aim of the theory is to generalise results obtained for the fundamental groups of closed compact hyperbolic manifolds to some larger class, where techniques similar to those used in studying Kleinian groups may be useful. The class includes most of the small cancellation groups which have been subject to much study by some group theorists, and many results from that theory hold for all hyperbolic groups. In this way many of the ideas follow on from Dehn's work around 1910 (see Dehn's papers translated into English [De]).

The authors were partially supported by N.S.F.grant 850-5550 administered through M.S.R.I. where the seminars were held. H. Short wishes to thank the Ecole Normale Superieure Lyon for their support while these notes were written up.

The aim of these notes is to give an accessible introduction to the ideas of hyperbolic groups, accentuating the group theoretic approach. Hopefully these notes can be read by a final year undergraduate or beginning graduate student without too much pain and work. We presuppose some basic knowledge of metric spaces and of groups given by generators and relations, though this goes little deeper than the triangle inequality and the definition of free groups and group presentations.

There has been some discussion about the proper way to refer to this class of groups. In the first preprint version of these notes, we referred to 'negatively curved groups', following the suggestion of Epstein, Thurston, Cannon, Rips and Cooper. However subsequent literature seems to favour the term 'word hyperbolic' or 'hyperbolic'.

Of course it was possible to work through only a small portion of the 200 or so pages of [G], and we have concentrated on establishing the basic definitions. At many points of [G], details are omitted or left to the reader; we have tried to complete some of these.

We did not cover the important concepts of quasiisometry and geometric properties. We simply did not have time to cover this very basic idea, which is more than adequately covered in [GH, Chapter 1] and [Gh].

While the seminar in MSRI was being held, we benefitted from access to early versions of the notes being produced at the time by three other groups, usually in hand written form. These were by:

W. Ballman, E. Ghys, A. Haefliger, P. de la Harpe, E. Salem, R. Strebel and M. Troyanov [GH], notes of a series held at Berne, edited by E. Ghys and P. de la Harpe and published as a book recently by Birkhäuser;

B. Bowditch at Warwick [Bow], to appear elsewhere in this volume;

M. Coornaert, T. Delzant and A. Papadopoulos [CDP] at Strasbourg, to appear shortly as a book (Springer-Verlag).

While we did indeed benefit from the use of the above notes (especially in Chapter 4), many of the proofs here are original. Partly this is because we did not have the now complete versions of these notes, partly because we were interested in the group-theoretic, rather than the metric space aspect. Some other articles known to us at the moment on the subject are:

D. Cooper's preprint [C] on automorphisms of hyperbolic groups;

F. Paulin's work [P1], [P2];

J. Alonso's article 'Combings of groups' [A] which grew out of his talk on the Rips complex (section 4) (this contains another definition of a hyperbolic group which is not discussed in these notes—we refer the interested reader to [A] for this definition and the proof of its equivalence to those given here); in [A2] Alonso shows that the type of isoperimetric inequality satisfied by a group is invariant under quasiisometry.

J.W. Cannon's notes from the Trieste 1989 meeting [Can2].

E. Ghys' Bourbaki seminar on hyperbolic groups contains, as well as the excellent main text, an extensive bibliography [Gh].

M. Bestvina and G. Mess' paper on the boundary of hyperbolic groups [BM].

M. Bestvina and M. Feighn's paper on obtaining hyperbolic groups from amalgamating two hyperbolic groups along a subgroup [BF].

S.M. Gersten and H. Short's article [GS] contains as an appendix the proof that the linear isoperimetric inequality implies that a group is hyperbolic, (included here as 2.4–2.7). I. G. Lysenok's article [L] also contains a proof of this fact and of the equivalence of these definitions with the existence of a Dehn algorithm.

G. Baumslag, S.M. Gersten, M. Shapiro and H.Short use properties of hyperbolic groups to show that the free product of two hyperbolic groups amalgamated along a cyclic subgroup is automatic.

The paper is divided up as follows:

The first chapter consists of a collection of alternative definitions, both of hyperbolic metric spaces and of hyperbolic groups including Gromov's inner product, slim and thin triangles, Cooper's diverging geodesics, the linear isoperimetric inequality, and Dehn's algorithm. The next chapter consists of proofs of the equivalence of these definitions of a hyperbolic metric space, and of a hyperbolic group. We also show here that the Dehn's algorithm definition gives immediately that there only a finite number of conjugacy classes of torsion elements, and also provides a time efficient algorithm for solving the word problem (a result originally due to Domanski and Anshel [DA]).

Some properties of quasigeodesics are developed in Chapter 3. These are used to establish the fact that the centralizer of an element of a hyperbolic group is cyclic-by-finite, and that thus there are no $\mathbb{Z}\times\mathbb{Z}$ subgroups in a hyperbolic group. We define the boundary of a hyperbolic metric space in Chapter 4, though we do not make much use of the construction to establish properties of hyperbolic groups, as is done say in [GH]. We finally build the Rips complex to show that a hyperbolic group is FP_∞. This gives another proof that there only a finite number of conjugacy classes of torsion elements.

Where possible we have tried to give references to original statements in Gromov's paper and to treatment of the topics in in [CDP] and in [GH].

Main differences between this version and the earlier MSRI preprint version of these notes:

Terminology : *negatively curved* has now become *hyperbolic* or *δ-hyperbolic.*

thin triangle has now become *slim* triangle. This is because we decided to return to Gromov's use of the term thin triangle—A. Haefliger suggested the 'slim' terminology. Thus *fine* triangle has become *thin* triangle.

There is much more work to be done before Gromov's work is properly understood; we hope that these notes be of some help to others working in this area. The various authors would like to also thank the other participants in the seminars for their contributions to the development of these notes. We would also like to thank T. Delzant for his talk in the series, though he wished to be absent from the list of authors.

Misprints and remaining mistakes in this written report on the activities of the seminar series are (mostly) due, of course, to

the editor, Hamish Short

Table of Contents

Chapter 0 Some definitions and notation

We assume that the reader knows what a free group is. If X is a finite set of generators for the group G, then there is a natural surjection $\mu : F(X) \to G$, where $F(X)$ is the free group on X.

We can construct a (geodesic) metric space called the Cayley graph $\Gamma_X(G)$ of G with respect to the generating set X (Dehn also called this the '*Gruppenbild*'). This graph has a vertex for each element of G, and an oriented edge labelled x from g to gx for each element $g \in G$ and each $x \in X$. The group G acts on $\Gamma_X(G)$ by multiplication on the left: the element $g \in G$ defines a map ϕ_g, which maps a vertex $h \in \Gamma_X(G)$ to the vertex gh; the endpoints of an edge go to the endpoints of an edge. Notice that in general multiplication on the right does not define an automorphism of the graph, as the endpoints of an edge are not in general sent to the endpoints of an edge. For more about Cayley graphs see for instance [MKS, 1.6].

A metric is defined by assigning unit length to each edge, and defining the distance between two points to be the minimum length of paths joining them (the space is clearly arc-connected).

With this metric, the left-action of G on $\Gamma_X(G)$ is by isometries.

We define the *length* of an element g of G with respect to the generators X, written $|g|_X$, to be the length of the shortest word in $F(X)$ representing g: i.e. $|g|_X = \min\{\ell(w) \mid w \in F(X),\ \mu(w) = g\}$. The distance between two vertices corresponding to elements $h, h' \in G$ is then $d(h, h') = |h^{-1}h'|_X$; this is called a *word metric* on G. The fact that the left action is by isometries can now be seen by noticing that $d(gh, gh') = |(gh)^{-1}(gh')|_X = |h^{-1}h'|_X$.

Examples

The Cayley graph of a free group with respect to a free basis is a tree.

The Cayley graph of $\mathbb{Z} \times \mathbb{Z}$ with respect to the standard pair of generators x, y is the square grid of horizontal and vertical lines in the plane.

The Cayley graph of the fundamental group of a closed, orientable surface of genus g greater than 1 can be embedded in the hyperbolic plane in an natural way. Take a convex fundamental domain consisting of a regular polygon of $4g$ sides and corner angle $\pi/2g$. The group is generated by reflections in the sides of the polygon, and repeatedly reflecting in the sides fills out the hyperbolic plane. The dual graph to the tiling is the Cayley graph with respect to these generators.

The same phenomenon occurs in higher dimensional manifolds. For more about this and quasiisometries see the first chapter of [GH], and other articles elsewhere in this volume.

Chapter 1. Some Notions of Hyperbolicity

We consider a path-connected metric space with distance function d. Always in mind is the example of the Cayley graph of a finitely generated group. We wish also to be able to talk about geodesic (i.e distance minimizing) paths between two points of the space, and in particular to be able to affirm their existence. We list a number of definitions which will later be shown to be equivalent. As usual different definitions will be useful in different contexts.

We say that a metric space X is a *geodesic* metric space if for all points x, y in X there is an isometric map from the interval $[0, d(x,y)]$ to a path in X joining x and y; that is, there is a path between the point x and y realising the distance $d(x,y)$. We denote an image of such an isometry by $[xy]$, and we use $d(w,[xy])$ to denote the distance of the point w from a geodesic arc $[xy]$ (notice that such a path is not necessarily unique—consider the Cayley graph of $\mathbb{Z} \times \mathbb{Z}$). For any path $\alpha : [0,n] \to X$ such that $\alpha : [0,n] \to \alpha([0,n])$ is an isometry, we call n the *length* of α, denoted by $\ell(\alpha)$. Thus for instance $\ell([xy]) = d(x,y)$ for geodesics $[xy]$.

Examples

A locally finite connected graph where the metric is induced by giving each edge unit length, is a geodesic metric space. The Euclidean plane with the usual metric, but the origin omitted is not a geodesic metric space.

We shall be particularly interested in the case when X is a Cayley graph of a finitely generated group (with respect to a finite generating set).

Definition 1.1 Inner product (Gromov [G, 1.1]) Given a base point $w \in X$, we define an *inner product* on X by

$$(x.y)_w = \frac{1}{2}(d(x,w) + d(w,y) - d(x,y)) .$$

If there is a constant $\delta \geq 0$ such that

$$\forall x, y, z \in X, \quad (x.y)_w \geq \min\{(x.z)_w, (z.y)_w\} \; - \; \delta.$$

we say that the inner product is *(δ) hyperbolic*.

Remark 1.2

(1) If X is a tree, we may take $\delta = 0$ (in fact this characterises an $\mathbb{R}$-tree (see for instance [GH, 1.6,7,8]).

(2) If w lies on a geodesic $[xy]$, then $(x.y)_w = 0$.

(3) Let $t \in [xy]$ such that $d(t, w) = d(w, [xy])$. Then

$$d(w,t) + d(t,x) \geq d(w,x) \text{ and } d(w,t) + d(t,y) \geq d(w,y).$$

As $d(x,t) + d(t,y) = d(x,y)$, adding these inequalities gives

$$d(w,[xy]) = d(w,t) \geq (x.y)_w .$$

We shall show that this definition is independent of base point; i.e. if the inner product is (δ) hyperbolic with respect to one base point, then it is (2δ) hyperbolic with respect to any base point.

In the standard hyperbolic plane $\mathbb{H}^2$, triangles do not have the same properties as triangles in the Euclidean plane. For instance in the Euclidean plane, in a large isosceles Euclidean triangle, the mid-point of the hypotenuse is far away from the other two sides. This cannot happen in hyperbolic space. This property gives rise to the following definition (the use of the word 'slim' is suggested by A. Haefliger):

Definition 1.3 Slim Triangles (attributed to Rips) Given any three points x, y, z in X, we say that a triangle xyz of geodesics joining these points is *δ-slim* if for any point w on $[xy]$ we have that $\min(d(w,[xz]), d(w,[yz])) \leq \delta$. We say that *triangles are slim* in X if there is a constant δ such that all geodesic triangles in X are δ-slim.

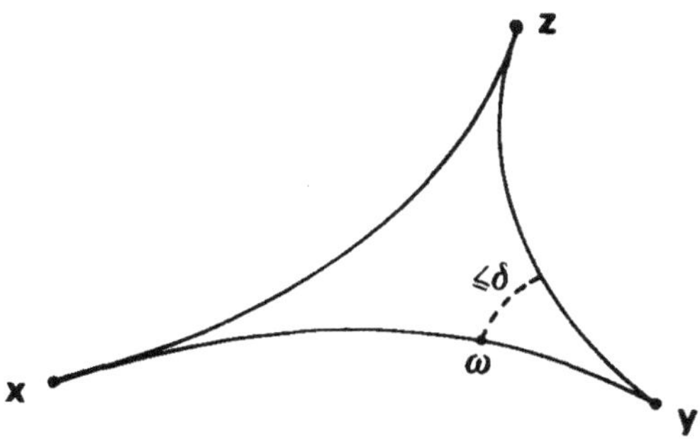

FIGURE 1.1

Now consider a slim triangle defined by the points x, y, z. Let

$$N^+ = \{p \in [xz] \text{ such that } d(p,[xy]) \leq \delta\}$$

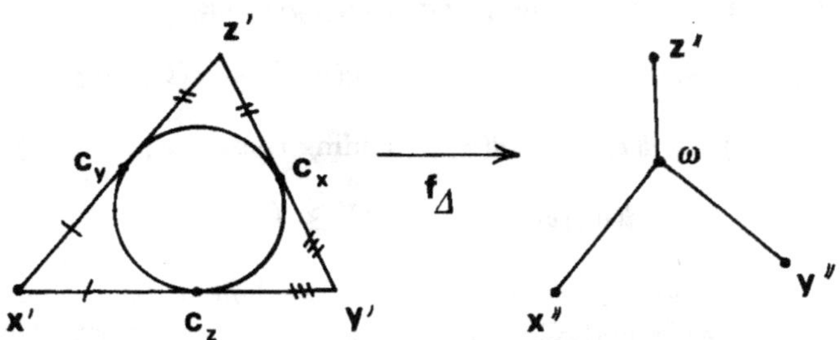

FIGURE 1.2

and let

$$N^- = \{q \in [xz] \text{ such that } d(q,[zy]) \leq \delta\}\ .$$

These two closed sets cover $[xz]$, so there is some point $y' \in N^+ \cap N^-$. Then there are points $z' \in [xy]$, $x' \in [yz]$ such that $d(y',z') \leq \delta$ and $d(y',z') \leq \delta$. Thus the set $\{x', y', z'\}$ has diameter at most 2δ. This suggests the following definition:

Definition 1.4 Minsize ('taille minimale' [CDP, 1.3] , [GH, 2.18]) Let xyz be a geodesic triangle and let x', y', z' be points on xyz (x' on the side opposite vertex x etc.). Define the *minsize* of the triangle to be

$$\text{minsize}(xyz) = \inf \text{diam}\{x', y', z'\}$$

where the infimum is taken over all triples of points $\{x', y', z'\}$.

Thus, if all geodesic triangles are δ-slim, then all geodesic triangles have minsize $\leq 2\delta$. We shall establish the converse below.

Definition 1.5 Thin triangles ([G, 6.3], 'fins' [CDP, 1.3], [GH, 2.16]) Given a geodesic triangle $\Delta = xyz$ in X, let $\Delta' = x'y'z'$ be a Euclidean comparison triangle with sides of the same lengths (i.e. $d_E(x',y') = d_X(x,y)$ etc., where d_E is the standard Euclidean metric). There is a natural identification map $f : \Delta \to \Delta'$. The maximum inscribed circle in Δ' meets the side $[x'y']$ (respectively $[x'z']$, $[y'z']$) in a point c_z (resp. c_y, c_x) such that

$$d(x', c_z) = d(x', c_y)\,, d(y', c_x) = d(y', c_z)\,, d(z', c_y) = d(z', c_x)\ .$$

We call the points c_x, c_y, c_z the *internal points* of xyz (here we are identifying c_x with $f^{-1}(c_x)$ etc.).

Notice that

$$d(x', c_z) = \frac{1}{2}(d(x', c_z) + d(x', c_y)) = \frac{1}{2}(d(x', z) + d(x', y) - d(z', y)) .$$

There is a unique isometry t_Δ of the triangle Δ' onto a *tripod* T_Δ, a tree with one vertex w of degree 3, and vertices x'', y'', z'' each of degree one, such that $d(w, z'') = d(z, c_y) = d(z, c_x)$ etc. Let f_Δ be the composite map $f_\Delta = t_\Delta \circ f : \Delta \to T_\Delta$.

We say that xyz is *δ-thin* if the fibres of f_Δ have diameter at most δ in X. In other words, for all p, q in Δ,

$$f_\Delta(p) = f_\Delta(q) \Rightarrow d_X(p, q) \leq \delta .$$

We say that *triangles are thin* if there is a constant δ such that all geodesic triangles in X are δ-thin.

Definition 1.6 insize ([G, 6.5], 'taille interne' [CDP, 1.3], [GH, 2.18])

We define the *insize* of xyz to be

$$\text{insize}(xyz) = \text{diam}\{c_x, c_y, c_z\} .$$

Remark

(1) It is immediately clear that minsize(xyz) $\leq$ insize(xyz).
(2) If a triangle is δ-thin then its insize is $\leq \delta$.

Another way of characterising hyperbolic geometry is by the way in which infinite rays emmanating from a point diverge. In euclidean space, rays diverge linearly, while in hyperbolic space, rays diverge exponentially. To make this precise requires a rather complicated-looking definition. Consider two people walking along two geodesic rays at unit speed, starting at the same point. The distance between them at time t is at most $2t$ in any metric space, by the triangle inequality. But what we are interested in is the distance between them by following a path which is outside of the ball of radius t around the start point. What characterises a hyperbolic space is that once the distance between the two travellers crosses a certain threshold, the length of the path outside the ball of radius t grows exponentially in t. Here is the detailed definition:

For $\rho > 0$ and $x \in X$, let $B_\rho(x)$ denote the ball of radius ρ about the point x in X. Recall that we consider a path α in X of length $\ell(\alpha)$ as a local isometry $\alpha : [0, \ell(\alpha)] \to X$.

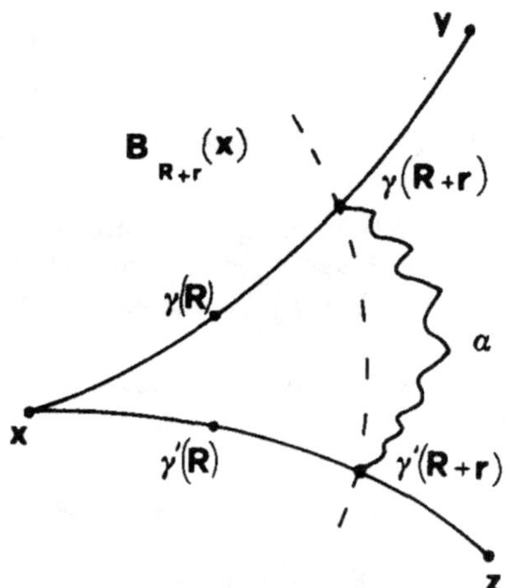

FIGURE 1.3

Definition 1.7 Geodesics Diverge We say that $e : \mathbb{N} \to \mathbb{R}$ is a *divergence function* for X if for all points $x \in X$, and all geodesics $\gamma = [xy], \gamma' = [xz]$, the function e satisfies the following condition.

For all $R, r \in \mathbb{N}$ such that $R + r < \min(\ell([xy]), \ell([xz]))$, if $d(\gamma(R), \gamma'(R)) > e(0)$, and α is a path in $\overline{X - B_{R+r}(x)}$ from $\gamma(R+r)$ to $\gamma'(R+r)$, then we have $\ell(\alpha) > e(r)$.

We say that *geodesics diverge exponentially* if there is an exponential divergence function.

Notice that in the Euclidean plane, there is a linear divergence function. However, it does not have an exponential divergence function.

D. Cooper has suggested a variation of this, where we say that *geodesics diverge supralinearly* if there is a divergence function $e(r)$ such that $\lim_{r \to \infty} e(r)/r = \infty$. We shall establish the rather surprising fact that supralinear divergence is equivalent to exponential divergence, and we shall just say that *geodesics diverge.*

Brian Bowditch has an interesting variant on this definition ([Bow]).

Some Definitions Relevant to Word Hyperbolic Groups

The case in which we are interested here is when the geodesic metric space under consideration is the Cayley graph $\Gamma_X(G)$ of a group G with respect to a finite generating set X.

Definition 1.8 Word hyperbolic groups We say that a group G is ***word hyperbolic*** (often abbreviated to ***hyperbolic***) if it has a finite set of generators X such that the corresponding Cayley graph $\Gamma_X(G)$ is a geodesic metric space with a δ-hyperbolic inner product, for some δ.

It follows from the definition that hyperbolic groups are finitely presented: this is shown in 2.18.

Yet another difference between hyperbolic and Euclidean geometries is the ratio of area to circumference of a circle (or polygon); in the Euclidean plane the area is a quadratic function of the circumference, whereas in the hyperbolic plane it is a linear function. This gives a further characterisation of a hyperbolic group, once we formulate a concept of area in a group.

Definition 1.9 Linear Isoperimetric Inequality (see [G, 2.3], [CDP, chap. 6]) Let $\langle X; R\rangle$ be a finite presentation of the group G with X finite. If w is a freely reduced word w in $F(X)$ of length $\ell(w)$, the free group on X, and $\overline{w} = 1$ in G, then there are words $p_i \in F(X)$, relators $r_i \in R$, and $\epsilon = \pm 1$ such that

$$w = \prod_{i=1}^{N} p_i r_i^{\epsilon_i} p_i^{-1} \quad \text{in} \quad F(X).$$

If there is a constant K such that for all such words w, $N < K.\ell(w)$, we say that G *satisfies a linear isoperimetric inequality* . (Compare [G].)

The reason for the restriction to the finitely presented case is that otherwise one could just throw in as relators of the presentation all words in $F(X)$ which represent the trivial element of G. This would give a very uninteresting linear isoperimetric inequality for any group.

Isoperimetric inequalities are more fully discussed in 2.4–2.7, where it is shown that a finitely presented group which satisfies a linear isoperimetric inequality is word hyperbolic. S.M. Gersten has developped a more general study of isoperimetric inequalities in [Ge]. If some finite presentation of a group satisfies a linear isopermetric inequality, then all finite presentations do, as can be seen by applying Tietze transformations (see [Ge]). (Care is required: notice that a free group satisfies a zero isoperimetric inequality with respect to a free basis, but a linear term is added when the basis is changed.) This shows that the definition is independent of generating set. More generally, Alonso has shown [A2] that the type of isoperimetric inequality satisfied by a group is invariant under quasiisometry.

It is clear that a free group has a linear isoperimetric inequality.

It is pointed out in [BGSS] that it follows from Newman's spelling theorem (see e.g. [LS]) that one relator groups with torsion are word hyperbolic.

One of the reasons Gromov gives for his study of hyperbolic groups is a desire to generalize small cancellation theory ([G, 0.4]). This latter theory has its origins in Dehn's solution of the word problem for the fundamental groups of surfaces, which was generalized by Greendlinger and Lyndon in the 1960s (see [LS, Chapter V] or R. Strebel's appendix to [GH] for more details). We give the conditions used by this theory here. Their main utility resides in the fact that they give easily checked conditions on a presentation which ensure hyperbolicity. Unfortunately the class of groups so defined is somewhat limited: a torsion-free C(7) small cancellation group has cohomological dimesion 2. Gromov indicates in [G, 0.2] that the class of hyperbolic groups is much larger than the class of C(7) small cancellation groups, while at the same time stating that hyperbolic presentations are generic. It would be interesting to formalise and understand some idea of the genericity of hyperbolic groups sketched by Gromov in these statements.

Definition 1.10 Small Cancellation Conditions Given a finite presentation $\mathcal{P} = \langle X; R\rangle$, let $\mathcal{R}$ denote the cyclic closure of R, i.e. the set of cyclic conjugates of elements of R and their inverses. A *piece* is a non-trivial word $v \in F(X)$ such that there are two different relators $r_1, r_2 \in \mathcal{R}$ such that $r_1 = vr_1'$ and $r_2 = vr_2'$.

We say that $\mathcal{P}$ satisfies the $C(p)$ condition if no element of $\mathcal{R}$ is a product of fewer than p pieces. We say that $\mathcal{P}$ satisfies the $C'(1/p)$ if for each piece v occurring in the relator r, $p\ell(v) < \ell(r)$. Thus if the $\mathrm{C}'(1/p)$ condition holds, then so does the $\mathrm{C}(p+1)$ condition.

Example The surface group of genus $g > 1$ has a presentation

$$\langle a_1, \dots, a_g, b_g, \dots b_g \mid \prod_{i=1}^{i=g} a_i b_i a_i^{-1} b_i^{-1}\rangle.$$

A maximal piece consists of a single letter, and so this presentation satisfies the condition $\mathrm{C}(4g)$, and also the condition $\mathrm{C}'(\frac{1}{4g-1})$.

The presentation satisfies the $\mathrm{T}(q)$ condition if for any sequence $r_1, r_2, \dots, r_k$ of elements of $\mathcal{R}$ with $k < q$, such that

$$r_1 = a_1 r_1' a_2^{-1}, r_2 = a_2 r_2' a_3^{-1}, \dots, r_k = a_k r_k' a_1^{-1},$$

where $a_i \in X \cup X^{-1}$, for some $j \leq k$, we have that $r_j = r_{j+1}^{-1}$ in $F(X)$ (suffices considered mod k). Notice that the $\mathrm{T}(3)$ condition is void.

Example The presentation $\langle x, y \mid xyx^{-1}y^{-1}\rangle$ of the group $\mathbb{Z} \times \mathbb{Z}$ satisfies the conditions $C(4) - T(4)$.

It is "well known" that a group which satisfies one of the C(7), C(5)–T(4), C(4)–T(5), C(3)–T(7) small cancellation conditions satisfies a linear isoperimetric inequality (see for instance [GS]). In the metric case (i.e. when for instance C′(1/6), C′(1/4)–T(4) or C′(1/4)–T(7) are satisfied), the main lemma of small cancellation theory states that a word $w \in F(X)$ which represents the trivial element of the group contains a subword which is more than half of some cyclic conjugate of a relator.

It follows that fundmental groups of compact surfaces of genus greater than 1 are hyperbolic.

In a series of papers studying the word problem for surface groups at the beginning of this century, Max Dehn studied the connection between hyperbolic geometry and surface groups (see [De]). There he gave a solution to the word problem, which we shall generalise here:

Definition 1.11 Dehn's Algorithm A *Dehn presentation* for the group G is a finite presentation $\langle X; R\rangle$ such that any non-trivial word in $F(X)$ which represents the identity element of G contains more than half of some word in R. That is, if $w \in F(X)$ is a reduced word, and $\mu(w) = 1$ in G, then there is a relation $r = r_1 r_2 \in R$ with $\ell(r_1) > \ell(r_2)$, such that $w = w_1 r_1 w_2$.

A group is said to have a Dehn's algorithm if it has a Dehn presentation.

It is clear that a group with a Dehn's algorithm satisfies a linear isoperimetric inequality (with multiplicative constant 1).

We shall show (2.12) that a group is hyperbolic if and only if it has a Dehn's algorithm. This was also established by Lysenok [L] and Cannon (see [Can2]).

Chapter 2. The Equivalence of the definitions

We now show the equivalence of several of the initial definitions concerning δ-hyperbolic metric spaces. We begin by showing the equivalence of some of the properties of geodesic triangles. Most of the proofs here are elementary and based on pictures of triangles. We have not tried to optimize the different values of δ involved.

Proposition 2.1. *The following are equivalent for a geodesic metric space X.*

(1) *Triangles are slim.*
(2) *Triangles are thin.*
(3) *There is a global bound on the insize of geodesic triangles.*
(4) *There is a global bound on the minsize of geodesic triangles.*
(5) *The inner product on X is hyperbolic with any choice of base point.*

Definition We say that a geodesic metric space is ***hyperbolic*** if it satisfies one of the above equivalent conditions.

Proof. It is clear that (2) implies (3) which in turn implies (4). (Also (2) immediately implies (1).)

(1) implies (3)

Suppose that all geodesic triangles in X are δ-slim. Let xyz be a geodesic triangle, and let c_x, c_y, c_z be the internal points.

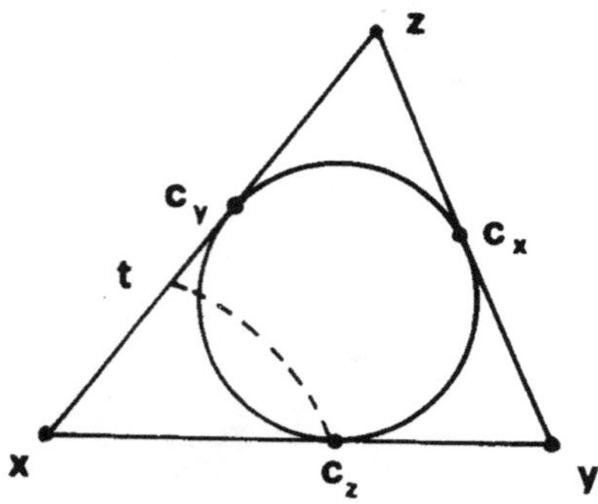

FIGURE 2.1

Consider the point c_z on $[xy]$; there is a point t on $[xz] \cup [yz]$ such that $d(c_z, t) \leq \delta$. Without loss of generality, suppose that t lies on $[xz]$. Then

$$d(x,t) + \delta \geq d(c_z, x) = d(c_y, x) \quad \text{and} \quad d(x,t) \leq d(x, c_z) + \delta$$

and so $d(t, c_y) \leq \delta$, and $d(c_z, c_y) \leq 2\delta$.

A similar argument shows that c_x is at distance not more than 2δ from one of c_z and c_y. It follows that $\operatorname{diam}\{c_x, c_y, c_z\} \leq 4\delta$, and (3) holds.

(1) implies (2)

Let u be a point on $[xc_y]$ and v a point on $[xc_z]$ such that $d(u, x) = d(v, x)$. As geodesic triangles are δ-slim,

$$d(u, [xc_z] \cup [c_y c_z]) \leq \delta \ .$$

If there is a point $t \in [xc_z]$ such that $d(u, t) \leq \delta$, then $d(t, v) \leq \delta$, so that $d(u, v) \leq 2\delta$. Thus if $d(u, v) > 2\delta$, it follows that there are points $t_u, t_v \in [c_y c_z]$ such that $d(u, t_u) \leq \delta$ and $d(v, t_v) \leq \delta$, and $d(u, v) \leq 6\delta$, and (2) holds.

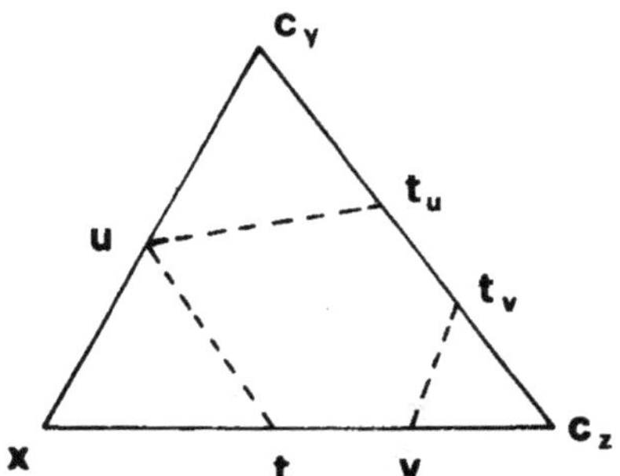

FIGURE 2.2

(4) implies (1)

Let x', y', z' be points on $[yz], [xz], [xy]$ such that $\operatorname{diam}\{x', y', z'\} \leq \delta$. This reduces the problem to studying three geodesic triangles, each with base $\leq \delta$.

Suppose that there is a point $t \in [xz']$ such that $d(t, [xy']) > 2\delta$.

Let u be the point in $[z't]$ nearest to t such that $d(u, u') = 2\delta$ for some point $u' \in [xy']$. Now consider the geodesic triangle $uu'x$. The bound on the minsize of triangles implies that there are points a, b, c on the three sides of $uu'x$ such that $\operatorname{diam}\{a, b, c\} \leq \delta$. The point a on $[xu]$ does not lie in $[tu]$ by supposition, and $d(u, a) \leq 3\delta$, so $d(t, u') \leq 3\delta$ or $d(t, c) \leq 3\delta$.

(2) implies (5)

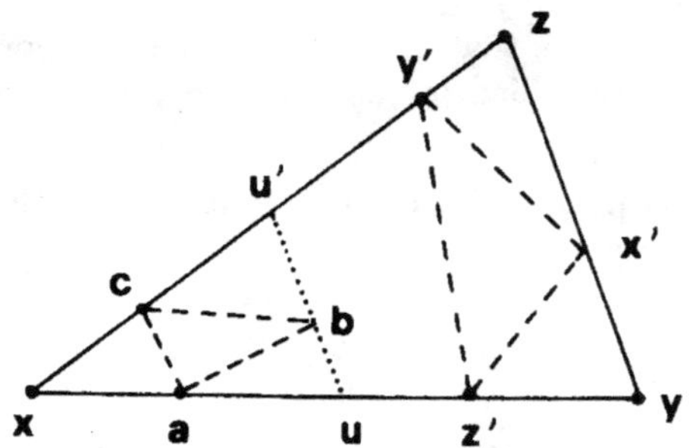

FIGURE 2.3

Consider a geodesic triangle wxy, with internal points c_x, c_y, c_w. The inner product with base point w is

$$(x.y)_w = \frac{1}{2}(d(w,x) + d(w,y) - d(x,y))$$

and we must show that $\forall z \in X$, $(x.y)_w \geq \min((x.z)_w, (y.z)_w) - \delta$.

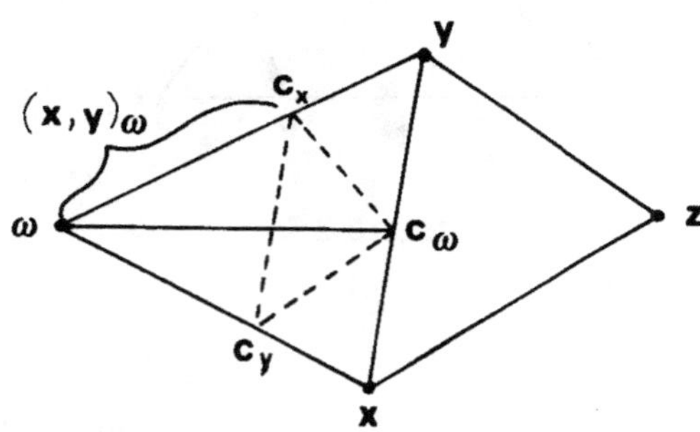

FIGURE 2.4

Recall that $d(w, c_x) = (x.y)_w$, and notice that

(1) $d(w, c_w) \leq d(w, c_x) + d(c_x, c_w) \leq (x.y)_w + \delta$

(2) $(x.y)_w \leq d(w, [xy])$.

Let z be another point in X. Then

$$(x.y)_w + 2\delta \geq d(w, c_w) + \delta \geq \min(d(w, [xz]), d(w, [yz]))$$

But $d(w,[xz]) \geq (x.z)_w$ so that

$$(x.y)_w + 2\delta \geq d(w,c_w) + \delta \geq \min((x.z)_w,(y.z)_w)$$

and the inner product condition holds.

(5) implies (1)

We first show:

Claim: If the inner product with base point w is hyperbolic, then for any geodesic triangle wxy,

$$(x.y)_w \leq d(w,[xy]) \leq (x.y)_w + 2\delta\,.$$

The left hand inequality follows immediately as in the remarks after definition 1.1. Now let c_w, c_x, c_y be the internal points of the triangle wxy on the side $[xy]$. Now consider the internal points d_x, d_w, d_y (resp. e_y, e_w, e_x) of geodesic triangles xwc_w (resp. ywc_w), where $d_y \in [wx]$ and $e_x \in [wy]$.

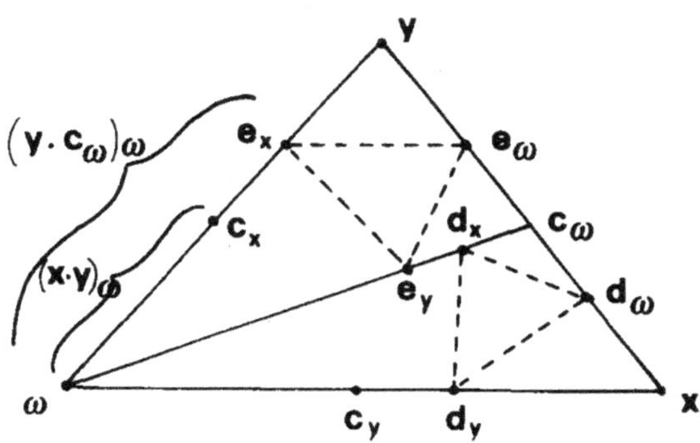

FIGURE 2.5

As d_w lies in $[xc_w]$, $d(x,d_y) = d(x,d_w) \leq d(x,c_y)$. It follows that $d(w,d_y) > d(w,c_y)$. Similarly $d(w,_x) > d(w,c_x)$. Thus

$$(x.c_w)_w \geq (x.y)_w \quad \text{and} \quad (c_w.y)_w \geq (x.y)_w\,.$$

Without loss of generality suppose that $(c_w.y)_w \leq (x.c_w)_w$; then by (5) (with c_w in place of z):

$$\delta \geq (c_w.y)_w - (x.y)_w = d(c_x,e_x) = d(e_w,c_w)\,.$$

But

$$d(w,c_w) = d(w,e_y) + d(e_y,c_w) = d(w,e_x) + d(c_w,e_w)$$
$$= d(w,c_x) + 2d(c_x,e_x) \leq 2\delta + (x.y)_w$$

and it follows that

$$d(w,[xy]) \leq d(w,c_w) \leq (x.y)_w + 2\delta \; .$$

This completes the proof of the claim.

Now suppose that the inner product is δ-hyperbolic with respect to any base point. Let xyz be a geodesic triangle, and let w be a point on the side $[xy]$. By the above,

$$0 = (x.y)_w \geq \min\{(x.z)_w, (z.y)_w\} - \delta \; .$$

Without loss of generality suppose that $(x.z)_w \leq (z.y)_w$. Then

$$\delta \geq (x.z)_w \geq d(w,[xz]) - 2\delta$$

and so

$$d(w,[xz]) \leq 3\delta$$

and the triangle xyz is 3δ-slim, as required.

This completes the proof of proposition 2.1. □

We now give Gromov's direct proof that for inner products, the property of being hyperbolic is independant of base point.

Proposition 2.2. ([G, 1.1B]) *If X is δ-hyperbolic with inner product based at w and $t \in X$ then X is 2δ-hyperbolic with inner product based at t.*

This means that the reference to base point can be removed from the definition: we say that X is hyperbolic if there is a positive constant δ such that $\forall w, x, y, z \in X$,

$$(x.y)_w \geq \min((x.z)_w, (z.y)_w) - \delta \; .$$

To prove the proposition we first show:

Lemma 2.3. ([G, 1.1A])

$$(x.y)_w + (z.t)_w \geq \min((x.z)_w + (y.t)_w, \; (t.x)_w + (y.z)_w) - 2\delta$$

Proof. We remove reference to the base point w.

$$\begin{aligned} x.y + z.t &\geq \min(x.t, t.y) + z.t - \delta \\ &= \min(x.t + z.t , t.y + z.t) - \delta \\ &\geq \min(x.t + \min(z.y, y.t) , t.y + \min(z.x, x.t)) - 2\delta \\ &= \min(x.t + z.y , x.t + y.t , t.y + z.x) - 2\delta \end{aligned}$$

and this achieves a unique minimum value of $x.t + y.t$ if and only if $y.t < z.y$ and $x.t < z.x$.

Similarly

$$\begin{aligned} x.y + z.t &\geq \min(x.z, z.y) + z.t - \delta \\ &= \min(x.z + z.t , z.y + z.t) - \delta \\ &\geq \min(x.z + \min(z.y, y.t) , z.y + \min(z.x, x.t)) - 2\delta \\ &= \min(x.z + z.y , x.z + y.t , z.y + x.t) - 2\delta. \end{aligned}$$

This achieve a unique minimum value of $x.z + z.y$ if and only if $z.y < y.t$ and $x.z < x.t$. Both of these cannot be true at the same time, so the result holds. □

Corollary 2.4. *If X is δ-hyperbolic, then for all $t, x, y, z \in X$,*

$$d(x, y) + d(z, t) \leq \max\{d(x, z) + d(y, t), d(x, t) + d(y, z)\} + 2\delta$$

Proof of Proposition 2.2. We wish to establish a lower bound for

$$\min\{(x.z)_t, (z.y)_t\} \;-\; (x.y)_t$$

$$\begin{aligned} = \min\{d(x,t) \;+\; d(z,t) \;-\; d(x,z)\,,\, d(z,t) \;+\; d(y,t) \;-d(z,y)\}\\ +\; d(x,y) \;-\; d(x,t) \;-\; d(y,t)\end{aligned}$$

$$= \min\{-d(y,t) \;-\; d(x,z), -d(x,t) \;-\; d(z,y)\} \;+\; d(x,y) \;+\; d(z,t)\;.$$

Adding $d(x,w) + d(y,w) + d(z,w) + d(t,w)$ inside and subtracting from the outside of the minimum gives

$$\min\{(y,t)_w \;+\; (x.z)_w\;,\; (x.t)_w \;+\; (z.y)_w\} \;-\; (z.t)_w \;-\; (x.y)_w$$

it follows , by Lemma 2.3 that

$$\min\{(x.z)_t\;,\; (z.y)_t\} \;-\; (x.y)_x \;\leq\; 2\delta$$

□

Linear isoperimetric inequality implies hyperbolic

We shall now show that a finitely presented group which satisfies a linear isoperimetric inequality is hyperbolic. The converse will be shown in 2.10, and again, using different methods in 2.12. We need first to develop some of the language of disc diagrams.

Singular Disc Diagrams

Let $F(X)$ denotes the free group on X. When R is a finite set of cyclically reduced words in $F(X)$, $\mathcal{P} = \langle X; R\rangle$ denotes a finite presentation of a group G; we use $\mathcal{R}$ to denote the cyclic closure of R, consisting of all elements r in R, their inverses, and all cyclic conjugates of $r^{\pm}$.

We form a 2-complex $K(\mathcal{P}) = K$ whose fundamental group is G in the standard way: K has one vertex, one labelled, directed edge for each generator, and one 2-cell for each relator. The 2-cell D_i corresponding to the relator r_i is glued to the 1-skeleton $K^{(1)}$ via a continuous map which identifies the boundary ∂D_i with a loop representing the word r_i.

A freely reduced word w in the generators is equal to the identity in G if an only if there is a continuous map from a disc $(D, \partial D)$ to $(K, K^{(1)})$ taking the boundary to a loop representing the word w. After a homotopy, the cell decomposition of K induces a cell decomposition of a simply connected complex, which we also call D, consisting of a set of discs joined by arcs or vertices. The vertices map to the vertex of K, the interiors of the 1-cells of D, called *edges*, map homeomorphically to the interiors of the 1-cells of K, and the interiors of the 2-cells, called *regions*, map homeomorphically to the interiors of the discs D_i of K. We can orient and label the edges according to the generating loop in $K^{(1)}$ to which they map, in such a way that reading the labels on the edges around the boundary of D gives the word w. The complex D we call a *singular disc diagram* (or Van Kampen or Dehn diagram) for $w = 1$ in G (for more details see [**LS**], chapter V). Regarding the singular disc diagram as a topological space, each component of the interior is a topological disc.

A singular disc diagram is *unreduced* if there are two regions R_1 and R_2 in D whose boundaries have an edge e in common, such that the labels on their boundaries, reading around from the edge e, clockwise on R_1 and anti-clockwise on R_2, are the same. It is not hard to see how to remove two such neighbouring regions in an unreduced singular disc diagram without changing the boundary label, so that we may concentrate our attention on reduced singular disc diagrams; *from now on all diagrams are assumed to be reduced.*

Let $\mathcal{P} = \langle X; R\rangle$ be a finite presentation of the group G. Suppose that $f : \mathbb{N} \to \mathbb{R}$ is a function with the property that if w is a freely reduced word of length n in the free group $F(X)$ and $\overline{w} = 1$ in the group G, then there is a singular disc diagram for w with at most $f(n)$ regions. Following S.M. Gersten [Ge], we say that f is a *Dehn function* for $\mathcal{P}$.

Notice that as $w = 1$ in G, there are words p_i in $F(X)$, and r_i in R, such that

$$w = \prod_{i=1}^{N} p_i r_i^{\epsilon_i} p_i^{-1}\ , \ \epsilon_i = \pm 1.$$

The Dehn function tells us that there is such a product with $N \leq f(n)$ relators. In addition, as we have a bound on the total length of the 1-skeleton of the diagram, we may take

$$\ell(p_i) < n + f(n)M \quad \text{where} \quad M = \max_{r\in R}(\ell(r))\,.$$

Tietze transformations transform lengths by scalar multiples together with the addition of new trivial words, and so transform Dehn functions by scalar multiples and the addition of linear terms [Ge]. Thus if a group G has a presentation with a linear (quadratic, exponential) Dehn function, then we say that G *satisfies a linear (quadratic, exponential) isoperimetric inequality* .

If X is a finite set of generators, and R is a recursive set of relations, then finding a recursive Dehn function for $\mathcal{P}$ solves the word problem for $\mathcal{P}$ [Ge].

Theorem 2.5. *If G is a finitely presented group satisfying a linear isoperimetric inequality, then there is a constant δ such that all geodesic triangles in the Cayley graph Γ are δ-slim.*

Proof. We argue by contradiction: suppose that there is no constant δ such all geodesic triangles are δ-slim.
Let K be the constant associated to the linear Dehn function for the finite presentation $P = \langle X; R \rangle$ for the group G, and let ρ be the maximum length of a relator in R. We can suppose that $\rho > 1$, else G is free and so hyperbolic (with radius of curvature 0). We can also assume that $K \geq 1$.
Then for each $r > 0$, there is a geodesic triangle xyz in Γ and a point $w \in [xy]$ such that

$$(\star) \qquad \min(d(w,[yz]),\, d(w,[xz])) \;>\; 2r\,.$$

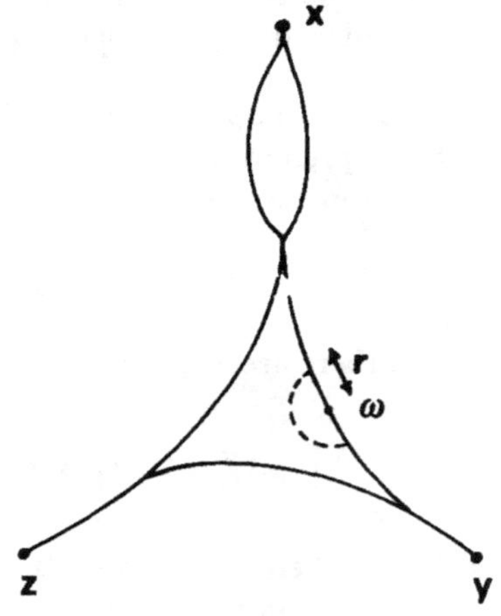

FIGURE 2.6

If xyz is degenerate (i.e. if $[xy] \cap [yz] - \{y\} \neq \emptyset$ or $y = z$ say) then xyz contains a non-degenerate geodesic triangle (or bigon) where $(\star)$ holds. So it suffices to consider non-degenerate geodesic triangles and bigons.
Let B_r be the ball of radius r (in Γ) with centre w.

Let ϵ be a constant, and suppose that $r > 6\epsilon$. We first cut off the corners of xyz such that the remaining segments are all at distance at least 4ϵ from each other, and the cut-off arcs are of length exactly 4ϵ. This will give one of three cases:

(1) a non-degenerate hexagon H with three sides of length 4ϵ.

(2) a non-degenerate quadrilateral with two sides of length 4ϵ.
(3) a degenerate hexagon.

As $r > 6\epsilon$, the length α of the side $[x'y']$ containing w is at least $2r$. Without loss of generality, we suppose that the cases are as shown below (figure 2.7).

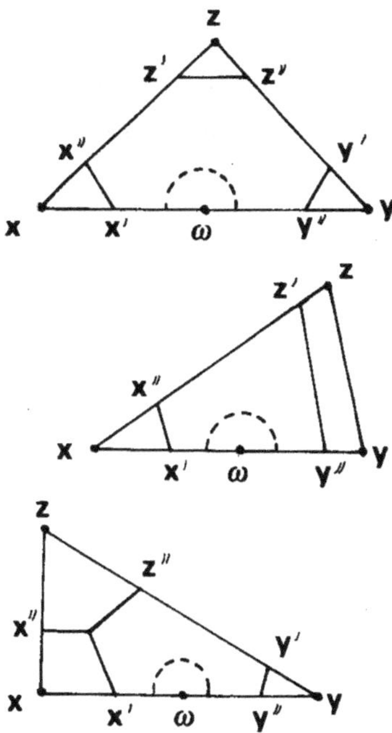

FIGURE 2.7

Consider first case (1), with $H = x'x''z'z''y''y'$ and let α, β and γ be the number of edges in the segments $[x'y']$, $[x''z']$ and $[y''z'']$.

Let D be a minimal disc diagram for the word represented by the hexagon H. We shall consider D both as a cell complex and as the underlying topological space, and we identify ∂D with H in Γ. As H is a simple closed curve in Γ, the diagram D is a topological disc, and each 1-cell which is not contained in ∂D is on the boundary of two 2-cells in D. If T is a subcomplex of D, we define $\text{star}_D(T)$ to be the set of all cells which intersect T. If θ is one of the geodesic arcs $[x'y']$, $[x''z']$, $[z''y'']$ we use $N(\theta)$ to denote the subcomplex of D obtained by iterating the star operation $[\epsilon/\rho]+1$ times ($[\epsilon/\rho]$ denotes integer part), starting from the arc θ in D. Let $\ell(\theta)$ denote the number of 1-cells in θ; thus $\ell([x'y'])=\alpha$. We need the following 2 lemmas to complete the proof. Their proofs are deferred.

Lemma 2.6. *If $\epsilon > \rho$, then there is a constant C_1 depending solely on ϵ, such that the number of 2-cells in $N(\theta)$ is at least $\ell(\theta)\epsilon/\rho^2 - C_1$.*

Let $A(D)$ be the number of 2-cells in the diagram D.

Lemma 2.7. *If $\epsilon > \rho$, there is a constant C_2 depending solely on ϵ such that*

$$A(D) > (\alpha+\beta+\gamma)\epsilon/\rho^2 - C_2 + 2r/\rho$$

We now use this last result to complete the proof of the main theorem. The linear isoperimetric inequality implies that

$$A(D) \leq (\alpha + \beta + \gamma)K + 12K\epsilon$$

Combining this inequality, and the result of Lemma 2.7, we have

$$(\alpha+\beta+\gamma)\epsilon/\rho^2 - C_2 + 2r/\rho \leq (\alpha + \beta + \gamma)K + 12K\epsilon\ .$$

Now set $\epsilon = K\rho^2$ (as $\rho > 1$ and $K \geq 1$, it follows that $\epsilon > \rho$). We thus obtain

$$2r/\rho - C_2 \leq 12K^2\rho^2$$

which is clearly a contradiction for sufficiently large values of r.

This completes the proof in case (1).

It remains to prove the lemmas. First notice that the map $\partial D \to H$ extends naturally to a map $h : D^{(1)} \to \Gamma$ from the 1-skeleton of D to the Cayley graph. The 1-cells of D are the inverse images of the edges of Γ, so there may well be vertices of degree 2 in the interior of D.

Proof of Lemma 2.6. Suppose that $\theta = [x'y']$; the other cases follow exactly analogously. Let γ_0 denote the segment θ in ∂D, and let $N_1 = \text{star}_D(\gamma_0)$.

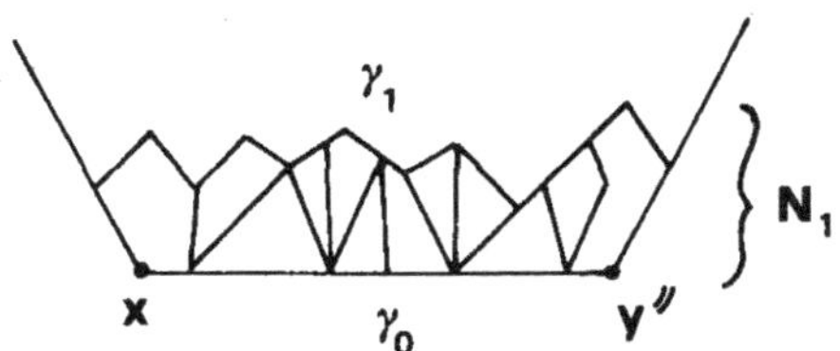

FIGURE 2.8

Then γ_0 contains α 1-cells, each of which lies in the boundary of a 2-cell. Each 2-cell has at most ρ 1-cells in its boundary, so there are at least $\ell(\theta)/\rho$ 2-cells in N_1. Now $N_1 \cap D^{(1)} \cap \overline{(\mathrm{int}(D) - N_1)}$ is a path γ_1 from the segment $[x'x'']$ to $[y'y'']$, and maps to a path in Γ lying in a ρ-neighbourhood of $[x'y']$.

It follows that γ_1 contains at least $(\ell(\theta) - 2\rho)$ 1-cells. Continue this process $[\epsilon/\rho] + 1$ times; i.e. let

$$N_i = \mathrm{star}(N_{i-1}) \text{ and let } \gamma_i = N_i \cap D^{(1)} \cap \overline{(\mathrm{int}(D) - N_i)} .$$

The number of 2-cells in $N_i - N_{i-1}$ is at least $(\ell(\theta) - 2(i-1)\rho)/\rho$. Repeating $[\epsilon/\rho] + 1$ times gives at least

$$\alpha\epsilon/\rho^2 \; - \; \epsilon(\epsilon + \rho)/\rho^2$$

2-cells in $N([x'y'])$.

This concludes the proof of Lemma 2.6, setting $C_1 = \epsilon(\epsilon + \rho)/\rho^2$. □

Proof of Lemma 2.7. We first show that

$$\begin{aligned} N([x'y']) \cap N([y''z'']) &= N([x'y']) \cap N([x''z']) \\ &= N([x''z']) \cap N([y''z'']) = \emptyset . \end{aligned}$$

In the construction of the proof of A.1, we have that γ_i maps to a path in Γ which lies in a $(i\rho)$-neighbourhood of $[x'y']$. It follows that $N([x'y'])^{(1)}$ maps into a $(\epsilon + \rho) < 2\epsilon$ neighbourhood of $[x'y']$. By the construction of H, $p \in [x'y']$ and $q \in [x''z']$ implies that $d(p, q) \geq 4\epsilon$. Hence the 2ϵ neighbourhoods of $[x'y']$

and of $[x''z']$ in Γ do not intersect, and so $N([x'y']) \cap N([x''z']) = \emptyset$. The other cases follow analogously.

Let ϕ' be the set of 1-cells in $N([x'y']) \cap \overline{(\mathrm{int}(D) - N([x'y']))}$, and let ϕ be the subset of ϕ' which maps into B_r (via the map h). A point $p \in \phi$, maps to a point at distance at most r from w, and so lies at least $r - 2\epsilon > 0$ from $N([x''z'] \cup N([y''z'']))$. Also $h(\phi)$ contains an arc of length at least $2r - 4\epsilon$, so that ϕ contains at least $(2r - 4\epsilon - 2)$ 1-cells. Each of these 1-cells lies in the boundary of a 2-cell which does not lie in $N([x'y'])$, so there are at least $(2r - 4\epsilon - 2)/\rho$ 2-cells in D which do not lie in $N([x'y']) \cup N([y''z'']) \cup N([x''z'])$.

So using Lemma 2.6, there are at least

$$(\alpha + \beta + \gamma)\epsilon/\rho^2 - 3C_1 - (4\epsilon + 2)/\rho + 2r/\rho$$

2-cells in D.
This concludes the proof of Lemma 2.7, setting $C_2 = 3C_1 + (4\epsilon + 2)/\rho$. □

The cases (2) and (3) of Theorem 2.5 remain to be considered
In case (2) Lemma 2.6 holds for the arcs $[x'y']$, $[x''z']$. Lemma 2.7 now holds for a minimal disc bounded by the quadrilateral $Q = x'x''z'y'$, with $\gamma = 0$ (and with $C_2 = 2C_1 + (4\epsilon + 2)/\rho$). The proof of the theorem is concluded by obtaining a contradiction as before.
In (3), look at the simple closed curve $P = x'\overline{z}z''y''y'$. Lemma 2.6 holds as before for the arcs $[x'y']$, $[y''z'']$, and Lemma 2.7 follows with $\beta = 0$ and $C_2 = 2C_1 + (4\epsilon + 2)/\rho$. Note that the sum of the lengths of the segments $[x'\overline{z}]$ and $[\overline{z}z'']$ lies between 4ϵ and 8ϵ. The side $[y'y'']$ has length 4ϵ, and the side $[y''z'']$ has length $\gamma > 2r - 12\epsilon > 0$.

The proof is concluded by obtaining a contradiction as before.

We shall give now give two proofs that a group with a linear isoperimetric inequality has slim (or thin) triangles. The first follows Gromov [**Gr**, 1.7C](see also [CDP, §5.3.1]). A second proof is given in theorems 2.15, 2.16.

A presentation of a group is said to be *triangular* if each relation has length three.

Every finitely presented group has a triangular presentation, and to such a presentation there is an associated simply-connected, locally finite 2 dimensional simplicial complex X, where each cell is isometric to a chosen standard 2-cell in $\mathbb{R}^2$. For a simplicial curve $\mathcal{C}$ in X, let $L(\mathcal{C})$ denote the length of $\mathcal{C}$, i.e. the number of 1-cells in $\mathcal{C}$, and let $A(\mathcal{C})$ denote the area bounded by $\mathcal{C}$, i.e. the minimal number of 2-cells in a singular disc embedded in X bounded by $\mathcal{C}$.

Proposition 2.10. *If geodesic triangles in X are slim then X satisfies a linear isoperimetric inequality.*

Proof. Let $d = 10\delta$ and let

$$A_0 = \max\{A(\mathcal{C}); L(\mathcal{C}) \leq 3d\}.$$

Let $N(\mathcal{C})$ denote $\text{int}(L(\mathcal{C}))/d)+1$. We shall show by induction on N that $A(\mathcal{C}) \leq 3N(\mathcal{C})A_0$.

The result clearly holds for $N = 1$.

Now suppose that the result holds for all curves $\mathcal{C}'$ such that $N(\mathcal{C}') \leq n$.

Let $\mathcal{C}$ be curve such that $N(\mathcal{C}) = n+1$ and choose a base point x on $\mathcal{C}$. Choose y_0 on the curve $\mathcal{C}$ farthest from x, and choose points y_1, y_2 on $\mathcal{C}$ at distance d from y.

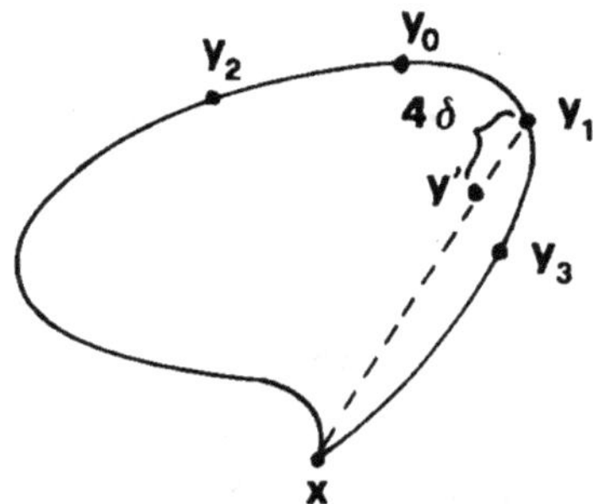

FIGURE 2.9

If $d(y_1, y_2) \leq d$ then the result holds, by applying the induction hypothesis to the curve obtained from $\mathcal{C}$ by omitting the segment between y_1 and y_2 which contains y_0, and replacing it by a geodesic $[y_1y_2]$. This curve has length at most $L(\mathcal{C}) - d$, and the induction hypothesis applies. The curve $\mathcal{C}$ then bounds a disc obtained from this one by adding on a disc bounded by the curve $y_1y_0y_2$ which has length at most $3d$ and the induction argument is complete in this case.

We are thus left with the case that $d(y_1, y_2) > d$; it follows that $d(x, y_0) > 5\delta$.

Without loss of generality, suppose that $d(x, y_1) \geq d(x, y_2)$; it follows that $d(x, y_1) > 5\delta$.

Let y' be a point on the geodesic arc $[xy_1]$ at distance 4δ from y_1, and let y_3 be a point on $\mathcal{C}$ at distance d from y_1, in the segment from x to y_1 not containing y_0.

To complete the argument, we use the following:

Lemma 2.11. *For $i = 0, 1, 2, 3$, we have $d(y', y_i) \leq d$.*

Given this, we see that the argument is complete, as then we have a disc bounded by C, made up of a disc bounded by a curve which is shorter than C by at least d, and three other discs, each bounded by curves of length at most $3d$.

Proof of Lemma 2.11. We apply Corollary 2.4 three times; firstly to the points x, y_0, y_1, y_2:

$$d(x, y_0) + d(y_1, y_2) \leq \max\{d(x, y_1) + d(y_0, y_2),\, d(x, y_2) + d(y_0, y_1)\} + 2\delta\,.$$

But $d(y_0, y_i) \leq d$ for $i = 1, 2$, and $d(y_1, y_2) > d$ by assumption, and $d(x, y_2) \leq d(x, y_1)$, so we get

$$(\star) \qquad d(x, y_0) \leq d(x, y_1) + 2\delta\,.$$

Now consider the points x, y_1, y', y_i where $i = 0, 1, 3$:

$$d(x, y_1) + d(y', y_i) \leq \max\{d(x, y') + d(y_1, y_i),\, d(x, y_i) + d(y_1, y')\} + 2\delta\,.$$

But by $(\star)$, and the definition of y', we get

$$d(x, y_1) + d(y', y_i) \leq \max\{d(x, y_1) - 4\delta + d,\, d(x, y_i) + 4\delta\} + 2\delta\,.$$

But $d(x, y_i) \leq d(x, y_0) \leq d(x, y_1) + 2\delta$, so we get

$$d(y', y_i) \leq 8\delta, \quad \text{for} \quad i = 0, 1, 3$$

and the lemma holds for these three points. It remains to show that $d(y', y_2) \leq 10\delta$. Consider the points x, y_0, y', y_2. We have

$$d(x, y_0) + d(y', y_2) \leq \max\{d(y', x) + d(y_0, y_2),\, d(x, y_2) + d(y', y_0)\} + 2\delta\,.$$

But $d(y', x) \leq d(y_0, x) - 4\delta$, and $d(y', y_0) \leq 8\delta$, so

$$d(y', y_2) \leq 10\delta\,.$$

This completes the proof of the Lemma. □

We now give an alternative proof of the fact that if triangles are thin in the Cayley graph of a group G, then G satisfies a the linear isoperimetric inequality. The proof will also show that G is finitely presented. Our plan follows the basic outline of Cannon's paper [Can] on co-compact hyperbolic groups; we subsequently discovered that a similar proof is given by Cannon in [Can2]. We first define ***local geodesics***, and show that these follow near their corresponding geodesics and are comparable to them in length. From this, we are able to show the existence of a ***Dehn's algorithm*** (see Definition 1.11), that is, a finite collection of relators such that any word which represents the trivial element contains more than half of one of the relators, and so may be shortened by use of one relator from this list. The linear isoperimetric inequality then follows immediately.

Let G be a group with finite generating set X. As usual, we regard a word u as a path $[0, \ell(u)] \to \Gamma_X(G)$.

We shall now show that:

Theorem 2.12. *Let G be hyperbolic, in the sense that geodesic triangles in the Cayley graph $\Gamma_X(G)$ are δ-thin. Let*

$$R = \{w \in F(X) \mid \ell(w) \leq 8\delta \text{ and } \mu(w) = 1\}.$$

Then $\langle X \mid R\rangle$ is a Dehn presentation for G.

Corollary. *Hyperbolic groups are finitely presented.*

When $\delta = 0$, G is a free group and X is a set of free generators, the theorem follows immediately. In what follows we assume $\delta \geq 1$, and that δ is an integer, so that the points in the Cayley graph which are considered are all vertices.
Definition 2.13 We will say that a path p is a *k-local geodesic* if each sub-path u of p of length at most k is geodesic.

Thus paths which are not k-local geodesics can be shortened locally.

Lemma 2.14. *Let $k = 4\delta$, and let u be a k-local geodesic, let v be a geodesic with $\mu(u) = \mu(v)$. Assume that each of these has length at least 2δ. Let r and s be the points in $\Gamma_X(G)$ on u and v at distance 2δ from $\mu(u)$ and $\mu(v)$. Then $d(r,s) \leq \delta$.*

Proof. Assume inductively that this is true when $\ell(u) \leq N$. We now take u, v such that $N \leq \ell(u) \leq N + k$. Let p be the vertex at distance k from $\mu(u)$ along u (recall k is integral). Let w be a geodesic from 1 to $\mu(w) = p$. Let q, t, and s be the points on u and w at distance 2δ from p as shown.

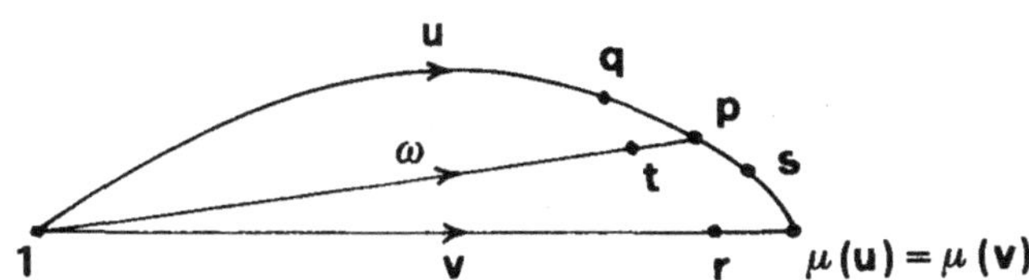

FIGURE 2.10

The segment of u from q to s has length k and thus is a geodesic, so that $d(q,s) = k = 4\delta$. By the induction hypothesis, $d(q,t) \leq \delta$. Hence, $d(t,s) \geq 3\delta$. But w, v, and the final length k segment of u form a geodesic triangle, and hence t lies close to v, and, in fact, is within δ of the point of v distance 2δ from $\mu(v)$. □

Theorem 2.15. *If u is a k-local geodesic, and v a geodesic with $\mu(u) = \mu(v)$, then u lies in a 3δ-neighbourhood of v.*

Proof. Let z be a point on u at least distance 2δ from the ends of u. (Otherwise there is nothing to prove.) Let x and y be geodesics from 1 to z and from z to $\mu(u)$. Let a, b be the points at distance 2δ from z along u, and let r and s be the points at distance 2δ from z along x and y, all as shown.

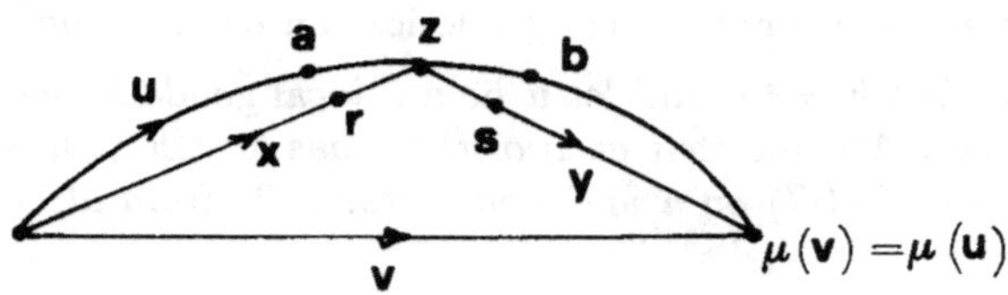

FIGURE 2.11

Since u is a k-local geodesic, $d(a,b) = 4\delta$, while by the lemma, $d(a,r) \leq \delta$ and $d(b,s) \leq \delta$. Consequently, $d(r,s) \geq 2\delta$. But x, y, and v form a geodesic triangle, so that $d(r,v) \leq \delta$. Hence $d(z,v) \leq 3\delta$. □

Proof of Theorem 2.12. We must show that if $w \in F(X)$ and $\mu(w) = 1$ then w may be shortened by using a single relator from R, that is, that w contains a subword r_1, and R contains a relator r_1r_2 with $\ell(r_1) > \ell(r_2)$. Thus $w = w_1r_1w_2$ and $\mu(w) = \mu(w_1)\mu(r_2)^{-1}\mu(w_2)$.

If w is not a k-local geodesic, then w has a sub-path p of length at most k which is not geodesic. This means that w can be shortened as required.

Suppose now that w is a k-local geodesic. Then w stays within distance 3δ of any geodesic for w. But since $\mu(w)$ is the identity element of G, a geodesic for $\mu(w)$ is the empty word. Thus the path w lies in the 3δ-neighborhood of the identity element in $\Gamma_X(G)$. But then w must be of length at most 3δ, for if w has an initial subpath of length $3\delta + 1$, then this subpath is geodesic and hence strays distance $3\delta + 1$ from the identity, which is a contradiction. In particular, $w \in R$. □

The proof of the following is immediate.

Theorem 2.16. *If G has a Dehn's algorithm, then G satisfies a linear isoperimetric inequality.*

The existence of a Dehn presentation has a simple consequence concerning elements of finite order in a hyperbolic group:

Corollary 2.17. ([GH,3.13]) *In a hyperbolic group, there are only finitely many conjugacy classes of elements of finite order.*

Proof. Consider the presentation giving a Dehn's algorithm. Let g be an element of finite order, and let $[g]$ be the set of all conjugates of g in G. Let w be a word in $F(X)$, chosen to be shortest over all $v \in F(X)$ such that $\mu(v) \in [g]$. Let n be the order of $\mu(w)$. As $\mu(w^n) = 1$, there must be some subword of w^n which is more than half of a word $r \in R$. This implies that $\ell(w) < \ell(r)$, else w, or some cyclic conjugate of w, can be shortened. (In fact w is contained in a cyclic conjugate of a relator.) Thus the number of conjugacy classes of elements of finite order is less than the number of elements of length at most $\max_{r \in R} \ell(r)$. □

This result contrasts with Gromov's result that in a hyperbolic group which is not a finite extension of a cylic group, there are an infinite number of conjugacy classes of prime (i.e. not proper powers) non-torsion elements [G, 5.1.B]. In particular an infinite torsion group is not hyperbolic (an alternative proof of this is to be found in [GS2]).

Solving the Word Problem in Linear Time

B. Domanski and M. Anshel have shown [DA] that if a group has a Dehn presentation, then there is an algorithm for solving the word problem in a length of time bounded linearly in terms of the length of the input word. To accomplish this requires a Turing machine with more than one tape. The way the machine works is to read through a given word, and on finding a place wher the word can be shortened, it does so. This may affect introduce a shortening in the part of the word already read, but not too far away. Here are the details.

Since the longest relator in our Dehn presentation has length $2k = 8\delta$, we know that we may shorten any word w representing the trivial element by replacing subwords of w of length at most $2k$ by shorter words of length at most k. Since each such replacement reduces the length of w by at least one, at most $\ell(w)$ replacements are required. We carry this procedure out on a Turing machine with an input tape, T and two pushdown stacks, S and S'. We also assume that there is enough internal memory to remember the final $2k$ letters of S at each step. We start with the word $w = a_1, \ldots, a_n$ written on tape T, and the two stacks S and S' empty.

Assume now that the first j letters from T have been read, and the contents of S and S' are respectively $x_1 \ldots x_p$ and $y_1 \ldots y_q$.

First we consider the case when a shortening of the last $2k$ letters is possible. Suppose that $1 \leq p - r + 1 \leq 2k$ and the final $p - r + 1$ letters $x_r \ldots x_p$ of S can be shortened to $v = v_1 \ldots v_s$. We now read v onto S' in reverse order, together with letters from S , until $2k$ letters have been transferred (if possible). The contents of S and S' are now $x_1 \ldots x_{t-1}$ and $y_1 \ldots y_q v_s \ldots v_1 x_{r-1} x_t$ where $t = \max\{1, r - 2k - 1\}$.

Now suppose that the final $\max\{p, 2k\}$ letters of S do not form a word which can be shortened. According to the values of p, q, j, one of the following happens. If $q \geq 1$, read the last letter of S' onto S, so that their new contents are $x_1 \ldots x_p y_q$ and $y_1 \ldots y_{q-1}$. If $q = 0$ (i.e. the S' tape is empty) and $j < n$, one letter is read from T onto S, so that $j + 1$ letters have now been read from T, and the word on S is $x_1 \ldots x_p a_{j+1}$; S' remains empty. If $q = 0$ and $j = n$ but $p \neq 0$, then all of T has been read, but w does not reduce to the empty word, and thus w is rejected by the machine as it cannot represent the trivial element of the group. Finally if $p = q = 0$ and $j = n$, the word w is accepted by the machine as we have reduced w to the empty word, so that $\mu(w) = 1$.

It is easy to see that this procedure halts after a length of time proportional to $n = \ell(w)$. To see this, notice that we read each letter of w once from T and move at most $4k$ letters from S to S' and back again for each replacement we make. Since each replacement reduces length, we make at most n replacements.

One might ask when the stack S' can be replaced by a finite amount of memory. In this case, the word problem can actually be solved by a pushdown automaton. This is equivalent to saying that the word problem for the group is a *context free grammar*. That is to say, the collection of all words representing the identity element of the group may be generated by a simple set of replacement rules (for more about languages and automata see for instance [HU]). It is a result of Muller and Schupp [MS] (together with Dunwoody's accessibility result) that this can be done if and only if the group is virtually free.

Now it is easy to see how to extend this procedure to a group G which is a direct product of finitely many hyperbolic groups, say $G_1, \ldots, G_m$. One simply takes a Turing machine with a tape T and m pairs of pushdown stacks, $S_1, S_1', \ldots, S_m, S_m'$. Then as each letter a_i is read off of T, one may rewrite it in terms of the generators of $\{G_j\}$ and sort these out to the appropriate stacks. Since this rewriting requires only a linear amount of time, this procedure solves the word problem in G in linear time.

Finally, if H is a finitely generated subgroup of G, the restriction of this procedure to elements of H solves the word problem in H in linear time. We have shown the following

Theorem 2.18. *Let H be a finitely generated subgroup of a direct product of hyperbolic groups. Then the word problem in H is soluble in linear time.*

Such groups can be fairly complicated. They include, for example, the finitely generated subgroups of the direct product of two free groups of rank 2. Many of these are not finitely presented. In fact, it is a theorem of G. Baumslag and J. Roseblade that these groups are finitely presented if and only if they contain a subgroup of finite index which is itself a direct product [BR].

Any problem which can be solved in linear time is also solved in linear space. Such problems correspond to the so-called *context sensitive grammars*. Such languages are also characterized by a set of rules for generating the words of the language (see [HU]). Clearly the class of groups with context sensitive word problem is much larger than the class of groups with context free word problem! In fact, though we will not show it here, the automatic groups of [**CEHPT**]all have context sensitive word problem, and hence so do their finitely generated subgroups.

Divergence of geodesics

Here we shall show (2.20) that in a geodesic metric space, non-linear divergence of geodesics implies that the space is hyperbolic. (The definition of divergence functions is given in 1.7.) We first show the more elementary result (2.19) that in a hyperbolic space geodesics diverge exponentially, leading to the remarkable fact that exponential and non-linear divergence are equivalent in geodesic metric spaces.

Theorem 2.19. *In a hyperbolic metric space geodesics diverge exponentially.*

Proof. Suppose that all geodesic triangles are δ-thin. Let γ and γ' be geodesics of length $R+r$ beginning at the point x such that $d(\gamma(R), \gamma'(R)) > \delta$. We thus set $e(0) = \delta$. Let p be a path from $\gamma(R+r)$ to $\gamma'(R+r)$ lying in the closure of the complement of $B_{R+r}(x)$ We will show there is an exponential function $e(r)$ independent of the choice of γ and γ' such that $\ell(p) \geq e(r)$.

Let α be a geodesic from $\gamma(R+r)$ to $\gamma'(R+r)$. Let b be a binary sequence of length s (possibly 0), and suppose α_b to have been chosen (the geodesic α thus corresponds to the empty sequence). Let m_b be the midpoint of the segment of p between the ends of α_b. We choose α_{b0} to be a geodesic from $\alpha_b(0)$ to m_b, and α_{b1} to be a geodesic from m_b to $\alpha_b(1)$.

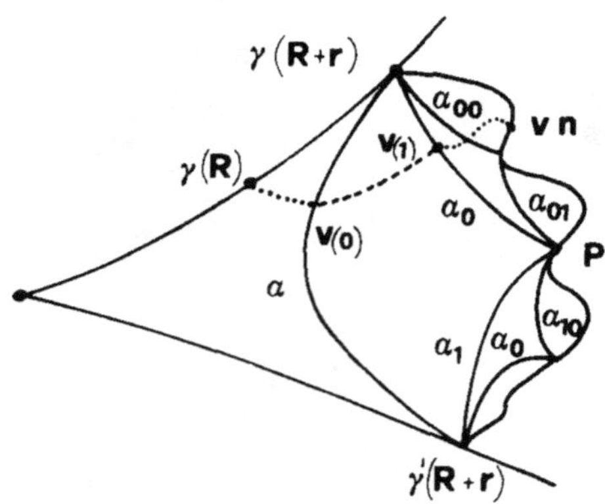

FIGURE 2.12

We continue this subdivision until each α_b in our final subdivision has length between $\frac{1}{2}$ and 1. This ensures that these last α_b's are contained in p. This also means that we have divided the original path α into n pieces, where

$$\log_2(\ell(p)) \leq n \leq \log_2(\ell(p)) + 1 .$$

Notice that for each b, the segments α_b, α_{b0} and α_{b1} form a geodesic triangle.

As $d(\gamma(R), \gamma'(R)) > \delta$, there is a point $v(0)$ on α wth the property that $(d(v, \gamma(R)) \leq \delta$.

For each i, if a point $v(i)$ lies on α_b, then as the triangle with sides $\alpha_b, \alpha_{b0}, \alpha_{b1}$ is δ-thin, there is a point $v(i+1)$ on α_{b0} or α_{b1} such that $d(v(i), v(i+1)) < \delta$.

Thus we may find a path from x to a point $v(n)$ on p whose length is at most $R + \delta(\log_2(\ell(p)) + 2)$. But as the path p lies outside of the ball of radius $R + r$, $d(x, v(n)) \geq R + r$, so that

$$R + \delta(\log_2(\ell(p)) + 2)) > R + r$$

$$\Leftrightarrow \ell(p) > 2^{\frac{r}{\delta} - 2},$$

and we see that e is an exponential function as required. □

We now establish the opposite direction of the equivalence.

Proposition 2.20. *If X is a geodesic metric space with a non-linear divergence function, then geodesic triangles are δ-slim for some δ.*

Proof. Let e be a divergence function for X, and let xyz be a geodesic triangle in X. Consider the edge $[xy]$ (resp. $[xz]$) as an isometric embedding α_1 (resp α_2) : $[0, n] \to X$ based at x.

Let T be the maximum value of $t \in [0, n]$ such that

$$\forall t \in [0, T],\ d(\alpha_1(t), \alpha_2(t)) \leq e(0),$$

and let $x_1 = \alpha_1(T)$, $x_2 = \alpha_2(T)$. Similarly define points z_1, z_2, y_1, y_2.
Claim 1 If $[xx_1] \cap [y_2y] \neq \emptyset$, then there is a bound on $\max\{d(z_2, y_1), d(z_1, x_2)\}$, and so the triangle is δ-slim with $\delta = \frac{K}{2} + 2e(0)$.

If $[xx_1] \cap [y_2y] \neq \emptyset$ then there are points $x_3 \in [xx_2]$ and $y_3 \in [yy_1]$ such that $d(x_3, y_3) < 2e(0)$. Applying the divergence function to $[zy_3]$ and $[zx_3]$ bounds the lengths of $[z_1x_3]$ and $[z_2y_3]$, and hence the lengths of $[z_1x_2]$ and $[z_2y_1]$.

Hence we assume that there are no such intersections. Let L_3, L_2, L_1 be the lengths $d(z_2, y_1), d(x_2, z_1), d(y_2, x_1)$, and suppose that $L_1 = \max\{L_3, L_2, L_1\}$. It suffices to show that L_1 is bounded by some constant K, for then xyz will be $(K/2 + e(0)$-thin).

Let t be the midpoint of $[x_1y_2]$.

Let $a = d(x, x_1)$ and $b = d(y, y_1)$; then $t \in B_{a+L_1/2}(x) \cap B_{b+L_1/2}(y)$, though these balls (call them B_1, B_2) have disjoint interiors.

Without loss of generality say $L_3 \geq L_2$ (i.e. $d(x_2, z) \leq d(y_1, z)$).
Claim 2 $[x_2z] \cap \mathrm{int}(B_2) = \emptyset$.

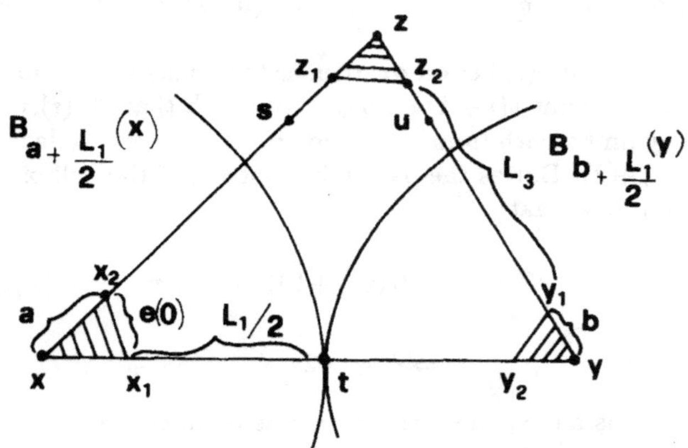

FIGURE 2.13

Suppose not, and let $s \in [x_2z] \cap \mathrm{int}(B_2)$. As $s \notin B_1$, it follows that $d(s, x_2) \geq L_1/2$. Since $L_3 \geq L_2$, there is a point $u \in [y, z]$ such that $d(u, z) = d(s, z)$. It follows that

$$\begin{aligned} L_1/2 \leq d(s, x_2) &= d(x_2, z) - d(z, s) \\ &= d(x_2, z_1) + d(z_1, z) - d(z, s) \\ &\leq d(z_2, y_1) + d(z_1, z) - d(z, u) \\ &= d(z, y_1) - d(z, u) = d(u, y_1) \end{aligned}$$

Hence $u \notin \mathrm{int}(B_2)$; but

$$d(z, y) = d(z, u) + d(u, y) \leq d(z, s) + d(s, y)$$

and so

$$b + L_1/2 \leq d(u, y) \leq d(s, y) < L_1/2 + b.$$

This contradiction establishes claim 2.

Let v be a point on the edge $[yz]$ such that $d(y, v) = b + L_1/2$. There is a path from t to v in the complement of B_2 of length at most

$$d(t, x_1) + 3e_0 + L_3 + 3e_0 + d(z_2, v) \leq L_1/2 + 6e_0 + L_1 + L_1/2.$$

Hence $e(L_1/2) \leq 2L_1 + 6e_0$, giving the required bound for L_1. $\square$

Chapter 3 Quasigeodesics

In this chapter we shall study the infinite cyclic subgroup $\langle g\rangle$ generated by a non-torsion element g in a word hyperbolic group. We shall see that the set of vertices in the Cayley graph $\Gamma_X(G)$ which correspond to $\langle g\rangle$ form something like a geodesic (a *quasigeodesic*). This in turn is used to obtain various results which are analogous to existing results about groups acting discretely and cocompactly on hyperbolic space. For instance, we show that an abelian subgroup of a hyperbolic group is a finite extension of a cyclic group (i.e. there are no $\mathbb{Z}\times\mathbb{Z}$ subgroups in a hyperbolic group). Most of the treatment here is due to Mihalik and Lustig.

In a geodesic metric space, we can define the length of an arc α between two points as $d_\alpha(x,y)=\sup\sum d(x_i,x_{i+1})$ over finite sets of points x_i on α between the points x,y.

Or, working in the domain, in order to give a precise sense to 'between' :

let $\alpha:[0,1]\to X$ be a path from x to y in X. We define the length of the arc to be $d_\alpha(x,y)=\lim_{n\to\infty}\sum_{i=0}^{n-1} d(\alpha(\frac{i}{n}),\alpha(\frac{i+1}{n}))$.

Definition 3.1 An arc α in a geodesic metric space X is called a (λ,ϵ)-*quasigeodesic* if there are positive constants $\lambda>1, \epsilon\geq 0$ such that for all points x,y on α,

$$d_\alpha(x,y)\leq \lambda d(x,y)+\epsilon.$$

We must first prove the following technical result, which is essential to all that follows.

Proposition 3.2. ([G, 8.1.D], [GH,8.21]) *Let g be an element of infinite order in a hyperbolic group G, and let Γ be the Cayley graph with respect to some finite generating set. Let α be a path from the vertex corresponding to the identity element to the vertex corresponding to g. Then the bi-infinite path*

$$(\ldots,g^{-1}\alpha,\alpha,g\alpha,\ldots)$$

is a quasigeodesic.

Proof. Suppose that the positive integer R is given, and choose k such that $d(g^k,1)>8R+2\delta$, Let β be the geodesic from 1 to g^k, and let y be the midpoint of β. Let I be the subinterval of β of length R centered on y. Recall that $B_r(z)$ denotes the ball of radius r about the point z in Γ. In what follows, by 'midpoint' of a geodesic we mean a vertex at distance at most $1/2$ from the actual midpoint of the arc.

Claim 1 If $p\in B_R(1)$ and $q\in B_R(g^k)$, then the midpoint m_1 of the geodesic arc $[pq]$ is in $N_{2\delta}(I)$, i.e. $d(m_1,I)<2\delta$.

Note that if $a \in B_R(1)$ and $b \in B_r(g^k)$ then the midpoint of $[ab]$ is at least distance R from balls of radius $R+\delta$ about 1 and g^k, by the choice of k.

If m_2 is the midpoint of $[p, g^k]$ then as $|\ell([pg^k]) - \ell([pq])| < R$, we have $|d(m_1,p) - d(m_2,p)| < R/2$.

By assumption geodesic triangles are δ-thin in Γ. Considering the geodesic triangle pqg^k, we see that the internal points are all inside the ball $B_{R+\delta}(g^k)$, so that there is a point $m_1' \in [pg^k]$ such that

(1) $d(m_1,p) = d(m_1',p)$,
(2) $d(m_1',m_2) < R/2$,
(3) $d(m_1,m_1') \leq \delta$,
(4) $d(m_1',1) > 3/2R + \delta$.

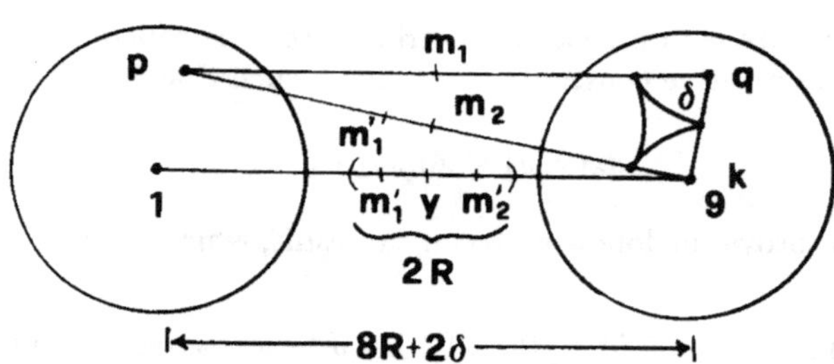

FIGURE 3.1

Similarly the internal points of the geodesic triangle $1pg^k$ are all inside the ball $B_{R+\delta}(1)$, and there are points m_2', m_1'' on $[1g^k]$ satisfying

(1) $d(m_2',g^k) = d(m_2,g^k)$ and $d(m_1'',g^k) = d(m_1',g^k)$,
(2) $d(m_2',m_2) \leq \delta$ and $d(m_1'',m_1') \leq \delta$,
(3) $d(m_1'',m_2') = d(m_1',m_2)$,
(4) $d(m_2',y) < R/2$

It follows from (2) that $d(m_1'',m_1) \leq 2\delta$, and from (3) and (4) that $d(y,m_1'') < R$. The claim is thus established.

Let N be the number of distinct vertices of Γ in the ball $B_{2\delta}(1)$; the 2δ neighbourhood of the interval I then contains at most RN vertices. Now consider the midpoints of each translate of the arc $\alpha = [1g^k]$ by each of the elements $1, g, g^2, \dots g^{NR}$. These midpoints are all distinct (else some power of g would act on Γ fixing a point, and thus would have finite order), and there are $1 + NR$ of them. Hence there is a number $p(R) \leq NR$ such that $g^{p(R)} \notin B_R(1)$ (and so $g^{k+p(R)} \notin B_R(g^k)$). Notice that $p(R) \geq R/\ell(g)$.

Claim 2 $\ell(g^{NR}) \geq R$ for all R.

Suppose that the claim is false, and thus that there is some R_0 with $\ell(g^{NR_0}) < R_0$. For all $s > NR_0$, let $s = nNR_0 + R_1$, with $n \in \mathbb{Z}^+$ and $0 \leq R_1 < NR_0$ Then

$$\ell(g^s) \leq \ell(g^{nNR_0}) + \ell(g^{R_1}) \leq n\ell(g^{NR_0}) + \ell(g^{R_1}) < n(R_0 - \epsilon) + \ell(g^{R_1})$$

$$< nR_0 \quad \text{when} \quad n\epsilon > \ell(g^{R_1})$$

But this means that for all sufficiently large values of n, we have that $\ell(g^s) < nR_0$. If we choose a value of R such that $p(R) > NR_0$, then by claim 1, we have that $\ell(g^{p(R)}) \geq R$. But the above says that $\ell(g^{p(R)}) < p(R)/N \leq R$, giving the required contradiction.

We now show that the bi-infinite arc $\beta = (\dots, g^{-1}\alpha, \alpha, g\alpha, \dots)$ is a quasi-geodesic. Let γ be a geodesic arc between two points x, y on β. By construction, the initial and final points of γ are within $N\ell(g)$ of vertices g^{aN} and g^{bN} for some integers a, b, and so the length $d_\beta(x, y)$ of the subarc of β between the points x, y is at most $(|b-a|+2)\ell(g)N$. But by the above, $d(g^{aN}, d^{bN}) \geq |b-a|$, and so

$$d(x, y) \geq |b-a| - 2\ell(g)N.$$

This means that

$$d_\beta(x, y) \leq |b-a|\ell(g)N + 2\ell(g)N \leq \ell(g)Nd(x, y) + 2\ell(g)^2N^2 + 2\ell(g)N$$

and it follows that β is a (λ, ϵ)-quasigeodesic for

$$\lambda = \ell(g)N \quad \text{and} \quad \epsilon = 2\ell(g)^2N^2 + 2\ell(g)N. \quad \square$$

We now show that geodesic arcs between points on a quasigeodesic lie near the quasigeodesic. More precisely, let $N_r(U)$ denote the r neighborhood of the subset U of X.

Proposition 3.3. ([G, 7.2.A], [CDP,3.1.3], [GH, 5.6, 5.11]) *Let x, y be points in the hyperbolic metric space X. If α is a (λ, ϵ) quasigeodesic between the points x, y, there are integers $L(\lambda, \epsilon)$, $M(\lambda, \epsilon)$ such that if γ is any geodesic $[xy]$, then $\gamma \subset N_L(\alpha)$ and $\alpha \subset N_M(\gamma)$.*

Proof. Let $e : \mathbb{N} \to \mathbb{R}^+$ be an exponential divergence function for X. We first show the existence of the bound L.

Let $D = \sup_{x \in \gamma}\{d(x, \alpha)\}$, and choose a point $p \in \gamma$ where this supremum is reached. Then

$$\operatorname{int} B_D(p) \cap \alpha = \emptyset$$

Let a, b be points on γ at distance D from p, and let a', b' be points on γ at distance $2D$ from p (or the points x or y if these are at distance $\leq 2D$ from p). There are points u, v on α such that $d(a', u) \leq D$, $d(b', v) \leq D$. Notice that $([a', u] \cup [b', v]) \cap \operatorname{int} B_D(p) = \emptyset$. Following a path via a', p and b', we see that $d(u, v) \leq 6D$, and as α is a (λ, ϵ) quasigeodesic, we have $d_\alpha(u, v) \leq 6\lambda D + \epsilon$. Hence there is a path of length $\leq 4D + 6\lambda D + \epsilon$ from a to b which does not meet $\operatorname{int} B_D(p)$. But the divergence function e says that the length of such a path is exponential in $D - (e(0)/2)$. This gives us the bound $L(\lambda, \epsilon)$ for D. (Notice again that we have only used the fact that the divergence function is nonlinear.)

Now suppose that $\alpha \not\subset N_L(\gamma)$. Then a component of $cl(\alpha - N_L(\gamma))$ is a path ξ with endpoints u, v at distance L from points a, b say on γ.

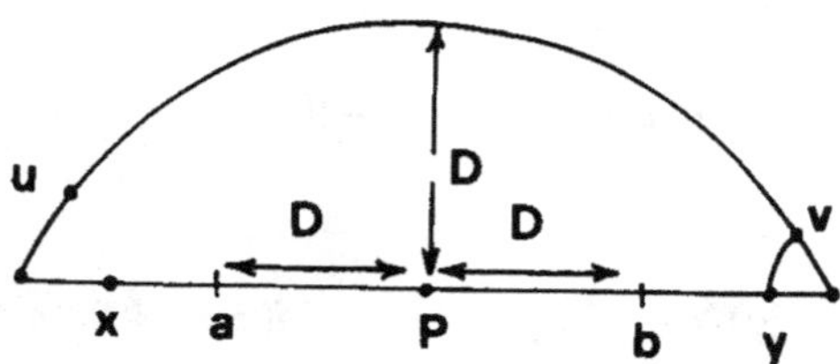

FIGURE 3.2

If α is an arc $[0, 1] \to X$, and $\xi = \alpha([t_1, t_2])$, then each point on γ between a and b is within distance L of some point of $\alpha([0, t_1]) \cup \alpha([t_2, 1]) = \alpha_1 \cup \alpha_2$ by the first part of the proof. Then there is some point z on γ between a and b which is at distance $\leq L$ from a point u_1 on α_1 and from a point u_2 on α_2. But then $d(u_1, u_2) \leq 2L$, so $d_\alpha(u_1, u_2) \leq 2\lambda L + \epsilon$, bounding the length of the arc ξ. It thus follows that every point on ξ is at distance at most $L + \lambda L + \epsilon/2$ from γ, and the proposition holds. □

Notice that when the geodesic space under consideration is the Cayley graph of a hyperbolic group, the above results say

Corollary 3.4. *If g is an element of infinite order in a hyperbolic group, then there is a constant L such that for any point x on a geodesic arc $[g^i g^j]$ there is an integer k such that $d(x, g^k) < L$.*

We now use these results on quasigeodesics to obtain information about subgroups of a hyperbolic group. In particular we shall show that an abelian subgroup of a hyperbolic group is cyclic-by-finite. This follows from:

Proposition 3.5. ([CDP, 10.7.2], [GH, 8.35]) *Let H be a hyperbolic group, and let g be an element of G of infinite order. Let $C(g)$ denote the centralizer of g, and let $\langle g \rangle$ denote the infinite cyclic group generated by g.*

Then $C(g)/\langle g \rangle$ is finite.

Proof. Let Γ be the Cayley graph of G with respect to some finite generating set, and suppose that geodesic triangles are δ thin in Γ. Let L be the constant (guaranteed by the above proposition) such that the geodesic $[1g^n]$ lies in a L-neighborhood of the set $\{1, g, g^2, \ldots, g^n\}$. Let $s \in C(g)$, and choose m such that $d(1, g^m) > 2\ell(s) + 2\delta$ (it is here that we use the fact that g has infinite order). Consider the geodesic 4-gon (parallelogram) with vertices $1, g^m, s, sg^m$ and sides $[s(sg^m)] = s[1g^m]$, $[g^m(sg^m)] = g^m[1s]$, $[1g^m]$ and $[1s]$.

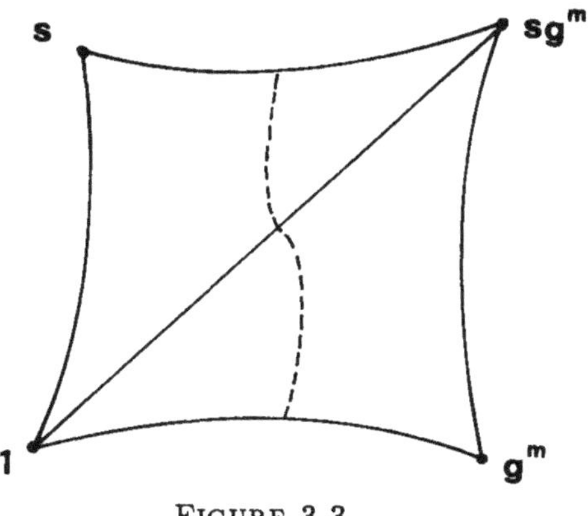

FIGURE 3.3

By considering the two (δ-thin) geodesic triangles with vertices $1, g^m, sg^m$ and $1, s, sg^m$ we see that either some point of $[1s]$ is at distance $\leq 2\delta$ from some point of $[g^m(sg^m)]$, or there is a point on $[1g^m]$ which is at distance $\leq 2\delta$ from some point of $[s(sg^m)]$. By the choice of m, the former cannot occur, else there is a path of length at most $2\ell(s)+2\delta$ from 1 to g^m. So let $a \in [1g^m]$ and $b \in [s(sg^m)]$ such that $d(a,b) \leq 2\delta$. But there are integers $i,j < m$ such that $d(a,g^i) \leq L$ and $d(b,sg^j) \leq L$, so $d(g^i,sg^j) \leq 2L+2\delta$. But then $d(1,sg^{j-i}) \leq 2L+2\delta$, and so every coset $s\langle g\rangle$ in $C(g)$ intersects the ball of radius $2L+2\delta$ about the identity. □

Corollary 3.6. *An abelian subgroup of a hyperbolic group which contains an element of infinite order is finite-by-cyclic.*

Definition 3.7 ([G, 5.3 page 139, and 7.3 page 191], [CDP, chap. 10]) If X is a geodesic metric space then a subset A is *ϵ-quasiconvex* if for all geodesics $[ab]$ with endpoints $a,b \in A$, $[ab] \subset N_\epsilon(A)$. A subgroup of a finitely generated group is said to be quasiconvex if the vertices in the subgroup form a quasiconvex subset of the Cayley graph.

It is not hard to see that this last definition is independent of the finite set of generators chosen (for more about quasiconvexity see [S]).

It follows from 3.3 that an infinite cyclic subgroup of a hyperbolic group defines a quasiconvex subset of the Cayley graph.

Also, a subgroup of finite index is quasiconvex. A finitely generated subgroup of a finitely generated free group is quasiconvex, as can be seen by thinking of the Cayley graph as a tree.

Rips has shown, [R], that a small cancellation group may have finitely generated subgroups which are infinitely related. This cannot happen for quasiconvex subgroups:

Lemma 3.8. ([CDP, 10.4.2]) *A quasiconvex subgroup H of a hyperbolic group G is finitely generated, and in fact is also hyperbolic .*

Proof. Let X be finite set of generators for G. Let $w = a_1 \ldots a_n$, $a_i \in X \cup X^{-1}$ be a geodesic in the Cayley graph $\Gamma_X(G)$ which ends at a vertex which lies in the subgroup H. The quasiconvexity condition implies that there for each $i = 1,\ldots,n$ there is a word $v(i) \in F(X)$ of length at most ϵ such that $a_1 \ldots a_i v(i)$ represents an element of H. Thus $w = \prod_{i=1}^n v(i-1)^{-1} a_i v(i)$ where $v(0) = v(n) =$ the empty word, and the set Y of words $v \in F(X)$ of length at most $2\epsilon+1$ is a finite set of generators for H.

Now let $d_Y(h,h')$ denote the distance between the vertices $h,h' \in H$ in $\Gamma_Y(H)$, and $d_X(g,g')$ denote distance between the vertices $g,g' \in G$ in $\Gamma_X(G)$, and $d_{X\cup Y}(g,g')$ denote the distance between the vertices g,g' in $\Gamma_{X\cup Y}(G)$.

Then by rewriting the elements of Y as words in $F(X)$, we see that

$$(2\epsilon+1)d_Y(h,h') \geq d_X(h,h') \geq d_{X\cup Y}(h,h').$$

Also, the way the generators Y were found shows that $d_Y(h,h') \le d_X(h,h')$. Again by rewriting words in $F(X \cup Y)$ as words in $F(X)$ we see that $(2\epsilon + 1)d_{X\cup Y}(g,g') \ge d_X(g,g')$ Thus

$$\frac{1}{2\epsilon+1} d_{X\cup Y}(h,h') \le d_Y(h,h') \le (2\epsilon+1) d_{X\cup Y}(h,h')$$

and a geodesic word for an element $h \in H$ (in terms of the generators Y) is a $(2\epsilon+1)$-quasigeodesic for h in terms of the generators $X \cup Y$. Thus a geodesic triangle in $\Gamma_Y(H)$ has quasigeodesic sides when regarded as a triangle in $\Gamma_{X\cup Y}(G)$, and these quasigeodesics lie close to geodesics, which are close in $\Gamma_{X\cup Y}(G)$, as G is hyperbolic. Thus the original geodesic triangle was slim. □

Proposition 3.9. *If $1 \to N \to G \to B \to 1$ is a short exact sequence of infinite groups, with G hyperbolic, then N is not quasiconvex.*

Proof. Let Γ be the Cayley graph of G with respect to some finite generating set, and suppose that N is quasiconvex in Γ. Let $p : \Gamma \to \Lambda$ be the quotient map of the action of N on Γ. When $x, y \in \Lambda$ and α is a geodesic of length n from x to y, for any preimages $a \in p^{-1}(x)$ and $b \in p^{-1}(y)$, we have that $d(a,b) \ge n$. Since B is infinite and finitely generated, Λ is a locally finite, infinite 1-complex. Choose a geodesic β of length $K > 2\epsilon + 2\delta$ with initial point $p(1)$ and endpoint b, and let $\tilde\beta$ be a lift of β based at 1, with endpoint c. Note that $\tilde\beta$ is also geodesic of length K. Choose $u \in N$ such that $\ell(u) > 2K + 2\delta$ and consider the geodesic 4-gon in Γ with sides $[1u]$, $\tilde\beta$, $u\tilde\beta$, $[c(uc)]$.

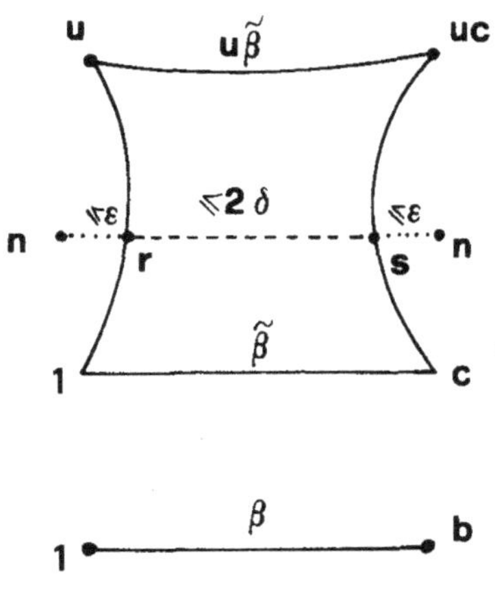

FIGURE 3.4

As we choose u to be long, no point of $u\tilde{\beta}$ is within 2δ of a point of $\tilde{\beta}$. As in the proof of proposition 3.5, some point r of $[1u]$ is within 2δ of a point s of $[c(uc)]$. As N is normal, $c^{-1}uc \in N$ so $uc = c(c^{-1}uc) \in cN = p^{-1}(b)$; also we have that $1, u \in N = p^{-1}(1)$. Since N is ϵ-quasiconvex for some ϵ, some point n_1 of N is within ϵ of r and some point n_2 of cN is within ϵ of s. But then n_1 is within $2\epsilon + 2\delta$ of n_2, contradicting the restriction on the length of β. □

S. M. Gersten has pointed out that the above proof shows:

Corollary 3.10. *If G is a hyperbolic group, and H is an infinite quasiconvex subgroup of G, then each element of the factor group $N_G(H)/H$ is of finite order.*

In general, it does not immediately follow that $N_G(H)/H$ is finite.
Remark M. Mihalik and W. Towle have since then used the methods of the above proof to show that if H is a quasiconvex subgroup of a hyperbolic group, then H has finite index in $N_G(H)$, the normalizer of H in G, and this index is bounded by a function depending on δ, ϵ and the number of generators for G. Furthermore they show that if H is a quasiconvex subgroup of the hyperbolic group G and $x \in G$ such that xHx^{-1} is contained in H, then $xHx^{-1} = H$.

When G is a finitely generated free group, all finitely generated subgroups are quasiconvex, so that 3.8 reduces to the result that a non-trivial finitely generated normal subgroup of a finitely generated free group is of finite index (due originally to Schreier).

Using results from [BGSS], it is shown in [GS2] that $C_G(H)/H$ is finite.

We finish off this chapter by relating quasigeodesics with the existence of a Dehn presentation and local geodesics (see Definition 1.11 and 2.12–2.17). Recall that a path is a k-local geodesic if all subpaths of length at most k are geodesic.

Theorem 3.11. *(cfr. [Can2]) Let X be a geodesic metric space where all geodesic triangles are δ-thin. If u is a 10δ-local geodesic, then u is a $(12, 7\delta)$-quasigeodesic.*

If v is a geodesic path in X, and $x \in X$, let $p(x)$ be a point on v at minimum distance from x.

Lemma 3.12. *Let u be a 10δ-local geodesic, and v a corresponding geodesic. If $d(p(u(t)), p(u(t+a))) \leq \delta/2$ then $|a| \leq 13\delta/2$.*

Proof of Lemma 3.12. Recall that Theorem 2.15 shows that a 4δ-local geodesic lies in a 3δ-neighborhood of a corresponding geodesic. It follows that $d(u(t), p(u(t))) \leq 3\delta$, and $d(u(t+a), p(u(t+a))) \leq 3\delta$, so that $d(u(t), u(t+a)) \leq 13\delta/2$.

Let u_0 be the segment of u between $u(t)$ and $u(t+a)$, and let v_0 be a geodesic between the same points. As before, u_0 lies in a 3δ neighborhood of v_0 and moreover v_0 lies in a $13\delta/2$ neighborhood of $u(t)$. It follows that u_0 lies in a $19\delta/2$ neighborhood of $u(t)$. If $\ell(u_0) \geq 10\delta$, then an initial 10δ segment is geodesic, so leaves the ball of radius $19\delta/2$, which is not possible. Thus $\ell(u_0) < 10\delta$ and, as it is geodesic, $\ell(u_0) \leq 13\delta/2$, and $|a| \leq 13\delta/2$. □

Returning to the proof of the theorem, let v be a geodesic corresponding to u (i.e. a geodesic between the endpoints of u). Divide v into subintervals $v_1, v_2, \ldots, v_n$ where $\ell(v_i) \leq \delta/2$ and $n \leq 1 + 2\ell(v)/\delta$.

Let $u_i = p^{-1}(v_i) \cap u$; then $\operatorname{diam}(u_i) \leq 13\delta/2$ by the lemma. It follows that

$$\ell(u) \leq \sum_{i=1}^{n} \operatorname{diam}(u_i) \leq n\frac{13\delta}{2} \leq \frac{13\delta}{2}(\frac{2\ell(v)}{\delta} + 1) = 13\ell(v) + \frac{13\delta}{2}$$

and u is therefore a $(13, 7\delta)$-quasigeodesic. □

By more careful bookkeeping, we can improve this to say that an 8δ-local geodesic is a $(8, 7\delta)$-quasigeodesic.

Chapter 4 The boundary

The aim of this section is to define a boundary ∂X for a hyperbolic metric space X, which gives a compactification $\hat{X} = X \cup \partial X$ when X is complete and locally compact. We follow the plan indicated by Gromov in his section 1.8, [G], using some of the ideas of [CDP]; the structure of the boundary is much more extensively developped in [CDP] and in [GH].

As usual we use $(x.y)$ to denote the inner product on the metric space X with respect to some basepoint w, i.e.

$$(x.y) = \frac{1}{2}\{d(x,w) + d(y,w) - d(x,y)\}.$$

The points of ∂X are defined to be equivalence classes of sequences in X as follows.

Definition 4.1: A sequence $\{a_i\}$ of points in X is said to *converge to infinity* if

$$\lim_{i,j\to\infty} (a_i.a_j) = \infty.$$

Note:

(1) This definition is independent of the choice of basepoint since $|(x.y)_w - (x.y)_{w'}| \le d(w,w')$.

(2) If $\{a_i\}$ converges to infinity then

$$\lim_{i\to\infty} d(a_i, w) = \infty$$

since $(x.y) \le \min\{d(x,w), d(y,w)\}$.

Let $S_\infty(X)$ denote the set of all sequences convergent to infinity, and define the relation

$$\{a_i\}R\{b_i\} \text{ iff } \lim_{i\to\infty} (a_i.b_i) = \infty.$$

Note that a sequence which converges to infinity is related to all of its subsequences.

The relation is symmetric and reflexive and independent of the choice of basepoint. For a general metric space R is not transitive as the following example shows.

Example. Let X be the Cayley graph of the group

$$\mathbb{Z} \times \mathbb{Z} = < x, y \mid xy = yx >$$

(with respect to the generators x and y) and let w be the vertex corresponding to the identity element, $e = x^0y^0$. Define $a_n = x^n$, $b_n = y^n$ and $c_n = x^ny^n$.

Then the sequences $\{a_i\}$, $\{b_i\}$ and $\{c_i\}$ all converge to infinity and $(a_n.c_n) = (b_n.c_n) = n$ while $(a_n.b_n) = 0$. Thus $\{a_i\}R\{c_i\}$ and $\{b_i\}R\{c_i\}$, but $\{a_i\}$ and $\{b_i\}$ are not related.

Example. With X as in the last example, consider the sequence defined by

$$p_n = \begin{cases} x^m, & \text{if } n = 2m \\ x^m y^m, & \text{if } n = 2m+1. \end{cases}$$

Then $\{p_n\} \in S_\infty(X)$ but the two subsequences consisting of odd and even terms are tending to infinity in different directions.

Recall that we say X is δ-hyperbolic if for all $x, y, z \in X$

$$(x.y) \geq \min\{(x.z), (y.z)\} - \delta.$$

Lemma 4.2. *When X is δ-hyperbolic the relation R is transitive.*

Proof. If $\{a_i\}, \{b_i\}, \{c_i\} \in S_\infty(X)$ with $\{a_i\}R\{b_i\}$ and $\{b_i\}R\{c_i\}$ then $\{a_i\}R\{c_i\}$ since $(a_i.c_i) \geq \min\{(a_i.b_i), (b_i.c_i)\} - \delta$. □

From now on we assume that X is a δ-hyperbolic metric space.

Observe that for $\{a_i\}, \{b_i\} \in S_\infty(X)$ we have

$$\{a_i\}R\{b_i\} \text{ iff } \lim_{i,j\to\infty} (a_i.b_j) = \infty.$$

Definition 4.3 The *boundary of* X is $\partial X = S_\infty(X)/R$. We say that $\{a_i\} \in S_\infty(X)$ *converges to* $x \in \partial X$ if $x = [\{a_i\}]$ and we write $a_i \to x$.

Note. If $x \in X$ then '$x_i \to x$' has the usual meaning of convergence in the metric sense.

Example. Take $X = \mathbb{R}$ with the usual metric and 0 as the basepoint. If $\{a_i\} \in S_\infty(X)$ then either $a_i > 0$ for almost all values of i or $a_i < 0$ for almost all values of i. This defines two distinct equivalence classes of sequences which we call $+\infty$ and $-\infty$ respectively and $\partial X = \{-\infty, +\infty\}$ as expected. Similarily for $\mathbb{R}$-trees the boundary is the set of ends.

In order to put a topology on the set $\hat{X} = X \cup \partial X$ we first extend the inner product to the boundary.

<u>Definition 4.4</u>: For $x, y \in \hat{X}$,

$$(x.y)_S = \inf\{\liminf_i (x_i.y_i)\},$$

where the infimum is taken over all pairs of sequences $x_i \to x$ and $y_i \to y$.

Note. This definition may seem unduly complicated. The next example illustrates the need for this complexity even in a very simple metric space.

Example. Let X be the Cayley graph (with respect to the generators x and y) of

$$\mathbb{Z} \times \mathbb{Z}_2 =< x, y \mid y^2 = 1,\ xy = yx > .$$

This space, like $\mathbb{R}$ above, has a boundary consisting of two points which we will also call $+\infty$ and $-\infty$. Let

$$a_n = x^n,\ b_n = x^{-n},\ c_n = yx^n,\ d_n = yx^{-n}$$

$$z_m = \begin{cases} x^{2n}, & \text{if } m = 2n \\ yx^{2n+1} & \text{m = 2n+1.} \end{cases}$$

Then a_n, c_n and z_n all tend towards $+\infty$, while b_n and d_n tend towards $-\infty$.

Observe that $(d_n.z_n)$ is 0 if n is even, and 1 if n is odd. Thus we use lim inf rather than lim in the definition. Furthermore $(a_n.b_n) = 0$, while $(c_n.d_n) = 1$. Thus we take the infimum over all sequences in the definition.

Some properties of the inner product on $\hat{X}$ are given in the following lemmas.

Lemma 4.5.

(1) *If* $x \in X$ *and* $y \in \hat{X}$ *then* $(x.y)_S = \inf\{\liminf_i (x.y_i)\}$ *where the infimum is taken over sequences* $y_i \to y$, *i.e. it suffices to consider the constant sequence at* x.

(2) *For* $x, y \in X$, $(x.y)_S = (x.y)$, *i.e.* $(\,.\,)_S$ *restricts to* $(\,.\,)$ *away from the boundary.*

Proof of (1). Let $x_i \to x$ and $y_i \to y$ be any pair of sequences. Since $x, x_i, y_i \in X$

$$\begin{aligned}
|(x_i.y_i) - (x.y_i)| =& \frac{1}{2}|d(x_i, w) + d(y_i, w) - d(x_i, y_i) \\
& - d(x, w) - d(y_i, w) + d(x, y_i)| \\
=& \frac{1}{2}|d(x_i, w) - d(x, w) + d(x, y_i) - d(x, x_i)| \\
\leq& \frac{1}{2}\{|d(x_i, w) - d(x, w)| + |d(x, y_i) - d(x_i, y_i)|\} \\
\leq& \frac{1}{2}\{d(x_i, x) + d(x, x_i)\} \\
=& d(x_i, x)
\end{aligned}$$

$$\Rightarrow \liminf_i (x_i.y_i) = \liminf_i (x.y_i).$$

Proof of (2). This follows from 4.5.1. □

Lemma 4.6.

(1) *If* $x, y \in \hat{X}$ *then*

$$(x.y)_S = \infty \Leftrightarrow x, y \in \partial X \text{ and } x = y.$$

(2) *If* $x \in \partial X$ *and* $\{x_i\}$ *is any sequence of points in* X *then*

$$(x_i.x)_S \to \infty \Leftrightarrow \{x_i\} \in S_\infty(X) \text{ and } x_i \to x.$$

(3) *If* $x, y \in \hat{X}$ *there are sequences* $\bar{x}_i \to x$, $\bar{y}_i \to y$ *with*

$$\lim_{i\to\infty} (\bar{x}_i.\bar{y}_i) = (x.y)_S$$

and if x *or* y *lies in* X *then the corresponding sequence can be chosen to be the constant sequence.*

(4) *If* $x, y \in \partial X$ *and* $x_i \to x$, $y_i \to y$ *then*

$$(x.y)_S \le \liminf_i (x_i.y_i) \le (x.y)_S + 2\delta.$$

(5) *If* $x, y, z \in \hat{X}$ *then*

$$(x.y)_S \ge \min\{(x.z)_S, (z.y)_S\} - \delta.$$

(6) *Let* $x, y \in \hat{X}$ *and* $y_i \to y$; *then*

$$\liminf_i (x.y_i)_S \ge (x.y)_S.$$

Proof of (1).

($\Leftarrow$): This follows from the definition of $(x.y)_S$.

($\Rightarrow$): First we show that $x, y \in \partial X$. Suppose, for example, that $x \in X$; then

$$(x.y)_S = \inf\{\liminf_i (x.y_i)\}$$

where the infimum is taken over sequences $y_i \to y$. But $(x.y_i) \le d(w, x) < \infty$. Thus $x \in X \Rightarrow (x.y)_S < \infty$.

Now suppose that $x, y \in \partial X$ with $a_i \to x$ and $b_i \to y$. Then

$$\liminf_i (a_i.b_i) \geq \inf_{x_i \to x,\, y_i \to y} \{\liminf_i (x_i.y_i)\} = \infty.$$

Thus

$$\lim_i (a_i.b_i) = \infty \text{ and } x = y.$$

Proof of (2). ($\Leftarrow$): Let $\{x_i\} \in S_\infty(X)$ with $x_i \to x$ and suppose that

$$\lim_i (x_i.x)_S \neq \infty.$$

Then the sequence of real numbers, $\{(x_i.x)_S\}$, has a bounded subsequence. Since $\{x_i\}$ is related to all its subsequences we can assume, by passage to a subsequence, that for some M, $(x_i.x)_S < M$, $\forall i$. Thus

$$\inf\{\liminf_j (x_i.y_j)\} < M,$$

where the infimum is taken over all sequences $y_j \to x$. Thus for each x_i, we can find a sequence $y_j^{(i)} \to x$ as $j \to \infty$ satisfying $(x_i.y_j^{(i)}) < M$, $\forall j$. Since X is δ-hyperbolic we also have

$$(x_i.y_j^{(i)}) \geq \min\{(x_i.x_k), (x_k.y_j^{(i)})\} - \delta.$$

This gives a contradiction because the quantity $(x_i.x_k)$ can be made arbitrarily large since $\{x_i\} \in S_\infty(X)$, and the quantity $(x_k.y_j^{(i)})$ can be made arbitrarily large since $\{x_k\}R\{y_j^{(i)}\}$.

($\Rightarrow$): First we show that $\{x_i\} \in S_\infty(X)$. Suppose that $(x_i.x)_S \to \infty$. Then there is a sequence $\{n_i\}$ of real numbers with

$$\lim_{i \to \infty} n_i = \infty$$

such that, for all i,

$$\inf\{\liminf_j (x_i.y_j)\} > n_i.$$

So, for each i, there is a sequence $y_j^{(i)} \to x$ as $j \to \infty$ satisfying $(y_j^{(i)}.x_i) > n_i$ for each j. Thus

$$\begin{aligned}
(x_i.x_j) &\geq \min\{(x_i.y_k^{(i)}), (y_k^{(i)}.x_j)\} - \delta \\
&\geq \min\{(x_i.y_k^{(i)}), (y_k^{(i)}.y_l^{(j)}), (y_l^{(j)}.x_j)\} - 2\delta \\
&\geq \min\{n_i, (y_k^{(i)}.y_l^{(j)}), n_j\} - 2\delta \\
&\geq \min\{n_i, n_j\} - 2\delta
\end{aligned}$$

for k and l large enough since $\{y_k^{(i)}\}R\{y_l^{(j)}\}$. Thus

$$\lim_{i,j\to\infty}(x_i.x_j)=\infty$$

and $\{x_i\}\in S_\infty(X)$.

Finally we show that $x_i\to x$. Assume $\{x_i\}\in S_\infty(X)$ with $(x_i.x)\to\infty$. Let $y_j^{(i)}\to x$ be as above and fix a specific sequence $z_i\to x$. Then

$$\begin{aligned}(x_i.z_i)&\geq\min\{(x_i.y_j^{(i)}),(y_j^{(i)}.z_i)\}-\delta\\&\geq\min\{(x_i.y_j^{(i)}),(y_j^{(i)}.z_j),(z_j.z_i)\}-2\delta\\&\geq\min\{(x_i.y_j^{(i)}),(z_j.z_i)\}-2\delta\end{aligned}$$

for j large enough since $\{y_j^{(i)}\}R\{z_j\}$. So

$$\lim_{i\to\infty}(x_i.z_i)=\infty$$

and $x_i\to x$.

Proof of (3). Let $x,y\in\partial X$ and let

$$A=(x.y)_S=\inf\{\liminf_i(x_i.y_i)\}.$$

Then for each $n\in\mathbb{Z}_+$ there are sequences $x_i^{(n)}\to x$ and $y_i^{(n)}\to y$ satisfying

$$A\leq\liminf_i(x_i^{(n)}.y_i^{(n)})\leq A+\frac{1}{n}.$$

By passing to subsequences we can assume that, for each i,

$$A\leq(x_i^{(n)}.y_i^{(n)})\leq A+\frac{1}{n}.$$

Now $x_i^{(n)}\to x$ and $y_i^{(n)}\to y$, so, by 4.6.2,

$$(x_i^{(n)}.x)\to\infty\text{ and }(y_i^{(n)}.y)\to\infty\text{ as }i\to\infty.$$

So, for each n, there is an i_n such that $(x_{i_n}^{(n)}.x)>n$ and $(y_{i_n}^{(n)}.y)>n$. Define $\bar{x}_i=x_{i_n}^{(n)}$ and $\bar{y}_n=y_{i_n}^{(n)}$. By 4.6.2, $\bar{x}_n\to x$ and $\bar{y}_n\to y$ and since $(\bar{x}_n.\bar{y}_n)=(x_{i_n}^{(n)}.y_{i_n}^{(n)})$

$$A\leq(\bar{x}_n.\bar{y}_n)\leq A+\frac{1}{n}.$$

If one of the points, say x, actually lies in X, then the proof is similar but uses 4.5.1 and a constant sequence for x. If both x and y lie in X constant sequences can be used for both.

Proof of (4). It follows from the definition of $(x.y)_S$ that $(x.y)_S \leq \liminf(x_i.y_i)$.

To get the other inequality we observe that, by 4.6.3, there are sequences $\bar{x}_i \to x$ and $\bar{y}_i \to y$ satisfying

$$\lim_{i\to\infty}(\bar{x}_i.\bar{y}_i) = (x.y)_S$$

so that

$$(\bar{x}_i.\bar{y}_i) \geq \min\{(\bar{x}_i.x_i), (x_i.y_i), (y_i.\bar{y}_i)\} - 2\delta.$$

Now $(\bar{x}_i.x_i) \to \infty$ and $(y_i.\bar{y}_i) \to \infty$, so, taking lim inf of both sides gives

$$(x.y)_S = \liminf(\bar{x}_i.\bar{y}_i) \geq \liminf(x_i.y_i) - 2\delta.$$

Proof of (5). By 4.6.3 there are sequences $x_i \to x$ and $y_i \to y$ satisfying

$$\lim_{i\to\infty}(x_i.y_i) = (x.y)_S.$$

Now, if $z_i \to z$ is any sequence converging to z,

$$(x_i.y_i) \geq \min\{(x_i.z_i), (z_i.y_i)\} - \delta.$$

Taking the lim inf of both sides gives

$$\begin{aligned}(x.y)_S &\geq \min\{\liminf(x_i.z_i), \liminf(z_i.y_i)\} - \delta \\ &\geq \min\{(x.z)_S, (z.y)_S\} - \delta\end{aligned}$$

by 4.6.4.

Proof of (6). Suppose that $y_i \to y$ satisfies

$$\liminf_i(x.y_i)_S < (x.y)_S - \epsilon < (x.y)_S.$$

By passing to a subsequence we can assume that $(x.y_i)_S < (x.y)_S - \epsilon$. Then for each i, there is a sequence $x_j^{(i)} \to x$ with

$$\liminf_j(x_j^{(i)}.y_i) < (x.y)_S - \epsilon$$

Again passing to subsequences, we can assume that

$$(x_j^{(i)}.y_i) < (x.y)_S - \epsilon.$$

Since $x_j^{(i)} \to x$ we can choose, for each i, a j_i satisfying $(x_{j_i}^{(i)}.x) > i$. This allows us to define a sequence $\bar{x}_i = x_{j_i}^{(i)}$. Then $\bar{x}_i \to x$ by 4.6.2 and

$$\liminf_i (\bar{x}_i.y_i) \le (x.y)_S - \epsilon < (x.y)_S$$

contradicting 4.6.4. □

Note: As the extended inner product $(\,,\,)_S$, has all these properties, we shall henceforth drop the suffix, S.

We now propose a basis of open sets for a topology on $\hat{X}$.

Definition 4.7: Let $\mathcal{B}$ be the collection of subsets of $\hat{X}$ consisting of

(1) the usual basis for the metric topology on X, i.e. open neighbourhoods $B_r(x) = \{y \in X, d(x,y) < r\}$, for each $x \in X$ and $r > 0$, and
(2) all sets of the form $N_{x,k} = \{y \in \hat{X} \mid (x.y) > k\}$, for each $x \in \partial X$ and $k > 0$.

Proposition 4.8. *The set $\mathcal{B}$ is a basis for a topology.*

Proof. We need to show that $\mathcal{B}$ satisfies the two requirements for a basis, i.e. that the elements of $\mathcal{B}$ form a cover of $\hat{X}$ and that if $B_1, B_2 \in \mathcal{B}$ with $y \in B_1 \cap B_2$ then there is a $B_3 \in \mathcal{B}$ with $y \in B_3 \subset B_1 \cap B_2$.

$\mathcal{B}$ has been chosen so that the first requirement is automatically satisfied. The proof that $\mathcal{B}$ satisfies the second condition breaks up into cases according to the types of the two basis elements.

Case 1: If B_1 and B_2 are both of type (1) the proof is the usual metric space proof.

Case 2: If one of the neighbourhoods is of type (1) and one is of type (2) then the neighbourhood B_3 will be of type (1). We need to show that if $y \in B(x,\epsilon) \cap N_{z,k}$ then there is an ϵ' satisfying $B(y,\epsilon') \subset B(x,\epsilon) \cap N_{z,k}$.

Since $y \in B(x,\epsilon)$ there is an ϵ_1 satisfying $B(y,\epsilon_1) \subset B(x,\epsilon)$. On the other hand, $y \in N_{z,k}$ means that $(z.y) > k$, so there is an ϵ_2 with $(z.y) > k + \epsilon_2 > k$. Set $\epsilon' = \min\{\epsilon_1, \epsilon_2\}$.

It follows that $B(y,\epsilon') \subset B(x,\epsilon)$. To see that $B(y,\epsilon') \subset N_{z,k}$ let $p \in B(y,\epsilon')$ and apply 4.6.3 to give a sequence $z_i \to z$ satisfying

$$\lim_{i\to\infty}(z_i.p) = (z.p).$$

As in the proof of 4.5.1,

$$|(z_i.p) - (z_i.y)| \le d(p,y) < \epsilon' \le \epsilon_2,$$

so that

$$-\epsilon_2 < (z_i.p) - (z_i.y) < \epsilon_2.$$

Applying lim inf to both sides gives

$$-\epsilon_2 \le \liminf_i (z_i.p) - \liminf_i (z_i.y) \le \epsilon_2$$

$$\Rightarrow -\epsilon_2 \le (z.p) - \liminf_i (z_i.y) \le \epsilon_2.$$

So by 4.5.1,

$$-\epsilon_2 \le (z.p) - (z.y).$$

$$\begin{aligned}\Rightarrow (z.p) &= (z.p) - (z.y) + (z.y)\\ &> (z.p) - (z.y) + k + \epsilon_2\\ &\ge -\epsilon_2 + k + \epsilon_2\\ &= k.\end{aligned}$$

Case 3: If both neighbourhoods are of type (2), we need to show that if

$$y \in N_{x,k} \cap N_{x',k'}$$

then there is a $B_3 \in \mathcal{B}$ with

$$y \in B_3 \subset N_{x,k} \cap N_{x',k'}.$$

If $y \in X$ then this B_3 will be of type (1). By Case 2, there is an ϵ_1 with $B(y,\epsilon_1) \subset N_{x,k}$ and an ϵ_2 with $B(y,\epsilon_2) \subset N_{x',k'}$. Letting $\epsilon = \min\{\epsilon_1,\epsilon_2\}$ we get

$$B(y,\epsilon) \subset N_{x,k} \cap N_{x',k'}.$$

If $y \in \partial X$ then the neighbourhood B_3 will be of type (2). Since $N_{y,r_1} \subset N_{y,r_2}$ for $r_1 > r_2$ it suffices to show that, if $x \in \partial X$ and $y \in \partial X \cap N_{x,k}$, with $y \neq x$, then there is an m satisfying $N_{y,m} \subset N_{x,k}$.

Suppose that this is not the case. Then for each m there is a

$$y_m \in (N_{y,m}) \cap (\hat{X} - N_{x,k}),$$

i.e. $(y.y_m) > m$ while $(x.y_m) \leq k$. We consider two subcases:

3a) A subsequence of the y_m's actually lies in X. By passing to this subsequence we get $(y_m.y) > m$, $(y_m.x) \leq k$ and $y_m \in X$. Thus by 4.6.2, $y_m \to y$, giving a contradiction since

$$k < (x.y) \leq \liminf_m (x.y_m) \leq k,$$

where the second inequality follows from 4.6.6.

3b) Suppose only a finite number of the y_m's lie in X. By passing to a subsequence we can assume all the y_m's lie on ∂X.

There is a k' such that

$$k < k' < (x.y) < \infty, \text{ while } (x.y_m) \leq k < k'$$

and $(y_m.y) > m$. By 4.6.3 we can choose, for each m, a pair of sequences $x_i^{(m)} \to x$ and $y_i^{(m)} \to y_m$ satisfying

$$\lim_i (x_i^{(m)}.y_i^{(m)}) = (x.y_m) \leq k < k'.$$

By excluding a finite number of terms, we can further assume that, for each m and for each i, $(x_i^{(m)}.y_i^{(m)}) < k'$. Since $x_i^{(m)} \to x$ we can require that

$$(*) \qquad (x_i^{(m)}.x) > m,$$

for each i and m.

Similarily, since, by 4.6.6,

$$\liminf_i (y_i^{(m)}.y) \geq (y_m.y) > m$$

we can require that

$$(**) \qquad (y_i^{(m)}.y) > m,$$

for each i and m. Now we define $\bar{x}_i = x_i^{(i)}$ and $\bar{y}_i = y_i^{(i)}$. Equations $(*)$ and $(**)$ together with 4.6.2 give $\bar{x}_i \to x$ and $\bar{y}_i \to y$. This gives a contradiction since

$$\begin{aligned} k' < (x.y) &= \inf\{\liminf_i (x_i.y_i)\} \\ &\leq \liminf_i (\bar{x}_i.\bar{y}_i) \\ &\leq k'. \quad \square \end{aligned}$$

<u>Note</u>: With this topology on $\hat{X}$, the inner product

$$(\, . \,) : \hat{X} \times \hat{X} \to \mathbb{R}$$

is not continuous. For suppose that $x, y \in \partial X$ and we have a pair of sequences $x_i \to x$ and $y_i \to y$. Then, by 4.6.2, $(x_i.x) \to \infty$ so that

$$\{x_i\} \cap N_{x,k} \neq \emptyset \text{ for all } k.$$

Thus $\{x_i\}$ converges to x in the topological sense. Similarily, $\{y_i\}$ converges to y in the topological sense. If (.) were continuous then we would have

$$(x_i.y_i) \to (x.y).$$

However, as we have seen in the examples above, $\{(x_i.y_i)\}_i$ need not converge and even $\liminf(x_i.y_i)$ may not be $(x.y)$.

The Rips complex

In this section we describe Rips' construction of a complex on which a word hyperbolic group acts in an specially nice way, allowing one to deduce properties of the group. More details on the construction, and some extensions of the methods to automatic groups are given in J. Alonso's preprint [A] (see also [CDP, chap.5]).

Theorem 4.11. (Rips) *A word hyperbolic group G acts simplicially on a simplicial complex P satisfying:*

(1) *P is contractible, locally finite and finite dimensional;*
(2) *on the vertices of P, G acts freely and transitively;*
(3) *the quotient complex P/G is compact.*

Before giving the construction required to prove the theorem, we state several corollaries:

Corollary 4.12. *Let G be a word hyperbolic group.*

(1) *G is finitely presented and is of type FP_∞ (i.e. $\mathbb{Z}$ has a free $\mathbb{Z}G$ resolution of finite type).*
(2) *The fact the action is free on the set of vertices means that the stabilizer of each simplex is finite.*
(3) *If G is torsion-free, then P/G is a finite $K(G,1)$, and G is of type FL, and has finite cohomological dimension (written $cd(G) < \infty$).*
(4) *If G is virtually torsion-free, then the virtual cohomological dimension is finite ($vcd(G) < \infty$).*
(5) *$H_\star(G;\mathbb{Q})$ and $H^\star(G;\mathbb{Q})$ are finite dimensional.*

Proof of Corollary. (1)
To prove this, we quote the following theorem

Theorem. (K. Brown, [Br2]) *Let X be a contractible G-complex such that the stabilizer of each cell is finitely presented and of type FP_∞. Suppose that X has a filtration by G-equivariant subcomplexes $\{X_j\}_{j\geq 1}$ such that each X_j is finite mod G. If in addition, for each j there is a j' such that the inclusion maps induce trivial maps $\pi_1(X_j) \to \pi_1(X_{j'})$ and $\widetilde{H}_i(X_j) \to \widetilde{H}_i(X_{j'})$, then G is finitely presented and of type FP_∞.*

The filtration used in this context is the sequence of complexes P_d defined below, and a variation of the slight variation of the argument given in the proof of 4.14 shows that every finite subcomplex of P_d collapses in some $P_{d'}$. (The details are given in [A].)

(2) Let S be the stabilizer of the simplex $\{g_1, \ldots, g_n\}$. Define a map from S to the set of permutations of n letters by simply recording the action of each $s \in S$ on the g_i. By (2) in Rips' Theorem, this map is injective so that S is finite.
(3) and (4) This follows from (1). Serre has shown that if G is virtually torsion-free, then $vcd(G) < \infty$ if and only if G acts on some contractible, finite dimensional CW complex with finite stabilizers.
(5) A spectral sequence argument shows that $H_*(G;\mathbb{Q}) \equiv H_*(P/G;\mathbb{Q})$ and $H^*(G;\mathbb{Q}) \equiv H^*(P/G;\mathbb{Q})$. □

We now build the complex P_d. The standard resolution of a group G is obtained by taking Y to be the simplex spanned by G. This has a vertex for each element of G, and a simplex for each finite subset of G.

Definition 4.13 Let X be a metric space, and let d be a positive real. We define the simplicial complex $P_d(X)$ to have a vertex for each point in X, and a simplex for each finite subset of X which has diameter at most d.

Proposition 4.14. *Let G be a finitely generated group and let Γ be the Cayley gaph of G with respect to some finite set of generators, regarded as a metric space in the usual way. If Γ is a δ-hyperbolic metric space, then $P_d(\Gamma)$ is contractible for $d \geq 4\delta + 1$.*

Proof. As $P_d(\Gamma)$ is simplicial, it is contractible if and only if $\pi_i(P_d(\Gamma)) = 0$ for all $i \geq 0$. It is sufficient to prove that every finite subcomplex K of $P_d(\Gamma)$ is contractible. Choose the identity element x_0 as base point for G.

Case (i) $\max\limits_{y \in K^0} d(x_0, y) \leq d/2$.

Then K lies inside a simplex of $P_d(\Gamma)$, and so is contractible.

Case (ii) $\max\limits_{y \in K^0} d(x_0, y) > d/2$.

Let y_0 be the point in K^0 furthest from x_0. Let y_0' be the point on a geodesic from x_0 to y_0 such that $d(y_0', x_0) = d(y_0, x_0) - [d/2]$, where $[r]$ denotes the integral part of r. We now define a function

$$f : K^0 \to P_d(\Gamma) \text{ by } f(y_0) = y_0', \; f(y) = y, \; y \in K^0 - \{y_0\}.$$

Claim: f can be extended to a simplicial map $K \to P_d(\Gamma)$.

We must show that whenever σ is a simplex in K, $f(\sigma)$ is a simplex in $P_d(\Gamma)$. But simplices of $P_d(\Gamma)$ consist of sets of elements of G of diameter at most d, so we must show that

$$(\star) \qquad \forall y \in K^0, \; d(y, y_0) \leq d \Rightarrow d(y, y_0') \leq d$$

Recall that (see Corollary 2.4) Γ is δ-hyperbolic if and only if for all $x, y, z, t \in \Gamma$,

$$d(x,y) + d(z,t) \leq \max\{d(x,z) + d(y,t), d(x,t) + d(y,z)\} + 2\delta.$$

Replacing (x, y, z, t) by (y, y_0', y_0, x_0), we get

$$d(y,y_0') + d(x_0,y_0) \leq \max\{d(y,y_0) + d(x_0,y_0'), d(y_0,y_0') + d(x_0,y)\} + 2\delta$$

$$\begin{aligned}
&\Leftrightarrow d(y,y_0') \leq \\
&\max\{d(y,y_0) + d(x_0,y_0') - d(x_0,y_0), d(y_0',y_0) + d(y,x_0) - d(x_0,y_0)\} + 2\delta \\
&\leq \max\{d - [d/2], d/2\} + 2\delta \leq d - [d/2] + 2\delta.
\end{aligned}$$

This is $\leq d$ when $d \geq 4\delta + 1$, as required.

It remains to show that f is homotopic to the inclusion map; but this follows immediately by noticing that $f(\sigma) \cup \sigma$ is contained in a simplex of $P_d(\Gamma)$, as $d(y_0, y_0') \leq d$.

References

[A] J. M. Alonso, *Combings of groups*, MSRI preprint # 04623–89, Proceedings of workshop on algorithmic problems, MSRI, January 1989, Springer-Verlag, MSRI series, edited by G. Baumslag and C.F. Miller III.

[A2] J. M. Alonso, *Inégalités isopérimétriques et quasi-isométries*, to appear C.R.A.S..

[BF] M. Bestvina and M. Feighn, *A combination theorem for negatively curved groups*, preprint Jan 1990.

[BM] M. Bestvina and G. Mess, *Local connectivity and the boundary tively curved groups*, preprint.

[BGSS] G. Baumslag, S. M. Gersten, M. Shapiro and H. Short, *Automatic groups and amalgams*, in preparation.

[BR] G. Baumslag and J. Roseblade, *Subgroups of direct products of free groups*, J. Lond. Math. Soc. **30** (1984), 44–52.

[Bow] B. Bowditch, *Notes on Gromov's hyperbolicity criterion*, this volume.

[Br] K. S. Brown, *Cohomology of groups*, Springer-Verlag, 1989.

[Br2] K.S. Brown, *Finiteness properties of groups*, J. Pure Appl. Algebra **44** (1987), 45–75.

[Can] J. W. Cannon, *The combinatorial structure of cocompact discrete hyperbolic groups*, Geom Dedicata **16** (1984), 123–148.

[Can2] J. W. Cannon, *Lectures at Trieste*, Proceedings of ICTP meeting 1989, Trieste.

[C] D. Cooper, *Automorphisms of negatively curved groups*, preprint.

[CDP] M. Coornaert, T. Delzant and A. Papadopoulos, *Notes sur les groupes hyerboliques de Gromov*, Lecture Notes in Mathematics, #1441, Springer-Verlag, 1990.

[CEHPT] annon, D. B. A. Epstein, D. F. Holt, M. S. Paterson and W. P. Thurston, *Word processing and group theory*, preprint.

[DA] B. Domanski and M. Anshel, *The complexity of Dehn's algorithm for word problems in jour J. of Algebra.*

[De] M. Dehn, *Papers on Group Theory and Topology*, translated and introduced by J. Stillwell, Springer-Verlag, 1987.

[G] M. Gromov, *Hyperbolic Groups*, Essays in group theory, MSRI series vol. 8, edited by S. M. Gersten, Springer-Verlag, pp. 75–263.

[Ge] S.M. Gersten Dehn functions and ℓ_1-norms of finite presentations, Proceedings of workshop on algorithmic problems, MSRI, January 1989, Springer-Verlag, MSRI series, edited by G. Baumslag and C.F. Miller III.

[Gh] E. Ghys, *Les groupes hyperboliques*, Séminaire Bourbaki, 722, S.M.F..

[GH] E. Ghys and P. de la Harpe (editors), *Sur les groupes hyperboliques d'aprés Mikhael Gromov*, Birkhäuser, Progress in Mathematics series, vol. 83, 1990.

[GS] S. M. Gersten and H. Short, *Small cancellation theory and Automatic jour Invent. Math.*.

[GS2] S. M. Gersten and H. Short, *Rational subgroups of biautomatic groups*, to appear Annals of Math.

[HU] J. E. Hopcroft and J. D. Ullman, *Formal languages and their relation to automata*, Addison Wesley, 1969.

[L] I. G. Lysenok, *On some algorithmic properties oh hyperbolic groups*, Izv. Akad. Nauk. SSSR Ser. Math. **53 (4)** (1989); English transl. in Math. USSR Izv. **35** (1990), 145–163.

[LS] R. C. Lyndon and P. E. Schupp, *Combinatorial group theory*, Springer-Verlag, 1977.

[MKS] W. Magnus, A. Karrass and D. Solitar, *Combinatorial group theory*, Dover, 1976.

[MS] Muller and P. E. Schupp, *Groups, the theory of ends and context-free languages*, J. of Computer and System Science **26** (1983), 295–310.

[P1] F. Paulin, *Points fixes d'automorphismes de groupes hyperboliques*, Annales de l'Institut Fourier **39** (1989), 651–662.

[P2] F. Paulin, *Outer automorphisms of hyperbolic groups and small actions on* $\mathbb{R}$*-trees*, to appear in "Proceedings of the Workshop on Arboreal Group Theory, M.S.R.I, Sept. 1988", Springer-Verlag M.S.R.I. series.

[R] E. Rips, *Subgroups of small cancellation groups*, Bull. London Math. Soc. **14** (1982), 45–47.

[S] H. Short, *Quasiconvexity and a theorem of Howson's*, This volume.

[Se] J.-P. Serre, *Cohomologie des groupes discrets*, Prospects in Mathematics,, Annals of Math Studies No. 70, 1971, pp. 77–169.

J.M. Alonso : Stockholms Universitet, Matematiska Institutionen, Box 6701, S–113 85, Stockholm, Sweden; alonso@matematik.su.se.

T. Brady: Department of Mathematics, University of Michigan, Ann Arbor, Michigan 48109; tbrady@math.lsa.umich.edu

D. Cooper: Department of Mathematics, University of California at Santa Barbara, California 93106; cooper%ori@sbitp.ucsb.bitnet

V. Ferlini: Department of Mathematics, Union College, Schenectady, New York 12308; ferliniv@gar.union.edu

M. Lustig : Math. Institut, Ruhr Universität, (D-4630) Bochum, Germany; p151210@dborub01.bitnet

M. Mihalik : Department of Mathematics, Vanderbilt University, Nashville, Tennessee 37235; mihalikm@vuctrvax.bitnet

M. Shapiro : Department of Mathematics, Ohio State University, 231 West 18th Avenue, Columbus OH 43210; rev@function.mps.ohio-state.edu.

H. Short: Mathematics Department, City College, CUNY, Convent Avenue at 138th Street, New York 10031; mathhs@ccnyvme.bitnet.

Notes on Gromov's hyperbolicity criterion for path-metric spaces.

B. H. Bowditch
I.H.E.S., 35 route de Chartres,
91440 Bures-sur-Yvette, France.

Abstract.

We give a survey of various properties of path-metric spaces satisfying the hyperbolicity criterion defined by Gromov. In particular we show the equivalence of various formulations of this criterion.

0. Introduction.

In [G], Gromov describes the notion of a "δ-hyperbolic" (or what we shall call "almost-hyperbolic") path-metric space. With one simple axiom, (essentially property H1 described here,) he is able to capture a remarkable number of the global properties of a "negatively curved" space, or more specifically, a simply-connected Riemannian manifold, all of whose sectional curvatures are less than some negative constant.

In his paper, Gromov is primarily concerned with developing the properties of a "hyperbolic group", i.e. one with an almost-hyperbolic Cayley graph. In other words he deals primarily with almost-hyperbolic spaces which admit cocompact groups of isometries. However, the group structure is largely irrelevant to understanding the global geometry of these spaces. The aim of the present paper is to give a more detailed exposition of Gromov's criterion purely in the context of path-metric spaces. Of course much of Gromov's paper will not be touched upon here. Other expositions of Gromov's work are [ABCDFLMSS], [BGHHSST] and [CDP].

One of the main aims of this paper is to show the equivalence of a number of different characterisations of almost-hyperbolicity. This will be achieved in the first six chapters. We shall not always take the most direct route in this, but get involved in discussions of spanning trees, convexity, isoperimetric inequalities and pseudoisometries etc. All the arguments of this paper are elementary. In particular, we make no use of Riemannian geometry, except by way of example.

The structure of this paper, in outline, is as follows. Chapter 1 describes the main terms and conventions we will be using. In Chapter 2, we discuss in more detail our main definitions of almost hyperbolicity, H1–H5. In chapter 3, we describe the "treelike" nature of almost-hyperbolic spaces, and show the equivalence of the first two definitions. In Chapter 4, we define "almost-convex" sets, and develop a few of their properties. We give proofs of H1⇒H4 and H1⇒H5. We also show that almost-hyperbolicity is a pseudoisometric invariant. It is this fact that allows one to define the notion of an almost- hyperbolic group. Chapter 5 is devoted to developing a notion of "area" that seems appropriate to the context of path-metric spaces. That H3⇒H4 follows easily from this. In Chapter 6, we give the remaining proofs, H4⇒H2 and H5⇒H3. Chapter 7 explores further the treelike nature of almost-hyperbolic space. In Chapter 8, we give a proof that the property of almost-hyperbolicity "propagates", that is, a path-metric space which is simply connected (in some sense) and almost-hyperbolic on a large scale is globally almost-hyperbolic.

This work was initiated by a series of seminars given at Warwick, and I am indebted to the other participants (in particular, David Epstein, Oliver Goodman, Greg McShane, Caroline Series) for their contribution. The first six chapters were completed at Warwick with the support of an S.E.R.C. fellowship. The remainder of the paper was produced at I.H.E.S. under a Royal Society European Exchange Fellowship.

CHAPTER 1 : Definitions.

1.1. Geodesic spaces.

Let $(\mathcal{S}, d)$ be a metric space. To say that d is a *path metric* means that, given any two points $X, Y \in \mathcal{S}$ and $\epsilon > 0$, there is a rectifiable path joining X and Y of length at most $d(X,Y) + \epsilon$. We then call $(\mathcal{S}, d)$ a *path-metric space*. Such spaces seem to be the natural context in which to speak of almost-hyperbolicity. However, to save ourselves a few unnecessary complications, we shall usually make an additional assumption.

Definition : A *geodesic space* is a complete, locally-compact path metric space.

It is an exercise to show that any closed uniform ball $\{X \in \mathcal{S} \mid d(X,Y) \leq r\}$ in a geodesic space $(\mathcal{S}, d)$ is compact. The following are easy consequences.

(1) If $X, Y \in \mathcal{S}$, then there is at least one path from X to Y of length equal to $d(X,Y)$. Such a path is called a *geodesic*.

(2) If $X \in \mathcal{S}$, and $Q \subseteq \mathcal{S}$ is any non-empty closed set, then there is some $Y \in Q$ with $d(X,Y) = d(X,Q)$, where $d(X,Q) = \inf_{Z \in Q} d(X,Z)$. We shall write $\text{proj}_Q(X) = \{Y \in Q \mid d(X,Y) = d(X,Q)\}$. Each such Y is called a *projection* of X to Q.

Given any closed set $Q \subseteq \mathcal{S}$, we shall write $\breve{Q}$ for its topological interior, and $\partial Q = Q \setminus \breve{Q}$ for its topological boundary. We may check that if $X \notin \breve{Q}$, then $\text{proj}_Q X \subseteq \partial Q$. If $r \geq 0$, we write $N_r(Q) = \{X \in \mathcal{S} \mid d(X,Q) \leq r\}$ for the uniform r-neighbourhood of Q. Note that $N_{r+s}(Q) = N_r(N_s(Q))$. We write $\breve{N}_r(Q)$ for the topological interior of $N_r(Q)$, and $\partial N_r(Q) = N_r(Q) \setminus \breve{N}_r(Q)$. Thus

$$\{X \in \mathcal{S} \mid d(X,Q) < r\} \subseteq \breve{N}_r(Q)$$

and

$$\partial N_r(Q) \subseteq \{X \in \mathcal{S} \mid d(X,Q) = r\}.$$

Any closed subset of a geodesic space will itself be a geodesic space in the induced path-metric, provided that we allow for the possibility that two points be an infinite distance apart if they cannot br joined by a rectifiable path. Given a closed subset Q of $(\mathcal{S}, d)$, we shall write $d_{0,Q}$ for the induced path-metric on $\mathcal{S} \setminus \breve{Q}$. More generally, we shall write $d_{r,Q}$ for the induced path-metric on $\mathcal{S} \setminus \breve{N}_r(Q)$.

Given two points X, Y in a geodesic space $(\mathcal{S}, d)$, we shall use $[X, Y]$ to denote some choice of geodesic from X to Y. We shall only be using this notation in the case where $(\mathcal{S}, d)$ is almost-hyperbolic. For such spaces, any two geodesics with the same endpoints remain a bounded distance apart. Once this has been established, we can afford to be a little careless in the use of this notation. For example, we will speak as though $[X,Y]$ were a well-defined object, even though it implies making a choice.

The above discussion applies only when $(\mathcal{S},d)$ is a geodesic space. However all the results of this paper may be interpreted for general path-metric spaces, with simple modification. For example, we could interpret a "geodesic" from X to Y as a rectifiable path of length at most $d(X,Y)+1$, or $\text{proj}_Q(X)$ as the set of points Y in Q satisfying $d(X,Y) \leq d(X,Q)+1$, and so on. Working with such things, however, would only confuse the exposition.

1.2. Almost-hyperbolicity.

We are now in a position to define the notion of almost-hyperbolicity. We give five definitions: H1–H5. For ease of reference, we collect together these definitions below. We shall discuss them in more detail in Chapter 2.

Let $(\mathcal{S},d)$ be a geodesic space. Let $k_i, h_i \in [0,\infty)$.

Definition 1 : $(\mathcal{S},d)$ is k_1-H1 if:
Given any four points $X,Y,Z,W \in \mathcal{S}$, at least one of $XY:ZW$, $XZ:YW$ or $XW:YZ$ holds, where $AB:CD$ is the statement that

$$\begin{aligned} d(A,B)+d(C,D) &\leq \max\big(d(A,C)+d(B,D), d(A,D)+d(B,C)\big)+k_1 \\ &\leq \min\big(d(A,C)+d(B,D), d(A,D)+d(B,C)\big)+2k_1. \end{aligned}$$

Definition 2 : $(\mathcal{S},d)$ is k_2-H2 if the following holds:
Suppose $X_1,X_2,X_3 \in \mathcal{S}$. Suppose α_i is any geodesic joining X_i to X_{i+1} (taking subscripts mod 3). Then,

$$N_{k_2}(\alpha_1) \cap N_{k_2}(\alpha_2) \cap N_{k_2}(\alpha_3) \neq \emptyset.$$

Definition 3 : $(\mathcal{S},d)$ is (k_3,h_3)-H3 if the following holds:
Suppose that σ is a path-metric on the circle S^1, and that $\gamma : (S^1,\sigma) \longrightarrow (\mathcal{S},d)$ is a distance non-increasing map. Then, there exist,

(i) a cellulation P of the unit disc D,

(ii) a path-metric ρ on the 1-skeleton Σ of P,

(iii) a distance non-increasing map $f : (\Sigma,\rho) \longrightarrow (\mathcal{S},d)$, and

(iv) a distance non-increasing map $\partial f : (\partial D, \rho_{\partial D}) \longrightarrow (S^1,\sigma)$, where $\rho_{\partial D}$ is the path-metric on ∂D induced from ρ,

such that,

(i) $f|\partial D = \gamma \circ \partial f$,
(ii) ∂f has topological degree 1, and
(iii) we have

$$\sum_{c \in C_2(P)} (\rho(\partial c))^2 \leq k_3(\sigma(S^1) + h_3)$$

where $C_2(P)$ is the set of 2-cells of P, $\rho(\partial c)$ is the ρ-length of the boundary ∂c of c, and $\sigma(S^1)$ is the total σ-length of S^1.

Definition 4 : $(\mathcal{S}, d)$ is (k_4, h_4)-H4 if:
Given any geodesic segment $\alpha \subseteq \mathcal{S}$, and $X, Y \in \partial N_{k_4}(\alpha)$, we have

$$d_{k_4,\alpha}(X,Y) \geq 3d(X,Y) - h_4,$$

where $d_{k_4,\alpha}$ is the induced path-metric on $\mathcal{S} \backslash \breve{N}_{k_4}(\alpha)$.

Definition 5 : $(\mathcal{S}, d)$ is (k_5, h_5)-H5 if the following holds:
Given any $A, X, Y \in \mathcal{S}$, with $d(A,X) = d(A,Y)$ and $d(X,Y) \geq k_5$, then we have $d_{r,A}(X,Y) \geq d(X,Y) + h_5$, where $r = d(A,X) = d(A,Y)$, and $d_{r,A}$ is the induced path-metric on $\mathcal{S} \backslash \breve{N}_r(A)$.

We shall show that these definitions are equivalent in the following effective sense. Given $i, j \in \{1,2,3,4,5\}$, and $\underline{k} \in [0,\infty) \sqcup [0,\infty)^2$, there is some $\underline{k}' \in [0,\infty) \sqcup [0,\infty)^2$ such that if $(\mathcal{S}, d)$ is $\underline{k}$-H(i), then it is $\underline{k}'$-H(j).

The cycle of proofs will be:

We include H1 $\Rightarrow$ H2, H1 $\Rightarrow$ H3, and H1 $\Rightarrow$ H4, since they are much more direct than following the cycle.

Definition : We say that $(\mathcal{S}, d)$ is *almost-hyperbolic* if it is $\underline{k}$-H(i) for some $\underline{k}$ and i.

We shall call the constants k_1, k_2, k_3, h_3, k_4, h_4, k_5, h_5, appearing in the definitions, *parameters* of hyperbolicity. Note that all can be imagined as having the physical dimensions of length. Also, all except h_5 have the property that increasing them would weaken the definition. In fact, the quantities $-1/k_1^2$, $-1/k_2^2$, $-1/(\max(k_3, h_3))^2$ and $-1/(\max(k_4, h_4))^2$, can be thought of as a measure of the upper bound of the curvature, seen on a large scale. This can be made more precise for negatively curved

Riemannian manifolds. (A space of "infinite negative curvature" is a metric tree, see Chapter 3.)

There is also a sense in which the parameters measure the "coarseness" of a space, where coarseness may be due to local concentrations of positive curvature, or to topological holes, etc. This is only an intuitive picture, and we make no attempt to formally isolate these notions. (It is possible to give a more clumsy version of H5 which fits into this scheme, see Chapter 2.)

1.3. Pseudoisometries.

From the definition H3, it is immediate that the notion of almost-hyperbolicity is invariant under bilipschitz equivalence. However, the notion of bilipschitz equivalence is is too strong for these kinds of spaces. It demands, for example that the spaces under consideration be homeomorphic, which is an unnatural constraint. We would like, for example, that an almost hyperbolic space admitting a properly-discontinuous, cocompact isometry group, should be equivalent to the Cayley graph of that group. A more appropriate notion, therefore, is that of a "pseudoisometry" (elsewhere known as a "coarse quasiisometry").

Definition : Let $(\mathcal{S}, d)$ and $(\mathcal{S}', d')$ be path-metric spaces, and $\lambda_1 \geq 1$, $\lambda_2 \geq 0$. A (λ_1, λ_2)-*pseudoisometry* between $\mathcal{S}$ and $\mathcal{S}'$ is a relation $R \subseteq \mathcal{S} \times \mathcal{S}'$ such that

$$\forall (x \in \mathcal{S})\, \exists (x' \in \mathcal{S}')\, xRx'$$
$$\forall (x' \in \mathcal{S}')\, \exists (x \in \mathcal{S})\, xRx'$$

and if xRx' and yRy', then

$$\begin{aligned}\frac{1}{\lambda_1}\big(d(x,y) - \lambda_2\big) &\leq d'(x',y') \\ &\leq \lambda_1 d(x,y) + \lambda_2.\end{aligned}$$

We say that a relation is a *pseudoisometry* if it is a (λ_1, λ_2)-pseudoisometry for some λ_1 and λ_2. Pseudoisometry is thus an equivalence relation on path-metric spaces.

We shall show (Proposition 4.10) that almost-hyperbolicity is invariant under pseudoisometry.

Note that if we take two finite generating sets for the same group, then the corresponding Cayley graphs are pseudoisometric. It therefore makes sense to define a finitely-generated group as being almost-hyperbolic if its Cayley graph is almost hyperbolic, irrespective of the choice of generators. We see in fact that a group is almost hyperbolic if and only if it acts as a properly discontinuous cocompact isometry group on some almost hyperbolic space.

Another point to note is that any path-metric $(\mathcal{S},d)$ space is $(1+\epsilon,2)$-pseudoisometric to a 1-complex in which each edge has length 1. The construction is as follows. For $\delta>0$ sufficiently small, we take a maximal packing P of $\mathcal{S}$ by disjoint δ-balls. We form a metric 1-complex, (G,ρ) by taking the vertices to correspond to the balls of P, and joining two vertices by an edge of length 1, if the centres of the corresponding balls are distant at most 1 in $\mathcal{S}$. We define $R\subseteq\mathcal{S}\times G$ by $(x,y)\in R$ if and only if, for some vertex v of G, we have $d(x,p(v))\leq 2\delta$ and $\rho(y,v)\leq\frac{1}{2}$, where $p(v)\in\mathcal{S}$ is the centre of the ball corresponding to v. We may check that this is a $(1+\epsilon,2)$-pseudoisometry for $\epsilon=O(\delta)$. Since this construction applies to any path-metric space, we see that we see that our entire discussion of almost-hyperbolicity can, in principle, be given a purely combinatorial formulation.

1.4. Convention on inequalities.

We end this chapter by introducing a convention that will streamline our manipulation of inequalities, and, we hope, make our arguments conceptually simpler.

Suppose $K\geq 0$, and $x,y\in\mathbf{R}$. we shall write $x\simeq_K y$ to mean that $|x-y|\leq K$, and $x\preceq_K y$ to mean that $x\leq y+K$. Thus,

$$x\simeq_K y\simeq_K z\Rightarrow x\simeq_{2K} z$$

and

$$x\preceq_K y\preceq_K z\Rightarrow x\preceq_{2K} z.$$

Whenever we use this notation, K will be a funcion only of the parameters of hyperbolicity.

Usually, we shall drop the subscripts K, and behave as though the relations $\simeq$ and $\preceq$ were transitive. Thus, one should think of the constants involved as increasing with each application of the transitive law in an argument. We may always explicitly relate the constants produced at the end to the constants introduced at the beginning. Usually, however, keeping track in this way would only confuse the argument.

Given points X,Y in a path-metric space $(\mathcal{S},d)$, we shall write $X\sim Y$ to mean that $d(X,Y)\simeq 0$. Again, we shall act as though $\sim$ were an equivalence relation.

In some places (mainly Chapter 3) we shall abbreviate $d(X,Y)$ to XY.

CHAPTER 2 : Five definitions of almost-hyperbolicity.

We discuss in more detail the definitions of Section 1.2.

2.1. Definition 1.

With the conventions introduced in Section 1.4, we may rewrite Definition 1 in the following way. Given any four points $X, Y, Z, W \in \mathcal{S}$, we may partition them into two sets of two elements, without loss of generality $\{\{X,Y\},\{Z,W\}\}$, so that

$$d(X,Y) + d(Z,W) \preceq d(X,Z) + d(Y,W) \simeq d(Y,Z) + d(X,W).$$

We shall write $XY : ZW$ for this pair of inequalities. This definition is just a rephrasing of the central one given in [G].

Lemma 2.1.1, below, tells us that this hypothesis is equivalent to saying that the distances between X, Y, Z, W may be read off approximately (i.e. up to an additive constant) along a tree with a length assigned to each edge (Figure 2a).

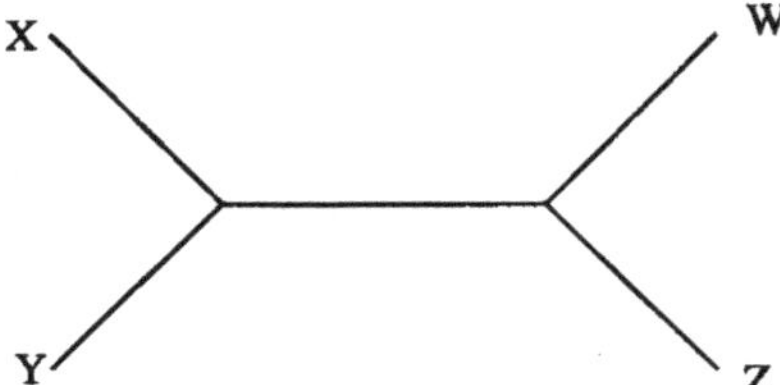

Figure 2a.

Note that there are three combinatorial possibilities for the tree, corresponding to the three possible partitions of $\{X, Y, Z, W\}$.

From this point of view, the axiom is analogous to the triangle inequalities of a metric space. Consider three points X, Y, Z in a metric space $(\mathcal{S}, d)$. The triangle inequalities tell us that we can find three non-negative numbers x, y, z such that

$$\begin{aligned} d(X,Y) &= x + y \\ d(Y,Z) &= y + z \\ d(Z,X) &= z + x, \end{aligned}$$

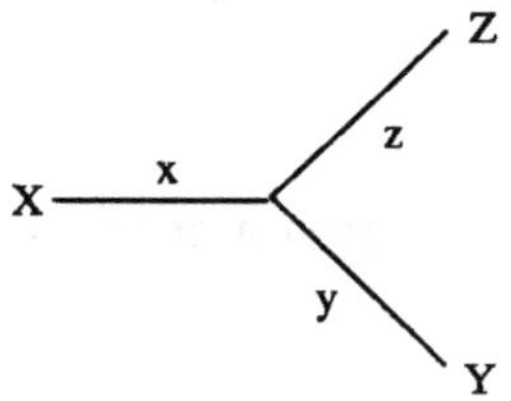

Figure 2b.

namely $x = \frac{1}{2}\big(d(Z,X) + d(X,Y) - d(Y,Z)\big)$ etc. In other words, the distances between X, Y and Z may be read off (precisely) from a tree with edges of length x, y and z. We shall write $(XYZ) \longleftrightarrow xyz$ to mean this (Figure 2b).

Lemma 2.1.1 : *Let X, Y, Z, W be four points in a metric space $(\mathcal{S}, d)$. Then, we have $XY : ZW$, that is*

$$d(X,Y) + d(Z,W) \preceq d(X,Z) + d(Y,W) \simeq d(Y,Z) + d(X,W),$$

if and only if there exist non-negative numbers x, y, z, w, u such that

$$\begin{aligned} d(X,Y) &\simeq x + y \\ d(Z,W) &\simeq z + w \\ d(X,W) &\simeq x + u + w \\ d(Y,Z) &\simeq y + u + z \\ d(X,Z) &\simeq x + u + z \\ d(Y,W) &\simeq y + u + w. \end{aligned}$$

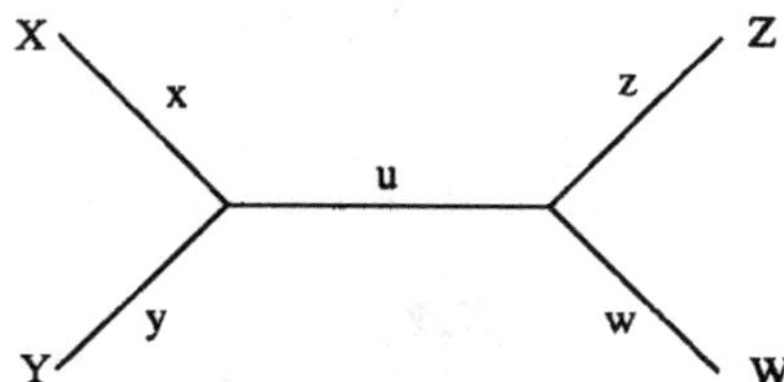

Figure 2c.

Proof : ($\Leftarrow$) is trivial. We prove ($\Rightarrow$).

Suppose we have $XY : ZW$. We find trees $(XYZ) \longleftrightarrow xya$ and $(YZW) \longleftrightarrow bzw$ as described above. (Figure 2d.)

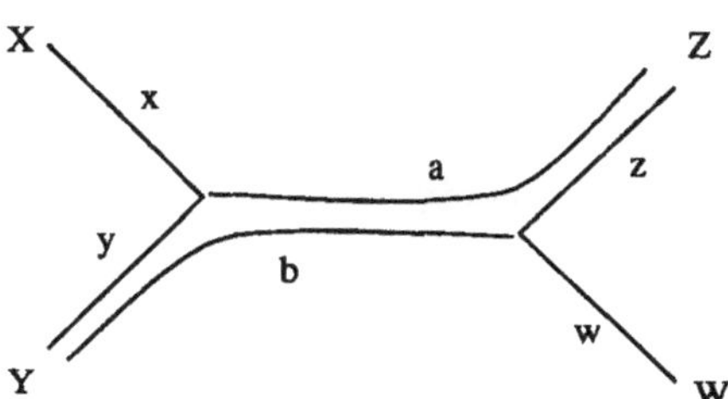

Figure 2d.

Thus,

$$d(Y,Z) = y + a = b + z.$$

Now,

$$\begin{aligned} d(X,Z) + d(Y,W) &\succeq d(X,Y) + d(Z,W) \\ (x+a) + (w+b) &\succeq (x+y) + (z+w) \\ a + b &\succeq y + z \end{aligned} .$$

But,

$$\begin{aligned} 2(a+y) &= (a+y) + (b+z) \\ &\succeq 2(y+z). \end{aligned}$$

Therefore,

$$a \succeq z.$$

Let $u = \max(a - z, 0)$. Then $u \simeq a - z = b - y$. Now,

$$\begin{aligned} d(X,W) &\simeq d(X,Z) + d(Y,W) - d(Y,Z) \\ &= (x+a) + (w+b) - (z+b) \\ &= x + (a-z) + w \\ &= x + u + w. \end{aligned}$$

$\diamond$

We shall write

$$XY : ZW \longleftrightarrow (xy)u(zw)$$

to express the situation described in Lemma 2.1.1. In fact, Lemma 3.1.7 tells us that, if $(\mathcal{S}, d)$ is almost-hyperbolic, then such a tree may be realised as a union of geodesic arcs in $\mathcal{S}$. (Though we may have to allow for a larger error of approximation than is obtained above.) If these geodesics are $[X, A]$, $[Y, A]$, $[A, B]$, $[B, Z]$, $[B, W]$ of lengths respectively x, y, u, z, w, then we shall write

$$(XY)AB(ZW) \longleftrightarrow (xy)u(zw).$$

In fact, this result may be extended to give spanning trees for any finite sets of points (see Section 3.3). Arguing further along these lines, one may show that any 0-H1 space is a metric tree, i.e. a path metric space which contains no topologically embedded circle (Section 3.4).

Finally, note that if we already know that the space $(\mathcal{S}, d)$ is H1, then the statement $XY : ZW$ becomes equivalent to $d(X, Z) + d(Y, W) \simeq d(Y, Z) + d(X, W)$.

2.2. Definition 2.

This is probably the conceptually simplest definition of almost-hyperbolicity. Given any triangle in $\mathcal{S}$, there is some point which lies within a bounded distance from all three edges. Such a point will be called a *centre* for the triangle. Lemma 3.1.5 shows that in an almost-hyperbolic space, any two centres for the same triangle are a bounded distance apart.

It should be stressed that in this definition, a "triangle" is taken to mean a set of three points, together with geodesic edges joining them. However, once it is known that a given space is almost-hyperbolic, it makes sense to speak of a centre of three points, without making an explicit choice of edges.

2.3. Definition 3.

This is the "isoperimetric inequality". Intuitively, it says that any closed curve (or *loop*) in our space "bounds a disc of area" at most a certain linear function of the curve's length. The main technical problem is in formulating what we mean by "bounding a disc" and "area" in an arbitrary path-metric space. The definition given in Section 1.2 gives one possibility, though the quantity representing the area in this case is perhaps more naturally termed "energy". We shall describe a few variations on these basic definitions below. Some will be more appropriate to certain contexts, for example when considering the Cayley graphs of almost-hyperbolic groups.

However we choose to define these terms, the essential property which we require of them may be summarised as follows.

Let (S,d) be a path-metric space. We define a *loop* in S as a distance non-increasing map $\gamma : (S^1,\sigma) \longrightarrow (S,d)$, where σ is a path-metric on the circle S^1. Suppose we represent S^1 as the union of four closed intervals, L_1, L_2, L_3, L_4, cyclically ordered, and intersecting only in their boundaries. Let $\alpha_i = \gamma|L_i$. We write $\gamma = \alpha_1 \cup \alpha_2 \cup \alpha_3 \cup \alpha_4$ for this, and we call $\alpha_1 \cup \alpha_2 \cup \alpha_3 \cup \alpha_4$ a *rectangle* in S. We shall want the following.

Rectangle Principle. (2.3.1) :

There are universal constants K_1 and K_2 such that the following holds.

Suppose that (S,d) is a path-metric space, and $\gamma = \alpha_1 \cup \alpha_2 \cup \alpha_3 \cup \alpha_4$ is a rectangle in S. Let $d_1 = d(\alpha_1,\alpha_3)$ and $d_2 = d(\alpha_2,\alpha_4)$. If γ "bounds a disc of area" A in S, then

$$A \geq K_1 d_1 d_2 - K_2(d_1 + d_2 + 1).$$

The constants K_1 and K_2 will depend only on the definition of area we choose. Chapter 5 is devoted to a discussion of these matters.

We now give the definitions we shall use in this paper.

Let G be any finite graph. To save on words, we shall identify G (thought of combinatorially) with its realisation as a topological 1-complex. We write $C_0(G)$ for the set of vertices of G, and $C_1(G)$ for the set of edges of G. Any path metric on G is determined, up to isotopy (rel $C_0(G)$), by a map $C_1(G) \longrightarrow (0,\infty)$. Any path-metric ρ, in turn determines a natural parameterisation on each edge, and thus a measure on G. Given any closed subset $G_0 \subseteq G$, we shall write $\rho(G_0)$ for the measure of G_0, which we shall refer to as the *ρ-length*, or just *length* of G_0. (Note that the length $\rho(e)$ of an edge $e \in C_1(G)$ may be greater than the distance between its endpoints.) We shall write ρ_{G_0} for the path-metric induced on G_0 from ρ. Thus $\rho_{G_0}(x,y) \geq \rho(x,y)$ for all $x,y \in G_0$.

Let D be the closed (unit) disc in $\mathbf{R}^2$.

Definition : A *cellulation*, P, of the disc D, is a presentation of D as a CW-complex, such that each 0-cell meets at least three 1-cells, and the boundary, ∂c, of any 2-cell, c, is an embedded circle.

Note that the conditions imply that the endpoints of each 1-cell are distinct.

We write $\Sigma(P)$ for the 1-skeleton of P, and $C_i(P)$ for the set of i-cells of P, $i = 0,1,2$. (Thus $C_0(P) = C_0(\Sigma(P))$ and $C_1(P) = C_1(\Sigma(P))$.) We shall use $A(P) = |C_2(P)|$ to denote the number of 2-cells of P.

We are also interested in the following special case of a cellulation.

Definition : A *triangulation*, P, of the disc D, is a presentation of the disc as a simplicial complex.

Thus any two triangles of P meet along an edge, or at a vertex, or not at all.

In this case, we shall write $A_T(P) = |C_1(P)|$ for the number of 1-cells of P. Lemma 5.6 tells us that we can always subdivide a cellulation P to give a triangulation P' satisfying

$$A_T(P') \leq 54A(P).$$

Definition : A *metric cellulation (triangulation)*, (P,ρ) is a cellulation (triangulation), P, of the disc D, together with a path-metric ρ on the 1-skeleton $\Sigma(P)$ of P.

Given any metric cellulation (P,ρ), we define the *mesh* of (P,ρ) as

$$m(P,\rho) = \max_{c\in C_2(P)} (\rho(\partial c)).$$

We define the *energy* as

$$I(P,\rho) = \sum_{c\in C_2(P)} (\rho(\partial c))^2.$$

We see that

$$I(P,\rho) \leq A(P)m(P,\rho)^2.$$

In the special case where P is a triangulation, we may define

$$m_T(P,\rho) = \max_{e\in C_1(P)} (\rho(e))$$

$$I_T(P,\rho) = \sum_{e\in C_1(P)} (\rho(e))^2.$$

Again,

$$I_T(P,\rho) \leq A_T(P)m_T(P,\rho)^2.$$

Comparing these formulations for the same metric triangulation, we get

$$m_T \leq m \leq 3m_T$$
$$A_T \leq 3A \leq 3A_T$$
$$I_T \leq I \leq 6I_T.$$

The last of these follow from the inequalities $x^2 + y^2 + z^2 \leq (x+y+z)^2 \leq 3(x^2 + y^2 + z^2)$, for $x, y, z \geq 0$.

Now, let $(\mathcal{S}, d)$ be any path-metric space.

Definition : A *cellular (simplicial) net*, (P,ρ,f), is a metric cellulation (triangulation), (P,ρ), of the disc D, together with a distance non-increasing map $f : (\Sigma(P),\rho) \longrightarrow (\mathcal{S}, d)$.

Recall that a *loop*, (γ, σ), is a distance non-increasing map $\gamma : (S^1, \sigma) \longrightarrow (S, d)$. We refer to $\sigma(S^1)$ as the *length* of γ.

Definition : A net (P, ρ, f) is said to *bound* a loop (γ, σ), if $f|\partial D$ factors through a distance non-increasing map $\partial f : (\partial D, \rho_{\partial D}) \longrightarrow (S^1, \sigma)$ of topological degree 1, i.e. $f|\partial D = \gamma \circ \partial f$.

One may mow give four versions of H3 as follows.

We may say that any loop (γ, σ), bounds a cellular (simplicial) net, (P, ρ, f), whose energy $I(P, \rho)$ $(I_T(P, \rho))$ is at most a certain linear function of $\sigma(S^1)$. This gives definitions H3ce (H3te).

Alternatively, we may say that any loop (γ, σ) bounds a cellular (simplicial) net whose mesh $m(P, \rho)$ $(m_T(P, \rho))$ is bounded by some fixed parameter, and for which the area $A(P)$ $(A_T(P))$ is at most a certain linear function of $\sigma(S^1)$. This gives definitions H3ca (H3ta).

Of these, the apparently weakest is H3ce, which we took as our main definition, and the apparently strongest is H3ta. That all these definitions amount to the same thing, should be apparent from the logical structure of Chapter 6. We prove, in fact, that H1 $\Rightarrow$ H3ta and that H3ce $\Rightarrow$ H4.

The formulations H3ca and H3ta seem best suited for combinatorial situations, for example where S is the Cayley graph of some finitely presented group. In this case, a loop is a word in the generators and their inverses, representing the trivial element. The area of a net bounding this loop is a measure of the number of applications of the relations we need to reduce this word to the trivial word.

The definition H3ta also makes the invariance of almost-hyperbolicity under pseudoisometries most apparent. Unfortunately, the details of the proof are a little messy, and instead, we give a different argument in Chapter 4 (Proposition 4.10).

Note that any metric on the 0-skeleton $C_0(P)$ of a cellulation P, determines (up to isotopy) a path-metric on the 1-skeleton $\Sigma(P)$. A metric cellulation (P, ρ) arises in this way precisely when the length of each edge equals the distance between its endpoints. If moreover, P is a triangulation, and the triangle inequalities for the three vertices of any triangle in $C_2(P)$ are strict, then (P, ρ) determines a singular-euclidean structure on the disc D, by gluing together euclidean triangles in the obvious way. This allows us to relate the energy $I_T(P, \rho)$ of such a triangulation to energy as defined in differential geometry as follows.

Recall that the energy of a map ϕ between two Riemannian manifolds is defined as

$$\frac{1}{2} \int \text{trace}(D\phi)^T (D\phi) d\mu,$$

where $d\mu$ is the volume element of domain, and the derivative $D\phi$ is expressed in terms of orthonormal coordinates. In dimension 2, this quantity depends only on the conformal structure of the domain. Now, suppose we have a euclidean triangle Δ, with sides of length a, b, c. We may imagine Δ as the image of an equilateral triangle under an affine map. Simple trigonometry shows that the energy of this map equals $\frac{1}{4\sqrt{3}}(a^2 + b^2 + c^2)$. We may think of a triangulation of the disc D as determining a conformal structure on D, by taking each triangle to be euclidean-equilateral of the same size. Suppose now, we have a path-metric ρ on $\Sigma(P)$ which happens to satisfy strict triangle inequalities for each triangle in $C_2(P)$. This determines a singular euclidean structure on the disc, as described above. The energy of the identity map is then equal to $\frac{1}{2\sqrt{3}}(I_T(P,\rho) - B)$, where B is the boundary correction $\frac{1}{2}\sum\{(\rho(e))^2 \mid e \in C_1(P), e \subseteq \partial D\}$.

Finally, we remark that Lemmas 6.1.2, and 6.1.4 provide another version of H3.

2.4. Definition 4.

This criterion is a simple consequence of the isoperimetric inequality and rectangle principle (see Proposition 5.12), and is a weak form of the pseudogeodesic property (Proposition 4.9).

2.5. Definition 5.

This may be thought of as expressing, in a weak sense, the fact that spheres, on a large scale, have extrinsic curvatures bounded away from 0. Compare with the statement that in a negatively curved Riemannian manifold, all of whose sectional curvatures are bounded away from 0, we have that all the principle curvatures of spheres are bounded away from 0.

In Chapter 6, we shall see that we may weaken the definition of H5 given in Chapter 1, by choosing only those X, Y which satisfy $k_5 \leq d(X,Y) \leq 2k_5 + 2h_5$. From Proposition 4.7, we may also give the stronger formulation, that for some k_5', h_5', we have that $d(X,Y) \geq k_5'$ implies that $h_5' d_{r,A}(X,Y) \geq d(X,Y)^2$. Now, the parameters k_5', h_5' have an interpretation in terms of the curvature/coarseness of S, as described in Section 1.2.

CHAPTER 3 : Centres and spanning trees.

In this chapter, we get the subject of almost-hyperbolicity off the ground, by showing the equivalence of definitions H1 and H2. Recall the notation $\simeq, \preceq, \sim$ introduced in Section 1.4. Also, in this chapter, we will find it useful to abbreviate $d(X,Y)$ to XY.

3.1. H1 $\Rightarrow$ H2.

Let $(\mathcal{S}, d)$ be a k_1-H1 geodesic space.

Lemma 3.3.1 : *Suppose that $X, Y, Z \in \mathcal{S}$, and that α is a geodesic joining X to Y. Then, there is some $A \in \alpha$ such that*

$$XZ \simeq XA + AZ$$
$$YZ \simeq YA + AZ.$$

(Figure 3a.)

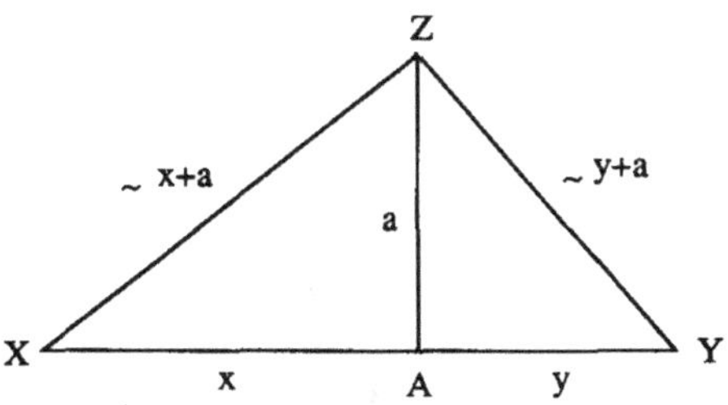

Figure 3a.

Proof : Let $XYZ \longleftrightarrow xyz$ so that $XY = x + y$, $YZ = y + z$ and $ZX = z + x$. Let $A \in \alpha$ be so that $XA = x$ and $AY = y$. Let $AZ = a$. (Figure 3b.)

Now

$$XZ \leq XA + AZ$$
$$x + z \leq x + a.$$

So,

$$z \leq a.$$

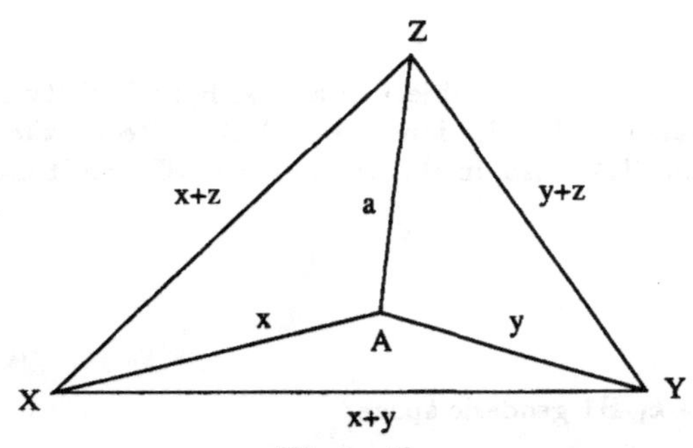

Figure 3b.

But,

$$XZ + AY = x + y + z$$
$$YZ + AY = x + y + z$$
$$XY + AZ = x + y + a.$$

Applying the hypothesis k_1-H1 to $\{X, Y, Z, A\}$, we see that we must have $z \simeq a$.
$\diamond$

Lemma 3.1.2 : *There is some constant K $(= K(k_1))$ so that: if α is any geodesic joining X to Y, $X, Y \in \mathcal{S}$, and β is any path from X to Y of length $\simeq d(X,Y)$, then $\beta \subseteq N_K(\alpha)$.*

Proof : Let $Z \in \beta$ be any point, and let $A \in \alpha$ be as in Lemma 3.1.1. Then,

$$\begin{aligned} XA + AY = XY &\simeq \text{length}\,\beta \\ &\geq XZ + ZY \\ &\simeq (XA + AZ) + (YA + AZ) \\ &= XA + AY + 2AZ. \end{aligned}$$

Thus

$$AZ \simeq 0,$$

i.e.

$$d(Z, [X, Y]) \simeq 0.$$

$\diamond$

In particular, this shows that any two geodesics joining the same pair of points remain a bounded distance apart.

Lemma 3.1.3 : *Suppose X, Y, Z are points of $\mathcal{S}$, and $[X,Y]$, $[Y,Z]$, $[Z,X]$ are any geodesics joining them. Then, there exists $A \in \mathcal{S}$ with*

$$d(A,[X,Y]) \simeq 0$$
$$d(A,[Y,Z]) \simeq 0$$
$$d(A,[Z,X]) \simeq 0.$$

Proof : Let $A \in [X,Y]$ be as in Lemma 3.1.1 (with $\alpha = [X,Y]$). Then $XA + AZ \simeq XY$. Applying Lemma 3.1.2, (with $\alpha = [X,Z]$ and $\beta = [X,A] \cup [A,Y]$), we see that $d(A,[Z,X]) \simeq 0$. Similarly, $d(A,[Y,Z]) \simeq 0$.
$\diamond$

Corollary 3.1.4, H1 $\Rightarrow$ H2 : $\forall k_1 \, \exists k_2$ *such that if $(\mathcal{S},d)$ is k_1-H1, then it is k_2-H2.*

In fact, we see that k_2 is at most a fixed (universal) multiple of k_1. This is because it is derived from k_1 by a certain number of applications of the transitive law to our approximate inequalities. In particular, we see that a 0-H1 space is also 0-H2.

Definition : Suppose that $(\mathcal{S},d)$ is k_1-H1, and that $X, Y, Z \in \mathcal{S}$. We call $A \in \mathcal{S}$ a *centre* of XYZ if $d(A,[X,Y]) \simeq 0$, $d(A,[Y,Z]) \simeq 0$ and $d(A,[Z,X]) \simeq 0$.

In view of Lemma 3.1.2, this definition makes sense, irrespective of the choice of geodesics $[X,Y]$, $[Y,Z]$ and $[Z,X]$.

Note that we may choose a centre to lie on any one of these three geodesic edges.

Lemma 3.1.5 : *Suppose that C and D are both centres for XYZ in $\mathcal{S}$. Then $C \sim D$.*

Proof : Choose C', D' in $[X,Y]$ with $C \sim C'$ and $D \sim D'$. Without loss of generality, C' is nearer X. Let x, u, v, a, y be as in Figure 3c.

Now, $d(C',[X,Z]) \simeq 0$, and $d(D',[X,Z]) \simeq 0$. So,

$$XZ \simeq x + u$$

and

$$XZ \simeq x + a + v.$$

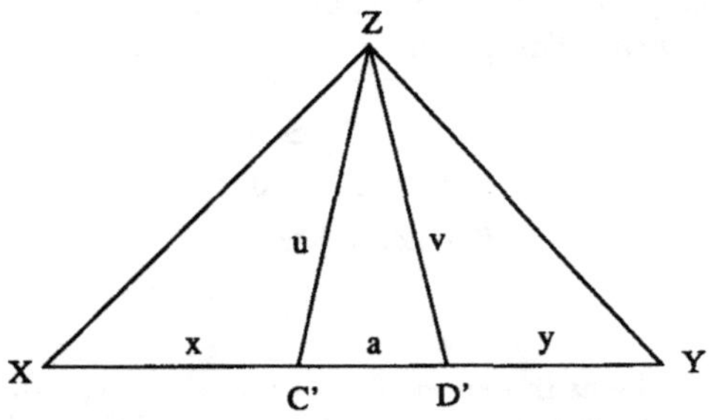

Figure 3c.

Thus,

$$u \simeq a + v.$$

Similarly.

$$v \simeq a + u.$$

Thus, $a \simeq 0$. So $C' \sim D'$ and $C \sim D$.
◇

Lemma 3.1.6 : *Given $X, Y, Z \in S$, let $A \in \mathrm{proj}_{[X,Y]} Z$ (i.e. A is a nearest point on $[X, Y]$ to Z). Then, A is a centre of XYZ.*

Proof : Let C be a centre of XYZ on $[X, Y]$. Without loss of generality, $XC \leq XA$. Let x, b, y, c, a be as in Figure 3d.

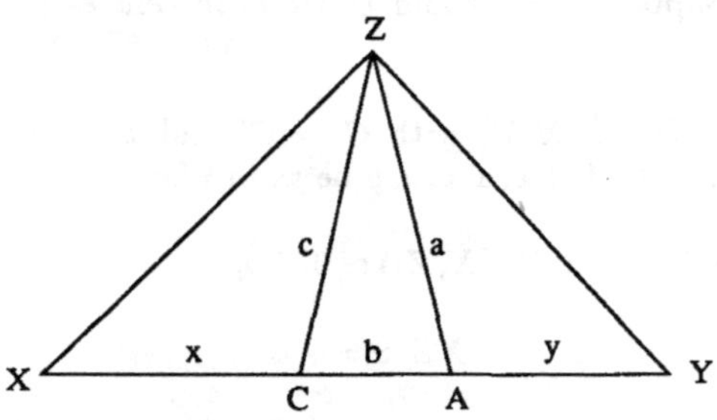

Figure 3d.

We have $a \le c$, thus

$$\begin{aligned} c + b + y &\simeq YZ \\ &\le a + y \\ &\le c + y. \end{aligned}$$

Thus,

$$b \simeq 0.$$

In other words, $A \sim C$, and so A is a centre.

◇

We stated in Section 2.1, that, in a H1 space, the statement $XZ + YW \simeq XW + YZ$ implies $XY + ZW \preceq XZ + YW$ and thus is equivalent to $XY : ZW$.

Lemma 3.1.7 : *Let $X, Y, Z, W \in \mathcal{S}$. Suppose we have*

$$XZ + YW \simeq XW + YZ.$$

Then, there exist $C, D \in \mathcal{S}$, with

$$d(C, [X_1, X_2]) \simeq 0 \text{ provided } \{X_1, X_2\} \ne \{Z, W\}$$

and

$$d(D, [Y_1, Y_2]) \simeq 0 \text{ provided } \{Y_1, Y_2\} \ne \{X, Y\},$$

where $X_i, Y_i \in \{X, Y, Z, W\}$, $X_1 \ne X_2$ and $Y_1 \ne Y_2$.

Moreover, we have

$$Z_1 Z_2 \simeq Z_1 C + CD + D Z_2,$$

whenever $Z_1 \in \{X, Y\}$ and $Z_2 \in \{Z, W\}$.

Proof : Let A, B be centres of XYZ, XYW respectively on $[X, Y]$. Without loss of generality, A is nearer X. Let x, a, u, y, b be as in Figure 3e.

We have

$$\begin{aligned} XZ + YW &\simeq XW + YZ \\ (x + a) + (y + b) &\simeq (x + u + b) + (y + u + a). \end{aligned}$$

Thus,

$$u \simeq 0.$$

Thus $A \sim B$, and so A is a centre for both XYZ and XYW. We take $C = A$.

Similarly, we find D on $[Z, W]$.

This proves the first part of the lemma.

Now, suppose that $Z_1 \in \{X, Y\}$ and $Z_2 \in \{Z, W\}$. We want to show that $Z_1 Z_2 \simeq Z_1 C + CD + D Z_2$. Without loss of generality, we can assume that $Z_1 = X$ and $Z_2 = Z$.

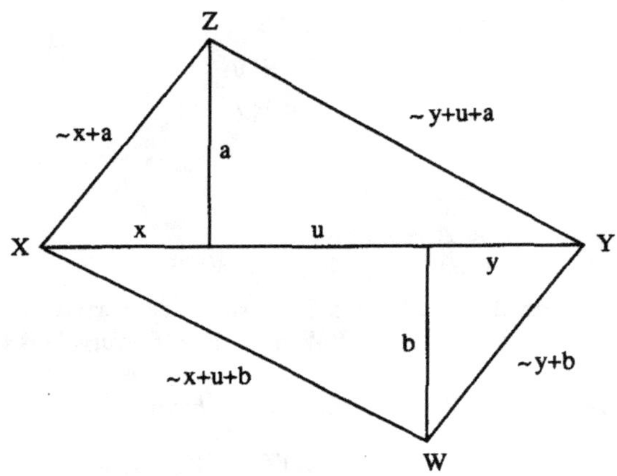

Figure 3e.

Now, there exist $C', D' \in [X, Z]$ with $C \sim C'$ and $D \sim D'$. If $XC' \leq XD'$, we have that

$$XZ = XC' + C'D' + D'Z \simeq XC + CD + DZ.$$

Thus, we may suppose that $XD' \preceq XC'$. Then,

$$XZ = XC' - C'D' + D'Z \simeq XC - CD + DZ.$$

But, $YW \leq YC + CD + DW$, and so

$$XZ + YW \simeq XC + YC + DZ + DW \simeq XY + ZW.$$

We noted, before the statement of the lemma that we must have $XZ + YW \succeq XY + ZW$, and so $XY + ZW \simeq XZ + YW$.

Applying the first half of the lemma to this case, we find a point $E \in \mathcal{S}$ so that $d(E, [W_1, W_2]) \simeq 0$, provided that $W_1, W_2 \in \{X, Y, Z, W\}$ are distinct, and $\{W_1, W_2\} \neq \{X, W\}$. Now, C and E are both centres of XYZ, and D and E are both centres of YZW. Applying Lemma 3.1.5, we find that $C \sim E \sim D$. Thus, $CD \simeq 0$, and so

$$XZ \simeq XC - CD + DZ \simeq XC + CD + DZ.$$

◇

Given the points C, D of Lemma 3.1.7, we may construct the "spanning tree"

$$[X, C] \cup [Y, C] \cup [C, D] \cup [D, Z] \cup [D, W],$$

which we shall write as $(XY)CD(ZW)$. In other words, from the points X, Y, Z and W of the hypotheses, the notation $(XY)CD(ZW)$ is intended to define the points C and D. Now, all the distances between the points of $\{X,Y,Z,W\}$ may be read off, up to an additive constant, along the tree (c.f. Lemma 2.1.1). In fact, the corresponding geodesics run within a bounded distance of the tree (see Lemma 3.3.1 or Lemma 4.1). If $XC = x$, $YC = y$, $CD = u$, $DZ = z$ and $DW = w$, we shall write

$$(XY)CD(ZW) \longleftrightarrow (xy)u(zw).$$

(Figure 3f.)

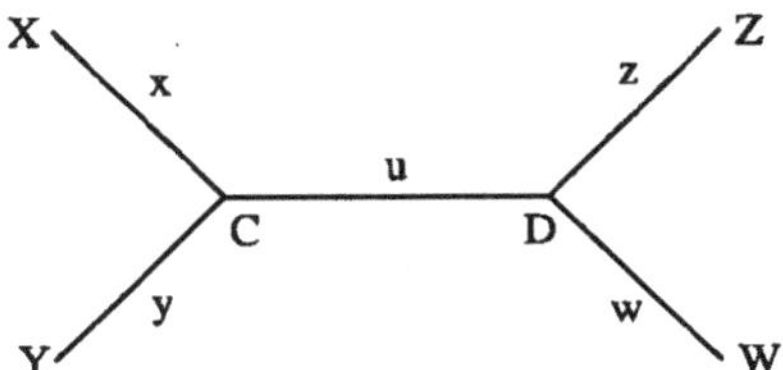

Figure 3f.

The above discussion may be generalised to define spanning trees for arbitrary finite sets of points (see Section 3.3). First, however, we give a proof of H2 $\Rightarrow$ H1.

3.2. H2 $\Rightarrow$ H1.

Let $(\mathcal{S},d)$ be k_2-H2.

Suppose that $X,Y,Z \in \mathcal{S}$, and that $[X,Y]$ is some geodesic from X to Y. If we choose geodesics α and β joining Z to X and Y respectively, then we may find some point $C \in [X,Y]$, with $d(C,\alpha) \leq 2k_2$ and $d(C,\beta) \leq 2k_2$. For all we know at the moment, the position of C on $[X,Y]$ might depend substantially on the choice of α and β. However, we must have

$$XZ \geq XC + CZ - 4k_2$$

and

$$YZ \geq YC + CZ - 4k_2.$$

Given X,Y,Z and $[X,Y]$ as above, we shall call any point $C \in [X,Y]$ a *near-projection* of Z to $[X,Y]$ if we have both $XZ \simeq XC + CZ$ and $YZ \simeq YC + CZ$. We shall deduce property H1 from the existence of such near-projections.

Lemma 3.2.1 : *Let $X, Y, Z \in S$ and $[X, Y]$ be any geodesic from X to Y. Suppose that $A \in [X, Y]$. Then either*

$$XZ \simeq XA + AZ$$

or

$$YZ \simeq YA + AZ.$$

Proof : Let C be a near-projection of Z to $[X, Y]$. Suppose $AX \leq AC$. Let x, u, a, v, b be as in Figure 3g.

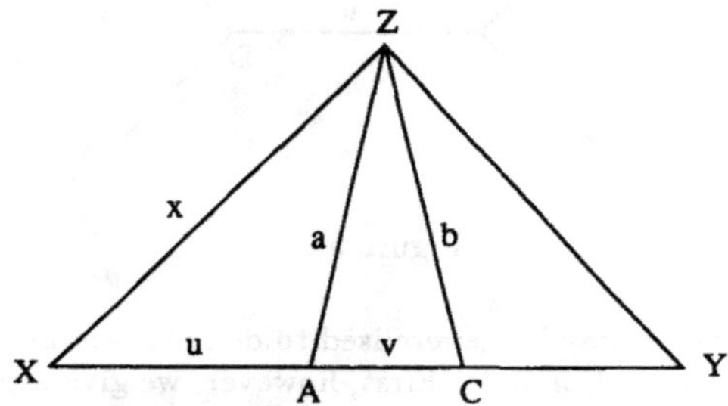

Figure 3g.

Thus $v + b \geq a$. Now,

$$\begin{aligned} XZ &\simeq XC + CZ \\ x &\simeq u + v + b \\ &\geq u + a. \end{aligned}$$

But $x \leq u + a$. Thus $x \simeq u + a$, i.e. $XZ \simeq XA + AZ$.

Similarly, if $AX \geq AC$, we have $YZ \simeq YA + AZ$.

◇

Lemma 3.2.2 : *Given any $X, Y, Z, W \in S$, we have*

$$XY + ZW \preceq \max(XW + YZ, XZ + YW).$$

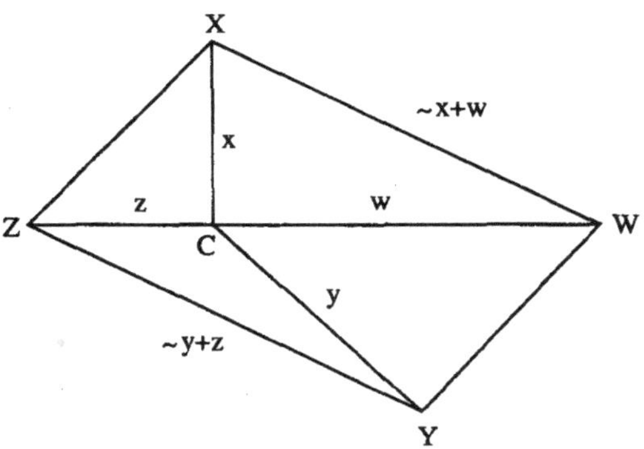

Figure 3h.

Proof : Choose any geodesic $[Z, W]$. Let C be any near-projection of X to $[Z, W]$ (Figure 3h).

From Lemma 3.2.1, we have, without loss of generality,

$$ZY \simeq CZ + CY.$$

But,

$$XW \simeq CX + CW.$$

Thus,

$$\begin{aligned} XW + YZ &\simeq (CX + CW) + (CZ + CY) \\ &= (CX + CY) + (CZ + CW) \\ &\geq XY + ZW. \end{aligned}$$

◇

Proposition 3.2.3 : *$\forall k_2\ \exists k_1$ such that if $(\mathcal{S}, d)$ is k_2-H2, it is k_1-H1.*

Proof : Suppose that $(\mathcal{S}, d)$ is H2. Given any $X, Y, Z, W \in \mathcal{S}$, we have, without loss of generality,

$$XY + ZW \geq \max(XW + YZ, XZ + YW).$$

By Lemma 3.2.2,

$$XY + ZW \preceq \max(XW + YZ, XZ + YW).$$

Thus, without loss of generality,

$$XZ + YW \leq XW + YZ \simeq XY + ZW,$$

i.e. $XZ : YW$. So, $(\mathcal{S}, d)$ is H1.
◇

Again k_1 is at most a certain fixed multiple of k_2. In particular, a 0-H2 space is also 0-H1.

3.3. Spanning trees.

Sections 3.3 and 3.4 are intended to make apparent the tree-like properties of almost-hyperbolic spaces.

In this section, we show that any finite set of points in an almost-hyperbolic space may be spanned by which allows us to measure distances up to an additive constant. We shall give a refinement of this result in Chapter 7.

We shall need the following stronger version of Lemma 3.1.2.

Suppose that $(\mathcal{S}, d)$ is k_1-H1.

Lemma 3.3.1 : *$\forall h \geq 0$ $\exists H = H(h, k_1)$ such that the following holds.*
If α is any geodesic joining X to Y in $\mathcal{S}$, and β is any path from X to Y of length at most $XY + h$, then we have both $\beta \subseteq N_H(\alpha)$ and $\alpha \subseteq N_H(\beta)$.

Proof : The statement $\beta \subseteq N_H(\alpha)$ is the same as Lemma 3.1.2, except that we are allowing h to be independent of k. This makes no essential difference to the proof.

The statement $\alpha \subseteq N_H(\beta)$ is new. There is a simple argument, see the proof of Proposition 4.9.
◇

Proposition 3.3.2 : *$\forall n \in \mathbf{N}$ $\exists K_n = K(n, k_1)$ such that the following holds.*
Let $V \subseteq \mathcal{S}$ be any set of $n+1$ points of $\mathcal{S}$. Then, there is an embedded tree $T_V \subseteq \mathcal{S}$ with geodesic edges, so that the distance XY between any two points $X, Y \in \mathcal{S}$ may be measured up to K_n along the arc $\beta(X, Y)$ in T_V joining X to Y, i.e.

$$XY \geq \text{length}\beta(X, Y) - K_n.$$

Proof : The proof is by induction on n. Suppose that V has $n+1$ elements, and that we have constructed T_V. Given $C, D \in T_V$, we write $\beta(C, D)$ for the arc in T_V joining C to D. For any $C, D \in T_V$, we may find $Z, W \in V$ such that

$\beta(C,D) \subseteq \beta(Z,W)$. Since, by hypothesis, length$\beta(Z,W) \leq ZW + K_n$, we may deduce that also length$\beta(C,D) \leq CD + K_n$.

Let $X \in \mathcal{S}$ be any $(n+2)^{\text{th}}$ point. We want to span $V \cup \{X\}$. Let $A \in \text{proj}_{T_V} X$ (i.e. A is a nearest point in T_V to X). Let $T_{V\cup\{X\}} = T_V \cup [A,X]$.

Suppose $Y \in V$, we want to measure XY. Let B be a centre for AXY on $[A,Y]$. Now, length$\beta(A,Y) \leq AY + K_n$. So, by Lemma 3.3.1, we have $d(B,\beta(A,Y)) \leq H = H(K_n,k_1)$. Now,

$$XB + BA \simeq XA = d(X,T_V) \leq d(X,\beta(A,Y)) \leq XB + H.$$

Thus, $BA \preceq H$. (The meanings of $\simeq$ and $\preceq$ depend on k_1, but not on n.) But,

$$\begin{aligned} XY &\simeq XB + BY \\ &\succeq XA + AY - 2H \\ &\geq (\text{length}\beta(A,Y) - K_n) + AX - 2H \\ &= \text{length}\big([X,A] \cup \beta(A,Y)\big) - (K_n + 2H). \end{aligned}$$

We may therefore take $K_{n+1} \simeq K_n + 2H(K_n,k_1)$.
$\diamondsuit$

3.4. Metric trees.

The main purpose of this section is to show that 0-H1 path metric spaces are precisely what we shall call "metric trees" (elsewhere known as "R-trees").

Definition : A *metric tree* is a path-metric space which contains no embedded rectifiable circle.

(In fact, we shall see that a metric tree can contain no topologically embedded circle.)

Note that we are not making any assumptions of completeness or local-compactness. However, it is true that in any metric tree, $(\mathcal{S},d)$, any two points may be joined by a (unique) geodesic. One may see this as follows.

First note that there is at most one rectifiable arc joining any two distinct points, $X,Y \in \mathcal{S}$. Suppose that α and β were two such arcs, with $\alpha \not\subseteq \beta$. Let β' be the closure of a component of $\beta\backslash\alpha$. Then β' meets α in two distinct points Z and W. If α' is the sub-arc of α lying between Z and W, then $\alpha' \cup \beta'$ is an embedded circle. This contradiction shows that $\alpha \subseteq \beta$. Similarly $\beta \subseteq \alpha$.

Now if γ is any rectifiable path joining X to Y, then the image of γ, being path-connected, contains an arc γ_0 with endpoints X and Y. (In this context, γ_0 may be obtained by applying Ascoli's theorem to a sequence of paths from X to Y lying in the image of γ, and whose lengths tend to the infinum for such paths.) Thus γ_0 is the unique rectifiable arc from X to Y.

Suppose that γ' is a path from X to Y of length at most $d(X,Y)+\epsilon$. Again, the image of γ' contains a rectifiable arc from X to Y which must therefore be γ_0. Thus $\text{length}\,\gamma_0 \leq \text{length}\,\gamma' \leq d(X,Y)+\epsilon$. Since ϵ is arbitrary, we have $\text{length}\,\gamma_0 = d(X,Y)$, i.e. γ_0 is a geodesic.

We have shown that any to points in a metric tree are joined by a unique rectifiable arc which is always a geodesic. (In fact, given that a metric tree contains no topological circle, we see that every closed arc in a metric tree is a geodesic.)

By a similar argument, we see that if we take any three points in a metric tree $(\mathcal{S},d)$, then the three geodesics joining them meet in a single point. Thus, $(\mathcal{S},d)$ is 0-H2.

Note that, in our discussion of geodesic spaces so far in this paper, we have made no use of the assumptions of local compactness or completeness, other than that there should exist a geodesic between any two points. (We have only used projection to compact sets.)

Proposition 3.4.1 : *Suppose that $(\mathcal{S},d)$ is a path-metric space in which any pair of points are joined by at least one geodesic. Then $(\mathcal{S},d)$ is 0-H1 if and only if it is 0-H2.*

Proof : See the remarks after Propositions 3.1.4 and 3.2.3.

$\diamond$

Suppose that $(\mathcal{S},d)$ is such a space (as described by Proposition 3.4.1), and that $V \subseteq \mathcal{S}$ is a finite set of points. Then, the construction of Proposition 3.3.2 gives an embedded tree $T_V \subseteq \mathcal{S}$ along which distances are measured precisely, i.e. $K(n,0)=0$ for all $n \in \mathbf{N}$. This is essentially because everywhere our approximate inequalities may be replaced by precise inequalities. Moreover, since the construction was inductive, we can take T_V to contain any previously chosen geodesic $[X,Y]$, with $X,Y \in V$. We leave the reader to check these statements.

Proposition 3.4.2 : *Let $(\mathcal{S},d)$ be a path-metric space. The following are equivalent:*

(1) $(\mathcal{S},d)$ is a metric tree,

(2) $(\mathcal{S},d)$ contains no topologically embedded circle,

(3) $(\mathcal{S},d)$ is 0-H1.

Proof : (2) $\Rightarrow$ (1) is trivial.

We argued above that any metric tree is 0-H2, and thus 0-H1 by Proposition 3.4.1. This proves (1) $\Rightarrow$ (3).

For (3) $\Rightarrow$ (2), suppose that $(\mathcal{S}, d)$ is a 0-H1 path-metric space. We will want to use the tree construction mentioned above, after Proposition 3.4.1. However, we do not yet know that every pair of points in $\mathcal{S}$ can be joined by a geodesic. We can get around this problem by taking the metric completion, $(\mathcal{S}_C, d)$, of $(\mathcal{S}, d)$. It is easily verified that $(\mathcal{S}_C, d)$ is also a 0-H1 path-metric space. (The extension of d to $\mathcal{S}_C$ is automatically a path-metric.) We claim that any pair of points, $A, B \in \mathcal{S}_C$, may be joined by a geodesic in $\mathcal{S}_C$. This may be seen as follows.

We can assume that $A \neq B$. Let $l = d(A, B)$, and suppose that $\alpha, \beta : [0, l] \longrightarrow \mathcal{S}_C$ are paths satisfying respectively $d(\alpha x, \alpha y) \leq |x-y|+\eta$ and $d(\beta x, \beta y) \leq |x-y|+\eta$ for all $x, y \in [0, l]$, and with $\alpha(0) = \beta(0) = A$ and $\alpha(l) = \beta(l) = B$. Given $x \in [0, l]$, let $E = \alpha x$ and $F = \beta x$. Thus, $d(A, E) \leq x+\eta$, $d(A, F) \leq x+\eta$, $d(B, E) \leq l-x+\eta$ and $d(B, F) \leq l - x + \eta$. Since $\mathcal{S}_C$ is 0-H1, we must have

$$\begin{aligned} d(A, B) + d(E, F) &\leq \max(d(A, E) + d(B, F), d(A, F) + d(B, E)) \\ &\leq l + 2\eta. \end{aligned}$$

Thus, $d(\alpha x, \beta x) \leq 2\eta$.

Now, for each $i \in \mathbf{N}$, choose a path $\gamma_i : [0, l] \longrightarrow \mathcal{S}_C$, with $\gamma_i(0) = A$, $\gamma_i(l) = B$, and $d(\gamma_i x, \gamma_i y) \leq |x - y| + 1/2^i$ for all $x, y \in [0, l]$. Thus if $x \in [0, d]$, and $j \geq i$, then $d(\gamma_i x, \gamma_j x) \leq 2/2^i$. Since $(\mathcal{S}_C, d)$ is complete, the paths γ_i converge uniformly to a path $\gamma : [0, d] \longrightarrow \mathcal{S}_C$. Clearly γ must be geodesic. This proves the claim.

Now, suppose (for contradiction) that $\Sigma \subseteq \mathcal{S} \subseteq \mathcal{S}_C$ is homeomorphic to a circle.

We need to find two points $Y, Z \in \Sigma$ with $[Y, Z] \not\subseteq \Sigma$. To do this, we take $W_1, W_2 \in \Sigma$ with $d(W_1, W_2)$ maximal. If $[W_1, W_2] \subseteq \Sigma$, then pick any $W_3 \in \Sigma \backslash [W_1, W_2]$. If $[W_1, W_3] \cup [W_3, W_2] \subseteq \Sigma$, then these three geodesics must cover Σ, and have disjoint interiors. Thus, $[W_1, W_2] \cap [W_2, W_3] \cap [W_3, W_1] = \emptyset$, contradicting the fact that $(\mathcal{S}_C, d)$ is 0-H1 and hence 0-H2.

Now, choose $C \in [Y, Z] \backslash \Sigma$, and let $\epsilon = d(C, \Sigma) > 0$. Since Σ is compact, we may find points $X_1, X_2, \ldots, X_p \in \Sigma$, cyclically ordered on Σ, so that $d(X_i, X_{i+1}) \leq \epsilon$. We can assume that $\{Y, Z\} \subseteq \{X_1, \ldots, X_p\} = V$. We may construct, in $\mathcal{S}_C$, a spanning tree T_V for V with $[Y, Z] \subseteq T_V$. Thus $C \in T_V$. Now C must must separate T_V, so we may write $T_V \backslash B = T_1 \cup T_2$, with T_i open in T_V. Let $V_i = V \cap T_i$. Thus $V = V_1 \cup V_2$ and $V_i \neq \emptyset$. We may find $D_1 \in V_1$ and $D_2 \in V_2$ adjacent on Σ. Thus $d(D_1, D_2) \leq \epsilon$. But the path β in T_V from D_1 to D_2 passes through C. Thus $d(D_1, D_2) = \text{length}\beta \geq 2\epsilon$.

We have contradicted the existence of Σ.

$\diamond$

CHAPTER 4. Convexity and pseudoisometries.

In this chapter, we define the notion of "almost convexity" for a subset of an almost-hyperbolic space $\mathcal{S}$. Examples, as we shall see, include "starlike" sets, and uniform neigbourhoods of other almost-convex sets. We define projections to almost-convex sets, and show that, in some sense, projections decrease distance by an exponential factor (Proposition 4.5). From this, we may deduce the exponential divergence of geodesic rays (Proposition 4.7), as well as the pseudogeodesic property (Proposition 4.9). We deduce properties H4 and H5, and conclude with a proof that almost-hyperbolicity is a pseudoisometric invariant (Proposition 4.10).

Let $(\mathcal{S}, d)$ be k_1-H1.

Lemma 4.1 : *There is some $h = h(k_1)$ such that for all $p \in \mathbf{N}$, the following holds.*

Suppose γ is a path in $\mathcal{S}$, consisting of at most p geodesic segments, which joins X to Y. Then, $[X,Y] \subseteq N_{ph}(\gamma)$, where $[X,Y]$ is any geodesic from X to Y.

Proof : Suppose $p = 2$, and $\gamma = [X,Z] \cup [Z,Y]$. Let A be a centre of XYZ on $[X,Y]$. Choose $B \in [X,Z]$ and $C \in [Y,Z]$ with $B \sim A$ and $C \sim A$. (Figure 4a.)

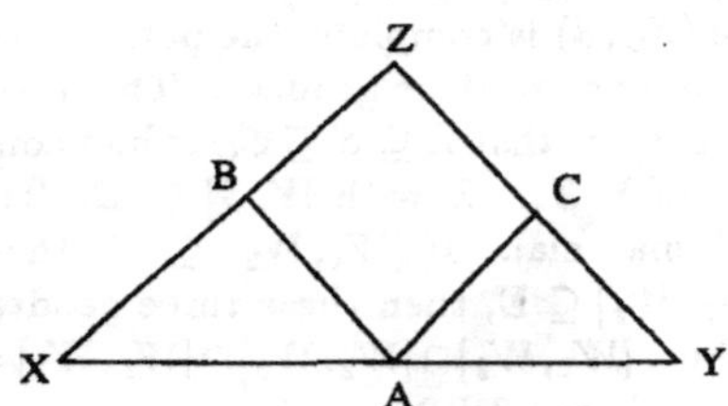

Figure 4a.

Lemma 3.1.2 now gives us a constant h such that $[X,A] \subseteq N_h([X,B])$ and $[A,Y] \subseteq N_h([Y,C])$. Thus, $[X,Y] \subseteq N_h([X,Z] \cup [Z,Y])$.

This also deals with the case $p = 1$: any two geodesics between the same points stay a distance at most h apart.

We now use induction on p. Suppose $\gamma = [X_0, X_1] \cup \cdots \cup [X_{p-1}, X_p]$. From the above,

$$\begin{aligned}[X_0, X_p] &\subseteq N_h\big([X_0, X_{p-1}] \cup [X_{p-1}, X_p]\big) \\ &\subseteq N_h\big(N_{(p-1)h}\big([X_0, X_1] \cup \cdots \cup [X_{p-2}, X_{p-1}]\big) \cup [X_{p-1}, X_p]\big) \\ &\subseteq N_{ph}(\gamma).\end{aligned}$$

$\diamond$

Let $\lambda \geq 0$, and let $Q \subseteq \mathcal{S}$ be a closed set.

Definition : The set $Q \subseteq \mathcal{S}$ is *λ-convex* if for all $X, Y \in Q$, and all geodesics $[X, Y]$, we have $[X, Y] \subseteq N_\lambda(Q)$.

We shall call a set *almost-convex* if it is λ-convex for some $\lambda \geq 0$.

Lemma 4.1 provides us with many examples of almost-convex sets. Any set in which any two points can be joined by a piecewise geodesic path with a bounded number of geodesic segments is almost-convex. This includes all starlike sets, and any uniform neighbourhood of a geodesic segment. We also have:

Lemma 4.2 : *There is some $\lambda_0 = \lambda_0(k_1)$ such that if $Q \subseteq \mathcal{S}$ is λ-convex, and $r \geq 0$, then $N_r(Q)$ is $(\lambda_0 + \max(\lambda - r, 0))$-convex.*

Proof : Let $\lambda_0 = 3h$, where h comes from Lemma 4.1. Suppose $X, Y \in N_r(Q)$. Then there exist $W, Z \in Q$ with $d(X, Z) \leq r$ and $d(Y, W) \leq r$. Now, $[Z, W] \subseteq N_\lambda(Q)$ and $[X, Z] \cup [Y, W] \subseteq N_r(Q)$. So, by Lemma 4.1,

$$\begin{aligned}[X, Z] &\subseteq N_{3h}([X, Z] \cup [Z, W] \cup [W, Y]) \\ &\subseteq N_{3h}(N_{\max(\lambda, r)}(Q)) \\ &= N_{\lambda_0 + \max(\lambda - r, 0)}(N_r(Q)).\end{aligned}$$

$\diamond$

Recall the definitions of $\check{Q}$, ∂Q, $\mathrm{proj}_Q X$ and $d(r, Q)$ from Section 1.1. Given any $M \subseteq \mathcal{S}$, we shall write $\mathrm{proj}_Q M = \bigcup_{X \in M} \mathrm{proj}_Q X$. Also, we write $\mathrm{diam} M = \sup\{d(X, Y) \mid X, Y \in M\}$ for the diameter of M.

Suppose that $Q \subseteq \mathcal{S}$ is λ-convex. Then, we claim that for any $X \in \mathcal{S}$, the quantity $\mathrm{diam}(\mathrm{proj}_Q X)$ may be bounded in terms of λ and the hyperbolicity parameter, k_1. In fact, we may make the more general statement:

Lemma 4.3 : *Suppose that $Q \subseteq \mathcal{S}$ is λ-convex, and $M \subseteq \mathcal{S}$ satisfies* $\mathrm{diam} M \leq 2d(Q, M)$, *then* $\mathrm{diam}(\mathrm{proj}_Q M) \leq J + 4\lambda$, *where J depends only on k_1.*

Proof : Let $\rho = d(Q, M)$. Suppose $X, Y \in M$, and $Z \in \mathrm{proj}_Q X$ and $W \in \mathrm{proj}_Q Y$ so that $d(X, Z) \leq 2\rho$, $d(X, Z) \geq \rho$ and $d(Y, W) \geq \rho$. We want to show that $d(Z, W) \preceq 4\lambda$.

Case (1), $XY : ZW$.

Let $(XY)AB(ZW)$ be a spanning tree, and let $(XY)AB(ZW) \longleftrightarrow (xy)u(zw)$. (Figure 4b. See the discussion after Lemma 3.1.7.)

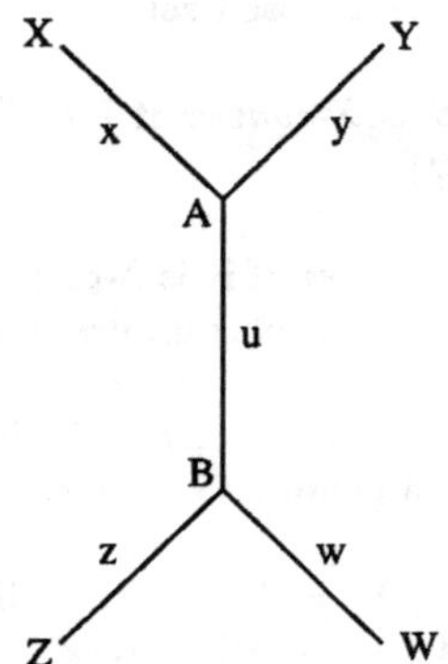

Figure 4b.

Now, $d(B,[Z,W]) \simeq 0$. Thus $d(B,Q) \preceq \lambda$. Now,

$$d(X,Z) = d(X,Q) \preceq d(X,B) + d(B,Q)$$
$$x + u + z \preceq (x+u) + \lambda.$$

Thus,

$$z \preceq \lambda.$$

Similarly,

$$w \preceq \lambda.$$

Thus,

$$d(Z,W) \simeq z + w \preceq 2\lambda.$$

Case (2), $XZ : YW$.

Let $(XZ)CD(YW) \longleftrightarrow (xz)u(yw)$ be a spanning tree (Figure 4c).

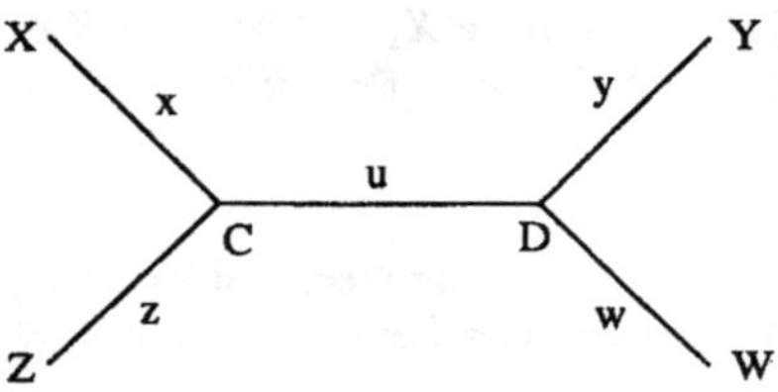

Figure 4c.

Similarly as in Case (1), we have $d(C,Q) \preceq \lambda$ and $d(D,Q) \preceq \lambda$. Thus,

$$\begin{aligned} d(X,Z) = d(X,Q) &\preceq d(X,C) + \lambda \\ x + z &\preceq x + \lambda. \end{aligned}$$

Thus,

$$z \preceq \lambda.$$

Similarly,

$$w \preceq \lambda.$$

From the hypotheses, we have $x + z \succeq \rho$, $y + w \succeq \rho$ and $x + y + u \preceq 2\rho$. Thus,

$$\begin{aligned} 2\rho + u &\preceq (x+z) + (y+w) + u \\ &= (x+y+u) + (z+w) \\ &\preceq 2\rho + 2\lambda. \end{aligned}$$

Thus,

$$u \preceq 2\lambda,$$

and so,

$$d(X,Z) \simeq w + z + u \preceq \lambda + \lambda + 2\lambda = 4\lambda.$$

Case (3), $XW : YZ$.

Let $(XW)FE(YZ) \longleftrightarrow (xw)u(yz)$ be a spanning tree (Figure 4d).

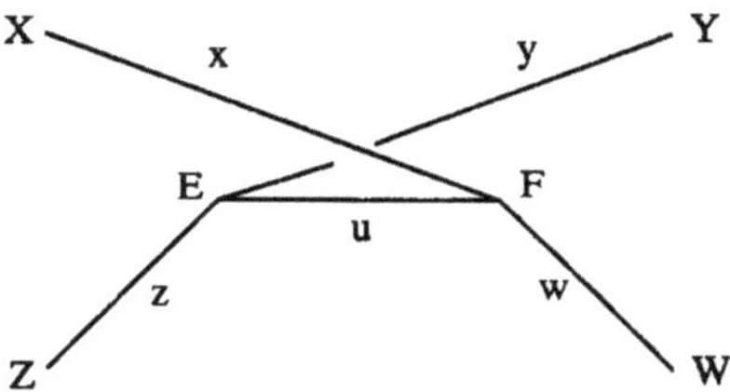

Figure 4d.

Now,

$$\begin{aligned} d(X,Z) = d(X,Q) &\le d(X,E) + d(E,Q) \\ x + u + z &\preceq x + \lambda. \end{aligned}$$

Thus,

$$u + z \preceq \lambda.$$

Similarly,

$$u + w \preceq \lambda.$$

Thus,

$$d(Z, W) \simeq z + w + u \leq (z + u) + (w + u) \preceq 2\lambda.$$

$\diamond$

Lemma 4.4 : *$\forall \lambda \, \exists r_0 = r_0(\lambda, k_1), K_0 = K_0(\lambda, k_1)$ such that the following holds.*

Suppose $Q \subseteq S$ is λ-convex, and $r \geq r_0$. Suppose $X, Y \in \mathcal{S} \setminus \check{N}_r(Q)$, $Z \in \text{proj}_Q X$ and $W \in \text{proj}_Q Y$. Let $d_1 = d_{0,Q}(Z, W)$ and $d_2 = d_{r,Q}(X, Y)$. Then $d_2 \geq 3d_1 - K_0$.

(Note that if $d_1 = \infty$, then clearly $d_2 = \infty$.)

Proof : Let λ_0 be the constant from Lemma 4.2. Let $h = J + 4\lambda_0$, where J comes from Lemma 4.3. Let $r_1 = \max(h/2, \lambda)$ and $r_2 = 3h/2$. Let $r_0 = r_1 + r_2$.

Suppose that $r \geq r_0$, and $X, Y \in \mathcal{S} \setminus \check{N}_r(Q)$. We join X to Y by a path α in $\mathcal{S} \setminus \check{N}_r(Q)$ of length d_2. Let $X = X_0, X_1, \ldots, X_p, X_{p+1} = Y$ be points on α which divide the path into p segments of length $3h$, and a remaining segment of length at most $3h$. Thus, $ph \leq d_2/3$.

Let $N = N_{r_1}(Q)$. Since $r_1 \geq \lambda$, by Lemma 4.2, we see that N is λ_0-convex.

For each $i = 1, \ldots, p$, we choose $Z_i \in \text{proj}_N X_i$. Let Z_0 be the point of $[X, Z]$ with $d(Z_0, Z) = r_1$, and let Z_{p+1} be the point of $[Y, W]$ with $d(Z_{p+1}, W) = r_1$. (Figure 4e.)

Thus, $Z_0 \in \text{proj}_N X$ and $Z_{p+1} \in \text{proj}_N Y$. Now, $d(X_i, X_{i+1}) \leq 3h$ and $d(X_i, N) \geq 3h/2$. Thus, by Lemma 4.3, $d(Z_i, Z_{i+1}) \leq J + 4\lambda_0 = h$ for each $i = 1, \ldots, p$. So, since $r_1 \geq h/2$, we have $[Z_i, Z_{i+1}] \cap \check{Q} = \emptyset$ for each i. We can therefore join Z to W by a path

$$[Z, Z_0] \cup [Z_0, Z_1] \cup \cdots \cup [Z_p, Z_{p+1}] \cup [Z_{p+1}, W]$$

in $\mathcal{S} \setminus \check{Q}$. Thus,

$$\begin{aligned} d_1 &\leq (p+1)h + 2r_1 \\ &= ph + (2r_1 + h) \\ &\leq \frac{d_2}{3} + (2r_1 + h). \end{aligned}$$

Thus,

$$d_2 \geq 3d_1 - (2r_1 + h).$$

$\diamond$

Proposition 4.5 : it $\forall \lambda \geq 0 \, \exists \zeta = \zeta(\lambda, k_1), L = L(\lambda, k_1)$ such that the following holds.

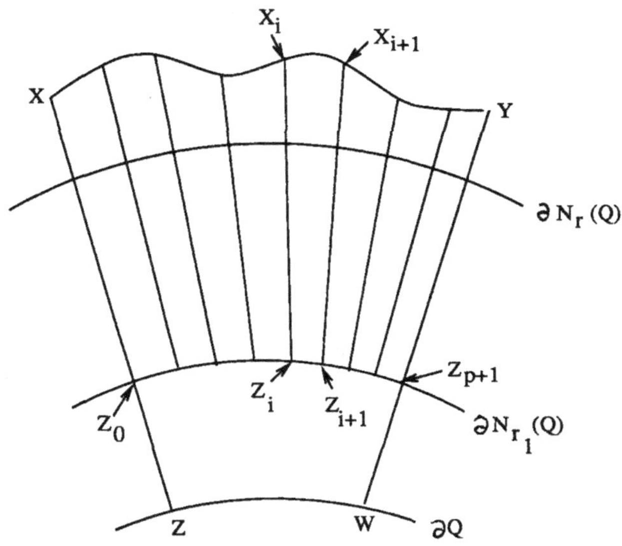

Figure 4e.

Let $Q \subseteq \mathcal{S}$ be λ-convex, and $X, Y \in \mathcal{S} \setminus \breve{N}_r(Q)$. Let $Z \in \text{proj}_Q X$ and $W \in \text{proj}_Q Y$. Set $d_1 = d_{0,Q}(Z,W)$ and $d_2 = d_{r,Q}(X,Y)$. If $r \geq L$ and $d_1 \geq L$, then

$$d_2 \geq e^{\zeta r} d_1.$$

Proof : By Lemma 4.2, there is some λ' so that any uniform neighbourhood of Q is λ'-convex. Let $r_0 = r_0(\lambda', k_1)$ and $K_0 = K_0(\lambda', k_1)$ be the constants given by Lemma 4.4. Let $L = \max(2r_0, K_0)$ and $\alpha = \frac{1}{2r_0} \log_e 2$.

Suppose that X, Y, Z, W, r, d_1, d_2 are as in the hypothesis. Let p be the integer part of (r/r_0). Let $Z = Z_0, Z_1, Z_2, \ldots, Z_p = X$ be points on $[Z, X]$ satisfying $d(Z_i, Z_{i+1}) = r_0$ for $i = 0, \ldots, p-2$. Let $W = W_0, W_1, W_2, \ldots, W_p = Y$ be similar points on $[W, Y]$. Thus, $Z_i \in \text{proj}_{N(i)} X$ and $W_i \in \text{proj}_{N(i)} Y$, where $N(i) = N_{ir_0}(Q)$. We write $\rho_i = d_{ir_0,Q}(Z_i, W_i)$. Thus, $d_1 = \rho_0$ and $d_2 = d_{r,Q}(X,Y) \geq d_{pr_0,Q}(X,Y) = \rho_p$, since $r \geq pr_0$.

By hypothesis, $\rho_0 = d_1 \geq L \geq K_0$, so by Lemma 4.4,

$$\rho_1 \geq 3\rho_0 - K_0 \geq 2\rho_0.$$

By induction,

$$\rho_i \geq 2\rho_{i-1} \geq 2^i \rho_0.$$

Thus,

$$d_2 \geq \rho_p \geq 2^p \rho_0 = 2^p d_1.$$

But $p \geq (r/r_0) - 1$, so

$$d_2 \geq (2^{1/r_0})^r \frac{d_1}{2} = \frac{1}{2} e^{2\zeta r} d_1.$$

Now, $r \geq L \geq 2r_0$, so $\zeta r \geq \log_e 2$. Thus,

$$d_2 \geq e^{\zeta r} d_1.$$

◇

Suppose that Q is λ-convex, and $X, Y \in \partial Q$. It is a fairly easy consequence of Proposition 4.5 (c.f. Proposition 4.5 below), that any shortest path from X to Y in $\mathcal{S} \backslash \mathring{Q}$ lies within a bounded distance of Q, depending on λ.

Another point to note is that we can arrange for the rate of expansion ζ to be chosen independently of λ:

Proposition 4.6 : $\forall k_1\ \exists \zeta_0, L_0$ *such that if* $Q, \lambda, r, X, Y, Z, W, d_1, d_2$ are as in *Proposition 4.5, with* $r \geq L_0 + \lambda$ *and* $d_1 \geq L_0 + 2\lambda$, *then*

$$d_2 \geq e^{\zeta_0 (r-\lambda)} (d_1 - 2\lambda).$$

Proof : Apply Proposition 4.5 to $N_\lambda(Q)$, which by Lemma 4.2 is λ_0-convex for fixed $\lambda_0 = \lambda_0(k_1)$.

◇

We make two applications of Proposition 4.5. The first is to the case where Q is a uniform ball. In this case Proposition 4.5 can be interpreted as the statement that geodesic rays diverge exponentially (Proposition 4.7). From this we deduce property H5 (Corollary 4.8).

The second application is to the case where Q is a geodesic segment. From this, we get the pseudogeodesic property (Proposition 4.9), and a direct proof of H4. We show that almost-hyperbolicity is a pseudoisometric invariant (Proposition 4.10).

Proposition 4.7 : $\forall k_1\ \exists \theta, M$ *such that the following holds. Suppose* $A, X, Y \in \mathcal{S}$ *with* $d(A, X) = d(A, Y) = r$. *Set* $d = d(X, Y)$ *and* $d' = d_{r,A}(X, Y)$. *If* $d \geq M$, *then* $d' \geq e^{\theta d}$.

Proof : By Lemma 4.2 (or Lemma 4.1), every ball in $\mathcal{S}$ is λ_0-convex for $\lambda_0 = \lambda_0(k_1)$. Let ζ and L be as in Proposition 4.5 for $\lambda = \lambda_0$. We can suppose that $L \geq 1$. Let $M = 3L$ and $\theta = \zeta/3$.

Now, suppose that $A, X, Y \in \mathcal{S}$ with $d(A,X) = d(A,Y) = r$, and $d = d(X,Y) \geq M$. By continuity of the distance function, there exist $Z \in [A,X]$ and $W \in [A,Y]$, with $d(A,Z) = d(A,W)$ and $d(Z,W) = L$. Set $d(X,Z) = d(Y,W) = l$. Now, $3L = M \leq d \leq L + 2l$, so $l \geq L$. Let $Q = N_{r-l}(A)$, so that $Z \in \mathrm{proj}_Q X$ and $W \in \mathrm{proj}_Q Y$. Also, $d_{0,Q}(Z,W) \geq d(Z,W) = L \geq 1$. Now, Q is λ_0-convex, so applying Proposition 4.5, we get

$$d' = d_{r,A}(X,Y) = d_{l,Q}(X,Y) \geq e^{\zeta l} d_{0,Q}(Z,W) \geq e^{\zeta l}.$$

But $d \leq L + 2l \leq \frac{d}{3} + 2l$. Thus $\frac{d}{3} \leq l$, and so $d' \geq e^{\zeta d/3} \geq \theta l$.
◇

Corollary 4.8 : *$\forall k_1, h_5\ \exists k_5$ such that if $\mathcal{S}$ is k_1-H1, then it is (k_5, h_5)-H5.*

Proof : Given θ and h_5, we have $e^{\theta d} \geq d + h_5$ for all sufficiently large d.
◇

We now go on to consider pseudogeodesics. We may think of a path $\gamma : [0,t] \longrightarrow \mathcal{S}$ as a distance non-increasing map, with respect to the standard metric, $\sigma(x,y) = |x - y|$, on $[0,t]$. We shall frequently use γ to denote both the map, and its image in $\mathcal{S}$.

Definition : We call γ a (ν_1, ν_2)-*pseudogeodesic*, if for all $x, y \in [0,t]$, we have

$$\sigma(x,y) \leq \nu_1 d(\gamma x, \gamma y) + \nu_2.$$

Proposition 4.9 : *$\forall k_1, \nu_1, \nu_2\ \exists l$ such that the following holds.*

Suppose $\gamma : [0,t] \longrightarrow \mathcal{S}$ is a (ν_1, ν_2)-pseudogeodesic. Let $X = \gamma(0)$, $Y = \gamma(t)$, and suppose that $[X,Y]$ is any geodesic joining X to Y. Then, $\gamma \subseteq N_l([X,Y])$ and $[X,Y] \subseteq N_l(\gamma)$.

Proof : We will first prove γ lies inside a uniform neighbourhood of $[X,Y]$. For this, we intend to apply Proposition 4.5 to the λ_0-convex set $Q = [X,Y]$.

Let ζ and L be the constants given by Proposition 4.5, for $\lambda = \lambda_0$. Let

$$\begin{aligned} l_1 &= \max\bigl(L, \frac{1}{\zeta} \log_e(1 + \nu_1)\bigr) \\ m &= \max(L, 2\nu_1 l_1 + \nu_2) \\ l_2 &= \frac{1}{2}\bigl(\nu_1(m + 2l_1) + \nu_2\bigr) \\ l_0 &= l_1 + l_2. \end{aligned}$$

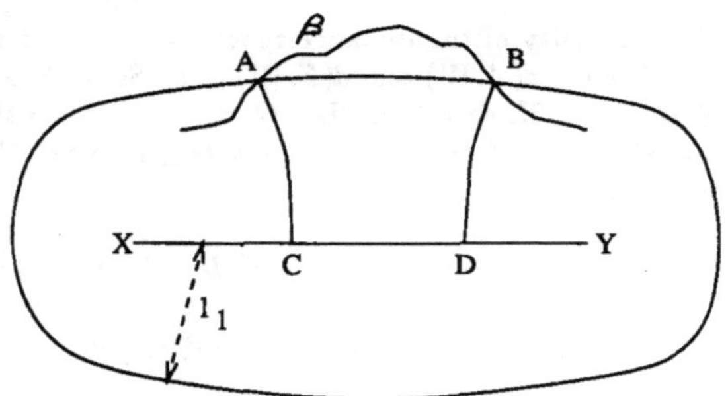

Figure 4f.

We claim that $\gamma \subseteq N_{l_0}([X,Y])$. If $\gamma \subseteq N_{l_1}([X,Y])$, we are done. If not, let β be a component of γ lying in $\mathcal{S}\backslash N_{l_1}([X,Y])$. Thus β is a (ν_1, ν_2)-pseudogeodesic joining two points A and B in $\partial N_{l_1}([X,Y])$. Let C and D be nearest points on $[X,Y]$ to A and B respectively. (Figure 4f.)

Set $d_1 = d(C,D)$ and $d_2 = d_{l_1,[X,Y]}(A,B)$. Thus $d_1 \leq d_{0,[X,Y]}(C,D)$. Now,

$$\begin{aligned} d_2 &\leq \text{length}\beta \leq \nu_1 d(A,B) + \nu_2 \\ &\leq \nu_1(d_1 + 2l_1) + \nu_2 \\ &= \nu_1 d_1 + (2l_1\nu_1 + \nu_2). \end{aligned}$$

Applying Proposition 4.5, we find that either

$$d_1 \leq L,$$

or

$$\begin{aligned} e^{\zeta l_1} &\leq d_2 \\ &\leq \nu_1 d_1 + (2l_1\nu_1 + \nu_2), \end{aligned}$$

so that

$$d_1 \leq (e^{\zeta l_1} - \nu_1)d_1 \leq 2l_1\nu_1 + \nu_2.$$

Either way,

$$d_1 \leq m.$$

Thus,

$$\begin{aligned} \text{length}\beta &\leq \nu_1(m + 2l_1) + \nu_2 \\ &= 2l_2 \end{aligned}$$

and so, $\beta \subseteq N_{l_2}\big(N_{l_1}([X,Y])\big) = N_{l_0}([X,Y])$ and the claim follows.

It remains to show that $[X,Y]$ lies in a uniform neighbourhood of γ. Let $l = 2l_0 + \frac{1}{2}$. We claim that $[X,Y] \subseteq N_l(\gamma)$.

To see this, we choose points $X = X_1, X_2, \ldots, X_p = Y$ on γ, so that $d(X_i, X_{i+1}) \leq 1$ for all i. Let $Y_i \in \mathrm{proj}_{[X,Y]} X_i$, so that $Y_0 = X$ and $Y_p = Y$. Now, $d(Y_i, Y_{i+1}) \leq l_0 + d(X_i, X_{i+1}) + l_0 \leq 2l_0 + 1$. We conclude that each point of $[X,Y]$ lies at most a distance $\frac{1}{2}(2l_0 + 1)$ from some point Y_i. But $Y_i \in N_{l_0}(\gamma)$. Thus, $[X,Y] \subseteq N_{l_0+\frac{1}{2}}(N_{l_0}(\gamma)) = N_l(\gamma)$.
$\diamond$

A simple corollary of Proposition 4.9 is that H1 $\Rightarrow$ H4. We shall deduce H3 $\Rightarrow$ H4 in the next chapter.

Our main application of Proposition 4.9 is the following. Recall the definition of pseudoisometry from Section 1.3.

Proposition 4.10 : *$\forall k, \mu_1, \mu_2\ \exists k'$ such that the following holds.*

Suppose that $(\mathcal{S}, d)$ and $(\mathcal{S}', d')$ are (μ_1, μ_2)-pseudoisometric geodesic spaces, and suppose that $(\mathcal{S}, d)$ is k-H2, then $\mathcal{S}'$ is k'-H2.

Proof : Let $R \subseteq \mathcal{S} \times \mathcal{S}'$ be a (μ_1, μ_2)-pseudoisometry. Suppose $X', Y', Z' \in \mathcal{S}'$. Choose $X, Y, Z \in \mathcal{S}$ with XRX', YRY' and ZRZ'. Let A be a centre for XYZ in $\mathcal{S}$. Choose $A' \in \mathcal{S}'$ with ARA'. Let α' be any geodesic in $\mathcal{S}'$ joining X' to Y'. We aim to show that $d(A', \alpha')$ is bounded in terms of k, μ_1 and μ_2.

Let $X' = X'_0, X'_1, \ldots, X'_p = Y'$ be points on α' so that $d'(X'_i, X'_{i+1}) = 1$ for $i \leq p-2$, and $d'(X'_{p-1}, Y') \leq 1$. For $i = 1, \ldots, p-1$, choose $X_i \in \mathcal{S}$ with $X_i R X'_i$. Let $X_0 = X$ and $X_p = Y$. Thus, $d(X_i, X_{i+1}) \leq \mu_1 + \mu_2$. Let $\alpha \subseteq \mathcal{S}$ be the piecewise geodesic path

$$[X_0, X_1] \cup [X_1, X_2] \cup \cdots \cup [X_{p-1}, X_p]$$

from X to Y (parametrised by arc-length).

Suppose that $C, D \in \alpha$. Then, without loss of generality, $C \in [X_i, X_{i+1}]$ and $D \in [X_{j-1}, X_j]$ with $i < j$. Write ρ for the distance from C to D measured along α. Then,

$$\rho \leq (j-i)(\mu_1 + \mu_2).$$

But

$$\begin{aligned}(j-i) &\leq d'(X'_i, X'_j) + 1 \\ &\leq \big(\mu_1 d(X_i, X_j) + \mu_2\big) + 1.\end{aligned}$$

Also,

$$d(X_i, X_j) \leq d(C, D) + 2(\mu_1 + \mu_2),$$

and so

$$\rho \leq (\mu_1 + \mu_2)\big(\mu_1 d(C, D) + (2\mu_1^2 + 2\mu_1\mu_2 + \mu_2 + 1)\big).$$

We see that α is a pseudogeodesic.

Now, Proposition 4.9 gives us a constant l such that $[X, Y] \subseteq N_l(\alpha)$. But A is a centre of XYZ, so there exists $B \in [X, Y]$ with $d(A, B) \leq k$. Now, $B \in N_l(\alpha)$, so there exists $i \in \{0, 1, \ldots, p\}$ such that

$$d(B, X_i) \leq l + \frac{1}{2}(\mu_1 + \mu_2).$$

Thus,

$$d(A, X_i) \leq k + l + \frac{1}{2}(\mu_1 + \mu_2),$$

and so,

$$\begin{aligned} d'(A', X_i') &\leq \mu_1\big(k + l + \frac{1}{2}(\mu_1 + \mu_2)\big) + \mu_2 \\ &= k'. \end{aligned}$$

We have shown that $d'(A', \alpha') \leq k'$.

The same argument applies to any geodesics β' from Y' to Z', and γ' from Z' to X'. Thus $d'(A', \beta') \leq k'$ and $d'(A', \gamma') \leq k'$. We conclude that

$$N_{k'}(\alpha') \cap N_{k'}(\beta') \cap N_{k'}(\gamma') \neq \emptyset.$$

CHAPTER 5. Area.

The purpose of this chapter is to describe a few notions of area appropriate to the setting of path-metric spaces. We shall begin with a discussion of the notion of degree for maps of the circle. This will enable us to give a description which is intrinsic to the disc. This gives a number of different formulations of the "rectangle principle" (5.7, 5.8, 5.11), of which the central one for our purposes is 5.7. We conclude with a proof that the rectangle principle, together with the isoperimetric inequality imply property H4.

We shall identify the circle S^1 with $\mathbf{R}/\mathbf{Z}$. Let J_1, J_2, J_3, J_4 be the quotients, under the $\mathbf{Z}$-action, of the intervals $\bar{J}_i = [\frac{i}{4} - \frac{3}{16}, \frac{i}{4} + \frac{3}{16}] \subseteq \mathbf{R}$ for $i = 1, 2, 3, 4$. Thus, $\{J_1, J_2, J_3, J_4\}$ is a covering of S^1 by four overlapping intervals satisfying $J_1 \cap J_2 = \emptyset$ and $J_2 \cap J_4 = \emptyset$.

Definition : A *4-link*, (F_1, F_2, F_3, F_4) is a (cyclically ordered) collection of four closed subsets F_1, F_2, F_3, F_4 of S^1 satisfying $F_1 \cup F_2 \cup F_3 \cup F_4 = S^1$, $F_1 \cap F_3 = \emptyset$ and $F_2 \cap F_4 = \emptyset$.

We shall take subscripts mod 4, so that $F_{i+4} = F_i$.

Definition : We say that a continuous map $f : S^1 \longrightarrow S^1$ is *associated* to a 4-link (F_1, F_2, F_3, F_4), if $f(F_i) \subseteq J_i$ for each $i = 1, 2, 3, 4$.

Lemma 5.1 : *Every 4-link has some map $f : S^1 \longrightarrow S^1$ associated to it. Moreover, any two maps associated to the same 4-link are homotopic.*

Proof : The second part is immediate: any two maps associated to the same 4-link are never antipodal, and are thus homotopic (by linear homotopy).

To prove the first part, let (F_1, F_2, F_3, F_4) be any 4-link. We can find another 4-link, (G_1, G_2, G_3, G_4), such that $F_i \subseteq G_i$ for each i, and $\{G_1, G_2, G_3, G_4\}$ is a collection of general-position 1-manifolds of S^1 (i.e. each G_i is a finite union of intervals, and $\partial G_i \cap \partial G_j = \emptyset$ if $i \neq j$). The sets G_i may be obtained by taking a small uniform neighbourhood of the F_i in the standard path-metric on S^1.

Now let $B_{ij} = G_i \cap \partial G_j$ for $i \neq j$, and $B = \bigcup_{i,j,i \neq j} B_{ij} = \bigcup_{i=1}^{4} \partial G_i$. Note that $B_{ij} = \emptyset$ if $j \equiv i + 2$. Thus, the B_{ij} partition B into eight subsets indexed by $\{(i,j) \mid i - j \text{ is odd}\}$.

Each complementary region e of $S^1 \backslash B$ is a connected component of one of the sets $E_i = \check{G}_i \backslash \bigcup_{j \neq i} G_j$ or $E_{ij} = \check{G}_i \cap \check{G}_j$. If e is bounded by $x \in B_{ij}$ and $y \in B_{kl}$, the possibilities are as follows:

(1) $i = k$, $j = l$, and either $e \subseteq E_i$ or $e \subseteq E_{ij}$,

(2) $i = k$, $j \equiv l + 2$, and $e \subseteq E_i$,

(3) $i = l$, $j = k$ and $e \subseteq E_{ij}$.

A similar discussion applies to the intervals J_i on S^1. This time, if $i-j$ is odd, then $J_i \cap \partial J_j$ consists of a single point x_{ij} ($= \frac{1}{16}(3i+j)$ if we take $0 \leq i,j \leq 3$).

We may now define $f : S^1 \longrightarrow S^1$ by sending each $v \in B_{ij}$ to x_{ij}, and mapping each complimentary region in linearly. From the above discussion, it is easily checked that $f(G_i) \subseteq J_i$. Thus $f(F_i) \subseteq J_i$.

◇

Lemma 5.1 allows us to define the degree $\deg(F_1, F_2, F_3, F_4)$ of any 4-link as the topological degree of any associated map. The following properties are easily deduced.

Proposition 5.2 : *Let (F_1, F_2, F_3, F_4) be any 4-link on S^1. Then,*
(1) $\deg(F_1, F_2, F_3, F_4) = \deg(F_2, F_3, F_4, F_1) = -\deg(F_4, F_3, F_2, F_1)$.
(2) If (F_1', F_2', F_3', F_4') is another 4-link with $F_i \subseteq F_i'$ for each i, then

$$\deg(F_1, F_2, F_3, F_4) = \deg(F_1', F_2', F_3', F_4').$$

(3) If $g : S^1 \longrightarrow S^1$ is any map, then $(g^{-1}F_1, g^{-1}F_2, g^{-1}F_3, g^{-1}F_4)$ is a 4-link, and

$$\deg(g^{-1}F_1, g^{-1}F_2, g^{-1}F_3, g^{-1}F_4) = \deg g \deg(F_1, F_2, F_3, F_4).$$

Proof :
(1) Compose an associated map with a rotation or reflection.
(2) Any map associated to (F_1', F_2', F_3', F_4') is also associated to (F_1, F_2, F_3, F_4).
(3) If the map f is associated to (F_1, F_2, F_3, F_4), then $f \circ g$ is associated to

$$(g^{-1}F_1, g^{-1}F_2, g^{-1}F_3, g^{-1}F_4).$$

◇

We will often abbreviate $\deg(F_1, F_2, F_3, F_4)$ to $\deg(\{F_i\})$.

Remark : We can clearly generalise the notion of 4-link to a "p-link" for any $p \geq 3$. The closed sets $F_1, \ldots, F_p$ form a p-link, if they cover S^1, and have the same intersection properties as p cyclically overlapping intervals on S^1, namely $F_i \cap F_j = \emptyset$ unless $j \equiv i-1, i, i+1 (\text{mod} p)$ for $p \geq 4$, or $F_1 \cap F_2 \cap F_3 = \emptyset$ if $p = 3$.

We can also give a combinatorial formulation of degree as follows.

Suppose that G is a finite graph. We write $C_0(G)$ for the set of vertices of G, and $C_1(G)$ for the set of edges of G.

Definition : A 4-*colouring*, (V_1, V_2, V_3, V_4) of G is a partition of $C_0(G)$ into four disjoint subsets, $C_0(G) = V_1 \cup V_2 \cup V_3 \cup V_4$, such that no vertex in V_1 is adjacent to any vertex in V_3, and no vertex in V_2 is adjacent to any vertex in V_4.

Now, any finite set of points $V \subseteq S^1$ gives us a representation of S^1 as a 1-complex with vertex set V. Any 4-colouring of this complex, $V = V_1 \cup V_2 \cup V_3 \cup V_4$ gives rise to a map $g : S^1 \longrightarrow S^1$, where $g(v) = i/4$ for all $v \in V_i$, and each edge of the 1-complex is mapped either to a point of $V^0 = \{\frac{1}{4}, \frac{2}{4}, \frac{3}{4}, \frac{4}{4}\}$, or to a component of $S^1 \setminus V^0$. This map is well defined up to homotopy on the edges, so we may define the degree of (V_1, V_2, V_3, V_4) as $\deg(V_1, V_2, V_3, V_4) = \deg g$.

The following lemma relates the degree of a 4-colouring on S^1 to the degree of a 4-link.

Lemma 5.3 : *Suppose that σ is a path-metric on S^1, and that (F_1, F_2, F_3, F_4) is a 4-link. Let $d = \min(\sigma(F_1, F_3), \sigma(F_2, F_4))$. Suppose that $V \subseteq S^1$ is a finite set of points, such that the σ-length of any of any component of $S^1 \setminus V$ is strictly less than d. Suppose that $\{V_1, V_2, V_3, V_4\}$ is a partition of V into disjoint subsets with $V_i \subseteq F_i$ for each i. Then, (V_1, V_2, V_3, V_4) is a 4-colouring, and* $\deg(V_1, V_2, V_3, V_4) = \deg(F_1, F_2, F_3, F_4)$.

Proof : That (V_1, V_2, V_3, V_4) is a 4-colouring, follows from the intersection properties of the F_i.

We associate a 4-link to (V_1, V_2, V_3, V_4) by constructing the "Dirichlet domains" about the points of V as follows. Let V' be the set of σ-midpoints of components of $S^1 \setminus V$. Given $v \in V$, let $D(v)$ be the closure of the component of $S^1 \setminus V'$ containing v. Thus, $D(v) \subseteq N_\delta(v)$, where 2δ is the maximum σ-length of any component of $S^1 \setminus V$. Let $D_i = \bigcup_{v \in V_i} D(v)$. Then, (D_1, D_2, D_3, D_4) is a 4-link. We define $f : S^1 \longrightarrow S^1$ associated to (D_1, D_2, D_3, D_4) as follows. If $v \in V_i$, we set $f(v) = \frac{i}{4}$. If $v' \in V'$, and the adjacent vertices of V lie in V_i and V_j respectively, with $|i - j| \leq 1$, we set $f(v') = \frac{i+j}{8}$. We map in linearly all the components of $S^1 \setminus (V \cup V')$. From the construction, f also defines the degree of (V_1, V_2, V_3, V_4). Thus $\deg(\{V_i\}) = \deg(\{D_i\})$.

Now, $D_i \subseteq N_\delta(V_i) \subseteq N_\delta(F_i)$. But $\delta < d/2$, and so we have that the 4-tuple $(N_\delta(F_1), N_\delta(F_2), N_\delta(F_3), N_\delta(F_4))$ is a 4-link. Making two applications of Proposition 5.2 (2), we find that $\deg(\{D_i\}) = \deg(\{N_\delta(F_i)\}) = \deg(\{F_i\})$.

◇

Let G be a finite graph, and let $W \subseteq C_0(G)$ be a subset of the vertices. We shall write span(W) for the subgraph of G whose vertex set is W, and whose edge set comprises those edges both of whose endpoints lie in W. By a "path in W", we mean a path in span(W), i.e. a path, all of whose vertices lie in W. An "arc" is a path which passes through no vertex more than once. A "component" of W is the intersection of $C_0(G)$ with a connected component of span(W). Any two points in the same component of W may be joined by an arc in W.

Lemma 5.4 : *Suppose that P is a triangulation of the disc D, and that $C_0(P) = V_1 \cup V_2 \cup V_3 \cup V_4$ is a 4-colouring of the 1-skeleton $\Sigma(P)$. Write $V_i^\partial = V_i \cap \partial D$. Then,* $\deg(V_1^\partial, V_2^\partial, V_3^\partial, V_4^\partial) = 0$.

Proof : Note that every triangle $c \in C_2(P)$ must have at least two vertices lying in the same V_i. We define $f : D \longrightarrow S^1$, by sending $v \in V_i$ to $\frac{i}{4}$, and mapping linearly triangles and edges. (Each triangle gets sent either to a vertex of $V^0 = \{\frac{1}{4}, \frac{2}{4}, \frac{3}{4}, \frac{4}{4}\}$, or to a component of $S^1 \setminus V^0$.) Thus $\deg(\{V_i^\partial\}) = \deg(f|\partial D) = 0$.

◇

Lemma 5.5 : *Suppose that P is a triangulation of the disc D, and that $C_0(P) \cap \partial D = V_1^\partial \cup V_2^\partial \cup V_3^\partial \cup V_4^\partial$ is a 4-colouring of ∂D. Suppose that $C_0(P) = W_1 \cup W_2$, with $W_1 \cap W_2 = \emptyset$, and $W_1 \cap \partial D = V_1^\partial \cup V_3^\partial$ and $W_2 \cap \partial D = V_2^\partial \cup V_4^\partial$. If $\deg(V_1^\partial, V_2^\partial, V_3^\partial, V_4^\partial) \neq 0$, then there is either an arc from V_1^∂ to V_3^∂ in W_1, or an arc from V_2^∂ to V_4^∂ in W_2.*

Proof : Suppose neither kind of arc exists. For $i = 1, 2$, let V_i be the set of vertices of P connected to V_i^∂ in W_i. Let $V_3 = W_1 \backslash V_1$ and $V_4 = W_2 \backslash V_2$. Thus (V_1, V_2, V_3, V_4) is a 4-colouring of $\Sigma(P)$, and $V_i \cap \partial D = V_i^\partial$ for each i. By Lemma 5.4, we have $\deg(V_1^\partial, V_2^\partial, V_3^\partial, V_4^\partial) = 0$.

◇

The following lemma will enable us to construct triangulations of the disc from cellulations. We demand that every vertex of a cellulation should meet at least three edges.

Lemma 5.6 : *Suppose P is a cellulation of the disc. Then, we can subdivide P to give a triangulation P' with at most $54|C_2(P)|$ edges.*

Proof : We construct the triangulation in two stages. First, we take a point in each 2-cell of P, and subdivide the 2-cell as a cone about this point (Figure 5a).

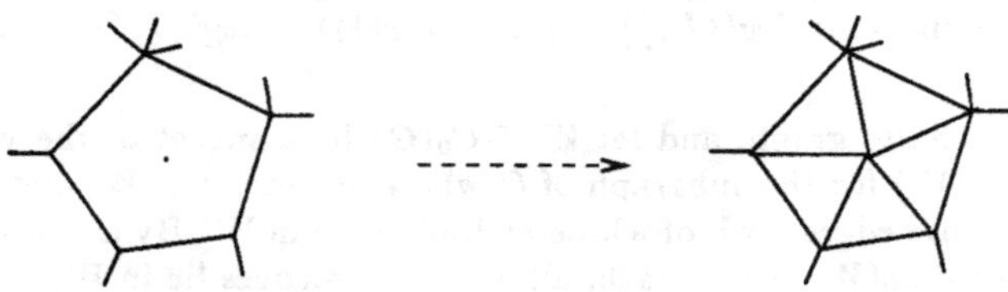

Figure 5a.

Figure 5b.

This gives a complex P^0 which might contain vertices of degree 2 (Figure 5b) To deal with this, we take the first barycentric subdivision P' of P^0 (Figure 5c).

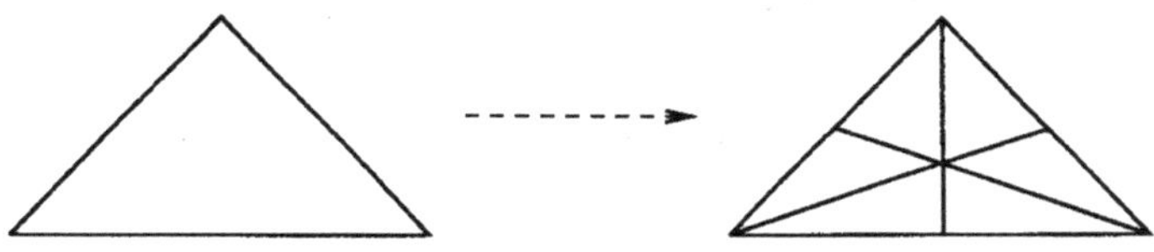

Figure 5c.

Let $t_i = |C_i(P)|$ be the number of i-cells of P. Similarly, let $t_i^0 = |C_i(P^0)|$ and $t_i' = |C_i(P')|$. We write $t_1 = t_1^I + t_1^B$, where $t_1^B = |\{e \in C_1(P) \mid e \subseteq \partial D\}|$ is the number of boundary edges of P, and t_1^I is the number of interior edges. Now $2t_1 \geq 3t_0$ since each vertex has degree 3, and $t_0 + t_2 - t_1 = 1$. Thus,

$$\begin{aligned} 3t_2 - t_1 &= 2t_1 + 3t_2 - 3t_1 \\ &\geq 3t_0 + 3t_2 - 3t_1 \geq 3, \end{aligned}$$

and so,

$$t_1 \leq 3t_2 - 3 \leq 3t_2.$$

Now,

$$t_2^0 = 2t_1^I + t_1^B \leq 2(t_1^I + t_1^B) = t_1 \leq 6t_2,$$

and

$$t_0 = t_1 + t_2^0 \leq 3t_2 + 6t_2 = 9t_2.$$

Thus,

$$\begin{aligned} t_1' &= 2t_1^0 + 6t_2^0 \\ &\leq 2(9t_2) + 6(6t_2) \\ &= 54t_2. \end{aligned}$$

◇

Now, recall the definitions of metric triangulations and cellulations from Section 2.3, as well as the notions of area, energy and mesh, A, A_T, I, I_T, m, m_T. We are now in a position to give the main formulation of the rectangle principle:

Proposition 5.7 : *There is some universal constant $\theta > 0$ such that the following holds.*

Suppose that (P, ρ) is a metric cellulation of the disc D, and that (F_1, F_2, F_3, F_4) is a 4-link on $S^1 = \partial D$. Let $d_1 = \rho(F_1, F_3) > 0$ and $d_2 = \rho(F_2, F_4) > 0$. Then,

$$I(P, \rho) \geq \theta d_1 d_2 |\deg(F_1, F_2, F_3, F_4)|.$$

Our proof will give $\theta = \frac{1}{216}$.

In fact, all we shall need in this paper is the result that, if $\deg(F_1, F_2, F_3, F_4) \neq 0$, then $I(P, \rho) \geq \theta d_1 d_2$. This much is easier to prove. However, for completeness, we shall outline the rest of the argument. Note that if P is a metric triangulation, then from the inequalities of Section 2.3, we get that

$$I_T(P, \rho) \geq \frac{1}{6} I(P, \rho) \geq (\frac{\theta}{6}) d_1 d_2 |\deg(F_1, F_2, F_3, F_4)|.$$

We shall begin by proving the analogous result for metric triangulations of bounded mesh. This would suffice for the corresponding formulation of H3 (H3ta, Section 2.3).

Proposition 5.8 : *Suppose that (P, ρ) is a metric triangulation, and that (F_1, F_2, F_3, F_4) is a 4-link on ∂D. Let $d_1 = \rho(F_1, F_3)$, $d_2 = \rho(F_2, F_4)$ and $n = \deg(F_1, F_2, F_3, F_4)$. Suppose that $m_T(P, \rho) < \min(d_1, d_2)$. Then,*

$$A_T(P)(m_T(P, \rho))^2 \geq |n| \big(d_1 d_2 - (m_T(P, \rho)) \min(d_1, d_2)\big).$$

Proof : Write $C_0^\partial(P) = C_0(P) \cap \partial D$. Write $m_T = m_T(P, \rho)$ and $A_T = A_T(P)$. Let m be any number strictly greater than m_T. Thus, $A_T m \geq A_T m_T \geq \rho(\Sigma)$ (where $\rho(\Sigma)$ is the ρ-length of the 1-skeleton Σ). Let q be the integer part of d_1/m, so that $(q+1)m \geq d_1$, hence $qm \geq d_1 - m$.

For $r \in \{1, 2, \ldots, q\}$, let $G_1^r = N_{(r-1)m}(F_1) \subseteq \Sigma$ and $G_3^r = \Sigma \setminus \mathring{N}_{rm}(F_1)$, where N_t denotes the uniform t-neighbourhood in (Σ, ρ). Thus, $\rho(G_1^r, G_3^r) = m > m_T$. (Figure 5d.)

Let $F_1^r = G_1^r \cap \partial D$ and $F_3^r = G_3^r \cap \partial D$. Thus $F_1 \subseteq F_1^r$, and, since $rm \leq qm \leq d_1 \leq \rho(F_1, F_3)$, we have $F_3 \subseteq F_3^r$. We see that (F_1^r, F_2, F_3^r, F_4) is a 4-link on ∂D. By Proposition 5.2 (2), we have

$$\deg(F_1^r, F_2, F_3^r, F_4) = \deg(F_1, F_2, F_3, F_4) = n.$$

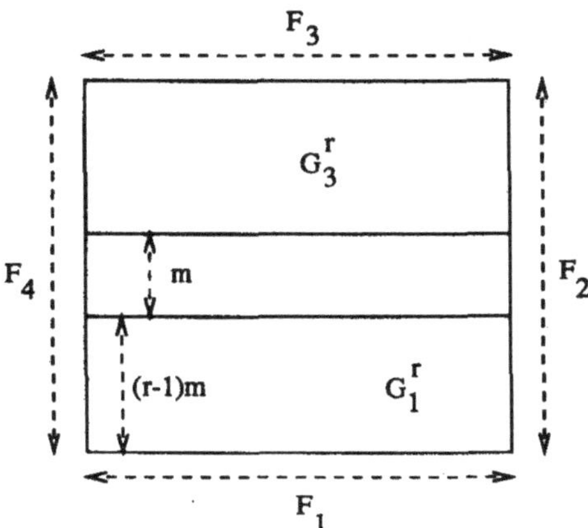

Figure 5d.

Now let

$$V_1^\theta = C_0^\theta(P) \cap F_1^r$$
$$V_3^\theta = C_0^\theta(P) \cap F_3^r$$
$$V_2^\theta = C_0^\theta(P) \cap F_2 \backslash (F_1^r \cup F_3^r)$$
$$V_4^\theta = C_0^\theta(P) \cap F_4 \backslash (F_1^r \cup F_3^r).$$

(Figure 5e.)

Since $m_T < \min(\rho(F_1^r, F_3^r), \rho(F_2, F_4))$, Lemma 5.3 tells us that the 4-tuple $(V_1^\theta, V_2^\theta, V_3^\theta, V_4^\theta)$ is a 4-colouring of ∂D, and that

$$\deg(V_1^\theta, V_2^\theta, V_3^\theta, V_4^\theta) = \deg(F_1^r, F_2, F_3^r, F_4) = n.$$

Let $W_1 = C_0^\theta(P) \cap (G_1^r \cup G_3^r)$ and $W_2 = C_0^\theta(P) \backslash W_1$, so that $W_1 \cap \partial D = V_1^\theta \cup V_2^\theta$ and $W_2 \cap \partial D = V_2^\theta \cup V_4^\theta$.

Since $m_T < \rho(G_1^r, G_3^r)$, there is no path from V_1^θ to V_3^θ in W_1.

Now, suppose that $n \neq 0$. Lemma 5.5 tells us that there must be an arc, α_r from V_2^θ to V_4^θ in W_2. Thus, α_r runs from F_2 to F_4, and lies entirely in $\Sigma \cap N_{rm}(F_1) \backslash N_{(r-1)m}(F_1)$. Such an arc must have ρ-length at least $\rho(F_2, F_4) = d_2$. Moreover, the α_r for $r = 1, 2, \ldots, q$ are all disjoint. Thus, $\rho(\Sigma) \geq \rho(\bigcup_{r=1}^q \alpha_r) \geq qd_2$, and so

$$\begin{aligned} A_T m^2 &\geq (\rho(\Sigma))m \\ &\geq (qd_2)m = d_2(qm) \\ &\geq d_2(d_1 - m) = d_1 d_2 - md_2. \end{aligned}$$

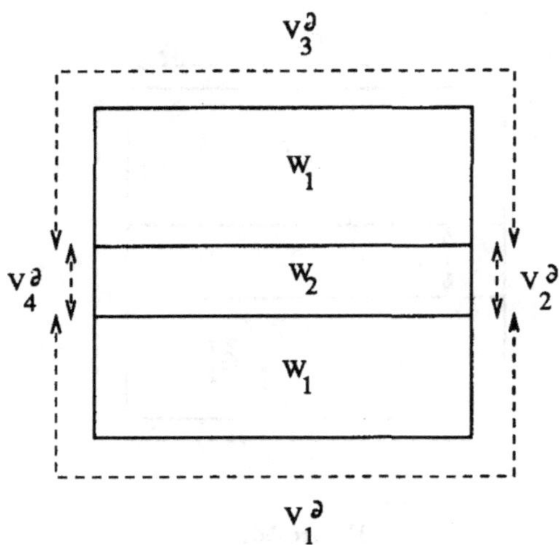

Figure 5e.

Now, without loss of generality, we have $d_2 \leq d_1$. Letting $m \longrightarrow m_T$, we have shown that, for $n \neq 0$,

$$A_T m_T^2 \geq d_1 d_2 - m_T \min(d_1, d_2).$$

As remarked above, this is all we shall need for this paper. However, we see that the full statement of Proposition 5.8 would follow if, in the above argument, we could show that there were $|n|$ edge-disjoint arcs from F_2 to F_4 lying in $\Sigma \cap N_{rm}(F_1) \backslash N_{(r-1)m}(F_1)$. This, in fact, is a consequence of a generalisation of Lemma 5.5, namely Lemma 5.9 below.

◇

Lemma 5.9 : *Suppose P is a triangulation of the disc D, and that $C_0^\partial(P) = C_0(P) \cap \partial D = V_1^\partial \cup V_2^\partial \cup V_3^\partial \cup V_4^\partial$ is a 4-colouring of ∂D. Suppose that $C_0(P) = W_1 \cup W_2$, with $W_1 \cap W_2 = \emptyset$, $W_1 \cap \partial D = V_1^\partial \cup V_3^\partial$ and $W_2 \cap \partial D = V_2^\partial \cup V_4^\partial$. Let $n = \deg(V_1^\partial, V_2^\partial, V_3^\partial, V_4^\partial)$. If there is no arc from V_1^∂ to V_3^∂ in W_1, then there is a set of $|n|$ edge-disjoint arcs from V_2^∂ to V_4^∂ in W_2.*

Sublemma 5.10 : *Suppose P is a triangulation of the disc. Suppose $C_0(P) = X_1 \cup X_2$ with $X_1 \cap X_2 = \emptyset$, and suppose $x, y \in X_2 \cap \partial D$. If there is no path from x to y in X_2, then there is a an arc α in X_1, which meets ∂D only at its endpoints, and which separates x from y.*

Proof of 5.10 : Let J, K be the two components of $\partial D \backslash \{x, y\}$, so that J, K are

open intervals. Let $Y_1^\theta = X_1 \cap J$ and $Y_3^\theta = X_1 \cap K$. Let Y_2^θ be the intersection, with ∂D, of the component of X_2 containing x. Let $Y_4^\theta = (X_2 \cap \partial D)\setminus Y_2^\theta$. Thus $y \in Y_4^\theta$. We check that $|\deg(Y_1^\theta, Y_2^\theta, Y_3^\theta, Y_4^\theta)| = 1$. Thus, by Lemma 5.5, there is an arc from Y_1^θ to Y_3^θ in X_1. Let α be such an arc with a minimal number of edges.
$\diamond$

Proof of 5.9 : (Sketch.) The proof is by induction over the number, c, of components of W_2.

Suppose first, that $c \geq 2$.

If there is some component X of W_2 with $X \cap \partial D = \emptyset$, then, by redefining $W_2' = W_2 \setminus X$ and $W_1' = W_1 \cup X$, we would reduce c. So, by induction, we could find $|n|$ edge-disjoint paths from V_2^θ to V_4^θ in $W_2' \subseteq W_2$.

Thus, we may assume that each component of W_2 meets ∂D. Let x, y lie in distinct components of W_2. By Sublemma 5.10, there is an arc α in W_1, which meets ∂D only in its endpoints w and z, and which separates x from y. Now, by hypothesis, there is no arc from V_1^θ to V_3^θ in W_1. Thus, without loss of generality, we have that $w, z \in V_1^\theta$.

Now, the arc α cuts D into two discs D^1 and D^2, with $\partial D^1 \cap \partial D^2 = \alpha$. For $j = 1, 2$, we may obtain 4-colourings $(V_1^{\theta(j)}, V_2^{\theta(j)}, V_3^{\theta(j)}, V_4^{\theta(j)})$ of ∂D^j, by $V_1^{\theta(j)} = (V_1^\theta \cap \partial D^j) \cup (\alpha \cap C_0(P))$ and $V_i^{\theta(j)} = V_i^\theta \cap \partial D^j$ for $i = 2, 3, 4$. In other words, we assign each vertex of α to $V_1^{\theta(j)}$. It is easily checked that

$$\deg(\{V_i^\theta\}) = \deg(\{V_i^{\theta(1)}\}) + \deg(\{V_i^{\theta(2)}\}).$$

For $i = 1, 2$ and $j = 1, 2$, define $W_i^j = W_i \cap D^j$. Now, each of W_2^1 and W_2^2 has fewer than c components. By induction, we may find $|n(j)|$ edge disjoint arcs from V_2^θ to V_4^θ in W_2^j, where $n(j) = \deg(\{V_i^{\theta(j)}\})$.

We are thus reduced to the case when $c = 0$ or 1, i.e. when W_2 is connected. We can assume (by interchanging the indices 1 and 3 if necessary) that $n = \deg(\{V_i^\theta\}) \geq 0$. Now, it is not difficult to find points $v_1, v_2, \ldots, v_{2n}$, cyclically arranged on ∂D, with $v_i \in V_2^\theta$ for i odd, and $v_i \in V_4^\theta$ for i even. (Note that if $n = 0$, there is nothing to prove.)

Let $T \subseteq \mathrm{span}(W_2) \subseteq \Sigma$ be a minimal subgraph which connects all the points of $\{v_1, v_2, \ldots, v_{2n}\}$. Thus, T is a planar tree whose endpoints are alternately labelled 2 and 4. We claim that such a tree contains n edge-disjoint arcs, each joining a vertex labelled 2 to one labelled 4.

We prove this claim by induction on n. Let T' be the subtree of T spanned by the set of interior nodes (i.e. vertices of degree at least 3 in T). Any extreme point w of T' will be "connected to" (at least) two consecutive vertices v_i and v_{i+1} in T. That is to say, the arc β from v_i to v_{i+1} in T, meets T' only in w. We now let T'' be the subtree of T spanned by $\{v_j \mid j \neq i, i+1\}$, so that $T'' \subseteq (T\setminus\beta) \cup \{w\}$, and apply induction.
$\diamond$

We next give a version of Proposition 5.8 for metric cellulations.

Proposition 5.11 : *Suppose that (P,ρ) is a metric cellulation, and suppose that (F_1,F_2,F_3,F_4) is a 4-link on ∂D. Let $d_1 = \rho(F_1,F_3)$, $d_2 = \rho(F_2,F_4)$ and $n = \deg(F_1,F_2,F_3,F_4)$. Suppose that $m(P,\rho) \leq \min(d_1,d_2)$. Then,*

$$A(P)(m(P,\rho))^2 \geq \frac{1}{54}|n|\big(d_1 d_2 - (m(P,\rho))\min(d_1 d_2)\big).$$

Proof : By Lemma 5.6, we can subdivide P to give a triangulation P' with

$$A_T(P') \leq 54A(P).$$

We put a path-metric ρ' on $\Sigma(P')$, by assigning each new edge a ρ'-length equal to $m = m(P,\rho)$. It is easily checked that ρ' restricted P agrees with ρ, i.e. we do not introduce any "short-cuts" across 2-cells of P. In particular, we have $\rho'(F_i,F_{i+2}) = \rho(F_i,F_{i+2}) = d_i$ for $i = 1,2$. Also $m_T(P',\rho') = m(P,\rho) = m$. Applying Proposition 5.8, we get

$$A_T(P')m^2 \geq |n|\big(d_1 d_2 - m\min(d_1,d_2)\big),$$

and so

$$A(P)m^2 \geq |n|\frac{1}{54}\big(d_1 d_2 - m\min(d_1,d_2)\big).$$

$\diamond$

From this, we may finally deduce the inequality of 5.7, namely that

$$I(P,\rho) \geq \frac{1}{216}|n|d_1 d_2.$$

Proof of 5.7 : Let (P,ρ), (F_1,F_2,F_3,F_4), d_1, d_2 be as in the hypotheses of Proposition 5.7. Let $n = \deg(F_1,F_2,F_3,F_4)$, and let $\delta = \min\{\rho(\partial c) \mid c \in C_2(P)\}$.

We subdivide (P,ρ) to give a new metric cellulation, (P',ρ'), as follows. We imagine each 2-cell $c \in C_2(P)$ as a euclidean square of side-length $\frac{1}{4}\rho(\partial c)$. We subdivide c into a grid of much smaller squares, with a path metric on the 1-skeleton induced from the euclidean metric, i.e. we assign to each 1-cell of this grid, a length equal to $\frac{\rho(\partial c)}{4h(c)}$, where $(h(c))^2$ is the number of subsquares. We perform such a construction for each 2-cell of P, and take ρ' to be the induced path-metric on the whole of $\Sigma(P')$. By taking the mesh of this subdivision much smaller than δ, we can can arrange that all the subsquares are about the same size. More precisely, given any $\epsilon > 0$, we can arrange that $\rho'(\partial c') \geq (1-\epsilon)m'$ for all $c' \in C_2(P')$, where $m' = m(P',\rho')$. Thus,

$$I(P',\rho') \geq (1-\epsilon)^2 A(P')(m')^2.$$

Note also that if $c \in C_2(P)$, then

$$\rho(\partial c)^2 = \sum\{\rho'(\partial c')^2 \mid c' \in C_2(P'), c' \subseteq c\},$$

since both sides are equal to 16 times the euclidean area of the subdivided square c. Thus,

$$I(P,\rho) = I(P',\rho').$$

Now, if $x, y \in \Sigma(P)$ are any two points, it is easily seen that $\rho'(x,y) \geq \frac{1}{2}\rho(x,y)$. In other words, taking short-cuts across 2-cells of P will shorten any path from x to y in $\Sigma(P)$ by a factor of at most 2. In particular, we have $\rho'(F_1, F_3) \geq d_1/2$ and $\rho'(F_2, F_4) \geq d_2/2$.

We may suppose that we have taken $m' < \frac{1}{2}\min(d_1, d_2)$, so can apply Proposition 5.11 to (P', ρ') to get

$$\begin{aligned} A(P')(m')^2 &\geq \frac{1}{54}|n|((\frac{d_1}{2})(\frac{d_2}{2}) - \frac{1}{2}m' \min(d_1, d_2)) \\ &= \frac{1}{216}|n|(d_1 d_2 - 2m' \min(d_1, d_2)). \end{aligned}$$

Thus,

$$\begin{aligned} I(P,\rho) = I(P',\rho') &\geq (1-\epsilon)^2 A(P')(m')^2 \\ &\geq \frac{1}{216}(1-\epsilon)^2|n|(d_1 d_2 - 2m' \min(d_1, d_2)). \end{aligned}$$

Now, let $\epsilon \longrightarrow 0$ and $m' \longrightarrow 0$. We conclude that

$$I(P,\rho) \geq \frac{1}{216}|n| d_1 d_2.$$

◇

Clearly, there is much room for improvement in the factor of $\frac{1}{216}$, particularly in the proof of Proposition 5.11.

We now use Proposition 5.7 to derive property H4 from H3.

Let (S, d) be a geodesic space. Suppose that $\gamma = \gamma_1 \cup \gamma_2 \cup \gamma_3 \cup \gamma_4$ is a rectangle (as defined in Section 2.3). Thus, $\gamma : (S^1, \sigma) \longrightarrow (S, d)$ is distance non-increasing, and we have $S^1 = L_1 \cup L_2 \cup L_3 \cup L_4$, with $\gamma_i = \gamma | L_i$. Let $d_i = d(\gamma(L_i), \gamma(L_{i+2}))$ for $i = 1, 2$. Suppose that γ bounds a cellular net (P, ρ, f) with $\partial f : (\partial D, \rho_{\partial D}) \longrightarrow (S^1, \sigma)$. Let $F_i = (\partial f)^{-1} L_i \subseteq \partial D$. Thus, $\deg(F_1, F_2, F_3, F_4) = \deg \partial f = 1$, and $f(F_i) = \gamma \circ \partial f(F_i) = \gamma(L_i)$. Since $f : (\Sigma, \rho) \longrightarrow (S, d)$ is distance non-increasing, we have $\rho(F_i, F_{i+2}) \geq d(\gamma(L_i), \gamma(L_{i+2})) = d_i$, for $i = 1, 2$. By Proposition 5.7, we get that

$$I(P,\rho) \geq \theta d_1 d_2.$$

Now, if $(\mathcal{S}, d)$ is (k_3, h_3)-H3, we can choose (P, ρ, f) so that

$$I(P,\rho) \le k_3(\sigma(S^1) + h_3),$$

and so

$$\sigma(S^1) \ge (\frac{\theta}{k_3})d_1 d_2 - h_3.$$

With the weaker version of the rectangle principle stated in Section 2.3, we get

$$k_3(\sigma(S^1) + h_3) \ge K_1 d_1 d_2 - K_2(d_1 + d_2 + 1).$$

This may be derived directly from Proposition 5.8 or 5.11 given the corresponding formulation of H3.

Lemma 5.12 : *$\forall k, l > 0\ \exists r, s$ such that the following holds.*

Suppose that $(\mathcal{S}, d)$ is a geodesic space, and suppose that for each rectangle $\gamma = \gamma_1 \cup \gamma_2 \cup \gamma_3 \cup \gamma_4$, we have

$$\text{length}\gamma \ge k d_1 d_2 - l(d_1 + d_2 + 1)$$

where $d_1 = d(\gamma_1, \gamma_3)$ and $d_2 = d(\gamma_2, \gamma_4)$. Then $(\mathcal{S}, d)$ is (r, s)-H4.

Proof : Let $r = \frac{4+l}{k}$ and $s = 2kr^2 + lr + l + 4r$.

Suppose that $\beta \subseteq \mathcal{S}$ is a geodesic segment. Let $X, Y \in \partial N_r(\beta)$, and write $d = d(X, Y)$ and $d' = d_{r,\beta}(X, Y)$. We join X to Y by a path, γ_3, of length d' lying in $\mathcal{S} \setminus \breve{N}_r(\beta)$. Let Z, W be nearest points to X, Y respectively, on β. Thus $d(X, Z) = d(Y, W) = r$. (Figure 5f.)

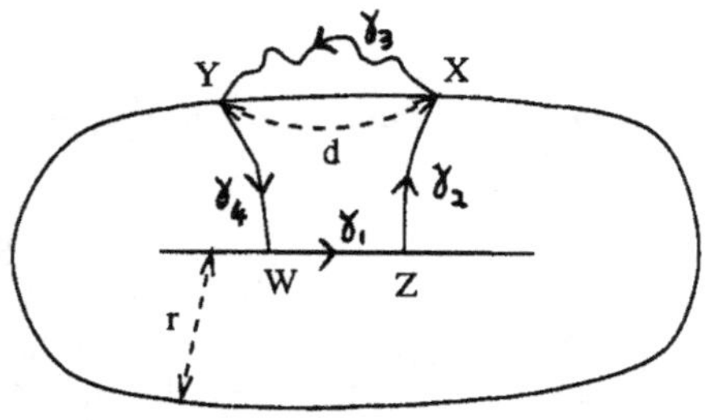

Figure 5f.

Let $\gamma_1 = [W, Z]$, $\gamma_2 = [Z, X]$ and $\gamma_4 = [Y, W]$. Let γ be the rectangle $\gamma_1 \cup \gamma_2 \cup \gamma_3 \cup \gamma_4$. We have $d_1 = d(\gamma_1, \gamma_3) = r$ and $d \ge d_2 = d(\gamma_2, \gamma_4) \ge d - 2r$. Also

$$\text{length}\gamma \le (d + 2r) + r + d' + r = d + d' + 4r.$$

So, by hypothesis,

$$\begin{aligned} d + d' + 4r &\geq kd_1d_2 - l(d_1 + d_2 + 1) \\ &\geq kr(d - 2r) - l(r + d + 1) \\ &= (kr - l)d - (2kr^2 + lr + l). \end{aligned}$$

Thus,

$$\begin{aligned} d' &\geq (kr - l - 1)d - (2kr^2 + lr + l + 4r) \\ &= 3d - s. \end{aligned}$$

◇

Proposition 5.13 : *$\forall k_3, h_3\ \exists k_4, h_4$ such that if (S, d) is (k_3, h_3)-H3, then it is (k_4, h_4)-H4.*

Proof : From Lemma 5.12 and previous discussion.

◇

CHAPTER 6. Remaining implications.

It remains to give proofs of H5 $\Rightarrow$ H3, and H4 $\Rightarrow$ H2. We prove the first implication in Section 6.1. We also include a proof of H1 $\Rightarrow$ H3, since it is much more direct than following the cycle. The final argument, $H4 \Rightarrow H2$, is given in Section 6.2.

6.1. H5 $\Rightarrow$ H3.

The idea of the proof is to show that any sufficiently long loop may be shortened by a definite amount, by replacing a portion of the loop by a geodesic segment. A sequence of these "short-cut" operations will give us a cellular net. The existence of such short-cuts is shown, for H1 spaces, in Lemma 6.1.4, and for H5 spaces, in Lemma 6.1.6. First, we get the technicalities out of the way.

Recall the definitions of Section 2.3.

Lemma 6.1.1 : *Let (S,d) be a path-metric space. Suppose that $\gamma = \gamma_1, \gamma_2, \ldots, \gamma_n$ is a sequence of loops in S, where $\gamma_i : (S^1, \sigma_i) \longrightarrow (S,d)$ is distance non-increasing. Suppose there are a sequence of arcs $J_i \subseteq S^1$, for $i = 1, \ldots, n-1$, such that γ_i agrees with γ_{i+1} on $S^1 \setminus J_i$ and σ_i agrees with σ_{i+1} on $S^1 \setminus J_i$. Suppose also that $\sigma_i(J_i) \leq K$, $\sigma_{i+1}(J_i) \leq K$, and $\sigma_n(S^1) \leq 2K$, where $K \in (0,\infty)$. Then, γ bounds a cellular net (P,ρ,f) with $A(P) = n$, and $m(P,\rho) \leq 2K$.*

Proof : We see (by induction on n) that the combinatorial structure of the arcs J_i determine a cellulation, P, of the disc D, with n 2-cells. The path-metrics, σ_i, determine a path-metric ρ on the 1-skeleton $\Sigma(P)$, so that for any 2-cell, $c \in C_2(P)$, we have $\rho(\partial c) \leq 2K$. The maps γ_i together give a distance non-increasing map $f : (\Sigma, \rho) \longrightarrow (S,d)$, such that $f|\partial D = \gamma$. Thus, $\partial f : (\partial D, \rho_{\partial D}) \longrightarrow (S^1, \sigma)$ is an isometry, and so (P,ρ,f) bounds γ.

◇

Lemma 6.1.2 : *$\forall K, b\, \exists k_3, h_3$ such that the following holds.*

Let (S,d) be a geodesic space, and suppose for that any loop $\gamma : (S^1,\sigma) \longrightarrow (S,d)$ with $\sigma(S^1) \geq 2K$, there is an arc $J \subseteq S^1$ satisfying $\sigma(J) \leq K$, and $d(\gamma(p),\gamma(q)) \leq \sigma(J) - b$, where $p, q \in S^1$ are the endpoints of J. Then, (S,d) is (k_3,h_3)-H3.

Proof : Let $\gamma : (S^1,\sigma) \longrightarrow (S,d)$ be a loop with $\sigma(S^1) \geq 2K$. Let $J_1 = J \subseteq S^1$ be the arc given by the hypotheses. Rescale the metric σ on J by a factor of $\frac{\sigma(J)-b}{\sigma(J)}$ to give a new path-metric σ_2 on S^1, with $\sigma_2(S^1) = \sigma(S^1) - b$. Define a new loop $\gamma_2 : (S^1,\sigma_2) \longrightarrow (S,d)$ by $\gamma_2|(S^1 \setminus J) = \gamma|(S^1 \setminus J)$, and taking $\gamma_2|J$ to map linearly along the geodesic $[\gamma(p), \gamma(q)]$. where $p, q \in S^1$ are the enpoints of J. Note that $\sigma_2(J) = \sigma(J) - b \geq d(\gamma(p), \gamma(q))$.

Continue by induction. After $n - 1 \leq \frac{\sigma(S^1)}{b}$ steps, we arrive at a loop $\gamma_n : (S^1, \sigma_n) \longrightarrow (S, d)$ with $\sigma_n(S^1) \leq 2K$. Applying Lemma 6.1.1, we get a cellular net (P, ρ, f) with $m(P, \rho) \leq 2K$, and $A(P) = n \leq \frac{1}{b}(\sigma(S^1) + b)$. This gives H3ca.

Now, the inequality $I(P, \rho) \leq A(P)(m(P, \rho))^2$ gives the main definition, H3ce. Finally, Lemma 6.1.3 below gives us the strongest definition H3ta.

◊

Lemma 6.1.3 : *Let (S, d) be a geodesic space. Suppose that is a loop in S bounding a cellular net (P, ρ, f). Then, γ bounds a simplicial net (P', ρ', f') with $m_T(P', \rho') \leq m(P, \rho)$, and $A_T(P') \leq 54A(P)$.*

Proof : Lemma 5.6 gives us P' combinatorially as a subdivision of P. We define ρ' by assigning, to each new edge $e \in C_1(P')$ with $e \not\subseteq \sigma(P)$, a length equal to $m(P, \rho)$.

Now, for each 2-cell $c \in C_2(P)$, we choose some $X_c \in f(\partial c)$, and map each vertex of $C_0(P') \cap (c \backslash \partial c)$ to X_c. We map in each edge $e \in C_1(P')$ with $e \subseteq c$ and $e \not\subseteq \partial c$, linearly along a geodesic segment. This defines $f' : (\Sigma(P'), \rho') \longrightarrow (S, d)$. We check that f' is distance non-increasing.

◊

Lemma 6.1.4 : *$\forall k_1, b \, \exists l$ such that the following holds.*

Suppose (S, d) is k_1-H1, and that $\gamma : (S^1, \sigma) \longrightarrow (S, d)$ is a loop with $\sigma(S^1) > l + b$. Then, there is an arc $J \subseteq S^1$ with $\sigma(J) = l + b$, and $d(\gamma(p), \gamma(p)) \leq l$ where p and q are the endpoints of J.

Proof : Lemma 2.1.1 tells us that the distances between any four points of S may be measured up to an additive constant $h = h(k_1)$ along a tree. Let $l = b + 8h$.

Suppose that γ is a loop with $\sigma(S^1) \geq l + b$. Choose any point $X \in S$, and let $Y = \gamma(t)$ be a point in $\gamma(S^1)$ furthest from X. Let J be a closed interval of σ-length $l + b$ centred at t. Let p, q be the endpoints of J. Thus, $\sigma(p, t) = \sigma(q, t) = \frac{1}{2}(l + b)$. Let $Z = \gamma(p)$ and $W = \gamma(q)$. Thus, $d(Z, Y) \leq \frac{1}{2}(l + b)$ and $d(W, Y) \leq \frac{1}{2}(l + b)$. We claim that $d(Z, W) \leq l$.

Case (1) : $XY : ZW \longleftrightarrow (xy)u(zw)$. (Figure 6a.)

Now, $d(X, W) \leq d(X, Y)$, thus

$$x + u + w \preceq_{2h} x + y$$
$$w \preceq_{2h} y,$$

and so,

$$d(W, Z) \simeq_h w + z \preceq_{2h} y + u + z \simeq_h d(Y, Z) \leq \frac{1}{2}(l + b),$$

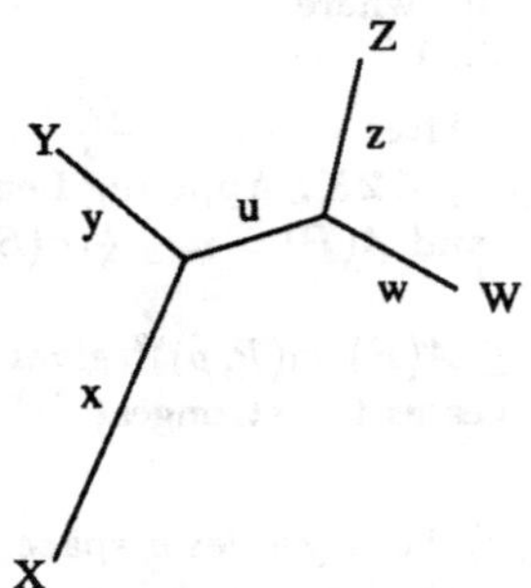

Figure 6a.

i.e.

$$d(W,Z) \leq \frac{1}{2}(l+b) + 4h$$
$$= l.$$

Case (2) : without loss of generality, $XZ : YW \longleftrightarrow (xz)u(yw)$. (Figure 6b.)

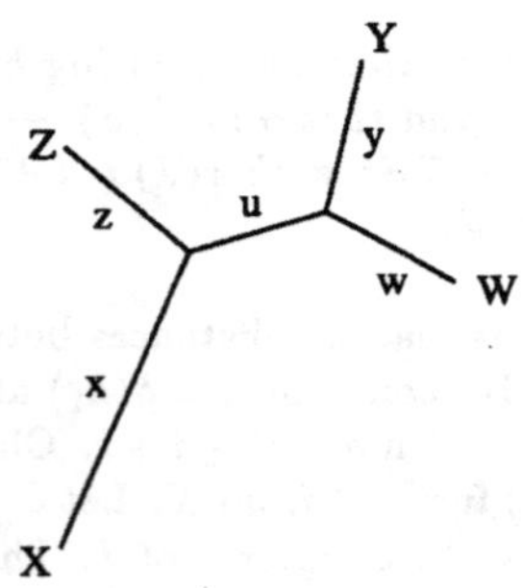

Figure 6b.

Again $d(X,W) \leq d(X,Y)$, thus

$$x+u+w \preceq_{2h} x+u+y$$
$$w \preceq_{2h} y,$$

and so

$$d(W,Z) \simeq_h w+u+z \preceq_{2h} y+u+z \preceq_h d(Y,Z) \leq \frac{1}{2}(l+b),$$

i.e.

$$\begin{aligned} d(W,Z) &\leq \frac{1}{2}(l+b)+4h \\ &= l. \end{aligned}$$

◇

Corollary 6.1.5 : *$\forall k_1 \exists k_3, h_3$ such that if $(\mathcal{S}, d)$ is k_1-H1, it is (k_3, h_3)-H3.*

Proof : Set $b = 1$. Apply Lemmas 6.1.4 and 6.1.2.

◇

Lemma 6.1.6 : *Suppose that $(\mathcal{S}, d)$ is (a, b)-H5, and $\gamma : (S^1, \sigma) \longrightarrow (\mathcal{S}, d)$ is a loop in $\mathcal{S}$, with $\sigma(S^1) > 2(a+b)$. Then, there is an arc $J \subseteq S^1$ with $\sigma(J) \leq 2(a+b)$ and $d(\gamma(p), \gamma(q)) \leq \sigma(J) - b$, where $p, q \in S^1$ are the endpoints of J.*

Proof : Choose any point $A \in \mathcal{S}$. Given any $r \geq 0$, let $C(r) = \gamma^{-1}(\mathcal{S} \setminus \check{N}_r(A))$. (Figure 6c.)

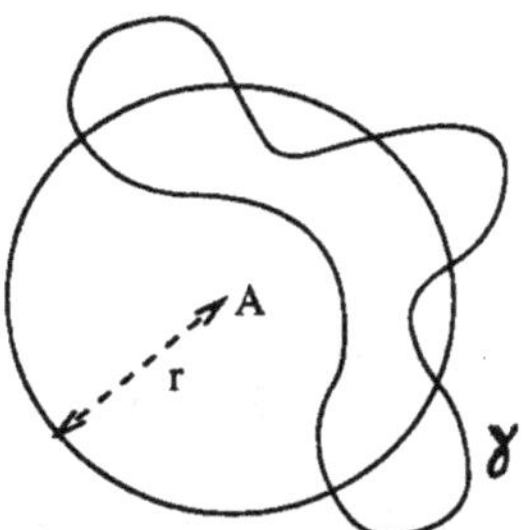

Figure 6c.

Thus, $C(r)$ is a closed subset of S^1, with $C(0) = S^1$ and $C(r_0) = \emptyset$ for some r_0. Define $g(r) \in [0, \infty)$ to be the largest σ-length of any connected component of $C(r)$. Now, $g : [0, r_0] \longrightarrow [0, \infty)$ in non-increasing and upper-semicontinuous in r. We have $g(0) = \sigma(S^1) > 2(a+b)$ and $g(r_0) = 0$. Let $R = \sup\{r > 0 \mid f(r) > a+b\}$. Thus $f(R) \geq a+b$, i.e. there is some component L of $C(R)$ with $\sigma(L) \geq a+b$. However, we see that each component of $\gamma^{-1}(\mathcal{S} \setminus N_R(A))$ has σ-length at most $a+b$. From this, it is easy to find an arc $J \subseteq L$ with $a+b \leq \sigma(J) \leq 2(a+b)$, and with $\gamma(p) \in \partial N_R(A)$ and $\gamma(q) \in \partial N_R(A)$, where $p, q \in S^1$ are the endpoints of J. Write $d = d(\gamma(p), \gamma(q))$ and $d' = d_{r,A}(\gamma(p), \gamma(q))$. (Figure 6d.)

If $d \leq a$, then $\sigma(J) \geq a+b \geq d+b$.

If $d \geq a$, then, since $(\mathcal{S}, d)$ is (a,b)-H5, we have $d' \geq d+b$. Thus, $\sigma(J) \geq d' \geq d+b$. Either way, $\sigma(J) \geq d+b$, i.e. $d(\gamma(p), \gamma(q)) \leq \sigma(J) - b$.

◇

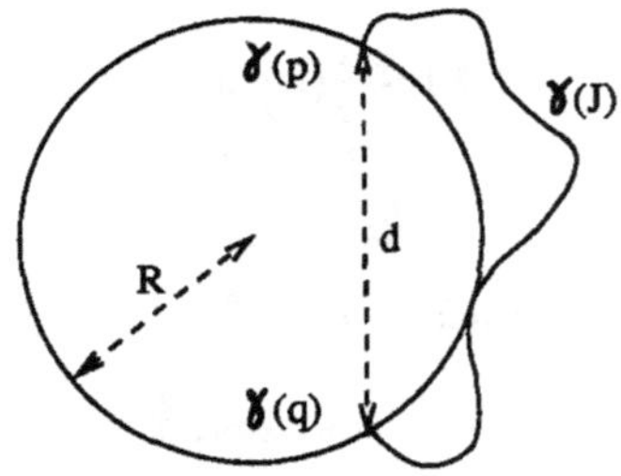

Figure 6d.

Proposition 6.1.7 : *$\forall k_5, h_5 \exists k_3, h_3$ such that if $(\mathcal{S}, d)$ is (k_5, h_5)-H5 it is (k_3, h_3)-H3.*

Proof : Lemmas 6.1.6 and 6.1.2.

◇

Remark : Note that the hypotheses of Lemma 6.1.2 serve as another definition of almost-hyperbolicity.

6.2. H4 ⇒ H2.

Let $(\mathcal{S}, d)$ be a geodesic space. Given $X, Y \in \mathcal{S}$, we write $[X, Y]$ for some choice of geodesic from X to Y. If Z, W in $[X, Y]$, we shall always take $[Z, W] \subseteq [X, Y]$.

Lemma 6.2.1 : *Suppose $(\mathcal{S}, d)$ is (r, L)-H4. Let $J = r + \frac{L}{4}$. Suppose $X, Y, Z \in \mathcal{S}$ with $d(Y, Z) \leq r$. Then $[X, Z] \subseteq N_J[X, Y]$ and $[X, Y] \subseteq N_J[X, Z]$, (that is, for any choice of geodesics $[X, Y]$ and $[X, Z]$).*

Proof : Let $[A, B] \subseteq [X, Z]$ be a component of $[X, Z] \setminus \breve{N}_r[X, Y]$ so that $A, B \in \partial N_r[X, Y]$. (Figure 6e.)

Since $(\mathcal{S}, d)$ is (r, L)-H4, we have $d(A, B) = d_{r,[X,Y]}(A, B) \geq 3d(A, B) - L$, and so $d = d(A, B) \leq \frac{L}{2}$.

Now, $[A, B] \subseteq N_{r+\frac{d}{2}}[X, Y]$, and so $[X, Z] \subseteq N_J[X, Y]$. Similarly $[X, Y] \subseteq N_J[X, Z]$.

◇

Lemma 6.2.2 : *Suppose $(\mathcal{S}, d)$ is (r, L)-H4. Let $R = J + 3r$ (where $J = r + \frac{L}{4}$ comes from Lemma 6.2.1). Suppose that $X, Y, Z \in \mathcal{S}$ and $[X, Y], [X, Z]$ are geodesics with $d(Y, [X, Z]) \geq R$ and $d(Z, [X, Y]) \geq R$. Then, there exist $A \in [X, Y]$ and $B \in [X, Z]$ such that*

$$d([A, Y], [X, Z]) \geq r$$

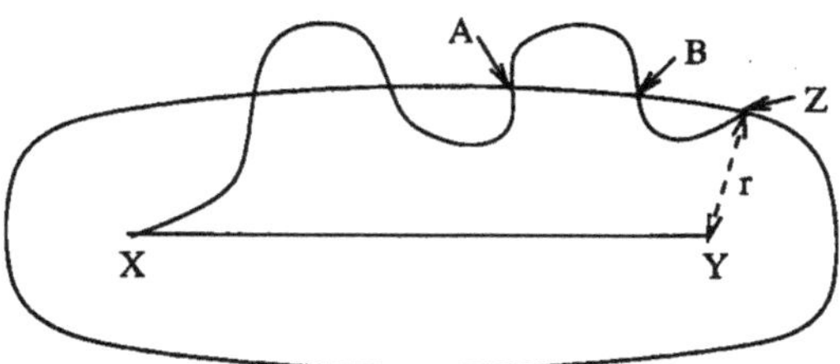

Figure 6e.

$$d([B,Z],[X,Y]) \geq r$$
$$[X,B] \subseteq N_R[X,A]$$
$$[X,A] \subseteq N_R[X,B].$$

(Note that this implies that $r \leq d(A,B) \leq R$.) (Figure 6f.)

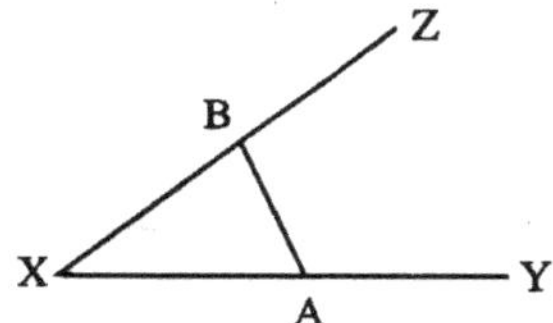

Figure 6f.

Proof : Let A be the point on $[X,Y] \cap N_r[X,Z]$ nearest Y. (By hypothesis, $Y \notin N_R[X,Z]$.) Let $C \in [X,Z]$ be a nearest point on $[X,Z]$ to A. Let $B \in [C,Z]$ be the point distant $J+2r$ from C. (B exists since $d(Z,[X,Y]) \geq R$. (Figure 6g.)

Now, $d(C,A) = r$, therefore by Lemma 6.2.1, $[X,A] \subseteq N_J[X,C]$, and so $[X,A] \subseteq N_R[X,B]$. Also by Lemma 6.2.1, $[X,C] \subseteq N_J[X,A]$. Since $[C,B] \subseteq N_{r+(J+2r)}(A)$, we have $[X,B] \subseteq N_R[X,A]$.

By construction, we have $d([A,Y],[X,Z]) \geq r$.

Finally, suppose for contradiction, that we could find $D \in [B,Z]$ and $E \in [X,Y]$ with $d(D,E) \leq r$. Since $d([A,Y],[X,Z]) \geq r$, we must have $E \in [X,A]$. Since $[X,A] \subseteq N_J[X,C]$, there is some $F \in [X,C]$ with $d(E,F) \leq J$. (Figure 6h.)

Now, $d(D,F) \leq d(D,E) + d(E,F) \leq J + r$. But $d(C,B) = J + 2r$, and $[C,B] \subseteq [F,D]$ so $d(D,F) \geq J + 2r$. We have contradicted the existence of D an E. Thus, $d([B,Z],[X,Y]) \geq r$.

◇

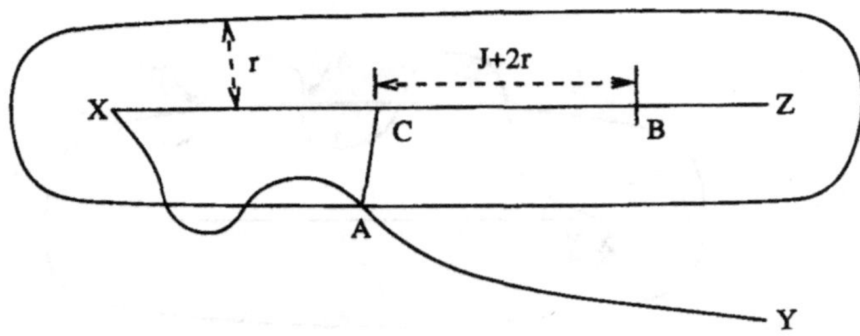

Figure 6g.

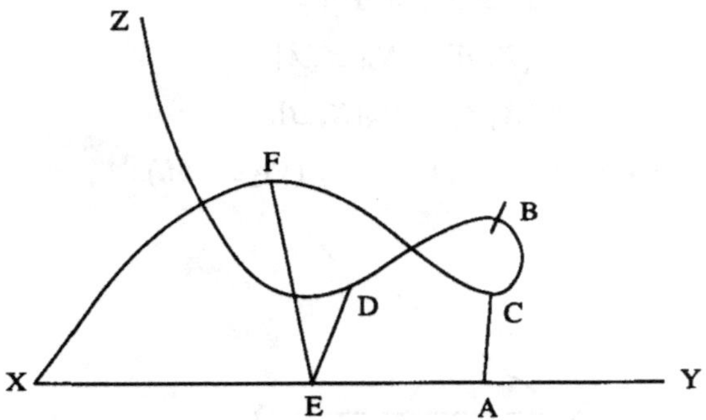

Figure 6h.

Lemma 6.2.3 : *Suppose* $(\mathcal{S}, d)$ *is* (r, L)*-H4. Let* $M = 10R + L$ *(where* $R = 4r + \frac{L}{4}$ *comes from Lemma 6.2.2). Then, given* $X, Y, Z \in \mathcal{S}$*, and geodesics* $[X, Y]$*,* $[Y, Z]$ *and* $[Z, X]$*, there is some* $W \in \mathcal{S}$ *such that*

$$d(W, [X, Y]) \leq M$$

$$d(W, [Y, Z]) \leq M$$

$$d(W, [Z, X]) \leq M.$$

We call such a point W, a "centre" for the triangle XYZ.

Proof : Suppose (for contradiction) that there is no centre for XYZ.

Then, in particular, $d(Y, [X, Z]) \geq M > R$ and $d(Z, [X, Y]) \geq M > R$. Let $A \in [X, Y]$ and $B \in [X, Z]$ be the points given by Lemma 6.2.2. Let C, D and E, F be similar points with respect to Y and Z respectively, so that $A, D \in [X, Y]$, $C, E \in [Y, Z]$ and $F, B \in [Z, X]$.

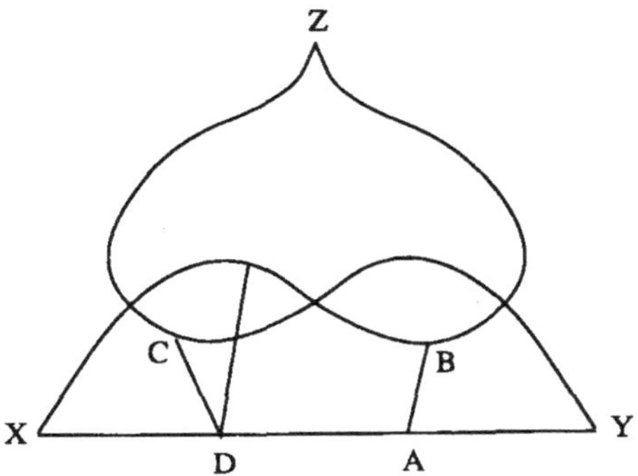

Figure 6i.

We claim that $A \in [X, D]$. Suppose (to contradict the claim) that $D \in [X, A]$. (Figure 6i.)

By hypothesis, (Lemma 6.2.2), $D \in N_R[X, B]$. Also $d(D, C) \leq R$. Thus, D is a centre for XYZ, contradicting the initial supposition. This proves the claim.

We can make similar statements about the order of points on $[Y, Z]$ and $[Z, X]$. Thus, the points are ordered cyclically $XADYCEZFBX$ about the triangle XYZ. (Figure 6j.)

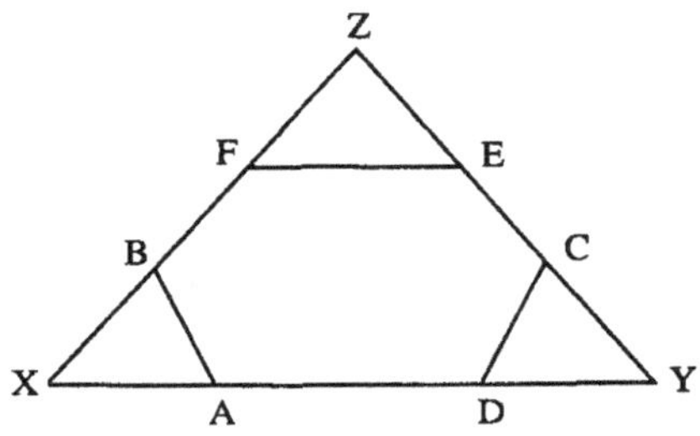

Figure 6j.

Now without loss of generality, we can assume that

$$d(A, D) \geq \max\big(d(C, E), d(F, B)\big).$$

Let $d = d(A, D)$. We know that $d(A, B) \geq r$ and $d(D, C) \geq r$. Let G be the point on $[A, B] \cap N_r[X, Y]$ nearest to B. Let H be the point on $[C, D] \cap N_r[X, Y]$ nearest to C. Thus $G, H \in \partial N_r[X, Y]$. (Figure 6k.)

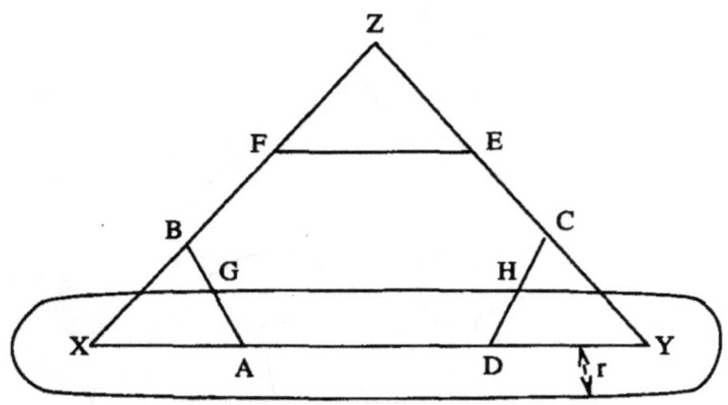

Figure 6k.

Let α be the path $[G,B]\cup[B,F]\cup[F,E]\cup[E,C]\cup[C,H]$, so that length$\alpha \leq R+d+R+d+R = 3R+2d$.

By hypothesis (Lemma 6.2.2), we have that $d([B,F],[X,Y]) \geq r$ and that $d([E,C],[X,Y]) \geq r$. Since $d(F,E) \leq R$, any point of $[F,E]\cap N_r[X,Y]$ would be a centre for XYZ. We see that $d([F,E],[X,Y]) \geq r$. Thus $\alpha \cap \breve{N}_r[X,Y] = \emptyset$. Since $\mathcal{S}$ is (r,L)-H4, we get

$$\begin{aligned}3R+2d &\geq \text{length}\alpha \\ &\geq d_{r,[X,Y]}(G,H) \\ &\geq 3d(G,H) - L \\ &\geq 3(d-2R) - L = 3d - 6R - L.\end{aligned}$$

Thus,

$$d \leq 9R + L.$$

Now,

$$\begin{aligned}d(A,[Y,Z]) &\leq d(A,D) + d(D,C) \\ &\leq d + R \\ &\leq 10R + L = M.\end{aligned}$$

Thus, A is a centre for XYZ.

This contradicts our initial supposition, and so proves the lemma.

$\diamond$

Proposition 6.2.4 : *$\forall k_4, h_4 \exists k_2$ such that if $(\mathcal{S}, d)$ is (k_4, h_4)-H4, then it is k_2-H2.*

Proof : Lemma 6.2.3. $\diamond$

CHAPTER 7 : More about trees.

7.1. Introduction.

In this chapter, we aim to explore further the "treelike" nature of hyperbolic spaces. The motivation for this chapter comes from [G, Chapter 6].

In Section 3.3, we showed that any finite subset of an almost-hyperbolic space may be spanned by a tree which measures distances up to a certain additive constant. The main aim of this chapter is to prove Theorem 7.6.1, which is a refinement of this result, though some results along the way seem to have some interest in their own right.

By a "tree", T, we mean a connected acyclic finite graph. As usual, we shall identify T, thought of combinatorially, with its realisation as a 1-complex.

Given any closed subset $P \subseteq T$, we write $\operatorname{span}_T P$, or just $\operatorname{span} P$ for the subtree of T spanned by P, i.e. the the smallest subtree containing P. Thus, in particular, if x, y are distinct points of T, then $\operatorname{span}(x, y)$ is the unique arc joining x to y.

Given any point $x \in T$, we write $\deg x$ for the degree of x, i.e. the number of connected components of $T \backslash \{x\}$. We write

$$\operatorname{node} T = \{x \in T \mid \deg x \geq 3\}$$

for the set of "nodes" of T, and

$$\operatorname{ext} T = \{x \in T \mid \deg x = 1\}$$

for the set of "extreme points" of T.

Another way to describe $\operatorname{ext} T$ is to say that it is the smallest subset, W, of T such that $T = \operatorname{span} W$.

Let V be a finite set. An *(abstract) spanning tree*, $\tau = (T, \sigma, f)$, for V consists of a tree T, together with a path-metric σ on T, and a map $f : V \longrightarrow T$ such that $T = \operatorname{span} f(V)$ (or equivalently $\operatorname{ext} T \subseteq f(V)$). Clearly (T, σ) is a geodesic space (as defined in Section 1.1). We can alternatively think of σ a finite 1-dimensional measure of full support on T. If $Q \subseteq T$, we write $\sigma(Q)$ for the measure, or "σ-length, of Q. We have $\sigma(x, y) = \sigma(\operatorname{span}(x, y))$ for all $x, y \in T$. Given $v, w \in V$, we write

$$\rho_\tau(v, w) = \sigma(f(v), f(w)).$$

Thus ρ_τ is a pseudometric V, i.e. we allow for the possibility of two distinct points being 0 distance apart. (This extra generality will be convenient in Section 7.3.) Corollary 7.3.2 gives a characterisation of which pseudometrics may be derived in this way.

We inserted the word "abstract" in the definition to distinguish it from from the following notion of "immersed" spanning tree. Let (S, d) be a geodesic space, and

$V \subseteq (\mathcal{S}, d)$ a finite set of points. An *immersed spanning tree*, (τ, g), for V, consists of an abstract spanning tree $\tau = (T, \sigma, f)$, together with a distance non-increasing map $g : (T, \sigma) \longrightarrow (\mathcal{S}, d)$ such that $g \circ f : V \longrightarrow (\mathcal{S}, d)$ is just the inclusion of V into $(\mathcal{S}, d)$. (Thus, in this case, f is injective and ρ_τ is a metric.) We call such a tree an *embedded spanning tree* if g is injective and σ is the path metric on T induced from d. In this case, we may identify T with $g(T) \subseteq \mathcal{S}$, so that the maps f and g become superfluous. In all the cases we deal with, the edges of an embedded spanning tree will be geodesic.

In Section 3.3, we showed that if $(\mathcal{S}, d)$ is k-H1, and $V \subseteq \mathcal{S}$ has $n+1$ elements, then there is an embedded spanning tree $T = T_V$ for V such that for all $v, w \in V$, we have

$$\rho_T(v, w) \leq d(v, w) + K(k, n),$$

where $K(k, n)$ depends only on k and n. In fact, we can write $K(k, n)$ in the form $kH(n)$, where $H : \mathbf{N} \longrightarrow [0, \infty)$ is a fixed function of n. This can be seen from a study of the proof, though in fact, it is a consequence of a more general principle, about which we shall say more in Section 7.2. With regard to the dependence on n, however, we were more careless. Our proof would give $H(n)$ exponential in n.

Definition : Given a function $f : \mathbf{N} \longrightarrow [0, \infty)$, and a set V of $n+1$ points in a k-H1 space $(\mathcal{S}, d)$, we shall call an immersed spanning tree, (τ, g), for V, an *$O(f(n))$-approximating tree*, or just *$O(f(n))$-tree*, if for all $v, w \in V$, we have

$$\rho_\tau(v, w) \leq d(v, w) + kO(f(n)).$$

Of course, this is really a property of a method of construction, rather than of a particular tree.

So far, we have shown the existence of embedded exponential approximating trees. Theorem 7.6.1. shows the existence of immersed $O(\log n)$ trees. In fact, this is the best order of growth one could hope for, as the following example shows.

Consider $n+1$ points equally spaced around a large circle in the hyperbolic plane. The best immersed tree in this case is obtained by joining each point to the centre of the circle by a geodesic segment. The angle between two adjacent segments is $O(1/n)$. We see that this gives us an $O(-\log(1/n)) = O(\log n)$-approximating tree.

We shall show (Proposition 7.5.2) that the construction of Section 3.3 gives, in fact, an $O(n)$-approximating tree. Another obvious way to construct an embedded spanning tree is to take a tree of minimal total length spanning the $n+1$ points of V. We call such a tree a *Steiner tree.* We shall show that this also gives an $O(n)$-approximating tree. In both these cases, this is, in general, the best result possible, as may seen from the following example.

Let $(\mathcal{S}, d)$ be the bi-infinite euclidean strip $\mathbf{R} \times [-1, 1] \subseteq \mathbf{R}^2$. For $i \in \mathbf{N}$ let X_i be the point $(4i, (-1)^i) \in \mathcal{S}$. Given $n \in \mathbf{N}$, let V_n be the set of points $\{X_i \mid 0 \leq i \leq n\}$.

The (unique) Steiner tree joining these points consists of the piecewise geodesic arc $\alpha = \bigcup_{i=1}^{n}[X_i, X_{i-1}]$. Clearly, for $n \geq 3$, we have length $\alpha \geq d(X_0, X_n) + \epsilon n$ for a fixed $\epsilon > 0$. The construction of Section 3.3 relied on a choice of ordering of the points of V. If we order the points in the obvious way, $X_0, X_2, \ldots, X_n$, then this construction also gives $T_V = \alpha$.

It seems quite likely that if we were to judiciously choose the ordering of the points of V, then the construction of Section 3.3 would always give an $O(\log n)$-approximating tree. However, I have been unable to prove this, and so Theorem 7.6.1 uses a slightly different construction (based on that of [G, Chapter 6]). This will in general give an immersed, rather than an embedded spanning tree. Perhaps the proof can be modified so as always to give an embedded tree. However, for general hyperbolic spaces, the assumption of embeddedness does not seem particularly natural (unless, for example, $\mathcal{S}$ happens to be a 2-manifold). Note that by a small perturbation, we can always arrange that a tree be embedded in $\mathcal{S} \times [0, \epsilon]$.

The main steps in the proof of Theorem 7.6.1. are as follows. We begin by giving a construction of abstract spanning trees which approximate a metric on a set V of $n + 1$ elements to $O(\log n)$ (Proposition 7.3.1). This gives us some hint as to the combinatorial structure of our desired immersed spanning tree when $V \subseteq \mathcal{S}$. However, we need to decide how to partition our tree into arcs, destined to become geodesics in $\mathcal{S}$. This is the purpose of the combinatorial result, Lemma 7.4.1. The final construction of our immersed spanning tree is based on a variant of that of Section 3.3. To show that this works, we need the result, stated above, that the construction of Section 3.3 is always $O(n)$. This is the only real geometric input. It can be viewed as a corollary of a result about piecewise geodesic paths in $\mathcal{S}$. This can in turn be interpreted as a statement about finite metric spaces (Proposition 7.3.4).

7.2. A note on parameters.

Throughout this paper, we have used the notation $x \simeq y$ to mean that $|x - y| \leq K$, where K has been assumed to be some function only of the parameter of hyperbolicity. Similarly, we have used $x \preceq y$ for $x \leq y + K$. If we look at Chapter 3, we see that all such numbers K arising in this way are the result of applying a transitivity law a certain finite number of times, starting with the parameter of hyperbolicity—either k_1 or k_2. We see that we can reinterpret the notation $x \simeq y$ to mean that $|x - y|$ is bounded by some universal multiple of k_1 (or k_2). Similarly for $\preceq$. We see that all the functions of the hyperbolicity parameter arising in Chapter 3 can be written in the form λk_1 for some $\lambda \in [0, \infty)$. This will apply equally well to this chapter, though we shall not always state this explicitly.

Another way to view this is to note that, if $k_1 \neq 0$, then after rescaling the metric by a factor of $1/k_1$, any k_1-H1 space becomes 1-H1. Now, any 0-H1 geodesic space is metric tree (Proposition 3.4.2), so we could restrict attention to 1-H1 spaces, and thus not worry about the dependence on k_1.

7.3. Abstract spanning trees.

Let V be a finite set. A *pseudometric* on V is a symmetric function $\rho : V \times V \longrightarrow [0,\infty)$ satisfying the triangle inequalities. Note that the definition H1 makes equally good sense applied to an arbitrary pseudometric space, in particular to (V,ρ). We may thus define

$$\operatorname{hyp}\rho = \min\{k \in [0,\infty) \mid (V,\rho) \text{ is } k-\text{H1}\}.$$

Equivalently,

$$\operatorname{hyp}\rho =$$
$$\max\{\rho(x,y)+\rho(z,w)-\max\bigl(\rho(x,w)+\rho(y,z),\rho(x,z)+\rho(y,w)\bigr) \mid x,y,z,w \in V\}.$$

The following result is copied from [G, Chapter 6], except that we have added more detail to the proof.

Proposition 7.3.1 : *Let V be a set of $n+1$ elements, and let ρ be a pseudometric on V. Then, there is an abstract spanning tree τ for V, such that*

$$\rho_\tau \leq \rho \leq \rho_\tau + (1+\log_2 n)\operatorname{hyp}\rho.$$

Proof : We construct $\tau = (T,\sigma,f)$ as follows.

First, choose any $v_0 \in V$, and write $V' = V\backslash\{v_0\}$. Let $\mathcal{I} = \{(t,v) \subseteq \mathbf{R}\times V' \mid 0 \leq t \leq \rho(v,v_0)\}$. Thus, we imagine $\mathcal{I}$ as a disjoint union $\mathcal{I} = \bigsqcup_{v\in V'} I_v$ of intervals I_v, where $I_v = [0,\rho(v,v_0)] \times \{v\}$. We topologise $\mathcal{I}$ accordingly.

Given $v,w \in V'$, write

$$\langle u,w\rangle = \frac{1}{2}\bigl(\rho(u,v_0)+\rho(w,v_0)-\rho(u,w)\bigr).$$

Thus, $\langle u,w\rangle \leq \min\bigl(\rho(u,v_0),\rho(w,v_0)\bigr)$. We define a relation

$$R_{u,w} \subseteq I_u \times I_w$$

by $\bigl((s,u),(t,w)\bigr) \in R_{u,w}$ if and only if $s = t \leq \langle u,w\rangle$. In other words, we identify the initial segments $[0,\langle u,w\rangle] \times \{u\}$ and $[0,\langle u,w\rangle] \times \{w\}$. Let $R \subseteq \mathcal{I}\times\mathcal{I}$ be the transitive closure of all the relations $R_{u,w}$ for $u, w \in V'$, and let $T = \mathcal{I}/R$ be the quotient.

One sees easily that T has the structure of a tree, with a path-metric σ induced from the parameterisation of the intervals I_v. Define $f : V \longrightarrow T$ as follows. For $v \in V'$, we let $f(v)$ be the projection of the point $(\rho(v, v_0), v) \in I_v$ to the quotient under R. We define $f(v_0)$ to be the projection of $(0, v) \in I_v$ for any $v \in V$, noting that all such points are identified under R. We see that $\operatorname{ext} T \subseteq f(V)$, so that $\tau = (T, \sigma, f)$ is a spanning tree for V.

By construction, we have

$$\rho_\tau(v, v_0) = \rho(v, v_0)$$

for all $v \in V'$.

Now, given $u, w \in V'$, write $\alpha(u, w)$ for the arc joining $f(u)$ to $f(w)$ in T. Let

$$\langle u, w\rangle_\tau = \frac{1}{2}\big(\rho_\tau(u, v_0) + \rho_\tau(w, v_0) - \rho_\tau(u, w)\big).$$

Thus, $\langle u, w\rangle_\tau$ is the distance, $\sigma(f(v_0), \alpha(u, w))$, from $f(v_0)$ to $\alpha(u, w)$.

From the construction, it is clear that

$$\langle u, w\rangle_\tau \geq \langle u, w\rangle$$

and thus,

$$\rho_\tau(u, w) \leq \rho(u, w).$$

To complete the proof, we need to show that

$$\langle u, w\rangle_\tau \leq \langle u, w\rangle + \frac{1}{2}k(1 + \log_2 n),$$

where $k = \operatorname{hyp} \rho$.

Now, from the definition of $k = \operatorname{hyp} \rho$, we find that for any $u, v, w \in V'$, we have

$$\rho(v, v_0) + \rho(u, w) \leq \max\big(\rho(u, v_0) + \rho(v, w), \rho(w, v_0) + \rho(u, v)\big) + k.$$

Rearranging the terms, we deduce that

$$\langle u, w\rangle \geq \min\big(\langle u, v\rangle, \langle v, w\rangle\big) - k/2.$$

More generally, therefore, if $v_1, v_2, \ldots, v_p$ is any sequence of p points of V', we have

$$\langle v_1, v_p\rangle \geq \min\{\langle v_i, v_{i+1}\rangle \mid 1 \leq i \leq p - 1\} - \frac{1}{2}k(1 + \log_2 p).$$

Thus, given $u, w \in V'$, it is enough to find some sequence $u = v_1, v_2, \ldots, v_p = w$ of distinct points of V' so that

$$\langle u, w\rangle_\tau \leq \min\{\langle v_i, v_{i+1}\rangle \mid 1 \leq i \leq p-1\}.$$

This may be accomplished as follows.

Let $x \in \alpha(u, w)$ be the nearest point of $\alpha(u, w)$ to $f(v_0)$. Let $d = \sigma(f(v_0), x) = \sigma(f(v_0), \alpha(u, w)) = \langle u, w\rangle_\tau$.

Recall that $T = \mathcal{I}/R$, where $\mathcal{I} = \bigsqcup_{v \in V'} I_v$. We see that the points $(d, u) \in I_u$ and $(d, w) \in I_w$ project to the same point $x \in T$. In other words, (d, u) and (d, w) are identified under R. From the definition of R as a transitive closure, this means that there are points $u = v_1, v_2, \ldots, v_p = w$ in V' with $\big((d, v_i), (d, v_{i+1})\big) \in R_{v_i, v_{i+1}}$ for each $i \in \{1, 2, \ldots, p-1\}$. Thus $d \leq \langle v_i, v_{i+1}\rangle$ for $i = 1, 2, \ldots, p-1$. In other words,

$$\langle u, w\rangle_\tau \leq \min\{\langle v_i, v_{i+1}\rangle \mid 1 \leq i \leq p-1\},$$

as required. (In fact we have equality in this expression.)

◇

Note that f might not be injective, even if ρ is a metric. (For example, take $V = \{1, 2, 3, 4\}$, and $\rho(1, 2) = \rho(2, 3) = \rho(3, 4) = \rho(4, 1) = 1$ and $\rho(2, 4) = \rho(1, 3) = 2$.)

Corollary 7.3.2 : *Suppose ρ is a pseudometric on a finite set V. Then,* $\operatorname{hyp} \rho = 0$ *if and only if there is some spanning tree τ for V such that $\rho = \rho_\tau$.*

Proof : It is a simple exercise that $\operatorname{hyp} \rho_\tau = 0$ for any spanning tree τ. The reverse implication is an immediate consequence of Proposition 7.3.1.

◇

By hypothesis, a spanning tree $\tau = (T, \sigma, f)$ for V satisfies $\operatorname{ext} T \subseteq f(V)$. It would be convenient if we could always take f to be a bijection to $\operatorname{ext} T$, so that we may identify V with $\operatorname{ext} T$. In fact, this can be arranged provided we allow σ to be a "path-pseudometric". If we think of a path-metric on T as assigning a positive length to each edge of T, then a *path-pseudometric* assigns a non-negative length to each edge. We do this as follows.

Given any spanning tree $\tau = (T, \sigma, f)$ for V, we define $\tau' = (T', \sigma', f')$ by attaching an arc of length 0 to $f(v)$ for each $v \in V$. More formally, define $T' = \big(([0,1] \times V) \sqcup T\big)/\sim$, where $(1, v) \sim f(v)$ for each $v \in V$. Let $f'(v)$ be the projection of $(0, v)$ to T'. Define σ' by $\sigma'|T = \sigma$ and $\sigma'([0,1] \times \{v\}) = 0$ for all v. Thus, f' is injective, $f'(V) = \operatorname{ext} T'$, and $\rho_{\tau'} = \rho_\tau$.

The remainder of this section is aimed at proving a result about finite metric spaces (Proposition 7.3.4), which may be interpreted, in the context of almost-hyperbolic geodesic spaces, as a statment about piecewise geodesic paths. The result is best motivated in that context (see the beginning of Section 7.5), though it fits more logically into this section.

Suppose that (V, ρ) is a pseudometric space. We introduce the following notation.

Given $x, y, z, w \in V$, write

$$xy \wedge_\rho zw = \frac{1}{2}\big(\rho(x,y) + \rho(z,w) - \rho(x,z) - \rho(y,w)\big).$$

We shall usually abbreviate $xy \wedge_\rho zw$ to $xy \wedge zw$.

We list the following properties of $\wedge$, though we shall only find explicit use for parts (1), (2) and (4).

Lemma 7.3.3 : *Suppose that x, y, z, w, u are any points in the pseudometric space (V, ρ), then*
(1) $xy \wedge zw = yx \wedge wz = -xz \wedge yw$,
(2) $xy \wedge zw + xw \wedge yz + xz \wedge wy = 0$,
(3) $xy \wedge zw \leq \min\big(\rho(x,y), \rho(z,w)\big)$,
(4) $xy \wedge zy \geq 0$,
(5) $xy \wedge xy = \rho(x,y)$,
(6) $uy \wedge wz + uw \wedge yx + wx \wedge zy = 0$.

Proof : Elementary.
◇

We remark that if V is finite, then we can define $\operatorname{hyp}\rho$ in terms of this notation, thus:

$$\operatorname{hyp}\rho = 2\max\{h(xy \wedge zw, xw \wedge yz, xz \wedge wy) \mid x, y, z, w \in V\},$$

where $h(a,b,c) \geq 0$ is defined for any $a, b, c \in \mathbf{R}$ with $a + b + c = 0$, as follows. If it happens that $a \geq \max(|b|, |c|)$, then we set $h(a,b,c) = -b$. From this, we may define $h(a,b,c)$ in general, by insisting that it have the symmetry $h(a,b,c) = h(b,c,a) = h(-a,-c,-b)$. It is readily checked that this agrees with the definition given earlier in this section.

Suppose that $\operatorname{hyp}\rho = 0$, so that $\rho = \rho_\tau$ for some spanning tree $\tau = (T, \sigma, f)$ (Corollary 7.3.2). Given $x, y \in V$, write $\alpha(x,y) = \operatorname{span}(f(x), f(y))$. In this case, if $xy \wedge zw \geq 0$, then $xy \wedge zw = \sigma(\alpha(x,y), \alpha(z,w)) = \sigma(\beta)$ (i.e. the σ-length of β), where $\beta = \alpha(x,y) \cap \alpha(z,w) = \alpha(x,w) \cap (y,z)$. Lemma 7.5.1 gives an interpretation of the quantity $xy \wedge zw$ in the context of almost-hyperbolic geodesic spaces.

Suppose now that $V = \{v_0, v_1, \ldots, v_n\}$ is a set of $n+1$ points with pseudometric ρ. We set

$$\Lambda(v_0, v_1, \ldots, v_n; \rho) = \max\{v_i v_{i+1} \wedge v_{j+1} v_j \mid 0 \leq i < j \leq n\}.$$

Thus, $\Lambda(v_0, v_1, \ldots, v_n; \rho) \geq 0$, by Lemma 7.3.3(4).

Proposition 7.3.4 : *Suppose that $V = \{v_0, v_1, \ldots, v_n\}$ is a set of $n+1$ points with $n \geq 2$, and that ρ is a pseudometric on V. Let $k = \operatorname{hyp}\rho$ and $l = \Lambda(v_0, v_1, \ldots, v_n; \rho)$. Then,*

$$\sum_{i=1}^{n} \rho(v_i, v_{i-1}) \leq \rho(v_0, v_n) + 2(2n-3)l + 2(3n-4)k.$$

The essential point is that the terms in k and l are both linear in n. In fact, the term in l is the best possible, though there is some room for improvement in the term in k.

The case $k = 0$ is not hard to deduce (set $k = 0$ in the argument presented below, and ignore most of the proof). Suppose we know this, and we are given a general pseudometric ρ on V. Then Proposition 7.3.1 gives us a ρ_τ, with $\operatorname{hyp}\rho_\tau = 0$, and

$$\rho_\tau \leq \rho \leq \rho_\tau + k(1 + \log_2 n).$$

We see that

$$l_\tau = \Lambda(v_0, v_1, \ldots, v_n; \rho_\tau) \leq l + k(1 + \log_2 n).$$

Thus,

$$\begin{aligned}
\sum_{i=1}^{n} \rho(v_i, v_{i-1}) &\leq \sum_{i=1}^{n} \rho_\tau(v_i, v_{i-1}) + kn(1 + \log_2 n) \\
&\leq \rho_\tau(v_0, v_n) + 2(2n-3)l_\tau + kn(1 + \log_2 n) \\
&\leq \rho(v_0, v_n) + 2(2n-3)l + (5n-6)(1 + \log_2 n)k.
\end{aligned}$$

So this gives the term in k to be $O(n \log n)$. If we want it $O(n)$, we need a more careful argument.

Proof of Proposition 7.3.4 : Let $V = \{v_0, v_1, \ldots, v_n\}, \rho, l, k$ be as in the hypothesis. We shall construct a spanning tree for V. We begin exactly as in the proof of Proposition 7.3.1. Let $V' = V \setminus \{v_0\}$, and let

$$\mathcal{I} = \{(t, v) \subseteq \mathbf{R} \times V' \mid 0 \leq t \leq \rho(v, v_0)\} = \bigsqcup_{v \in V'} I_v.$$

Write

$$\begin{aligned}
\langle v, w \rangle &= vv_0 \wedge wv_0 \\
&= \frac{1}{2}\big(\rho(v, v_0) + \rho(w, v_0) - \rho(v, w)\big),
\end{aligned}$$

so that, for $u, v, w \in V'$, we have

$$\langle u, w \rangle \geq \min(\langle u, v \rangle, \langle v, w \rangle) - k/2.$$

For $i \in \{1, 2, \ldots, n\}$, let R_i be the relation $R_{v_{i-1}, v_i} \subseteq I_{v_{i-1}} \times I_{v_i}$, i.e.

$$((s, v_{i-1}), (t, v_i)) \in R_i \Leftrightarrow s = t \leq \langle v_{i-1}, v_i \rangle.$$

So far, there has been no difference from Proposition 7.3.1. This time, however, we set $R \subseteq \mathcal{I} \times \mathcal{I}$ to be the equivalence relation generated by the R_i for $i \in \{1, 2, \ldots, n\}$ (rather than using all the $R_{v,w}$). Again, $T = \mathcal{I}/R$ is a tree with path-metric σ and a natural map $f : V \longrightarrow T$. Let $\tau = (T, \sigma, f)$. By attaching arcs of length 0, as described above, we can identify V with $\operatorname{ext} T$, and $\sigma|V$ with ρ_τ. We shall abbreviate $\rho(v_i, v_j)$, $\rho_\tau(v_i, v_j)$ and $\langle v_i, v_j \rangle$ respectively to $\rho(i, j)$, $\rho_\tau(i, j)$ and $\langle i, j \rangle$. We shall write $\alpha(i, j)$ for $\operatorname{span}(v_i, v_j)$. (Thus, $\rho_\tau(i, j)$ is the σ-length of $\alpha(i, j)$.)

From the construction of T, we have that $\rho_\tau(0, i) = \rho(0, i)$ and $\rho_\tau(i-1, i) = \rho(i-1, i)$ for each $i \in \{1, 2, \ldots, n\}$.

Given $x, y \in T$, we shall write $x \leq y$ to mean that $x \in \operatorname{span}(v_0, y)$. Thus, $\leq$ is a partial ordering on the points of T. We write $x < y$ if $x \leq y$ and $x \neq y$. Given $x \in T$, we write $V(x) = \{v \in V \mid x \leq v\}$. From the construction of T, we see that $V(x)$ necessarily consists of a consecutive sequence of points $\{v_{p(x)}, v_{p(x)+1}, \ldots, v_{q(x)}\}$ of points of V, where $0 \leq p(x) \leq q(x) \leq n$.

Suppose that $x \in T \backslash (\operatorname{ext} T \cup \operatorname{node} T)$. We see that $x \in \alpha(i-1, i)$ if and only if i equals either $p(x)$ or $q(x) + 1$, using the convention that $n + 1 \equiv 0$. Thus, the closed path $\bigcup_{i=1}^{n+1} \alpha(i-1, i)$ traverses each component of $T \backslash (\operatorname{ext} T \cup \operatorname{node} T)$ precisely twice. Thus,

$$\sum_{i=1}^{n} \rho_\tau(i-1, i) = \rho_\tau(0, n) + 2\sigma(T \backslash \alpha(0, n)).$$

Since $\rho_\tau(i-1, i) = \rho(i-1, i)$ and $\rho_\tau(0, n) = \rho(0, n)$, we see that we need to show that

$$\sigma(T \backslash \alpha(0, n)) \leq (2n-3)l + (3n-4)k.$$

We shall split this into two parts. To each node $x \in \operatorname{node} T$, we shall associate a subset $\beta(x) \subseteq T$ containing the point x, which is either a closed arc or a single point. Writing $\beta = \bigcup \{\beta(x) \mid x \in \operatorname{node} T\}$, we shall show that

$$\sigma(\beta) \leq (n-1)k,$$

and that

$$\sigma(T \backslash (\beta \cup \alpha(0, n))) \leq (2n-3)(l+k).$$

We shall begin with the first part. In fact, we shall show that for any $x \in \operatorname{node} T$, we have $\sigma(\gamma(x)) \leq (\deg x - 2)k$, where

$$\gamma(x) = \beta(x) \backslash \bigcup \{\beta(y) \mid y \in \operatorname{node} T \text{ and } x < y\}.$$

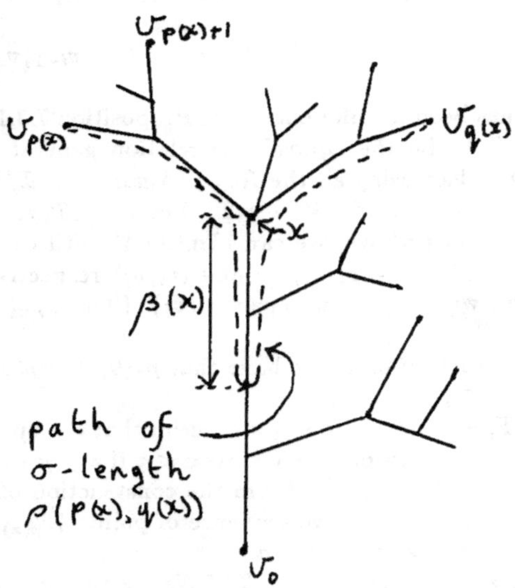

Figure 7a.

If $z \in \beta$, and we choose $x \in \operatorname{node} T$, maximal with respect to $<$ such that $z \in \beta(x)$, then $z \in \gamma(x)$. It follows that $\beta = \bigcup\{\gamma(x) \mid x \in \operatorname{node} T\}$. We may then deduce that

$$\sigma(\beta) \leq \sum_{x \in \operatorname{node} T} (\deg x - 2)k = (|\operatorname{ext} T| - 2)k = (n-1)k,$$

as required.

So, suppose that $x \in \operatorname{node} T$ so that $V(x) = \{v_{p(x)}, v_{p(x)+1}, \ldots, v_{q(x)}\}$. We define $\beta(x)$ as follows. If $\sigma(x, v_0) \leq \langle p(x), q(x) \rangle$, we set $\beta(x) = \{x\}$. If $\sigma(x, v_0) > \langle p(x), q(x) \rangle$, then we set $\beta = \operatorname{span}(x, y)$, where $y \in \sigma(x, v_0)$ is the point with $\sigma(v_0, y) = \langle p(x), q(x) \rangle$. (It is conceivable that there may be some ambiguity in the choice of y, arising from the fact that σ is only a pseudometric. Though this does not give us serious problem—all distances and lengths are well-defined.) We could alternatively define $\beta(x)$ as the closed arc (or point) in $\operatorname{span}(x, v_0)$, with one endpoint at x, and of σ-length equal to $\frac{1}{2}\max(0, \rho(p(x), q(x)) - \rho_\tau(p(x), q(x)))$. (See Figure 7a.) If $x \in \operatorname{ext} T$, we shall write $\beta(x) = \{x\}$. In each case, we have

$$\sigma(v_0, \beta(x)) = \min(\sigma(v_0, x), \langle p(x), q(x) \rangle).$$

For any $x \in \operatorname{node} T$, the set $V(x) = \{v_{p(x)}, v_{p(x)+1}, \ldots, v_{q(x)}\}$ has a naturally partition as $V(x) = \bigsqcup_{i=1}^{r} V(x_i)$, where $r = \deg x - 1$ and $x_1, x_2, \ldots, x_r \in \operatorname{node} T \cup \operatorname{ext} T$ are vertices of the tree T, adjacent to x. In fact, we can express the sequence $p(x), p(x)+1, \ldots, q(x)$ as the concatenation of the consecutive sequences $p(x_i), p(x_i)+1, \ldots, q(x_i)$ for $i = 1, 2, \ldots, r$. Thus $p(x_1) = p(x)$, $q(x_r) = q(x)$ and $q(x_i) + 1 = p(x_{i+1})$ for $i \in \{1, 2, \ldots, r-1\}$. (See Figure 7b.)

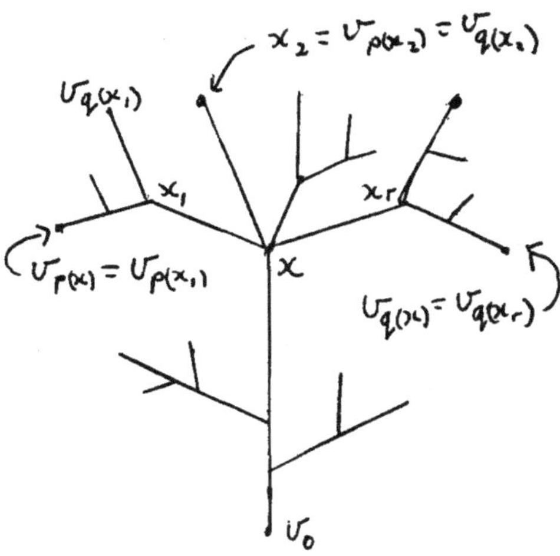

Figure 7b.

By the construction of T, we have that $\langle q(x_i), q(x_i)+1 \rangle = \sigma(v_0, \alpha(q(x_i), q(x_i)+1)) = \sigma(v_0, x)$ for each $i \in \{1, 2, \ldots, r-1\}$.

Now, for any $i \in \{1, 2, \ldots, r\}$, we see that

$$\begin{aligned}\sigma(\beta(x)\backslash\beta(x_i)) &= \max(0, \sigma(v_0, \beta(x_i)) - \sigma(v_0, \beta(x))) \\ &= \max(0, \langle p(x_i), q(x_i)\rangle - \langle p(x), q(x)\rangle).\end{aligned}$$

Also, $\sigma(\beta(x)) = \max(0, \sigma(v_0, x) - \langle p(x), q(x)\rangle)$. Since $\gamma(x) \subseteq \beta(x)\backslash\bigcup_{i=1}^{r}\beta(x_i)$, we have that

$$\sigma(\gamma(x)) \leq \max(0, \min(\{\langle p(x_i), q(x_i)\rangle \mid 1 \leq i \leq r\} \cup \{\sigma(v_0, x)\}) - \langle p(x), q(x)\rangle).$$

Now, we have that $\operatorname{hyp} \rho = k$. So, after $(2r-2)$ applications of the inequality $\langle u, w\rangle \geq \min(\langle u, v\rangle, \langle v, w\rangle) - k/2$ (to the sequence $p(x) = p(x_i), q(x_1), q(x_1)+1 = p(x_2), q(x_2), \ldots, p(x_r), q(x_r) = q(x)$), we find that

$$\langle p(x), q(x)\rangle \geq \min(\{\langle p(x_i), q(x_i)\rangle \mid 1 \leq i \leq r\} \cup \{\sigma(v_0, x)\}) - (2r-2)(k/2).$$

We conclude that

$$\sigma(\gamma(x)) \leq (2r-2)(k/2) = (r-1)k = (\deg x - 2)k,$$

as required.

We have shown that $\sigma(\beta) \leq (n-1)k$. It remains to show that $\sigma\big(T\setminus(\beta \cup \alpha(0,n))\big) \leq (2n-3)(l+k)$.

By an "edge" of T, we mean the closure of a component of $T\setminus \operatorname{node} T$. Now T has at most $2|\operatorname{ext} T|-3 = 2n-1$ edges. Thus, there are at most $2n-3$ edges not lying in $\alpha(0,n)$. Let $e = \operatorname{span}(x,y)$ be such an edge, where $y \in \operatorname{node} T$, $x \in \operatorname{node} T \cup \operatorname{ext} T$ and $y \in \operatorname{span}(v_0, x)$. We claim that $\sigma(e\setminus\beta(x)) \leq l+k$. The result then follows by summing over all such edges.

Suppose first that $x \in \operatorname{node} T$. Write $p = p(x)$ and $q = q(x)$. If $e \subseteq \beta(x)$, then there is nothing to prove, so we can assume that

$$\sigma(v_0, y) = \sigma(v_0, e) < \sigma(v_0, \beta(x)) \leq \langle p, q\rangle.$$

Now, $V(x) = \{v_p, v_{p+1}, \ldots, v_q\}$ is, by definition, the set of points of V separated from v_0 by x. Since y is the adjacent node to x in the direction of v_0, we must have that both $\alpha(p-1,p)$ and $\alpha(q,q+1)$ pass through y. Thus,

$$\langle p-1, p\rangle = \sigma(v_0, \alpha(p-1,p)) \leq \sigma(v_0, y) \leq \langle p, q\rangle.$$

Similarly,

$$\langle q, q+1\rangle \leq \langle p, q\rangle.$$

Writing

$$\langle i,j\rangle_\tau = \frac{1}{2}\big(\rho_\tau(0,i) + \rho_\tau(0,j) - \rho_\tau(i,j)\big) = \sigma(v_0, \alpha(i,j)),$$

(so that $\langle i, i+1\rangle_\tau = \langle i, i+1\rangle$ for each i) we see that

$$\begin{aligned}\langle p-1, q+1\rangle &\geq \min(\langle p-1,p\rangle, \langle p,q\rangle, \langle q,q+1\rangle) - 2(k/2)\\ &= \min(\langle p-1,p\rangle_\tau, \langle q,q+1\rangle_\tau) - k\\ &= \langle p-1, q+1\rangle_\tau - k.\end{aligned}$$

Thus,

$$\rho_\tau(p-1,q+1) \geq \rho(p-1,q+1) - 2k.$$

(Recall that $\rho_\tau(0,p-1) = \rho(0,p-1)$ and $\rho_\tau(0,q+1) = \rho(0,q+1)$.) From the definition of $\beta(x)$, we have that $\sigma(\beta(x)) = \frac{1}{2}\max\big(0, \rho(p,q) - \rho_\tau(p,q)\big)$, so that

$$\rho_\tau(p,q) + 2\sigma(\beta(x)) \geq \rho(p,q).$$

Also, we have $\rho_T(p-1,p) = \rho(p-1,p)$ and $\rho_T(q,q+1) = \rho(q,q+1)$. From the construction of T, we have that $e = \alpha(p-1,p) \cap \alpha(q,q+1)$, and so

$$\begin{aligned}
\sigma(e\backslash\beta(x)) &= \sigma(e) - \sigma(\beta(x)) \\
&= \frac{1}{2}(\rho_T(p-1,p) + \rho_T(q,q+1) - \rho_T(p,q) - \rho_T(p-1,q+1)) - \sigma(\beta(x)) \\
&\leq \frac{1}{2}(\rho(p-1,p) + \rho(q,q+1) - \rho(p,q) - (\rho(p-1,q+1) - 2k)) \\
&= v_{p-1}v_p \wedge v_{q+1}v_q + k \\
&\leq l + k,
\end{aligned}$$

as required.

The case where $x \in \operatorname{ext} T$ is similar, but simpler. We have, by definition, that $\beta(x) = \{x\}$. If we set $p = q$ so that $x = v_p = v_q$ then the argument goes through, more or less, as before.

◇

7.4. Pinnate Structures.

In this section, we describe a few combinatorial properties of trees.

Let T be a tree (i.e. a finite acyclic connected 1-complex). Let $V = \operatorname{ext} T$ be the set of extreme points of T. Suppose x, y, z, w are distinct points of V. We shall write $xy|zw$ to mean that $\operatorname{span}(x,y)$ meets $\operatorname{span}(z,w)$ in at most one point. (Figure 7c.)

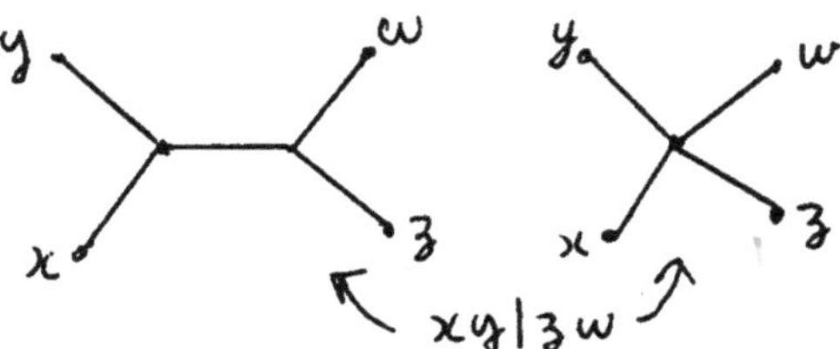

Figure 7c.

We have the following properties.

(1) $xy|zw \Leftrightarrow xy|wz \Leftrightarrow zw|xy$,

(2) $xy|zw$ and $xz|yw \Rightarrow xw|yz$.

Remark : Given a finite set V, we can regard the the quaternary relation $(..)|(..)$ as defining a structure on V equivalent to the notion of a combinatorial spanning tree for V, with V equal to the set of extreme points. As axioms for $(..)|(..)$, we can take properties (1) and (2) stated above, as well as an enumeration of the possibilities for the relation restricted to any set of five distinct points $\{a,b,c,d,e\} \subseteq V$. That is, we need to assume that the relation on $\{a,b,c,d,e\}$ is derived from one of the possible combinatorial spanning trees for five points. (There are twenty-six such, in total.) We leave as an exercise the observation that one can reconstruct uniquely a spanning tree for V from such data.

In this section, we want to describe an additional structure on V, which we shall call a "pinnate structure".

Let T be a tree with $V = \operatorname{ext} T$.

Definition : A *pinnate decomposition* of T is a partition of T into disjoint subsets indexed by V, written $T = \bigsqcup_{x \in V} \gamma(x)$, such that $x \in \gamma(x)$ for all x, and $\gamma(x)$ is homeomorphic to a half-open interval for each $x \in V$, except one point $x_0 \in V$ for which $\gamma(x_0) = \{x_0\}$.

We refer to x_0 as the *root*. We call $\gamma(x)$ for $x \in V \backslash \{x_0\}$ a *branch* of the pinnate decomposition.

If $x \in V$, then there is a unique $\phi x \in V$ such that the closure $\bar{\gamma}(x)$ of $\gamma(x)$ meets $\gamma(\phi x)$. Thus, $\phi x_0 = x_0$, otherwise $\phi x \neq x$. Also, since x_0 is an extreme point of T, there is a unique point $x_1 \in V$ with $\phi x_1 = x_0$. We call $\operatorname{span}(x_0, x_1) = \{x_0\} \cup \gamma(x_1)$ the *stem* of the pinnate decomposition.

Thus, we may imagine a tree with a pinnate decomposition as resembling a fern frond with one main stem, and a sequence of successive branching. (Figure 7d.)

Given T, we may recover the pinnate decomposition from the map ϕ by taking

$$\gamma(x) = \operatorname{span}(x, \phi x, \phi^2 x) \backslash \operatorname{span}(\phi x, \phi^2 x)$$

for $x \neq x_0$, and

$$\gamma(x_0) = \{x_0\},$$

where x_0 is the unique fixed point of ϕ.

We call a map $\phi : V \longrightarrow V$ arising in this way a *pinnate structure* on T (or a pinnate structure on V compatible with T).

Remark : We may give a set of axioms for a pinnate structure $\phi : V \longrightarrow V$ compatible with $(..)|(..)$ as follows.

(1) There is some $x_0 \in V$ with $\phi x_0 = x_0$.

(2) If $x, y \in V \backslash \{x_0\}$ and $\phi x = \phi y = x_0$, then $x = y$.

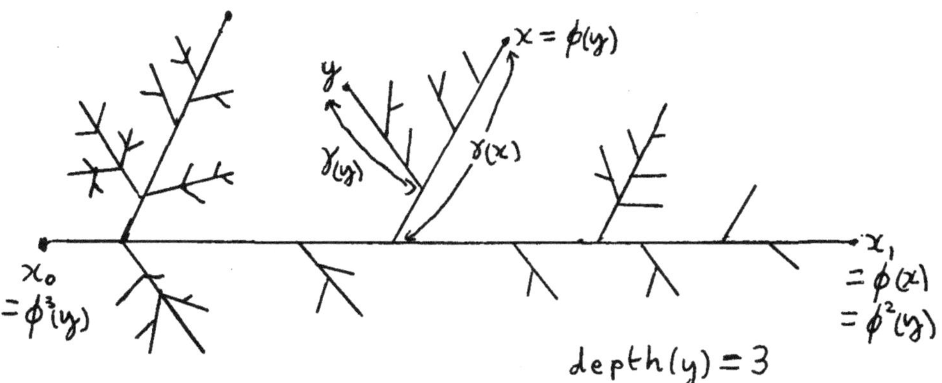

Figure 7d.

(3) If $\phi^2 x \neq x_0$, then $x, \phi x, \phi^2 x, \phi^3 x$ are all distinct.

(4) If $x, y, \phi x, \phi y$ are all distinct, then $x(\phi x)|y(\phi y)$.

(5) If $x, y, \phi x, \phi^2 x$ are all distinct, and $\phi x = \phi y$, then either $xy|(\phi x)(\phi^2 x)$ or $yx|(\phi x)(\phi^2 x)$.

We leave as an exercise that this data suffices to reconstruct a pinnate decomposition $\{\gamma(x) \mid x \in V\}$ as described above.

Let $\phi : V \longrightarrow V$ be a pinnate structure compatible with T. If $\phi^t x \neq x_0$, then it follows that $x, \phi x, \phi^2 x, \ldots, \phi^t x$ are all distinct. Thus, there is some r such that $\phi^r x = x_0$. We write $\text{depth}_\phi(x)$ (or just $\text{depth}(x)$) for the smallest such r. We write

$$\text{depth}\,\phi = \max_{x \in V}(\text{depth}_\phi(x)).$$

Proposition 7.4.1 : *Any tree T admits a pinnate structure of depth at most $1 + \log_2(\frac{m+1}{3})$, where $m = |\text{ext}\, T|$.*

The worst case is, in fact, a tree with 3.2^{p-2} extreme points, and each node of degree three, as shown in Figure 7e for $p = 5$. Any pinnate structure on such a tree must have depth at least p.

We shall first prove the following lemma (7.4.2). In fact, this lemma suffices to give a logrithmic bound on the depth of a pinnate structure, which is all we shall need for Theorem 7.6.1.

Lemma 7.4.2 : *Suppose T is a tree with at most 2^p extreme points, and that*

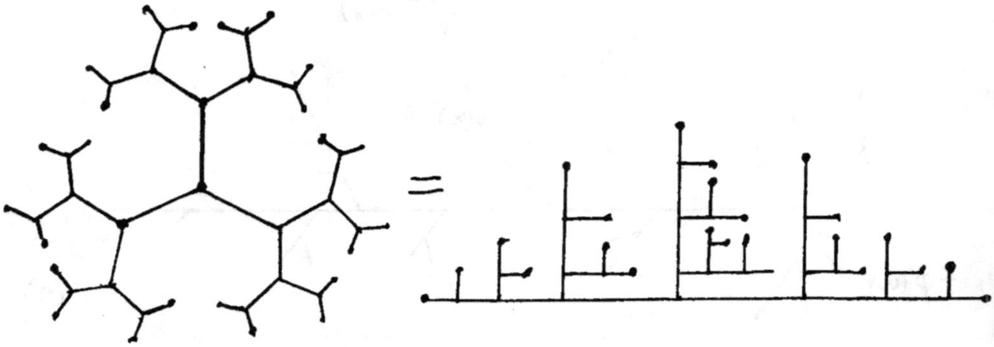

Figure 7e.

$x_0 \in \operatorname{ext} T$ is any extreme point. Then T admits a pinnate structure of depth at most p and with root x_0.

Proof : Let $V = \operatorname{ext} T$. We construct a pinnate decomposition for T. We begin by constructing its stem α as follows. We imagine moving along T, starting at x_0. Each time we come to a node $y \in \operatorname{node} T$, we follow the edge of T which separates a maximal number of extreme points from x_0. In other words, we move into a component C of $T \backslash \{y\}$ which maximises $|C \cap V|$ for $x_0 \notin C$. We continue until we reach an extreme point $x_1 \in V$. Let $\alpha = \operatorname{span}(x_0, x_1)$ be the path we have followed. Let $\gamma(x_1) = \alpha \backslash \{x_0\}$.

Now, if $D \subseteq T$ is a component of $T \backslash \alpha$, then by construction, $|D \cap V| \leq 2^{p-1} - 1$. The closure $\bar{D} \subseteq T$ is a subtree with $\operatorname{ext} \bar{D} = (D \cap V) \cup \{x_D\}$, where x_D is the point where $\bar{D}$ meets α. Thus, $|\operatorname{ext} \bar{D}| \leq 2^{p-1}$. By induction, $\bar{D}$ has a pinnate decomposition $\bar{D} = \bigsqcup_{x \in \operatorname{ext} \bar{D}} \gamma_D(x)$, with root x_D, and depth at most $p-1$. If $x \in D \cap V$, set $\gamma(x) = \gamma_D(x)$.

Doing this for each such component D, we arrive at a pinnate decomposition $\{\gamma(x) \mid x \in V\}$ of depth at most p, and root x_0.

◇

Proof of Proposition 7.4.1 : Let $V = \operatorname{ext} T$, and suppose that $|V| \leq 3.2^{p-1} - 1$ with $p \geq 2$. We want to construct a pinnate decomposition of depth at most p. (The case $p = 1$ is trivial.)

Suppose that y is a node of T. Let $C_1, C_2, \ldots, C_r$ $(r \geq 2)$ be the connected components of $T \backslash \{y\}$. Let $e_i = |C_i \cap V|$, so that $\sum_{i=1}^{r} e_i = |V|$. For each $j \in \{1, 2, \ldots, r\}$, we must have $e_j \leq \sum_{i \neq j} e_i$, otherwise we would do better choosing the node in C_j adjacent to y. It follows that

$$e_j \leq \frac{1}{2}\sum_{i=1}^{r} e_i \leq \frac{1}{2}(3.2^{p-1} - 1) \leq 2^p - 1.$$

Now, without loss of generality, we have $e_1 \geq e_2 \geq \cdots \geq e_r$. Since $\sum_{i=1}^{r} e_i \leq 3.2^{p-1} - 1$, we must have $e_i \leq 2^{p-1} - 1$ for $i \geq 3$.

Considering each closures $\bar{C}_i$ as a subtree of T, we have $\operatorname{ext} \bar{C}_i = (C_i \cap V) \cup \{y\}$. Thus, $|\operatorname{ext} \bar{C}_i| \leq 2^p$ for $i = 1, 2$, and $|\operatorname{ext} \bar{C}_i| \leq 2^{p-1}$ for $i \geq 3$. For each i, let $\{\gamma_i(x) \mid x \in \operatorname{ext} \bar{C}_i\}$ be the pinnate decomposition of $\bar{C}_i$, with root y, given by Lemma 7.4.2. Let $x_0 \in C_1 \cap V$ and $x_1 \in C_2 \cap V$ be, respectively, the unique points such that $\bar{\gamma}_1(x_0)$ and $\bar{\gamma}_2(x_1)$ meet y. (Thus span (y, x_0) and span (y, x_1) are, respectively, the stems for C_1 and C_2.)

Define $\gamma(x_0) = \{x_0\}$, $\gamma(x_1) = \operatorname{span}(x_0, x_1) \backslash \{x_0\}$, and $\gamma(x) = \gamma_i(x)$ for $x \in (C_i \cap V) \backslash \{x_0, x_1\}$. We check that $\{\gamma(x) \mid x \in V\}$ is a pinnate decomposition of T of depth at most p.

$\diamond$

Suppose that $\phi : V \longrightarrow V$ is a pinnate structure compatible with the tree T. Let $\{\gamma(x) | x \in V\}$ be the corresponding decomposition of T. For any $x \in V$, we have that $\gamma(z)$ meets span(x_0, x) if and only if $z = \phi^t x$ for some t with $0 \leq t \leq r = \operatorname{depth} x$. We call $x_0 = \phi^r x, \phi^{r-1} x, \ldots, \phi x, x$ the *path sequence* for x. (Figure 7f.)

Figure 7f.

More generally, suppose that $x, y \in V$ with $r = \operatorname{depth} x$ and $s = \operatorname{depth} y$. Let $t \geq 0$ be the largest integer such that $\phi^{r-t} x = \phi^{s-t} y$. Then, $\gamma(z)$ meets span(x, y) if and only if z belongs to the sequence

$$x, \phi x, \ldots, \phi^{r-t} x = \phi^{s-t} y, \ldots, \phi y, y.$$

We write $w = w(x, y) = \phi^{r-t} x = \phi^{s-t} y$. The path sequences for x and y agree on the first $t + 1$ terms, namely, $x_0 = \phi^t w, \ldots, \phi w, w$. Note that this is precisely the path sequence for w. We must have either $x x_0 | y w$ or $y x_0 | x w$. (Figure 7g.)

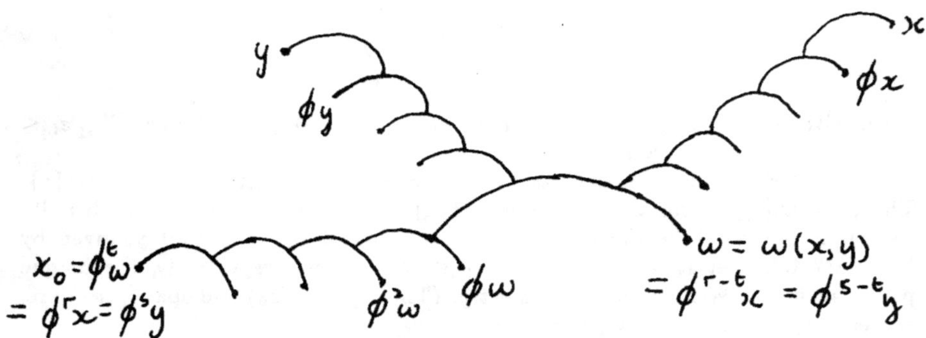

Figure 7g.

This dicussion will be relevant to Section 7.6.

7.5. Piecewise geodesic paths.

In this section, we return to the context of almost-hyperbolic geodesic spaces.

Suppose that V is a set of $n+1$ points in such a space. We have already stated (in Section 7.1) that both Steiner trees, and the construction of Section 3.3, give trees which approximate distances up to an additive term $kO(n)$. The property both trees have in common which gives rise to this linear estimate may be summarised as follows. Suppose $X, Y \in V$, and that β is the path in the tree joining X to Y. Then β consists of the union of at most n geodesic segments, no two of which run "almost parallel" over a long distance. The result can thus be rephrased as a property of such paths, namely that length $\beta \leq d(X,Y) + kO(n)$. We shall see that this is a simple consequence of Proposition 7.3.4. In fact, we are able to weaken the hypothesis, taking account of the direction of geodesic segments. Thus, we need only assume that the path β should not double back on itself over a large distance.

We introduce the following notation. Suppose that $(\mathcal{S}, d)$ is a (pseudo)-metric space. If $X_0, X_1, \ldots, X_p \in \mathcal{S}$, we shall write $\langle X_0 X_1 \cdots X_p \rangle$ to mean that

$$d(X_0, X_p) = \sum_{i=1}^{p} d(X_i, X_{i-1}).$$

If $(\mathcal{S}, d)$ happens to be a geodesic space, then we shall always choose geodesics so that $[X_i, X_j] \subseteq [X_0, X_p]$ for any $i, j \in \{0, 1, \ldots, p\}$. Thus we may imagine the sequence of points $X_0, X_1, \ldots, X_p$ occurring in order along $[X_0, X_p]$. If X, Y, Z, W are any points in $(\mathcal{S}, d)$, we have already defined (Section 7.3)

$$XY \wedge ZW = \frac{1}{2}(d(X,Y) + d(Z,W) - d(X,Z) - d(Y,W)).$$

To the properties (1)–(6) of Lemma 7.3.3, we may add that if X, Y, Z, W, X', Y', Z' and W' lie in $\mathcal{S}$ with $\langle XX'Y'Y\rangle$ and $\langle ZZ'W'W\rangle$, then $X'Y' \wedge Z'W' \leq XY \wedge ZW$. We have already observed that we can express property H1 in terms of this notation. In fact, we can write $XY : ZW$ as

$$XZ \wedge YW \simeq XW \wedge YZ \succeq 0.$$

If we already know that $(\mathcal{S}, d)$ is H1, then $XY : ZW$ becomes equivalent to the statement $XZ \wedge YW \simeq XW \wedge YZ$, or to the statement $XZ \wedge YW \simeq 0$.

Suppose that $(\mathcal{S}, d)$ is an almost hyperbolic space. Suppose that $XZ : YW$, and that $(XZ)AB(YW)$ is a spanning tree for X, Y, Z, W (Section 3.1). Then it is readily checked that

$$XY \wedge ZW \simeq XW \wedge ZY \simeq d(A,B).$$

Thus, the quantity $XY \wedge ZW$ may be thought of as measuring the distance over which the geodesics $[X,Y]$ and $[Z,W]$ run almost parallel.

The following lemma gives a more careful formulation of this idea.

Lemma 7.5.1 : *$\forall k\, \exists h$ such that the following holds.*

Suppose $(\mathcal{S}, d)$ is a k-H1 geodesic space, and that $X, Y, Z, W \in \mathcal{S}$ satisfy $XY \wedge ZW \geq 0$. Then, there exist $A, B, C, D \in \mathcal{S}$ with $\langle XABY\rangle$ and $\langle ZCDW\rangle$ and $d(A,B) = d(C,D) = XY \wedge ZW$. Also, if $E \in [A,B]$ and $F \in [C,D]$ satisfy $d(A,E) = d(C,F)$, then $d(E,F) \leq h$. (In particular, $d(A,C) \leq h$ and $d(B,D) \leq h$.) See Figure 7h.

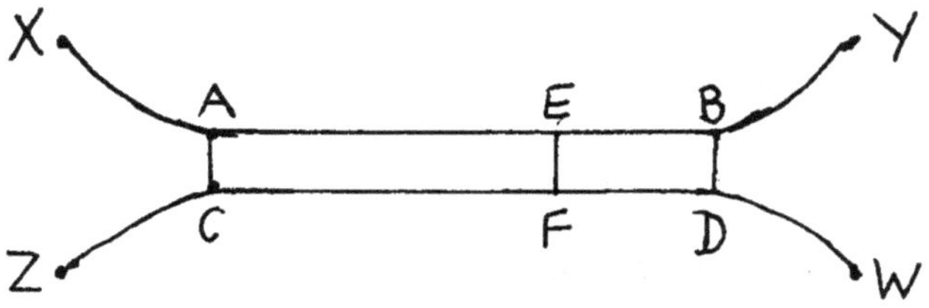

Figure 7h.

Proof :

Case (1) : $XW : YZ$ (or equivalently, $XY \wedge ZW \simeq 0$).

We have $d([X,Y],[Z,W]) \simeq 0$. Since $XY \wedge ZW \leq \min(d(X,Y), d(Z,W))$ (Lemma 7.3.3(3)), we can find $A \sim B \sim C \sim D$ with $\langle XABY \rangle$, $\langle ZCDW \rangle$ and $d(A,B) = d(C,D) = XY \wedge ZW$.

Case (2) : $XY : ZW$.

We have $XY \wedge ZW = -XZ \wedge YW \preceq 0$. So, $XY \wedge ZW \simeq 0$, and we are back in Case (1).

Case (3) : $XZ : YW$.

Let $(XZ)PQ(YW)$ be a spanning tree for X, Y, Z, W (Section 3.1). Thus, $d(P,Q) \simeq XY \wedge ZW$.

Now, there exist $A', B' \in [X,Y]$ with $A' \sim P$ and $B' \sim Q$, so that $d(A',B') \simeq d(P,Q) \simeq XY \wedge ZW$. Since $d(X,Y) \geq XY \wedge ZW$, we can find $A, B \in [X,Y]$ with $A \sim A'$ and $B \sim B'$ and $d(A,B) = XY \wedge ZW$. Now,

$$\begin{aligned} d(X,Y) &\simeq d(X,P) + d(P,Q) + d(Q,Y) \\ &\simeq d(X,A) + d(A,B) + d(B,Y). \end{aligned}$$

Thus if $\langle XBAY \rangle$, then $d(A,B) \simeq 0$, and so $XY \wedge ZW \simeq d(A,B) \simeq 0$, and we are back in Case (1). Therefore, we can assume that $\langle XABY \rangle$.

Similarly, we can find $C, D \in [Z,W]$ with $C \sim P$, $D \sim Q$ and $\langle ZCDW \rangle$ and $d(C,D) = XY \wedge ZW$.

Now, suppose that $E \in [A,B]$ and $F \in [C,D]$ satisfy $d(A,E) = d(C,F)$. Applying Lemma 3.1.2 twice, we find $F' \in [C,D]$ with $F' \sim E$. Now, $d(C,F') \simeq d(C,E) \simeq d(A,E) \simeq d(C,F)$, and so $F \sim F'$. Thus $E \sim F$.

◇

Suppose $X_0, X_1, \ldots, X_n$ is any sequence of points in the geodesic space $(\mathcal{S}, d)$. We shall write

$$[X_0, X_1, \ldots, X_n] = \bigcup_{i=1}^{n} [X_{i-1}, X_i]$$

for the piecewise geodesic path joining these points. If $\gamma = [X_0, X_1, \ldots, X_n]$, then, as in Section 7.3, we shall write

$$\Lambda(\gamma) = \Lambda(X_0, X_1, \ldots, X_n) = \max\{X_i X_{i+1} \wedge X_{j+1} X_j \mid 0 \leq i < j \leq n-1\}.$$

(Formally we should think of γ as consisting just of the $(n+1)$-tuple of points $(X_0, X_1, \ldots, X_n)$, though we are imagining it as a piecewise geodesic path, or as the image of this path in $\mathcal{S}$.) Note that if γ' is obtained by subdividing γ, i.e. by adding additional points along the geodesic segments of γ, then we have $\Lambda(\gamma') \leq \Lambda(\gamma)$.

If (S, d) is k-H1, then the quantity $\Lambda(\gamma)$ measures the maximum distance along which γ doubles back on itself (along a pair of geodesic segments). Proposition 7.3.4 tells us that the total length of γ is at most $d(X_0, X_n) + 2(2n-3)\Lambda(\gamma) + 2(3n-4)k$.

We shall apply this to the tree construction given in Section 3.3.

Let $V \subseteq S$ be a set with $(n+1)$ elements. A linear ordering on V can be thought of as a bijection $\underline{X} : \{0, 1, \ldots, n\} \longrightarrow V$. We shall write $X_i = \underline{X}(i)$ and $\underline{X} = (X_0, X_1, \ldots, X_n)$. We define $T_V = T_{\underline{X}} \subseteq S$ inductively as follows. We take $T_{(X_0,X_1)} = [X_0, X_1]$ and $T_{(X_0,X_1,\ldots,X_n)} = T_{(X_0,X_1,\ldots,X_{n-1})} \cup [X_n, Y_n]$, where Y_n is a nearest point to X_n on the tree $T_{(X_0,X_1,\ldots,X_{n-1})}$. We see that $T_{\underline{X}}$ an embedded tree in S, and that $\operatorname{ext} T_{\underline{X}} \subseteq V$. Thus, $T_{\underline{X}}$ is an embedded spanning tree for V in the sense described in Section 7.1.

Suppose that $i, j \in \{0, 1, \ldots, n\}$, with $i > j$. Let α be the arc in $T_{\underline{X}}$ joining X_i to X_j. Then α is piecewise geodesic with at most n segments. In fact, α has one of the following forms (Figure 7i).

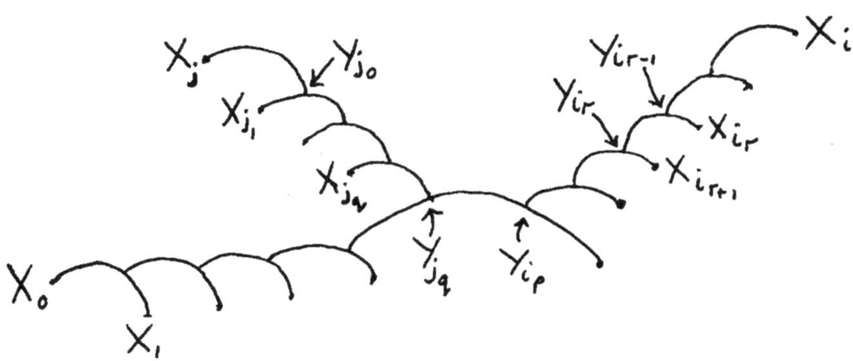

Figure 7i.

(1) $\alpha = [X_i, Y_{i_0}, Y_{i_1}, Y_{i_2}, \ldots, Y_{i_p}, Y_{j_q}, Y_{j_{q-1}}, \ldots, Y_{j_1}, Y_{j_0}, X_j]$ where p and q lie in $\{0, 1, \ldots, n\}$ and $i = i_0 > i_1 > \cdots > i_p$ and $j = j_0 > j_1 > \cdots > j_q$. Also, the i_r and j_s are all distinct. Moreover, we have $Y_{i_r} \in [X_{i_{r+1}}, Y_{i_{r+1}}]$ for $0 \leq r \leq p-1$, and $Y_{j_s} \in [X_{j_{s+1}}, Y_{j_{s+1}}]$ for $0 \leq s \leq q-1$.

(2) $\alpha = [X_i, Y_{i_0}, Y_{i_1}, Y_{i_2}, \ldots, Y_{i_p}, X_j]$ where $p \in \{0, 1, \ldots, n\}$ and $i = i_0 > i_1 > \cdots > i_p$. Also $Y_{i_r} \in [X_{i_{r+1}}, Y_{i_{r+1}}]$ for $0 \leq r \leq p-1$.

(3) $\alpha = [X_0, X_1]$.

Suppose that $[A, B]$ and $[C, D]$ are two distinct geodesic segments of α, so that A, B, C, D occur in this order along α. (Possibly $B = C$.)

By inspection of the form of α given above, we see that we must have either

$d(A,B) \le d(A,[C,D])$ or $d(D,C) \le d(D,[A,B])$. We claim that this implies that $AB \wedge DC \le h$, where h is the constant of Lemma 7.5.1.

We can assume, without loss of generality, that $d(A,B) \le d(A,[C,D])$, and (since $h \ge 0$) that $AB \wedge CD \ge 0$. Lemma 7.5.1 gives us points $A', B' \in [A,B]$ and $D' \in [C,D]$ with $\langle AA'B'B \rangle$, $d(A',B') = AB \wedge DC$ and $d(A',D') \le h$. Now,

$$\begin{aligned} d(A,A') + AB \wedge DC &= d(A,A') + d(A',B') \le d(A,B) \\ &\le d(A,[C,D]) \le d(A,A') + d(A',C') \\ &\le d(A,A') + h. \end{aligned}$$

Thus, $AB \wedge DC \le h$ as required.

Since this applies to any pair of geodesic segments of α, we conclude that

$$\Lambda(\alpha) \le h.$$

As remarked in Section 7.2, the constant h must have the form λk for some universal $\lambda \in \mathbf{R}$. We may now apply Proposition 7.3.4 to find that

$$\begin{aligned} \operatorname{length} \alpha &\le d(X_i, X_j) + 2(2n-3)(\lambda k) + 2(3n-4)k \\ &= d(X_i, X_j) + kH(n), \end{aligned}$$

where $H(n) = 2\lambda(2n-3) + 2(3n-4)$.

We have shown:

Proposition 7.5.2 : *There is a linear function $H : \mathbf{N} \longrightarrow \mathbf{R}$ such that the following holds.*

Suppose that $(\mathcal{S}, d)$ is a k-H1 geodesic space, and that $X_0, X_1, \ldots, X_n \in \mathcal{S}$. Let $T_{\underline{X}}$ be the embedded spanning tree as defined above. If $i, j \in \{0, 1, \ldots, n\}$ and α is the arc in $T_{\underline{X}}$ joining X_i to X_j, then

$$\operatorname{length} \alpha \le d(X_i, X_j) + kH(n).$$

This proposition is used in the proof of Theorem 7.6.1.

We conclude this section with a brief discussion of Steiner trees.

Suppose that $(\mathcal{S}, d)$ is a geodesic space, and $V \subseteq \mathcal{S}$ a set of $(n+1)$ points. Among all spanning trees (T, σ, f, g) for V, we choose one of minimal length $\sigma(T)$. It is an exercise to show that this minimum is always attained in geodesic spaces. Such a tree must be embedded, so we may identify T with its image, S_V, in $\mathcal{S}$. Clearly, each edge of S_V must be geodesic. We call S_V a *Steiner tree* for V.

Suppose that $(\mathcal{S}, d)$ is k-H1 and that α an arc in S_V joining two points of V. A simple application of Lemma 7.5.1 shows that, as in the previous case, $\Lambda(\alpha) \le h$

(otherwise we could construct a shorter tree). Applying Proposition 7.3.4 (as in the case of Proposition 7.3.2) we deduce:

Proposition 7.5.3 : *There is a linear function $J : \mathbf{N} \longrightarrow \mathbf{R}$ such that the following holds.*

Suppose that $(\mathcal{S}, d)$ is a k-H1 geodesic space, and that $V \subseteq \mathcal{S}$ is a set of $(n+1)$ points. Let S_V be a Steiner tree for V. If $X, Y \in V$, and α is the arc in S_V joining X to Y, then

$$\operatorname{length} \alpha \leq d(X,Y) + kJ(n).$$

7.6. A logrithmic spanning tree.

In this section we collect the various pieces together to show how to construct an $O(\log n)$-approximating tree.

Suppose $(\mathcal{S}, d)$ is a k-H1 geodesic space, and $V \subseteq \mathcal{S}$ is a set of $(n+1)$ points. We shall write ρ for the metric d restricted to V. Thus, $\operatorname{hyp} \rho \leq k$ (Section 7.3). Proposition 7.3.1 gives us an abstract spanning tree $\tau' = (T', \sigma', f')$ for V with

$$\rho_{\tau'} \leq \rho \leq \rho_{\tau'} + (1 + \log_2 n)k.$$

As remarked after the proof of Proposition 7.3.1, we can identify V with $\operatorname{ext} T$.

Now, Lemma 7.4.1 gives us a pinnate structure $\phi : V \longrightarrow V$ for T of depth p, where $p \leq 1 + \log_2 \left(\frac{n+2}{3}\right)$. Let X_0 be the root of this pinnate structure.

Suppose $X \in V$ has depth r. We defined, at the end of Section 7.4, the "path sequence" $\underline{X} = (X_0, X_1, \ldots, X_r)$ for X, where $X_i = \phi^{r-i} X$, so that $X = X_r$. If X' is some other point of V, of depth s, then the path sequence $\underline{X}' = (X_0, X_1', \ldots, X_s')$ for X' will agree with that for X precisely on some initial segment $(X_0, X_1, \ldots, X_t) = (X_0, X_1', \ldots, X_t')$. This initial segment is, itself, the path sequence, $\underline{W}$, for some point $W = W(X, X') \in V$ of depth t. (Possibly W is the same as X or X'.) In the tree T', we must have either $XX_0|X'W$ or $X'X_0|XW$ (see Section 7.4). We see that

$$\rho_{\tau'}(X,X') + \rho_{\tau'}(W,X_0) = \max\big(\rho_{\tau'}(X,W) + \rho_{\tau'}(X',X_0), \rho_{\tau'}(X',W) + \rho_{\tau'}(X,X_0)\big).$$

Thus,

$$\begin{aligned} &\rho(X,X') + 2k(1+\log_2 n) \\ &\quad \geq \max\big(\rho(X,W) + \rho(X',X_0) - \rho(W,X_0), \rho(X',W) + \rho(X,X_0) - \rho(W,X_0)\big). \end{aligned}$$

We shall now use these path sequences to construct our immersed tree $\tau = (T, \sigma, f, g)$ as follows.

For each $X \in V$, we construct the embedded tree $T(X) = T_{\underline{X}}$, as defined in Section 7.5, where $\underline{X}$ is the path sequence for X. Since the definition of $T_{\underline{X}}$ was inductive, we can do this consistently over initial segments, so that for any $X, X' \in V$, the trees $T(X)$ and $T(X')$ agree on the common subtree $T(W(X, X'))$. Thus the trees $T(X)$, as X varies in V, form a directed set under inclusion. This allows us to define T as the direct limit of the $T(X)$.

More explicitly, let $\tilde{T} = \{(X, Y) \in V \times \mathcal{S} \mid Y \in T(X)\}$. We write $(X, Y) \sim (X', Y')$ if $Y = Y' \in T(W(X, X'))$. We check that $\sim$ is an equivalence relation, and set $T = \tilde{T}/\sim$. Let ι_X be the natural inclusion of $T(X)$ in T. We define $f : V \longrightarrow T$ by $f(X) = \iota_X(X)$. We define $g : T \longrightarrow \mathcal{S}$ by demanding, for each $X \in V$, that $g \circ \iota_X$ be the inclusion of $T(X)$ in $\mathcal{S}$. Thus, $g \circ f$ is the identity on V. Finally, the path-metrics on the $T(X)$, induced from $\mathcal{S}$, themselves induce a path-metric σ on T. Clearly, f is distance non-increasing from (T, σ) to $(\mathcal{S}, d)$. This defines our spanning tree $\tau = (T, \sigma, f, g)$.

Note that if $Y, Y' \in V$ both lie in the path sequence for X, then $f(Y), f(Y') \in \iota_X T(X) \subseteq T$. Applying Proposition 7.5.2, we see that

$$\rho(Y, Y') \leq \rho_\tau(Y, Y') = \sigma(f(Y), f(Y')) \leq \rho(Y, Y') + kH(p),$$

where $p = \operatorname{depth} \phi$.

Now, suppose that $X, X' \in V$ are arbitrary. Let $W = W(X, X')$. Now, X_0 and W each lie in the path sequences of both X and X'. Thus,

$$\begin{aligned}
&\rho(X, X') + 2k(1 + \log_2 n) \\
&\geq \max\big(\rho(X, W) + \rho(X', X_0) - \rho(W, X_0), \rho(X', W) + \rho(X, X_0) - \rho(W, X_0)\big) \\
&\geq \max\big(\rho_\tau(X, W) + \rho_\tau(X', X_0) - \rho_\tau(W, X_0), \\
&\qquad \rho_\tau(X', W) + \rho_\tau(X, X_0) - \rho_\tau(W, X_0)\big) - 2kH(p).
\end{aligned}$$

But $\operatorname{hyp} \rho_\tau = 0$, and so

$$\rho_\tau(X, X') + \rho_\tau(W, X_0) \leq \max\big(\rho_\tau(X, W) + \rho_\tau(X', X_0), \rho_\tau(X', W) + \rho_\tau(X, X_0)\big).$$

We conclude that

$$\rho_\tau(X, X') \leq \rho(X, X') + 2k(1 + \log_2 n + H(p)).$$

Now $p \leq 1 + \log_2 \left(\frac{n+2}{3}\right)$, and H is linear, so

$$\rho_\tau(X, X') \leq \rho(X, X') + kF(n),$$

where

$$\begin{aligned}
F(n) &= 2\big(1 + \log_2 n + H(1 + \log_2 \left(\frac{n+2}{3}\right))\big) \\
&= O(\log n).
\end{aligned}$$

We have shown:

Theorem 7.6.1 : *There is a function $F : \mathbf{N} \longrightarrow \mathbf{R}$ such that the following holds.*

Suppose (S, d) is a k-H1 geodesic space, and $V \subseteq S$ is a set of $(n+1)$ points. Then, there is an immersed spanning tree τ for V in S, such that for any $X, Y \in V$, we have

$$\rho_\tau(X, Y) \leq d(X, Y) + kF(n).$$

Moreover, $F(n) = O(\log n)$.

CHAPTER 8 : Propagation of hyperbolicity.

8.1. Summary.

The aim of this chapter is to show that the property of almost-hyperbolicity "propagates". Thus, we shall show that a path-metric space is almost-hyperbolic if and only if it is "locally almost-hyperbolic" and "almost simply-connected", provided, of course, that we properly quantify the various parameters involved.

Let $(\mathcal{S}, d)$ be a path-metric space. Recall, from Section 2.3, the notion of a cellular net, (P, ρ, f), and of its mesh, $m(P, \rho)$, as well as the notion of a loop, (γ, σ). Given $X \in \mathcal{S}$ and $r \in [0, \infty)$, write $N_r(X)$ for the uniform ball $\{Y \in \mathcal{S} | d(X, Y) \leq r\}$.

Definition : The space $(\mathcal{S}, d)$ is *m-simply-connected* if every loop in $\mathcal{S}$ bounds a cellular net of mesh at most m.

Definition : Given $i \in \{1, 2, 3, 4, 5\}$, and $\underline{k} \in [0, \infty) \sqcup [0, \infty)^2$, and $r \in (0, \infty)$, we shall say that $(\mathcal{S}, d)$ is *r-locally* $\underline{k}$-H(i) if, for each $X \in \mathcal{S}$, the uniform ball $N_r(X)$ about X is $\underline{k}$-H(i) in the induced path metric.

Given any $Q \subseteq \mathcal{S}$, we shall write d_Q for the induced path-metric on Q. Applying the equivalence of definitions H(i) to the spaces (N, d_N) for $N = N_r(X)$ as X varies in $\mathcal{S}$, we see that if $(\mathcal{S}, d)$ is r-locally $\underline{k}$-H(i), then it is r-locally $\underline{k}'$-H(j) for $\underline{k}' = \underline{k}'(k, i, j)$. This is the primary reason for making the definitions in this way. We shall use mostly the hypothesis of locally H1. Clearly any space is r-locally k-H1 if $r \leq k/2$, so the hypothesis is only useful if r is large in relation to k.

Proposition 8.1.1 :

(1) $\forall k \; \exists m$ such that any k-H1 geodesic space is m-simply-connected.

(2) $\forall k \; \exists k' \; \forall r$ any k-H1 geodesic space is r-locally k'-H1.

Part (1) is an immediate consequence of property H3ca (Section 6.1). .

Part (2) is in not quite immediate, since we have defined local hyperbolicity in terms of the induced path metric on balls. However, it is a simple consequence of the uniform convexity of balls in an almost hyperbolic space, as we shall explain in Section 8.2.

The main result of this chapter is the following converse to Proposition 8.1.1.

Theorem 8.1.2 : *$\forall k \exists k' \forall m \exists R$ such that any m-simply-connected R-locally k-H1 geodesic space is k'-H1.*

As usual, we can take k' to be a certain universal multiple of k (see Section

8.7).

The necessity of some sort of simple-connectedness is easy to see. Consider, for example, the subset $W_R = \big([0,\infty) \times \{0\}\big) \cup \bigcup_{n=R}^{\infty} C_n$ of the euclidean plane $\mathbf{R}^2$, where $R \in \mathbf{N}$, and for each $n \in \mathbf{N}$, C_n is the circle of radius n about the origin $(0,0)$. If we take the induced path-metric on W_R, then it is R-locally 0-H1, but not k-H1 for any k.

The fact that simple-connectedness is central to this result suggests that we should make use of the linear isoperimetric inequality (property H3). The idea is roughly as follows.

Suppose that $(\mathcal{S}, d)$ is almost simply-connected, and locally almost-hyperbolic, for suitable parameters. Let γ be any loop in $\mathcal{S}$. We span γ by a "disc" of "minimal area". We may imagine this disc as having some kind of metric structure induced from $\mathcal{S}$. In this induced structure, the disc is, itself, locally almost-hyperbolic. The reason for this is as follows. Any loop in the disc of small diameter satisfies the linear isoperimetric inequality in $\mathcal{S}$ since $\mathcal{S}$ is locally H3, and thus intrinsically in the disc since it is area-minimising. We see that the disc is intrinsically locally H3.

We have essentially essentially reduced the problem to a 2-dimensional situation. In this context, we can imagine that the "curvature" of our disc is always negative when averaged on a certain scale. From this we might hope to deduce that the global average of the curvature of the disc is negative. This should then imply the global isoperimetric inequality, namely that the area of the disc is at most a certain linear function of the length of its boundary. This would show that $(\mathcal{S}, d)$ were H3.

There are many ways one might try to make sense of this argument. It seems that all lead into technical complications at some point or other. We shall try to keep these to a minimum by adopting a notion of spanning disc best suited to our purposes, namely what we shall call a "vertex net". We can thus contain the most unpleasant technicalities to relating this to our previous notion of cellular net (Lemma 8.5.1). The essential passage from local to global will take the form of a combinatorial lemma (8.4.1).

We shall give a more detailed outline of the proof in Section 8.6. For the moment, we return to a discussion of local hyperbolicity.

8.2. Local hyperbolicity.

Suppose $(\mathcal{S}, d)$ is a geodesic space. If Q is a closed subset of $\mathcal{S}$, and d_Q is the induced path-metric on Q, then (Q, d_Q) is also a geodesic space. (In general, we need to allow d_Q to take the value ∞ when two points are not joined by a rectifiable path in Q.)

Recall the discussion of convexity from Chapter 4.

Lemma 8.2.1 : *Suppose that $(\mathcal{S}, d)$ is a geodesic space, and that $Q \subseteq \mathcal{S}$ is closed and λ-convex. Let $N = N_\lambda(Q)$. Then, for all $X, Y \in N$, we have*

$$d_N(X,Y) \leq d(X,Y) + 4\lambda.$$

Proof : Let $Z \in \text{proj}_Q X$ and $W \in \text{proj}_Q Y$. Then $[Z, W] \subseteq N$, and the path $[X, Z, W, Y] = [X, Z] \cup [Z, W] \cup [W, Y] \subseteq N$ has length at most $d(X, Y) + 4\lambda$.
◇

We deduce Proposition 8.1.1(2) as follows.

Corollary 8.2.2 : *Suppose $(\mathcal{S}, d)$ is a k-H1 geodesic space, and $\lambda_0 = \lambda_0(k)$ is the constant of Lemma 4.2. For any $r \in [0, \infty)$ and $X \in \mathcal{S}$, we have that (N, d_N) is $(k + 8\lambda_0)$-H1, where $N = N_r(X)$.*

Proof : If $r \leq \lambda_0$, then clearly (N, d_N) is $(4\lambda_0)$-H1.

If $r > \lambda_0$, then $Q = N_{(r-\lambda_0)}(X)$ is λ_0-convex, by Lemma 4.2. Thus, $d \leq d_N \leq d + 4\lambda_0$ by Lemma 8.2.1. The result follows easily.
◇

Remark : There is an alternative way one might define local hyperbolicity. Given a geodesic space $(\mathcal{S}, d)$, we weaken the hypothesis of $\underline{k}$-H(i), for $i \in \{1, 2, 3, 4, 5\}$, by demanding that the set of points under consideration should have diameter at most some fixed number, which we call the *range*. Thus, for example, to say that $(\mathcal{S}, d)$ is locally k-H1 in this sense means that $XY : ZW$ or $XZ : WY$ or $XW : YZ$ for any $X, Y, Z, W \in \mathcal{S}$ satisfying $\text{diam}\{X, Y, Z, W\} \leq R$ for some fixed range R. The equivalence of these five definitions for large range follows from the arguments of previous chapters, although we cannot apply these results directly. Relating these notions to the definitions of Section 8.1 is easiest using property H1. Suppose we are given $k, r \in [0, \infty)$. Then if the range, $R = R(k, r)$, is large enough, and $(\mathcal{S}, d)$ is a locally k-H1 geodesic space, in the sense just defined, then, using the arguments of previous chapters, we see that any uniform r-ball in $\mathcal{S}$ is almost convex. It follows (as in 8.2.1 an 8.2.2) that $(\mathcal{S}, d)$ is r-locally k'-H1 in the original sense (i.e. that of Section 8.1), where k' depends only on k. The converse of this statement (similarly quantified) is a consequence of Lemma 8.2.3.

We conclude this section with the following trivial observation.

Lemma 8.2.3 : *Suppose that $(\mathcal{S}, d)$ is a geodesic space, and $N = N_r(X)$, where $X \in \mathcal{S}$ and $r \geq 0$. Then for any $Y, Z \in N_{r/2}(X)$, we have $d_N(Y, Z) = d(Y, Z)$.*

Proof : $[Y, Z] \subseteq N$.

◇

This means that if we know that $(\mathcal{S}, d)$ is r-locally almost hyperbolic, we can apply any result obtained for globally almost hyperbolic spaces, provided that we ensure that our constructions do not take us outside a set of diameter r.

8.3. A geometric lemma.

Suppose $(\mathcal{S}, d)$ is a geodesic space, and that $X, Y, Z, W, X_0, X_1, \ldots, X_n \in \mathcal{S}$. Recall, from Section 7.5, the notation $XY \wedge ZW$, $\Lambda(X_0, X_1, \ldots, X_n)$, $[X_0, X_1, \ldots, X_n]$ and $\langle X_0 X_1 \ldots X_n \rangle$, from Section 7.5. Note that if $X_0 = X_n$, then

$$\Lambda(X_0, X_1, \ldots, X_n) = \Lambda(X_1, X_2, \ldots, X_{n-1}, X_0, X_1).$$

Thus, if $\gamma = [X_0, X_1, \ldots, X_n]$ is a closed piecewise geodesic path, then $\Lambda(\gamma)$ is defined independently of the choice of basepoint. We shall write $\operatorname{length}\gamma$ for the quantity $\sum_{i=1}^{n} d(X_i, X_{i-1})$.

Lemma 8.3.1 : *There is a map $F : \mathbf{N} \longrightarrow \mathbf{R}$ and a fixed $\lambda \geq 0$ such that the following holds.*

Suppose that $(\mathcal{S}, d)$ is a k-H1 geodesic space, and that γ is a closed path in $\mathcal{S}$ consisting on n geodesic segments. Then,

$$\operatorname{length}\gamma \leq (\Lambda(\gamma) + \lambda k) F(n).$$

Proof : This is just a special case of Proposition 7.3.4 (which gives F linear in n).

Alternatively, one may give a short proof of Lemma 8.3.1 as follows.

Let $L = \operatorname{length}\gamma$. Without loss of generality, we can suppose that $d(X_{n-1}, X_0) \geq d(X_i, X_{i+1})$ for each $i \in \{0, 1, \ldots, n-1\}$. Thus, $d(X_{n-1}, X_0) \geq \frac{1}{n} L$. For $i = 0, 1, \ldots, n-1$, let Y_i be a nearest point in $[X_0, X_{n-1}]$ to X_i. Thus, $Y_0 = X_0$ and $Y_{n-1} = X_{n-1}$. Clearly there is some $i \in \{0, 1, \ldots, n-1\}$ with $\langle X_0 Y_i Y_{i+1} X_{n-1} \rangle$ and

$$d(Y_i, Y_{i+1}) \geq \frac{1}{(n-1)} d(X_{n-1}, X_0) \geq \frac{1}{n(n-1)} L.$$

Then, either $\frac{1}{n(n-1)} L \simeq 0$, or else one can show without difficulty that

$$\Lambda(\gamma) \geq X_0 X_{n-1} \wedge Y_i Y_{i+1} \succeq \frac{1}{n(n-1)} L.$$

(This argument gives F quadratic in n.)

◇

Lemma 8.3.2 : *$\forall k\, \exists a\, \forall H\, \exists L$ such that the following holds. Suppose $(\mathcal{S}, d)$ is a k-H1 geodesic space, and that $\gamma = [X_0, X_1, \ldots, X_n]$ is a closed piecewise geodesic path with at most 13 segments. (Thus, $X_0 = X_n$ and $n \leq 13$.) Then, either* $\text{length}\,\gamma \leq L$, *or else we can find $0 \leq i < j \leq n$, and $A, B \in [X_j, X_{j+1}]$, such that $\langle X_j A B X_{j+1} \rangle$, $d(A, B) \geq H$ and*

$$d(X_i, B) + d(B, A) + d(A, X_{i+1}) \leq d(X_i, X_{i+1}) + a.$$

Proof : Let $L = (H + \lambda k) \max\{F(r) \mid 1 \leq r \leq 13\}$.

If $\Lambda(\gamma)$, then by Lemma 8.3.1, $\text{length}\,\gamma \leq L$. So, suppose $\Lambda(\gamma) \geq H$. This means there exist i, j with $0 \leq i < j \leq n$ and with $X_i X_{i+1} \wedge X_{j+1} X_j \geq H$. Applying Lemma 7.5.1, we find $A, B \in [X_j, X_{j+1}]$ as required, with $a = 2h$ (h being the constant of Lemma 7.5.1).

$\diamond$

8.4. A combinatorial lemma.

Let P be a presentation of the disc D as a CW-complex. Write $C_i(P)$ for the set of i-cells of P, and $\Sigma(P)$ for the 1-skeleton of P. We say that Σ is *j-edge-connected*, for $j \in \mathbf{N}$, if no set of $j-1$ edges disconnects Σ. We say that Σ is *trivalent* if every vertex of Σ has degree 3.

Note that if Σ is trivalent and 2-edge-connected, then the boundary, ∂c, of every 2-cell $c \in C_2(P)$ is an embedded topological circle. Moreover, the two endpoints of any edge are distinct. Thus, in this case, P, is a cellulation in the sense of Section 2.3. If Σ is trivalent and 3-edge-connected, then any two distinct 2-cells of P meet either along a single common edge or not at all. Moreover, any 2-cell of P meets ∂D either in a single edge or not at all.

For $i = 0, 1$, let $C_i^\partial(P)$ be the set of i-cells lying in ∂D. Write $C_i^I(P) = C_i(P) \setminus C_i^\partial(P)$. Clearly, $|C_0^\partial(P)| = |C_1^\partial(P)|$. Given $c \in C_2(P)$, write $\nu(c) = |C_0(P) \cap \partial c|$ for the number of vertices in the boundary of c (or equivalently, the number of edges of c).

We call $c \in C_2(P)$ an *interior* 2-cell if $c \cap \partial D = \emptyset$. Write $C_2^I(P)$ for the set of interior 2-cells of P, and $C_2^\partial(P) = C_2(P) \setminus C_2^I(P)$.

Lemma 8.4.1 : *Suppose that P is a cellulation of the disc D, with $\Sigma(P)$ trivalent and 3-edge-connected. Suppose that for any two distinct interior 2-cells, $c_1, c_2 \in C_2^I(P)$, satisfying $\nu(c_1) \leq 13$ and $\nu(c_2) \leq 13$, we have that $c_1 \cap c_2 = \emptyset$. Then, it follows that*

$$|C_2(P)| \leq 15 |C_1^\partial(P)|.$$

Proof : We construct a new cellulation P' of the disc by contracting each interior 2-cell with fewer than 14 edges to a single point. This is a well-defined process, since, by hypothesis, no two such cells intersect. The set of interior 2-cells of P' corresponds bijectively to the the set of interior 2 cells of P with at least 14 edges. Now each such cell has lost at most half its edges through this process of contraction, and so each $c \in C_2^I(P')$ has at least 7 edges. Clearly, there is a bijective correspondence between $C_1^\theta(P)$ and $C_1^\theta(P')$, since ∂D is unchanged. Note also that P' is 3-edge-connected, and so $C_2^\theta(P')$ is also in bijective correspondence with $C_1^\theta(P)$. Also, each vertex of P' has degree at least 3.

For $i = 0, 1, 2$, write $\Lambda_i = |C_i(P)|$, $\Lambda_i^\theta = |C_i^\theta(P)|$, $\lambda_i = |C_i(P')|$, $\lambda_i^\theta = |C_i^\theta(P')|$ and $\lambda_i^I = |C_i^I(P')|$.

The following relations are easily verified.

(1) $\lambda_0 + \lambda_2 - \lambda_1 = 1$,

(2) $3\lambda_0 \leq 2\lambda_1$,

(3) $\lambda_1 = \lambda_1^\theta + \lambda_1^I$,

(4) $\lambda_2 = \lambda_2^\theta + \lambda_2^I$,

(5) $7\lambda_2^I \leq 2\lambda_1^I$,

(6) $\Lambda_1^\theta = \lambda_1^\theta = \lambda_2^\theta$,

(7) $\Lambda_2 \leq \lambda_0 + \lambda_2$.

Now, from (1) and (2), we obtain the following inequality (*),

$$\lambda_1 \leq 3(\lambda_2 - 1).$$

Applying (3), (4) and (6) to this, we get

$$\lambda_1^I \leq 3\lambda_2^I + 2\lambda_1^\theta - 3.$$

From (5), we get

$$7\lambda_2^I \leq 2(3\lambda_2^I + 2\lambda_1^\theta - 3),$$

and so

$$\lambda_2^I \leq 4\lambda_1^\theta - 6.$$

From (4) and (6),

$$\lambda_2 = \lambda_2^I + \lambda_1^\theta \leq (4\lambda_1^\theta - 6) + \lambda_1^\theta = 5\lambda_1^\theta - 6.$$

Using (*) again, we get

$$\lambda_1 \leq 3(\lambda_2 - 1) \leq 15\lambda_1^\theta - 21.$$

By (7) and (1),

$$\Lambda_2 \leq \lambda_0 + \lambda_2 = 1 + \lambda_1 \leq 15\lambda_1^\theta - 20.$$

Using (6),

$$\Lambda_2 \le 15\Lambda_1^\theta - 20.$$

So certainly,

$$|C_2(P)| \le 15|C_1^\theta(P)|.$$

$\diamond$

8.5. Vertex nets.

This is a technical section. We define the notion of a "vertex net", which is another formulation of the idea of a spanning disc. We relate this to the idea of a cellular net from Section 2.3.

Let G be a finite graph. We write $C_0(G)$ for the set of vertices of G, and $C_1(G)$ for the set of edges of G. By a ***subgraph***, G', of G, we mean a subset $C_1(G')$ of the edges of G, together with all their endpoints.

Suppose that $(\mathcal{S}, d)$ is metric space, and that $g : C_0(G) \longrightarrow \mathcal{S}$ is any map. If $e \in C_1(G)$, we shall write

$$\text{length}(e, g) = d(g(v), g(w)),$$

where $v, w \in C_0(G)$ are the endpoints of e. If G' is a subgraph of G, then we write

$$\text{length}(G', g) = \sum_{e \in C_1(G')} \text{length}(e, g)$$

and

$$\text{coarseness}(G', g) = \max_{e \in C_1(G')} \text{length}(e, g).$$

As an example, consider the circle S^1 as the unit circle in the complex plane. For $n \in \mathbf{N}$ and $j \in \mathbf{Z}$, let v_j^n be the point $e^{2\pi i(j/n)} \in S^1$. Let $V^n = \{v_1^n, v_2^n, \ldots, v_n^n\} \subseteq S^1$. For any n, we can represent S^1 as a graph with vertex set $C_1(S^1) = V^n$. Suppose $(\mathcal{S}, d)$ is a geodesic space, and $g : V^n \longrightarrow \mathcal{S}$ any map. Then, we can identify (S^1, g) with the closed piecewise geodesic path $\gamma = [X_0, X_1, \ldots, X_n]$. Thus, $\text{length}(S^1, g) = \text{length}\gamma = \sum_{i=1}^n d(X_i, X_{i-1})$. We shall call such an object a *cycle* of n points in $\mathcal{S}$.

Let P be a cellulation of the disc D. We call such a cellulation *good* if the 1-skeleton, $\Sigma = \Sigma(P)$, is trivalent and 2-edge-connected (see Section 8.4), and if each 2-cell of P meets ∂D either in a single edge or not at all.

We identify $S^1 \equiv \partial D$.

Definition : Let $(\mathcal{S}, d)$ be a geodesic space. A *vertex net*, (P, g), consists of a good cellulation P of the disc D, together with a map $g : C_0(P) \longrightarrow \mathcal{S}$.

We can thus define length(P, g) and coarseness(P, g) by putting $G' = G = \Sigma(P)$ in the formulae given above. We set

$$\text{mesh}(P) = \max_{c \in C_2(P)} \text{length}(\partial c, g).$$

Definition : Suppose that $\gamma \equiv (S^1, g_0)$ is a cycle of n points in $\mathcal{S}$. We say that γ *bounds* a vertex net (P, g), if $C_0^\partial(P) = V^n$, and $g|C_0^\partial(P) = g_0$. (Recall that $C_0^\partial(P) = C_0(P) \cap \partial D$.)

Clearly, coarseness $\gamma \leq$ coarseness$(P, g) \leq$ mesh(P, g).

We now relate all this to the definitions of Section 2.3.

Lemma 8.5.1 : *Let σ be a path-metric on the circle S^1. Suppose that the points $x_1, x_2, \ldots, x_n \in S^1$ are cyclically ordered, and cut S^1 into segments of σ-length at most m_0. Suppose that (S, d) is a geodesic space, and that $\beta : (S^1, \sigma) \longrightarrow (S, d)$ is distance non-increasing (i.e. β is a "loop" in the sense of Section 2.3). Let γ be the cycle $[X_0, X_1, \ldots, X_n]$, where $X_i = \beta(x_i)$ for $i \geq 1$, and $X_0 = X_n$. (Thus* coarseness $\gamma \leq m_0$.) *Then, we have the following.*

(1) Suppose that γ bounds a vertex net (P, g), then for any $\epsilon > 0$, β bounds a cellular net (P, ρ, f) with

$$m(P, \rho) \leq m_0 + \text{mesh}(P, g) + \epsilon.$$

(2) Suppose that β bounds a cellular net (P, ρ, f), then γ bounds a vertex net (P', g) with

$$\text{mesh}(P', g) \leq m_0 + m(P, \rho).$$

Proof :

(1) Write $C_1^\partial(P) \subseteq C_1(P)$ for the set of 1-cells lying in ∂D. Write $C_1^I(P) = C_1(P) \setminus C_1^\partial(P)$. By hypothesis, each 2-cell of P meets ∂D either in single edge, or not at all. Let $\eta = \epsilon / |C_1(P)|$.

We define a path-metric ρ on $\Sigma(P)$ as follows. If $e \in C_1^\partial(P)$, set $\rho(e) = m_0$. If $e \in C_1^I(P)$, set $\rho(e) = \max(\eta, \text{length}(e, g))$.

Now (by definition), $C_0^\partial(P) = C_0(P) \cap \partial D = V^n = \{v_1^n, v_2^n, \ldots, v_n^n\}$, and $g(v_i^n) = X_i$. We define $\partial f : (\partial D, \rho) \longrightarrow (S^1, \sigma)$ by setting $\partial f(v_i^n) = x_i$ for $i = 1, 2, \ldots, n$, and mapping each component of $\partial D \setminus C_1^\partial(P)$ onto the corresponding component of $S^1 \setminus \{x_1, x_2, \ldots, x_n\}$, linearly with respect to the parameterisations induced by ρ and σ respectively. Note that $\beta \circ \partial f | C_0^\partial(P) = g | C_1^\partial(P)$, and that ∂f is homotopic to the identification $\partial D \equiv S^1$.

We define $f : \Sigma(P) \longrightarrow (\mathcal{S}, d)$ as follows. We take $f|C_0(P) = g$ and $f|\partial D = \beta \circ \partial f$. (Note that this is consistent on $C_0^\partial(P)$.) Finally, if $e \in C_1^I(P)$, with endpoints $v, w \in C_0(P)$, then we map e (with parameterisation induced by ρ) linearly onto the geodesic $[g(v), g(w)]$ in $\mathcal{S}$.

We have defined (P, ρ, f). It is readily checked that $m(P, \rho) \leq m_0 + \text{mesh}(P, g) + \epsilon$. This completes the proof of part (1).

(2) We have that β bounds a cellular net (P, ρ, f). We can suppose that $m_0 > 0$. We shall construct (P', g) in a series of stages.

We identify $\partial D \equiv S^1$. Let D_0 be the topological disc $D_0 = D \cup_{\partial D} (S^1 \times [0, 2])$, where we identify $S^1 = \partial D$ with $S^1 \times \{0\}$ via $x \leftrightarrow (x, 0)$.

Let Q be any triangulation of the annulus $S^1 \times [0, 1]$. Let $h : S^1 \times [0, 1] \longrightarrow S^1$ be a homotopy from $h(., 0) = \partial f : (\partial D, \rho_{\partial D}) \longrightarrow (S^1, \sigma)$ to the identity $h(., 1) = 1_{S^1} : S^1 \longrightarrow S^1$. By subdividing Q enough times, we obtain a new triangulation Q_1 of $S^1 \times [0, 1]$ with the property that $\sigma(h(p), h(q)) \leq m/3$ for the endpoints, p and q, of any edge of Q_1. We define a cellulation, Q_2, of $S^1 \times [1, 2]$ (i.e. a presentation of $S^1 \times [1, 2]$ as a CW-complex) as follows. Set $x_0 = x_n$, and for each $i \in \{1, 2, \ldots, n\}$, write $[x_{i-1}, x_i] \subseteq S^1$ for the closed interval joining x_{i-1} to x_i in S^1 (in the positive direction). We take

$$C_0(Q_2) = \{x_1, x_2, \ldots, x_n\} \times \{1, 2\},$$

$$C_1(Q_2) = \{[x_{i-1}, x_i] \times \{j\} \mid 1 \leq i \leq n,\ j = 1, 2\} \cup \{\{x_i\} \times [1, 2] \mid 1 \leq i \leq n\},$$

and

$$C_2(Q_2) = \{[x_{i-1}, x_i] \times [1, 2] \mid 1 \leq i \leq n\}.$$

The cellulations P, Q_1 and Q_2, together give us a cellulation, P_0, of the disc D_0, after taking a common subdivision on the circles $\partial D \equiv S^1 \times \{0\}$ and $S^1 \times \{1\}$. By taking the maximal such subdivision, we will have that each 0-cell of P_0 has degree at least 3 in $\Sigma(P_0)$. Moreover each 0-cell in $\partial D_0 = S^1 \times \{2\}$ has degree exactly 3. We define $g_0 : C_0(P_0) \longrightarrow \mathcal{S}$ as follows.

$$g_0|C_0(P) = f|C_0(P),$$

$$g_0|C_0(Q_1) = \beta \circ h|C_0(Q_1),$$

and

$$g_0(x_i, j) = \beta(x_i)$$

for $i = 1, 2, \ldots, n$ and $j = 1, 2$. One can check that g_0 is well-defined on any 0-cells P, Q_1 or Q_2 may have in common. Moreover, one may check that $\text{mesh}(P_0, g_0) \leq m_0 + m(P, \rho)$.

Now let $\phi : D \longrightarrow D_0$ be any homeomorphism with $\phi(v_i^n) = (x_i, 2)$ for $i = 1, 2, \ldots, n$. Let P_1 be the pull-back of the cellulation P_0 to D. Let $g_1 = g_0 \circ \phi :$

$C_0(P_1) \longrightarrow \mathcal{S}$. Thus $g_0(v_i^n) = X_i = \beta(x_i)$. Clearly, $\text{mesh}(P_1, g_1) = \text{mesh}(P_0, g_0)$. Note that every vertex of $C_0^\theta(P_1)$ has degree 3, and that no 2-cell meets ∂D in more than one edge.

Finally, we define P' by arbitrarily splitting apart each vertex of $C_0(P_1)$, with degree at least 4, to give a tree in $\Sigma(P_1)$. Thus, we can arrange that each vertex of P' has degree 3. There is a natural map $\theta : C_0(P') \longrightarrow C_0(P_1)$ given by contracting all these trees back to points. We define $g = g_1 \circ \theta : C_0(P') \longrightarrow \mathcal{S}$. Clearly, $\text{mesh}(P', g) = \text{mesh}(P_1, g_1) \le m_0 + m(P, \rho)$.

◇

Corollary 8.5.2 : *In an m-simply-connected geodesic space, any cycle of coarseness m_0 bounds a vertex net of mesh at most $m + m_0$.*

Proof : Apply Lemma 8.5.1(2) to the closed piecewise geodesic path given by the cycle (i.e. by joining the points of the cycle by geodesic segments).

◇

Corollary 8.5.3 : *$\forall m, m_0, \lambda, \mu\ \exists k_3, h_3$ such that if $(\mathcal{S}, d)$ is a geodesic space in which every cycle of n points, with coarseness at most m_0, bounds a vertex net of mesh at most m and with at most $\lambda n + \mu$ 2-cells, then $(\mathcal{S}, d)$ is (k_3, h_3)-H3.*

Proof : Apply Lemma 8.5.1(1).

◇

8.6. Main Proof.

In this section, we give the proof that almost hyperbolicity propagates. It will remain in the final section (8.7) to refine this statement with regard to the parameters involved.

The idea of the proof is as follows. Suppose that $(\mathcal{S}, d)$ is an m-simply-connected r-locally k-H1 geodesic space, with r much larger than both m and k. Choose any cycle, $\gamma = [X_0, X_1, \ldots, X_n]$, in $\mathcal{S}$. Let $m_0 \ge \text{coarseness}\,\gamma = \max\{d(X_i, X_{i-1}) \mid 1 \le i \le n\}$. Corollary 8.5.2 tells us that γ bounds a vertex net of mesh at most $m + m_0$. We choose $M \ge m + m_0$ very large (but smaller than r). Among all vertex nets bounded by γ, and of mesh at most m, we select those of (close to) minimum length. Then, among these, we choose one, (P, g), with a minimal number of 2-cells. We aim to show that P satisfies the hypotheses of Lemma 8.4.1.

First, we note that P must be 3-edge-connected. This is a simple argument.

Next, we show that there is some upper bound, L, on $\text{length}(\partial c, g)$ for any interior 2-cell c with fewer than 14 edges. The reason is that, if length$(\partial c, g)$ were very large, we could use Lemma 8.3.2 to reduce length(P, g) by partially collapsing the 2-cell c. In doing this, however, we disturb one of the neighbouring cells, so it is conceivable that we may increase the mesh of the net above M. In this case, however, Lemma 6.1.4 allows us to cut the offending 2-cell in half, without increasing the total length of the net too much.

Finally, suppose that c_1 and c_2 are adjacent interior 2-cells, each with at most 13 edges. We have shown that length$(\partial c_i) \leq L$, and we can assume that $L \leq M/2$. Thus, if we delete the common edge of c_1 and c_2, we will not increase the mesh beyond M. However, we reduce the number of 2-cells. This contradicts the choice of (P, g).

We conclude that P satisfies the hypothesis of Lemma 8.4.1. Thus P has at most $15n$ 2-cells. We now apply Corollary 8.5.1 to deduce that $(\mathcal{S}, d)$ is almost hyperbolic.

To make this argument precise, we need to describe, more carefully, the modifications of vertex nets we use, and to quantify all the parameters involved.

Suppose that (P, g) is a vertex net. As before, write $C_1^\partial(P)$ for the set of 1-cells of P which lie in ∂D. Write $C_1^I = C_1(P) \setminus C_1^\partial(P)$. We describe the following ways to modify (P, g) to give a new net (P', g').

(1) Addition of an edge.

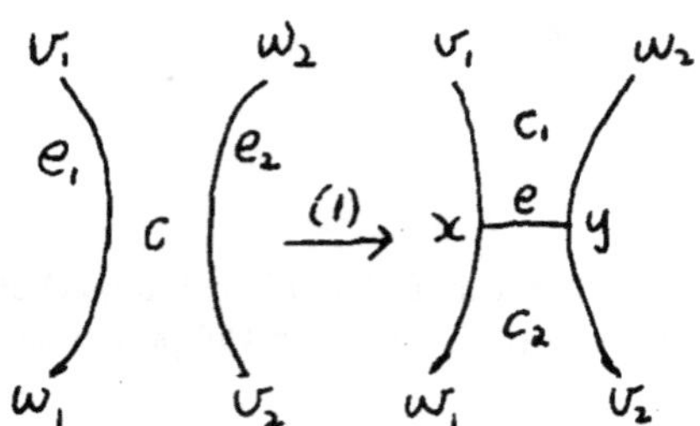

Figure 8a.

Suppose $c \in C_2(P)$. Suppose $e_1, e_2 \in C_1^I(P)$ are distinct edges of c, with endpoints v_1, w_1 and v_2, w_2 respectively. We assume that v_1, w_1, v_2, w_2 (not necessarily distinct) are cyclically ordered in this way around ∂c. (Figure 8a.) Suppose that $C \in \mathcal{S}$ lies in the geodesic segment $[g(v_1), g(w_1)]$ and that $D \in [g(v_2), g(w_2)]$. We form P' by adding a new edge, e, between the points $x \in e_1$ and $y \in e_2$. Thus, $C_0(P') = C_0(P) \cup \{x, y\}$. We define g' by $g'|C_0(P) = g$ and $g'(x) = C$ and

$g'(y) = D$. We have

$$\begin{aligned}\text{length}(P', g') &\le \text{length}(P, g) + d(C, D)\\ \text{mesh}(P', g') &\le \max\big(\text{mesh}(P, g), \text{length}(\partial c_1, g'), \text{length}(\partial c_2, g')\big),\end{aligned}$$

where $c_1, c_2 \in C_2(P')$ are the 2-cells meeting e.

(2) Partial contraction of a 2-cell.

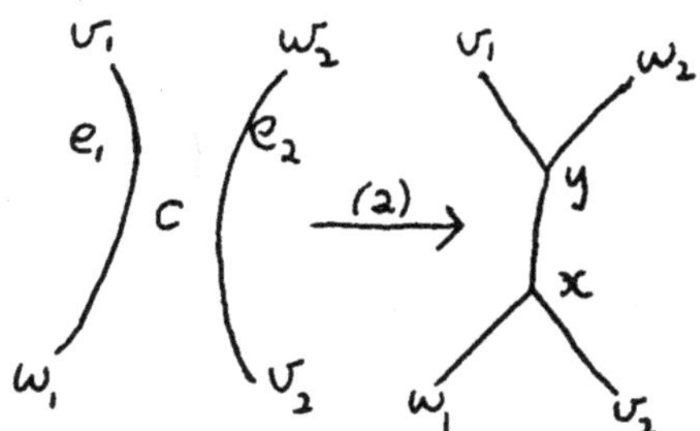

Figure 8b.

Let $c, e_1, e_2, v_1, w_1, v_2, w_2$ be as in modification (1). Suppose that $A, B \in \mathcal{S}$ with $\langle g(v_2)ABg(w_2)\rangle$. We form P' by contracting a strip joining e_1 to e_2 as shown in the Figure 8b. Let x and y be the new vertices introduced. Thus, $C_0(P') = C_0(P) \cup \{x, y\}$. Define g' by $g'|C_0(P) = g$ and $g'(x) = A$ and $g'(y) = B$. We have

$$\begin{aligned}\text{length}(P', g') &\le \text{length}(P, g) - d(A, B) + \delta\\ \text{mesh}(P', g') &\le \text{mesh}(P, g) + \delta,\end{aligned}$$

where $\delta = d(g(v_1), B) + d(B, A) + d(A, g(w_1)) - d(g(v_1), g(w_1))$.

(3) Removal of an edge.

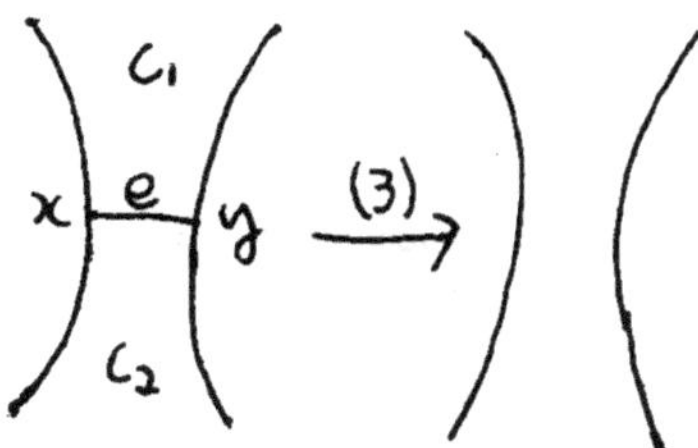

Figure 8c.

Suppose $e \in C_1(P)$ with endpoints $x, y \in C_0(P)$. Suppose that $x, y \notin \partial D$. Thus $C_0(P') = C_0 \backslash \{x, y\}$. Let $g' = g|C_0(P)$. We have

$$\begin{aligned} \text{length}(P', g') &\leq \text{length}(P, g) \\ \text{mesh}(P', g') &\leq \max\big(\text{mesh}(P, g), \text{length}(\partial c_1, g) + \text{length}(\partial c_2, g)\big), \end{aligned}$$

where $c_1, c_2 \in C_2(P)$ are the 2-cells of P meeting e.

(4) Removal of a subgraph.

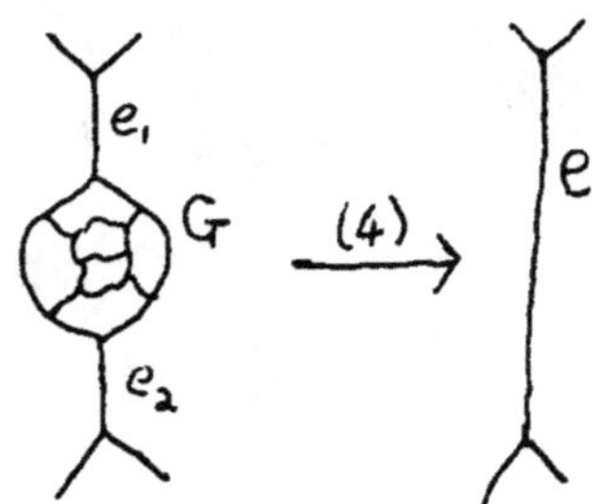

Figure 8d.

Suppose the edges e_1 and e_2 of P separate $\Sigma(P)$ into two components. Let G be the component not meeting ∂D. (Figure 8d.) We form P' by removing G and joining e_1 and e_2 to form a single edge e. Thus $C_0(P') = C_0(P) \backslash C_0(G)$. Let $g' = g|C_0(P')$. We have

$$\begin{aligned} \text{length}(P', g') &\leq \text{length}(P, g) \\ \text{mesh}(P', g') &\leq \text{mesh}(P, g). \end{aligned}$$

We now give the proof in detail. We make no essential use of the completeness or local compactness assumptions on $\mathcal{S}$.

Proposition 8.6.1 : *$\forall k, m\ \exists r, k'$ such that if $(\mathcal{S}, d)$ is an m-simply-connected r-locally k-H1 geodesic space, then it is k'-H1.*

Proof : We are given k, m.
Choose $\epsilon > 0$ and $m_0 > 0$ arbitrarily (for example take $\epsilon = m_0 = 1$).
Let a be the constant arising from Lemma 8.3.2, given k.

Let $b = a + 2m_0$.
Let l be the constant arising from Lemma 6.1.4, given $k_1 = k$ and b.
Let $H = a + l + m_0 + 2\epsilon$.
Let L be the constant arising from Lemma 8.3.2, given k and H.
Let $M = \max(2L, m + m_0, 2(l + b))$.
Let $r = M + a$.

Suppose that $(\mathcal{S}, d)$ is an m-simply-connected r-locally k-H1 geodesic space. Suppose $\gamma = [X_0, X_1, \ldots, X_n]$ is some cycle in $\mathcal{S}$, with coarseness $\gamma \leq m_0$.

Let $\mathcal{P}$ be the set of all vertex nets in $\mathcal{S}$ spanning γ, and with mesh at most M. Since $m + m_0 \leq M$, applying Corollary 8.5.2, we have that $\mathcal{P} \neq \emptyset$.

Let

$$\lambda = \inf\{\text{length}(P, g) \mid (P, g) \in \mathcal{P}\},$$

and

$$\mathcal{P}_0 = \{(P, g) \in \mathcal{P} \mid \text{length}(P, g) \leq \lambda + \epsilon\}.$$

Choose a fixed $(P, g) \in \mathcal{P}_0$ with minimal number of 2-cells. We shall show that (P, g) satisfies the hypothesis of Lemma 8.4.1. We can deduce from this that $|C_2(P)| \leq 15n$. The result then follows by applying Corollary 8.5.3 and the equivalence of H1 and H3.

First note that P is 3-edge-connected — otherwise, we could use modification (4) to strictly reduce the number of 2-cells, without increasing the length or mesh.

Now, suppose (for contradiction) that there is some $c \in C_2^I(P)$ with length(∂c, g) $> L$, and $\nu(c) \leq 13$. Let $y_1, y_2, \ldots, y_s$ be the set of vertices $\partial c \cap C_0(P)$, ordered cyclically around ∂c. Thus $s \leq 13$. Set $y_0 = y_s$, and let $Y_i = g(y_i) \in \mathcal{S}$ for each i. Let α be the closed piecewise geodesic path $[Y_0, Y_1, \ldots, Y_s]$. Now, $\text{diam}[Y_0, Y_1, \ldots, Y_s] \leq \text{mesh}(P, g) \leq m < r$. Thus, we apply Lemmas 8.3.2 and 8.2.3 to find $0 \leq i < j \leq n$ and $A, B \in [Y_j, Y_{j+1}]$ with $\langle Y_j A B Y_{j+1} \rangle$, $d(A, B) \geq H$ and $d(Y_i, B) + d(B, A) + d(A, Y_{i+1}) \leq d(Y_i, Y_{i+1}) + a$. We perform modification (2), with $v_1 = y_i$, $w_1 = y_{i+1}$, $v_2 = y_j$ and $w_2 = y_{j+1}$ (Figure 8e).

(Possibly $w_1 = v_2$ or $v_1 = w_2$.) This gives us a new net (P', g'). Let c_0 be the 2-cell of P, other than c, which meets the edge, e, between v_1 and w_1. Let c' be the corresponding cell of P'.

We have

$$\begin{aligned}\text{length}(P', g') &\leq \text{length}(P, g) - d(A, B) + a \\ &\leq \lambda + \epsilon - H + a.\end{aligned}$$

However we may have increased the mesh, so that, perhaps, (P', g') does not belong to $\mathcal{P}$. Since $d(A, B) \geq H > a$, the only cell of P' which may have boundary length greater than M is c'. We know, at least, that length($\partial c'$, g) $\leq M + a$.

If length($\partial c'$, g') $\leq M$, then mesh(P', g') $\leq M$. Since $H - a + \epsilon < 0$, we have have length(P', g') $< \lambda$, contradicting the definition of λ. Thus, we can assume that

$$M \leq \text{length}(\partial c', g') \leq M + a.$$

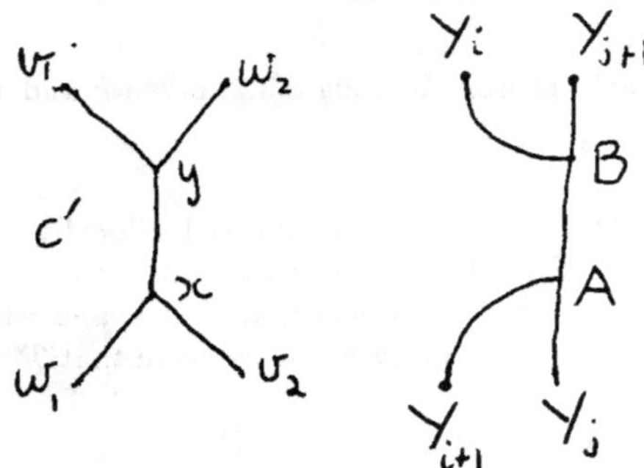

Figure 8e.

Let $z_1, z_2, \ldots, z_t$ be the vertices of $\partial c' \cap C_0(P')$ cyclically ordered around c. Let $z_0 = z_t$, and let $Z_i = g'(z_i)$ for $0 \leq i \leq t$. Let β be the closed piecewise geodesic path $[Z_0, Z_1, \ldots, Z_t]$. Now $\operatorname{diam} \beta \leq M + a \leq r$ and $\operatorname{length} \beta \geq M \geq 2(l + b)$. So, applying Lemmas 6.1.4 and 8.2.3, we find $C', D' \in \beta$ with $d(C', D') \leq l$, but with the distance between C' and D', measured along β, at least $l + b$. If $c' \in C_2^I$ (i.e. $c' \cap \partial D = \emptyset$), then set $C = C'$ and $D = D'$. If, on the other hand, $c' \cap \partial D \neq \emptyset$, then c' meets ∂D in precisely one edge, which, without loss of generality, we can take to have endpoints z_0 and z_1. Thus Z_0 and Z_1 must be consecutive points in our original cycle γ. It follows that $d(Z_0, Z_1) \leq m_0$. It is impossible that C' and D' lie in the same geodesic segment of β, so we can assume, without loss of generality, that $D' \neq [Z_0, Z_1]$. We set $D = D'$. If $C' \in [Z_0, Z_1]$, we set $C = Z_0$, otherwise, we set $C = C'$. Thus, in any case, $d(C, C') \leq m_0$. Now, $d(C, D) \leq l + m_0$, and the distance between C and D, measured along β is at least $l + b - m_0$. We have $C \in [Z_p, Z_{p+1}]$ and $D \in [Z_q, Z_{q+1}]$, where we can suppose that $p \neq 0$ and $q \neq 0$. Since $b - m_0 > m_0$, we must have $p \neq q$. Let e_1' be the edge in $\partial c'$ bounded by z_p and z_{p+1}, and let e_2' be the edge of $\partial c'$ bounded by z_{q+1}. We now perform modification (1), to give a new net (P'', g''). If $c_1'', c_2'' \in C_2(P'')$ are the 2-cells into which c' is split, then it is easily seen that, for $i = 1, 2$,

$$\operatorname{length}(\partial c_i'', g'') \leq \operatorname{length}(\partial c', g') - (b - 2m_0) \leq (M + a) - (b - 2m_0) = M.$$

Thus,

$$\operatorname{mesh}(P'', g'') \leq M,$$

and so $(P'', g'') \in \mathcal{P}$. However, we have

$$\begin{aligned}\operatorname{length}(P'', g'') &\leq \operatorname{length}(P', g') + (l + m_0) \\ &\leq (\lambda + \epsilon - H + a) + (l + m_0) \\ &= \lambda - \epsilon,\end{aligned}$$

which contradicts the definition of λ, and thus the existence of the original cell c.

We have shown that, for any 2-cell $c \in C_2^I(P)$ with $\nu(c) \leq 13$, we have $\text{length}(\partial c, g) \leq L$.

Finally, suppose (for contradiction) that there exist $c_1, c_2 \in C_2^I(P)$ satisfy $\nu(c_1) \leq 13$, $\nu(c_2) \leq 13$ and meet along some edge $e \in C_1(P)$. Thus, $e \cap \partial D = \emptyset$. We know that $\text{length}(\partial c_i, g) \leq L \leq M/2$. Thus, if we perform modification (3) to produce (P', g'), then we get

$$\text{mesh}(P', g') \leq M$$

and

$$\text{length}(P', g') \leq \text{length}(P, g) \leq \lambda + \epsilon.$$

Thus $(P', g') \in \mathcal{P}_0$. However, $|C_0(P')| \leq |C_0(P)| - 1$. This contradicts the assumption that P has a minimal number of 2-cells, and so such c_1 and c_2 cannot exist.

We have shown (P, g) satisfies the hypotheses of Lemma 8.4.1, as required.

◇

8.7. Refinement of parameters.

The argument of the last section does not give any sensible way of relating the final hyperbolicity parameter k' to k. In fact, however, one can always take k' to be a certain universal multiple, ξk of k (independent of the mesh m). The main theorem of this chapter (Theorem 8.1.2) follows from Proposition 8.6.1, and the following proposition.

Proposition 8.7.1 : *There is some universal $\xi \in [0, \infty)$, and a function $R : [0, \infty) \longrightarrow [0, \infty)$ such that the following holds.*

Suppose that (S, d) is a κ-H1 and $R(\kappa)$-locally k-H1 geodesic space. Then it is ξk-H1.

We shall need two lemmas.

Lemma 8.7.2 : *$\forall k \exists h$ such that (S, d) if a k-H1 geodesic space, and if $X, Y, Z, W \in S$ satisfy*

$$\min\big(d(X,Y), d(Z,W)\big) \geq \max\big(d(X,Z), d(Y,W)\big),$$

then $d([X,Y],[Z,W]) \leq h$.

Proof : $XY \wedge ZW \geq 0$. Apply Lemma 7.5.1.

◇

Lemma 8.7.3 : *$\forall k, h\, \exists K$ such that if $(\mathcal{S}, d)$ is a k-H1 geodesic space, and if the points $Y_1, Y_2, Y_3, Z_1, Z_2, Z_3 \in \mathcal{S}$ satisfy $d(Y_i, Z_i) \leq h$ for $i = 1,2,3$, then there is some $C \in \mathcal{S}$ with $d(C, [Y_{i+1}, Z_i]) \leq K$ for $i = 1,2,3$.*

Proof : Let C be a centre of $Y_1Y_2Y_3$. Apply Lemma 3.1.2.

◇

Proof of Proposition 8.7.1 : Suppose $(\mathcal{S}, d)$ is a κ-H1 geodesic space. Let X_1, X_2, X_3 be any three points of $\mathcal{S}$. Let A be a centre for $X_1X_2X_3$ (with respect to the parameter κ). For $i = 1,2,3$, let W_i be a nearest point on $[X_i, X_{i+1}]$ to A. (We take subscripts mod 3.) Applying Lemma 3.1.2, we can find $J = J(\kappa)$ such that $d(W_i, W_{i+1}) \leq J$ and $[X_i, W_i] \subseteq N_J[X_i, W_{i+1}]$ for $i = 1,2,3$. Let $R = 12J$. Thus R depends only on κ.

Suppose now that $(\mathcal{S}, d)$ is also R-locally k-H1. We want to find some point $C \in \mathcal{S}$ whose distance from each of the geodesics $[X_i, X_{i+1}]$ has a bound depending only on k.

Suppose $d(X_1, W_1) > 3J$. Then, let $E \in [W_1, X_1]$ be the point with $d(E, W_1) = 3J$. There is some $F \in [W_3, X_1]$ with $d(E, F) \leq J$. We must have $J \leq d(F, W_3) \leq 5J$. Thus, $\min(d(E, W_1), d(F, W_3)) \leq \max(d(E, F), d(W_1, W_3))$. Also, we have $\operatorname{diam}[W_1, E, F, W_3, W_1] \leq 5J < R$. Thus, applying Lemmas 8.7.2 and 8.2.3, we can find $Y_1 \in [W_3, F]$ and $Z_1 \in [W_1, E]$ with $d(Y_1, W_3) \leq 5J$, $d(Z_1, W_1) \leq 5J$ and $d(Y_1, Z_1) \leq \min(h, J)$, where $h = h(k)$ comes from Lemma 8.7.2.

In the case when $d(X_1, W_1) \leq 3J$, we take $Y_1 = Z_1 = X_1$.

We perform similar constructions with respect to X_2 and X_3. This gives points $Y_i \in [X_i, W_{i-1}]$ and $Z_i \in [X_i, W_i]$ with $d(Y_i, W_{i-1}) \leq 5J$, $d(Z_i, W_i) \leq 5J$ and $d(Y_i, Z_i) \leq \min(h, J)$.

We thus have $\operatorname{diam}[Y_1, Z_1, Y_2, Z_2, Y_3, Z_3, Y_1] \leq 12J \leq R$. Thus, applying Lemmas 8.7.2 and 8.2.3, we find $C \in \mathcal{S}$ with $d(C, [Y_{i+1}, Z_i]) \leq K$ for $i = 1,2,3$, where $K = K(k, h(k))$ depends only on k.

We have shown that $(\mathcal{S}, d)$ is K-H2. It is thus k'-H1 for some k' depending only on k. As discussed in Section 7.2, we can assume that k' has the form ξk for some universal $\xi > 0$.

◇

References.

[**ABCDFLMSS**] J.M.Alonso, T.Brady, D.Cooper, T.Delzant, V.Ferlini, M.Lustig, M. Mihalik, M.Shapiro, H.Short, *Notes on negatively curved groups* : preprint, M.S.R.I., Berkeley (1989).

[**BGHHSST**] W.Ballmann, E.Ghys, A.Haefliger, P.de la Harpe, E.Salem, R.Strebel, M.Troyanov, *Sur les groupes hyperboliques d'après Mikhael Gromov* : preprint, Geneva (1989).

[**CDP**] M.Coornaert, T.Delzant, A.Papadopoulos, *Notes sur les groupes hyperboliques de Gromov* : preprint, Université Louis Pasteur, Strasbourg (1989).

[**G**] M.Gromov, *Hyperbolic groups,* in "Essays in group theory" : ed. S.M.Gersten, M.S.R.I. Publications No. 8 (1988).

Quasiconvexity and a Theorem of Howson's

HAMISH SHORT

Chaire Louis Néel,
Unité de mathématiques pures et appliquées,
Ecole Normale Supérieure, Lyon 69364
France.

Abstract. This exposition aims to be an elementary introduction to the idea of quasiconvex subgroups of a group, as introduced by Gromov. We show that a subgroup of a free group is finitely generated group is quasiconvex if and only if it is quasiconvex, and use this to obtain a proof of Howson's theorem.

I shall present yet another idea of Gromov's connecting the geometry of the Cayley graph with properties of the group. Here I shall talk about a simple application of these ideas in the context of free groups, though I believe that the basic definition of quasiconvexity introduced by Gromov should have many applications elsewhere (see for instance recent work of S.M. Gersten (of University of Utah) and myself [**GS**]).

The aim of the definition is to enable one to say something about subgroups of a finitely generated group. The aim of the present talk is to illustrate the ideas involved by using them to give a leisurely proof of a relatively well-known theorem. We shall be concerned with the *"finitely generated intersection property" (f.g.i.p.),* which a group may or may not have: a group has the f.g.i.p. if and only if the intersection of any two finitely generated subgroups is a finitely generated subgroup. Not all groups have this property, as we shall see in §1, but we have:

HOWSON'S THEOREM. (1954, [**H**])
In a free group, the intersection of two finitely generated subgroups is finitely generated.

(It is at least worth mentioning that the intersection of two subgroups in a group is a subgroup, and that subgroups of a free group are free.)

Lots of proofs are available of this result (something can be said about the rank of the intersection (e.g. Hanna Neumann [**NH**]) — see final remarks). Some of the proofs are even topological, using covering space techniques (e.g. R.G. Burns [**B**], S.M. Gersten [**G**], W.D. Neumann [**NW**], and references therein). There is even a proof using regular languages and the theory of finite state automata (Anissimov and Seifert [**AS**]). It is shown in [**K**] that finite extensions of free groups also have the finitely generated intersection property, a result Krstić attributes to an unnamed Soviet mathematician. This can be further generalized to graphs of groups, where the vertex groups have the f.g.i.p. and the edge groups are finite [**C**, 8.33] (this class includes finite extensions of free groups, as was originally shown by Karass, Pietrowski and Solitar in the finitely generated case – see for instance [**C**, 8.55]).

I would like to thank S.J. Pride and J. Howie for correcting my example §1.iii), and S. Krstić for bring his article [**K**] to my attention. I am very grateful to the Fondation Scientifique de Lyon et du Sud–Est for their financial support, and to the members of the Laboratoire de Mathématiques de l'Ecole Normale Supérieure de Lyon for their hospitality during the academic year 1989–90.

§1 Infinitely generated subgroups

i) An infinitely generated subgroup in a finitely generated free group. Recall that in the free group $F(s,t) = \langle s,t \mid \ \rangle$ the subgroup generated by $\{s^n t s^{-n} \mid n \in \mathbf{Z}\}$ is infinitely generated. This can be seen as each element of the subgroup can be written uniquely as a reduced product of the given elements.

ii) Here is a simple example of a group without the f.g.i.p..
Let $\Gamma = F(a,b) \times \mathbf{Z} = \langle a,b; \rangle \times \langle z \rangle = \langle a,b,z \mid aza^{-1}z^{-1}, bzb^{-1}z^{-1} \rangle$. Take $A = Subgp_\Gamma\{a,b\}$, and $B = Subgp_\Gamma\{a,bz\}$. Then, as z commutes with a and with b, a word is in the intersection if and only if the exponent sum of z in w (the total number of occurrences of z counted according to sign) is zero. The exponent sum of z in a word is in fact the image of the element represented under the obvious map $\Gamma \to \mathbf{Z}$ (sending a,b to 0). But in a word written in terms of the generators of B, the exponent sum of z is the same as the exponent sum of b. Thus $A \cap B = Subgp_\Gamma\{b^n a b^{-n}\}$ which is again infinitely generated.

iii) A hyperbolic group without the f.g.i.p. (using Rips construction [**R**]).

Let $r_i(s,t) = s^i t^i \dots s^{i+5} t^{i+5}$.

$$\begin{aligned}\Gamma' = \langle a,b,z,s,t \mid & aza^{-1}z^{-1} = r_1(s,t), bzb^{-1}z^{-1} = r_7(s,t),\\ & asa^{-1} = r_{13}(s,t), a^{-1}sa = r_{19}(s,t), bsb^{-1} = r_{25}(s,t),\\ & b^{-1}sb = r_{31}(s,t), zsz^{-1} = r_{37}(s,t), z^{-1}sz = r_{43}(s,t)\\ & ata^{-1} = r_{49}(s,t), a^{-1}ta = r_{55}(s,t), btb^{-1} = r_{61}(s,t),\\ & b^{-1}tb = r_{67}(s,t), ztz^{-1} = r_{73}(s,t), z^{-1}tz = r_{79}(s,t)\rangle\end{aligned}$$

This group is hyperbolic (it satisfies the $C(7)$ small cancellation condition). There is an exact sequence of groups:

$$1 \to F(s,t) \to \Gamma' \to \Gamma \to 1\,.$$

The preimages of the subgroups A and B of (ii) are the subgroups generated by $\{a, b, s, t\}$ and $\{a, bz, s, t\}$ respectively. Their intersection is infinitely generated (as its image is infinitely generated in Γ). Notice that we have produced a two generator normal subgroup N in the hyperbolic group Γ'; in general this subgroup has no finite presentation: an example of a relation comes from

$$s = a^{-1} r_{13}(s,t) a = a r_{19}(s,t) a^{-1}$$

$$\Rightarrow r_{13}(r_{19}(s,t), r_{55}(s,t)) = r_{19}(r_{13}(s,t), r_{49}(s,t)).$$

§2 Geodesics for elements of a subgroup of a free group

Let A be a finitely generated subgroup of the free group $F(S) = F$. The Cayley graph $\mathcal{G}(F,S)$ is a tree, which we consider as a metric space via the word metric, as usual (for definitions see elsewhere in this volume, or for instance [**DG**]). Let $T = \{a_1, \ldots, a_k\}$ be a finite set of generators for A. The inclusion map $\alpha : A \to F$ assigns to each generator in T a reduced word $\alpha(a_j)$ in $F(S)$. Notice that it suffices to consider the finite subset of S consisting of those letters which occur in the words $\alpha(a_i)$. So we shall only consider finitely generated free groups, when the Cayley graph $\mathcal{G}(F,S)$ is a locally finite tree.

For each element $h \in A$ there is thus a unique geodesic path $w(h)$ in the Cayley graph $\mathcal{G}(F,S)$ from the identity vertex 1 to the vertex h.

We shall show that the path $w(h)$ lies close to the vertices in $\mathcal{G}(F,S)$ which correspond to elements of A.

Let $v = b_1 \ldots b_m$ be a reduced word in the generators T (i.e. each b_j is some $a_p^{\pm 1}$) representing the element h.

Then in F, $\alpha(v) = \alpha(b_1) \ldots \alpha(b_m)$ also represents the element h (as a word in the generators S). The path $\alpha(v)$ in $\mathcal{G}(F,S)$ beginning at the identity vertex ends at the vertex h. The image of the path $w(h)$ is contained in the image of the path $\alpha(v)$, as omitting a point of the path $w(h)$ disconnects $\mathcal{G}(F,S)$. (Omitting a point from the geodesic path w disconnects the tree.)

Thus each point c in the path w lies in the image $\alpha(a_j)$ of some generator a_j, and so c is at distance at most

$$\frac{1}{2}\max_j\{\ell(\alpha(a_j))\}$$

from some vertex in A. As A is finitely generated, this number is finite.

§3 QUASICONVEXITY

The above considerations lead us to the following definition:

Definition (Gromov [**Gr**, §5.3 p.139 and §7.3 p.191])

Let Γ be a finitely generated group, with a finite set of generators S. A subset A is *quasiconvex* with respect to S if there is a constant $C \geq 0$ such that:

every geodesic path w in $\mathcal{G}(\Gamma, S)$ which joins two vertices in A lies in a C-neighbourhood of A.

When A is a subgroup, as Γ acts on $\mathcal{G}$ on the left it suffices to consider geodesic paths which begin at the identity vertex and end at a vertex in A.

Examples

o) Any finite subset of a group is quasiconvex. The subgroup of even integers is quasiconvex in the additive group of integers – in fact $n\mathbf{Z}$ is quasiconvex in $\mathbf{Z}$.

i) A finitely generated subgroup of a free group is quasiconvex.

ii) Consider $\mathbf{Z} \times \mathbf{Z}$ generated as usual by $\{(1,0),(0,1)\}$.
There are essentially only 4 subgroups which are quasiconvex with respect to these generators: the whole group, the trivial subgroup, and subgroups of $\mathbf{Z} \times \{0\}$ and $\{0\} \times \mathbf{Z}$.

iii) A subgroup of finite index in a finitely generated group is quasiconvex with respect to any finite set of generators (see figure 1).

iv) Let S_1, S_2 be finite sets of generators for the groups Γ_1, Γ_2. Then Γ_1 and Γ_2 are quasiconvex subgroups of the free product $\Gamma_1 \star \Gamma_2$ (and of the direct product $\Gamma_1 \times \Gamma_2$) with respect to the finite generating set $S_1 \cup S_2$.

At last a proposition!

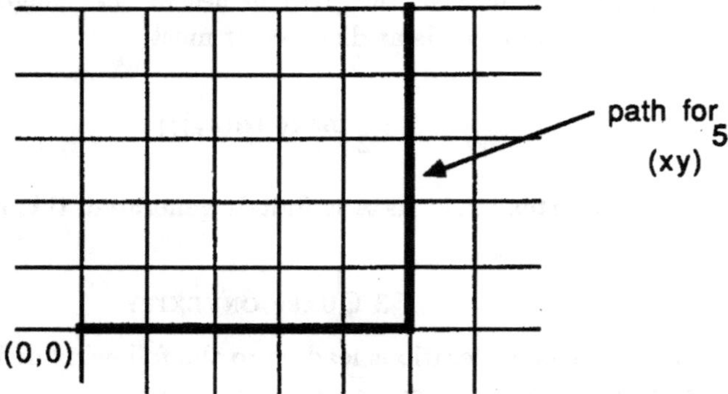

Figure 1 *Sbgp*$\{(1,1)\}$ is not quasiconvex.

PROPOSITION 1.

Suppose that the group Γ is generated by the finite set S, and that A is a subgroup of Γ.

If A is quasiconvex with respect to S, then A is finitely generated.

PROOF: Let $h \in A$, and let $w = a_1 \dots a_n$, $a_i \in S \cup S^{-1}$ be a geodesic path in $\mathcal{G}(\Gamma, S)$ from 1 to h. By assumption there is a constant K such that each vertex on the path w lies at distance at most C from a vertex in A.

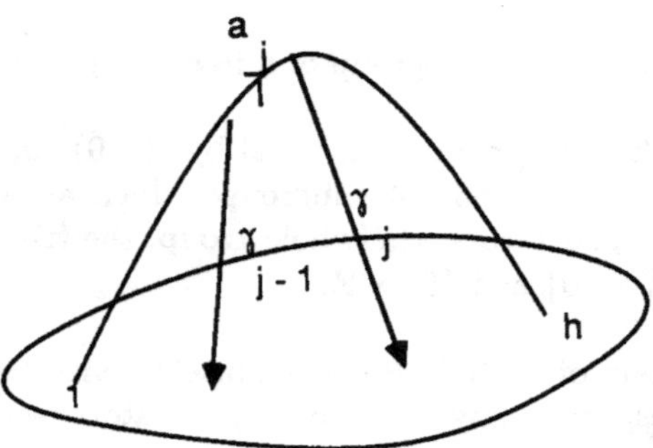

Figure 2 Generators for a quasiconvex subgroup

This means that there are words $\gamma_1, \dots \gamma_{n-1}$, $\ell(\gamma_j) \leq C$, such that the path $a_1 \dots a_j \gamma_j$ based at the identity vertex ends at a vertex in A.

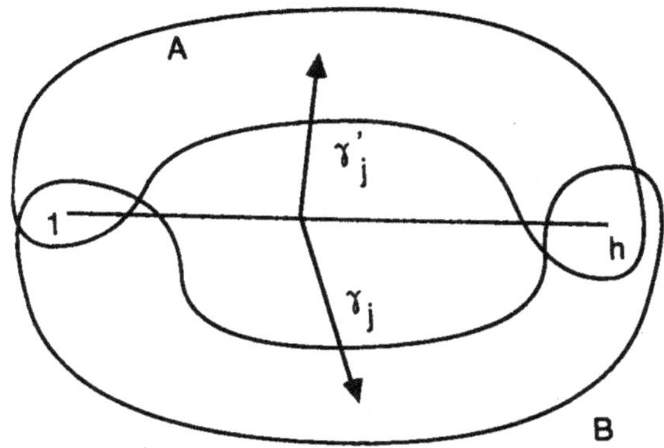

Figure 3

Rewrite the word w as (letting $\gamma_0 = \gamma_n$ denote the empty word)

$$(\gamma_0^{-1}a_1\gamma_1)(\gamma_1^{-1}a_2\gamma_2)\dots(\gamma_{n-1}^{-1}a_n\gamma_n).$$

Each word $(\gamma_i^{-1}a_{i+1}\gamma_{i+1})$ has length at most $2C+1$, and represents an element of A. It is now clear that the set

$$\{g \in \Gamma \mid g \text{ has a representative } v \in F(S), \ell(v) \leq 2C+1 \text{ and } g \in A\}$$

generates A. ∎

Combining this with the previous remarks about a finitely generated subgroup of a free group we obtain:

PROPOSITION 2.
Let Γ be the free group freely generated by S.
A subgroup A of Γ is quasiconvex with respect to S if and only if it is finitely generated.

In order to prove Howson's theorem, it suffices to show :

PROPOSITION 3.
Let Γ be a group generated by the finite set S. Let A, B be subgroups of Γ which are quasiconvex with respect to S.
Then $A \cap B$ is quasiconvex with respect to S.

PROOF: Let $w = a_1 \dots a_n$ be a geodesic word for an element $h \in A \cap B$. Let K_A, K_B be quasiconvex constants for the two subgroups. For each j there are words $\gamma_j, {\gamma'}_j$, such that $\ell(\gamma_j) \leq K_A$, $\ell({\gamma'}_j) \leq K_B$ such that $(a_1 \dots a_j\gamma_j)$ represents an element of A, and $(a_1 \dots a_j{\gamma'}_j)$ represents an element of B. Notice that $(\gamma_j^{-1}a_{j+1}\dots a_n)$ also represents an element of A, and $({\gamma'}_j^{-1}a_{j+1}\dots a_n)$ represents an element of B.

Let N be the number of different such pairs $(\gamma_j, \gamma'_j) \in \Gamma \times \Gamma$. If $\ell(w) > N$, then for some $1 \leq i < j < N$ we have $\gamma_i = \gamma_j$ and $\gamma'_i = \gamma'_j$. In this case,

$$\begin{aligned}(a_1 \dots a_i\gamma_i)(\gamma_j^{-1}a_{j+1} \dots a_n) &= a_1 \dots a_i a_{j+1} \dots a_n \\ &= (a_1 \dots a_i\gamma'_i)(\gamma'^{-1}_j a_{j+1} \dots a_n)\end{aligned}$$

and this is an element of $A \cap B$. Continuing in this way, we eventually obtain a word of length less than N, and we see that $A \cap B$ is quasiconvex with constant N.

Alternatively, for each pair (γ_i, γ'_i), let $\alpha(\gamma_i) \in A$, $\beta(\gamma'_i) \in B$ be shortest elements such that

$$\gamma_i\alpha(\gamma_i) = \gamma'_i\beta(\gamma'_i) = \delta(\gamma_i, \gamma'_i) \text{ in } \Gamma.$$

There is a bound on the number of such pairs, as γ_i, γ'_i have bounded length, and hence there is a bound C on the length of the elements $\delta(\gamma_i, \gamma'_i)$. But $a_1 \dots a_i\delta(\gamma_i, \gamma'_i)$ is in $A \cap B$. ∎

§4 Generalisations, Exercises, Questions etc.

i) (Open Question; [**NH**], [**B**], [**G**], [**NW**], ...) More about Howson's theorem:

Hanna Neumann showed [**NH**]that

$$rk(A \cap B) - 1 \leq 2(rk(A) - 1)(rk(B) - 1),$$

and asked whether

$$rk(A \cap B) - 1 \leq (rk(A) - 1)(rk(B) - 1)\ ?$$

This is now known as Hanna Neumann's Conjecture.
Burns has shown [**B**] that

$$rk(A \cap B) - 1 \leq 2(rk(A) - 1)(rk(B) - 1) - \min\{rk(A) - 1, rk(B) - 1\}\ .$$

Hanna Nemann's conjecture is still open.

Bounds on the rank obtained from the method given here seem to depend on the lengths of the generators of the subgroups when expressed as elements in the group F, and they are much much worse than any existing estimates.

Can at least Hanna Neumann's original estimate be recovered from this method?

ii) (Question [**GS**])

To what class of groups can Proposition 2 be extended? i.e. for what class of groups does the class of quasiconvex subgroups coincide with the class of finitely generated subgroups? What class of groups has the f.g.i.p.? As we remarked in the introduction, the class includes graphs of groups where the vertex groups already have the f.g.i.p. and the edge groups are finite [**C**, 8.33]. Can this result be proved using quasiconvexity methods? Notice that Rips (as seen in §1 iii) above) has shown that the class does not contain all small cancellation groups [**R**, Corollary (a)], and hence in general hyperbolic groups fail to have the property.

Is the class closed under quasiisometry?

iii) (Exercise) Hyperbolic groups.

Show that in a hyperbolic group, quasiconvexity is independent of generating set.

Show that a quasiconvex subgroup of a hyperbolic group is itself hyperbolic ("*by an easy argument* "says Gromov [**Gr**,p139]). Hint: show that the inclusion map is a quasiisometry (see also [**GS**]).

iv) (Harder exercise) Show that in a hyperbolic group the centralizer of an element is a quasiconvex subgroup. And thus, using iii) and one of the examples given in the text, that a hyperbolic group has no $\mathbf{Z} \times \mathbf{Z}$ subgroup. (see [**GS**])

v) (Exercise) In proposition 1, the fact that the words studied are geodesic was never really used. Generalise the definition of quasiconvexity with respect to a set of representatives $R \subset F(X)$.

References

[AS] A. Anissimov and F. Seifert, *Zur algebraischen Charakteristik der durch kontextfreie Sprachen definierten Gruppen*, Elektron. Inform. Verarb. u. Kybernetik **11** (1975), 695–702.

[B] R.G. Burns, *On the intersection of finitely generated subgroups of a free group*, Math. Z. **119** (1971), 121–130.

[C] D.E. Cohen, "Combinatorial group theory: a topological approach," L.M.S. student texts vol. 14, C.U.P., 1989.

[DG] P. de la Harpe and E. Ghys (editors), "Sur les groupes hyperboliques, d'après Mikhael Gromov," Birkhäuser, 1990.

[G] S.M. Gersten, ***Intersections of finitely generated subgroups of free groups and resolutions of graphs***, Invent. Math. **71** (1983), 567–591.

[Gr] M. Gromov, *Hyperbolic groups*, in "Essays in Group Theory," M.S.R.I. series vol. 8, edited by S.M. Gersten, Springer–Verlag, 1987.

[GS] S.M. Gersten and H. Short, ***Rational subgroups of biautomatic groups***, to appear in Annals of Math., M.S.R.I. preprint, MSRI-07923-89, (1989).

[H] A.G. Howson, ***On the intersection of finitely generated free groups***, J. London Math. Soc. **29** (1954), 428–434.

[K] S. Krstić, ***Fixed subgroups of automorphisms of free by finite groups: an extension of Cooper's proof***, Arch. Math. **48** (1987), 25–30.

[NH] Hanna Neumann, ***On the intersection of finitely generated free groups***, Publ. Math. Debrecen **4** (1955–56), 186–189; ***Addendum***, ibid. **5** (1957–58), 128.

[NW] W.D. Neumann, ***On the intersections of finitely generated subgroups of free groups***, Research report, Centre for Mathematical Analysis, The Australian National University, CMA-R59-89, (1989) .

[R] E. Rips, ***Subgroups of small cancellation groups***, Bull. London Math. Soc. **14** (1982), 45–47.

Sous-groupes à deux générateurs des groupes hyperboliques .

Thomas Delzant

Soit Γ un groupe hyperbolique ; on donne (lemme 1.2) une condition géométrique simple portant sur des éléments $f_1, ..., f_n$ de ce groupe permettant d'assurer que ceux-ci engendrent un groupe libre ; on en déduit une nouvelle preuve du théorème suivant annoncé par M. Gromov [Gr].

Théorème 1 [Gr] : *Soit Γ un groupe hyperbolique sans torsion . Γ ne contient qu'un nombre fini de classes de conjuguaison de sous-groupes non libres engendrés par deux éléments.*

Ce théorème est à rapprocher du théorème suivant concernant l'étude des sous-groupes à un générateur des groupes hyperboliques.

Théorème 2 [Gr] : *Un groupe hyperbolique ne contient qu'un nombre fini de classes de conjuguaison d'éléments de torsion.*

Notons que le théorème 1 était connu dans le cas des groupes à petite simplification C'(1/10) (Hill Pride et Vella [HPV]).

Plan de la preuve :
Après avoir étudié un critère de liberté (paragraphe 1) , on montre (paragraphe 2) qu'un sous-groupe non libre engendré par deux éléments contient un système de générateurs (u,v) ou u est conjugué à un élément de longueur inférieure à 18δ. Enfin on montre au paragraphe 3 que dans un groupe hyperbolique sans torsion, si un élément u est fixé, modulo tranlation à gauche ou à droite par u, il n'y a qu'un nombre fini de v tels que u et v engendrent un sous-groupe non libre.

Remerciements : C'est avec plaisir que je remercie H.Short et C. Champetier de m'avoir signalé une erreur dans une première version de ce texte.

§ 0 Notations et rappels de définitions:

Pour toutes les notions brièvement rappelées ici on renvoit le lecteur à [Gr] ou pour plus de détails [CDP] [GH] ou[Gh] .

Dans un espace métrique , on note $|x-y|$ la distance de deux points x et y.
Rappelons la définition de l'hyperbolicité : un espace métrique est dit δ-hyperbolique si pour tout quadruplet x, y, z, t de points on a l'inégalité suivante :

(Hyp) $$|x-y|+|z-t| \leq \max(|x-z|+|y-t|, |x-t|+|y-z|) + 2\delta$$

Rappellons qu'un groupe finiment engendré est dit hyperbolique si son graphe de Cayley muni de la métrique du mot invariante à gauche est hyperbolique. Dans le cas des groupes, on note $|g|$ la distance d'un élément du groupe considéré à l'origine notée e de ce groupe.
Une écriture $g=a_1a_2....a_n$ de g est dite géodésique si $|g| = |a_1|+..+|a_n|$; elle est dite géodésique à A près si l'identité précédente est remplacée par l'inégalité $|g| \geq |a_1|+..+|a_n|-A$ (l'inégalité $|g| \leq |a_1|+..+|a_n|$ est toujours satisfaite)
La restriction de la fonction distance au groupe ne prend que des valeurs entières, ainsi nous pouvons supposer que la constante δ de la définition de l'hyperbolicité est un entier, de plus avec cette convention, l'inégalité opposée à $|g| \leq 2\delta$ est $|g| \geq 2\delta+1$ (par exemple).
Nous nous servirons du fait que dans un espace hyperbolique géodésique (par exemple dans un groupe hyperbolique) tout ensemble fini de points s'approche par un arbre (voir [Gr] ou pour plus de détails [CDP] chap8) Par exemple soit ABC trois points d'un espace hyperbolique, a, b, c les trois géodésiques qui forment les cotés de ce triangle et considérons un tripode c'est à dire un arbre à trois sommets A*B*C* dont les longueurs des cotés sont a, b, c ; il existe une unique application notée f du triangle dans le tripode qui est une isométrie en restriction à chacun des cotés; *un résultat fondamental est que cette application est une isométrie à 2δ près, en particulier si deux points ont même image dans ce tripode ils sont distants d'au plus 2δ.* Ce tripode (ou plus généralement l'arbre si il y a plus de 3 points s'appelle le tripode de comparaison du triangle. Notons que le lemme d'approximation par les arbres ayant plus de trois sommets ne sera utilisé qu'à partir du troisième paragraphe.

§ 1 Un critère de liberté

Commençons par le lemme suivant qui est à comparer au lemme 7.2.C de [Gr] .

Lemme 1.1 : *Soit X un espace métrique δ-hyperbolique et x_n une suite de points de X vérifiant :*

$$|x_{n+2} - x_n| \geq \max(|x_{n+2} - x_{n+1}|, |x_n - x_{n+1}|) + 2\delta + a$$

alors $$|x_n - x_p| \geq |n-p|.a$$

Démonstration : On va voir par récurrence sur l'entier k que pour tout n, on a

$$|x_{n+k+1} - x_n| \geq |x_{n+k} - x_n| + a$$

ce qui entraine la propriété voulue.
Compte tenu de l'inégalité triangulaire et de l'hypothèse du lemme

$$\min (| x_{n+2} - x_{n+1} | , | x_n - x_{n+1} |) \geq a + 2\delta$$

Ce qui établi la propriété voulue pour k=0; notons que le cas k=1 est une conséquence immédiate de la même hypothèse. Supposons donc notre propriété satisfaite pour tout n jusqu'au rang k et appliquons alors la définition de l'hyperbolicité aux quatre points $x_n, x_{n+2}, x_{n+1}, x_{n+k+1}$.
Il vient :

$$|x_n - x_{n+2}| + |x_{n+1} - x_{n+1+k}| \leq \max(|x_n - x_{n+1}| + |x_{n+2} - x_{n+k+1}|, |x_n - x_{n+k+1}| + |x_{n+2} - x_{n+1}|) + 2\delta$$

Notons que $|x_n - x_{n+2}| + |x_{n+1} - x_{n+1+k}| \geq |x_n - x_{n+1}| + |x_{n+2} - x_{n+k+1}| + 2\delta + a$ puisque
$|x_{n+1} - x_{n+1+k}| \geq |x_n - x_{n+1}| + a$ par hypothèse de récurrence et que
$| x_{n+2} - x_n | \geq | x_n - x_{n+1} | + 2\delta + a$ par l'hypothèse du lemme.
Ainsi on a nécéssairement

$$|x_n - x_{n+2}| + |x_{n+1} - x_{n+1+k}| \leq |x_n - x_{n+k+1}| + |x_{n+2} - x_{n+1}| + 2\delta$$

Puis comme $| x_{n+2} - x_n | \geq | x_n - x_{n+1} | + 2\delta + a$, il vient
$a + |x_{n+1} - x_{n+1+k}| \leq |x_n - x_{n+k+1}|$ ce que nous souhaitions montrer.

La remarque essentielle de cet article est que ce lemme entraine immédiatement le critère de liberté suivant :

Lemme 1.2 : *Soit Γ un groupe δ-hyperbolique et* $f_1,\dots,f_n$ *n éléments de Γ ; on suppose que*
1°) Quel que soit i $\quad | f_i^2 | \geq | f_i | + 2\delta + 1$
2°) Quels que soient i et j distincts $\quad | f_i^{\pm 1} f_j^{\pm 1} | \geq \max(| f_i | ; |f_j|) + 2\delta + 1$
Alors le sous groupe engendré par les f_i est libre et ces éléments en forment une base.

Démonstration du lemme 1.2 : On considère un mot non trivial en les f_i et leurs inverses réduit dans le groupe libre c'est-à-dire ne contenant pas la suite de deux lettres $f_i f_i^{-1}$ ni $f_i^{-1} f_i$. On note x_n ce mot tronqué à la n-ième lettre ; la suite x_n satisfait les hypothèses du lemme 1.1 le mot x_n ne peut donc vérifier $|x_n| = 0$.

Nous allons maintenant analyser les conditions 1 et 2 du lemme1.2
Pour comprendre la condition 1°) remarquons le lemme suivant

Lemme 1.3 (voir [CDP] chap.9 lemme 3.5) *Si un élément u d'un groupe hyperbolique Γ satisfait*

$$| u^2 | \leq | u | + 2\delta$$

Alors u est conjugué à un élément de longueur inférieure à 6δ. Plus précisement soit ab une écriture géodésique de u telle que $|a|+|b|=|u|$ et $||a|-|b|| \leq 1$ alors $|ba| \leq 6\delta$.

Démonstration :

On considère le triangle e, u, u^2 que l'on dessine à coté de son tripode de comparaison (figure 1)

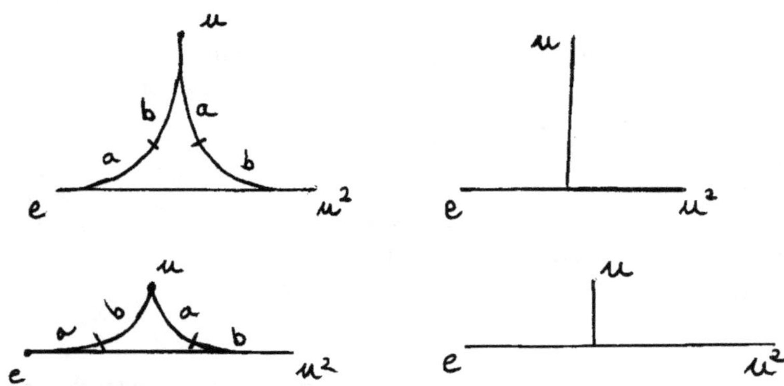

Distinguons comme sur la figure deux cas:

le milieu du segment eu est situé au dessus du centre du tripode auquel cas nous avons $|ba| \leq 2\delta$

le milieu du segment eu est situé au dessous du centre du tripode auquel cas $|ba| \leq |u^2| - |a| - |b| - 4\delta$; ainsi par hypothèse, $|ba| \leq 6\delta$, ce qu'il fallait montrer puisque ba est un conjugué de ab.

On déduit immédiatement du lemme 1.1 et du lemme1.3 le théorème 2; en effet si un élément u est de torsion, la suite x_n ne satisfait pas le lemme1.1 de sorte que $|u^2| \leq |u| + 2\delta$, et le lemme 1.3 montre que u est conjugué à un élément de longueur inférieure à 6δ.

La deuxième condition du lemme 1.2 peut s'analyser aussi comme une condition de simplification :

Lemme 1.4 Soient u et v deux élément satisfaisant $|u^{-1}v| \leq \max(|u|,|v|)+2\delta$ alors il existe trois éléments a, b, c vérifiant u =ab; v =ac , $|u| = |a|+|b|$; $|v| \geq |a| + |c| - 2\delta$ et $|a| \geq \min |b|, |c| - 4\delta$.

et de plus $|b^{-1}c| \geq |b|+|c| - 2\delta$.

Dessinons le triangle e (élément neutre), u , v ainsi que son tripode d'approximation :

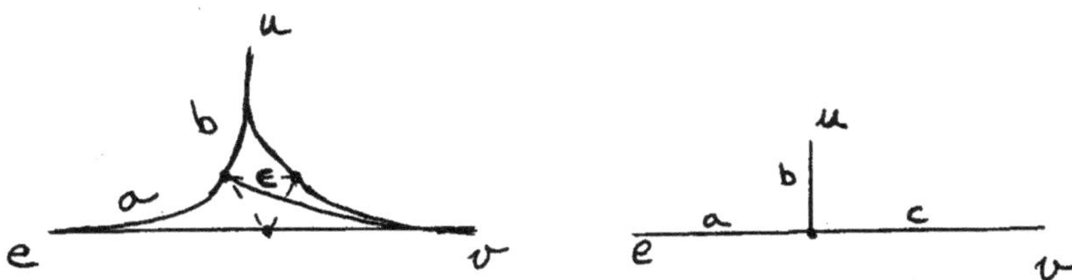

La finesse de ce triangle (c'est-à-dire la définition de l'hyperbolicité) montre que l'on peut trouver des éléments a, b, c dans Γ vérifiant :

$$u = ab \quad v = ac \qquad \text{avec}$$
$$|a| + |b| = |u| \text{ et } |a| + |c| -2\delta \leq |v| \leq |a| +|c|$$

Comme $|u^{-1}v| \leq \max(|u|,|v|)+2\delta$, on a bien $|a| \geq \min |b|, |c| - 4\delta$.

Nous pouvons maintenant préciser un peu le lemme 1.3

Lemme 1.5 : *Un élément u de Γ admet au moins l'une des formes normales suivantes* :

a) *u est conjugué à un élément de longueur inférieure à* 20δ.

b) $u = mvm^{-1}$ *avec* $|m^{\pm}v^{\pm}| \geq \max(|m|,|v|) +2\delta+1$ et $|u| \geq |v| +2|m| -4\delta$

c) $u= u_1u_2$ *avec* $|u| = |u_1| + |u_2|$ et $|u_2u_1| \geq \max(|u_1| , |u_2|) +2\delta+1$

Démonstration : Soit $u = u_1u_2$ une éciture géodésique de u; supposons $u_1$et u_2 sensiblement de même longueur . Si ces deux mots ne satisfont pas à la condition c du lemme, le lemme 1.3 montre que l'on peut écrire $u_1 =ab$; $u_2 = ca^{-1}$ de sorte que $u = abca^{-1}$et cette écriture est presque géodésique (à 4δ près) de plus $|a| \geq \min |b|, |c| - 4\delta$. Si u n'est conjugué à un élément de longueur inférieure à 20δ, bc est de longueur au moins 20δ et donc a est de longueur au moins égale à 2δ; mais alors $|abc|$ est au moins égale à $|a|+|bc| -4\delta$ puisque l'écriture de u est presque géodésique ainsi $\min(|a|,|bc|)\geq 20\delta \geq 6\delta+1$et donc $|abc| \geq \max(|a|,|bc|) + 2\delta + 1$. On pose a=m et v=bc pour avoir la seconde forme normale.

§2 Première réduction .

On se fixe dans tout ce paragraphe un groupe δ-hyperbolique Γ (on ne suppose pas que Γ est sans torsion); notre but est de montrer l'énoncé intermédiaire suivant intéressant en soi :

Théorème 2.1 : *Si un sous groupe Γ_0 de Γ à deux générateurs est non libre, il est conjugué à un sous-groupe à deux générateurs u et v tels que* $|u| \leq 18\delta$.

Démonstration : Parmi tous les conjugués de Γ_0 l'un possède un système (u,v) de générateurs vérifiant la propriété de minimalité suivante :
Pour tout conjugué de Γ_0 et tout système (x,y) de générateurs de ce conjugué, on a :

$$|u| < |x| \quad \text{ou} \quad |u| = |x| \text{ et } |v| \leq |y|$$

Nous fixerons donc un tel conjugué de Γ_0 et un tel système de générateurs.

D'après le lemme 1.3, nous avons $|u^2| \geq |u| + 2\delta + 1$ et $|v^2| \geq |v| + 2\delta + 1$ sauf si l'un de ces éléments est conjugué à un élément de longueur inférieure à 6δ , auquel cas il n'y a rien à montrer. Comme ces deux inégalités sont satisfaites et que le groupe engendré par u et v n'est pas libre, l'une des inégalités $|u^{\pm 1}v^{\pm 1}| \geq \max(|u|;|v|) + 2\delta + 1$ n'est pas satisfaite . Quitte à changer u en u^{-1} ou v en v^{-1}, nous pouvons supposer :
2.1 $|u^{-1}v| \leq \max(|u|;|v|) + 2\delta$
D'après le lemme 1.4 nous pouvons écrire

2.2 $$u = ab \quad v = ac \quad \text{avec}$$
$$|a| + |b| = |u| \text{ et } |a| + |c| - 2\delta \leq |v| \leq |a| + |c|$$

La propriété de minimalité de (u,v) montre que :
2.3 $ba = a^{-1}(ab)a$ vérifie $|a| + |b| = |ba| = |ab|$
2.4 De plus $|v| \geq |u|$ entraine $|c| \geq |b|$
Evidement la longueur de $u^{-1}v$ est au moins égale à la longueur de v et à celle de u, ainsi
2.5 $|b^{-1}c| = |u^{-1}v| \geq |a| + |c| - 2\delta$ et $|b^{-1}c| = |u^{-1}v| \geq |a| + |b|$

Notons que $bc^{-1} = a^{-1}uv^{-1}a$, ainsi bc^{-1} et $ba = a^{-1}ua$ engendrent un conjugué de Γ_0 ; la propriété de minimalité entraine donc :

2.6 $|bc^{-1}| = |v| \geq |a| + |c| -2\delta$ et $|bc^{-1}| = |u| \geq |a| + |b|$

Par l'inégalité triangulaire nous obtenons $|b| \geq |a| -2\delta$

Enfin $|ca| \geq |v|$ sinon on pourrait remplacer (u,v) par $a^{-1}ua$ et $a^{-1}va$; ainsi

2.7 $|ca| = |a^{-1}c^{-1}| \geq |a| + |c| -2\delta$

Afin de pouvoir appliquer le lemme 1.1, nous préférons réécrire les inégalités 2.1 à 2.7 de la façon suivante :

2.8 $|ab| = |ba| = |a^{-1}b^{-1}| = |b^{-1}a^{-1}| \geq \max(|a|, |b|) + |a| -2\delta$

$|c^{-1}a^{-1}| = |ac| \geq \max(|a|, |c|) + |a| -4\delta$

$|c^{-1}b| = |b^{-1}c| \geq \max(|b|, |c|) + |a| -2\delta$

$|cb^{-1}| = |bc^{-1}| \geq \max(|b|, |c|) + |a| -2\delta$

On a alors le lemme suivant :

Lemme 2.2 $|a| \leq 6\delta$

Démonstration : Soit $u^{\alpha_1}v^{\beta_1} \ldots\ldots u^{\alpha_n}v^{\beta_n}$ un mot en u et v dont l'image dans Γ est nulle ; écrivons ce mot en fonction de a, b et c ; on a donc $(ab)^{\alpha_1}(ac)^{\beta_1}\ldots\ldots(ab)^{\alpha_n}(ac)^{\beta_n} = e$. Le mot réduit associé après simplification des séquences de la forme aa^{-1} ou $a^{-1}a$ est un mot en a,b et c dont les séquences de deux lettres consécutives sont de la forme ab , ba , $a^{-1}b^{-1}$, $b^{-1}a^{-1}$, ac, ca , $c^{-1}a^{-1}$, $a^{-1}c^{-1}$, $b^{-1}c$ bc^{-1}, cb^{-1} et bc^{-1}.

D'après le lemme 1.1 et compte tenu des inégalités 2.8 ce mot ne peut être trivial que si $|a| -4\delta \leq 2\delta$ ce qu'il fallait voir.

En regardant à nouveau e, u, v et son tripode d'approximation (figure 1) nous remarquons que

$|c^{-1}b| = |v^{-1}u| \geq |b|+ |c| \;\; -4\delta \geq \max(|u| , |v|) + |b| -|a| \;\; -4\delta$.

Ainsi l'inégalité 2.1 entaine à son tour $|b| -|a| \;\; -4\delta \leq 2\delta$ et donc $|b| \leq 12\delta$.

Il en résulte que $|u| \leq 18\delta$ ce que nous souhaitions prouver.

§3 Deuxième partie de la preuve

Le but de ce paragraphe est d'étudier pour un élément u fixé l'ensemble des v tels que u et v n'engendrent pas un goupe libre. Pour simplifier cette étude, nous supposerons que *le groupe Γ est* *sans torsion* .

Rappellons qu'un élément u est dit primitif, si pour tout élément v , v commute à une puissance non triviale de u entraine que v est une puissance de u. L'énoncé suivant nous sera utile

Proposition 3.1 : *Dans un groupe hyperbolique sans torsion tout élément non nul est la puissance d'un élément primitif.*

Démonstration : Rappellons ([Gr] ou pour une preuve détaillée [CDP] chapite10) que le centralisateur C(u) d'un élément non nul contient $\mathbb{Z}$ comme sous groupe d'indice fini, de sorte qu'il est isomorphe à $\mathbb{Z}$ si Γ est sans torsion; la proposition en résulte aisément.

On définit aussi la norme relative d'un élément par rapport à un élément u de la façon suivante :
$|g|_u = \min_{k,l} (|u^l g| ; |gu^k|)$

Le résultat central de ce paragraphe est le

Théorème 3.2 *Soient u un élément fixé; il existe une constante K telle que* $|v|_u > K$ *implique que u et v engendrent un groupe libre.*

Notons que ce théorème entraine immédiatement le théorème 1 compte tenu du théorème 2.1.
On peut supposer que |u| est minimale parmi tous les conjugués de u. Cela entraine immédiatement que si $u=u_1u_2$ est une écriture géodésique de u, alors u_2u_1 est donc un conjugué de u de même longueur que u .

Le premier lemme utile est le suivant :

Lemme 3.3 : *Soit u un élément primitif ; il existe une constante C et une constante A ne dépendant que de u telles que*
$|g|_u = |g| > C$ implique $|g^{-1}ug| \geq |g^{-1}u| + |g| - 2|u| - 10\delta - A \geq \max(|g^{-1}u| , |g|) + 10\delta$;
$|g^{-1}u^{-1}g| \geq |g^{-1}u^{-1}| + |g| - 2|u| - 10\delta - A \geq \max(|g^{-1}u^{-1}| , |g|) + 10\delta$;
De plus k étant un nombre entier fixé, il existe une constante C_k *et une constante* A_k *ne dépendant que de u et k telles que*
$|g|_u = |g| > C_k$ implique $|g^{-1}u^kg| \geq |g^{-1}u^k| + |g| - 2|u^k| - 10\delta - A_k \geq \max(|g^{-1}u^k| , |g|) + 10\delta$;
$|g^{-1}u^{-k}g| \geq |g^{-1}u^{-k}| + |g| - 2|u^k| - 10\delta - A_k \geq \geq \max(|g^{-1}u^{-k}| , |g|) + 10\delta$;

Nous ne montrons que la première assertion, la seconde résultant d'un raisonement analogue. Considérons l'arbre d'approximation du quadruple e, g, u, ug (voir [Gr] ou pour plus de détails [CDP] chap.8) Celui-ci est l'un des deux arbres dessinés ci-dessous.

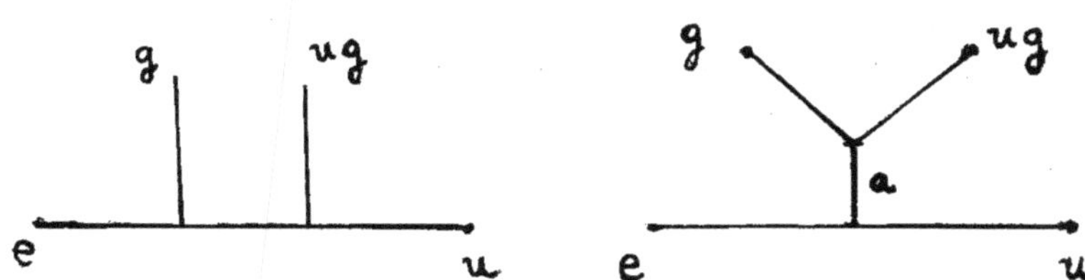

Dans le premier cas :

3.1 $|ug-g| \geq |g-u| + |g| - 2|u| - 10\delta$. (ou 10δ est la constante d'approximation pour un arbre à 4 sommets)

Dans le second cas montrons que la distance notée a sur le dessin ne peut être arbitrairement grande. Si $|a| \geq |u|$ soit g_1 n'importe quel point situé entre e et g au dessous du point A . On a alors $|ug_1-g_1| \leq |u| + 10\,\delta$ et est bornée.

En particulier si la distance a est arbitrairement grande, on peut écrire $g=g_1g_2$ avec g_1 arbitrairement grand et $|g_1^{-1}ug_1| \leq |u| + 10\,\delta$ et $g_1=xy$, $|x|<|g|$ et $g_1^{-1}ug_1 = x^{-1}ux$.

Mais alors xg_1^{-1} commute à u et est donc une puissance de u; ainsi $g_1=u^kx^{-1}$ et $g = u^kx^{-1}g_2$ en particulier $|u^{-k}g|<|g|$ ce qui contredit l'hypothèse $|g| =|g|_u$.

On note A le plus grand des a ainsi obtenus

Ainsi en regardant à nouveau l'arbre d'approximation de la figure 1, nous voyons que

3.2 $|g-ug| \geq |g-u|+|g| - A - 2|u| - 10\delta$.

On repète l'argument précédent en remplaçant u par u^{-1}; la conjonction de 3.2 et 3.3 est précisement le lemme 3.3.

Lemme 3.4 *Pour tout u fixé il existe un entier k_1 tel que si* $l \geq k \geq 0$ $|u^{k_1+l}| \geq \max(|u^{k_1}|, |u^l|)+2\delta+1$

Démonstration : il résulte de la classification des isométries d'un espace hyperbolique (voir [Gr] ou pour plus de détails [CDP] chap9) que la suite u^n est une quasi-géodésique c'est-à-dire que $|u^n| \geq na - c$ pour deux constante positives a et c.

D'autre part on sait (mêmes références) que toute quasi-géodésique reste à une distance bornée d'une géodésique ; la suite u^n reste donc à une distance inférieure à A (grande devant δ mais ne dépendant que de u) d'une certaine géodésique . Choisissons k_1 de sorte que ka-c>10A; le dessin suivant montre que si $l \geq k_1$ alors $|u^{2l}| \geq 2|u^l| - 4A \geq |u^l| + 6A \geq |u^l| + 2\delta + 1$ si A était choisi suffisament grand, ce qui est loisible.On en déduit que si l est au moins égal à k_1,

3.3 $\quad |u^{l+k_1}| \geq \max (|u^l|, |u^{k_1}|) + 2\delta + 1$

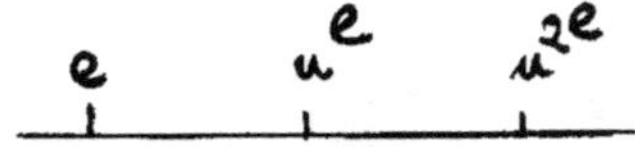

On a enfin :

Lemme 3.5 : *Pour tout u fixé il existe un entier k tel que si* $l \geq k_2 \geq 0$ *et si* $|v|_u = |v|$ *est suffisament grand alors*

$$|v^{\pm} u^{\pm l}| \geq \max(|v|, |u^l|) + 2\delta + 1$$

Soit G la géodésique décrite dans la démonstation du lemme 3.4 si $|v|_u = |v|$ le point de G situé à une distance minimale de v est à une distance bornée (inférieure à 2A) de e. Si un point p est situé sur cette géodésique on a (par finesse des triangles) $|p-v| \geq |p| + |v| - 4A - 10\delta$ (voir le dessin ci-dessous) Par suite comme la distance de u^l à cette géodésique est inférieure à A, on a $|vu^{\pm l}| \geq |v| + |u^{\pm l}| - 4A - 10\delta$; d'ou l'inégalité recherchée pourvu que $|v|_u$ et $|u^l|$ soient au moins égales à $4A + 12\delta + 1$, ce qui est le cas si $l \geq k_2$ pour un certain k_2 .

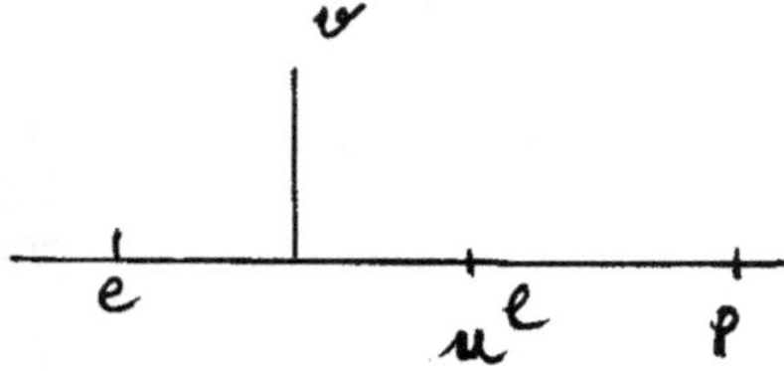

Pour achever la démonstration du théorème il nous faut distinguer trois cas suivant que v admet la forme normale a, b, ou c du lemme 1.5.

1°)On note k la plus grande des deux constantes k_1 et k_2 trouvées au lemmes 3.4 et 3.5.
Supposons que l'on puisse écrire $v=mwm^{-1}$ avec

3.4 $|m^{\pm}w^{\pm}| \geq \max(|m|,|w|) +2\delta+1$ et $|v| \geq |w| +2|m| -4\delta$

et supposons de plus que:

3.5 $|w|\geq \max_{n\leq k}|u^n| +100\delta$

Comme v n'est pas conjugué à un élément de longueur $\leq 6\delta$ on a :

3.6 $|w^2|\geq|w|+2\delta+1$

Comme l'écriture $v=mwm^{-1}$ est presque géodésique, nous pouvons supposer que $|m|_u$ est arbitrairement grand. De plus en vertu du lemme 3.5 on a

3. 7 $|m^{-1}u^{\pm k}|\geq \max(|m|, |u^k|)+10\delta$

Notons enfin que l'inégalité 3.5 montre que :

3.8 $|u^n m.w| \geq \max(|u^n m|, |w|) +2\delta$ pourvu que $|m|$ soit assez grand

Soit $u^{\alpha_1}v^{\beta_1}\ldots\ldots u^{\alpha_n}v^{\beta_n}$ un mot en u et v dont l'image dans Γ est nulle ; écrivons ce mot en fonction de m, u et w : il s'écrit $u^{\alpha_1}mw^{\beta_1}m^{-1}\ldots\ldots m^{-1}u^{\alpha_n}mw^{\beta_n}m^{-1}$;
ainsi le mot $m^{-1}u^{\alpha_1}mw^{\beta_1}m^{-1}\ldots\ldots m^{-1}u^{\alpha_n}mw^{\beta_n}$ est trivial

Montrons que cela impose à $|m|_u$ d'être plus petit que $C_{2k}(u)$ ou C_{2k} est la constante trouvée au lemme 3.4 en remplaçant k par 2k dans l'énoncé.
De fait il suffit de découper le mot étudié en des séquences de la forme $m^{-1}.u^n m$ avec $-2k\leq n\leq 2k$ puis $m.u^k$ puis $u^k.u^m$ avec $m \geq k$ puis $u^k.m$ puis $m.w$ ou $u^n m.w$ puis $w^{\pm 2}$ pour se rendre compte que cette suite satisfait au critère du lemme 1.1 (grâce aux inégalités 3.5 à 3.8) et donc ne peut définir un mot trivial .

2°) v est conjugué à un mot de longueur inférieure à 20δ.
On peut donc écrire $v=mwm^{-1}$ avec $|w|\leq 20\delta$ et il n'y a qu'un nombre fini de tels w.
Pour chacun on détermine le nombre entier k que l'on avait obtenu pour u au lemme 3.4 et 3.5. On coupe m en deux mots de longueur sensiblement égale $m=m_1m_2$.
En raisonnant comme précédement on voit qu'une relation entre u et v entraine que $|m_1|_u$ ou $|m|_{2w}$ est inférieur à une constante qui ne dépend que de u ou w, c'est-à-dire que de u .Ainsi $|m|$ est inférieure au

double de cette constante.

3°) Ecrivons $v=v_1v_2$ comme au lemme 1.4 ; soit $u^{\alpha_1}v^{\beta_1}\ldots\ldots u^{\alpha_n}v^{\beta_n}$ un mot en u et v dont l'image dans Γ est nulle ; cette relation entraine que le mot $v^{\beta_1}\ldots\ldots u^{\alpha_n}v^{\beta_n}u^{\alpha_1}$ est trivial ; en découpant ce mot en v^2 , $vu^{\pm l}$ avec $l\geq k$ puis vu^lv^{-1} ou $v^{-1}u^lv$ puis $v^{-1}u^{\pm l}$ puis vu^lv, ou $v^{-1}u^lv^{-1}$,on se rend compte que cette suite satisfait aux hypothèses du lemme1.1 sauf si l'un des deux mots vu^lv, ou $v^{-1}u^lv^{-1}$,ne satisfait pas $|vu^lv| \geq \max(|vu^l||v|) +2\delta +1$ ou bien $|v^{-1}u^lv^{-1}| \geq \max(|v^{-1}u^l|,|v|)+2\delta +1$;
On peut donc supposer par exemple que :

3.8 $|vu^lv| \leq \max(|vu^l|,|v|)+2\delta$
Montrons que cela implique :

3.9 $|v_2u^lv_1| \leq C$ ou C est une certaine constante qui ne dépend que de u.
En effet dessinons l'arbre d'approximation des quatre points e, v^{-1} u^l et u^lv :

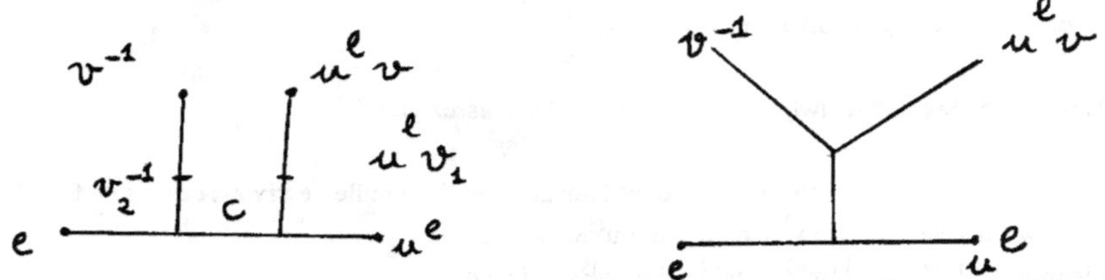

Si on suppose que l'arbre est donné par la première figure et que la distance C est grande cela contredit 3.8; on peut donc supposer que celui-ci est donné par la seconde figure. Afin de préciser nos notations redessinons cet arbre :

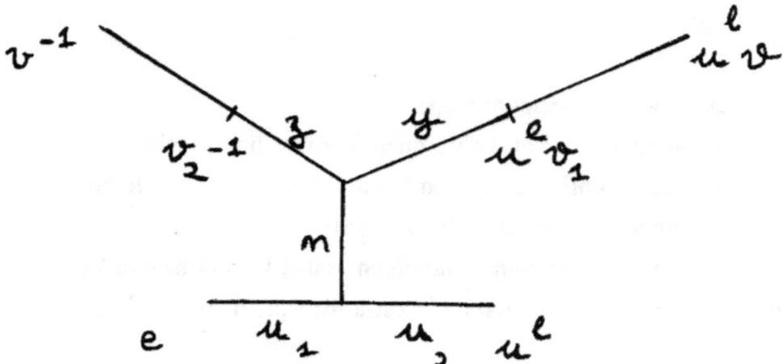

Ainsi on peut écrire :

$u^1 = u_1 u_2$ (écriture géodésique)

$v_2^{-1} = u_1 mz$ (écriture géodésique à 10δ près)

$v_1 = u_2^{-1} my$ (écriture géodésique à 10δ près)

De sorte que $v = u_2^{-1} myz^{-1} m^{-1} u_1^{-1}$; on pose $x = yz^{-1}$ et $|x|$ est majoré par une constante qui ne dépend que de u ($|x| \leq |u^1| + 10\delta$)

Les deux éléments $u' = u_2 u_1 = u_2 u u_2^{-1}$ et $v' = mxm^{-1} = = u_2 v u_1 = u_2 v u_2^{-1} u'$ engendrent donc un sous-groupe non libre .

Notons aussi que $|m|_{u'}$ est du même ordre que $|v|_u$. De plus, u n'admet qu'un nombre fini de conjugués de cette sorte , puisque k est fixé. On est donc ramené à remplacer u par u' dans la discussion des deux premier cas pour conclure.

Bibliographie :

[CDP] M. Coornaert, T. Delzant, A. Papadopoulos : Notes sur les groupes hyperboliques Preprint Irma, à paraitre dans Lecture Notes

[Gh] E. Ghys : Les groupes hyperboliques, Séminaire Bourbaki, preprint 1990

[G-H] E.Ghys P. de la Harpe Sur les groupes hyperboliques d'après M. Gromov Birkauser1990

[Gr] : M. Gromov : Hyperbolic groups in Essay in group theory (Gersten Editor), Publication of MSRI, 1985

[HPV] H. Hill, S. Pride & A. Vella, Subgroups of small cancellation groups J. fur die r. und a. Math. 1984

Irma, 7 rue R. Descartes
67084 Strasbourg Cedex
France

II. COXETER GROUPS AND BUILDINGS

AN INVITATION TO COXETER GROUPS

PIERRE DE LA HARPE
Section de Mathématiques
C.P. 240
CH-1211 Genève 24

ABSTRACT

These notes provide an introduction to Coxeter groups based on examples. They include discussions of symmetric groups, of regular polytopes in Euclidean spaces, of the 27 lines on a cubic surface, of Poincaré theorem for fundamental polygons (the reflection case), and some hints for the classifications of finite and Euclidean Coxeter groups.

Introduction

Coxeter groups, also called reflection groups, offer a very rich mixture of geometry and algebra. The aim of the present notes is to indicate some of the fascinating connections disclosed by their study.

The subject has historical roots going back to the regular solids of Pythagoras, the tesselations of arabic tradition and of Kepler, or the high-dimensional visions of L. Schläfli around 1850, to mention but a few. It is also a cornerstone of the theory of Lie groups and Lie algebras, both classical (W. Killing, E. Cartan, H. Weyl) and more recent (see e.g., [Car], [Kac], [SeP]). Indeed, it is difficult to find a subject without meaningful connections to the Coxeter world (see [Arn] and [HSV]).

The contribution of Harold Scott Macdonald Coxeter to this subject cannot be overestimated. He published his first mathematical paper in 1928, at a time geometry was somehow unfashionable (see [DGS] for a listing of his books and papers before 1980 and [Log] for an interview of Coxeter). It is striking how Coxeter has been both remarkably faithful to the 19th century tradition and successful in opening new fruitful domains of research. Today, he claims that he is impressed by other people's work and adds: "I am pleased to see these new developments and extensions of my old ideas" [Co7].

Most of what follows is a description of examples. Chapter 1 is about finite symmetric groups. Chapter 2 deals with other examples of finite Coxeter groups related to regular polytopes or appearing in algebraic geometry. Chapter 3 is mainly devoted to a theorem of Poincaré showing how algebraic presentations of (reflection) groups can be read out of appropriate geometric data; applications of this result provide examples of infinite Coxeter groups. The last chapter introduces a theorem of Tits on linear representations of Coxeter groups, and includes a proof of the classification of irreducible finite Coxeter groups.

For other expositions of Coxeter groups, see [BeG], [Bli], [Bro], [Co6], [Hil], [Ko2] and [Vi2].

These notes are based on lectures organised by the "troisième cycle romand de mathématiques" in November and December 1988, and on lectures in Trieste in March and April 1990. I am grateful to all those who have helped me with discussions, suggestions and encouragement, and particularly to Herbert Abels, Roland Bacher, Fokko du Cloux, Etienne Ghys, Fred Goodman, André Haefliger, Vaughan Jones, Michel Kervaire, Gus Lehrer, Luis Paris, Rudolf Scharlau and Thierry Vust.

Chapter 1. Symmetric Groups

Symmetric groups are prototypes of finite Coxeter groups. We denote here by σ_n the symmetric group of n objects ($n \geq 2$).

1.1. A Geometric Approach

The group σ_n acts by isometries on the usual Euclidean space $\mathbb{R}^n$ and permutes the vectors of the canonical orthonormal basis $\{e_1, e_2, \ldots, e_n\}$. Thus σ_n acts also

- on the affine hyperplane $\mathbb{E}^{n-1} = \{(x_1, x_2, \ldots, x_n) \in \mathbb{R}^n \ : \ x_1 + \ldots + x_n = 1\}$,
- on the unit sphere $\mathbb{S}^{n-1} = \{(x_1, x_2, \ldots, x_n) \in \mathbb{R}^n \ : \ x_1^2 + \ldots + x_n^2 = 1\}$,
- on the intersection $\mathbb{E}^{n-1} \cap \mathbb{S}^{n-1}$, often denoted by $\mathbb{S}^{n-2}$.

1. Reflections. A *reflection* of $\mathbb{R}^n$ is an isometry of order 2 which is the identity on an hyperplane. Similarly, a reflection of $\mathbb{E}^{n-1}$ [respectively of $\mathbb{S}^{n-1}$] is an isometry of order 2 which is the identity on an affine hyperplane [respectively on an equator (namely on a great circle in case $n-1=2$)]. For example, given any pair (i,j) of distinct integers in $\{1, 2, \ldots, n\}$, the corresponding transposition of σ_n acts on $\mathbb{R}^n$ as the reflection with fixed points

$$H_{i,j} = \{(x_1, x_2, \ldots, x_n) \in \mathbb{R}^n \ : \ x_i = x_j\} \ .$$

In particular, σ_n (in the action on $\mathbb{R}^n$ considered here) is *generated by reflections.* Let us illustrate the small values of n.

2. Case $n=2$. The element $w \neq 1$ in σ_2 acts on R^2 as the reflection with respect to the diagonal, on the affine line E^1 as a half-turn, and on $\mathsf{S}^0=\{e_1,e_2\}$ as the unique permutation distinct from the identity.

3. Case $n=3$. There are three transpositions in σ_3 which are $s_1=(1,2)$, $s_2=(2,3)$ and $(1,3)=s_1s_2s_1=s_2s_1s_2$. For graphical convenience, let us consider the action of σ_3 on E^2 (rather than on R^3). The three transpositions of σ_3 act by reflections with respect to three lines dividing regularly E^2 in six wedges which are called *chambers*, as in Figure 1. (These chambers are in bijection with the 1-cells of the barycentric subdivision of the boundary of a triangle.) The resulting action of σ_3 on the set of these six chambers is simply transitive. Consequently, once a so-called *fundamental chamber* has been chosen — which is C_0 in Figure 1 — one may label the other chambers by the elements of σ_3.

In some obvious sense, C_0 is at distance 0 from itself, there are 2 chambers at distance 1 from C_0, and 2 chambers at distance 2, and one chamber at distance 3. We encode this information in a *counting function*

$$\sum t^{d(C_0,C)} = 1 + 2t + 2t^2 + t^3 = (1+t)(1+t+t^2)$$

(where the summation is over all chambers) which does not depend on the choice of C_0.

All this can be repeated for the action of σ_3 on the circle S^1 in E^2; chambers in S^1 are arcs of circle, and Figure 1 shows also the six boundary points written in terms of the canonical basis $\{e_1,e_2,e_3\}$ of R^3.

4. Case $n=4$. We shall illustrate the action of σ_4 on the 2-sphere $\mathsf{S}^2 = \mathsf{S}^3 \cap \mathsf{E}^3 \subset \mathsf{R}^4$, represented via stereographic projection as $\mathsf{R}^2 \cup \{\infty\}$. There are six transpositions in σ_4, including

$$s_1 = (1,2) \qquad s_2 = (2,3) \qquad s_3 = (3,4)\ .$$

The transpositions of σ_4 act as reflections with respect to six great circles of S^2, three of them being straight lines in the stereographic projection chosen for Figure 2 (namely s_1, s_2 and $(2,4)=s_2s_3s_2=s_3s_2s_3$). These six great circles divide S^2 into 24 chambers, which are spherical triangles with angles $(\pi/2,\pi/3,\pi/3)$, and σ_4 act simply transitively on the set of these 24 chambers. (The angles make sense because the stereographical projection is conformal.) The basic vectors e_1,e_2,e_3,e_4 of R^4 are the vertices of a regular tetrahedron inscribed in S^2; these vertices are marked by a thick dot in Figure 2. The chambers are the 2-cells of the *barycentric subdivision of the boundary of this tetrahedron* (or of the dual tetrahedron, with three vertices marked by 0 in Figure 2 and the fourth vertex at infinity). Caution: the element $(1,2)(3,4)=s_1s_3$ is of order 2 in σ_4; however, it is not a reflection on S^2, but rather a half-turn with two fixed points.

5. Exercise. Choose a fundamental chamber C_0 such as that indicated in Figure 2, check that the counting function $\sum t^{d(C_0,C)}$ (summation over all

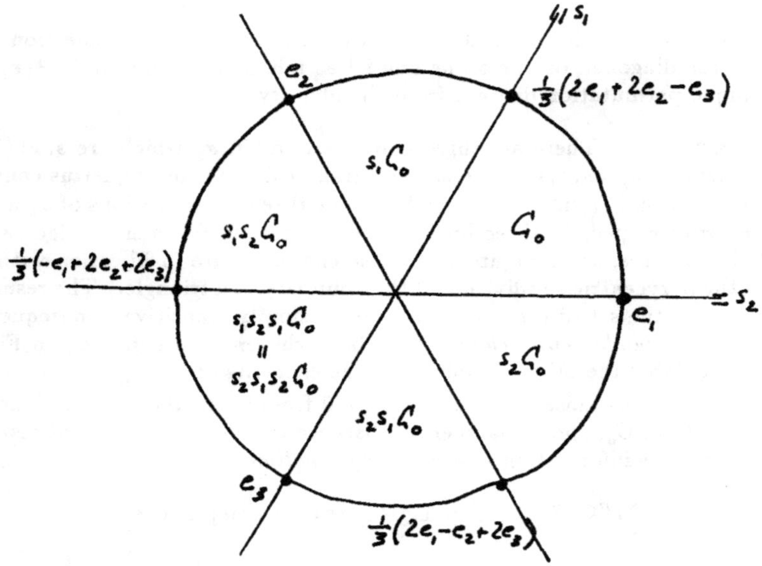

Figure 1.

chambers) is given by

$$1+3t+5t^2+6t^3+5t^4+3t^5+t^6=(1+t)^2(1+t^2)(1+t+t^2)$$

and observe that all roots of this polynomial are on the unit circle of the complex plane.

6. Case $n=5$. The group σ_5 acts on $\mathsf{S}^3=\mathsf{R}^3\cup\{\infty\}$. The corresponding figure should show the five basic vectors of R^5, namely the five vertices of a 4-simplex in E^4, appearing in $\mathsf{S}^3=\mathsf{R}^3\cup\{\infty\}$ as the four vertices of a tetrahedron together with its center. These points should be completed by six planes and four spheres limiting altogether 120 chambers in S^3.

1.2. A Presentation of the Symmetric Group

Let us write $s_1=(1,2)$, $s_2=(2,3),\ldots,s_{n-1}=(n-1,n)$. These transpositions generate σ_n and satisfy

$$\begin{aligned} &s_j^2=1 \\ &s_js_ks_j=s_ks_js_k \qquad \text{if} \quad |j-k|=1 \\ &s_js_k=s_ks_j \qquad \text{if} \quad |j-k|\geq 2 \end{aligned}$$

where $j,k\in\{1,2,\ldots,n-1\}$.

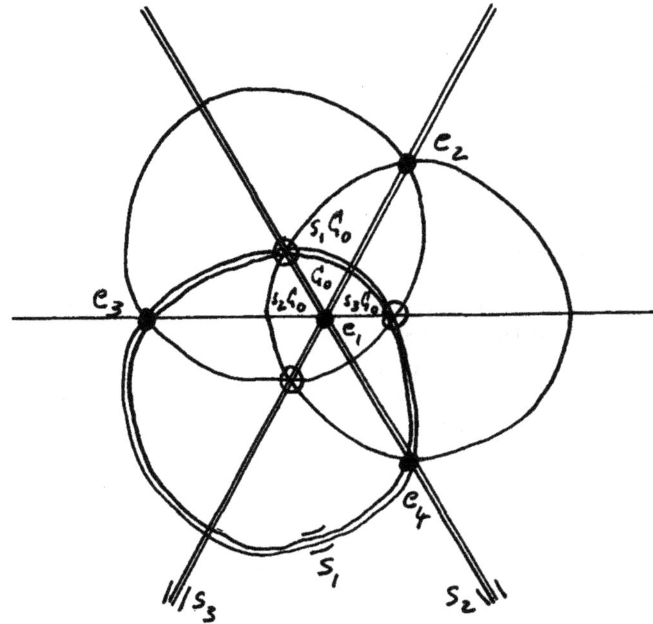

Figure 2.

7. Presentation of a group. Let us introduce an abstract group $W(A_{n-1})$ having a presentation with generators

$$\widehat{s}_1, \widehat{s}_2, \ldots, \widehat{s}_{n-1}$$

and relations

$$\begin{aligned} &\widehat{s}_j^2 = 1 \\ &\widehat{s}_j \widehat{s}_k \widehat{s}_j = \widehat{s}_k \widehat{s}_j \widehat{s}_k \qquad \text{if} \quad |j-k| = 1 \\ &\widehat{s}_j \widehat{s}_k = \widehat{s}_k \widehat{s}_j \qquad \text{if} \quad |j-k| \geq 2 \end{aligned}$$

for $j, k \in \{1, 2, \ldots, n-1\}$, as well as the homomorphism $\varphi : W(A_{n-1}) \longrightarrow \sigma_n$ defined by $\varphi(\widehat{s}_j) = s_j$ for $j = 1, 2, \ldots, n-1$. (Later, the complicated notation $W(A_{n-1})$ will fit in a larger scheme.)

8. Proposition. *The homomorphism φ defined above is an isomorphism from $W(A_{n-1})$ onto σ_n.*

Proof. In this proof, let us write $\widehat{W}$ for $W(A_{n-1})$. It is obvious that $\varphi : \widehat{W} \longrightarrow \sigma_n$ is onto, and it is thus sufficient to show that $|\widehat{W}| \leq n!$. Set $S = \{\widehat{s}_1, \widehat{s}_2, \ldots, \widehat{s}_{n-1}\}$. For each $k \in \{0, 1, \ldots, n-1\}$, let $\widehat{t}_{n-1,k}$ be the product of the first k letters of the sequence $(\widehat{s}_{n-1}, \widehat{s}_{n-2}, \ldots, \widehat{s}_1)$; in particular $\widehat{t}_{n-1,0} = e$ and $\widehat{t}_{n-1,n-1} = \widehat{s}_{n-1}\widehat{s}_{n-2}\ldots\widehat{s}_1$. We shall check by induction on n the two following claims:

(i) Any $w \in \widehat{W}$ can be written as a word in the letters of S containing $\widehat{s}_{n-1}$ at most once.

(ii) Any $w \in \widehat{W}$ can be written as $w = w'\widehat{t}_{n-1,k}$ for some $k \in \{0, 1, \ldots, n-1\}$, where w' is a word in $\widehat{s}_1, \widehat{s}_2, \ldots, \widehat{s}_{n-2}$.

The claims are obvious for $n = 2$. We assume from now on that they hold for some $n \geq 2$, and we show that they hold also for $n+1$.

(i) Consider $w \in W(A_n)$ written as a word in $\widehat{s}_1, \widehat{s}_2, \ldots, \widehat{s}_n$. Assume first that this word contains $\widehat{s}_n$ at least twice. By the induction hypothesis, one may write $w = a\widehat{s}_n b\widehat{s}_n c$ where b, c are words in $\widehat{s}_1, \ldots, \widehat{s}_{n-1}$; moreover $b = b'\widehat{t}_{n-1,k}$ for some $k \in \{0, 1, \ldots, n-1\}$, where b' is a word in $\widehat{s}_1, \ldots, \widehat{s}_{n-2}$; observe that b' commutes with $\widehat{s}_n$. If $k = 0$ then $b = b'$ and $w = abc$. If $k > 0$ then $\widehat{t}_{n-1,k} = \widehat{s}_{n-1}\widehat{t}_{n-2,k-1}$; as $\widehat{t}_{n-2,k-1}$ commutes with $\widehat{s}_n$ one has

$$w = ab'\widehat{s}_n\widehat{s}_{n-1}\widehat{s}_n\widehat{t}_{n-2,k-1}c = ab'\widehat{s}_{n-1}\widehat{s}_n\widehat{s}_{n-1}\widehat{t}_{n-2,k-1}c \ .$$

Thus w can also be written as a word containing $\widehat{s}_n$ once less as in the previous form. This proves Claim (i).

(ii) If w can be written as a word which does not contain $\widehat{s}_n$, Claim (ii) for $n+1$ is obvious (with $k = 0$). Otherwise Claim (i) for $n+1$ and the induction hypothesis show that one can write

$$w = a\widehat{s}_n b\widehat{t}_{n-1,k}$$

for some $k \in \{0, 1, \ldots, n-1\}$, where a is a word in $\widehat{s}_1, \ldots, \widehat{s}_{n-1}$ and where b is a word in $\widehat{s}_1, \ldots, \widehat{s}_{n-2}$. If we set $w' = ab$, one has also

$$w = w'\widehat{s}_n\widehat{t}_{n-1,k} = w'\widehat{t}_{n,k+1} \ .$$

This proves Claim (ii).

It follows that any $w \in \widehat{W} = W(A_{n-1})$ can be written as $w = \widehat{t}_{1,k_1}\widehat{t}_{2,k_2}\ldots\widehat{t}_{n-1,k_{n-1}}$ for some $k_1 \in \{0, 1\}, k_2 \in \{0, 1, 2\}, \ldots, k_{n-1} \in \{0, 1, \ldots, n-1\}$, and in particular that $|\widehat{W}| \leq n!$ In view of Number 12 below, we observe that the re-writings of the word w in the proof above change it in a word which is not longer than the initial one. □

A proof of the kind given above was already known to L.E. Dickson at the end of last century [Di1]. (The result itself is due to E.H. Moore.) More recently, it has been popularized by V. Jones and A. Ocneanu in their work on Hecke algebras and link invariants (see §4 in [HKW]). See Number 71 below for another proof.

From now on, we identify the abstract group given by the Presentation 7 and the symmetric group σ_n.

9. Exercise. For any subset X of $S=\{s_1,s_2,\ldots,s_{n-1}\}$, let R_X be the set of those relations appearing in the Presentation 7 in which all letters belong to X. Let W_X be the abstract group defined by the presentation $\langle X : R_X\rangle$. By a minor modification of the proof of Proposition 8, check that the homomorphism $W_X \longrightarrow W(A_{n-1})$ induced by the inclusion $X \subset S$ is injective, so that one may (and does) identify W_X to a subgroup of $W(A_{n-1})$. Check moreover that W_X is isomorphic to a direct product of the form $\sigma_p\times\sigma_q\times\ldots\times\sigma_r$ where $p+q+\ldots+r\leq n$.

10. Exercise. Consider the Cayley graph $\mathcal{G}_n$ of the group σ_n with respect to the generating set $\{s_1,s_2,\ldots,s_{n-1}\}$. Check that $\mathcal{G}_n$ is a hexagon for $n=3$, and a truncated octahedron for $n=4$ (namely the dual graph, in a sense to be precised, of the graph of Figure 2).

11. Counting functions. Let W be a group generated by a set S. For each $w\in W$, the *length* of w with respect to S is the minimum $\ell(w)$, or $\ell_S(w)$ if necessary, of all integers q such that w is a product of q elements in $S\cup S^{-1}$. A *reduced decomposition* of w with respect to S is a sequence $(s_1,s_2,\ldots,s_q)$ of elements in $S\cup S^{-1}$ such that $w=s_1s_2\ldots s_q$ and $q=\ell(w)$. We assume from now on that S is finite. For each integer $q\geq 0$, set

$$b_q = \text{ cardinal } \{w\in W \ : \ \ell(w)=q\} \ .$$

The *counting function* is defined to be

$$W(t)=\sum_{q\geq 0} b_q t^q \ .$$

In fact, $W(t)$ is a polynomial if W is finite and a formal power series with coefficients in $\mathbb{Z}$ if W is infinite.

For example, if W is free on a set S of cardinal ℓ, it is easy to check that

$$b_0=1 \qquad b_q=2\ell(2\ell-1)^{q-1} \quad \text{for} \quad q\geq 1$$

so that

$$W(t)=1+2\ell\sum_{q\geq 1}(2\ell-1)^{q-1}t^q=\frac{1+t}{1-(2\ell-1)t} \ .$$

For any group W with a generating set S of cardinal ℓ, one has $b_0=1$, and it follows from the free case that $b_q\leq 2\ell(2\ell-1)^{q-1}$ for $q\geq 1$.

12. Counting function of $\sigma_n=W(A_{n-1})$. Consider again the symmetric group σ_n given by the Presentation 7. From the proof of Proposition 8, we know that each $w\in W$ can be written as a product $\widehat{t}_{1,k_1}\widehat{t}_{2,k_2}\ldots\widehat{t}_{n-1,k_{n-1}}$, and indeed that the length of such a product is $k_1+k_2+\ldots+k_{n-1}$. If b_q^n denotes the number of words of length q in σ_n, one has consequently

$$b_q^n=b_q^{n-1}+b_{q-1}^{n-1}+\ldots+b_{q-(n-1)}^{n-1}$$

(where we set $b_r^{n-1}=0$ if $r<0$), and this can also be written

$$\sigma_n(t) = (1+t+\ldots+t^{n-1})\sigma_{n-1}(t) .$$

(According to Comtet [Com], this result goes back to 1871 and is due to Bourget.) By induction on n, one has consequently

$$\sigma_n(t) = \prod_{j=1}^{n-1}(1+t+\ldots+t^j)$$

For example, the *longest word* $w_0 \in \sigma_n$ is of length $1+2+\ldots+(n-1)=\frac{n(n-1)}{2}$. It is easy to check that w_0 is the order-reversing permutation:

$$w_0 = (1,n)(2,n-1)\ldots(n/2,1+n/2) \qquad \text{if } n \text{ is even} \ ,$$
$$w_0 = (1,n)(2,n-1)\ldots\left(\frac{n-1}{2},\frac{n+3}{2}\right) \qquad \text{if } n \text{ is odd} \ .$$

For $w \in \sigma_n$, there are in general many reduced decompositions other than the sequence suggested by the product $\hat{t}_{1,k_1}\hat{t}_{2,k_2}\ldots\hat{t}_{n-1,k_{n-1}}$. In particular, it has been shown [St2] that the number of reduced decompositions of the longest word $w_0 \in \sigma_n$ is

$$\frac{\left(\frac{n(n-1)}{2}\right)!}{1^{n-1}\,3^{n-2}\,5^{n-3}\ldots(2n-3)} .$$

For more counting results in σ_n and their interpretation in algebraic geometry, see [St1].

Let for example $n=4$. The counting function is

$$\sigma_4(t) = (1+t)(1+t+t^2)(1+t+t^2+t^3)$$

(see Exercise 5) and the longest word

$$w_0 = s_1s_2s_1s_3s_2s_1 = (1,3)(2,4)$$

is of length 6. There are altogether 16 reduced decompositions for this w_0.

1.3. Some Coxeter Terminology

13. Definitions. Let S be a finite set. A *Coxeter matrix* of type S is a symmetric square matrix $M=(m_{s,t})_{s,t\in S}$ in which the entries are either integers or the symbol ∞, such that

$$m_{s,s} = 1 \qquad \text{for all} \quad s \in S \ ,$$
$$m_{s,t} \geq 2 \qquad \text{for all} \quad s,t \in S \qquad \text{such that } s \neq t \ .$$

Such a matrix is encoded in the corresponding *Coxeter graph* $\mathcal{G}_M$: it is a weighted graph with set of vertices S, with set of edges the pairs $\{s,t\} \subset S$ such that

$m_{s,t} \geq 3$, where each edge is marked by its *weight* $m_{s,t}$. An edge of weight m is represented by

$$\overset{m}{\bullet\!\!-\!\!-\!\!-\!\!\bullet}$$

or also by $\bullet\!\!-\!\!-\!\!-\!\!\bullet$ in case $m=3$.

The *Coxeter system* defined by a Coxeter matrix M is the pair (W, S) where W is an abstract group defined by the presentation with set of generators S and with set of relations $(st)^{m_{s,t}} = 1$ (for pairs with $m_{s,t} < \infty$). In particular:

$$\begin{aligned} s^2 &= 1 && \text{for all} \quad s \in S \ , \\ st &= ts && \text{whenever } \{s,t\} \text{ is not an edge of } \mathcal{G}_M \ , \\ sts &= tst && \text{whenever } \{s,t\} \text{ is an edge of weight } 3 \ . \end{aligned}$$

If S is empty, it is convenient to define $W = \{1\}$. Here are a few other examples.

14. Example. If $S = \{s\}$ then $M = (1)$ and $\mathcal{G}_M$ is reduced to a point. The corresponding group $W = \langle s : s^2 = 1 \rangle$ is the group of order two.

15. Example. If $S = \{s, t\}$, then there exists $m = \{2, 3, \ldots, \infty\}$ such that

$$M = \begin{pmatrix} 1 & m \\ m & 1 \end{pmatrix} \ .$$

The graph $\mathcal{G}_M$ is

$$\begin{aligned} &\bullet \quad \bullet && \text{if } m = 2 \ , \\ &\bullet\!\!-\!\!-\!\!-\!\!\bullet && \text{if } m = 3 \ , \\ &\overset{m}{\bullet\!\!-\!\!-\!\!-\!\!\bullet} && \text{if } m \geq 3 \ , \end{aligned}$$

The corresponding group is

the Vierergruppe $C_2 \times C_2$ if $m=2$,

the dihedral group $D_{2m} = C_m \rtimes C_2$ if $3 \leq m < \infty$,

the infinite dihedral group $D_\infty = C_\infty \rtimes C_2 = C_2 * C_2$ if $m = \infty$.

A few words about notations: the cyclic group of order m is denoted by C_m. There is a standard action of C_2 on C_m, in which the element $s \neq 1$ in C_2 acts by sending any element of C_m to its inverse; the resulting semi-direct product $C_m \rtimes C_2$ is the dihedral group of order $2m$, denoted here by D_{2m} (and in some places by D_m!). If $m=2$, the action of C_2 on C_2 is trivial and $D_4 = C_2 \times C_2$. If $m=3$, the group D_6 is isomorphic to σ_3. If $m=\infty$, the group D_∞ is isomorphic to the free product of two copies of C_2; the subgroup C_∞ of $C_\infty \rtimes C_2$ corresponds to reduced words of even length in the free product $C_2 * C_2$.

16. Example. If the Coxeter graph $\mathcal{G}_M$ is a segment with n vertices

$$\bullet\!\!-\!\!-\!\!-\!\!\bullet\!\!-\!\!-\!\!-\!\!\bullet\!\!-\!\!-\!\!-\ \cdots\cdots\ -\!\!-\!\!-\!\!\bullet\!\!-\!\!-\!\!-\!\!\bullet$$

and thus with $n-1$ edges, all with weight 3, then W is the symmetric group σ_{n+1}, as shown by Proposition 8. The standard notation for this graph is A_n.

17. Example. If $\mathcal{G}_M$ has two connected components $\mathcal{G}_1$ and $\mathcal{G}_2$, the corresponding group is the direct product of the groups W_1 and W_2 associated to $\mathcal{G}_1$ and $\mathcal{G}_2$. Thus the graph

m

defines the group $C_2 \times D_{2m}$

18. Remark. If m is odd, this group $C_2 \times D_{2m}$ is isomorphic to the dihedral group $D_{4m} = C_{2m} \rtimes C_2$ associated to the graph

$2m$

This shows that there exist distinct systems (W', S'), (W'', S''), with isomorphic groups $W' \approx W''$. We shall also see in the end of Section 2.1 that the graphs

give rise to isomorphic groups. However, it is customary (and abusive!) to use the words *Coxeter group* for a group W appearing in a Coxeter system (W, S).

19. Definition. A Coxeter system (W, S) is *irreducible* if the corresponding graph is connected, and reducible otherwise. By reduction to factors of direct products, it is often enough to consider irreducible systems.

20. Example. Let $\mathcal{G}_1, \mathcal{G}_2$ be two Coxeter graphs and let W_1, W_2 be the corresponding groups. Let $\mathcal{G}_1 * \mathcal{G}_2$ be the Coxeter graph obtained from the disjoint union of $\mathcal{G}_1$ and $\mathcal{G}_2$ by adding an edge of weight ∞ connecting s_1 and s_2 for each pair of a vertex s_1 of $\mathcal{G}_2$ and a vertex s_2 of $\mathcal{G}_2$. The group corresponding to $\mathcal{G}_1 * \mathcal{G}_2$ is the free product $W_1 * W_2$.

21. Exercise. For each finite integer $m \geq 2$, the dihedral group $D_{2m} = \langle s, t : s^2 = t^2 = (st)^m = 1 \rangle$ acts on the Euclidean plane R^2 in such a way that s and t act as reflections with respect to lines making an angle of π/m. Extend to any m the considerations of Section 1.1 about $D_6 = \sigma_3$; what is the counting function?

Check also that the infinite dihedral group $D_\infty = \langle s, t : s^2 = t^2 = 1 \rangle$ acts on the *affine* line $\mathsf{E}^1 \approx \mathsf{R}$ in such a way that s [respectively t] acts as a half-turn around 0 [resp. around 1/2]. Observe that the counting "function" (a priori a formal power series) is in fact a *rational* function.

Chapter 2. Regular Polytopes and Other Finite Coxeter Groups

We introduce in this chapter some groups which share several properties with symmetric groups.

2.1. The Groups of the Standard Examples

In this section, V denotes an affine Euclidean space of some dimension $n \geq 1$.

22. Definitions. A *convex polyhedron* in V is a subset of V which is the intersection of a finite number of closed affine half-spaces. A *polytope* in V is a convex polyhedron P which is compact and which has a non empty interior. The *group* of such a polytope is the group $G(P)$ of all isometries of V by which P is invariant.

23. Faces. Let us choose some origin in V. If H is any affine hyperplane of V which does not contain this origin, we denote by H^+ the closed half-plane of V limited by H and containing the origin. Let P be a polytope in V of which the interior contains the origin and let $\{H_i\}_{i \in I}$ be a family of affine hyperplanes such that

$$P = \bigcap_{i \in I} H_i^+ \ ,$$

$$P \subsetneqq \bigcap_{j \in J} H_j^+ \qquad \text{for any proper subset } J \text{ of } I \ .$$

One shows that

the set $\{H_i\}_{i \in I}$ is finite and is determined by P,

$H_i \cap P$ is a polytope in H_i for each $i \in I$.

The $H_i \cap P$'s are the *faces*, or $(n-1)$-*facettes*, of P. One defines similarly the $(n-2)$-facettes of P (the faces of its faces), the $(n-3)$-facettes, ... The 0-facettes and 1-facettes are respectively the *vertices* and *edges* of P. One shows also that

$$P \text{ is the convex hull of its vertices .}$$

A *flag* in P is a nested sequence $F_0 \subset F_1 \subset \ldots \subset F_{n-1}$ where each F_j is a j-facette of P.

The previous generalities have a good intuitive content, but it is not completely straightforward to prove them in general (see any of [Ber], Chapter 12, or [Brø], Chapter 2, or [Gru], Chapter 3). However, they are quite easy to check in many of the examples below.

24. Regularity. The polytope P in V is said to be *regular* if $G(P)$ acts transitively on the set $\mathsf{F}(P)$ of flags of P. If P is regular, $G(P)$ acts simply transitively on $\mathsf{F}(P)$, so that the cardinal of $G(P)$ coincides with that of $\mathsf{F}(P)$. (There are other equivalent definitions for "regular": see for example n°7.5 in [Co6].)

25. Example. Let $\{e_1, e_2, \ldots, e_{n+1}\}$ be the standard basis of the Euclidean space $\mathbf{R}^{n+1}$. The *standard simplex* of dimension n is the convex hull Simp_n of this basis; it is a polytope in the affine space

$$\mathsf{E} = \left\{(x_1, x_2, \ldots, x_{n+1}) \in \mathbf{R}^{n+1} \ : \ \sum x_i = 1\right\}$$

of dimension n. (The natural choice of $\frac{1}{n+1}(e_1 + \ldots + e_{n+1})$ as an origin for E makes it possible to view E as a vector space.) To each permutation w of $\{1, 2, \ldots, n+1\}$ corresponds a flag of Simp_n with vertex $e_{w(1)}$, with edge limited by $e_{w(1)}$ and $e_{w(2)}$, and so on. The group of Simp_n is the symmetric group σ_{n+1}.

26. Example. The *standard cube* Cub_n of dimension n is the convex hull of the 2^n vertices $\sum_{1 \leq i \leq n} \varepsilon_i e_i$ of $\mathbf{R}^n$, where $\varepsilon_i \in \{1, -1\}$ and where $\{e_1, e_2, \ldots e_n\}$ is now the standard basis of $\mathbf{R}^n$.

27. Proposition. *The group $G = G(\mathrm{Cub}_n)$ is the semi-direct product $C_2^n \rtimes \sigma_n$ defined by the natural action of the symmetric group σ_n on the direct product of n copies of the cyclic group of order 2. Moreover G is generated by reflections of $\mathbf{R}^n$.*

Proof. The group G contains the reflections with respect to the coordinate hyperplanes in $\mathbf{R}^n$, which generate a normal subgroup C_2^n of G. In particular G acts transitively on the vertices of Cub_n.

Let A be a vertex of Cub_n. The vertices connected to A by an edge constitute a standard simplex Simp_{n-1}. The flags of Cub_n with vertex A are then in bijection with the flags in this Simp_{n-1}. As moreover $G(\mathrm{Simp}_{n-1}) \subset G(\mathrm{Cub}_n)$, this shows that Cub_n is regular. This shows also that the order of G is $2^n n!$, and more precisely that G is the "obvious" semi-direct product $C_2^n \rtimes G(\mathrm{Simp}_{n-1})$.

As both $G(\mathrm{Simp}_{n-1})$, viewed as a group acting in $\mathbf{R}^n$, and C_2^n are generated by reflections, the same holds for G. □

28. Remark. The barycenters of the faces of Cub_n are the $2n$ points $\varepsilon_i e_i$, where $\varepsilon_i \in \{1, -1\}$ and where $i \in \{1, 2, \ldots, n\}$. The *dual* of the cube, namely the convex hull of these $2n$ points, is the cross polytope Cross_n of dimension n (or the regular octahedron if $n = 3$). The polytope Cross_n is also regular and its group is obviously the same as the group of Cub_n.

29. The case $n = 3$. The group $W = C_2^3 \rtimes \sigma_3$ of the standard cube in $\mathbf{R}^3$ acts on the unit sphere $\mathbf{S}^2$ of $\mathbf{R}^3$. It is easy to check that there are nine reflections, namely

(a) three reflections with respect to the coordinate planes,

(b) three pairs of reflections, one for each pair of antipodal faces of the cube, the planes fixed by the reflections in one pair intersecting diagonally the corresponding faces.

In a suitable stereographic projection, the nine reflections in W correspond to nine lines and circles organised as in Figure 3. This shows 48 chambers which are triangles with angles $(\pi/2, \pi/3, \pi/4)$, as well as a regular cube (with vertices marked by thick dots) and the dual regular octahedron (with five vertices marked 0 and one vertex at infinity). The chambers are the 2-cells of the barycentric subdivision of the cube, and thus also the 2-cells of the barycentric subdivision of the octahedron. Figure 3 shows the three basis vectors e_1, e_2, e_3 and the chamber of points $(x_1, x_2, x_3) \in S^2$ such that $x_3 \geq x_2 \geq x_1 \geq 0$.

30. Exercise. Check that the counting function corresponding to the group $C_2^3 \rtimes \sigma_3$ of previous number is

$$\begin{aligned} &1 + 3t + 5t^2 + 7t^3 + 8t^4 + 8t^5 + 7t^6 + 5t^7 + 3t^8 + t^9 \\ &= (1+t)^3(1+t^2)(1+t^2 + t^4) \, . \end{aligned}$$

Compare with Exercise 5, and also with Corollary 35 below.

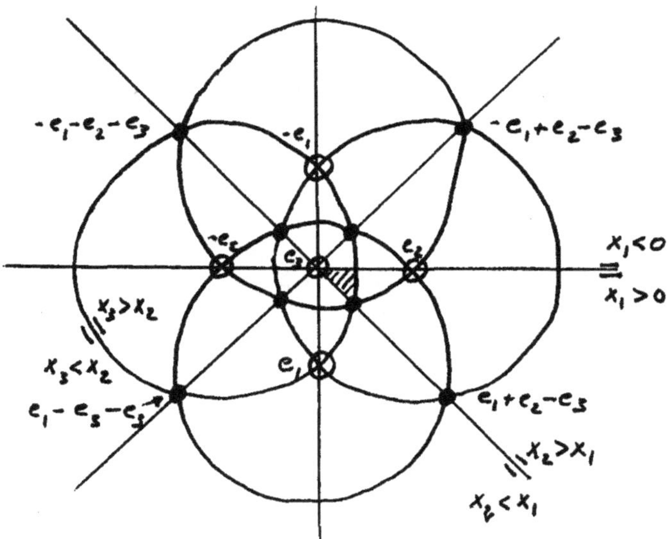

Figure 3.

31. Remark. Let P be the *half-measure polytope* of dimension n, namely the convex hull of the vertices

$$\left\{ \sum_{1\le i\le n} \varepsilon_i e_i \ : \ \varepsilon_i \in \{1,-1\} \quad \text{and} \quad \prod_{1\le i\le n} \varepsilon_i = 1 \right\}$$

of the standard cube. Then $G(P)$ contains the subgroup W of the group $G(\mathrm{Cub}_n) \approx C_2^n \rtimes \sigma_n$ generated by the permutation of the basic vectors, namely by σ_n, and by the products of even numbers of sign changes of coordinates. Thus W is a subgroup of index 2 in $C_2^n \rtimes \sigma_n$, and W is more precisely the semi-direct product $C_2^{n-1} \rtimes \sigma_n$, where C_2^{n-1} is identified to

$$\{(\varepsilon_1, \varepsilon_2, \ldots, \varepsilon_n) \in C_2^n \ : \ \prod_{1\le i\le n} \varepsilon_i = 1\} \, .$$

Observe that W is generated by reflections, for example by the reflection s_1 of $\mathbb{R}^n$ defined by

$$s_1(e_j) = \begin{cases} -e_2 & \text{if } j=1, \\ -e_1 & \text{if } j=2, \\ e_j & \text{otherwise} \end{cases}$$

and by the transpositions

$$s_2 = (1,2) \qquad s_3 = (2,3) \quad \ldots \quad s_n = (n-1, n)$$

of σ_n. The polytope P is a regular tetrahedron for $n=3$, a regular crosspolytope for $n=4$, and is not regular for $n \ge 5$. I believe that $G(P) = W$ for $n \ge 5$.

For $n=3$, observe that $G(\mathrm{Cub}_3)$ is the direct product of $G(P) \approx \sigma_4$ and of the group C_2 generated by (minus the identity).

2.2. Presentations

Proposition 27 shows that the group $G(\mathrm{Cub}_n) = C_2^n \rtimes \sigma_n$ is generated by the sign change s_1 of the first coordinate and by the transpositions $s_2 = (1,2), \ldots, s_n = (n-1, n)$. These reflections satisfy

$$\begin{aligned} s_j^2 &= 1 & & \\ (s_1 s_2)^4 &= 1 & & \\ (s_j s_{j+1})^3 &= 1 & \text{if} \quad & j \ge 2 \\ s_j s_k &= s_k s_j & \text{if} \quad & |j-k| \ge 2 \end{aligned}$$

where $j, k \in \{1, 2, \ldots, n\}$.

32. The Coxeter system of type B_n. For each integer $n \ge 2$, we introduce the Coxeter graph of type B_n

with n vertices. The corresponding Coxeter system is

$$(W(B_n), S) \quad \text{with} \quad S = \{\widehat{s}_1, \widehat{s}_2, \ldots, \widehat{s}_n\} \ ,$$

and there is an obvious surjective homomorphism $\varphi : W(B_n) \longrightarrow C_2^n \rtimes \sigma_n$ defined by $\varphi(\widehat{s}_j) = s_j$ for $j = 1, \ldots, n$. (Compare with Section 1.2.)

33. Digression. If $n = 2$, we know from Example 15 that φ is an isomorphism onto, because $W(B_2)$ is the dihedral group of order 8. Let us enumerate three presentations of this group. The first one is the one above, namely (if we write s for s_1 and t for s_2):

$$\langle s, t \ : \ s^2 = t^2 = (st)^4 = 1 \rangle \ .$$

The second one shows how the group contains a Vierergruppe (generated by the reflections of $\mathbb{R}^2$ with respect to the two axes):

$$\langle s, \widetilde{s}, t \ : \ s^2 = \widetilde{s}^2 = t^2 = 1, \quad s\widetilde{s} = \widetilde{s}s \ , \quad tst = \widetilde{s} \rangle \ .$$

The third one shows how this group contains the group of rotations of the square (generated by $\varrho = st$):

$$\langle s, \varrho \ : \ s^2 = \varrho^4 = 1 \ , \quad s\varrho s^{-1} = \varrho^{-1} \rangle \ .$$

The situation is pictured in Figure 4.

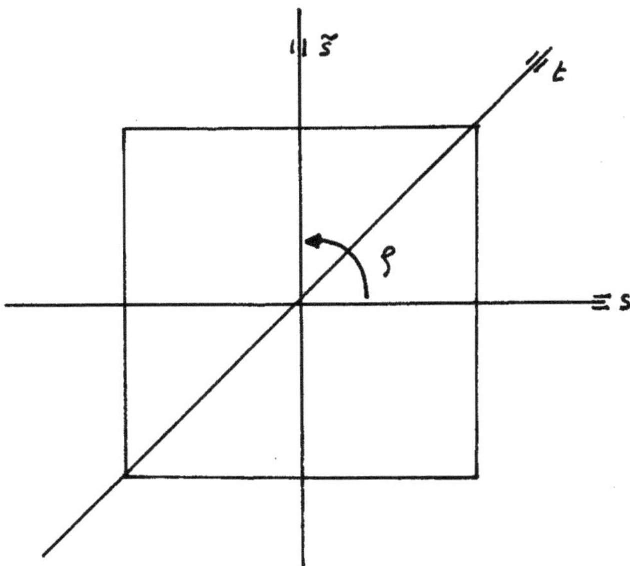

Figure 4.

34. Proposition. *With the notations of Number 32, the homomorphism*

$$\varphi \; : \; W(B_n) \to C_2^n \rtimes \sigma_n$$

is an isomorphism onto.

Proof. (I am grateful to F. du Cloux and G. Lehrer who have indicated to me the following argument. See Number 72 below for another proof.) In this proof, we write $\widehat{W}_n$ or $\widehat{W}$ for $W(B_n)$ and we denote by $\widehat{s}_1, \ldots, \widehat{s}_n$ the generators of $S \subset \widehat{W}$. If $\sigma = (\widehat{s}_{j_1}, \ldots, \widehat{s}_{j_p})$ is a sequence of generators, we denote by $\langle\sigma\rangle$ the product $\widehat{s}_{j_1} \ldots \widehat{s}_{j_p} \in W$. For each $k \in \{0, 1, \ldots, 2n-1\}$, let $\tau_{n,k}$ be the sequence of the k first letters of the sequence

$$\tau_{n,2n-1} = (\widehat{s}_n, \widehat{s}_{n-1}, \ldots, \widehat{s}_2, \widehat{s}_1, \widehat{s}_2, \ldots, \widehat{s}_{n-1}, \widehat{s}_n) \; .$$

In particular $\langle\tau_{n,0}\rangle = e$, and one may check that $\varphi(\langle\tau_{n,2n-1}\rangle)$ is the sign change in the last coordinate of $\mathbb{R}^n$. We shall check by induction on n the two following claims.

(i) Any $w \in \widehat{W}$ can be written as a word in the letters of S which contains $\widehat{s}_n$ either at most once, or exactly twice within a subword $\langle\tau_{n,2n-1}\rangle$.

(ii) Any $w \in \widehat{W}$ can be written as $w = w'\langle\tau_{n,k}\rangle$ for some $k \in \{0, 1, \ldots, 2n-1\}$, where w' is a word in $\widehat{s}_1, \widehat{s}_2, \ldots, \widehat{s}_{n-1}$.

From the facts recalled in Digression 33, it is straightforward to check the claims for $n=2$. We assume from now on that they hold for some $n \geq 2$, and we show that they hold also for $n+1$.

(i) Consider $w \in \widehat{W}_{n+1}$ written as a word in $\widehat{s}_1, \ldots, \widehat{s}_{n+1}$, and assume that this word contains $\widehat{s}_{n+1}$ at least twice. By the induction hypothesis, one may write

$$w = \langle\alpha\rangle \; \widehat{s}_{n+1} \; \langle\beta'\rangle \; \langle\tau_{n,k}\rangle \; \widehat{s}_{n+1} \; \langle\gamma\rangle$$

where

α is a sequence in $\widehat{s}_1, \ldots, \widehat{s}_n$,
β' is a sequence in $\widehat{s}_1, \ldots, \widehat{s}_{n-1}$,
$k \in \{0, 1, \ldots, 2n-1\}$,
γ is a sequence in $\widehat{s}_1, \ldots, \widehat{s}_{n+1}$.

If $k=0$, then $w = \langle\alpha\rangle\langle\beta'\rangle\langle\gamma\rangle$. If $1 \leq k \leq 2n-2$ then $\langle\tau_{n,k}\rangle = \widehat{s}_n\langle\tau_{n-1,k-1}\rangle$ and

$$\begin{aligned} w &= \langle\alpha\rangle \, \langle\beta'\rangle \widehat{s}_{n+1} \widehat{s}_n \widehat{s}_{n+1} \langle\tau_{n-1,k-1}\rangle \, \langle\gamma\rangle \\ &= \langle\alpha\rangle \, \langle\beta'\rangle \widehat{s}_n \widehat{s}_{n+1} \widehat{s}_n \langle\tau_{n-1,k-1}\rangle \, \langle\gamma\rangle \; . \end{aligned}$$

Thus, if $k \leq 2n-2$, one may write w as a word containing $\widehat{s}_{n+1}$ once less as in the previous form.

If $k=2n-1$, then

$$\begin{aligned} w &= \langle\alpha\rangle\,\langle\beta'\rangle\widehat{s}_{n+1}\langle\tau_{n,2n-1}\rangle\widehat{s}_{n+1}\langle\gamma\rangle \\ &= \langle\alpha\rangle\,\langle\beta'\rangle\langle\tau_{n+1,2n+1}\rangle\,\langle\gamma\rangle\ . \end{aligned}$$

If the sequence γ does not contain $\widehat{s}_{n+1}$, there is nothing left to prove. Otherwise, one may write $\gamma=(\gamma',\widehat{s}_{n+1},\gamma'')$ where γ' is a sequence in $\widehat{s}_1,\ldots,\widehat{s}_n$. Now $\langle\tau_{n+1,2n+1}\rangle$ commutes with any of $\widehat{s}_1,\ldots,\widehat{s}_n$ (this is easy to check, and we leave the details to the reader), and one has also

$$\begin{aligned} w &= \langle\alpha\rangle\,\langle\beta'\rangle\,\langle\gamma'\rangle\,\langle\tau_{n+1,2n+1}\rangle\widehat{s}_{n+1}\langle\gamma''\rangle \\ &= \langle\alpha\rangle\,\langle\beta'\rangle\,\langle\gamma'\rangle\,\langle\tau_{n+1,2n}\rangle\langle\gamma''\rangle \end{aligned}$$

so that one may again write w as a word containing $\widehat{s}_{n+1}$ twice less as in the previous form.

This proves Claim (i).

(ii) The proof of Claim (ii) is now straightforward. See the proof of Proposition 8.

Claim (ii) shows firstly that $|W(B_n)|\leq 2^n n!$, and thus secondly that φ is an isomorphism onto. □

35. Corollary. *The counting function for the Coxeter system* (W,S) *of type* B_n *is*

$$W(t)=\prod_{j=1}^{n}\left(1+t+t^2+\ldots+t^{2j-1}\right)\ .$$

Proof. The previous proof shows the following: if $w\in W(B_n)$ has a reduced decomposition of some length p, then it has also a reduced decomposition $(s_{j_1},\ldots,s_{j_p})$ which ends by one of the sequences $\tau_{n,k}$ and which does not contain s_n outside this subsequence. Hence any $w\in W$ can be written uniquely as $w=w_1w_2\ldots w_n$ with

$$w_j=\langle\tau_{j,k_j}\rangle \ \text{ for some } k_j\in\{0,1,\ldots,2j-1\}\ ,$$

$$k_1+k_2+\ldots+k_n=\ \text{length}(w)\ .$$

In particular

$$W(B_n)(t)=\left(1+t+\ldots+t^{2n-1}\right)W(B_{n-1})(t)$$

and the claim follows.

36. Remarks. The longest word in a Coxeter system of type B_n is of length

$$1+3+5+\ldots+2n-1=n^2\ .$$

We have just shown that $(\tau_{1,1},\tau_{1,3},\ldots,\tau_{n,2n-1})$ is *one* reduced decomposition for this word. It has been shown [Kra] that the number of reduced decompositions of this word is

$$(n^2)!\,\frac{1!2!\dots(n-1)!}{n!\,(n+1)!\,(n+2)!\dots(2n-1)!}\,.$$

37. Exercise. Consider the Coxeter graph D_n

and the associated Coxeter system $(W(D_n), S)$.

(i) Define a homomorphism φ from $W(D_n)$ onto the group $C_2^{n-1} \rtimes \sigma_n$, sending the generators $\hat{s}_1, \dots, \hat{s}_n$ in S to the reflections denoted by $s_1, \dots, s_n$ in Remark 31.

(ii) For each $k \in \{0, 1, \dots, 2n-2\}$, let $\tau_{n,k}$ be the sequence of the k first letters of the sequence

$$\tau_{n,2n-2} = (\hat{s}_n, \hat{s}_{n-1}, \dots, \hat{s}_4, \hat{s}_3, \hat{s}_1, \hat{s}_2, \hat{s}_3, \hat{s}_4, \dots, \hat{s}_{n-1}, \hat{s}_n)$$

and set

$$\tau'_{n,n-1} = (\hat{s}_n, \hat{s}_{n-1}, \dots, \hat{s}_4, \hat{s}_3, \hat{s}_2)\,.$$

Describe the transformations $\varphi(\tau_{n,2n-2})$ and $\varphi(\tau'_{n,n-1})$ of $\mathbf{R}^n$.

(iii) For $n \geq 3$, show that any $W(D_n)$ can be written either as $w = w'\tau_{n,k}$ or as $w = w'\tau'_{n,n-1}$, with $w' \in W(D_{n-1})$.

(iv) Deduce from (iii) that $|W(D_n)| \leq 2^{n-1}!$ and that φ is an isomorphism.

(v) Check that the counting function of $W(D_n)$ is given by

$$\begin{aligned} W(D_n)(t) &= \left(1 + t + \dots + t^{n-2} + 2t^{n-1} + t^n + \dots + t^{2n-2}\right) W(D_{n-1})(t) \\ &= \frac{(1+t+\dots+t^{2n-3})(1+t+\dots+t^{n-1})}{1+t+\dots+t^{n-2}}\, W(D_{n-1})(t) \\ &= (1+t+\dots+t^{n-1}) \prod_{j=1}^{n-1} (1+t+\dots+t^{2j-1}) \end{aligned}$$

2.3. Lists of Schläfli and Coxeter

38. Definitions. Consider a regular polytope P in $\mathbb{E}^n$, where $n \geq 2$. Let A be a vertex of P. Vertices of P joined to A by an edge of P are vertices of a regular polytope Q_A in an affine hyperplane of $\mathbb{E}^n$; this Q_A is independent (up to isometry) of the choice of A and is called the *vertex figure* of P. One associates to P its *Schläfli symbol*, inductively defined as follows:

if $n = 2$, the symbol of a regular p-gon is $\{p\}$,

if $n \geq 3$, the symbol of P is $\{p, q, \ldots, r\}$ if the 2-facettes of P are regular p-gons and if the vertex figure of P has symbol $\{q, \ldots, r\}$.

Two regular polytopes are *similar* if one may be obtained from the other by the composition of an isometry and of a dilation. It is obvious that two similar regular polytopes have the same Schläfli symbol, and it is easy to prove (by induction on n) that the converse holds (see e.g., §12.6 in [Ber]).

39. Standard examples. The symbol of Simp_n is $\{3, 3, \ldots, 3\}$, also written $\{3^{n-1}\}$, that of Cub_n is $\{4, 3, \ldots, 3\} = \{4, 3^{n-2}\}$ and that of Cross_n is $\{3^{n-2}, 4\}$.

40. Exceptional regular polytopes in $\mathbb{E}^3$. There exist two "exceptional" regular polytopes in 3-space which are the dodecahedron $\{5, 3\}$ and its dual icosahedron $\{3, 5\}$, where "dual" is defined as in Remark 28. Their group has order 120 (count the flags in either $\{5, 3\}$ or $\{3, 5\}$) and contains 15 reflections. The picture for $\{5, 3\}$ and $\{3, 5\}$ which is analogous to Figure 2 for $\{3, 3\}$ and to Figure 3 for $\{4, 3\}$ and $\{3, 4\}$ is given as Figure 5. One may show that the group of $\{5, 3\}$ and of $\{3, 5\}$ is the Coxeter group associated to the Coxeter graph

$$\bullet \overset{5}{\text{———}} \bullet \text{———} \bullet \qquad H_3$$

and that it is isomorphic to $C_2 \times \mathcal{A}_5$, where $\mathcal{A}_5$ denotes the alternating group of five objects. (See chapter 3 below, or Exercise II in §6.4 of [Bli].)

41. A self-dual regular polytope in $\mathbb{E}^4$. There exists in $\mathbb{E}^4$ a regular polytope which is similar to its dual. Its Schläfli symbol is $\{3, 4, 3\}$, and it has

24	vertices ,
96	edges ,
96	triangular 2-facettes ,
24	octahedral 3-faces .

Its group is associated (see §1.3) to the Coxeter graph

$$\bullet \text{———} \bullet \overset{4}{\text{———}} \bullet \text{———} \bullet \qquad F_4$$

and it is a solvable group of order $1152 = 2^7 3^2$.

Here is one model for this polytope (see [Co6] for others). One identifies $\mathbb{R}^4$ to the space $\mathbb{H}$ of Hamilton quaternions, the universal covering group of $SO(3)$ to

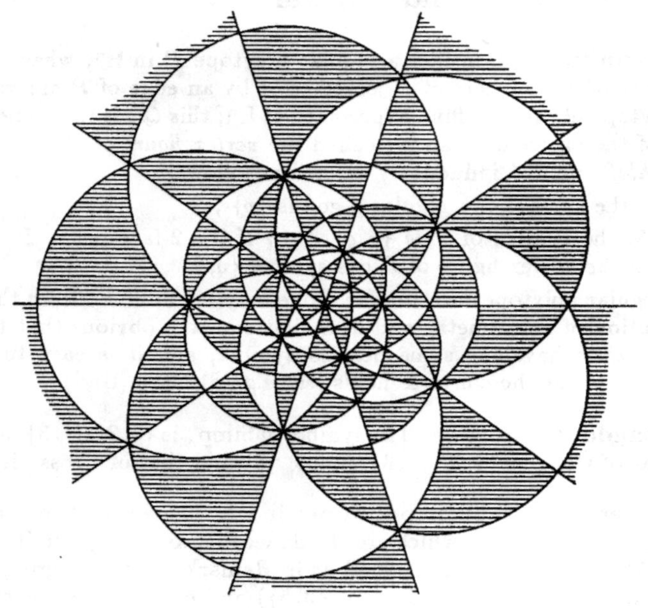

Figure 5.

$Sp(1)=\{q \in \mathsf{H} : |q|=1\}$, and the covering map $\pi : Sp(1) \longrightarrow SO(3)$ is defined by $\pi(q)(\xi)=q\xi\overline{q}$ for all $\xi \in \mathsf{R}^3$, where R^3 is identified to $\{q \in \mathsf{H} : \mathrm{Re}\,(q)=0\}$. Then $\{3,4,3\}$ is the convex hull in H of $\pi^{-1}(T)$, where T is the group of rotations of a regular tetrahedron in R^3; thus T is a subgroup of $SO(3)$ isomorphic to the alternating group $\mathcal{A}_4$. (See e.g., II.3 in [Lam].)

42. Two more regular polytopes in E^4. There are two regular polytopes in E^4 which are dual to each other and which have respectively

120 vertices	600 vertices
720 edges	1200 edges
1200 triangular 2-facettes	720 pentagonal 2-facettes
600 tetrahedral 3-faces	120 dodecahedral 3-faces
Schläfli symbol $\{3,3,5\}$,	Schläfli symbol $\{5,3,3\}$.

Their common group is associated to the Coxeter graph

$$\bullet\overset{5}{\text{---}}\bullet\text{---}\bullet\text{---}\bullet \qquad H_4$$

and has order $14400 = 2^6 3^2 5^2$. With the notations of the previous number, one model for the polytope with 120 vertices is $\pi^{-1}(\mathcal{A}_5)$, where $\mathcal{A}_5$ denotes here the group of rotations of a regular dodecahedron in $\mathbb{R}^3$ (of course this group is isomorphic to the alternating group of five objects).

It may be shown that the finite subgroups of $SO(3)$ for which the procedure used above provides regular polytopes in dimension four are precisely the dihedral group D_4 (which provides the cross polytope $\{3,3,4\}$), the alternating group $\mathcal{A}_4$ (which provides $\{3,4,3\}$ as mentioned in the previous number), and $\mathcal{A}_5$ which provides $\{3,3,5\}$ as above.

43. Theorem (Schläfli, around 1850). *In affine Euclidean spaces, the regular polytopes are up to similarity exactly those which are mentioned above, namely those of the following table. In particular, there is not any exceptional regular polytope in dimension* $n \geq 5$.

n	Symbol	Coxeter graph	Type
2	$\{p\}$ with $p \geq 3$	$\bullet\overset{p}{\text{---}}\bullet$	$I_2(p)$
3	$\{3,3\}$	$\bullet\text{---}\bullet\text{---}\bullet$	A_3
	$\{4,3\}$ and $\{3,4\}$	$\bullet\overset{4}{\text{---}}\bullet\text{---}\bullet$	B_3
	$\{5,3\}$ and $\{3,5\}$	$\bullet\overset{5}{\text{---}}\bullet\text{---}\bullet$	H_3
4	$\{3,3,3\}$	$\bullet\text{---}\bullet\text{---}\bullet\text{---}\bullet$	A_4
	$\{4,3,3\}$ and $\{3,3,4\}$	$\bullet\overset{4}{\text{---}}\bullet\text{---}\bullet\text{---}\bullet$	B_4
	$\{3,4,3\}$	$\bullet\text{---}\bullet\overset{4}{\text{---}}\bullet\text{---}\bullet$	F_4
	$\{5,3,3\}$ and $\{3,3,5\}$	$\bullet\overset{5}{\text{---}}\bullet\text{---}\bullet\text{---}\bullet$	H_4
$n \geq 5$	$\{3^{n-1}\}$	$\bullet\text{---}\bullet\text{---}\ \cdots\ \text{---}\bullet\text{---}\bullet$	A_n
	$\{4,3^{n-2}\}$ and $\{3^{n-2},4\}$	$\bullet\overset{4}{\text{---}}\bullet\text{---}\bullet\text{---}\ \cdots\ \text{---}\bullet\text{---}\bullet$	B_n

To keep the table small, the type of $\bullet\overset{p}{\text{---}}\bullet$ has been called $I_2(p)$. This is usual for $p=5$ and for $p \geq 7$, but the types for $p=3,4,6$ are rather called A_2, B_2 and G_2 respectively.

44. Comment on the proof. Given a regular polytope P with Schläfli symbol $\{p,q,\ldots,r\}$, denote by $\ell(P)$ the length of an edge of P, by $r(P)$ the radius of a sphere containing all vertices of P, and set $\varrho(P) = \frac{1}{4}\ell(P)^2 r(P)^{-2}$. If

Q denotes the vertex figure of P, an easy computation shows that

$$\varrho(P) = 1 - \frac{\cos^2(\pi/p)}{\varrho(Q)} .$$

In particular $\varrho(P)$ depends only on $p, q, \ldots, r$.

As $p \geq 3$ one has $\cos^2(\pi/p) \geq \frac{1}{4}$. If one had $\varrho(Q) \leq \frac{1}{4}$, the formula above would imply $\varrho(P) \leq 0$ and this is absurd. Thus $\varrho(Q) > \frac{1}{4}$, and this is a severe limitation. For example, if P is of dimension 3 and if its symbol is $\{p, q\}$, one has $\varrho(Q) = \frac{1}{\sin^2(\pi/q)}$ and the limitation implies $q \leq 5$. Details may be found for example in [Ber] and in [Lam].

45. Theorem (Coxeter, 1935). *Any connected Coxeter graph for which the associated Coxeter group is finite appears in the following list.*

A_n — •—•— ... —•—• $(n \geq 1)$ $\Rightarrow$ group σ_{n+1} of order $n!$

B_n — •$\overset{4}{—}$•—•— ... —•—• $(n \geq 2)$ $\Rightarrow$ group $C_2^n \rtimes \sigma_n$ of order $2^n\, n!$

D_n — >•—•— ... —• $(n \geq 4)$ $\Rightarrow$ group $C_2^{n-1} \rtimes \sigma_n$ of order $2^{n-1}\, n!$

E_6 $\Rightarrow$ group of order $51840 = 2^7\, 3^4\, 5$

E_7 $\Rightarrow$ group of order $2903040 = 2^{10}\, 3^4\, 5\, 7$

E_8 $\Rightarrow$ group of order $696729600 = 2^{14}\, 3^5\, 5^2\, 7$

F_4 — •—•$\overset{4}{—}$•—• $\Rightarrow$ solvable group of order $1152 = 2^7\, 3^2$

G_2 — •$\overset{6}{—}$• $\Rightarrow$ dihedral group $C_6 \rtimes C_2$ of order 12

H_3 — •$\overset{5}{—}$•—• $\Rightarrow$ group $\mathcal{A}_5 \times C_2$ of order $120 = 2^3\, 3\, 5$

H_4 — •$\overset{5}{—}$•—•—• $\Rightarrow$ group of order $14400 = 2^6\, 3^2\, 5^2$

$I_2(p)$ — •$\overset{p}{—}$• $(p = 5$ or $p \geq 7)$ $\Rightarrow$ dihedral group $C_p \rtimes C_2$ of order $2p$

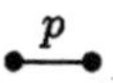

Proof. See [Co2] and [Co3], or Chapter 4 below. □

2.4. A Digression in Algebraic Geometry

We discuss first the 27 lines on a cubic surface and the Coxeter group of type E_6. I am grateful to T. Vust for tutorials on this subject.

46. Cubic surfaces. Surfaces here are subsets of the complex projective space $\mathbf{P}^3_{\mathbf{C}}$, with homogeneous coordinates $[z_0 : z_1 : z_2 : z_3]$. A *cubic form* is a homogeneous polynomial

$$F(Z) = \sum c_\alpha Z_0^{\alpha_0} Z_1^{\alpha_1} Z_2^{\alpha_2} Z_3^{\alpha_3}$$

where the summation is over quadruples $\alpha = (\alpha_0, \ldots, \alpha_3)$ of non negative integers such that $|\alpha| = 3$, where $|\alpha| = \alpha_0 + \alpha_1 + \alpha_2 + \alpha_3$.

To a non zero cubic form F is associated the *cubic surface* X_F, which is the subset of $\mathbf{P}^3_{\mathbf{C}}$ of points $[z_0 : z_1 : z_2 : z_3]$ such that $\sum c_\alpha z_0^{\alpha_0} z_1^{\alpha_1} z_2^{\alpha_2} z_3^{\alpha_3} = 0$. Two non zero forms which are proportional define the same surface. As there are 20 coefficients c_α, it is natural to introduce the space $\mathbf{P}^{19}_{\mathbf{C}}$, with homogeneous coordinates c_α, of all these forms up to proportionality; this $\mathbf{P}^{19}_{\mathbf{C}}$ may also be viewed as the space of cubic surfaces. A cubic form F defines a surface X_F which is *smooth* if and only if F and the four partial derivatives $\frac{\partial F}{\partial Z_j}$ do not have any common zero in $\mathbf{C}^4 - \{0\}$.

For example, the *Fermat cubic* X_{Fer} is the smooth surface of equation

$$z_0^3 + z_1^3 + z_2^3 + z_3^3 = 0$$

in $\mathbf{P}^3_{\mathbf{C}}$. The subset of $\mathbf{P}^{19}_{\mathbf{C}}$ of smooth cubic surfaces is non empty, open and dense.

47. Proposition. *The Fermat cubic contains 27 lines.*

Proof. Let ℓ be a line in X_{Fer}. Possibly after a permutation of the homogeneous coordinates of $\mathbf{P}^3_{\mathbf{C}}$, the equations of ℓ may be written

$$z_0 = az_2 + bz_3$$
$$z_1 = cz_2 + dz_3$$

with $a, b, c, d \in \mathbf{C}$. By substitution in the equation of X_{Fer}, we find

$$a^3 + c^3 + 1 = 0 \qquad a^2 b + c^2 d = 0$$
$$ab^2 + cd^2 = 0 \qquad b^3 + d^3 + 1 = 0 .$$

If $abcd \neq 0$, these equations do not have any solution. Indeed, set $\alpha = a^3, \ldots, \delta = d^3$; one finds firstly the relation $\alpha = \beta = \gamma = \delta = -\frac{1}{2}$. Set then $a = -\varepsilon_a 2^{-1/3}, \ldots, d = -\varepsilon_d 2^{-1/3}$, with $\varepsilon_a, \ldots, \varepsilon_d$ cubic roots of 1; one finds secondly the relation $\varepsilon_a^2 \varepsilon_b + \varepsilon_c^2 \varepsilon_d = 0$, which does not have any solution.

If $a = 0$, then one has also $d = 0$, and one finds 9 lines of equations

$$z_0 + \omega^j z_3 = 0 = z_1 + \omega^k z_2 \qquad j, k = 1, 2, 3$$

where $\omega = \exp(2i\pi/3)$. Similarly for

$$z_0 + \omega^j z_2 = 0 = z_1 + \omega^k z_3$$
$$z_0 + \omega^j z_1 = 0 = z_2 + \omega^k z_3$$

so that one has 27 lines. □

In Claim (2) of the following theorem, we say that 6 points in $\mathbb{P}^2_{\mathbb{C}}$ are in *general position* if no three of them are on one line, and if they are not all six lying on a plane conic. (A plane conic is the zero set of a non zero homogeneous polynomial of degree 2 in the homogeneous coordinates $[z_0 : z_1 : z_2]$ of $\mathbb{P}^2_{\mathbb{C}}$.) In Claim (3), the line in $\mathbb{P}^2_{\mathbb{C}}$ through two distinct points P_j, P_k is denoted by P_jP_k, and the bar indicates closure. We denote by C_1 the (uniquely defined) plane conic containing $P_2, \ldots, P_6$, and by C_1° the complement of these 5 points in C_1; similarly for $C_2^\circ, \ldots, C_6^\circ$, by circular permutation.

48. Theorem. *Let X be a smooth cubic surface in $\mathbb{P}^3_{\mathbb{C}}$.*

(1) The surface X contains exactly 27 lines.

(2) There are six points $P_1, \ldots, P_6$ in general position in $\mathbb{P}^2_{\mathbb{C}}$ and a morphism $\pi : X \longrightarrow \mathbb{P}^2_{\mathbb{C}}$ such that:
(i) the restriction of π to $X - \pi^{-1}(P_1, \ldots, P_6)$ is an isomorphism,
(ii) the $E_j = \pi^{-1}(P_j)$ are disjoint lines in X $(j = 1, \ldots, 6)$.

(3) The 27 lines of X are
(i) the 6 lines $E_i (1 \le i \le 6)$,
(ii) the 15 lines $F_{j,k} = \overline{\pi^{-1}(P_jP_k - \{P_j, P_k\})}$ $\quad (1 \le j < k \le 6)$
(iii) the 6 lines $G_i = \overline{\pi^{-1}(C_i^\circ)}$ $\quad (1 \le i \le 6)$.

(4) The incidence relations between pairs of lines in X are as follows:

$$E_i \cap E_j = \emptyset \quad \text{if } i \neq j \ ,$$
$$E_i \text{ meets } F_{j,k} \text{ if and only if } i \in \{j, k\} \ ,$$
$$E_i \text{ meets } G_j \text{ if and only if } i \neq j \ ,$$
$$F_{i,j} \text{ meets } F_{k,\ell} \text{ if and only if } i, j, k, \ell \text{ are distinct } \ ,$$
$$F_{i,j} \text{ meets } G_k \text{ if and only if } \{i, j\} \in k \ ,$$
$$G_j \cap G_k = \emptyset \text{ if } j \neq k \ .$$

Moreover, any 6 lines among these 27 which are pairwise skew play the role of $\{E_1, \ldots, E_6\}$ for a suitable morphism $X \longrightarrow \mathbb{P}^2_{\mathbb{C}}$.

Proof. We refer to §8.D in [Mum] and to §V.4 (including the exercises) in [Har] for the details, but we give nevertheless the following ideas.

If $F(Z) = \sum c_\alpha Z_0^{\alpha_0} Z_1^{\alpha_1} Z_2^{\alpha_2} Z_3^{\alpha_3}$ is a non-zero cubic form, denote now by X_c the corresponding cubic surface, with homogeneous coordinates $[c_\alpha] \in \mathbb{P}^{19}_{\mathbb{C}}$.

Define

$$\mathcal{H} = \{([z_0 : z_1 : z_2 : z_3] \times [c_\alpha]) \in \mathbf{P}^3_{\mathbf{C}} \times \mathbf{P}^{19}_{\mathbf{C}} \ : \ [z_0 : z_1 : z_2 : z_3] \in X_c\}$$

and let $\pi_2 : \mathcal{H} \longrightarrow \mathbf{P}^{19}_{\mathbf{C}}$ be the restriction of the second projection of $\mathbf{P}^3_{\mathbf{C}} \times \mathbf{P}^{19}_{\mathbf{C}}$. Define also

$$\Delta = \{[c_\alpha] \in \mathbf{P}^{19}_{\mathbf{C}} \ : \ X_c \quad \text{is not smooth}\}$$
$$S = \pi_2^{-1}(\Delta) \subset \mathcal{H}$$

so that $\mathcal{H} - S$ is the "union" of all smooth cubics.

It is a classical fact that the set of lines in $\mathbf{P}^3_{\mathbf{C}}$ can be identified with the set of points of a hypersurface G in $\mathbf{P}^4_{\mathbf{C}}$ (where G holds for "Grassmannian"). Define

$$I = \{(\ell, [c_\alpha]) \in G \times \mathbf{P}^{19}_{\mathbf{C}} \ : \ \ell \subset X_c\}$$

and let $p_2 : I \longrightarrow \mathbf{P}^{19}_{\mathbf{C}}$ be the restriction of the second projection of $G \times \mathbf{P}^{19}_{\mathbf{C}}$.

The main step of the proof of (1) is to show that the restriction of p_2 to $I - p_2^{-1}(\Delta)$ is a covering. Thus any smooth cubic surface has exactly the same number of lines as the Fermat cubic X_{Fer} of Proposition 47. One may show in a somehow similar way that the incidence relations between lines is the same for X_c and for X_{Fer}.

We refer to [Har] for the other claims; arguments make a crucial use of the notion of "blowing up" in algebraic geometry. □

For the next result, let $\ell_1, \ldots, \ell_{27}$ be some enumeration of the 27 lines of some smooth cubic X. We denote by G the subgroup of the symmetric group σ_{27} of those permutations w of $\{1, \ldots, 27\}$ such that the lines $\ell_{w(i)}$ and $\ell_{w(j)}$ meet if and only if ℓ_i and ℓ_j meet $(1 \leq i, j \leq 27)$. Recall also that E_6 is the Coxeter graph

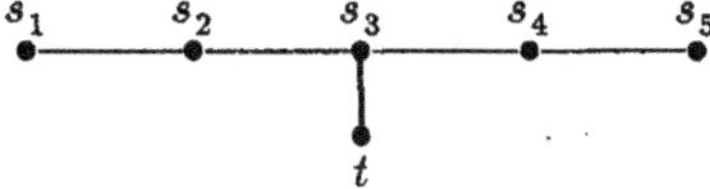

and observe that we denote here by $s_1, \ldots, s_5, t$ the canonical generators of the Coxeter group $W(E_6)$.

49. Theorem. *We keep the notations above, including those of Theorem 48.*

(1) There is an isomorphism $\varphi : W(E_6) \longrightarrow G$ with the following properties:

$$\varphi(s_i) \quad \textit{exchanges } E_i \textit{ and } E_{i+1}, \textit{ and } \varphi(s_i)(E_k) = E_k \textit{ for } k \neq i, i+1,$$

$$\varphi(t) \quad \begin{cases} \textit{permutes } E_i \textit{ and } F_{j,k} \textit{ if } \{i,j,k\} = \{1,2,3\} \\ \textit{fixes } E_i \textit{ if } 4 \leq i \leq 6 \ . \end{cases}$$

(2) The commutator subgroup of G is of index two in G, and it is the simple group of order $25920 = 2^6 3^4 5$.

Proof. See [Har]. For the unicity of the simple group of order 25920, see [Con]. □

50. On the Picard group of a cubic surface. Let X be a smooth cubic surface. Consider the cohomology group $H^2(X, \mathbf{Z})$ together with the intersection form

$$\beta \; : \; H^2(X,\mathbf{Z}) \times H^2(X,\mathbf{Z}) \longrightarrow H^4(X,\mathbf{Z}) \approx \mathbf{Z}$$

and the class $w \in H^2(X, \mathbf{Z})$ of the "canonical divisor". It is known that the group of automorphisms of $H^2(X, \mathbf{Z})$ preserving β and w is the Coxeter group $W(E_6)$.

Each line ℓ in X defines a class $[\ell]$ in $H^2(X, \mathbf{Z})$ and two distinct lines ℓ, ℓ' intersect in X if and only if $\beta([\ell], [\ell']) \neq 0$. Thus the automorphism group of $(H^2(X, \mathbf{Z}), \beta, w)$ acts on the configuration of the 27 lines in X, and one may give a proof of Theorem 49 in these terms.

For this and much more on the cohomology of X, see Manin's book [Man]. For other examples of interplay between Coxeter groups and algebraic surfaces, see [Nik], [Mat] and [CoD].

51. Remarks. (1) It is a natural question to ask whether the group G above acts as a group of transformations of a cubic surface. The answer is no. What I understand from §100 in [Seg] implies that the group of automorphisms of X_c, which depends strongly on $c \in \mathbf{P}^{19}_{\mathbf{C}} - \Delta$, is the trivial group for a generic point c, and is otherwise a finite group of one of the following orders: 2, 4, 6, 12, 24, 54, 108, 120, 648.

(2) One may associate to the 27 lines of a cubic surface the 27 points in $\mathbf{R}^6 = \mathbf{C}^3$ of coordinates

$$\left.\begin{array}{l}(0,\omega^j,-\omega^k)\\(\omega^j,-\omega^k,0)\\(-\omega^k,0,\omega^j)\end{array}\right\} \qquad j,k = 1,2,3\,; \quad \omega = \exp(2i\pi/3)\;.$$

Let P be the convex hull of these points in $\mathbf{R}^6$. The automorphism group of P is again $W(E_6)$; see [Co2] and [Fra].

(3) The 27 lines and the group G appear to have been a must of mathematical education some time ago. For example, G is quite a natural example in the first books on groups: see §III.V in [Jor], Chapter XIV in [Dic], Note H in [Bur] or Chapter XIX in [MBD].

52. Centers and commutator subgroups of Coxeter groups. Let X be a connected Coxeter graph and let $(W(X), S)$ be the corresponding irreducible Coxeter system.

If $W(X)$ is infinite or if X is one of

$$A_n (n \geq 2)\;, \qquad D_{2m+1}\;(m \geq 2)\;, \quad E_6$$

then the center of $W(X)$ is $\{1\}$. In all other cases, the center of $W(X)$ is cyclic of order 2 and is generated by the "longest word" w_0. (See for example Number 12 and Remark 36; for the proof of these facts, see [Bli], Chap. V, §4, Exercise 3.)

Let $W(X)^+$ be the subgroup of $W(X)$ of elements which are products of an even number of generators. If all edges in X have odd weights and in particular if X is one of

$$A_n(n \geq 1) , \qquad D_n(n \geq 4) , \qquad E_n(6 \leq 7 \leq 8)$$

then $W(X)^+$ is the subgroup of $W(X)$ generated by all commutators $w_1 w_2 w_1^{-1} w_2^{-1}$. (See [Bli], Chap. IV, §1, Exercise 9.) For example, by Theorem 49, the group $W(E_6)^+$ is the simple group of order 25920.

53. On E_7. The group $W(E_7)$ is the direct product of its center, which is cyclic of order 2, and of $W(E_7)^+$, which is the unique simple group of order

$$1451520 = 4 \cdot 9! = 2^9 \cdot 3^4 \cdot 5 \cdot 7 .$$

This $W(E_7)^+$ is known to be "the group of automorphisms of the 28 bitangents of the general quartic curve in the projective plane" (quoted from Section 9.6 in [CoM]). It is also known that $W(E_7)$ is the group of automorphisms of the polytope P in $\mathbb{R}^7$, defined as the convex hull of the following 56 points: $(\pm 1, 0, \pm 1, \pm 1, 0, 0, 0)$ and circular permutations; see [Col] and [Fra].

54. On E_8. The center of $W(E_8)$ is again cyclic of order 2, but is now contained in $W(E_8)^+$. The quotient of $W(E_8)^+$ by this group of order 2 is the simple group of order

$$17418200 = 48 \cdot 10! = 2^{12} \cdot 3^5 \cdot 5^2 \cdot 7$$

and is "the group of automorphisms of the 120 tritangent planes of the sextic curve intersection of a cubic surface and a quadric cone" (quoted again from Section 9.6 in [CoM]).

Let us indicate one more appearance of $W(E_8)$, related with a so-called "Del Pezzo surface of degree 1" in the same way that $W(E_6)$ is related to cubic surfaces. (See Number 50 above, and [Man].) Consider the Euclidean space $\mathbb{R}^8$ together with its canonical basis $e_1, \ldots, e_8$. Let Γ_8 be the lattice generated in $\mathbb{R}^8$ by

$$e_i + e_j \quad (1 \leq i, j \leq 8) \quad \text{and} \quad \frac{1}{2}(e_1 + \ldots + e_8)$$

which is the "lattice of Korkine and Zolotareff". The scalar product of $\mathbb{R}^8$ restricts to a bilinear form $\beta : \Gamma_8 \times \Gamma_8 \longrightarrow \mathbb{Z}$ which has the following properties:

it is unimodular, namely the determinant of $(\beta(\alpha_i, \alpha_j))_{1 \leq i,j \leq 8}$ is $\pm$ for any $\mathbb{Z}$-basis $\alpha_1, \ldots, \alpha_8$ of Γ_8,

it is definite, namely $\beta(\xi, \xi) > 0$ for any $\xi \in \Gamma_8 - \{0\}$,

it is of type II, namely $\beta(\xi, \xi)$ is always even.

Moreover, any form β with these three properties on a free abelian group of some rank $n < 16$ is isomorphic to this one (so that in particular $n = 8$). In this context, a *root* is a vector $\xi \in \Gamma_8$ such that $\beta(\xi, \xi) = 2$, and one may show that Γ_8 contains 240 roots.

Let G be the group of automorphisms of the group Γ_8 which preserve β. Then G acts transitively on the 240 roots, and G is isomorphic to $W(E_8)$. The subgroup of G of orientation preserving automorphisms is isomorphic to $W(E_8)^+$. (See Chap. V in [Ser].)

Chapter 3. Poincaré Theorem for Reflection Groups

In 1882, Poincaré found a geometric criterium for a finite set T of isometries of the hyperbolic plane $\mathbb{H}^2$ to generate a so-called Fuchsian group [Po1]. One year later [Po2], he extended this to dimension 3. There are several expositions of these results, either in particular cases such as triangle groups or surface groups (see §2 in [Mi1] or Section 3.9 in [Sie]), or in more general cases (see [Rha], [Ma1], [Sei], [MaS], or Chapter 9 in [Bea]). Section IV.F in Maskit's book [Ma2] contains a proof which is explicitely written for any dimension $n \geq 2$.

The purpose of this Chapter is to expose Poincaré theorem in the particular case where the given set of generators (T above) is a set S of reflections. The resulting groups provide numerous examples of Coxeter groups, finite as well as infinite, acting properly discontinuously and by isometries on familiar spaces. The exposition below has been strongly influenced by A. Haefliger; it follows partly [Ma2]. Among many good introductions to hyperbolic geometry, we quote [Ree], for dimension 2, Chapter 19 of [Ber], for all dimensions, and [Mi2], for some history.

3.1. Statement and Proof of the Theorem

55. Reflections. We denote by X a complete smooth Riemannian manifold of some dimension $n \geq 2$, from the following list:

the Euclidean space $\mathbb{E}^n$,

the unit sphere $\mathbb{S}^n$ in $\mathbb{E}^{n+1}$,

the hyperbolic space $\mathbb{H}^n$.

There is a natural notion of *hyperplane* in X, and thus also of open and closed *half-spaces*. A *reflection* of X is an isometry r of X of order 2 such that the set X^r of its fixed points is a hyperplane (see Number 1). There is a bijection $r \longleftrightarrow X^r$ between the set of reflections of X and the set of hyperplanes in X.

56. Angles. Let s, t be two distinct reflections of X such that $X^s \cap X^t \neq \emptyset$; it follows that $X^s \cap X^t$ is of codimension 2 in X. Let X^s_+ [respectively X^t_+] be one of the two open half-spaces limited by X^s [resp. X^t]. Consider a point $x \in X^s \cap X^t$, a non zero tangent vector $\mathcal{V}^s_x$ at x which is normal to X^s and points towards X^s_+, and similarly a non zero vector $\mathcal{V}^t_x$ normal to X^t pointing towards X^t_+. Then the angle β of $\mathcal{V}^s_x$ and $\mathcal{V}^t_x$ is independent of the choice of x in $X^s \cap X^t$; one has $0 < \beta < \pi$. The *angle* of X^s_+ and X^t_+ is defined to be $\alpha = \pi - \beta$, as in Figure 6.

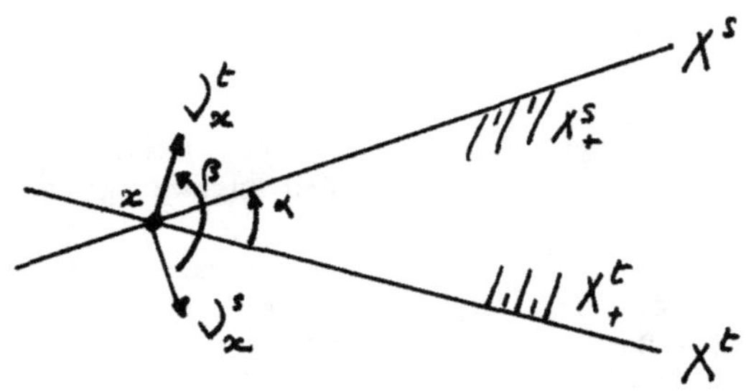

Figure 6.

57. Reflection data. Let S be a countable family of reflections of X, each $s \in S$ being given together with a distinguished open half-space X^s_+ limited by X^s. We say that S and the family $(X^s_+)_{s\in S}$ constitute *reflection data* in X if the four following conditions hold

(1) The family $(X^s)_{s\in S}$ is *locally finite.* This means that, for any compact subset K of X, there are only finitely many $s \in S$ such that $K \cap X^s \neq \emptyset$. (In most applications below, the set S is indeed finite.)

(2) The so-called *fundamental chamber* $C = \bigcap_{s\in S} X^s_+$ is non empty. (It follows from (1) and (2) that C is open in X.)

(3) For each $s \in S$, the intersection $\overline{C} \cap X^s$ has a non empty interior in X^s. This is equivalent (exercise) to the condition $C \subsetneqq \bigcap_{s\in S'} X^s_+$ for any proper subset S' of S.

(4) For each pair $(s,t) \in S\times S$ such that codim $(\overline{C} \cap X^s \cap X^t) = 2$, there is an integer $m_{s,t} \geq 2$ such that the angle of X^s_+ and X^t_+ is $\frac{\pi}{m_{s,t}}$.

One sets moreover $m_{s,s} = 1$ for each $s \in S$, and $m_{s,t} = \infty$ for each $(s,t) \in S{\times}S$ such that $\overline{C} \cap X^s \cap X^t$ is either empty or of codimension at least 3. The *Coxeter matrix* associated to reflection data as above is the matrix $M = (m_{s,t})_{s,t\in S}$. Even though the set S need not be finite, one associates to M the *Coxeter graph* $\mathcal{G}_M$ and the corresponding *Coxeter system* as in Number 13.

Let (s,t) be as in Condition (4) and let $x \in \overline{C} \cap X^s \cap X^t$. Then s and t induce in the tangent space $T_xX \approx \mathsf{E}^n$ two reflections with respect to hyperplanes making an angle $\frac{\pi}{m_{s,t}}$. It follows that st is an isometry of order $m_{s,t}$ exactly, and that s, t generate a dihedral group of order $2m_{s,t}$ (see Exercise 21).

58. The developing map. Let S and $(X_+^s)_{s \in S}$ be reflection data in X, and let M be the corresponding Coxeter matrix. Let $\widehat{S}$ be a set given together with a bijection $\widehat{s} \mapsto s$ from $\widehat{S}$ onto S. We define successively:

(1) The abstract Coxeter system $(\widehat{W}, \widehat{S})$ associated to M.

(2) The group W of isometries of X generated by S.

(3) The surjective homomorphism $\varphi : \widehat{W} \to W$ such that $\varphi(\widehat{s}) = s$ for all $\widehat{s} \in \widehat{S}$.

(4) The topological space $\widehat{W} \times \overline{C}$, with the product of the discrete topology on $\widehat{W}$ and of the topology induced on $\overline{C}$ by X, and the second projection $pr : \widehat{W} \times \overline{C} \to \overline{C}$.

(5) The equivalence relation on $\widehat{W} \times \overline{C}$ generated by $(\widehat{w}, x) \sim (\widehat{w}\widehat{s}, x)$ for $\widehat{w} \in \widehat{W}$, $\widehat{s} \in \widehat{S}$ and $x \in \overline{C} \cap X^s$. The quotient space (with the quotient topology) is denoted by $\widehat{X}$ and the canonical map by

$$r \begin{cases} \widehat{W} \times \overline{C} & \longrightarrow & \widehat{X} \\ (\widehat{w}, x) & \longmapsto & [\widehat{w}, x] \end{cases} .$$

The projection pr factors via a map denoted by

$$p \begin{cases} \widehat{X} & \longrightarrow & \overline{C} \\ [\widehat{w}, x] & \longmapsto & x \end{cases} .$$

(6) The developing map

$$D \begin{cases} \widehat{X} & \longrightarrow & X \\ [\widehat{w}, x] & \longmapsto & \varphi(\widehat{w})x \end{cases} .$$

All this is summarized in the following diagram.

$$\begin{array}{ccccc} & & \widehat{W} & \xrightarrow{\varphi} & W \\ \widehat{W} \times \overline{C} & \xrightarrow{r} & \widehat{X} & \xrightarrow{D} & X \\ & \searrow^{pr} & \downarrow p & & \\ & & \overline{C} & & \end{array}$$

Observe that $\widehat{W}$ acts on the left by homeomorphisms on $\widehat{X}$, by $\widehat{w}'[\widehat{w}, x] = [\widehat{w}'\widehat{w}, x]$, that W acts by isometries on X, and that D is φ-equivariant.

59. Theorem. *Let S and $(X_+^s)_{s\in S}$ be reflection data on X. With the notations above, one has:*

the developing map D is a homeomorphism,

the homomorphism φ is an isomorphism.

The first claim implies that the open chamber C is a *fundamental domain* for the action of W on X: this means here that $wC \cap w'C = \emptyset$ for distinct elements $w, w' \in W$ and that $X = \bigcup_{w\in W} w\overline{C}$; in particular, the action of W on X is *properly discontinuous*. (These statements hold for W and X because they are obvious for $\widehat{W}$ and $\widehat{X}$.) The second claim implies that W is a Coxeter group. Consequently, Theorem 59 implies the following:

60. Poincaré theorem for reflection groups. *Let S and $(X_+^s)_{s\in S}$ be reflection data on X. Then, with the notations as above:*

(i) The group W generated by S acts properly discontinuously on X, and the chamber $C = \bigcap_{s\in S} X_+^s$ is a fundamental domain for this action.

(ii) (W, S) is a Coxeter system associated to the Coxeter matrix $(m_{s,t})_{s,t\in S}$.

The proof of Theorem 59, below as well as in [Ma2], goes by induction on the dimension n of X. For each dimension $n \geq 3$, the proof uses the result for $\mathbf{S}^{n-1}$; in particular, the proof for the hyperbolic case requires a proof which holds for the spherical case.

61. Facettes of the fundamental chamber. Consider an integer $k \in \{1, \ldots, n\}$ and a subset $\{s_1, \ldots, s_k\}$ of S of cardinal k such that

$$\operatorname{codim}\left(\overline{C^{s_1,\ldots,s_k}}\right) = k \quad \text{where} \quad \overline{C^{s_1,\ldots,s_k}} = \overline{C} \cap X^{s_1} \cap \ldots \cap X^{s_k} .$$

The interior $C^{s_1,\ldots,s_k}$ of $\overline{C^{s_1,\ldots,s_k}}$ in $X^{s_1} \cap \ldots \cap X^{s_k}$ is a $(n-k)$-*facette* of C; it is homeomorphic to an open disc of dimension $n-k$ if $1 \leq k \leq n-1$, and it is reduced to a point called a *vertex* of C if $k = n$. A $(n-1)$-*facette* is also called a *side* or a *face*, and a $(n-2)$-*facette* an *edge*. It is useful to consider C as an n-*facette* of itself. The closed chamber $\overline{C}$ is then the union of its k-*facettes* for $k \in \{0, 1, \ldots, n\}$. (See Number 23.)

To each facette $C^{s_1,\ldots,s_k}$ of C one may associate the subgroup $W_{\{s_1,\ldots,s_k\}}$ of W generated by $\{s_1, \ldots, s_k\}$ and the subgroup $\widehat{W}_{\{s_1,\ldots,s_k\}}$ of $\widehat{W}$ generated by $\{\widehat{s}_1, \ldots, \widehat{s}_k\}$. Observe that $\varphi : \widehat{W} \to W$ maps $\widehat{W}_{\{s_1,\ldots,s_k\}}$ onto $W_{\{s_1,\ldots,s_k\}}$, and that elements of $W_{\{s_1,\ldots,s_k\}}$ act as the identity on points of $C^{s_1,\ldots,s_k}$. In case $k = 0$, both $\widehat{W}_\emptyset$ and $W_\emptyset$ are reduced to one element; if $k = 1$, both $\widehat{W}_{\{s\}}$ and $W_{\{s\}}$

are of order 2, for each $s \in S$.

62. Statement of four intermediate claims. Let $x \in \overline{C}$. During the proof, we shall need a small enough neighbourhood U_x of x in $\overline{C}$; for definiteness, we choose it as follows.

Let k be the integer in $\{0, 1, \ldots, n\}$ and let F be the k-facette of C such that $x \in F$. Let ϱ_x be the smallest distance (for the Riemannian metric on X) between x and a point x' in some k'-facette F' on C, where $k' \leq k$ and $F' \neq F$. The conditions of Number 57 imply that $\varrho_x > 0$. We define

$$U_x = \{y \in \overline{C} \ : \ d(x,y) < \frac{1}{2}\varrho_x\}$$

to be the open ball in $\overline{C}$ (*not* in X !) of center x and of radius $\frac{1}{2}\varrho_x$.

If $x \in F = C^{s_1,\ldots,s_k}$, we write $\widehat{W}_x$ for $\widehat{W}_{\{s_1,\ldots,s_k\}}$ and W_x for $W_{\{s_1,\ldots,s_k\}}$. We denote by

$$\varphi_x \ : \ \widehat{W}_x \longrightarrow W_x$$

the restriction of φ to $\widehat{W}_x$.

One has obviously $pr^{-1}(U_x) = \widehat{W} \times U_x$. Thus $p^{-1}(U_x) = r(\widehat{W} \times U_x)$. This subset $p^{-1}(U_x)$ of $\widehat{X}$ is not connected: it has one distinguished connected component which is $r(\widehat{W}_x \times U_x)$; the other ones are obtained by the action of $\widehat{W}$ on $\widehat{X}$, and are consequently naturally indexed by the coset space $\widehat{W}/\widehat{W}_x$.

Let $W_x U_x$ denote the set of all points in X of the form wy for some $w \in W_x$ and $y \in U_x$. We denote by

$$D_x \ : \ r(\widehat{W}_x \times U_x) \longrightarrow W_x U_x$$

the restriction of D to $r(\widehat{W}_x \times U_x)$.

Claim (i): φ_x is an isomorphism of groups, for all $x \in \overline{C}$.

Claim (ii): D_x is a homeomorphism between two open n-discs, for all $x \in \overline{C}$.

It follows from Claim (ii) that $\widehat{X}$ is locally homeomorphic to X (namely to $\mathbf{R}^n$). Thus $\widehat{X}$ is a manifold with charts of the form

$$\varphi(\widehat{w}) D_x \widehat{w}^{-1} \ : \ r(\widehat{w}\widehat{W}_x \times U_x) \longrightarrow \varphi(\widehat{w}) W_x U_x$$

and the transition maps are appropriate restrictions of elements of W, namely of isometries of X. Thus $\widehat{X}$ inherits via the D_x 's a Riemannian structure from that of X.

Claim (iii): The Riemannian manifold $\widehat{X}$ is Hausdorff, connected and complete.

Claim (iv): The developing map $D : \widehat{X} \to X$ is a covering.

Observe that the first claim of Theorem 59, according to which D is a homeomorphism, follows from Claim (iv) because X is simply connected.

For $x \in C$, we have already observed that both $\widehat{W}_x$ and W_x are reduced to one element. Thus the connected components of $p^{-1}(U_x)$ are $r(\{\widehat{w}\} \times U_x)$, for $\widehat{w} \in \widehat{W}$. As the homeomorphism D restricts to a φ-equivariant bijection from $p^{-1}(x)$ to Wx, it follows that the homomorphism $\varphi : \widehat{W} \to W$ is indeed an isomorphism, and also that the map $w \mapsto wx$ from W to X is injective.

Thus Theorem 59 follows from Claims (i) to (iv) above.

Consider again $x \in C$. Claim (i) is obvious because $\widehat{W}_x \approx W_x \approx \{1\}$. As U_x is an open n-disc, Claim (ii) is also obvious in this case.

Consider now a generator $s \in S$ and a point x in the corresponding face C^s. Both $\widehat{W}_x$ and W_x are of order 2 and Claim (i) is again straightforward. The neighbourhood U_x is a closed half of an open n-disc limited by an hyperplane through the center of this disc. Both $r(\widehat{W}_x \times U_x)$ und W_xU_x are open discs obtained by gluing two copies of U_x along their boundaries, and D_x is again clearly a homeomorphism.

Thus, for Claims (i) and (ii), it is sufficient hereafter to consider points of $\overline{C}$ which are in facettes of codimensions at least 2.

63. Claims (i) and (ii) in dimension $n=2$. The closed chamber $\overline{C}$ in dimension 2 is a disjoint union of C, of its one-dimensional sides indexed by S and of its vertices (=edges) indexed by pairs (s,t) with $m_{s,t} < \infty$. (These may be absent if $|S|=1$ or more generally if $X^s \cap X^t = \emptyset$ whenever $s \neq t$.)

We have already observed (end of Number 57) that W_x is a dihedral group of order $2m_{s,t}$ for any vertex x of C, and it is straightforward to check that $|\widehat{W}_x| \leq 2m_{s,t}$. Claim (i) follows.

The neighbourhood U_x of the vertex x in $\overline{C}$ is a closed sector of angle $\pi/m_{s,t}$ in an open disc. Both $r(\widehat{W}_x \times U_x)$ and W_xU_x are open discs obtained by gluing $2m_{s,t}$ of these sectors. Thus Claim (ii) is also straightforward, because it reduces to the situation of two reflections in $\mathbb{E}^2$.

64. Claim (ii) implies Claim (iii). Let us first show that $\widehat{X}$ is Hausdorff. Let $\widehat{x}_1, \widehat{x}_2$ be two distinct points in $\widehat{X}$. Set $x_j = p(\widehat{x}_j) \in \overline{C}$ and choose $\widehat{w}_j \in \widehat{W}$ such that $\widehat{x}_j = r(\widehat{w}_j, x_j)$, for $j=1,2$.

If $x_1 = x_2 = x$ (say), set $\widehat{U}_j = r(\widehat{w}_j\widehat{W}_x \times U_x)$ and observe that $\widehat{x}_j \in \widehat{U}_j$ for

$j=1,2$. As $\widehat{w}_1\widehat{W}_x \cap \widehat{w}_2\widehat{W}_x = \emptyset$ (because $\widehat{x}_1 \neq \widehat{x}_2$), one has $\widehat{U}_1 \cap \widehat{U}_2 = \emptyset$. If $x_1 \neq x_2$, let d be the Riemannian distance between x_1 and x_2 in $\overline{C}$. Let U_j be the open ball in $\overline{C}$ of center x_j and of radius $\frac{1}{2}d$, let $\widehat{U}_j = p^{-1}(U_j)$, and observe that $\widehat{x}_j \in \widehat{U}_j$, for $j=1,2$. As $U_1 \cap U_2 = \emptyset$ one has $\widehat{U}_1 \cap \widehat{U}_2 = \emptyset$.

The connectedness of $\widehat{X} = r(\widehat{W} \times \overline{C})$ follows from the connectedness of $\overline{C}$ and from the fact that $\widehat{S}$ generates $\widehat{W}$.

Let us finally show that the Riemannian manifold $\widehat{X}$ is complete. Let $(\widehat{x}_j)_{j\geq 1}$ be a Cauchy sequence in $\widehat{X}$. For each $j \geq 1$, set $x_j = p(\widehat{x}_j) \in \overline{C}$. As p contracts distances and as $\overline{C}$ is complete, $(x_j)_{j\geq 1}$ converges to some point of $\overline{C}$, say x. For j large enough, one has $x_j \in U_x$. As each component of $p^{-1}(U_x)$ is of the form $r(\widehat{w}\widehat{W}_x \times U_x)$ for some $\widehat{w} \in \widehat{W}$, there exists $\widehat{w}_j \in \widehat{W}$ such that $\widehat{x}_j \in r(\widehat{w}_j\widehat{W}_x \times U_x)$. Using again the fact that $(\widehat{x}_j)_{j\geq 1}$ is a Cauchy sequence, we see that there exists $\widehat{w} \in \widehat{W}$ such that $\widehat{w}_j\widehat{W}_x = \widehat{w}\widehat{W}_x$, and one may write $\widehat{x}_j = r(\widehat{w}, x_j)$, for j large enough. As $D_x\widehat{w}^{-1}$ is an isometry from $r(\widehat{w}\widehat{W}_x \times U_x)$ onto W_xU_x by Claim (ii), it is sufficient to check that the sequence of general term $D_x\widehat{w}^{-1}r(\widehat{w}, x_j) = x_j$ converges, but this is obvious. Thus the sequence $(\widehat{x}_j)_{j\geq 1}$ is convergent.

65. Lemma. *Let $\widehat{M}, M$ be two Riemannian manifolds of the same dimension and let $D : \widehat{M} \to M$ be a local isometry. If $\widehat{M}$ is complete and if M is connected, then D is a covering.*

Proof. The first step is to observe that geodesics in M can be lifted to $\widehat{M}$. Indeed, let $\gamma : [0,\ell] \to M$ be a geodesic parametrized by arc length and let $\widehat{m} \in \widehat{M}$ be such that $D(\widehat{m}) = \gamma(0)$. As D is a local isometry around $\widehat{m}$, we can lift some initial part of γ to a geodesic $\widehat{\gamma} : [0,\widehat{\ell}[\to \widehat{M}$ such that $\widehat{\gamma}(0) = \widehat{m}$, for $\widehat{\ell}$ small enough in $]0,\ell]$. Choose $\widehat{\ell}$ maximal with these properties. As $\widehat{M}$ is complete, one can also define $\widehat{\gamma}(\widehat{\ell}) = \lim_{t\to\widehat{\ell}}\widehat{\gamma}(t)$ and thus extend $\widehat{\gamma}$ to a geodesic $[0,\widehat{\ell}] \to \widehat{M}$. Now $\widehat{\ell} = \ell$ and $\widehat{\gamma}$ is a lift of γ itself. (If $\widehat{\ell} < \ell$, as D is a local isometry around $\widehat{\gamma}(\widehat{\ell})$, one could extend $\widehat{\gamma}$ to some $[0,\widehat{\ell}+\varepsilon[$, and this is impossible by maximality of $\widehat{\ell}$.)

Though it is useless for the proof, one may observe here that the previous argument shows that M is necessarily complete.

The second step is to show that $D(\widehat{M}) = M$. As D is a local homeomorphism, $D(\widehat{M})$ is open in M. It remains to check that $D(\widehat{M})$ is also closed. Let $x \in \overline{D(\widehat{M})}$. Let B_x be an open ball around x in M such that any pair of points in B_x can be joined by a geodesic segment (see e.g., Theorem I.9.9 in [Hel]). Let $y \in B_x \cap D(\widehat{M})$ and let γ be a geodesic segment from y to x. Choose $\widehat{y} \in \widehat{M}$ such that $D(\widehat{y}) = y$, let $\widehat{\gamma}$ be the geodesic lift of γ in $\widehat{M}$ starting from $\widehat{y}$, and denote by $\widehat{x}$ the end of

$\widehat{\gamma}$. Then $x = D(\widehat{x}) \in D(\widehat{M})$, and $D(\widehat{M})$ is indeed closed.

Consider finally some point $x \in M$. Choose again an open ball B_x as above; then B_x is the image by the exponential map at x of some open ball A_x around the origin in the tangent space $T_x M$ to M at x. As D is a local homeomorphism, $D^{-1}(x)$ is countable and we may write $D^{-1}(x) = (\widehat{x}_j)_{j \in J}$ for some countable index set J.

For each $j \in J$, any geodesic segment from x to some point $y \in B_x$ can be lifted to a geodesic segment from $\widehat{x}_j$ to some point $\widehat{y}_j \in \widehat{M}$. When y describes B_x, these $\widehat{y}_j$ describe a subset $\widehat{B}_{x,j}$ of $\widehat{M}$ containing $\widehat{x}_j$. As $\widehat{B}_{x,j}$ is the image by the exponential map at $\widehat{x}_j$ of the ball $(T_{\widehat{x}_j} D)^{-1}(A_x)$ of $T_{\widehat{x}_j} \widehat{M}$, this $\widehat{B}_{x,j}$ is an open ball in $\widehat{M}$, and the map D provides by restriction a map $D_{x,j} : \widehat{B}_{x,j} \to B_x$ which is a homeomorphism with inverse $y \mapsto \widehat{y}_j$.

Thus $D^{-1}(B_x) = \bigcup_{j \in J} B_{x,j}$, and it remains to check that the $B_{x,j}$ are pairwise disjoint. Let $j, k \in J$ and set

$$E = \{y \in B_x \ : \ \widehat{y}_j = \widehat{y}_k\} .$$

As the lifts

$$\lambda_j \ : \ \begin{cases} B_x & \longrightarrow & B_{x,j} \\ y & \longmapsto & \widehat{y}_j \end{cases} \qquad \text{and} \qquad \lambda_k \ : \ \begin{cases} B_x & \longrightarrow & B_{x,k} \\ y & \longmapsto & \widehat{y}_k \end{cases}$$

are continuous, E is closed in B_x. As $D\lambda_j = D\lambda_k$ and as D is a local homeomorphism, E is also open in B_x. Thus, as B_x is connected, either $E = B_x$, in which case $B_{x,j} = B_{x,k}$ and $j = k$, or $E = \emptyset$, in which case $B_{x,j} \cap B_{x,k} = \emptyset$. □

66. Remark. The proof of Theorem 59 (and thus also Theorem 60) is complete for the dimension $n = 2$. Assume from now on that $n \geq 3$. We have yet to prove Claims (i) and (ii) of Number 62 for n, and we assume inductively that Theorem 59 holds for dimension $n-1$.

67. The induction step. Let $x \in \overline{C}$. Notations being as in Number 62, the point x is in a facette $C^{s_1,\dots,s_k}$ for some $k \in \{0, 1, \dots, n\}$. Consider a positive number ϱ such that $0 \leq \varrho < \varrho_x$ as well as

the sphere X'_ϱ in X of center x and radius ϱ,

the set $S' = \{s_1, \dots, s_k\}$ of generators of W_x,

the half-space $X'^s_+ = X^s_+ \cap X'_\varrho$ of X' for each $s' \in S'$.

Then S' and $(X'^s_+)_{s \in S'}$ constitute (for $\varrho > 0$) reflection data in X'_ϱ. The corresponding fundamental chamber $C'_\varrho = \bigcap_{s \in S'} X'^s_+$ is equal to $C \cap X'_\varrho$, and the Coxeter matrix M' is equal to the submatrix $(m_{s,t})_{s,t,\in S'}$ of $M=(m_{s,t})_{s,t\in S}$. One may define as in Number 58

the product $\widehat{W}_x \times \overline{C'_\varrho}$, which is a subspace of $\widehat{W} \times \overline{C}$,

the quotient $\widehat{X'_\varrho} = r'(\widehat{W}_x \times \overline{C'_\varrho})$, a subspace of $\widehat{X}$,

the developing map $D'_\varrho : \widehat{X}'_\varrho \to X'_\varrho$, a restriction of D.

Theorem 59 shows that $\varphi_x : \widehat{W}_x \to W_x$ is an isomorphism, and this implies Claim (i) of Number 62. (It implies also the following: the canonical map from the abstract Coxeter group defined by the matrix M' to the abstract Coxeter group $\widehat{W}$ defined by M is an injection.) Theorem 59 shows also that D'_ϱ is a homeomorphism from $\widehat{X'_\varrho}$ onto the sphere X'_ϱ. It follows that

$$r(\widehat{W}_x \times U_x) = \bigcup_{0\leq\varrho<\varrho_x} \widehat{X'_\varrho} \qquad \text{and} \qquad W_x U_x = \bigcup_{0\leq\varrho<\varrho_x} X'_\varrho$$

are open discs, in $\widehat{X}$ and X respectively, and that

$$D_x \ : \ r(\widehat{W}_x \times U_x) \longrightarrow W_x U_x$$

is a homeomorphism. This is Claim (ii) of Number 62.

The proofs of Theorems 59 and 60 are now complete.

68. Remarks. (i) As an exercise, the reader should state and prove analogues of Theorems 59 and 60 for $X=\mathbb{E}^1$ and $X=\mathbb{S}^1$.

(ii) One should carefully distinguish Poincaré theorem from other results on groups on transformations where some group generated by "symmetries" is *assumed* to act properly on some manifold: see e.g., [Kol] and [Gut].

(iii) In condition (4) for reflection data (Number 57), it is definitely *not* sufficient to assume that the angle between two faces X^s_+, X^t_+ is a rational multiple of π, as easy examples in $\mathbb{H}^2$ show (see e.g., Propositions III.6 and III.8 in [GHJ]).

(iv) The fact that the closed chamber $\overline{C}$ is complete is (obvious and) crucial above: see e.g., Number 64. In the general form of Poincaré theorem, when sides of C are paired by isometries which are not necessarily reflections, it is a delicate point to find good criteria for the completion of the appropriate quotient of $\overline{C}$.

69. Corollary. *Let S and $(X^s_+)_{s\in S}$ be reflection data in X, and assume that $s,t \in S$ are such that* $\operatorname{codim}(\overline{C}\cap X^s \cap X^t)\geq 3$. *Then* $X^s \cap X^t = \emptyset$.

Proof. It follows from Poincaré theorem that st is of infinite order, so that X^s and X^t cannot meet inside X. □

Corollary 69 shows for example that the fundamental chamber cannot be combinatorially equivalent to a pyramid with quadrilateral basis.

70. Problem. Consider the situation of Poincaré theorem. Let TX be the tangent bundle of X. For each reflection $r \in W$, one identifies $T(X^r)$ to a submanifold of TX of codimension 2. Set $\Sigma = \bigcup T(X^r)$, where the union is over all reflections in W. Show that the fundamental group B_W of $TX - \Sigma$ has a presentation with set of generators S and set of relations

$$\left\{(st)^{m(s,t)} \quad : \; s,t \in S \text{ with } s \neq t \text{ and } m_{s,t} < \infty\right\} .$$

The group B_W is the *braid group* of W; see [Mag] and [Lin].

3.2. Examples

The most elementary illustration of Poincaré theorem is that where S is reduced to one generator; see Example 14. The next case is that where $S = \{s,t\}$ has two generators, so that W is a dihedral group. Finite dihedral groups are found acting on any of $\mathbb{S}^n (n \geq 1)$, $\mathbb{E}^n$ or $\mathbb{H}^n$ $(n \geq 2)$, and the corresponding Coxeter graphs are the graphs $A_2, B_2, I_2(5), G_2, I_2(p)$ for $p \geq 7$, of Theorem 45. The infinite dihedral group is found acting on any of $\mathbb{E}^n (n \geq 1)$ or $\mathbb{H}^n (n \geq 2)$ and the corresponding Coxeter graph is $\bullet\overset{\infty}{\text{———}}\bullet$. See Example 15. The next three numbers give more interesting examples of finite groups.

71. Type $A_n (n \geq 1)$. Consider in $X = \mathbb{E}^{n+1}$ the fundamental chamber

$$C = \left\{(x_1, \ldots, x_{n+1}) \in \mathbb{E}^{n+1} \; : \; x_{n+1} \geq x_n \geq \ldots \geq x_2 \geq x_1\right\}$$

of which the intersection with the hyperplane of equation $x_1 + \ldots + x_{n+1} = 1$ appears already in Section 1.1. This C has n faces of equations $x_j = x_{j+1}$ corresponding to the transpositions $s_j = (j, j+1)$ of the symmetric group σ_{n+1}, for $j = 1, \ldots, n$. Thus Poincaré theorem provides a new proof of the isomorphism

$$W(A_n) \longrightarrow \sigma_{n+1}$$

of Proposition 8.

72. Type $B_n (n \geq 2)$ and type $D_n (n \geq 4)$. Consider first $X = \mathbb{E}^n (n \geq 2)$ and the fundamental chamber

$$C = \left\{(x_1, \ldots, x_n) \in \mathbb{E}^n \; : \; x_n \geq x_{n-1} \geq \cdots \geq x_1 \geq 0\right\} .$$

(For $n = 3$, the intersection $C \cap \mathbb{S}^2$ appears already in Figure 3; see Number 29.) Then C has n faces of equations

$$x_1 = 0 \qquad \text{and} \qquad x_j = x_{j+1} \;\; (j = 2, \ldots, n-1) .$$

The corresponding reflection data give rise to the Coxeter graph B_n, so that Poincaré theorem provides a new proof of the isomorphism

$$W(B_n) \longrightarrow C_2^n \rtimes \sigma_n$$

of Proposition 34.

Similarly, for $n \geq 4$, the chamber

$$C = \left\{(x_1, \dots, x_n) \in \mathsf{E}^n \ : \ x_n \geq x_{n-1} \geq \dots \geq x_2 \geq x_1 \quad \text{and} \quad x_2 + x_1 \geq 0\right\}$$

corresponds to the Coxeter graph D_n, and Poincaré theorem provides a proof of the isomorphism

$$W(D_n) \longrightarrow C_2^{n-1} \rtimes \sigma_n$$

of Exercise 37.

73. Exercise (see the tables in [Bli]). In $X = \mathsf{E}^8$, show that the fundamental chamber defined by the equations

$$\begin{aligned} &x_1 \leq x_2 \leq x_3 \leq x_4 \leq x_5 \\ &x_1 + x_2 \geq 0 \\ &x_1 + x_8 \geq x_2 + x_3 + x_4 + x_5 + x_6 + x_7 \end{aligned}$$

gives rise to a Coxeter group of type E_6. Also, the fundamental chamber defined by these equations and $x_5 \leq x_6$ [respectively $x_5 \leq x_6 \leq x_7$] gives rise to a Coxeter group of type E_7 [resp. E_8].

In $X = \mathsf{E}^4$, the fundamental chamber defined by the equations

$$x_2 \geq x_3 \geq x_4 \geq 0 \qquad \text{and} \qquad x_1 \geq x_2 + x_3 + x_4$$

gives rise to a Coxeter group of type F_4.

74. Euclidean groups in dimension 2. Let $\mathcal{P}$ be a regular tesselation of the plane E^2 by squares. Choose one tesselating square, consider the three lines ℓ_0, ℓ_1, ℓ_2 indicated in Figure 7, and let s_0, s_1, s_2 be the corresponding reflections. Poincaré theorem shows that s_0, s_1, s_2 generate a group $G(\mathcal{P})$ of isometries of E^2, with fundamental domain the triangular chamber shown in Figure 7, and which is the Coxeter group defined by the Coxeter graph $\bullet\overset{4}{—}\bullet\overset{4}{—}\bullet$ of type $\widetilde{B}_2$.

Similarly, let $\mathcal{Q}$ be a regular tesselation of E^2 by equilateral triangles. The reflections t_0, t_1, t_2 with respect to the lines m_0, m_1, m_2 shown in Figure 8 generate a group $G(\mathcal{Q})$ associated to the Coxeter graph $\bullet\overset{6}{—}\bullet—\bullet$ of type $\widetilde{G}_2$. The reflections with respect to the three sides of one triangle of $\mathcal{Q}$ generate a group of index 6 in $G(\mathcal{Q})$ which is associated to the Coxeter graph (triangle) of type $\widetilde{A}_2$.

75. Exercise. The notations are as in the previous number.

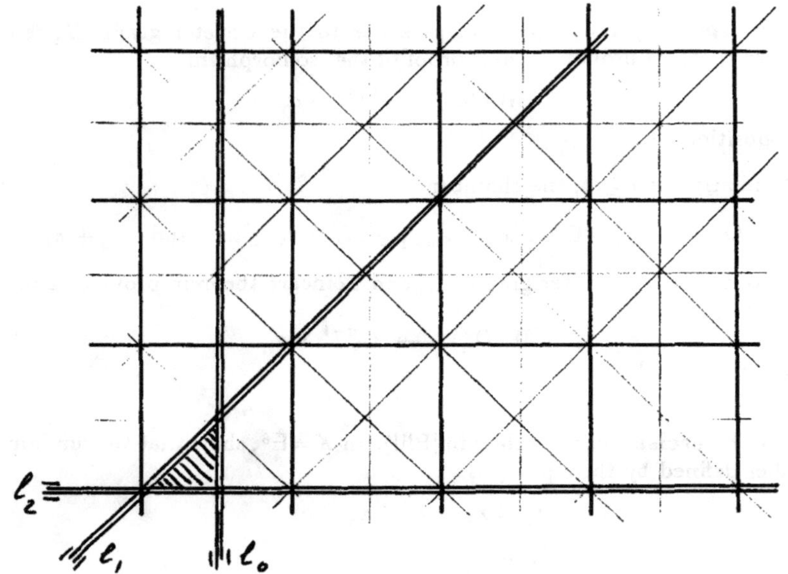

Figure 7.

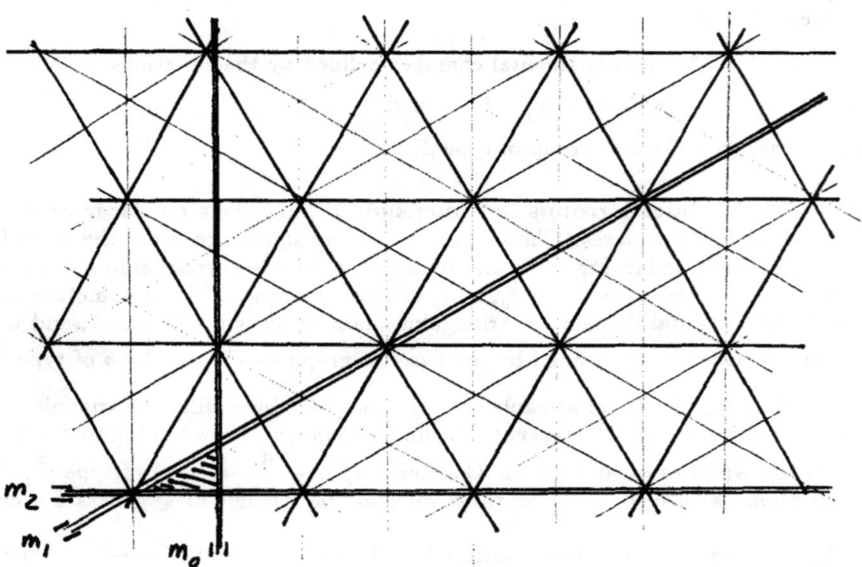

Figure 8.

(i) Show that $G(\mathcal{P})$ contains all the isometries of E^2 which leave the tessela-tion $\mathcal{P}$ invariant.

(ii) Check that the counting function for the triangular chambers of Figure 7 is given by

$$\sum t^{d(C,C_0)} = 1 + 3t + 5t^2 + 8t^3 + 11t^4 + 13t^5 + 16t^6 + \dots$$
$$= \frac{1+t}{1-t} \; \frac{1+t+t^2+t^3}{1-t^3}$$

where the summation is over all chambers C of $\mathcal{P}$, where C_0 is some fundamental chamber, and where $d(C, C_0)$ denotes the combinatorial distance between C_0 and C, which is also the minimum number of generators needed to write that element $w \in G(\mathcal{P})$ such that $wC_0 = C$. This is also the growth series of $W(\widetilde{B}_2)$; see [Pa2].

(iii) Show that $G(\mathcal{Q})$ contains all the isometries of E^2 which leave the tesse-lation $\mathcal{Q}$ invariant.

(iv) Check that the counting function for the equilateral triangles of $\mathcal{Q}$ is given by

$$1 + 3\sum_{k=1}^{\infty} k\, t^k = \frac{1+t}{1-t} \; \frac{1+t+t^2}{1-t^2} \; .$$

This is also the growth series of $W(\widetilde{A}_2)$; see [Pa2].

(v) Check that the counting function for the "small" triangular chambers of Figure 8 is given by

$$1 + 3t + 5t^2 + 7t^3 + 9t^4 + 12t^5 + 15t^6 + 17t^7 + 19t^8 \dots$$
$$= \frac{1+t}{1-t} \; \frac{1+t+t^2+t^3+t^4+t^5}{1-t^5}$$

This is also the growth series of $W(\widetilde{G}_2)$; see [Pa2].

76. Euclidean groups in dimension 3. Let now $\mathcal{P}$ denote a regular tesselation of E^3 by cubes. Choose such a cube P_0, one tetrahedron C_0 of the barycentric subdivision of this cube, and consider the group $G(\mathcal{P})$ generated by the reflections with respect to the four sides of C_0. Let s_i denote the reflection whose fixed points contain the face of C_0 opposed to the barycenter $A_i \in C_0$ of a facette of dimension i in the cube $(i = 0, 1, 2, 3)$. Then $G(\mathcal{P})$ is the Coxeter group associated to the diagram

s_0 —4— s_1 —— s_2 —4— s_3

of type $\widetilde{C}_3$. It can be checked that $G(\mathcal{P})$ is the group of all isometries of E^3 leaving $\mathcal{P}$ invariant.

Set $A_0' = s_0(A_0)$; thus A_0' is the vertex of P_0 such that A_1 is the middle of A_0A_0'. Then A_0, A_0', A_2, A_3 are the vertices of a tetrahedron giving rise to a subgroup of index two in $G(\mathcal{P})$, which is the Coxeter group associated to the diagram

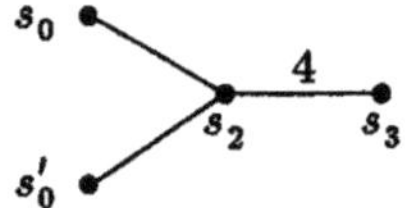

of type $\widetilde{B}_3$. Set finally $A_3' = s_3(A_3)$. Then A_0, A_0', A_3, A_3' are the vertices of a third tetrahedron giving rise to a subgroup of index two in the previous group (thus of index four in $G(\mathcal{P})$), which is associated to the diagram

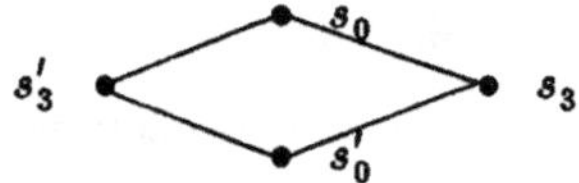

of type $\widetilde{A}_3$.

77. Other regular Euclidean tesselations. A tesselation of E^n is here a family $\mathcal{P}=(P_i)_{i\in I}$ of polytopes which cover E^n without overlapping, each $(n-1)$-face of one P_i being a face of exactly one other P_j. As for polytopes, there is a notion of *flag* for such a tesselation (see Number 23). Then $\mathcal{P}$ is defined to be *regular* if the group $G(\mathcal{P})$ of all isometries of E^n preserving $\mathcal{P}$ is transitive on the set of flags of $\mathcal{P}$.

The standard examples are given by n-dimensional cubes paving E^n. In this case, $G(\mathcal{P})$ is the Coxeter group of

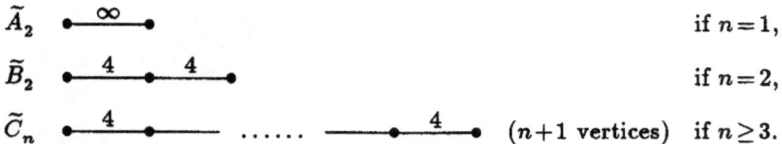

$\widetilde{A}_2$ •—∞—• if $n=1$,

$\widetilde{B}_2$ •—4—•—4—• if $n=2$,

$\widetilde{C}_n$ •—4—•—— ...... ——•—4—• $(n+1$ vertices$)$ if $n\geq 3$.

The only non standard examples exist in dimension 2 (triangles, hexagons, see Number 74) and in dimension 4. An exceptional tesselation in dimension 4 give rise to the Coxeter group of

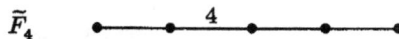

$\widetilde{F}_4$ •——•——4——•——•——•

(see n° 12.6.10.4 in [Ber]).

78. Hyperbolic triangle groups. In the hyperbolic plane H^2, consider a triangle C with angles $\pi/p, \pi/q, \pi/r$ where p, q, r are "integers" in $\{2, 3, \ldots, \infty\}$ such that

$$\frac{1}{p} + \frac{1}{q} + \frac{1}{r} < 1 ;$$

zero angles correspond to ideal vertices of C. Such a triangle is uniquely defined by p, q, r, up to isometry. By Poincaré theorem, the reflections with respect to the sides of such a triangle generate a Coxeter group with graph

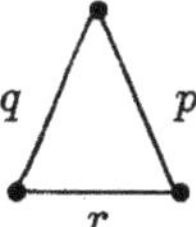

(if some of p, q, r is 2, the corresponding side does not occur). These groups are important for the study of automorphism groups of closed Riemann surfaces [Ma1].

A remarkable case is given by $(p,q,r) = (2,3,\infty)$, because the Coxeter group is then isomorphic to $PGL_2(\mathbb{Z})$. Coxeter generators are given by

$$s_1 = \begin{bmatrix} 0 & 1 \\ 1 & 0 \end{bmatrix} \qquad s_2 = \begin{bmatrix} -1 & 1 \\ 0 & 1 \end{bmatrix} \qquad s_3 = \begin{bmatrix} -1 & 0 \\ 0 & 1 \end{bmatrix}$$

where $\begin{bmatrix} a & b \\ c & d \end{bmatrix}$ denotes the class modulo $\pm \begin{pmatrix} 1 & 0 \\ 0 & 1 \end{pmatrix}$ of the matrix $\begin{pmatrix} a & b \\ c & d \end{pmatrix}$. The relations are $s_j^2 = 1$ for $j = 1,2,3$ as well as $(s_1 s_2)^3 = 1$ and $(s_3 s_1)^2 = 1$. More on this in Section II.2.C of [Bro].

79. Other Coxeter groups acting on $\mathbb{H}^2$. For any real number $d > 0$, there exists a quadrilateral ABCD in $\mathbb{H}^2$ as in Figure 9.

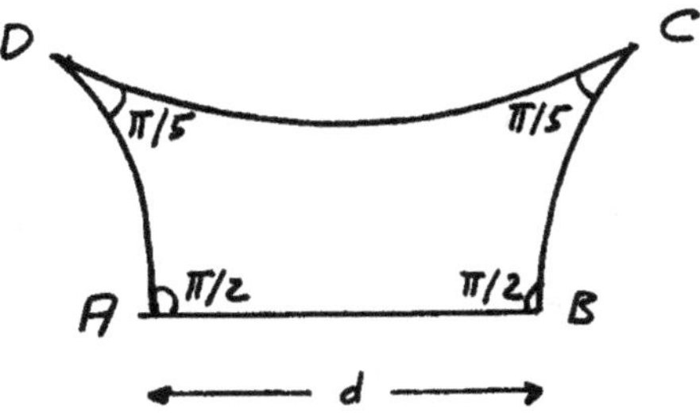

Figure 9.

Two such quadrilaterals for d_1, d_2 are *not* isometric for $d_1 \neq d_2$. However, the reflections with respect to the sides of such a quadrilateral generate a group isomorphic to the Coxeter group of

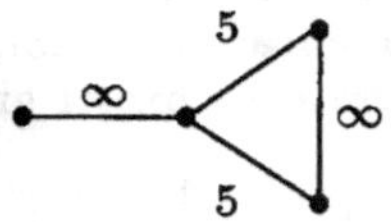

for any $d > 0$. This gives an example of a family of groups, indexed by $d > 0$, which are pairwise isomorphic, which are not conjugate in the group of isometries of $\mathbf{H}^2$, and which have fundamental regions of finite volume. It is a deep rigidity result of Mostow that similar families cannot occur in $\mathbf{H}^n$ for $n \geq 3$. See [Mo1], [Mo2] and [Pra].

More generally, let $p_1, p_2, \ldots, p_k$ be a sequence of "integers" with $k \geq 3$ and $p_j \in \{2, 3, \ldots, \infty\}$ for each j. One may check that there exists in $\mathbf{H}^2$ a polygon with angles $\pi/p_1, \pi/p_2, \ldots, \pi/p_k$ if and only if $\sum p_j < k-2$. When this condition holds, these polygons are described (up to isometry) by a family of k–3 independent parameters. Each of these polygons gives rise to a Coxeter group of isometries of $\mathbf{H}^2$.

One may also consider a region in $\mathbf{H}^2$ limited by an infinite number of arcs and giving rise to a Coxeter system (W, S) with $|S| = \infty$.

80. Coxeter groups and integral quadratic forms. (We follow the beginning of [Men].) Consider an integral quadratic form of rank $n \geq 2$

$$f(x) = \sum_{1 \leq i,j \leq n} b_{i,j} x_i x_j = x^t B x$$

where $x \in \mathbf{R}^n$ is a column vector with coordinates $x_1, \ldots, x_n$, where x^t denotes its transpose and where $B = (b_{i,j})_{1 \leq i,j \leq n}$ is a symmetric matrix with coefficients in $\mathbf{Z}$. We define the orthogonal group

$$O(f, \mathbf{Z}) = \left\{ \gamma \in GL_n(\mathbf{Z}) \ : \ \gamma^t B \gamma = B \right\} .$$

If f is positive definite or negative definite, the group $\{\gamma \in GL_n(\mathbf{R}) : \gamma^t B \gamma = B\}$ is isomorphic to $O(n)$ and is compact; then $O(f, \mathbf{Z})$ is finite, because it is the intersection of this compact group with the discrete subgroup $GL_n(\mathbf{Z})$ of $GL_n(\mathbf{R})$. Let us turn to examples of indefinite forms.

Set first $f(x, y) = x^2 - 2y^2$ (with x, y instead of x_1, x_2). Observe that $P = \begin{pmatrix} 3 & 4 \\ 2 & 3 \end{pmatrix}$ and $Q = \begin{pmatrix} -1 & 0 \\ 0 & 1 \end{pmatrix}$ are in $O(f, \mathbf{Z})$. Let $\gamma = \begin{pmatrix} a & b \\ c & d \end{pmatrix}$ be an integral matrix. Then

$$f\left(\begin{pmatrix} a & b \\ c & d \end{pmatrix} \begin{pmatrix} x \\ y \end{pmatrix} \right) = (a^2 - 2c^2)x^2 + 2(ab - 2cd)xy + (b^2 - 2d^2)y^2 \ .$$

Thus $\gamma \in O(f, \mathbf{Z})$ if and only if

$$a^2 - 2c^2 = 1 \qquad ab - 2cd = 0 \qquad b^2 - 2d^2 = -2 \ .$$

The first of these equations is an example of Pell's equation, and its integral solutions are given by

$$a + c\sqrt{2} = \pm(1+\sqrt{2})^{2n} \qquad n \in \mathbb{Z}$$

(see e.g., Theorem 244 in [HaW]). From this, it follows easily that $O(f,\mathbb{Z})$ is generated by $\pm\begin{pmatrix}1 & 0\\ 0 & 1\end{pmatrix}$, by P and by Q. Thus, up to its center $\pm\begin{pmatrix}1 & 0\\ 0 & 1\end{pmatrix}$, the group $O(f,\mathbb{Z})$ is also generated by PQ and Q, which are both of order 2. Indeed, $O(f,\mathbb{Z})\Big/\pm\begin{pmatrix}1 & 0\\ 0 & 1\end{pmatrix}$ is isomorphic to the infinite dihedral group, namely to the group of the Coxeter graph $\bullet\overset{\infty}{\text{—}}\bullet$.

One may show that the form $f(x,y,z) = x^2 - 2y^2 - 3z^2$ gives rise the same way to the group of the Coxeter graph

$$\bullet\overset{\infty}{\text{———}}\bullet\overset{4}{\text{———}}\bullet\overset{\infty}{\text{———}}\bullet$$

and that $f(x,y,z,u) = x^2 - 2y^2 - 3z^2 - 4u^2$ gives rise to the group of the Coxeter graph of Figure 10. There are many more examples in [Vi3], [Vi4].

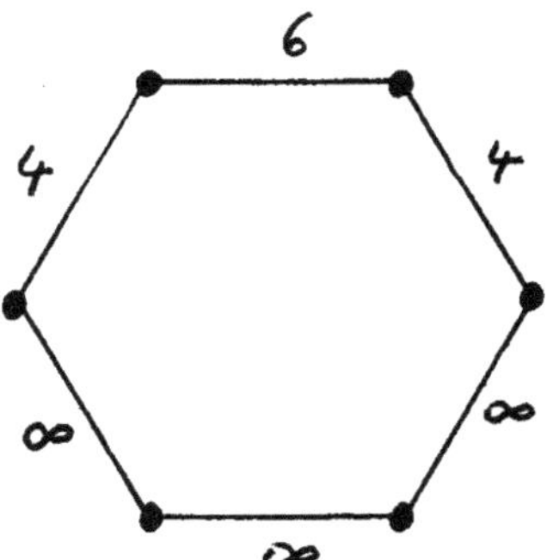

Figure 10.

81. Coxeter groups on hyperbolic spaces of large dimensions. The subject of Coxeter groups acting on H^n for $n \geq 3$ and on other manifolds is beyond the scope of these notes. (See [Vi5], [Vi6], [Dav]). Let us however mention the few following facts.

There are regular tesselations in H^n with compact tiles for $n \leq 4$ and with tiles of finite volume for $n \leq 5$. They have been classified, respectively by Schlegel

(1883) and Coxeter [Co5]. Each of these provides an example of Coxeter group, as in Number 74.

It is a very strong condition for a simplex of finite volume in H^n to fulfill the hypothesis of Poincaré theorem, and thus to provide an example of Coxeter group acting on this H^n. In particular, one must have $n \leq 9$, and even $n \leq 4$ for compact simplices. These simplices are classified up to isometry: see [Lan], [Ko2], [Che], [Vi1], and the exercises of §V.4 in [Bli]. The corresponding groups are sometimes named the "hyperbolic Coxeter groups" ([Ko2], [Bli]), but should be confused neither with the "hyperbolic Coxeter groups" of Vinberg [Vi6], nor with the Coxeter groups which are hyperbolic in the sense of Gromov [Mou], nor with the "hyperbolic type" of Cossec and Dolgachev (see Remark 2.4.1 of [CoD]; see also [Max]).

Some appropriately truncated simplices in H^n fulfilling the hypothesis of Poincaré theorem have been classified by Kaplinskaya [Kap] and by Im Hof [Imh].

Let W be a Coxeter group acting by isometries on H^n. Suppose first that W has a compact fundamental domain. Results due to Nikulin, Vinberg and Prokhorov show that $n \leq 29$; see [Vi7] and [Pro]. Examples are known up to $n = 8$, and are due to Bugaenko (see the indications of [Vi6] for $n = 6$ and $n = 7$). Experts tend to believe that the maximum dimension is close to 8 (and thus far from 29).

Suppose then that W has a fundamental domain of finite volume which is not compact. In this case, Vinberg and Prokhorov have shown that $n \leq 996$; again, this is believed to be far from a sharp estimate. Known examples for large n are arithmetic. (We shall not define this term here; see for example [Vi6].) There is an example of Borcherd [Bor] for $n = 21$. Recent results of Frank Esselman [Ess] show that noncompact arithmetic groups can act on H^n if and only if either $n \leq 19$ or $n = 21$, and that there is a unique example on H^{21} which is that constructed by Borcherd. I am grateful to R. Scharlau [Sch] for communicating these results to me.

Chapter 4. The Geometric Representation of a Coxeter Group

In this Chapter, S denotes a *finite* set, $M=(m_{s,t})_{s,t\in S}$ a given Coxeter matrix of type S, and (W,S) the associated Coxeter system (see Section 1.3). We follow §V.4 of [Bli].

4.1. The Geometric Representation

82. The bilinear form B. Let E denote the real vector space of functions from S to $\mathbb{R}$ and let $(e_s)_{s\in S}$ be its canonical basis. The symmetric bilinear form $B : E\times E\to\mathbb{R}$ is defined by

$$B(e_s,e_t) = -\cos(\pi/m_{s,t}) \qquad s,t\in S\ .$$

In particular

$$\begin{aligned}
&B(e_s,e_s)=1 && \text{for all} \quad s\in S \\
&B(e_s,e_t)=0 && \Longleftrightarrow \quad m_{s,t}=2 \quad \Longleftrightarrow \quad st=ts \\
&B(e_s,e_t)=-\frac{1}{2} && \Longleftrightarrow \quad m_{s,t}=3 \quad \Longleftrightarrow \quad sts=tst \\
&B(e_s,e_t)=-1 && \Longleftrightarrow \quad m_{s,t}=\infty\ .
\end{aligned}$$

For all $s\in S$, let $\sigma_s\in \mathrm{End}_{\mathbb{R}}E$ be defined by

$$\sigma_s(x)=x-2B(e_s,x)e_s\ .$$

One has

$$\begin{aligned}
&\sigma_s(e_s)=-e_s && \sigma_s^2=\mathrm{id}_E \\
&\mathrm{codim}\{x\in E:\sigma_s(x)=x\}=1 && \mathrm{Im}(1-\sigma_s)\subset\mathbb{R}e_s
\end{aligned}$$

and σ_s is a *reflection* of E. This reflection leaves B invariant:

$$B\big(\sigma_s(x),\sigma_s(y)\big)=B(x,y)$$

for all $x,y\in E$. Otherwise said, σ_s belongs to the *orthogonal group*

$$O(B)=\{g\in GL(E)\ :\ B(gx,gy)=B(x,y) \quad \text{for all} \quad x,y\in E\}$$

of B. (For a smart way to view the σ_s, see [Moz] and [AKP].)

83. The case $|S|=2$. In this case, we write $S=\{s,t\}$ and $m=m_{s,t}$. Set $r=st$, so that $srs=trt=r^{-1}$. Each element in W can be written either as r^j or as r^js for some $j\in\mathbb{Z}$.

Suppose first that $m < \infty$. The form B is positive and non degenerate, because

$$B(xe_s + ye_t, xe_s + ye_t) = \left(x - y\cos\frac{\pi}{m}\right)^2 + \left(y\sin\frac{\pi}{m}\right)^2 > 0$$

for all $(x,y) \neq (0,0)$ in $\mathbb{R}^2$. Then r is a rotation of order m with respect to the scalar product B, and W is the dihedral group of order $2m$. The map sending s to σ_s and t to σ_t extends to a faithful representation of W in the vector space E which is irreducible as soon as $m \geq 3$.

Suppose now that $m = \infty$. The form B is positive and degenerate, because

$$B(xe_s + ye_t, xe_s + ye_t) = (x-y)^2 .$$

The kernel E^0 of B is generated by the vector $e^0 = e_s + e_t$, which is fixed by σ_s and by σ_t. Set $f = e_s - e_t$; then $\sigma_s\sigma_t f = 4e^0 + f$, so that $\sigma_s\sigma_t$ has infinite order. The map sending s to σ_s and t ot σ_t extends again to a faithful representation of W in E. Observe that this representation is reducible and indecomposable; compare with Exercise 21. (Recall that a representation of W on E is *reducible* if E has a non trivial W-invariant subspace, and *decomposable* if E has two complementary non trivial W-invariant subspaces.)

84. Proposition. *Let (W, S) be a Coxeter system and let $E, B, (\sigma_s)_{s\in S}$ be as above. Then there exists a homomorphism*

$$\sigma \ : \ W \longrightarrow O(B) \subset GL(E)$$

such that $\sigma(s) = \sigma_s$ for all $s \in S$. Moreover $\sigma(st) = \sigma_s\sigma_t$ is exactly of order $m_{s,t}$.

Proof. One may assume $|S| \geq 2$ and choose two distinct elements $s, t \in S$. It is sufficient to show that $\sigma(st)$ is of order $m_{s,t}$. Set $E_{s,t} = \mathbb{R}e_s \oplus \mathbb{R}_t$ and let $F_{s,t}$ be its orthogonal in E with respect to B. It is straightforward to check that $E_{s,t}$ is invariant by σ_s and σ_t and that each vector of $F_{s,t}$ is fixed by σ_s and σ_t.

If $m_{s,t} < \infty$, the restriction of B to $E_{s,t}$ is non degenerate, so that $E = E_{s,t} \oplus F_{s,t}$. It follows from Number 83 that σ_s and σ_t generate in $GL(E)$ a dihedral group of order $2m$.

If $m_{s,t} = \infty$, it follows from Number 83 that the restriction of $\sigma_s\sigma_t$ to $E_{s,t}$ is of infinite order. Hence $\sigma_s\sigma_t$ itself is of infinite order. □

85. On reflections. Let s be a reflection of a real vector space E. Set $E^s = \ker(s-1)$ and $E^{-s} = \ker(s+1)$. Here, "reflection" means by definition that E^{-s} has dimension one.

Observe that, for any subspace E' of E invariant by s, one has either $E' \subset E^s$ or $E' \supset E^{-s}$. Indeed, if there exists $x \in E'$ such that $sx - x \neq 0$, then E' contains $\mathbb{R}(sx-x) = E^{-s}$.

Lemma. *Let G be a group, let E be a finite dimensional real vector space and let $\varrho : G \to GL(E)$ be an irreducible representation of which the image contains one reflection. Then ϱ is absolutely irreducible (this means that $\varrho \otimes \mathrm{id}_{\mathbf{C}}$ is still irreducible on $E \otimes_{\mathbf{R}} \mathbf{C}$) and there exists at most one bilinear form B on E invariant by $\varrho(G)$ (up to multiplication by a real constant).*

Proof. Let $s \in G$ be such that $\varrho(s)$ is a reflection. To show that ϱ is absolutely irreducible, it is enough to show (Schur's lemma) that the commutant

$$\varrho(G)' = \{u \in \mathrm{End}(E) \ : \ u\varrho(g) = \varrho(g)u \quad \text{for all} \quad g \in G\}$$

is reduced to $\mathbf{R}$. Let $u \in \varrho(G)'$. As $u\varrho(s) = \varrho(s)u$, the line $E^{-\varrho(s)}$ is invariant by u and there exits $\alpha \in \mathbf{R}$ such that u acts on $E^{-\varrho(s)}$ as multiplication by α. As $\ker(u-\alpha)$ is not $\{0\}$ and is invariant by $\varrho(G)$, one has $\ker(u-\alpha) = E$. Thus u acts on E as multiplication by α.

Suppose moreover that there exists a non zero bilinear form B on E which is invariant by $\varrho(G)$. The kernels

$$\{x \in E \ : \ B(x, E) = 0\} \quad \text{and} \quad \{y \in E \ : \ B(E, y) = 0\}$$

are invariant by $\varrho(G)$, and are consequently reduced to $\{0\}$, so that B is non degenerate. Let B' be another bilinear form on E invariant by $\varrho(G)$. As B is non degenerate, there exists $u \in \mathrm{End}(E)$ such that

$$B'(x, y) = B\big(u(x), y\big) \quad \text{for all} \quad x, y \in E\ .$$

Moreover $u \in \varrho(G)'$ by invariance of B'. The first part of the proof implies now that B' is a scalar multiple of B. □

86. Proposition. *Let the Coxeter system (W, S) be irreducible (Definition 19), let E, B and σ be as above, and let E^0 denote the kernel of B. Then*

(i) W acts on E^0 as the identity,

(ii) any proper W-invariant subspace of E is contained in E^0.

In particular the representation $\sigma : W \to GL(E)$ is indecomposable.

Proof. Claim (i) follows from the definitions. For Claim (ii), consider a W-invariant subspace E' of E. If there exists $s, t \in S$ such that $e_s \in E'$ and $m_{s,t} \geq 3$, then

$$\sigma_t(e_s) = e_s - 2B(e_t, e_s)e_t \in E'$$

and $e_t \in E'$ because $B(e_t, e_s) \neq 0$. There are now two cases.

If there exists $s \in S$ such that $e_s \in E'$, then $e_t \in E'$ for all $t \in S$, because the Coxeter graph of (W, S) is connected, and consequently $E' = E$. Otherwise, the first observation of Number 85 shows that $E' \subset E^{\sigma(s)}$ for all $s \in S$, so that E' is a subspace of the intersection E^0 of these $E^{\sigma(s)}$. □

87. Corollary. *Let (W,S) be a Coxeter system. If W is finite, the form B is positive and non degenerate.*

Proof. It is enough to prove the corollary when (W,S) is irreducible. If W is finite, there exists clearly a $\sigma(W)$-invariant scalar product on E. By Lemma 85, the form B is proportional to this scalar product. □

For the converse to Corollary 87, see Corollary 99.

4.2. Classification of Finite and Euclidean Groups

88. Perron-Frobenius vectors. Say that a (real or complex) matrix $A=(A_{i,j})_{1\leq i,j\leq \ell}$ is *decomposable* if there exists a non trivial partition $\{1,\ldots,\ell\}=I\amalg J$ such that $A_{i,j}=0$ for all $i\in I$ and $j\in J$. In practice, one may associate to A a graph with vertex set $\{1,\ldots,\ell\}$ and with an edge between i and j whenever $A_{i,j}A_{j,i}\neq 0$; if the graph is connected, then A is indecomposable (and the converse holds for symmetric matrices).

Recall that the *spectral radius* $r(A)$ of the square matrix A is the maximum of the absolute values $|\lambda|$ of the eigenvalues λ of A.

Assume now that A has real non negative entries: $A_{i,j}\geq 0$ for all $i,j\in\{1,\ldots,\ell\}$. A *Perron-Frobenius vector* for A is an eigenvector $x=(x_i)_{1\leq i\leq \ell}\in\mathbb{R}^\ell$ of A such that $x_i\geq 0$ for all $i\in\{1,\ldots,\ell\}$.

89. Theorem. *Let $A=(A_{i,j})_{1\leq i,j\leq \ell}$ be an indecomposable matrix with non negative entries. Then*

(i) There exists a Perron-Frobenius vector x for A, unique up to multiplication by a positive constant; moreover $x_i>0$ for all $i\in\{1,\ldots,\ell\}$.

(ii) The eigenvalue corresponding to x is the spectral radius $r(A)$; it is a simple root of the characteristic polynomial of A.

(iii) Let $A'=(A'_{i,j})_{1\leq i,j\leq \ell}$ be a matrix such that $0\leq A'_{i,j}\leq A_{i,j}$ for all $i,j\in\{1,\ldots,\ell\}$, and such that $A'\neq A$. Then $r(A')<r(A)$.

Proof: see e.g., [Gan] and [Sam]. □

90. Corollary. *Let $B=(B_{i,j})_{1\leq i,j,\leq \ell}$ be an indecomposable symmetric real matrix such that $B_{i,i}=1$ and $B_{i,j}\leq 0$ for $i\neq j$, and let K be the kernel of B. If the bilinear form associated to B is positive, then $\dim_{\mathbb{R}}(K)\leq 1$.*

Proof. One may write $B=1-\frac{1}{2}A$ with A as in the previous theorem. Let $\lambda_1\geq\ldots\geq\lambda_\ell$ be the eigenvalues of A. As λ_1 is simple, $\lambda_1>\lambda_2$. The eigenvalues of B are

$$1-\frac{1}{2}\lambda_1 < 1-\frac{1}{2}\lambda_2 \leq \ldots \leq 1-\frac{1}{2}\lambda_\ell\,.$$

If B defines a positive bilinear form, then $1-\frac{1}{2}\lambda_1 \geq 0$, and B has at most one eigenvalue, namely $1-\frac{1}{2}\lambda_1$, which is zero. □

91. The Perron-Frobenius vector on a Coxeter graph. Consider a Coxeter matrix $M=(m_{s,t})_{s,t\in S}$ as in Section 4.1, together with the associated graph $\mathcal{G}_M$, the space $E=\mathbf{R}^S$ and the bilinear form B on E. Then $A=2-2B$ can be seen as a symmetric matrix with nonnegative entries acting on the space E. Assume from now on that A is indecomposable, namely that $\mathcal{G}_M$ is connected.

Let $x=(x_s)_{s\in S}\in E$ be a Perron-Frobenius vector for A (see Theorem 89.i). As the vectors of the canonical basis $(e_s)_{s\in S}$ of E are in bijection with the vertices of $\mathcal{G}_M$, it is convenient to represent x by the *labelled graph* consisting of $\mathcal{G}_M$ together with the label x_s on the vertex s, for each $s\in S$. Let λ be the corresponding eigenvalue, so that $Ax=\lambda x$ and $Bx=(1-\frac{1}{2}\lambda)x$. The labelled graph is such that

$$x_s=\lambda\sum\left(2\cos\frac{\pi}{m_{s,t}}\right)x_t \qquad \text{for all} \quad s\in S$$

where the summation is over the neighbours t of s in $\mathcal{G}_M$. Let us illustrate this by the graphs of type D_4 and B_4: the labelled graphs

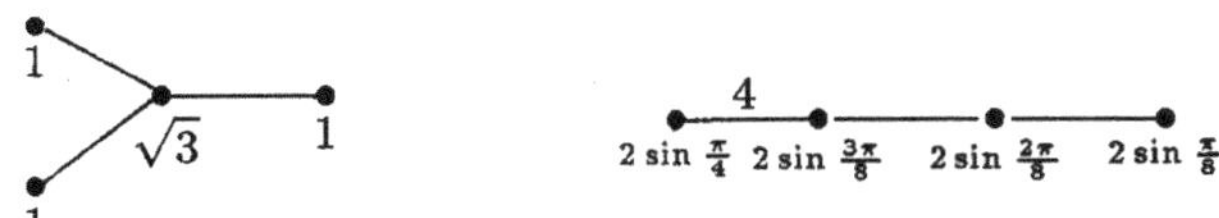

encode respectively the eigenvalue equations

$$\begin{pmatrix}0&0&1&0\\0&0&1&0\\1&1&0&1\\0&0&1&0\end{pmatrix}\begin{pmatrix}1\\1\\\sqrt{3}\\1\end{pmatrix}=\sqrt{3}\begin{pmatrix}1\\1\\\sqrt{3}\\1\end{pmatrix}$$

(recall that $2\cos\pi/3=1$) and

$$\begin{pmatrix}0&\sqrt{2}&0&0\\\sqrt{2}&0&1&0\\0&1&0&1\\0&0&1&0\end{pmatrix}\begin{pmatrix}\sqrt{2}\\\sqrt{2+\sqrt{2}}\\\sqrt{2}\\\sqrt{2-\sqrt{2}}\end{pmatrix}=\sqrt{2+\sqrt{2}}\begin{pmatrix}\sqrt{2}\\\sqrt{2+\sqrt{2}}\\\sqrt{2}\\\sqrt{2-\sqrt{2}}\end{pmatrix}$$

(recall that $2\cos\pi/4=\sqrt{2}$). The list of labelled graphs associated to finite irreducible Coxeter systems is given in Table 1.4.8 of [GHJ] (where the label on the vertex of order 3 in the graph of type D_ℓ has to be corrected as $\sin[(\ell-2)\pi/(2\ell-2)]$).

92. Euclidean graphs. An irreducible Coxeter system (W, S) and the corresponding Coxeter graph are said to be *Euclidean* if the associated bilinear form B is positive and degenerate. With the notations of the previous number, this is equivalent to saying that the Perron-Frobenius eigenvalue of $A = 2 - 2B$ satisfies the equation $\lambda = 2$. The proof of the next result is left to the contemplation of the reader (see also Number VI.4.3 in [Bli]).

93. A list of Euclidean graphs. See Table I; note that the type of a graph has an index ℓ when the graph has $\ell+1$ vertices.

94. Classification of finite irreducible Coxeter groups. Let S be a finite set and let $M = (m_{s,t})_{s,t\in S}$ be a Coxeter matrix. Let $\mathcal{G}$ be the Coxeter graph and let (W, S) be the Coxeter system associated to M. Consider also a set S', a Coxeter matrix $M' = (m'_{s,t})_{s,t\in S'}$, the graph $\mathcal{G}'$ and the system (W', S') associated to M'. Assume that S' is a subset of S. We say that $\mathcal{G}$ *dominates* $\mathcal{G}'$ and that (W, S) dominates (W', S') if $m_{s,t} \geq m'_{s,t}$ for all $s, t \in S'$. For example the graph

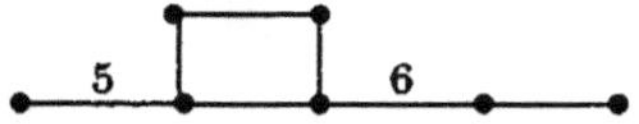

dominates the graph

Let B be the bilinear form on $\mathbf{R}^S$ associated to M and set $A = 2 - 2B$, viewed as a matrix over S. Consider similarly B' and $A' = 2 - 2B'$; one may view A' as a matrix over S, with $A'_{s,t} = 0$ if at least one of s, t is in $S - S'$. The definitions imply that $\mathcal{G}$ dominates $\mathcal{G}'$ if and only if $0 \leq A'_{s,t} \leq A_{s,t}$ for all $s, t \in S$.

Assume now that the group W is finite. Corollary 87 shows that B is positive non degenerate, and thus that the spectral radius of A is strictly smaller than 2. It follows from Theorem 89.iii and from Definition 92 that $\mathcal{G}$ cannot dominate any Euclidean graph, and a fortiori any graph of Table I.

Suppose first for simplicity that $m_{s,t} \in \{2, 3\}$ for all pairs (s, t) of distinct elements in S, so that each edge in $\mathcal{G}$ is of weight 3, and each entry in A is either 0 or 1. Then $\mathcal{G}$ contains

no cycle, otherwise $\mathcal{G}$ would dominate some $\widetilde{A}_\ell$,

no vertex of degree 4, otherwise $\mathcal{G}$ would dominate $\widetilde{D}_4$,

at most one vertex of degree 3, otherwise $\mathcal{G}$ would dominate some $\widetilde{D}_\ell$.

Thus $\mathcal{G}$ is either a segment, namely a graph of type A_ℓ for some $\ell \geq 1$ (see Theorem 45), or a tripod; as $\mathcal{G}$ cannot dominate $\widetilde{E}_\ell$ for $\ell \in \{6, 7, 8\}$, it is straightforward to check that $\mathcal{G}$ is either of type D_ℓ for some $\ell \geq 4$ or of type E_ℓ for some $\ell \in \{6, 7, 8\}$.

Suppose now that at least one of the $m_{s,t}$ is strictly larger than 3. Then there exists exactly one pair (s, t) with $m_{s,t} > 3$, otherwise $\mathcal{G}$ would dominate one

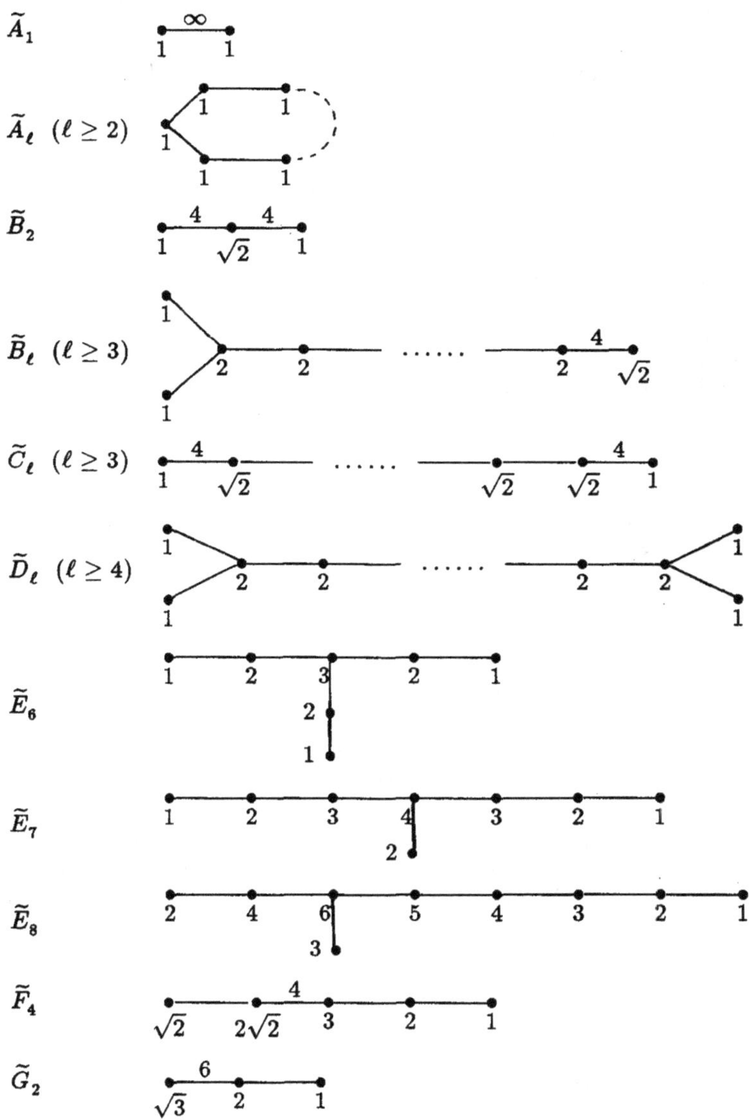
$\widetilde{A}_1$
∞
$\widetilde{A}_\ell$ $(\ell \geq 2)$
$\widetilde{B}_2$
$\widetilde{B}_\ell$ $(\ell \geq 3)$
$\widetilde{C}_\ell$ $(\ell \geq 3)$
$\widetilde{D}_\ell$ $(\ell \geq 4)$
$\widetilde{E}_6$
$\widetilde{E}_7$
$\widetilde{E}_8$
$\widetilde{F}_4$
$\widetilde{G}_2$

Table I.

of $\widetilde{B}_2$ or $\widetilde{C}_\ell$ for $\ell \geq 3$, and the graph underlying $\mathcal{G}$ is a segment, otherwise $\mathcal{G}$ would dominate $\widetilde{B}_\ell$ for some $\ell \geq 3$. Let us suppose that $|S| \geq 3$. Then the maximal $m_{s,t}$ is either 4 or 5, otherwise $\mathcal{G}$ would dominate $\widetilde{G}_2$. (See Table I.) We leave the following steps as an exercise for the reader, for a graph $\mathcal{G}$ with $|S| \geq 3$ (the notations are those of Theorem 45):

(i) $\mathcal{G}$ cannot dominate

(ii) if $\mathcal{G}$ has an edge of weight 5 then $\mathcal{G}$ is either H_3 or H_4

(iii) if $\mathcal{G}$ has an edge of weight 4 then $\mathcal{G}$ is either F_4 or B_ℓ for some $\ell \geq 3$.

The previous arguments show that the list of Theorem 45 is the *complete* list of Coxeter graphs with B positive and non degenerate, and Corollary 87 shows that this list contains all the graphs defining finite irreducible Coxeter groups. Conversely, it is true that all groups appearing in this list are finite: this follows either from a lengthy case by case checking (see the previous chapters) or from Section 4.3 below.

95. Exercise. Show that *all* Euclidean graphs appear in Table I. (Hint. An Euclidean graph cannot dominate *strictly* a graph of the list. In particular, either it is a cycle $\widetilde{A}_\ell$, or it does not contain any cycle. In the latter case, either it is a cross $\widetilde{D}_4$, or it does not contain any vertex of degree 4. And so on.)

96. Remarks. (i) The organisation of the arguments in Number 94 appears in many places, including [Smi] and [CGS].

(ii) It is known that an irreducible Coxeter system (W, S) is Euclidean if and only if there exists in W a subgroup of finite index which is isomorphic to $\mathbb{Z}^n$ for some $n \geq 1$ (see Number VI.2.1 in [Bli]). A Coxeter group which is neither finite nor Euclidean contains non abelian free subgroups [Hae].

(iii) The two previous numbers complete the classification of connected Coxeter graphs such that the associated form B is positive, either non degenerate (94) or degenerate (95). For other signatures of the form B, the corresponding classifications are still incomplete: see [Max] and [Pal].

4.3. A Theorem of Tits

97. The Tits representation. Notations being as in Section 4.1, consider moreover the dual space E^* of $E = \mathbb{R}^S$. The Tits representation

$$\sigma^* : W \longrightarrow GL(E^*)$$

is the contragredient of the geometric representation; this means that $\sigma^*(w)$ is the transposed map of $\sigma(w^{-1})$ for all $w \in W$. We introduce also the open simplicial cone

$$C = \{x^* \in E^* \ : \ x^*(e_s) > 0 \quad \text{for all} \quad s \in S\}$$

and the *Tits cone*

$$\Omega = \bigcup_{w \in W} w\overline{C}$$

where $w\overline{C}$ denotes the image by $\sigma^*(w)$ of the closure of C.

98. Theorem. *The notations are as above.*

(i) Let $w \in W$ be such that $wC \cap C \neq \emptyset$. Then $w = 1$. In particular σ^* is injective with discrete image in $GL(E^*)$ and the same holds for σ. The group W operates simply transitively on the set $\{wC\}_{w \in W}$.

(ii) The action of W on Ω is properly discontinuous, and Ω is a convex cone in E^*.

(iii) Let (W, S) be irreducible. Then the Tits cone coincides with E^* if and only if the group W is finite, it is an open half-space if and only if W is Euclidean (see Number 92), and its closure does not contain any line in the other cases.

Proof. see §V.4 in [Bli], or [Vi2]. □

99. Corollary. *Let (W, S) be a Coxeter system and let B be the associated bilinear form on $E = \mathbb{R}^S$. Then W is finite if and only if B is positive non degenerate.*

Proof. If B is positive and non degenerate, then $\sigma(W)$ is a discrete subset of the compact group $O(B)$. It follows that W is finite. For the converse, see Corollary 87. □

Bibliography

[AKP] N. Alon, I. Krasikov and Y. Peres, Reflection sequences, *Amer. Math. Monthly* **96** (1989), 820-823.

[Arn] V. Arnold, Problem VIII, the A-D-E classifications, *Mathematical developments arising from Hilbert problems,* Proc. Symp. Pure Math. **28** (Amer. Math. Soc. 1976), p. 46.

[Bea] A.F. Beardon, *The geometry of discrete groups,* Springer, 1983.

[BeG] C.T. Benson and L.C. Grove, *Finite reflection groups,* Bogden and Quigley, 1971. Second edition, Springer, 1985.

[Ber] M. Berger, *Géométrie, volumes 1-5,* Cedic/F. Nathan, 1977-1978.

[Bli] N. Bourbaki, *Groupes et algèbres de Lie, chapitres IV-VI,* Hermann, 1968.

[Bor] R. Borcherds, Automorphism groups of Lorentzian lattices, *J. of Algebra* **111** (1987), 133-153.

[Brø] A. Brøndsted, *An introduction to convex polytopes,* Springer, 1983.

[Bro] K.S. Brown, *Buildings,* Springer, 1989.

[Bur] W. Burnside, *Theory of groups of finite order, second edition,* Cambridge Univ. Press, 1911.

[CGS] P.J. Cameron, J.M. Goethals, J.J. Seidel and E.E. Shult, Line graphs, root systems, and elliptic geometry, *J. of Algebra* **43** (1976), 305-327.

[Car] R. Carter, *Simple groups of Lie type,* J. Wiley, 1972.

[Che] M. Chein, Recherche des graphes des matrices de Coxeter hyperboliques d'ordre ≤ 10, *Rev. Française Informat. Recherche Opérationnelle* **3** (1969), Sér. R-3, 3-16.

[Com] L. Comtet, *Advanced combinatorics,* Reidel, 1974.

[Con] J.H. Conway et al., *Atlas of finite groups,* Clarendon Press, 1985.

[CoD] F.R. Cossec and I.V. Dolgachev, *Enriques surfaces I,* Birkhäuser, 1989.

[Co1] H.S.M. Coxeter, The polytope with regular-prismatic vertex figures, Part 2, *Proc. London Math. Soc.* **34** (1931), 126-189.

[Co2] H.S.M. Coxeter, Discrete subgroups generated by reflections, *Ann. of Math.* **35** (1934), 588-621.

[Co3] H.S.M. Coxeter, The complete enumeration of finite groups of the form $R_i^2 = (R_i R_j)^{k_{i,j}} = 1$, *Proc. London Math. Soc.* **10** (1935), 21-25.

[Co4] H.S.M. Coxeter, The polytope 2_{21}, whose twenty-seven vertices correspond to the lines on the general cubic surface, *Amer. J. Math.* **62** (1940), 457-486.

[Co5] H.S.M. Coxeter, Regular honbeycombs in hyperbolic space, *Proc. Internat. Congr. Math., Amsterdam, 1954,* volume III, 155-169.

[Co6] H.S.M. Coxeter, *Regular polytopes, 3rd ed.,* Dover, 1973.

[Co7] H.S.M. Coxeter, letter to L. Paris, 7th of March 1990.

[CoM] H.S.M. Coxeter and W.O.J. Moser, *Generators and relations for discrete groups, 4th ed.,* Springer 1980.

[DGS] C. Davis, B. Grunbaum and F.A. Sherk (ed.), *The geometric vein, The Coxeter Festschrift,* Springer, 1981.

[Dav] M.W. Davis, Groups generated by reflections and aspherical manifolds not covered by Euclidean space, *Ann. of Math.* **117** (1983), 293-324.

[Di1] L.E. Dickson, The abstract group isomorphic with the symmetric group on k letters (1899), *Collected papers IV,* 401-403. See also E.H. Moore, Concerning the abstract group of order $k!$ and $\frac{1}{2}k!$ holohedrically isomorphic with the symmetric and the alternating substitution-groups on k letters, *Proc. London Math. Soc.* **28** (1897), 357-366.

[Di2] L.E. Dickson, *Linear groups with an exposition of the Galois field theory,* Teubner, 1901.

[Ess] F. Esselmann, *Ueber die maximale Dimension von Lorentz Gittern mit coendlicher Spiegelungsgruppe*, in preparation, University of Bielefeld, 1990.

[Fra] J.S. Frame, The classes and representations of the groups of 27 lines and 28 bitangents, *Ann. Mat. Pura Appl.* **32** (1951), 83-119.

[Gan] F.R. Gantmacher, *The theory of matrices, volume two,* Chelsea 1959.

[GHJ] F.M. Goodman, P. de la Harpe and V.F.R. Jones, *Coxeter graphs and towers of algebras,* Springer, 1989.

[Gru] B. Grünbaum, *Convex polytopes,* Interscience, 1967.

[Gut] E. Gutkin, Geometry and combinatorics of groups generated by reflections, *Enseign. Math.* **32** (1986), 95-110.

[HaW] G.H. Hardy and E.M. Wright, *An introduction to the theory of numbers,* fifth ed., Oxford Univ. Press, 1979.

[Hae] P. de la Harpe, Groupes de Coxeter infinis non affines, *Expo. Math.* **5** (1987), 91-96.

[HKW] P. de la Harpe, M. Kervaire and C. Weber, On the Jones polynomial, *Enseign. Math.* **32** (1986), 271-335.

[Har] R. Hartshorne, *Algebraic geometry,* Springer, 1977.

[HSV] M. Hazewinkel, W. Hesselink, D.Siersma and F.D. Veldkamp, The ubiquity of Coxeter-Dynkin diagrams (an introduction to the A-D-E problem), *Nieuw Arch. Wiskunde, III. Ser.* **25** (1977), 257-307.

[Hel] S. Helgason, *Differential geometry and symmetric spaces,* Academic Press, 1962.

[Hil] H. Hiller, *Geometry of Coxeter groups,* Pitman, 1982.

[Imh] H-C. Im Hof, A class of hyperbolic Coxeter groups, *Expo. Math.* **3** (1985), 179-186.

[Jor] C. Jordan, *Traité des substitutions et des équations algébriques,* Gauthier-Villars, 1870.

[Kac] V.G. Kac, *Infinite dimensional Lie algebras,* Birkhäuser, 1983.

[Kap] I.M. Kaplinskaya: Discrete groups generated by reflections in the faces of simplicial prisms in Lobachevskian spaces, *Math. Notes* **15** (1974), 88-91.

[Ko1] J.L. Koszul, *Lectures on groups of transformations,* Tata Institute, Bombay 1965.

[Ko2] J.L. Koszul, *Lectures on hyperbolic Coxeter groups,* University of Notre Dame, 1967.

[Kra] W. Kraskiewicz, Reduced decompositions in hyperoctahedral groups, *C.R. Acad. Sci. Paris,* **309,** Ser. I (1989), 903-907.

[Lam] K. Lamotke, *Regular solids and isolated singularities,* Vieweg, 1986.

[Lan] F. Lannér, On complexes with transitive groups of automorphisms, *Lunds Univ. Math. Sem.* **11** (1950), 71pp.

[Lin] V. Ya. Lin, Artin braids and the groups and spaces connected with them, *J. Soviet Math.* **18**-5 (1982), 736-788.

[Log] D. Logothetti, Interview of Coxeter, *Mathematical people, profiles and interviews,* D.J. Albers and G.L. Alexanderson, ed., Birkhäuser, 1985, 51-63.

[MaS] A.M. Macbeath and D. Singerman, Spaces of subgroups and Teichmuller space, *Proc. London Math. Soc.,* **31** (1975), 211-256.

[Ma1] W. Magnus, *Noneuclidean tesselations and their subgroups,* Academic Press, 1974.

[Ma2] W. Magnus, Braid groups : a survey. *Lecture Notes in Math.* **372** (Springer 1974), 463-487.

[Man] Yu.I. Manin, *Cubic forms, algebra, geometry, arithmetic,* North Holland, 1974.

[Ma1] B. Maskit, On Poincaré's Theorem for fundamental polygons, *Adv. in Math.* **7** (1971), 219-230.

[Ma2] B. Maskit, *Kleinian groups,* Springer, 1988.

[Mat] J. Matsuzawa, Monoidal transformations of Hirzebruch surfaces and Weyl groups of type C, *J. Fac. Sci. Univ. Tokyo Sect. IA, Math.* **35** (1988), 425-429.

[Max] G. Maxwell, Hyperbolic trees, *J. of Algebra* **54** (1978), 46-49.

[Men] J. Mennicke, Eine Pflasterung des dreidimensionalen hyperbolischen Raumes, *Math. Phys. Semesterberichte* **27** (1980), 55-68.

[MBD] G.A. Miller, H.F. Blichfeldt and L.E. Dickson, *Theory and applications of finite groups,* J. Wiley, 1916.

[Mi1] J. Milnor, On the 3-dimensional Brieskorn manifolds $M(p,q,r)$, in *Knots, groups, and 3-manifolds,* Papers dedicated to the memory of R.H. Fox, L.P. Neuwirth ed., Princeton Univ. Press, 1975.

[Mi2] J. Milnor, Hyperbolic geometry : the first 150 years, *Bull. Amer. Math. Soc.* **6** (1982), 9-24.

[Mo1] G.D. Mostow, Quasi-conformal mappings in n-space and the rigidity of hyperbolic space forms, *Publ. Math. IHES* **34** (1967), 53-104.

[Mo2] G.D. Mostow, *Strong rigidity of locally symmetric spaces,* Annals of Math. Studies **78** , Princeton Univ. Press, 1973. For [Mo2] and [Pra], see the reviews by M.S. Raghunathan, M.R. **53** ♯ 5874-5.

[Mou] G. Moussong, *Hyperbolic Coxeter groups,* Thesis, Ohio State University, 1988.

[Moz] S. Mozes, Reflection processes on graphs and Weyl groups, *J. Combin. Theory, Ser. A* **53** (1990), 128-142.

[Mum] D. Mumford, *Algebraic geometry I, complex projective varieties,* Springer, 1976.

[Nik] V.V. Nikulin, Discrete reflection groups in Lobachevsky spaces and algebraic surfaces, *Proc. Internat. Congr. Math., Berkeley, 1986,* Volume **1** , 654-671.

[Pa1] L. Paris, *Minimal non standard Coxeter trees,* preprint, Université de Genève, February 1990.

[Pa2] L. Paris, *Growth series of Coxeter groups,* this volume.

[Po1] H. Poincaré, Sur la théorie des fonctions fuchsiennes (1881) *Oeuvres II,* 75-91; Théorie des groupes fuchsiens (1882), *Oeuvres II,* 108-168.

[Po2] H. Poincaré, Mémoire sur les groupes kleinéens (1883), *Oeuvres II,* 258-

299.

[Pra] G. Prasad, Strong rigidity of Q-rank 1 lattices, *Invent. Math.* **21** (1973), 255-286.

[PrS] A. Pressley and G. Segal, *Loop groups,* Clarendon Press, 1986.

[Pro] M.N. Prokhorov, The absence of discrete reflection groups with noncompact fundamental polyhedron of finite volume in Lobachevsky space of large dimension, *Math. USSR Izvestija* **28** (1987), no 2, 401-411.

[Ree] E.G. Rees, *Notes on geometry,* Springer, 1983.

[Rha] G. de Rham, Sur les polygones générateurs de groupes fuchsiens, *Enseign. Math.* **17** (1971), 49-61.

[Sam] H. Samelson, On the Perron-Frobenius theorem, *Michigan Math. J.* **4** (1956), 57-59.

[Sch] R. Scharlau, *On the classification of arithmetic reflection groups on hyperbolic 3-space*, preprint, University of Bielefeld, 1990.

[Seg] B. Segre, *The non-singular cubic surfaces,* Oxford Univ. Press, 1942.

[Sei] H. Seifert, Komplexe mit Seitenzuordnung, *Nachr. Akad. Wiss. Göttingen Math.-Phys. Kl. II* **1975** , no. 6, 49-80.

[Ser] J-P. Serre, *Cours d'arithmétique,* P.U.F., 1970.

[Sie] C.L. Siegel, *Topics in complex function theory, volume II, automorphic functions and abelian integrals,* J. Wiley, 1971.

[Smi] J.H. Smith, Some properties of the spectrum of a graph, *Combinatorial structures and their applications (Proc. Calgary Internat. Cong., Calgary, 1969),* 403-406.

[St1] R.P. Stanley, Weyl groups, the hard Lefschetz theorem, and the Sperner property, *SIAM J. Algebraic Disccrete Methods* **1** (1980), 168-184.

[St2] R.P. Stanley, On the number of reduced decompositions of elements of Coxeter group, *European J. Combin.* **5** (1984), 359-372.

[Vi1] E.B. Vinberg, Discrete groups generated by reflections in Lobacevskii spaces, *Math. USSR Sbornik* **1** (1967) no 3, 429-444.

[Vi2] E.B. Vinberg, Discrete linear groups generated by reflections, *Math. USSR Izvestija* **5** (1971) no 5, 1083-1119.

[Vi3] E.B. Vinberg, On groups of unit elements of certain quadratic forms, *Math. USSR Sbornik* **16**-1 (1972), 17-35.

[Vi4] E.B. Vinberg, Some arithmetical discrete groups in Lobacevskii spaces, *Discrete subgroups of Lie groups and applications to moduli, Bombay Colloquium 1973,* Oxford Univ. Press, 1975.

[Vi5] E.B. Vinberg, Discrete reflection groups in Lobachevsky spaces, *Proc. Internat. Congr. Math., Warsaw, 1983,* Volume **1** , 593-601.

[Vi6] E.B. Vinberg, Hyperbolic reflection groups, ***Russian Math. Surveys*** **40** (1985) no 1, 31-75.

[Vi7] E.B. Vinberg, The absence of crystallographic groups of reflections in Lobachevsky spaces of large dimension, *Trans. Moscow Math. Soc.* **47** (1985), 75-112.

After completion of this work, I became aware of J.E. Humphreys' book on Coxeter groups. Here is the precise reference, and a few other ones.

R. Charney and M.W. Davis, *Singular metrics of nonpositive curvature on branched covers of Riemannian manifolds,* preprint, Ohio State University, 1990.

R. Charney and M.W. Davis, *Reciprocity of growth functions of Coxeter groups,* preprint, Ohio State University, 1990.

M.W. Davis and M. Shapiro, *Coxeter groups are almost convex,* preprint, Ohio State University, 1990.

M. Dyer, Reflection subgroups of Coxeter systems, *J. of Algebra* **135** (1990), 57-73.

J.E. Humphreys, *Reflection groups and Coxeter groups,* Cambridge Univ. Press 1990.

L. Paris, *Complex growth series of Coxeter systems,* preprint, University of Wisconsin, 1990.

J.A. de la Pena and M. Takane, Spectral properties of Coxeter transformations and applications, *Arch. Math.***55** (1990), 120-134.

M.K.Shaieev, Reflective subgroups in Bianchi groups, *Selecta Math. Sov.* **9** (1990), 315-322.

O.P. Shcherbak, Wawevronts and reflection groups, *Russian math. surveys* **43**-3 (1988), 149-194.

O.V. Shvartsman, Reflective subgroups of Bianchi groups, *Selecta Math. Sov.* **9** (1990), 323-329.

E.B. Vinberg, Reflective subgroups in Bianchi groups, *Selecta Math. Sov.* **9** (1990), 309-314.

FIVE LECTURES ON BUILDINGS

KENNETH S. BROWN

Cornell University

These lectures are intended to provide an introduction to J. Tits's theory of buildings. The point of view is generally the same as that of my book [10], which will be the main reference in what follows. For a different point of view and much more information about buildings, see [18]. See also the conference proceedings [17] and [19] for some examples of recent research.

I have tried to minimize the prerequisites, especially for the first lecture. But from the second lecture on, readers will need to be familiar with Coxeter groups and Coxeter complexes. Appendix A below provides a brief review of some of this material. Readers can refer to it as necessary.

LECTURE 1. DEFINITION AND EXAMPLES

Buildings are simplicial complexes satisfying certain axioms that will be given below. In order to motivate these axioms, we begin with two examples.

1. Examples

Example 1. Let k be a field and let V be the n-dimensional vector space k^n. Let Δ be the simplicial complex associated to the poset of proper non-zero subspaces of V, ordered by inclusion. Thus the vertices of Δ are the proper non-zero subspaces of V, and the simplices are the chains

$$V_1 < \cdots < V_q$$

of such subspaces. The maximal simplices, called *chambers*, are those with $\dim V_i = i$ for all i. They have rank $n-1$ (i.e., they have $n-1$ vertices) and hence dimension $n-2$. Note that, for a suitable choice of basis $e_1, \ldots, e_n$, such a simplex corresponds to the chain

$$[e_1] < [e_1, e_2] < \cdots < [e_1, \ldots, e_{n-1}],$$

Partially supported by a grant from the National Science Foundation.

where the square brackets denote the subspace spanned by a set of vectors. We call Δ the ***building*** associated to V and we write $\Delta = \Delta(V)$.

If $n = 3$, for instance, then $\dim \Delta = 1$, and Δ is simply the incidence graph of lines and planes in k^3 passing through the origin. Thus there is a vertex for each line L, a vertex for each plane P, and an edge joining these vertices whenever $L \subset P$. [Equivalently, Δ is the incidence graph of points and lines in the projective plane over k.]

The picture below shows this graph in case k is the field $\mathbf{F}_2$ with 2 elements. It has 14 vertices and 21 edges. The vertices are shown in two different "colors" to indicate the two possible types of proper non-zero subspaces of k^3 (lines and planes). Note that each vertex has degree 3, i.e., has exactly 3 edges coming out of it. This corresponds to the fact that, over $\mathbf{F}_2$, each plane contains exactly 3 lines through the origin and each line is contained in exactly 3 planes.

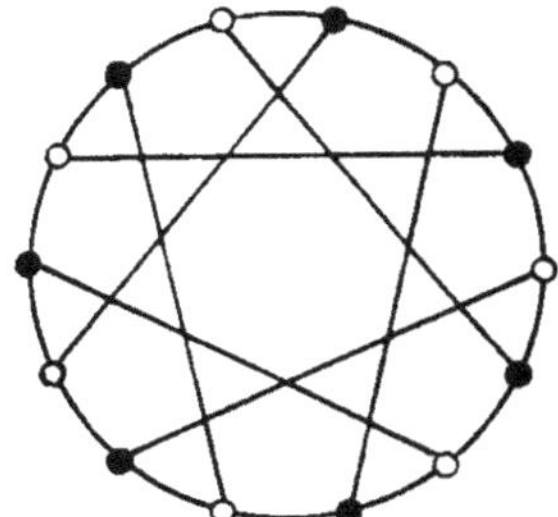

Continuing with the case $n = 3$, note that Δ contains lots of hexagons. Indeed, each basis e_1, e_2, e_3 for V yields a configuration

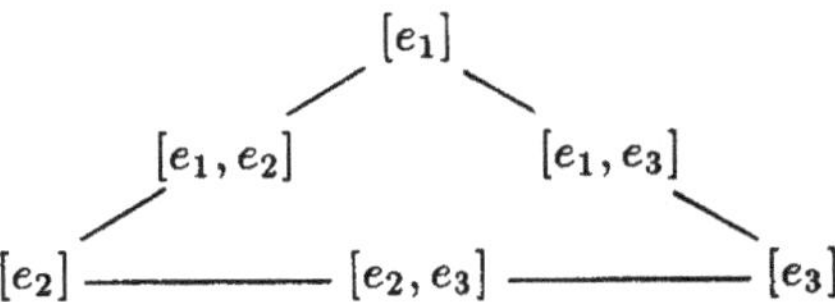

in Δ. These hexagons are called ***apartments***. There is one for each unordered triple of independent 1-dimensional subspaces of k^3 [or, equivalently, for each triangle in the projective plane]. In the example with $k = \mathbf{F}_2$, there are 28 apartments. The reader should locate a few of them in the picture.

The phrase "lots of hexagons" above can be made more precise. For example, one can easily find an apartment containing any given chamber. And with only slightly more effort, one can even find an apartment containing any two given chambers. (Try a few examples of this in the picture above.)

Returning now to the case of arbitrary n, it is still true that every basis $e_1, \ldots, e_n$ determines a subcomplex $\Sigma \subset \Delta$, called an *apartment*. It consists of the simplices which can be constructed by using subspaces spanned by subsets of $\{e_1, \ldots, e_n\}$. One can check that Σ is isomorphic to the barycentric subdivision of the boundary of an $(n-1)$-simplex. In particular, Σ is topologically an $(n-2)$-sphere.

And it is still true that any two chambers are contained in an apartment, but this is a more substantial exercise than in the case $n = 3$. See [10], Exercise 2 of §IV.2, where this assertion is deduced from the proof of the Jordan–Hölder theorem. See also Abels ([1], [2]) for generalizations.

Example 2. For our second example of a building, let Δ be a tree in which every vertex has degree ≥ 3. For example, Δ could be the tree pictured below, in which every vertex has degree exactly 3. As in Example 1, the vertices have been drawn in two colors, so that each chamber (edge) has one vertex of each color. Vertices of the same color will be said to have the same *type*. This equivalence relation is intrinsic to Δ; in fact, two vertices have the same type if and only if the distance between them is even.

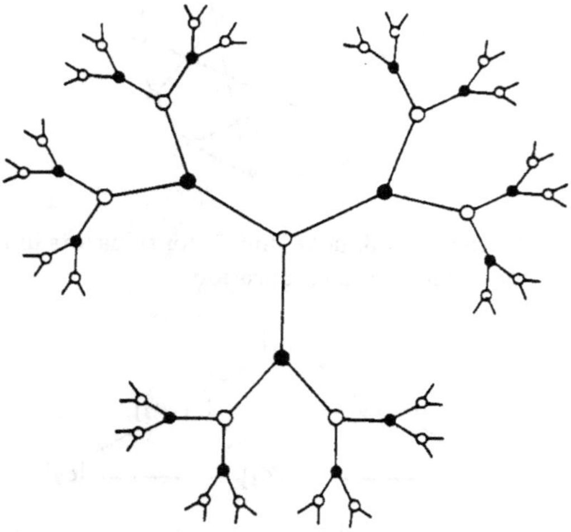

By an *apartment* in Δ we will mean any subcomplex Σ which is isomorphic to a triangulated line (infinite in both directions). There are uncountably many apartments. Once again, it is easy to see that any two chambers are contained in an apartment.

2. The definition

We need to recall some terminology. Let Δ be a finite-dimensional simplicial complex in which all maximal simplices have the same dimension. Call the maximal simplices *chambers*. Two chambers C, D are *adjacent* if they have a common codimension 1 face. A *gallery* from C to D is a sequence of chambers

$$C = C_0, C_1, \dots, C_l = D$$

such that C_{i-1} and C_i are adjacent for $i = 1, \dots, l$. The gallery is said to have *length* l. We call Δ a *chamber complex* if any two chambers can be connected by a gallery. For example, every triangulated manifold is a chamber complex.

A chamber complex is *thin* if every simplex of codimension 1 is a face of exactly two chambers. Thus a thin chamber complex is precisely what combinatorial topologists call a *pseudomanifold without boundary*. In particular, every triangulated manifold without boundary is a thin chamber complex. Finally, a chamber complex is *thick* if every simplex of codimension 1 is a face of at least 3 chambers.

Note that a 1-dimensional chamber complex is nothing but a connected simplicial graph; it is thick if and only if every vertex has degree ≥ 3, and it is thin if and only if every vertex has degree 2 [in which case it is a line or a polygon].

We are ready now for the main definition. The theory is somewhat simpler if we require our buildings to be thick, so we will do that for the moment. At the end of this lecture we will remove that restriction.

Definition (Tits, 1965). Let Δ be a finite-dimensional simplicial complex. We call Δ a (thick) *building* if it can be expressed as the union of a family of subcomplexes Σ, called *apartments*, satisfying the following axioms:

(B0) *Each apartment is a thin chamber complex of the same dimension as* Δ.

(B1) *Any two simplices of* Δ *are contained in an apartment.* [Hence Δ is a chamber complex.]

(B2) *Given two apartments* Σ, Σ' *with a common chamber, there is an isomorphism* $\Sigma \xrightarrow{\approx} \Sigma'$ *fixing* $\Sigma \cap \Sigma'$ *pointwise.*

(B3) Δ *is thick.*

It should be clear from our discussion of Examples 1 and 2 that those examples do in fact satisfy **(B0)**, **(B1)**, and **(B3)**. To see that **(B2)** holds in Example 2, note that the intersection of two lines is a convex subset of each of them, hence an interval (possibly unbounded); it is now a simple matter to construct the desired isomorphism. It is also easy to verify **(B2)** in Example 1 with $n = 3$; one need only think about the possibilities for the intersection of two of our hexagonal apartments. The general case of Example 1 takes a little more work; see [10], Exercise 2 of §IV.2.

One can best get a feeling for these axioms by looking at some consequences of them.

3. Consequences of the axioms

We will omit most of the proofs in this section; they can all be found in Chapter IV of [10]. We begin with an easy observation:

Proposition 1. *All apartments are isomorphic.*

Proof. Given two apartments Σ, Σ', choose by **(B1)** an apartment Σ'' containing a chamber of Σ and a chamber of Σ'. Then **(B2)** yields isomorphisms $\Sigma \xrightarrow{\approx} \Sigma'' \xrightarrow{\approx} \Sigma'$. □

Much less obviously, one can show:

Proposition 2. *The isomorphism type of the apartments is determined by* Δ.

The content of this is the following: Let $\mathcal{A}$ be a system of apartments, i.e., a family of subcomplexes Σ satisfying the axioms. Then for any other system of apartments $\mathcal{A}'$, the complexes $\Sigma \in \mathcal{A}'$ are isomorphic to those in $\mathcal{A}$. If, for instance, Δ has an apartment system consisting of lines (resp. hexagons) then every apartment system consists of lines (resp. hexagons). We will explain why this is true in Lecture 2.

Proposition 3. *If the apartments are finite complexes, then* Δ *admits a unique system of apartments.*

For example, if there is an apartment system $\mathcal{A}$ consisting of hexagons, then $\mathcal{A}$ is the only possible system of apartments in Δ. [One can even show in this case that $\mathcal{A}$ necessarily consists of all of the hexagons in Δ.]

Proposition 3′. *In the general case, there is a unique maximal system of apartments.*

Another way to say this is that any two apartment systems are compatible with one another, in the sense that their union is again an apartment system. [What has to be shown here is that the union satisfies **(B2)**.] In the tree case, for instance, the maximal apartment system is the one we described in Example 2 above, consisting of all lines. But there are also smaller apartment systems, containing only some of the lines; one need only take enough lines to satisfy **(B1)**.

Proposition 4. *The apartments are Coxeter complexes.*

The reader can refer to Appendix A for a review of what this means. For our present purposes, however, it suffices to recall that Coxeter complexes are certain very special thin chamber complexes, associated to "generalized reflection groups". For example, every finite Coxeter complex is topologically a sphere, with a triangulation induced by a finite reflection group. A special case of the proposition, then, is that no closed manifold other than a sphere can ever occur as an apartment in a building. This may seem surprising at first, and it illustrates the force of the axioms.

The idea of the proof of Proposition 4 is the following: Given an apartment Σ, one combines various isomorphisms given by **(B2)** in order to construct automorphisms of Σ which behave like reflections. These "reflections" generate a group W whose action on Σ resembles the action of a finite reflection group on Euclidean space (or on the unit sphere in that space); one deduces that W is a Coxeter group and that Σ is the associated Coxeter complex. The proof uses the thickness axiom **(B3)** in order to construct "enough" reflections. [It is clear *a priori* that thickness has to be used; for if we were to drop **(B3)**, then any thin chamber complex Σ could occur as an apartment: We could simply take $\Delta = \Sigma$, which would then be a building with a single apartment.]

Looking back at the examples, one can check directly that the apartments are Coxeter complexes. In Example 1, the associated Coxeter group is the symmetric group on n letters (or the dihedral group of order 6 if $n = 3$). And in Example 2 it is the infinite dihedral group.

The next result is not about the apartments themselves, but rather about how they sit as subcomplexes of Δ:

Proposition 5. *Every apartment Σ is a retract of Δ and is convex in Δ.*

The word "convex" here is used in a combinatorial sense: For any two chambers of Σ, any gallery of minimal length joining them in Δ is entirely contained in Σ. [Intuitively, a minimal gallery is something like a geodesic, whence the term "convex".]

Sketch of Proof. To construct a retraction $\rho : \Delta \to \Sigma$, choose a chamber C of Σ, which will be fixed throughout the construction. Given a simplex σ of Δ, we can find an apartment Σ' containing C and σ. By **(B2)** there is then an isomorphism $\phi : \Sigma' \to \Sigma$ fixing C pointwise, and we set $\rho(\sigma) = \phi(\sigma)$:

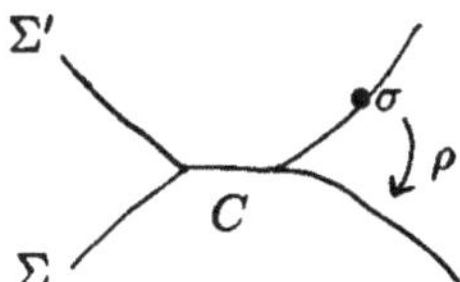

It is not hard to check that ρ is a well-defined simplicial map. And it is clearly a retraction, since we can take $\Sigma' = \Sigma$ and $\phi = \mathrm{id}_\Sigma$ if $\sigma \in \Sigma$.

Note, for future reference, that ρ depends on the choice of C but is canonical otherwise. [This assertion requires a little thought, but it is not difficult to check.] We denote ρ by $\rho_{\Sigma,C}$. The retractions $\rho_{\Sigma,C}$ are extremely useful technical tools. They are used in the proofs of many of the results that we have stated above without proof. We will also use them now, to prove the convexity of apartments.

Suppose $C_0, \ldots, C_l$ is a minimal gallery from C_0 to C_l in Δ, with extremities $C_0, C_l \in \Sigma$. Suppose the gallery does not stay in Σ, say $C_{i-1} \in \Sigma$ but $C_i \notin \Sigma$. Let C be the chamber of Σ which is adjacent to C_{i-1} along the same face as C_i, and let ρ be the retraction $\rho_{\Sigma,C}$ constructed above. From the definition of ρ, one can check that $\rho(C_i) = C_{i-1}$:

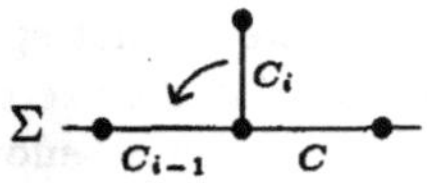

But then the image of our gallery under ρ is a gallery which has the same extremities but "stutters" [it repeats a chamber]. Removing the repetition, we obtain a shorter gallery from C_0 to C_l, contradicting minimality of the original gallery. □

Here is another application of retractions:

Proposition 6. *Δ is labellable.*

This means that it is possible to partition the vertices into n "types", where $n = \operatorname{rank} \Delta$, in such a way that each chamber has exactly one vertex of each type. (See §3 of Appendix A for more details.) In Example 1, for instance, a vertex is a subspace V' of a vector space V, and its type is determined by its dimension. The proof of labellability is immediate: One knows that Coxeter complexes are labellable (cf. Appendix A, §3), so we need only label one apartment Σ and then extend the labelling to Δ by using a retraction $\rho : \Delta \to \Sigma$.

Finally, we state the ***Solomon–Tits theorem***, which shows that the axioms severely limit how complicated the algebraic topology of a building can be:

Proposition 7. *If the apartments are finite (hence $(n-1)$-spheres, where $n = \operatorname{rank} \Delta$), then Δ has the homotopy type of a bouquet of $(n-1)$-spheres. If the apartments are infinite, then Δ is contractible.*

4. The role of thickness

Although thick buildings as defined above are the most important ones for applications, it is sometimes convenient to have a more general notion of "building". It turns out that there is a perfectly satisfactory theory without the thickness axiom, provided we add the assumption that the apartments are Coxeter complexes. In other words, we drop **(B3)** but then take the result of Proposition 4 as a new axiom. All of the results above remain valid if we do this. [The point is that thickness is used only to prove Proposition 4, but the latter is used in many of the other proofs.]

Here, then, is our revised definition:

Definition. Let Δ be a finite-dimensional simplicial complex. We call Δ a *building* if it can be expressed as the union of a family of subcomplexes Σ, called *apartments*, satisfying the following axioms:

(B0) *Each apartment is a Coxeter complex of the same dimension as* Δ.

(B1) *Any two simplices of* Δ *are contained in an apartment.*

(B2) *Given two apartments* Σ, Σ' *with a common chamber, there is an isomorphism* $\Sigma \xrightarrow{\approx} \Sigma'$ *fixing* $\Sigma \cap \Sigma'$ *pointwise.*

Note, for instance, that a building in our new sense can even be thin; indeed, a thin building is precisely the same thing as a Coxeter complex. Note also that we can now generalize Example 2: Any tree in which every vertex has degree ≥ 2 is a building.

LECTURE 2. GALLERIES, ETC.

Let Δ be a building and let $\mathcal{C}$ be the set of chambers of Δ. One can view $\mathcal{C}$ as a metric space with the gallery metric

$$d : \mathcal{C} \times \mathcal{C} \to \mathbf{Z},$$

where $d(C, D)$ is the minimal length of a gallery from C to D. In this lecture we will take a close look at minimal galleries, and we will see that we can associate to a pair (C, D) much more than just the number $l = d(C, D)$. The upshot of our investigation will be a new way of axiomatizing buildings, which has become quite important in recent research. We need to begin by explaining how to associate a matrix to Δ which captures its "type", i.e., the isomorphism type of the apartments.

1. The Coxeter matrix of a building

Consider first the case where Δ is the Coxeter complex $\Sigma = \Sigma(W, S)$ associated to a Coxeter group (W, S). [See Appendix A for the notation.] Let $M = (m(s,t))_{s,t \in S}$ be the Coxeter matrix of (W, S); thus $m(s,t)$ is the order of st. This has a geometric interpretation which is explained in §5 of Appendix A, and which we review briefly here.

Recall that Σ is labellable. In fact, there is a canonical labelling, with S as the set of labels, which is defined in §3 of Appendix A. An example of this is illustrated below, where Σ is the plane tiled by equilateral triangles, and W is generated by reflections s, t, u with respect to the sides of one "fundamental" triangle. In the picture on the left, a labelling is indicated by the use of three "colors" for the vertices; in addition, the stabilizers of the faces of the fundamental chamber are shown. The picture on the right shows the canonical labelling, with s, t, u as labels.

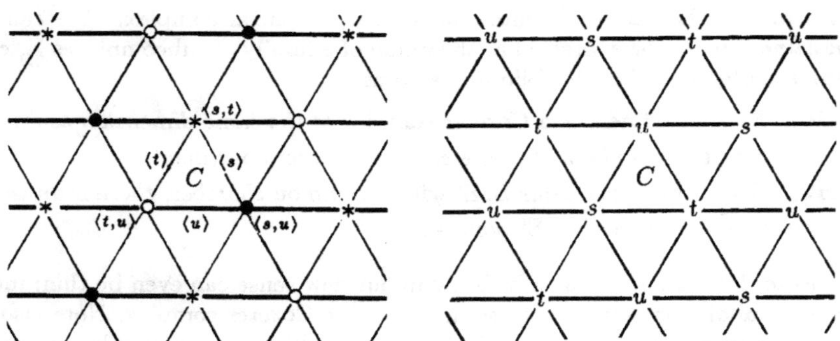

Returning to the general case, every simplex $\sigma \in \Sigma$ has a well-defined *type*, which is a subset of S. In particular, a codimension 2 simplex σ has type $S - \{s,t\}$ for some $s, t \in S$ with $s \neq t$. The geometric interpretation of the matrix M, then, is that the link of σ is a $2m$-gon, where $m = m(s,t)$. (Note that m could be ∞, in which case an ∞-gon is to be interpreted as a line.) In the example above, for instance, all the numbers m are equal to 3, and the link of any vertex is a hexagon.

Note that a $2m$-gon has diameter m, so we can also write

$$m(s,t) = \operatorname{diam}(\operatorname{lk} \sigma).$$

One consequence of this is the following: Suppose we are given a Coxeter complex Σ as an abstract simplicial complex, but we are not told what Coxeter group W it comes from. Then we can reconstruct W from Σ by looking at the links of the various types of codimension 2 simplices, where "type" makes sense because Σ is labellable. Here is a convenient way to state this result:

Proposition 1. *Let Σ be a Coxeter complex, labelled by a set I. Then there is a well-defined matrix $M = (m_{ij})_{i,j \in I}$ with $m_{ii} = 1$ and, for $i \neq j$,*

$$m_{ij} = \operatorname{diam}(\operatorname{lk} \sigma)$$

for any simplex σ of type $I - \{i,j\}$. Let W_M be the Coxeter group defined by M, with generating set $S = \{s_i\}_{i \in I}$ and relations $(s_i s_j)^{m_{ij}} = 1$, and let Σ_M be the associated Coxeter complex $\Sigma(W_M, S)$. Then $\Sigma \approx \Sigma_M$.

Remark. There is some ambiguity in our use of the word "diameter" above. On the one hand, the diameter of a chamber complex is defined as the supremum of the gallery distances $d(C,D)$, where C and D range over all chambers. On the other hand, the links discussed above are graphs, for which one usually defines diameter in terms of lengths of edge paths between vertices. This ambiguity is harmless, since a $2m$-gon has diameter m in both senses.

It is now easy to generalize to buildings. Let Δ be an arbitrary building, with an arbitrary system of apartments. Recall that Δ is labellable (Lecture 1, §3, Proposition 6). Choose a fixed labelling by a set I. In view of the essential uniqueness of labellings (Appendix A, §3), nothing we do will depend on this choice in any significant way.

Proposition 2. *Let Δ be a labelled building as above. There is a well-defined matrix $M = (m_{ij})_{i,j \in I}$ with $m_{ii} = 1$ and, for $i \neq j$,*

$$m_{ij} = \operatorname{diam}(\operatorname{lk} \sigma)$$

for any simplex σ of type $I - \{i, j\}$. Moreover, every apartment Σ is isomorphic to Σ_M.

This is a fairly easy consequence of Proposition 1, once one checks two things: (a) For any simplex $\sigma \in \Delta$, its link $\operatorname{lk}_\Delta \sigma$ is a building with apartments $\operatorname{lk}_\Sigma \sigma$, where Σ ranges over the apartments of Δ. (b) Any building has the same diameter as its apartments. The proof of (a) is easy, right from the definitions. And (b) follows from the convexity of apartments (Lecture 1, §3, Proposition 5). For more details see [10], §§IV.1 and IV.3.

The reader might find it an instructive exercise to verify the conclusions of Proposition 2 directly in case $\Delta = \Delta(k^n)$.

Remark. Proposition 2 shows that the isomorphism type of the apartments depends only on Δ; this was stated without proof in Lecture 1 (§3, Proposition 2).

The matrix M is called the *Coxeter matrix* of Δ, and the associated Coxeter group $W = W_M$ is called the *Weyl group* of Δ.

2. Galleries and words

We continue with the notation just established: Δ is a labelled building, M is its Coxeter matrix, and W is its Weyl group, with distinguished generating set S in 1-1 correspondence with the set I of labels. It is convenient to identify I with S and thereby to regard S as the set of labels. This double use of S (as a set of group elements and a set of labels) is potentially confusing, but it turns out to be quite convenient.

We now focus on the set $\mathcal{C}$ of chambers of Δ. For the moment, we view $\mathcal{C}$ as a set with a relation (adjacency). Recall that this relation enabled us to define the gallery metric d, making $\mathcal{C}$ a metric space. Using the labelling, we will refine the adjacency relation and obtain from this a refined distance function.

Given a label $s \in S$, we say that two chambers C, D are *s-adjacent*, and we write $C \overset{s}{\sim} D$, if C and D have the same face of type $S - \{s\}$. Note, then, that any two distinct adjacent chambers are s-adjacent for a unique $s \in S$. Consequently, a non-stuttering gallery $C_0, \ldots, C_l$ has a well-defined *type* $\mathbf{s} = (s_1, \ldots, s_l)$, such that $C_{i-1} \overset{s_i}{\sim} C_i$ for $i = 1, \ldots, l$. For example, the solid line in the picture below indicates a gallery of type (s, t, s, u, t) between two chambers C and D. If, on the other hand, we follow the broken line instead of the solid line at the beginning, then we obtain a gallery of type (t, s, t, u, t) between the same two chambers. Similarly, there is a gallery of type (t, s, u, t, u) from C to D.

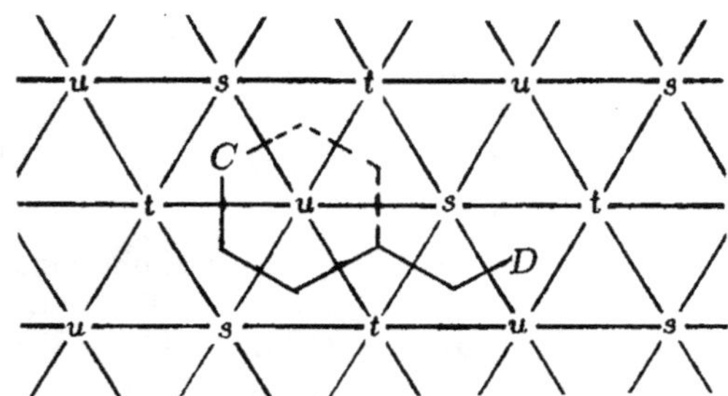

Switching now to the other role played by S (as a set of generators of W), note that the type $\mathbf{s}$ of a gallery can be viewed as a *word*. Such a word represents an element of W. For instance, the type (s, t, s, u, t) which arose in our example above represents the element $w = stsut \in W$. Note, in this example, that the other two galleries from C to D that we mentioned have types that represent the same element w. Indeed, the relations $(st)^3 = 1$ and $(tu)^3 = 1$ imply that

$$stsut = tstut = tsutu$$

in W. Thus we seem to have a well-defined element $w \in W$ associated to the ordered pair (C, D). This illustrates a general principle, valid in any building:

Theorem. *There is a unique function $\delta : \mathcal{C} \times \mathcal{C} \to W$ with the following property: For any $C, D \in \mathcal{C}$, if there is a minimal gallery from C to D in Δ of type $\mathbf{s} = (s_1, \ldots, s_l)$, then $\delta(C, D)$ is the element $w = s_1 \cdots s_l$ represented by $\mathbf{s}$. Moreover, $d(C, D) = l(\delta(C, D))$, where $l : W \to \mathbf{Z}$ is the length function l_S.*

Thus δ is the promised refinement of the distance function d:

$$\begin{array}{ccc} & \overset{\delta}{\nearrow} & W \\ & & \downarrow l \\ \mathcal{C} \times \mathcal{C} & \xrightarrow{d} & \mathbf{Z} \end{array}$$

I like to think of $w = \delta(C, D)$ as something like a "vector" pointing from C to D. It has a "magnitude" $l(w)$, which is the distance from C to D, but it contains a great deal more information; for example, reduced decompositions of w are related to "geodesics" from C to D.

The proof of the theorem is not difficult. First, by the convexity of apartments, every minimal gallery between two given chambers is contained in any apartment containing those chambers. So we reduce easily to the case where Δ is a single apartment, which may be assumed to be $\Sigma(W, S)$ with its canonical labelling. The proof in this case is then an easy consequence of the well-known correspondence between galleries in $\Sigma(W, S)$ and S-words; see Appendix A, §6, for more details. Incidentally, that proof also shows that δ in this special case can be identified with the "difference function"

$$W \times W \to W,$$

given by $(w, w') \mapsto w^{-1}w'$. Thus the last assertion of the theorem reduces to the familiar formula $d(w, w') = l(w^{-1}w')$ for the word metric.

3. A new axiomatization of buildings

We have just seen that a building gives rise to a pair $(\mathcal{C}, \delta)$ consisting of a set with a W-valued distance function. It is not hard to see that this assignment is essentially 1-1, i.e., that one can recover Δ up to canonical isomorphism from $(\mathcal{C}, \delta)$. (The point here is that we can recover the s-adjacency relations from δ, since $C \overset{s}{\sim} D \iff \delta(C, D) \in \{1, s\}$; it is then easy to reconstruct Δ, cf. §D of the appendix to Chapter I of [10].) The obvious question, then, is which pairs $(\mathcal{C}, \delta)$ arise from buildings? In other words, what are the appropriate axioms for "W-metric spaces", in order that they correspond precisely to buildings?

Tits [27] has recently given the following beautiful answer to this question:

Theorem. *Let (W, S) be a Coxeter system with S finite. Given a set $\mathcal{C}$ and a function $\delta : \mathcal{C} \times \mathcal{C} \to W$, the pair $(\mathcal{C}, \delta)$ arises from a building if and only if it satisfies the following four axioms:*

(1) *$\delta(C, D) = 1$ if and only if $C = D$.*
(2) *$\delta(D, C) = \delta(C, D)^{-1}$.*
(3) *If $\delta(C', C) = s \in S$ and $\delta(C, D) = w$, then $\delta(C', D) = sw$ or w. If, in addition, $l(sw) = l(w) + 1$, then $\delta(C', D) = sw$.*
(4) *If $\delta(C, D) = w$, then for any $s \in S$ there is a C' such that $\delta(C', C) = s$ and $\delta(C', D) = sw$. If $l(sw) = l(w) - 1$, then there is a unique such C'.*

Remarks. 1. The first three axioms resemble the three axioms for (ordinary) metric spaces. Axiom (3), for example, or at least the first part of (3), is something like the triangle inequality. The second part of (3), from this point of view, gives a "collinearity" condition under which equality holds in the triangle inequality. Axiom (4), on the other hand, is probably harder to grasp intuitively; our discussion of the tree case below should help.

2. One can actually get by with a simpler set of axioms. For example, it is not hard to show that (1), (3), and the first part of (4) imply (2) and the second part of (4).

3. Perhaps the most interesting aspect of the theorem is that it provides an axiomatization of buildings that does not assume the existence of anything resembling apartments. The heart of the proof of the theorem is the construction of apartments, in the guise of "strong isometries" from W into $\mathcal{C}$. A *strong isometry* is a function that preserves the W-valued distance function, where W is equipped with the distance function $(w, w') \mapsto w^{-1}w'$ as in §2 above.

See Appendix B for a sketch of the proof of the theorem. See also Tits [25] for an earlier axiomatization in the same spirit.

We close this lecture by explaining why (3) and (4) are true in the tree case. Assume, then, that Δ is a tree (in which every vertex has degree ≥ 2). The Weyl group W in this case is the infinite dihedral group, and S consists of two generators of order 2. If one carefully checks the definition of the adjacency relations, one finds that two adjacent edges are s-adjacent ($s \in S$) if and only if their common vertex does *not* have label s.

Now fix two chambers C, D, and let $\delta(C, D) = w$. To avoid uninteresting cases, assume $w \neq 1$. Suppose $\delta(C', C) = s \in S$, and let t be the other element of S. Thus the common vertex of C and C' has label t. Assume first that $l(sw) = l(w) + 1$, i.e., that the (unique) S-word representing w starts with t. Let $\Gamma = (C_0, C_1, \ldots, C_l)$ be the (unique) minimal gallery from C to D. Then we have $C = C_0 \overset{t}{\sim} C_1$, so C and C_1 have a common vertex with label s. We can therefore picture Γ as the following edgepath:

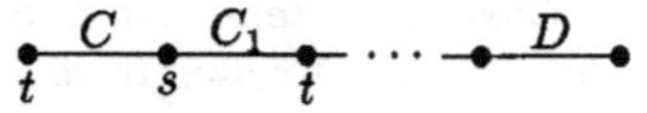

Putting C' in front of Γ, we obtain the path

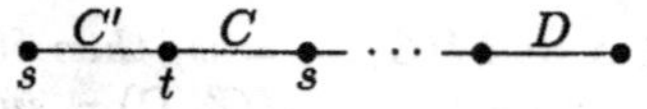

This path is a geodesic since $C' \neq C$. The composite gallery $(C', C_0, C_1, \ldots, C_l)$ is therefore minimal, whence $\delta(C', D) = sw$. This proves (3) and the first part of (4) when $l(sw) = l(w) + 1$.

Suppose now that $l(sw) = l(w) - 1$, i.e., that w starts with s. The labels on Γ are then reversed from those above, but C and C' still share a vertex with label t. The picture therefore takes the form

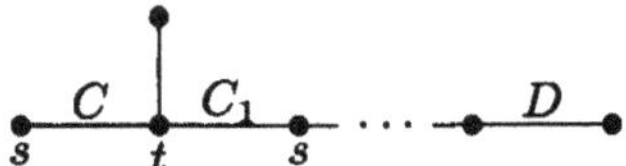

where the vertical edge is C'. Now it is possible that $C' = C_1$; in this case we have the minimal gallery $C' = C_1, \ldots, D$ whose type is obtained from that of Γ by removing the initial s. Hence $\delta(C', D) = sw$ for this particular C'. For all other choices of C', we have a minimal gallery $C', C_1, \ldots, D$ of the same type as Γ, whence $\delta(C', D) = w$. All assertions in (3) and (4) should now be clear.

Lecture 3. Buildings and Groups

Recall that every Coxeter system (W, S) has an associated simplicial complex $\Sigma = \Sigma(W, S)$, on which W acts as a group of type-preserving simplicial automorphisms. [In fact, one can show that W is the full group of type-preserving simplicial automorphisms of Σ.] The complexes Σ which correspond to Coxeter groups in this way are precisely the thin buildings.

It is natural to try to generalize this construction. Thus we seek a class of groups G to which we can associate a building Δ. By analogy with the special case of Coxeter groups, we expect G to act on Δ by type-preserving automorphisms, and we expect the action to be transitive on chambers. One might also hope that the action would be transitive on apartments. It turns out that the most satisfactory theory is obtained by demanding a transitivity property which is strong enough to imply both of those just mentioned.

1. Strongly transitive group actions

Let Δ be a building and $\mathcal{A}$ a system of apartments. Let a group G act on Δ by type-preserving automorphisms leaving $\mathcal{A}$ invariant. The action is said to be *strongly transitive* if G acts transitively on the set of pairs (Σ, C) with $\Sigma \in \mathcal{A}$ and $C \in \operatorname{Ch}\Sigma$, where the latter is the set of chambers of Σ.

Note that the action is strongly transitive if and only if G is transitive on $\mathcal{A}$ and the stabilizer of some $\Sigma \in \mathcal{A}$ is transitive on $\operatorname{Ch}\Sigma$. Alternatively, the action is strongly transitive if and only if G is transitive on $\operatorname{Ch}\Delta$ and the stabilizer of some $C \in \operatorname{Ch}\Delta$ is transitive on the set of apartments containing C.

Our goal is to discover a class of groups G for which we can construct a building with a strongly transitive G-action. We will do this by working the problem backwards: We assume that we have a strongly transitive action, and we will see what structure this imposes on G. Fix an apartment $\Sigma \in \mathcal{A}$ (the "fundamental apartment") and a chamber $C \in \operatorname{Ch}\Sigma$ (the "fundamental chamber"). We need some notation for stabilizers, etc.:

Let B be the stabilizer of C; it acts transitively on the set of apartments containing C.

Let N be the stabilizer of Σ; it leaves Σ invariant and acts transitively on $\mathrm{Ch}\,\Sigma$. One can deduce that N surjects onto the group $\mathrm{Aut}_0\,\Sigma$ of type-preserving automorphisms of Σ. [Given $\phi \in \mathrm{Aut}_0\,\Sigma$, there is an $n \in N$ such that $nC = \phi(C)$. Then ϕ and n agree pointwise on C since they are both type-preserving; a standard argument based on the thinness of Σ now shows that ϕ and n agree on all of Σ. For more details, look at the "standard uniqueness argument" in [10].]

Let T be the fixer of Σ, i.e., the (normal) subgroup of N consisting of those elements that fix Σ pointwise. In other words,

$$T = \ker\{N \twoheadrightarrow \mathrm{Aut}_0\,\Sigma\}.$$

Note that we can also describe T as $N \cap B$; this follows from the same standard uniqueness argument used above.

Let W be the quotient group N/T. It is isomorphic to $\mathrm{Aut}_0\,\Sigma$, hence it is a Coxeter group whose associated Coxeter complex is isomorphic to Σ. More precisely, the action of W on Σ yields a set S of "fundamental reflections", these being the non-trivial elements of W which fix a codimension 1 face of the fundamental chamber C, and (W,S) is a Coxeter system such that $\Sigma(W,S)$ is canonically isomorphic to Σ. [The point here is that Σ is known to be a Coxeter complex, so the assertions just made about W are known to be true about the group $\mathrm{Aut}_0\,\Sigma$, which is canonically isomorphic to W.]

The notation that has been introduced so far is summarized in the following diagram:

$$\begin{array}{ccc} & G & \\ \diagup & & \diagdown \\ B & & N \twoheadrightarrow W = \langle S \rangle \\ \diagdown & & \diagup \\ & T & \end{array}$$

We now give Σ its canonical labelling with S as the set of labels; this labelling is characterized by the property that $wC \overset{s}{\sim} wsC$ for all $w \in W$ and $s \in S$, where C is still the fundamental chamber. Extend this labelling to a labelling of Δ. Consideration of types of minimal galleries then yields, as in Lecture 2, a function

$$\delta : \mathrm{Ch}\,\Delta \times \mathrm{Ch}\,\Delta \to W.$$

Since the action of G is label-preserving, δ is G-invariant:

$$\delta(gC_1, gC_2) = \delta(C_1, C_2)$$

for all $C_1, C_2 \in \mathrm{Ch}\,\Delta$ and all $g \in G$.

Before proceeding further, let's look at an example.

2. Example

Let $\Delta = \Delta(k^n)$ as in Lecture 1 (§1, Example 1). Let G be the general linear group $\mathrm{GL}_n(k)$. It acts on k^n and permutes the subspaces, hence it acts on Δ. The action is easily seen to be type-preserving and strongly transitive. As fundamental chamber we take the "standard flag"

$$[e_1] < [e_1, e_2] < \cdots < [e_1, \ldots, e_{n-1}]$$

constructed from the standard basis vectors. And as fundamental apartment, we take the apartment determined by the standard basis vectors; its chambers are the flags

$$[e_{\pi(1)}] < [e_{\pi(1)}, e_{\pi(2)}] < \cdots < [e_{\pi(1)}, \ldots, e_{\pi(n-1)}],$$

where π ranges over the permutations of $\{1, \ldots, n\}$.

The stabilizer B of C is the group of upper triangular matrices, called the *Borel* subgroup of G; this explains the use of the letter "B" in the general theory. The stabilizer N of Σ is the monomial group, i.e., the group of matrices with exactly one non-zero entry in each row and in each column. The intersection $T = N \cap B$ is the group of diagonal matrices. And the quotient $W = N/T$ can be identified with the symmetric group on n letters. The letters "T", "N", and "W" are reminders that T is a maximal *torus* in G, N is the *normalizer* of T, and W is the *Weyl group*.

Finally, the reader might enjoy trying to guess what δ is in this example. It associates to any two maximal flags in k^n a certain permutation of $\{1, \ldots, n\}$. This turns out to be the so-called *Jordan–Hölder permutation*. See [10], §IV.2, Exercise 2; see also [1] and [2].

3. The structure of G

We return to the general setup of §1. In order to illustrate the ideas of Lecture 2, we will focus our attention on the set $\mathrm{Ch}\,\Delta$, together with the function $\delta : \mathrm{Ch}\,\Delta \times \mathrm{Ch}\,\Delta \to W$. We will identify $\mathrm{Ch}\,\Delta$ with the set G/B of left cosets of B in G, interpret δ group-theoretically, and then deduce results about G. See Chapter V of [10] for a different way of deriving the same results.

Recall first that δ is G-invariant, so that

$$\delta(gC, hC) = \delta(C, g^{-1}hC)$$

for any $g, h \in G$, where C, as always in this lecture, is the fundamental chamber. Thus δ is completely known as soon as we know $\delta(C, gC)$ for all g. Now strong transitivity implies, for any $g \in G$, that there is a $b \in B$ such that $bgC \in \Sigma$. [Choose an apartment containing C and gC, and use the transitivity of B on the set of apartments containing C.] Since W acts transitively on $\mathrm{Ch}\,\Sigma$, we can write $bgC = wC$ for some $w \in W$. Using the invariance of δ under the action of b, we now obtain

$$\begin{aligned}\delta(C, gC) &= \delta(C, bgC)\\ &= \delta(C, wC)\\ &= w.\end{aligned}$$

The last equation here follows from the correspondence between galleries in $\Sigma(W, S)$ and S-words, cf. Appendix A, §1.

We now use the fact that chambers are in 1-1 correspondence with cosets of B, so that the equation $bgC = wC$ above can also be written as $bgB = wB$. [Here wB denotes the coset nB for any representative $n \in N$ of w; this is independent of the choice of n.] Consequently, $gB = b^{-1}wB$, whence $g \in BwB$ ($= BnB$ for any representative n as above). Conversely, if $g \in BwB$ then $bgC = wC$ for some $b \in B$. The upshot of the previous paragraph, then, is that every $g \in G$ is in some double coset BwB, and that one then has $\delta(C, gC) = w$.

An immediate consequence of this is that g is in a *unique* double coset BwB. In other words:

Theorem 1. $G = BWB \overset{\text{def}}{=} \bigcup_{w\in W} BwB = \coprod_{w\in W} BwB.$

For historical reasons that will be explained in the next lecture, this result is called the *Bruhat decomposition* of G. Note that it implies, in particular, that G is generated by B and N. The reader might find it helpful at this point to think about what the Bruhat decomposition means when $G = \mathrm{GL}_n(k)$ as in §2; there is a concrete interpretation in terms of row and column operations.

Returning to the study of δ, let's now view δ as a function $G/B \times G/B \to W$. Our calculation above then says that $\delta(B, gB)$ is the unique $w \in W$ such that $BgB = BwB$. Since $\delta(gB, hB) = \delta(B, g^{-1}hB)$, we conclude that $\delta(gB, hB)$ is the element w which represents the double coset $Bg^{-1}hB = (gB)^{-1}(hB)$. In other words, we have arrived at a group-theoretic description of δ as the composite

$$G/B \times G/B \to B\backslash G/B \to W,$$

where the second arrow is given by the Bruhat decomposition and the first arrow is the "difference map" $(gB, hB) \mapsto (gB)^{-1}(hB)$. Viewed in this way, δ appears as a very natural generalization of the analogous function on W itself which arose in Lecture 2.

We close this section by giving the group-theoretic translation of properties (3) and (4) of δ stated in Lecture 2. This translation involves products of double cosets, about which one can ordinarily say practically nothing.

Theorem 2. *For any $s \in S$ and $w \in W$, we have:*

(a) *If $l(sw) = l(w) + 1$, then $BsB \cdot BwB = BswB$.*

(b) *If $l(sw) = l(w) - 1$, then $BsB \cdot BwB \subseteq BswB \cup BwB$. Equality holds if Δ is thick.*

Proof. Take $g \in BsB$ and $h \in BwB$. Then the double coset containing the product gh is what arises when one computes $\delta(g^{-1}B, hB)$. Since $\delta(g^{-1}B, B) = \delta(B, gB) = s$ and $\delta(B, hB) = w$, we can apply the "triangle inequality" (i.e., axiom (3)) to $g^{-1}B, B, hB$; we conclude that $\delta(g^{-1}B, hB) = sw$ or w, with the first case occurring if $l(sw) = l(w) + 1$. Hence $gh \in BswB$ or BwB, with the first case occurring if $l(sw) = l(w) + 1$. This proves (a) and the first part of (b).

Assume now that $l(sw) = l(w) - 1$, and fix $h \in BwB$. By axiom (4) there is a unique coset $g_0^{-1}B$ with $\delta(g_0^{-1}B, B) = s$ and $\delta(g_0^{-1}B, hB) = sw$; for all other cosets $g^{-1}B$ with $\delta(g^{-1}B, B) = s$, we have $\delta(g^{-1}B, hB) = w$. Now if Δ is thick, then there must in fact exist at least one such g. [In other words, the common face of $g_0^{-1}C$ and C is contained in at least one other chamber.] We then have $g \in BsB$ and $gh \in BwB$. Thus $BsB \cdot BwB$ meets BwB, and the second part of (b) follows easily. □

Note that we can take $w = s$ in (b). If Δ is thick, we conclude that $BsB \cdot BsB \not\subseteq B$, or, equivalently, that $sBs \not\subseteq B$. Since s has order 2, we can also write this as

$$sBs^{-1} \not\subseteq B.$$

In conclusion, we have seen in this section that a strongly transitive action of a group on a building gives rise to a pair of subgroups B, N with some very special properties. We now reverse the procedure and show that a "BN-pair" in an abstract group is sufficient for the construction of a building. For simplicity, we will only consider the thick case, which suffices for most applications.

4. BN-pairs

We say that a pair of subgroups B and N of a group G is a *BN-pair* if B and N generate G, the intersection $T = B \cap N$ is normal in N, and the quotient $W = N/T$ admits a set of generators S such that the following two conditions hold for all $s \in S$ and $w \in W$:

(BN1) $$BsB \cdot BwB \subseteq BswB \cup BwB.$$

(BN2) $$sBs^{-1} \not\subseteq B.$$

One also says, in this situation, that the quadruple (G, B, N, S) is a *Tits system*.

The reader may be surprised at how little is assumed here, given how much more is known to hold in the setup of §3. For example, the definition does not require W to be a Coxeter group. It does not even require the elements of S to be of order 2. The reason for not assuming more is that everything one needs turns out to be a consequence of our two axioms. Indeed, one derives, by fairly short group-theoretic arguments, all of the following results about BN-pairs (cf. [10], §V.2):

(1) $G = \coprod_{w \in W} BwB$.

(2) $BsB \cdot BwB = \begin{cases} BswB & \text{if } l(sw) = l(w) + 1 \\ BswB \cup BwB & \text{if } l(sw) = l(w) - 1. \end{cases}$

(3) S consists of elements of order 2, and (W, S) is a Coxeter system.

(4) S is uniquely determined by B and N. [This explains why we define "BN-pairs" instead of "BNS-triples".]

(5) The subgroups P with $B \subseteq P \subseteq G$ are in 1-1 correspondence with the special subgroups $W' = \langle S' \rangle$ of W (where S' ranges over all subsets of S) via $W' \leftrightarrow BW'B$. Moreover, these subgroups P are mutually non-conjugate, and each is equal to its own normalizer.

The subgroups P in (5) were not mentioned explicitly in §3, but their significance in that context is that they are the stabilizers of the faces of the fundamental chamber C.

Finally, we state the result that we have been aiming for:

Theorem. *Given a BN-pair with S finite, one can construct a building Δ with G-action, together with a G-invariant system of apartments $\mathcal{A}$, with the following properties:*

(a) *The G-action is strongly transitive.*

(b) *There is a chamber C whose stabilizer is B.*

(c) *There is an apartment Σ containing C and stabilized by N.*

The proof is straightforward: One just directly constructs the simplices of the desired Δ by using cosets of the subgroups P discussed above (cf. [10], §V.3). Alternatively: Set $\mathcal{C} = G/B$ and define $\delta : \mathcal{C} \times \mathcal{C} \to W$ by letting $\delta(gB, hB)$ be the unique $w \in W$ such that $(gB)^{-1}(hB) = BwB$. It is not hard to verify that $(\mathcal{C}, \delta)$ satisfies the axioms stated in Lecture 2, and we therefore get a building [with G-action, since G acts on $(\mathcal{C}, \delta)$] by applying the main theorem of that lecture. Incidentally, we do not need to appeal to the hard part of the proof of the theorem just cited, which is the construction of apartments (in the form of strong isometries $W \to \mathcal{C}$). For we can use the canonical map

$$W = N/T \to G/B = \mathcal{C}$$

to get one apartment (the "fundamental" one), and we can get enough other apartments by applying the G-action.

Remark. N is not necessarily the full stabilizer of the fundamental apartment. In order to achieve this, one needs to add an extra axiom (called **(BN3)** in [10]) which says that N is "big enough".

There are two obvious questions that suggest themselves now. First, where do BN-pairs occur naturally? Second, what good is the building associated to a BN-pair, i.e., what can we discover about G once we have a building that G acts on? The remaining two lectures will be devoted to these two questions.

LECTURE 4. BN-PAIRS AND ALGEBRAIC GROUPS

We can best understand where BN-pairs arise "in nature" by looking at the history of the subject.

1. History

We begin with a paper of Bruhat [11] published in 1954. Bruhat was interested in representations of complex Lie groups, with emphasis on the classical matrix groups, such as $\mathrm{SL}_n(\mathbf{C})$. At the time of Bruhat's work, it had been known for a long time how to associate to such a group G a finite reflection group W, called the Weyl group of G. It is given by $W = N/T$, where T is a maximal "torus" and N is its normalizer. And people were becoming aware of the importance of a certain subgroup $B \subset G$, which eventually became known as the Borel subgroup of G as a result of the fundamental work of Borel [4]. What was not yet known, however, was the connection between B and W provided by the Bruhat decomposition $G = \coprod_{w \in W} BwB$.

Bruhat discovered this while studying so-called "induced representations". Questions about these led him to ask whether the set $B\backslash G/B$ of double cosets was finite. He was apparently surprised to discover, by a separate analysis for each of the families of classical groups, that the set of double cosets was not only finite but was in 1-1 correspondence with W.

Chevalley [13] immediately realized the importance of this result, and it became a basic tool in his work on algebraic matrix groups. He replaced Bruhat's case-by-case proof by a unified proof that applied not only to the classical groups but also to the five exceptional groups E_6, E_7, E_8, F_4, and G_2. Moreover, he worked over an arbitrary field, not just $\mathbf{C}$. In particular, his work included the construction of analogues of these exceptional groups over any field k. By letting k range over the finite fields, one obtained five new families of finite simple groups.

Meanwhile, Tits had been trying for some time to give geometric interpretations of the exceptional (complex) simple Lie groups. He thought E_6, for instance, should be the automorphism group of some sort of "geometry", in the same way that $\mathrm{SL}_n(\mathbf{C})$ is essentially the automorphism group of $(n-1)$-dimensional complex projective space. [More precisely, we should replace SL_n by the projective linear group here; and it is not really the full automorphism group of projective space, but it is a subgroup easily characterized in geometric terms.] By the time of the work of Bruhat and Chevalley, he had succeeded in doing this for some but not all of the exceptional groups G.

One motivation for Tits's project was that if one could describe G geometrically, then one should be able to construct analogues of G over an arbitrary field. This motivation disappeared as a result of Chevalley's work. But the question of finding geometric interpretations was still of intrinsic interest, and, after Chevalley, one could phrase that question as follows: Now that we know that the exceptional groups exist over any field, can we use the groups (or perhaps Chevalley's method of constructing them) to construct the geometries that we have been seeking?

So Tits studied Chevalley's methods, succeeded in constructing geometries for the "Chevalley groups", and extracted the axioms **(BN1)** and **(BN2)** as the essential properties that made this work. At more-or-less the same time, he wrote down axioms satisfied by his geometries, cf. [21]. These axioms are almost identical to the axioms for buildings that we stated in Lecture 1, except that they are stated in terms of incidence geometries instead of simplicial complexes. A reformulation of the axioms in terms of simplicial complexes appeared a few years later [22]. This reformulation was quite natural, since Tits had made extensive use of flags in [21], and these flags form the simplices of a simplicial complex.

[For the reader not familiar with this terminology, we remark that an incidence geometry involves "subspaces" of various dimensions, together with a relation called "incidence"; a flag is then a finite set of pairwise incident subspaces. For example, the building $\Delta(k^n)$ can be described as the flag complex of $(n-1)$-dimensional projective space.]

2. Algebraic groups and spherical buildings

It is clear now where we should expect to find BN-pairs. Namely, we should look at algebraic matrix groups. For example, we have already seen, in a somewhat roundabout way, how to construct a BN-pair in $G = \mathrm{GL}_n(k)$ for any field k. Our approach was to deduce this from the strongly transitive action of G on $\Delta(k^n)$; but, in fact, it is completely elementary to verify the BN-pair axioms by direct matrix computations. [It then follows that $\Delta(k^n)$ is in fact a building, which we stated without proof in Lecture 1; see [10], §V.5.] One can similarly treat $\mathrm{SL}_n(k)$, $\mathrm{Sp}_{2n}(k)$, and other matrix groups.

For the benefit of readers familiar with the language of algebraic groups, we state the general result: Every reductive algebraic group (over any field) gives rise to a BN-pair with finite Weyl group W, from which one obtains a spherical building (i.e., a building in which the apartments are finite and hence spheres). The result in this generality is due to Borel and Tits [7]; it is a vast generalization of the existence of BN-pairs in Chevalley groups.

Tits [23] has proven the remarkable fact that all thick spherical buildings of rank ≥ 3 (dimension ≥ 2) arise in this way. Thus we have a very good correspondence between spherical buildings and algebraic groups, except in dimension 1.

3. Algebraic groups and Euclidean buildings

It turns out that many of the same algebraic groups mentioned in §2 admit a *second* BN-pair, whenever the ground field comes equipped with a discrete valuation. This was first discovered by Iwahori and Matsumoto [16] for Chevalley groups and was greatly generalized by Bruhat and Tits [12]. The second BN-pair has an infinite Euclidean reflection group as its Weyl group, and the resulting building therefore has Euclidean spaces as apartments. Such a building is said to be of *Euclidean* (or *affine*) type. See §4 below for the precise definition. Tits [26] has shown that in rank ≥ 4 (dimension ≥ 3) every thick Euclidean building arises from an algebraic group over a field with discrete valuation.

As an example, we will describe the second BN-pair for $G = \mathrm{SL}_n(K)$, where K is a field with a discrete valuation $v : K \to \mathbf{Z} \cup \{\infty\}$. We will confine ourselves to a sketch; further details can be found in [10], §V.8.

Let A be the valuation ring associated to v, i.e.,

$$A = \{ a \in K : v(a) \geq 0 \}.$$

It has a unique maximal ideal, which is the principal ideal πA, where $\pi \in A$ is any element with $v(\pi) = 1$. Let k be the residue field $A/\pi A$. Our point of view is that we are studying things (namely, matrix groups) defined over K, and we wish to "reduce" to a simpler field k as an aid in this study; a discrete valuation makes this possible by providing us with a nice ring A to serve as intermediary between K and k:

$$\begin{array}{ccc} A & \hookrightarrow & K \\ \downarrow & & \\ k & & \end{array}$$

In particular, we obtain our new B in G as the inverse image in $\mathrm{SL}_n(A)$ of the "ordinary B" in $\mathrm{SL}_n(k)$, i.e., the upper triangular subgroup of $\mathrm{SL}_n(k)$. On the other hand, we take N to be the same N we used before, namely, the monomial subgroup of G. [It would not make sense to also obtain N by lifting a subgroup of $\mathrm{SL}_n(k)$, since then B and N would both be subgroups of $\mathrm{SL}_n(A)$ and hence would not generate G.]

It is somewhat tedious (but not conceptually difficult) to verify that our new B and N do indeed form a BN-pair. To understand what type of building we get, we need to compute the Weyl group $W = N/T$.

Note first that $T = N \cap B$ is the group of monomial matrices in $\mathrm{SL}_n(A)$ which are upper triangular mod π. This is precisely the group $D(A)$ of diagonal matrices in $\mathrm{SL}_n(A)$. Now if we form $N/D(K)$ instead of $N/D(A)$ (where $D(K)$ is the diagonal subgroup of $G = \mathrm{SL}_n(K)$), then the quotient $\overline{W}$ is the "ordinary" Weyl group of G, i.e., the symmetric group on n letters. So there is a surjection $W \twoheadrightarrow \overline{W}$ whose kernel L is given by

$$L = D(K)/D(A) \approx (K^*/A^*)^{n-1} \approx \mathbf{Z}^{n-1}.$$

The asterisk here indicates the group of invertible elements of a ring, and the last isomorphism is induced by the valuation v. We therefore have an extension

$$1 \to L \to W \to \overline{W} \to 1.$$

It is not hard to show that the extension splits, so that

$$W \approx L \rtimes \overline{W}.$$

The action of $\overline{W}$ on L here can be described as follows: Let $V = \{ (x_1, \ldots, x_n) \in \mathbf{R}^n : \sum_{i=1}^n x_i = 0 \}$. Then $\overline{W}$ acts on V by permuting the coordinates, and we may identify L with the $\overline{W}$-invariant lattice $\mathbf{Z}^n \cap V$.

Using this description of W, one can see that it is a Euclidean reflection group acting on $V \approx \mathbf{R}^{n-1}$, with L acting as a lattice of translations. The corresponding Coxeter complex can then be identified with V itself (suitably triangulated). The building associated to the BN-pair is therefore a union of triangulated Euclidean spaces of dimension $n-1$. One can profitably think of it as an $(n-1)$-dimensional analogue of a tree. When $n = 2$, it really is a tree; in fact, it is the same as the tree described in Shalen's lectures [this volume] in terms of classes of lattices. One can give a similar description for arbitrary n.

Remarks. 1. We now have two buildings that $G = \mathrm{SL}_n(K)$ acts on. On the one hand, there is the spherical building $\Delta(K^n)$ that we get by forgetting that we have a valuation. Its apartments are spheres of dimension $n - 2$. On the other hand, we have the Euclidean building just constructed, in which the apartments are Euclidean spaces of dimension $n-1$. Moreover, there is a 1-1 correspondence between the apartments in the first building and those in the second; for both sets of apartments are in 1-1 correspondence with G/N. [Recall that we used the same N in our two BN-pairs.] As we will see in Lecture 5, there is a geometric explanation for this: Roughly speaking, the spherical building is the "boundary" of the Euclidean building; it is obtained by adjoining a "sphere at infinity" to each apartment. Thus our two buildings fit together to form a single geometric object on which G acts.

2. There is a third building that comes to mind, namely, the (spherical) building $\Delta(k^n)$, on which $\mathrm{SL}_n(k)$ operates. This too shows up in our Euclidean building: It is isomorphic to the link of any vertex.

3. We have not assumed that the field K is complete with respect to the metric defined by the valuation v. What happens if K is incomplete and we pass to its completion $\hat{K}$? For example, we might have $K = \mathbf{Q}$ and $v = v_p$ (p-adic valuation) for some prime p, so that $\hat{K}$ is the field $\mathbf{Q}_p$ of p-adic numbers. The answer is that the Euclidean building associated to $\mathrm{SL}_n(\hat{K})$ is the *same* as the one associated to $\mathrm{SL}_n(K)$, but the apartment system is bigger. In fact, the apartment system that one gets from $\mathrm{SL}_n(\hat{K})$ is the complete one, i.e., the unique maximal one (which exists by Proposition 3′ of §3 of Lecture 1), whereas the apartment system that one gets from SL_n over an incomplete field is not (cf. [10], §VI.9F). Thus some of the geometry of Δ is hidden if K is incomplete.

We close this lecture by making some general remarks about Euclidean buildings, in preparation for the applications to be given in Lecture 5.

4. Introduction to Euclidean buildings

A general reference for this section is [10], Chapter VI.

By a *Euclidean reflection group* we will mean an essential, infinite, irreducible, affine reflection group W acting on Euclidean space $\mathbf{R}^n$ for some n. [Readers not familiar with this terminology will lose little by just thinking about a typical example, such as the group of isometries of the plane generated by the reflections with respect to the sides of an equilateral triangle.] The (affine) hyperplanes whose reflections are in W decompose $\mathbf{R}^n$ into chambers, which turn out to be simplices. Any choice of "fundamental" chamber C determines a set S of $n+1$ "fundamental" reflections, and there is then a canonical homeomorphism $|\Sigma(W,S)| \approx \mathbf{R}^n$. The vertical bars here indicate the geometric realization of an abstract simplicial complex.

The Euclidean reflection groups were classified by Witt: They are precisely the "affine Weyl groups" of the root systems associated to the simple complex Lie algebras, cf. [8]. Thus there is one Euclidean reflection group for each of the types $\mathrm{A}_n, \mathrm{B}_n, \ldots, \mathrm{G}_2$ in the usual list of simple Lie algebras. And it is these same affine Weyl groups that arise as the groups $W = N/T$ associated to the BN-pairs mentioned in §3 and illustrated with the example of $\mathrm{SL}_n(K)$. The buildings we are interested in, then, have apartments $\Sigma \approx \Sigma(W,S)$ for some (W,S) as in the previous paragraph. We will call such a building Δ a *Euclidean* building, although the more common term in the literature is *building of affine type*.

In discussing Euclidean buildings, we will shift point of view slightly and work with geometric realizations $|\Delta|$ instead of abstract simplicial complexes Δ. But it will be convenient to keep using the same terminology that we have been using up to now. Thus, for instance, we will now call $X = |\Delta|$ a building and we will call the subspaces $E = |\Sigma|$ ($\Sigma \in \mathcal{A}$) apartments. Similarly, a chamber will be viewed as a geometric (open) n-simplex $C \subset X$; the corresponding closed simplex (i.e., the topological closure of C in X) will be denoted $\overline{C}$.

It is not hard to show that each apartment E carries a canonical Euclidean metric d_E and that the isomorphisms $E \xrightarrow{\approx} E'$ that come from axiom **(B2)** are isometries. It follows that the metrics d_E extend to a well-defined function $d : X \times X \to \mathbf{R}$. The following proposition lays out the basic properties of this distance function.

Proposition.

(1) *d is a metric.*
(2) *X is complete with respect to d.*
(3) *For every apartment E and chamber $C \subset E$, the retraction $\rho = \rho_{E,C} : X \to E$ is distance-decreasing, i.e.,*

$$d(\rho(x), \rho(y)) \leq d(x, y).$$

Equality holds for $x \in \overline{C}$.
(4) *Any two points $x, y \in X$ are the endpoints of a unique geodesic segment (i.e., subset isometric to a closed interval of real numbers). It is contained in every apartment containing x and y, hence the apartments are convex.*
(5) *X is contractible and satisfies the CAT(0) inequality.*

Here CAT(0) is the comparison condition discussed in Paulin's lectures [this volume]; see also [15]. Roughly speaking, it says that X is a space of non-positive curvature.

We will sketch the proof of (1), (3), (4), and (5). All omitted details, as well as the proof of (2), can be found in [10], §VI.3, although the "CAT" terminology is not used there.

The first step is to prove (3), which makes sense even before we know that d is a metric. It follows from the definition of ρ (cf. Lecture 1) that ρ maps every apartment containing C isometrically onto E. This immediately yields the second assertion of (3), and it also shows that ρ maps every closed chamber isometrically onto its image. To prove the first assertion of (3), consider the line segment $[x, y]$ joining x and y in some apartment. One can subdivide $[x, y]$ so that each piece lies in a chamber; applying ρ, we get a polygonal path in E from $\rho(x)$ to $\rho(y)$, and the desired inequality now follows from the triangle inequality in E.

It is now easy to prove (1), the content of which is that d satisfies the triangle inequality: Given $x, y, z \in X$, choose an apartment E containing x and y, and let $\rho = \rho_{E,C}$ for some chamber C of E. Using the triangle inequality in E and the first part of (3), we find

$$d(x,y) \leq d(x,\rho(z)) + d(\rho(z),y) \leq d(x,z) + d(z,y),$$

as required.

We turn next to (4). The crucial observation is that the triangle inequality just proved is strict unless z is in the line segment $[x,y]_E$ joining x and y in the Euclidean space E. For suppose $d(x,y) = d(x,z)+d(z,y)$, and let z' be the unique point of $[x,y]_E$ with $d(x,z') = d(x,z)$ and $d(z',y) = d(z,y)$. Using the chain of inequalities above (which must be equalities under our present hypothesis), one easily concludes that $\rho(z) = z'$. Recall now that ρ was defined with respect to an arbitrary chamber C in E; in particular, we could have taken C with $z' \in \overline{C}$. The second assertion of (3) now implies that $z = z'$, which proves our claim that $z \in [x,y]_E$. We now know that

$$[x,y]_E = \{\, z \in X : d(x,y) = d(x,z) + d(z,y)\,\},$$

whence (4).

Finally, we prove CAT(0). (Contractibility is a formal consequence of this; it is also proven directly in [10], §VI.3.) A formulation of CAT(0) which will be convenient for us is the following: Given $w,x,y \in X$ and $t \in [0,1]$, let $z = (1-t)x+ty \in [x,y]$, the latter being the unique geodesic segment joining x and y; then

$$d^2(w,z) \leq (1-t)d^2(w,x) + td^2(w,y) - t(1-t)d^2(x,y). \qquad (*)$$

To see why this is equivalent to CAT(0), consider the geodesic triangle with vertices w,x,y, and choose a comparison triangle in the Euclidean plane with vertices $\bar{w},\bar{x},\bar{y}$. Let $\bar{z} = (1-t)\bar{x} + t\bar{y} \in [\bar{x},\bar{y}]$. According to criterion C for CAT(0), one has to show $d(w,z) \leq d(\bar{w},\bar{z})$, cf. Chapter 3 of [15]. Now one can compute $d(\bar{w},\bar{z})$ by Euclidean geometry, and one finds that its square is precisely the right-hand side of $(*)$. Hence $(*)$ is indeed equivalent to CAT(0).

We now prove $(*)$. Choose an apartment E containing x and y, and let C be a chamber of E with $z \in \overline{C}$. Let ρ be the retraction $\rho_{E,C} : X \to E$, and consider the triangle in E with vertices $\rho(w),x,y$. Using (3), and applying the formula from Euclidean geometry mentioned in the previous paragraph, we find

$$\begin{aligned} d^2(w,z) &= d^2(\rho(w),z) \\ &= (1-t)d^2(\rho(w),x) + td^2(\rho(w),y) - t(1-t)d^2(x,y) \\ &\leq (1-t)d^2(w,x) + td^2(w,y) - t(1-t)d^2(x,y), \end{aligned}$$

as required.

We now have some idea what sorts of groups have BN-pairs and hence act on buildings. Our next goal is to see what this is good for.

A good example to keep in mind is $SL_n(\mathbf{Q}_p)$ and its associated Euclidean building X of dimension $n-1$. Recall that X is a complete geodesic metric space satisfying CAT(0). We can get immediate applications of this by using a suitable fixed-point theorem.

1. The Bruhat-Tits fixed-point theorem and applications

Recall that a finite group acting on a tree always has a fixed point (cf. [20], §I.4.3, or Shalen's lectures in this volume). One also has the classical theorem of E. Cartan that a compact group of isometries of a complete simply-connected Riemannian manifold of non-positive curvature always has a fixed point. The following result of Bruhat and Tits [12] simultaneously generalizes these two theorems.

Theorem. *Let X be a complete geodesic metric space satisfying CAT(0). If G is a group of isometries of X with a bounded orbit, then G has a fixed point.*

The idea of the proof is extremely simple: One associates to every non-empty bounded set $A \subset X$ a "center" $c(A)$. The construction depends only on the metric on X and hence is compatible with isometries. Consequently, we can take A to be any non-empty G-invariant bounded set, such as a bounded orbit, and $c(A)$ is then fixed by G.

It remains to say how $c(A)$ is defined. The original definition given in [12] was somewhat awkward, but Serre later observed that the classical notion of "circumcenter" would do the job, i.e., that one could take $c(A)$ to be the center of the smallest closed ball containing A. More precisely, let $B_r(x)$ be the closed ball of radius r centered at x, and define the *circumradius* of A, denoted $r(A)$, to be the infimum of the numbers r such that $A \subseteq B_r(x)$ for some $x \in X$. Call x a *circumcenter* of A if this infimum is achieved at x, i.e., if $A \subseteq B_r(x)$ with $r = r(A)$. The claim is that A admits a unique circumcenter, which we can then take as our definition of $c(A)$.

This result turns out to be an easy consequence of the inequality $(*)$ that we used in Lecture 4 as our formulation of CAT(0). To prove uniqueness, for instance, suppose x and y are two distinct circumcenters. Then

$$A \subseteq B_r(x) \cap B_r(y),$$

where $r = r(A)$. On the other hand, if we set $z = (1-t)x + ty$ for any t with $0 < t < 1$, then there is an $s < r$ such that

$$B_r(x) \cap B_r(y) \subseteq B_s(z);$$

for we can apply $(*)$ with $w \in B_r(x) \cap B_r(y)$ to obtain

$$d^2(w, z) \leq r^2 - t(1-t)d^2(x, y),$$

so our assertion holds with $s = \sqrt{r^2 - t(1-t)d^2(x, y)}$. Thus $A \subseteq B_s(z)$, contradicting the minimality of r. This proves that there cannot be two distinct circumcenters.

Similar ideas are used to prove existence of the circumcenter. One shows that two "approximate circumcenters" x and y must be close together, so that a sequence of better and better approximations is a Cauchy sequence and hence converges to a circumcenter. Details can be found in [10], §VI.4.

As a sample application, we indicate how one can analyze the maximal compact subgroups of a group like $G = \mathrm{SL}_n(\mathbf{Q}_p)$. [This is a p-adic Lie group. What we are about to do is analogous to a classical application of the Cartan fixed-point theorem, in which one uses the latter to prove that a real Lie group has a unique conjugacy class of maximal compact subgroups.] Let X be the Euclidean building on which G acts. The fundamental chamber C has n vertices, whose stabilizers can be computed explicitly. One of them turns out to be $\mathrm{SL}_n(\mathbf{Z}_p)$, where $\mathbf{Z}_p$ is the ring of p-adic integers, i.e., it is the stabilizer in G of the standard lattice $\mathbf{Z}_p^n \subset \mathbf{Q}_p^n$. The other $n-1$ are stabilizers of other $\mathbf{Z}_p$-lattices.

All n of these subgroups are compact. In fact, they are maximal compact subgroups, since they are even maximal proper subgroups; moreover, they are mutually non-conjugate. [The general situation when one has a BN-pair is that the stabilizers of C and its non-empty faces are the proper subgroups of G containing B. No two of these are conjugate to one another, and the maximal ones are the stabilizers of the minimal faces of C, i.e., the vertex stabilizers. These are therefore maximal proper subgroups.] The Bruhat-Tits fixed-point theorem allows us to say more:

Corollary 1. *G has exactly n conjugacy classes of maximal compact subgroups, represented by the stabilizers of the vertices of C.*

Proof. In view of what has been said above, it suffices to prove that every compact subgroup $H \subset G$ fixes a vertex of X. Now the fixed-point theorem says that H fixes a point of X, hence H stabilizes a simplex of X; but then H fixes the vertices of that simplex, because the H-action is type-preserving. □

Similarly, one can use the fixed-point theorem to say things about torsion in discrete subgroups of G. For example:

Corollary 2. *If Γ is a discrete co-compact subgroup of G, then Γ has only finitely many conjugacy classes of finite subgroups.*

Sketch of proof. Γ acts on X with finite stabilizers and compact quotient. Let D be a compact subset of X that meets every Γ-orbit. Then Γ has only finitely many distinct stabilizers in D, and the fixed-point theorem implies that every finite subgroup of Γ is conjugate to a subgroup of one of them. □

Readers familiar with the cohomology theory of discrete groups will note that the action of Γ on X also leads to cohomological information. For example, Γ has virtual cohomological dimension $\leq n-1$, and its homology and cohomology groups are finitely generated.

As our next application of the theory of buildings, we would like to indicate how to get much sharper cohomological results. For this we need to know more about Euclidean buildings.

2. Some facts about Euclidean buildings

Throughout this section, X denotes an arbitrary Euclidean building, and d is its dimension. For example, X could be the building associated to $\mathrm{SL}_n(\mathbf{Q}_p)$, in which case $d = n-1$. In an effort to give the reader some feeling for the nature of Euclidean buildings, I will state more in this section than is strictly necessary for the application to group cohomology.

Theorem 1. *The complete apartment system consists of all subsets of X which are isometric to $\mathbf{R}^d$.*

The content of this statement is that every such subset is the geometric realization of a subcomplex, and that the family $\mathcal{A}$ of these subcomplexes is a system of apartments. One can get a feeling for this by thinking about the tree case, where the result is easy. One should also compare Theorem 1 to a combinatorial fact about arbitrary buildings, hinted at in §3 of Lecture 2 and essentially proved in Appendix B: If we view a building as a set $\mathcal{C}$ with a W-valued distance function, then the complete apartment system consists of all subsets of $\mathcal{C}$ strongly isometric to W. The proof of Theorem 1 is actually quite similar to the proof of this combinatorial fact; see [10], §VI.7.

The next result requires the notion of "sector". Let E be an apartment in X. Thus E is a Euclidean space divided up into simplicial chambers by walls, the latter being the reflecting hyperplanes associated to a Euclidean reflection group W. Associated to W is a canonical finite quotient $\overline{W}$, which is a finite reflection group. We can visualize it by choosing an origin in E and translating all the W-walls so that they pass through the origin; these translated hyperplanes are the $\overline{W}$-walls. Suppose, for example, that W is the group generated by the reflections with respect to the sides of an equilateral triangle. The apartment E is shown in the first picture below, with the three heavy lines being the $\overline{W}$-walls relative to one particular choice of origin. In this case $\overline{W}$ is the dihedral group of order 6.

The $\overline{W}$-walls divide E up into chambers which are simplicial cones. Each of these chambers is called a *sector* in E. More generally, in order to free the definition from our arbitrary choice of origin, we will also call any translate of a $\overline{W}$-chamber a sector. Finally, a subset of X is called a *sector* if it is a sector in some apartment (in the complete apartment system). The second picture shows some sectors. Note that the cone point of a sector need not be at a vertex. Note also that one can translate a sector into itself, thereby obtaining a *subsector.*

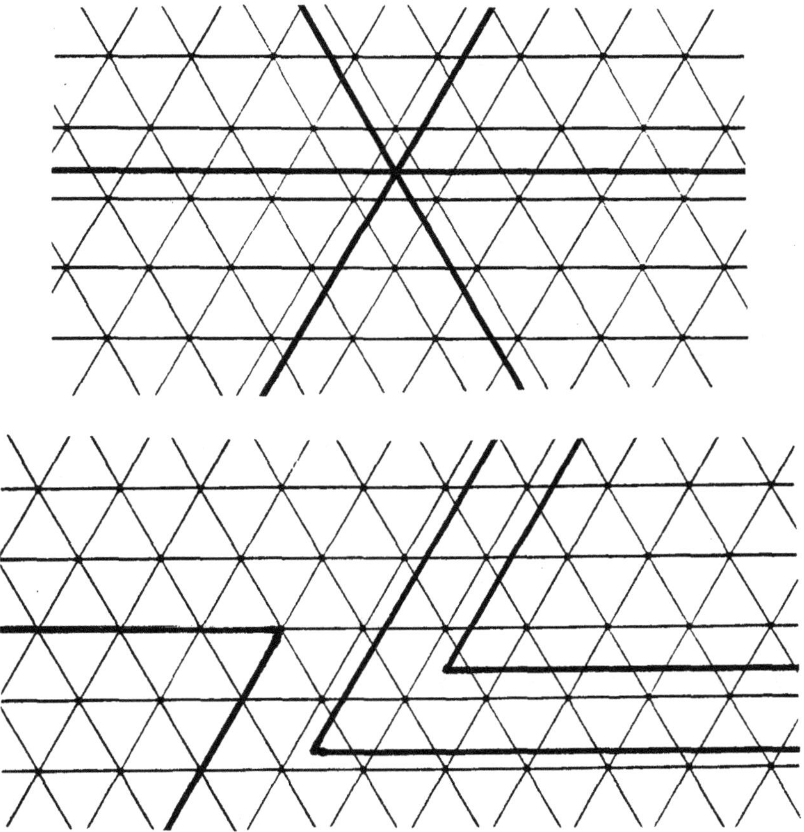

We can now state:

Theorem 2. *Given any two sectors* $\mathfrak{C}, \mathfrak{D} \subset X$, *there are subsectors* $\mathfrak{C}' \subseteq \mathfrak{C}$ *and* $\mathfrak{D}' \subseteq \mathfrak{D}$ *such that* $\mathfrak{C}'$ *and* $\mathfrak{D}'$ *are contained in some apartment.*

In the tree case, for instance, a sector is a ray, or half-line, and an apartment is a line. The reader should draw some pictures to see why any two rays have subrays which are contained in some line. The proof of Theorem 2 in the general case is more difficult and is given in [10], §VI.8.

We turn next to the ***boundary*** ∂X of X, or the set of points "at infinity". One can construct such a boundary for fairly general CAT(0) spaces, and the reader may be familiar with one or more special cases. The boundary of a tree, for instance, is its set of ends. And the boundary of a complete simply-connected Riemannian manifold of non-positive curvature is well-known in differential geometry, cf. [3] or [14]. In the present case, where X is a Euclidean building, the theory goes roughly as follows (cf. [10], §VI.9):

A point $e \in \partial X$ is represented by a ray (subset isometric to $[0, \infty)$), with two rays representing the same boundary point e if and only if they are at finite Hausdorff distance from one another.

One can organize the boundary points into simplices. For example, each sector $\mathfrak{C} \subset X$ determines a $(d-1)$-simplex (or "chamber") in ∂X, which can be visualized as the face of $\mathfrak{C}$ at infinity. [Recall that $\mathfrak{C}$ is a d-dimensional simplicial cone. The rays in $\mathfrak{C}$ starting at the cone point are parametrized by a $(d-1)$-simplex.] And the boundary points represented by rays in a given apartment E form a sphere (the "sphere at infinity"), decomposed into simplices by the sectors in E and their faces. This decomposition makes the sphere a Coxeter complex; the corresponding Coxeter group is the finite reflection group $\overline{W}$ mentioned above. Note that Theorem 2 implies that any two chambers in ∂X are contained in one of these spheres at infinity. All of this suggests the following result (cf. [10], §VI.9B):

Theorem 3. *There is a $(d-1)$-dimensional spherical building whose geometric realization is in 1-1 correspondence with ∂X.*

In case X is the Euclidean building associated to $\mathrm{SL}_n(\mathbf{Q}_p)$, for instance, the spherical building in Theorem 3 is just the usual spherical building $\Delta(\mathbf{Q}_p^n)$ that one gets by forgetting that $\mathbf{Q}_p$ has a valuation. This is proved in [10], §VI.9F.

Up to now we have been working with ∂X just as a set, with no topology. But one can impose a useful topology on ∂X, at least if X is locally compact. [And X will in fact be locally compact if it comes from an algebraic group over a field K with discrete valuation, provided the residue field k is finite.] One way to describe this topology is to choose a basepoint $x \in X$ and identify ∂X with the set of rays starting at x; we can view this set of rays as a set of maps $[0, \infty) \to X$ and give it the topology of uniform convergence on compact subintervals. It is not hard to show that the topology is independent of the choice of basepoint; moreover, ∂X, with this topology, is compact (Ascoli's theorem).

Using similar ideas, we can compactify X by adjoining ∂X "at infinity" (still assuming X is locally compact). Namely, we identify $X \cup \partial X$ with a space of maps $[0, \infty) \to X$, with points of X corresponding to maps which are isometries on a compact subinterval and constant thereafter. For future reference, we note that the compact space $X \cup \partial X$ is contractible.

Remarks. 1. The construction just sketched is well-known to work for any complete locally compact CAT(0) space, but I am not aware of any reference where the details are given in this generality.

2. The 1-1 correspondence in Theorem 3 is generally not a homeomorphism. In the tree case, for instance, ∂X might be homeomorphic to the Cantor set, whereas the spherical building in Theorem 3 is 0-dimensional and hence has a discrete geometric realization.

The final result we wish to state makes use of the topology on ∂X. Recall the Solomon–Tits theorem (Lecture 1, §3, Proposition 7), which implies that the reduced homology of a spherical building is non-trivial only in the top dimension, where it is free abelian. Borel and Serre ([6], 2.6) prove the analogue of this for the reduced cohomology $\tilde{H}^*(\partial X)$ of the compact space ∂X. [Here H^* denotes some reasonable cohomology theory for compact spaces, such as Alexander–Spanier cohomology.] Borel and Serre assume that X is the Euclidean building associated to an algebraic group over a complete field with discrete valuation and finite residue field, and they prove:

Theorem 4. *$\tilde{H}^i(\partial X)$ is trivial for $i \neq d-1$ and is free abelian for $i = d-1$.*

Remark. It seems certain that Theorem 4 remains valid for an arbitrary locally compact Euclidean building, but I have not checked this.

Since $X \cup \partial X$ is compact and contractible, one can use the long exact cohomology sequence of the pair $(X \cup \partial X, \partial X)$ to restate Theorem 4 in terms of the cohomology of X with compact supports:

Corollary. *$H_c^i(X)$ is trivial for $i \neq d$ and free abelian for $i = d$.*

3. Applications to the cohomology of discrete groups

The purpose of this short section is to give the reader a brief glimpse at one possible application of the theory of buildings. For a more detailed survey of this and related applications, see Chapter VII of [10].

The most obvious application, given what we have just done in §2, concerns discrete subgroups of p-adic Lie groups like $\mathrm{SL}_n(\mathbf{Q}_p)$. Let $\Gamma \subset \mathrm{SL}_n(\mathbf{Q}_p)$ be discrete and co-compact. For simplicity, assume further that Γ is torsion-free. Then Γ acts freely and co-compactly on the $(n-1)$-dimensional Euclidean building associated to $\mathrm{SL}_n(\mathbf{Q}_p)$. One can now use standard cohomological arguments to deduce from the corollary above that Γ has cohomological dimension exactly $n-1$ and that Γ satisfies a homological duality condition analogous to Poincaré duality ("Bieri–Eckmann duality").

More surprisingly, Borel and Serre [6] have applied the results of §2 to certain subgroups which are neither discrete nor co-compact. We will illustrate this by discussing the group $\mathrm{SL}_n(\mathbf{Z}[1/p])$ and its torsion-free subgroups of finite index.

We can view the ring $\mathbf{Z}[1/p]$ as a subring of both $\mathbf{R}$ and $\mathbf{Q}_p$. It is non-discrete in both cases, but for opposite reasons: In $\mathbf{R}$, we have $p^n \to \infty$ as $n \to +\infty$, and $p^n \to 0$ as $n \to -\infty$; in $\mathbf{Q}_p$, it is the other way around. One deduces that $\mathbf{Z}[1/p]$ embeds as a discrete subring of $\mathbf{R} \times \mathbf{Q}_p$. Consequently, we can view $\mathrm{SL}_n(\mathbf{Z}[1/p])$ as a discrete subgroup of $\mathrm{SL}_n(\mathbf{R}) \times \mathrm{SL}_n(\mathbf{Q}_p)$.

Now each of these factors has an associated contractible space on which it acts properly. In the case of $\mathrm{SL}_n(\mathbf{R})$, it is the usual symmetric space $X_\infty = \mathrm{SL}_n(\mathbf{R})/\mathrm{SO}_n(\mathbf{R})$, which is a manifold. In the case of $\mathrm{SL}_n(\mathbf{Q}_p)$, it is the Euclidean building which we have been discussing and which we now denote by X_p. This yields a proper action of $\mathrm{SL}_n(\mathbf{R}) \times \mathrm{SL}_n(\mathbf{Q}_p)$ on the product $X = X_\infty \times X_p$, and hence a properly discontinuous action of the discrete subgroup $\mathrm{SL}_n(\mathbf{Z}[1/p])$ on X.

If $\Gamma \subset \mathrm{SL}_n(\mathbf{Z}[1/p])$ is a torsion-free subgroup of finite index, then the action of Γ on X is free. It turns out, however, that the quotient $\Gamma\backslash X$ is not compact, so further work needs to be done before we can hope to draw the usual sorts of cohomological consequences.

This further work is done in an earlier paper of Borel and Serre [5]. There they attach a boundary to X_∞, obtaining a manifold with boundary $\bar{X}_\infty$. The construction is canonical enough that the action of Γ extends to the boundary. Moreover, if we replace X_∞ by $\bar{X}_\infty$ in the product above, then the action of Γ is still properly discontinuous, but now it is also co-compact.

The next step is to compute $H_c^*(\bar{X}_\infty \times X_p)$, for which it remains only to compute $H_c^*(\bar{X}_\infty)$. This computation, in turn, is reduced by standard duality theorems of algebraic topology to the computation of the homology of the boundary of $\bar{X}_\infty$. Here we find another application of buildings: Borel and Serre [5] show that the boundary they have attached to X_∞ has the homotopy type of the Tits building $\Delta(\mathbf{Q}^n)$ associated to $\mathrm{SL}_n(\mathbf{Q})$. Hence its reduced homology is non-trivial in only one dimension, where it is free abelian.

One can now deduce that Γ has finitely generated homology and satisfies Bieri–Eckmann duality; moreover, one gets an explicit calculation of its cohomological dimension.

4. Final remarks

In this lecture I have not even been able to hint at the range of applications of buildings. Many more applications are discussed in Tits's Vancouver lecture [24]. See also [17] and [19] for recent developments, with emphasis on connections with finite geometries and finite group theory.

Appendix A. Review of Coxeter Complexes

Most of what is summarized here can be found in P. de la Harpe's lectures [this volume] and/or in Chapters I–III of [10].

1. The complex associated to a Coxeter group

Let (W, S) be a Coxeter system of rank n (i.e., S has n elements). Recall that there is an associated simplicial complex $\Sigma = \Sigma(W, S)$ of rank n (dimension $n-1$), characterized by the following properties:

(1) W acts on Σ by simplicial automorphisms.
(2) There is a distinguished simplex C which is a fundamental domain for the action, in the following sense: Every simplex of Σ is W-equivalent to a unique face of C.
(3) The stabilizers of the faces of C (including C itself and the empty face) are the special subgroups of W, i.e., the subgroups generated by subsets of S, including S itself and the empty set.

In particular, the stabilizer of C is the trivial subgroup, so that W acts simply-transitively on the chambers (maximal simplices) of Σ. And the stabilizers of the codimension 1 faces of C are the subgroups $\langle s \rangle$ of order 2 generated by the elements of S. It follows that the set $\mathcal{C}$ of chambers of Σ can be identified with W (via $wC \leftrightarrow w$), and that two distinct elements $w, w' \in W$ correspond to adjacent chambers if and only if $w' = ws$ for some $s \in S$. [Another way to say this is that chambers correspond to vertices of the Cayley graph of (W, S), and pairs of distinct adjacent chambers correspond to edges of the Cayley graph.]

Recall that a *gallery* is a sequence of chambers $C_0, C_1, \ldots, C_l$ such that C_{i-1} and C_i are adjacent for $i = 1, \ldots, l$. It is said to *stutter* if $C_{i-1} = C_i$ for some i. In view of the previous paragraph, non-stuttering galleries in Σ starting at C are in 1-1 correspondence with S-words (and hence with paths starting at 1 in the Cayley graph). Under this correspondence, minimal galleries correspond to reduced words (and hence to geodesics in the Cayley graph). It follows that the metric spaces $\mathcal{C}$ and W are isometric, where the former has the gallery metric and the latter has the word metric.

The picture below illustrates the case where W is the group generated by the reflections in the sides of an equilateral triangle. Thus Σ is the Euclidean plane, tiled by equilateral triangles. The Cayley graph has been superimposed on Σ; it is the 1-skeleton of the dual tiling. Note also that the vertices of Σ have been drawn in three different "colors", to indicate the three W-orbits (or, in the language to be introduced in §3, the three *types* of vertices). The reader is advised to choose some words at random and to trace out the corresponding galleries in Σ. [One first has to choose a fundamental chamber in Σ or, equivalently, an origin in the Cayley graph.]

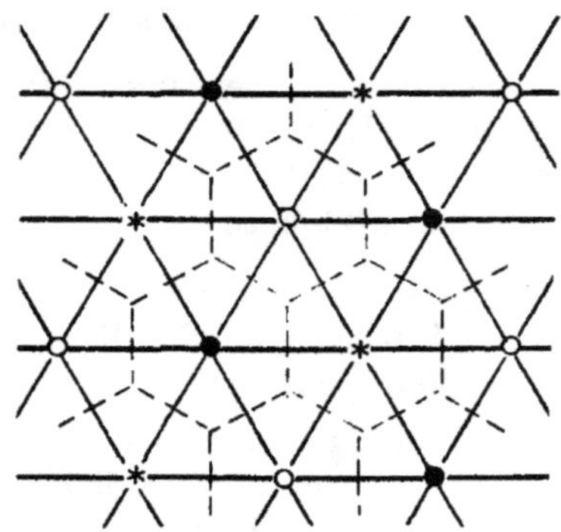

2. Coxeter complexes

By a *Coxeter complex* we will mean an abstract simplicial complex Σ which is isomorphic to $\Sigma(W,S)$ for some Coxeter system (W,S) (with S finite).

Examples. (a) The only 0-dimensional Coxeter complex is the 0-sphere S^0.

(b) The 1-dimensional Coxeter complexes are the $2m$-gons, $2 \le m \le \infty$.

(c) If X is the boundary of a regular convex polytope in $\mathbf{R}^n$, then the barycentric subdivision of X is a Coxeter complex; its simplices are the chains $F_0 < \cdots < F_q$ of non-empty faces of X.

(d) For any triple $\{p,q,r\}$ of integers ≥ 2, there is a well-known tiling of the "plane" (spherical, Euclidean, or hyperbolic, depending on the sign of $\frac{1}{p}+\frac{1}{q}+\frac{1}{r}-1$) by triangles with angles $\frac{\pi}{p}$, $\frac{\pi}{q}$, and $\frac{\pi}{r}$. The abstract simplicial complex underlying this tiling is a Coxeter complex. The case $p=q=r=3$ was drawn above.

(e) If Σ is a Coxeter complex and $\sigma \in \Sigma$ is a simplex of codimension i, then the link of σ in Σ, denoted $\operatorname{lk}\sigma$ or $\operatorname{lk}_\Sigma \sigma$, is a Coxeter complex of rank i (dimension $i-1$). This assertion will be used later, so we explain it briefly: We may assume that $\Sigma = \Sigma(W,S)$ and that σ is a face of the fundamental chamber C. In this case the stabilizer of σ is a special subgroup $W' = \langle S' \rangle$, and I claim that $\operatorname{lk}\sigma \approx \Sigma(W',S')$. To prove the claim, one need only observe that W' acts on $\operatorname{lk}\sigma$ with a simplex σ' as fundamental domain and with the special subgroups of W' as the stabilizers of the faces of σ'; this is easy to check. Alternatively, one can deduce the claim from the definition of $\Sigma(W,S)$ in terms of "special cosets"; see [10], §III.2.

(f) If Σ_1 and Σ_2 are Coxeter complexes, then so is their join $\Sigma_1 * \Sigma_2$.

Remarks. 1. Tits has characterized Coxeter complexes in terms of the existence of enough "half-spaces"; see [10], §III.4.

2. Note that the Coxeter complexes in Examples (a)–(d) above are topologically either spheres or Euclidean spaces. The general fact is that every finite Coxeter complex is homeomorphic to a sphere and that every infinite Coxeter complex is contractible (but not necessarily homeomorphic to Euclidean space). See [10], §IV.6.

3. Labellings

Let Σ be a chamber complex, i.e., a finite-dimensional simplicial complex in which all the maximal simplices (called *chambers*) have the same dimension, and any two chambers can be connected by a gallery. Let n be the rank of Σ and let I be a set of cardinality n. A *labelling* of Σ by I is a function $\lambda : V(\Sigma) \to I$ (where $V(\Sigma)$ is the set of vertices of Σ), such that the vertices of any chamber are mapped bijectively onto I. If Σ can be labelled, then the labelling is essentially unique: Any two labellings (say by sets I and I') differ by a bijection $I \approx I'$. (To see this, just note that if the labelling is known on a chamber C, then it is determined on any chamber adjacent to C.)

It is sometimes helpful to think of a labelling as a "coloring" of the vertices. The number of colors used is required to be the rank of Σ, and joinable vertices are required to have different colors.

We often call $\lambda(v)$ the *type* of v. More generally, the *type* of a simplex σ is the subset of I consisting of the labels $\lambda(v)$ of the vertices of σ. This notion of "type" depends on λ, but the induced equivalence relation (where two simplices are equivalent if they have the same type) is independent of λ.

Not every chamber complex is labellable. (For example, a 1-dimensional complex is labellable if and only if it contains no closed polygon with an odd number of sides.) But Coxeter complexes are always labellable. To see this, we may assume $\Sigma = \Sigma(W, S)$; then the W-action partitions the vertices into n orbits, and we can label Σ by associating one label $i \in I$ to each orbit. The *canonical labelling* of $\Sigma(W, S)$ is obtained by taking $I = S$ and defining λ as follows: If v is a vertex of the fundamental chamber C, then the face of C opposite v has stabilizer $\langle s \rangle$ for some $s \in S$; set $\lambda(v) = s$. This defines λ on the vertices of C, and we use the W-action to extend λ to all the vertices.

For a second situation where there is a canonical labelling, suppose Σ is the barycentric subdivision of a cell complex X, as in Example (c), for instance. Then Σ has one vertex v for every cell F of X, and we set $\lambda(v) = \dim F$. Thus $I = \{0, \dots, n-1\}$ in this case.

4. Types of galleries

Let Σ be a labelled chamber complex. Then we can use the labelling to refine the adjacency relation on the set $\mathcal{C}$ of chambers: Let I be the set of labels. Then any codimension 1 simplex of Σ has type $I - \{i\}$ for some $i \in I$. Given $i \in I$, two chambers of Σ will be called *i-adjacent* if they have the same face of type $I - \{i\}$. This is an equivalence relation, unlike the ordinary adjacency relation.

Note that two distinct adjacent chambers are i-adjacent for a unique i. Hence a non-stuttering gallery $\Gamma = (C_0, \ldots, C_l)$ has a well-defined *type* $\mathbf{i} = (i_1, \ldots, i_l)$, such that C_{j-1} and C_j are i_j-adjacent for $j = 1, \ldots, l$. We will call $\mathbf{i}$ a *word*, or an *I-word*, even though there is no group under discussion at the moment. In case Σ is thin (i.e., every codimension 1 simplex is a face of exactly two chambers), the assignment $\Gamma \mapsto \mathbf{i}$ yields a bijection from the set of non-stuttering galleries starting at a given chamber C to the set of I-words.

For a familiar example of this, suppose $\Sigma = \Sigma(W, S)$ with its canonical labelling, and let C be the fundamental chamber. One can easily check that the bijection just described coincides with the one mentioned in §1.

5. The Coxeter matrix

Let Σ be a Coxeter complex with an arbitrary labelling $\lambda : V(\Sigma) \to I$. Given $i, j \in I$ with $i \neq j$, choose a simplex σ of type $I - \{i, j\}$. Then the link of σ in Σ is a rank 2 Coxeter complex, hence it is a $2m$-gon for some m $(2 \leq m \leq \infty)$. This number m depends only on $\{i, j\}$, and not on the choice of σ, since the group of type-preserving automorphisms of Σ is transitive on the simplices of any given type. Hence we may write

$$m = m_{ij}.$$

If we further set $m_{ii} = 1$, then the resulting matrix $M = (m_{ij})$ is called the *Coxeter matrix* of Σ (with respect to λ). This terminology is explained by the following result:

Proposition. *Let $\Sigma = \Sigma(W, S)$ with its canonical labelling. Then its Coxeter matrix M is the same as the Coxeter matrix of (W, S) in the usual sense, i.e., $m_{s,t}$ is the order of st.*

Proof. To compute the Coxeter matrix, it suffices to consider codimension 2 faces of the fundamental chamber C. Let σ be such a face, and let its type be $S - \{s, t\}$. We must show that the link of σ is a $2m$-gon, where m is the order of st. Now we know that the link of σ is the Coxeter complex associated to the stabilizer W' of σ (cf. §1, Example (e)); and one easily checks from the definition of the canonical labelling that W' is the dihedral subgroup $\langle s, t \rangle$, of order $2m$. So the proposition follows from the fact that the Coxeter complex of a dihedral group of order $2m$ is a $2m$-gon. □

For an arbitrary Σ and λ, the proposition motivates us to introduce the Coxeter group $W = W_M$ associated to M, with generating set $S = \{s_i\}$ in 1-1 correspondence with I, and relations $(s_i s_j)^{m_{ij}} = 1$. It should be reasonably clear to the reader that we will then have $\Sigma \approx \Sigma(W, S)$, as stated in Lecture 2 (§1,

Proposition 1). But our real reason for introducing W_M is that it provides the right framework for understanding types of galleries. This is the subject of the next section.

6. The W-valued distance function

Let Σ be a labelled Coxeter complex. The set $\mathcal{C}$ of chambers of Σ can be viewed as a metric space (with distance defined by galleries) or as a set with a family of equivalence relations (i-adjacency). We close this "review" by showing that there is a single structure on $\mathcal{C}$ from which one can obtain both the metric and the adjacency relations.

Proposition. *Let Σ be a labelled Coxeter complex, let M be its Coxeter matrix, and let W_M be the associated Coxeter group.*

(1) *There is a unique function $\delta : \mathcal{C} \times \mathcal{C} \to W_M$ with the following property: Let $C_0, \dots, C_l$ be a non-stuttering gallery and let $\mathbf{i} = (i_1, \dots, i_l)$ be its type. Then $\delta(C_0, C_l)$ is the element $s_{i_1} \cdots s_{i_l}$ of W_M represented by $\mathbf{i}$.*

(2) *Suppose $\Sigma = \Sigma(W, S)$ with its canonical labelling. If we identify both $\mathcal{C}$ and W_M with W in the usual way, then $\delta : W \times W \to W$ is given by*

$$\delta(w, w') = w^{-1} w'.$$

Proof. Uniqueness in (1) is obvious. To prove existence, we may assume $\Sigma = \Sigma(W, S)$ with its canonical labelling. We then take the formula in (2) as a definition. A non-stuttering gallery of type $\mathbf{s} = (s_1, \dots, s_l)$ from w to w' in $\Sigma(W, S)$ has the form

$$w, ws_1, ws_1s_2, \dots, ws_1 \cdots s_l = w',$$

whence $\delta(w, w') = s_1 \cdots s_l$. Thus δ satisfies the condition in (1). □

Note that we recover the gallery metric d from δ by

$$d(C, D) = l(\delta(C, D)),$$

where $l : W_M \to \mathbf{Z}$ is the length function with respect to the generating set $S = \{ s_i : i \in I \}$. And we recover the adjacency relations by

$$C \overset{i}{\sim} C' \iff \delta(C, C') \in \langle s_i \rangle.$$

Appendix B. Buildings as W-Metric Spaces

The purpose of this appendix is to sketch a proof of the theorem stated in §3 of Lecture 2. We will base our sketch on the exercises in §IV.4 of [10], which contain all the necessary ideas. The reader who has worked through those exercises should be able to fill in the missing details.

For ease of reference, we restate the theorem to be proved:

Theorem. *Let (W,S) be a Coxeter system with S finite. Given a set $\mathcal{C}$ and a function $\delta : \mathcal{C} \times \mathcal{C} \to W$, the pair $(\mathcal{C},\delta)$ arises from a building if and only if it satisfies the following four axioms:*

(1) $\delta(C,D) = 1$ *if and only if* $C = D$.
(2) $\delta(D,C) = \delta(C,D)^{-1}$.
(3) *If* $\delta(C',C) = s \in S$ *and* $\delta(C,D) = w$, *then* $\delta(C',D) = sw$ *or* w. *If, in addition,* $l(sw) = l(w)+1$, *then* $\delta(C',D) = sw$.
(4) *If* $\delta(C,D) = w$, *then for any* $s \in S$ *there is a* C' *such that* $\delta(C',C) = s$ *and* $\delta(C',D) = sw$. *If* $l(sw) = l(w) - 1$, *then there is a unique such* C'.

1. Proof of the "only if" part

The notation here is as in Lecture 2: Δ is a labelled building with Weyl group W; S is the distinguished generating set of W and is also identified with the set of labels of Δ; $\mathcal{C}$ is the set of chambers of Δ; and $\delta : \mathcal{C} \times \mathcal{C} \to W$ is the "distance function" obtained by considering types of minimal galleries. We need to prove that δ satisfies axioms (1)–(4) above.

(1) and (2) are immediate. To prove (3) and (4), let $\delta(C,D) = w$, let $\mathbf{s} = (s_1,\ldots,s_l)$ be a reduced decomposition of w, and choose a minimal gallery $C = C_0,\ldots,C_l = D$ of type $\mathbf{s}$; this is possible by Exercise 3(c) in the set of exercises cited above. Suppose first that $l(sw) = l(w)+1$. Then for any C' with $\delta(C',C) = s$, the composite gallery $C', C_0,\ldots,C_l$ has reduced type $\mathbf{s}' = (s,s_1,\ldots,s_l)$ so it is minimal (*loc. cit.*, Exercise 1) and $\delta(C',D) = sw$. Both (3) and (4) follow in this case. Now suppose $l(sw) = s(w) - 1$. Then we can choose $\mathbf{s}$ so that $s_1 = s$. We then have $\delta(C,C_1) = s$ and $\delta(C_1,D) = sw$. And if C' is any chamber distinct from C_1 and s-adjacent to C, then the gallery $C', C_1,\ldots,C_l$ is minimal [being non-stuttering and of reduced type $\mathbf{s}$], hence $\delta(C',D) = w$. □

The reader who has done the exercises cited above will note that we have just repeated an argument used in the solution of Exercise 4(b), as outlined in the hint to that exercise. It turns out that the rest of Exercise 4, which was concerned with the construction of apartments in the building Δ, can be done as a formal consequence of the properties of δ. In other words, it depends only on (1)–(4), and not on the fact that $(\mathcal{C},\delta)$ comes from a building. We will indicate how this is done in the next section. We will then be able to prove the "if" part of the theorem in the following section.

2. The construction of apartments

In this section we assume only that (W,S) is a Coxeter system and that $\mathcal{C}$ is a set with a function $\delta : \mathcal{C} \times \mathcal{C} \to W$ satisfying (1)–(4). If we set $d(C,D) = l(\delta(C,D))$, then it is easy to see that d is a $\mathbf{Z}$-valued metric in the usual sense.

Note also that we have for each $s \in S$ an equivalence relation of ***s-adjacency*** on $\mathcal{C}$, defined by

$$C \overset{s}{\sim} C' \iff \delta(C, C') \in \langle s \rangle.$$

Note next that, in view of (2), there are analogues of (3) and (4) obtained by interchanging the roles of C and D. The analogue of the first part of (3), for example, is:

(3′) If $\delta(C, D) = w$ and $\delta(D, D') = s \in S$, then $\delta(C, D') = ws$ or w.

This has the following interpretation: Fix $C \in \mathcal{C}$ and let $\rho : \mathcal{C} \to W$ be the function $\delta(C, -)$; then ρ preserves s-adjacency for all $s \in S$. [In case $(\mathcal{C}, \delta)$ comes from a building, one can explain this fact conceptually by choosing an apartment Σ containing C and relating ρ to the retraction $\rho_{\Sigma,C} : \Delta \to \Sigma$. Details are left to the interested reader.]

We turn now to the main goal of this section, which is the construction of strong isometries from W into $\mathcal{C}$. [Recall that a strong isometry is a function that preserves δ, where δ is defined on W by $\delta(w, w') = w^{-1}w'$.] Even though we do not yet know that $(\mathcal{C}, \delta)$ comes from a building, one should think of the image of a strong isometry $W \to \mathcal{C}$ as an apartment. So the following theorem, in effect, proves the existence of apartments containing given subsets of $\mathcal{C}$.

Theorem. *For any subset $\mathcal{D} \subset W$, any strong isometry $\alpha : \mathcal{D} \to \mathcal{C}$ extends to a strong isometry $W \to \mathcal{C}$.*

We confine ourselves to an outline of the proof. The missing details are not difficult and can be found in the hints to Exercise 4 (*loc. cit.*), or the paper of Tits [25], or Ronan's book [18].

The crucial step is to show that if w_1 and w_2 are adjacent, with $w_1 \in \mathcal{D}$ and $w_2 \notin \mathcal{D}$, then α extends to $\mathcal{D} \cup \{w_2\}$. We may assume that $w_1 = 1$ (the "fundamental chamber"), in which case $w_2 = s$ for some $s \in S$. Let $\alpha(1) = C$. We wish to extend α by setting $\alpha(s) = C'$ for a suitable $C' \in \mathcal{C}$ with $\delta(C, C') = s$. The extension will be a strong isometry provided

$$\delta(C', \alpha(w)) = sw$$

for all $w \in \mathcal{D}$. Equivalently, if we define $f : \mathcal{D} \to W$ by $f(w) = s\delta(C', \alpha(w))$, then we want $f(w) = w$ for all w.

[Note, for future reference, that f is distance-decreasing, i.e.,

$$d(f(w), f(w')) \le d(w, w').$$

One sees this by writing $f(w) = s\rho(\alpha(w))$, where $\rho = \delta(C', -) : \mathcal{C} \to W$, and using the fact that ρ preserves the adjacency relations. See the discussion following (3′) above.]

Now we know, since α is a strong isometry, that $\delta(C, \alpha(w)) = w$ for all $w \in \mathcal{D}$. So the second part of axiom (3) implies that, for any choice of C', we will have $f(w) = w$ for all $w \in \mathcal{D}$ such that $l(sw) = l(w) + 1$. Moreover, axiom (4) says that for any $w \in \mathcal{D}$ with $l(sw) = l(w) - 1$, we can find a C' such that the resulting f will satisfy $f(w) = w$ for that particular w. What we must show, then, is that the (unique) C' which works for one such w works for all of them.

Let $\mathcal{E} = \{ w \in \mathcal{D} : l(sw) = l(w) - 1 \}$, and choose C' as above so that f satisfies $f(w) = w$ for at least one $w \in \mathcal{E}$ (unless $\mathcal{E} = \varnothing$, in which case we are already done). Then f is distance-decreasing and satisfies $f(w) = w$ or sw for any $w \in \mathcal{E}$. An easy geometric argument (cf. *loc. cit.*, last part of the hint to Exercise 4(b)) now shows that in fact $f(w) = w$ for all $w \in \mathcal{E}$. □

3. Proof of the "if" part of the theorem

Using the theorem just proved, we can complete the sketch of the proof of our main theorem. Assume, as in §2, that $(\mathcal{C}, \delta)$ satisfies (1)–(4); our task is to construct a building with $(\mathcal{C}, \delta)$ as the associated "W-metric space". We already know what the chambers of the desired Δ should be, and we already know what the adjacency relations should be. To construct the rest of the simplices, fix a subset $S' \subseteq S$, let $W' = \langle S - S' \rangle$, and consider the following relation on $\mathcal{C}$:

$$C \sim D \iff \delta(C, D) \in W'.$$

This is an equivalence relation, and the equivalence classes will be called *simplices of type* S'. [Motivation: We want to specify a simplex σ by giving the set of chambers of which σ is a face.] If $S'' \subseteq S'$, then any simplex σ of type S', viewed as a subset of $\mathcal{C}$, is contained in a unique simplex τ of type S''; we say that τ is the *face* of σ of type S''. It is now straightforward to verify that the set of simplices, with the face relation just defined, is indeed the set of simplices of a simplicial complex Δ.

For any strong isometry $\alpha : W \to \mathcal{C}$, the image of α generates a subcomplex Σ of Δ, called an apartment. One now verifies the building axioms **(B0)** and **(B2)** without much difficulty. [For **(B2)**, use the maps ρ introduced in §2.] And **(B1)** follows from the theorem of §2, since any two-element subset of $\mathcal{C}$ is strongly isometric to a subset of W. □

References

1. H. Abels, *The gallery distance of flags*, preprint, Bielefeld, 1989.
2. H. Abels, *The geometry of the chamber system of a semimodular lattice*, preprint, Bielefeld, 1990.
3. W. Ballmann, M. Gromov, and V. Schroeder, *Manifolds of nonpositive curvature*, Progress in Mathematics, vol. 61, Birkhäuser, Boston, 1985.
4. A. Borel, *Groupes linéaires algébriques*, Ann. of Math. **64** (1956), 20–80.

5. A. Borel and J-P. Serre, *Corners and arithmetic groups*, Comment. Math. Helv. **48** (1974), 244–297.
6. A. Borel and J-P. Serre, *Cohomologie d'immeubles et de groupes S-arithmétiques*, Topology **15** (1976), 211–232.
7. A. Borel and J. Tits, *Groupes réductifs*, Inst. Hautes Études Sci. Publ. Math. **41** (1972), 5–251.
8. N. Bourbaki, *Groupes et Algèbres de Lie*, Chapitres 4–6, Masson, Paris, 1981.
9. K. S. Brown, *Cohomology of Groups*, Graduate Texts in Mathematics 87, Springer-Verlag, New York, 1982.
10. K. S. Brown, *Buildings*, Springer-Verlag, New York, 1989.
11. F. Bruhat, *Représentations des groupes de Lie semi-simples complexes*, C. R. Acad. Sci. Paris **238** (1954), 437–439.
12. F. Bruhat and J. Tits, *Groupes réductifs sur un corps local. I. Données radicielles valuées*, Inst. Hautes Études Sci. Publ. Math. **41** (1972), 5–251.
13. C. Chevalley, *Sur certains groupes simples*, Tohôku Math. J. (2) **7** (1955), 14–66.
14. P. Eberlein and B. O'Neill, *Visibility manifolds*, Pacific J. Math. **46** (1973), 45–109.
15. E. Ghys and P. de la Harpe (eds.), *Sur les groupes hyperboliques d'après Mikhael Gromov*, Progress in Mathematics, vol. 83, Birkhäuser, Boston, 1990.
16. N. Iwahori and H. Matsumoto, *On some Bruhat decomposition and the structure of the Hecke rings of p-adic Chevalley groups*, Inst. Hautes Études Sci. Publ. Math. **25** (1965), 5–48.
17. W. M. Kantor, R. A. Liebler, S. E. Payne, and E. E. Shult (eds.), *Finite Geometries, Buildings, and Related Topics*, Oxford University Press, Oxford, 1990.
18. M. Ronan, *Lectures on Buildings*, Academic Press, Boston, 1989.
19. L. A. Rosati (ed.), *Buildings and the Geometry of Diagrams*, Proc. CIME Como 1984, Lecture Notes in Mathematics 1181, Springer-Verlag, Berlin, 1986.
20. J-P. Serre, *Trees*, Springer-Verlag, Berlin, 1980, translation of "Arbres, Amalgames, SL_2," Astérisque **46**, 1977.
21. J. Tits, *Géométries polyédriques et groupes simples*, Atti della II Riunione del Groupement de Mathématiciens d'Expression Latine, Florence (1961), Edizioni Cremonese, Rome, 1963, pp. 66–88.
22. J. Tits, *Structures et groupes de Weyl*, Séminaire Bourbaki 1964-65, exposé no. 288, February 1965.
23. J. Tits, *Buildings of Spherical Type and Finite BN-Pairs*, Lecture Notes in Mathematics 386, Springer-Verlag, Berlin, 1974.
24. J. Tits, *On buildings and their applications*, Proc. International Congress of Mathematicians, Vancouver 1974, vol. 1, Canad. Math. Congress, Montreal, 1975, pp. 209–220.
25. J. Tits, *A local approach to buildings*, The Geometric Vein (C. Davis, B. Grünbaum, and F. A. Sherk, eds.), Coxeter Festschrift, Springer-Verlag, New York, 1981, pp. 519–547.
26. J. Tits, *Immeubles de type affine*, Buildings and the Geometry of Diagrams (L. A. Rosati, ed.), Proc. CIME Como 1984, Lecture Notes in Mathematics 1181, Springer-Verlag, Berlin, 1986, pp. 159–190. *Note*: The author has written a correction to this paper, which is not yet published, in which he has modified axiom (A5).
27. J. Tits, *Théorie des groupes*, Résumé du cours, Annuaire du Collège de France, 1988–89, pp. 81–96.

DEPARTMENT OF MATHEMATICS, WHITE HALL, CORNELL UNIVERSITY, ITHACA, NY 14853

E-mail: kbrown@mssun7.msi.cornell.edu

The group of values of a W–distance is a Coxeter group

Herbert Abels

Recently, Tits [T] has given a characterization of buildings by what he called Weyl group valued distance functions. It is shown here that one need not suppose that the group of values of this distance is a Coxeter group, this follows from the properties of this distance. So there are no buildings modeled on other groups than Coxeter groups. The result may be considered as a non group theoretical version of the theorem that the Weyl group of a Tits system is a Coxeter group. This result answers a question raised by K. Brown and P. de la Harpe at the workshop on geometries and groups in Trieste in April 1990.

W–valued distances

We fix a pair (W, S) consisting of a group W and a subset S of W which generates W. Let $\ell : W \longrightarrow \mathbf{Z}$ be the corresponding *length*, so $\ell(w)$ is the infimum of numbers $q \geq 0$ such that $w = s_1^{\varepsilon_1} \ldots s_q^{\varepsilon_q}$ with $s_i \in S$ and $\varepsilon_i \in \{\pm 1\}$. Let $\mathfrak{C}$ be a set. A *W-valued distance function* on $\mathfrak{C}$, or *W–distance* for short, is a map

$$\delta : \mathfrak{C} \times \mathfrak{C} \longrightarrow W$$

with the following properties:

(D0) $\delta(C, D) = 1$ *iff* $C = D$.

(D1) *If* $\delta(C, D) = w$ *and* $\delta(D, D') = s \in S$, *then* $\delta(C, D') \in \{w, ws\}$. *If furthermore* $\ell(ws) = \ell(w) + 1$, *then* $\delta(C, D') = ws$.

(D2) *Given* C *and* D *in* $\mathfrak{C}$ *and* $s \in S$ *there is a* $D' \in \mathfrak{C}$ *such that* $\delta(D, D') = s$ *and* $\delta(C, D') = \delta(C, D) \cdot s$.

The elements of $\mathfrak{C}$ are called *chambers*. $\mathfrak{C}$ with W–distance δ is called *thick* if for every $s \in S$ and every chamber C there are at least two chambers C' with $\delta(C, C') = s$.

The content of this note is the following

Proposition *If δ is a W-valued distance on $\mathfrak{C}$ and $\mathfrak{C}$ is thick, then (W,S) is a Coxeter system.*

The hypothesis that $\mathfrak{C}$ be thick is indispensable since for every group W and every subset S of generators of W the function

$$\delta : W \times W \longrightarrow W$$
$$\delta(w_1, w_2) = w_1 w_2^{-1}$$

is a W-distance.

Proof

We fix a set $\mathfrak{C}$ with W-distance δ and suppose that $\mathfrak{C}$ is thick.

2.1 *For every $s \in S$ we have $s \neq 1$ and $s^2 = 1$.*

Proof Fix a chamber C and two distinct chambers D and E with $\delta(C,D) = \delta(C,E) = s$. At least one of them is different from C, hence $s \neq 1$ by D0. Note that hence C, D, E are all different.

We next show that $\delta(D,C) = s^{-1}$. Put $w = \delta(D,C)$. Then $1 = \delta(D,D) \in \{w, ws\}$ by D1, hence $ws = 1$ since $w = \delta(D,C) \neq 1$.

It follows that $\delta(D,E) = s^{-1}$, since $\delta(D,E) \in \{s^{-1}, s^{-1} \cdot s\}$ by D1 and $\delta(D,E) \neq 1$ by D0. Also $\delta(E,D) = s^{-1}$ by interchanging the roles of D and E. It thus remains to prove that $\delta(D,E) = s$. By D2 there is a chamber F such that $\delta(E,F) = s$ and $\delta(D,F) = s^{-1} \cdot s = 1$, hence $D = F$ and $\delta(E,D) = s$.

We obtain a number of corollaries of 2.1.

2.2 *For every $s \in S$ the relation "$C \underset{s}{\sim} D$ iff $\delta(C,D) \in \{1, s\}$" is an equivalence relation, called s-adjacency.*

Proof Use D1.

As usual, a sequence $(C_0, \ldots, C_p)$ of chambers is called a *gallery* of type $\sigma = (s_1, \ldots s_p)$, $s_i \in S$, if $C_{i-1} \underset{s_i}{\sim} C_i$ for every i from 1 to p . Now D1 implies

2.3 *If $(C_0, \ldots, C_p)$ is a gallery of type $\sigma = (s_1, \ldots, s_p)$ then there is a subsequence $(t_1, \ldots, t_q)$ of σ such that $\delta(C_0, C_p) = t_1 \ldots t_q$. If furthermore $\ell(w) = p$ for $w = s_1 \ldots s_p$ then $\delta(C_0, C_p) = w$.*

And D2 implies

2.4 *Let C be a chamber, let $\sigma = (s_1, \ldots, s_p)$ be a sequence in S and put $w = s_1 \ldots s_p$. Then there is a chamber D with $\delta(C, D) = w$ and for every such D there is a gallery $(C_0, \ldots, C_p)$ of type σ with $C_0 = C$ and $C_p = D$.*

2.5 *For any two chambers C and D we have*

$$\delta(C, D) = \delta(D, C)^{-1} .$$

Proof Take a sequence $\sigma = (s_1, \ldots, s_p)$ such that $w = \delta(C, D)$ is the product $s_1 \ldots s_p$ and $\ell(w) = p$. There is a gallery as in 2.4. Read this gallery backwards and apply 2.3.

2.6 D1 *and* D2 *also hold with respect to the first variable C . I.e. if $\delta(C, D) = w$ and $\delta(C', C) = s \in S$ then $\delta(C', D) \in \{w, sw\}$, etc.*

Proof Use 2.5 and 2.1.

The next step is to prove that $\ell(w) \neq \ell(sw)$.

2.7 *$\ell(sw) = \ell(w) \pm 1$ for $w \in W$ and $s \in S$.*

Proof We have to prove that $\ell(sw) \neq \ell(w)$ and do this by induction on $\ell(w) = q$. The case $q = 0$ is in 2.1. So suppose 2.7 holds for every $w \in W$ of length $< q$ and every $s \in S$. The main step is to prove that the following stronger form of D1 holds, which we state for the first variable.

Claim If $\delta(C', C) = s \in S$ and $\delta(C, D) = w$ and $\ell(sw) \geq \ell(w) = q$, then $\delta(C', D) = sw$.

The case $\ell(sw) > \ell(w)$ is D1 for the first variable. To prove the claim, take elements $v \in W$ and $t \in S$ such that $vt = w$ and $\ell(v) = q - 1$. We first show that

$$\ell(sv) = q .$$

We have $q - 1 \leq \ell(sv) \leq q$, the first inequality holds since $\ell(svt) = \ell(sw) = q$, the second one since $\ell(v) = q \dot{-} 1$. But $\ell(sv) = q - 1 = \ell(v)$ contradicts our inductive hypothesis.

We now prove the claim. There is a chamber E such that $\delta(C, E) = v$ and $\delta(E, D) = t$, by D2. Then $\delta(C', E) = sv$ by D1, since $\delta(C, E) = v$ has length $q - 1 < \ell(sv)$.

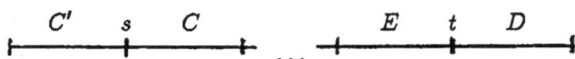

We obtain

$$\delta(C', D) \in \{sv\,,\, svt\}$$
$$\text{and } \delta(C', D) \in \{vt\,,\, svt\}\,,$$

the first inclusion by D1 for C', E, D and the second one by D1 for the first variable for the triple C', C, D with $\delta(C, D) = w = vt$. It follows that $\delta(C', D) = svt = sw$, since if $\delta(C', D) \neq svt$, then $\delta(C', E)$ must be equal to both sv and vt, but $sv = vt$ implies $sw = svt = vt^2 = v$, contradicting $\ell(sw) = q > q - 1 = \ell(v)$.

The claim now implies 2.7 for $\ell(w) = q$ as follows. Suppose $\ell(sw) = \ell(w)$. Take chambers C and D with $\delta(C, D) = w$, using 2.4. By thickness there are two chambers C' and C'' with $\delta(C', C) = \delta(C'', C) = s$. Then $\delta(C', D) = \delta(C'', D) = sw$, by the claim. But we can apply the claim also for C'', C', D, since $\delta(C', D) = sw$ and $\ell(s(sw)) = \ell(sw) = q$, hence $\delta(C'', D) = s(sw) = w$, a contradiction.

2.8 Now it is not hard to finish the proof using the following characterization of Coxeter groups [B IV §1 n° 7] by systems of half spaces. Let W be a group and

S a subset of W generating W such that $1 \notin S$ and $s^2 = 1$ for every $s \in S$. Let P_s, $s \in S$, be a family of subsets of W with the following properties.

(A′) $1 \in P_s$ for every $s \in S$.

(B′) P_s and sP_s are disjoint.

(C) Suppose s and s' are in S and $w \in W$. If $w \in P_s$ and $ws' \notin P_s$ then $sw = ws'$.

Then (W, S) is a Coxeter system and $P_s = \{w \in W \mid \ell(sw) > \ell(w)\}$.

2.9 Let us return to our thick $\mathfrak{C}$ with W-distance δ. Define

$$P_s = \{w \in W \mid \ell(sw) > \ell(w)\} .$$

2.10 If $\delta(C, D) = w$ and $w \notin P_s$ the chamber C' of D2 such that $\delta(C', C) = s$ and $\delta(C', D) = sw$ is unique.

Proof If C' and C'' are two different such chambers then $\delta(C', C'') = s$ by 2.2 and hence $\delta(C', D) = w$ by D1 for the triple (C', C'', D) since $\ell(w) = \ell(s(sw)) > \ell(sw)$ as $w \notin P_s$.

2.11 We finish our proof by checking that for the family P_s, $s \in S$, defined in 2.9 property (C) holds — (A') and (B') are obvious. Suppose s, s' and w are as in (C). Let C and D be chambers such that $\delta(C, D) = w$. There is a chamber D' such that $\delta(D, D') = s'$ and $\delta(C, D') = ws'$, by D2. There is a chamber C' such that $\delta(C', C) = s$ and $\delta(C', D') = ws'$, by thickness and 2.10 since $ws' \notin P_s$. On the other hand $\delta(C', D) = sw$, by D1 since $\delta(C', C) = s$, $\delta(C, D) = w$ and $w \in P_s$. Hence $\delta(C', D')$ is sw or sws'. We know already that $\delta(C', D') = ws'$, which is different from sws'. Hence $sw = ws'$, q.e.d.

2.12 Remark Note that the proof uses only the following weak version of thickness: For every $s \in S$ there is a chamber C and two chambers C' with $\delta(C, C') = s$.

References

[B] Bourbaki, N. *Groupes et algèbres de Lie.* Chap IV–VI

[T] Tits, J. *Théorie des groupes. Immeubles jumelés.* Résumé de cours. Collège de France 1988/89

Herbert Abels
Fakultät für Mathematik
Universität Bielefeld
4800 Bielefeld 1
FRG

GROWTH SERIES OF COXETER GROUPS

LUIS PARIS
Université de Genève, Section de Mathématique,
2–4 rue du Lièvre, CH-1211 Genève 24.

1. Introduction

Let W be a finitely generated group and S be a finite symmetric generating set of W (i.e S is finite, generates W and, if $s \in S$, then $s^{-1} \in S$). The set S determines a length on W, called *word length*. It is defined by

$$|w| = |w|_S = \min\{r | w = s_1 \dots s_r,\ s_i \in S\},$$

for $w \in W$. The *growth series* of W with respect to S is the formal series

$$W_S(t) = \sum_{w \in W} t^{|w|}.$$

We know that $W_S(t)$ is rational for some families of groups, like Coxeter groups [**6**, §4.1, exercise 26], virtually-abelian groups [**3**], [**4**], Gromov hyperbolic groups [**8**], [**10**], [**14**] and groups of alternating link complements [**11**, p. 58]. There are finitely presented groups with non rational growth series [**28**, p. 131].

After two independent papers [**20**], [**26**], there have been many studies on asymptotic properties of the coefficients of the growth series [**2**], [**12**], [**13**], [**21**], [**27**], [**29**]. The case of Coxeter groups is solved in [**15**]; if a Coxeter group W is finite or affine, then W has polynomial growth, otherwise W has exponential growth.

Another problem on growth series is to understand the relation between the Euler characteristic of a group and its growth series [**9**], [**11**], [**18**], [**22**], [**23**], [**24**]. It is often the case that

$$W_S(1) = \frac{1}{\chi(W)},$$

where $\chi(W)$ is the Euler characteristic of W. For example, this equality holds when (W, S) is a Coxeter system [**23**, p. 112].

We refer to [**28**] for a good exposition on growth series, discussing finitely generated groups in general as well as Coxeter groups.

Our aim here is to prove the *recursive formulas* on growth series of Coxeter groups (the Main Theorem below). These formulas are stated in [**6**, §4.1, exercise 26] and proved in exercises 15 to 26 of [**6**, §4.1]. Remark that one of these recursive formulas (the one about finite groups) was proved earlier by Solomon [**25**].

An immediate corollary of these formulas will be, for a Coxeter system (W, S):

1) $W_S(t)$ is rational,

2) the zeros of $W_S(t)$ are roots of the unity,

3) $1/W_S(+\infty)$ is an integer,

4) $1/W_S(t^{-1})$ is a formal series with integer coefficients.

Solomon used in [**25**] the recursive formula for finite Coxeter groups to state explicitly the growth series of finite Coxeter groups; he proved that, for a finite Coxeter system (W, S),

$$W_S(t) = \prod_{i=1}^{l}(1 + t + \cdots + t^{m_i}),$$

where $m_1, \ldots, m_l$ are the exponents of W (see also [**16**]). There are also explicit formulas for affine Coxeter groups [**5**], [**17**], [**6**, §6.4, exercise 10] and cocompact hyperbolic Coxeter groups [**7**]. Serre used in [**23**] the recursive formulas to prove the following results:

1) If S is the standard generating set of a Coxeter group W, then

$$W_S(1) = \frac{1}{\chi(W)},$$

where $\chi(W)$ is the Euler characteristic of W.

2) $W_S(t^{-1}) = \pm W_S(t)$, for W either an (infinite) affine Coxeter group or a cocompact hyperbolic Coxeter group.

In Section 2 we define a Coxeter system, we state the Exchange Lemma and the Main Theorem (recursive formulas) and we prove the Main Theorem. The Exchange Lemma is a characterisation of Coxeter groups given by Matsumoto [**19**]. It is essential in the proof of the Main Theorem.

2. Main Theorem

Let W be a group and let S be a finite generating set of W. We say that (W, S) is a *Coxeter system* iff W admits the following presentation:

$$W = \langle s \in S \mid (st)^{m_{s,t}} = 1, \text{ for } s, t \in S \rangle,$$

where $M = (m_{s,t})_{s,t\in S}$ is a symmetric matrix with the properties
1) $m_{s,s} = 1$, $\forall s \in S$,
2) $m_{s,t} \in \{2, 3, \ldots, +\infty\}$ if $s \neq t$.

EXCHANGE LEMMA (MATSUMOTO [**19**]). *Let W be a group and let S be a finite generating set of W, with $s^2 = 1$ for all $s \in S$. The system (W, S) is a Coxeter system if and only if it satisfies the following condition (E):*

(E) For all $w \in W$, if $w = s_1 \ldots s_r$ with $s_i \in S$ and $|w| < r$, then there exist two indices $i < j$ such that $w = s_1 \ldots \hat{s}_i \ldots \hat{s}_j \ldots s_r$.

COROLLARY 1. *Let (W, S) be a Coxeter system. If $w \in W$ can be written $w = s_1 \ldots s_r$, then there exists a subsequence $s_{i_1}, \ldots, s_{i_l}$ of $s_1, \ldots, s_r$, such that $w = s_{i_1} \ldots s_{i_l}$ is a reduced form (i.e. $l = |w|$).*

For $X \subseteq S$ a subset, we denote by W_X the subgroup of W generated by X.

COROLLARY 2. *Let (W, S) be a Coxeter system and let $X \subseteq S$ be a subset.*
1) $|w|_X = |w|_S$, for all $w \in W_X$,
2) (W_X, X) is a Coxeter system.

PROOF: 1) Let $w \in W_X$. Write $w = x_1 \ldots x_r$, with $x_i \in X$. By corollary 1 (applied to (W, S)), there exists a subsequence $x_{i_1}, \ldots, x_{i_l}$ of $x_1, \ldots, x_r$ such that $w = x_{i_1} \ldots x_{i_l}$ is a reduced form (with respect to S). We have then $|w|_X = |w|_S = l$.

2) If (W, S) satisfies (E) then, as in the proof of 1), (W_X, X) satisfies (E) too. Thus it is a Coxeter system. □

MAIN THEOREM (BOURBAKI [**6**], SOLOMON [**25**]). *Let (W, S) be a Coxeter system.*

1) If W is finite, then

$$\frac{t^m + (-1)^{|S|+1}}{W_S(t)} = \sum_{X \subsetneq S} \frac{(-1)^{|X|}}{W_X(t)}, \tag{1}$$

where

$$m = \max_{w \in W} |w|.$$

2) If W is infinite, then

$$\frac{(-1)^{|S|+1}}{W_S(t)} = \sum_{X \subsetneq S} \frac{(-1)^{|X|}}{W_X(t)}. \tag{2}$$

COROLLARY. *Let (W,S) be a Coxeter system:*
1) $W_S(t)$ is rational,
2) the zeros of $W_S(t)$ are roots of unity,
3) $1/W_S(+\infty)$ is an integer,
4) $1/W_S(t^{-1})$ is a formal series with integer coefficients.

This corollary can be proved without difficulties by induction on the cardinality of S.

STEPS OF THE PROOF OF THE MAIN THEOREM: Let (W,S) be a Coxeter system. For a subset $A \subseteq W$ we set

$$A(t) = \sum_{w \in A} t^{|w|}.$$

Let $X \subseteq S$. We say that $v \in W$ is *X-minimal* iff v is of minimal length in $W_X v$, namely

$$|v| = \min_{w \in W_X v} |w|.$$

We put

$$A_X = \{v \in W \mid v \text{ is } X\text{-minimal}\}$$
$$B_X = \{v \in W \mid v \text{ is } X\text{-minimal and is not } Y\text{-minimal}, \quad \text{for any } Y \text{ with } X \subsetneq Y \subset S\}.$$

The proof of Main Theorem has three steps: Lemmas 1, 2 and 3. The Corollary of Lemma 1 gives the equality

$$W_S(t) = A_X(t) W_X(t) \tag{3}$$

for all $X \subseteq S$. Lemma 2 shows that

$$B_X(t) = \sum_{Y \supseteq X} (-1)^{|Y-X|} \frac{W_S(t)}{W_Y(t)} \tag{4}$$

for all $X \subseteq S$. The Corollary of Lemma 3 shows that $B_\emptyset = \emptyset$ if W is infinite and $B_\emptyset = \{w_0\}$ if W is finite, where w_0 is the longest element of W. This implies $B_\emptyset(t) = 0$ if W is infinite, and $B_\emptyset(t) = t^m$ if W is finite, m being the maximal length in W. We replace $B_\emptyset(t)$ in (4) by its value and we obtain the equalities (1) and (2) of the Main Theorem. □

Let (W,S) be a Coxeter system and let X be a subset of S.

LEMMA 1. *Let $v \in W$ represent a class $W_X v \in W_X \setminus W$, assume that v is X-minimal (i.e. of minimal length in $W_X v$). Then*

1) v is the unique X-minimal element of $W_X v$,
2) for every $w = uv \in W_X v$, with $u = wv^{-1} \in W_X$, we have $|w| = |u| + |v|$.

PROOF: Let $u \in W_X$ and consider reduced forms

$$\begin{aligned} u &= x_1 \dots x_r, \quad (x_i \in X), \\ v &= s_1 \dots s_t, \quad (s_j \in S), \end{aligned}$$

of u and v. If $uv = x_1 \dots x_r s_1 \dots s_t$ is not a reduced form of uv, then, by the Exchange Lemma, we have

(a) either $uv = x_1 \dots \hat{x}_i \dots \hat{x}_j \dots x_r s_1 \dots s_t$,

(b) or $uv = x_1 \dots \hat{x}_i \dots x_r s_1 \dots \hat{s}_j \dots s_t$,

(c) or $uv = x_1 \dots x_r s_1 \dots \hat{s}_i \dots \hat{s}_j \dots s_t$.

If (a) holds then $u = x_1 \dots \hat{x}_i \dots \hat{x}_j \dots x_r$. This is not possible because we have supposed that $u = x_1 \dots x_r$ is a reduced form. If (b) holds then $s_1 \dots \hat{s}_j \dots s_t \in W_X v$. This is not possible because we have supposed that $v = s_1 \dots s_t$ is of minimal length in $W_X v$. If (c) holds, then $v = s_1 \dots \hat{s}_i \dots \hat{s}_j \dots s_t$. This is not possible because we have supposed that $v = s_1 \dots s_t$ is a reduced form. It follows that $uv = x_1 \dots x_r s_1 \dots s_t$ is a reduced form and $|u| + |v| = |uv|$. This shows the second assertion and the first is a straightforward consequence. □

COROLLARY. *We have*

$$W_S(t) = A_X(t) W_X(t). \tag{5}$$

PROOF:

$$\begin{aligned} W_S(t) &= \sum_{w \in W} t^{|w|} \\ &= \sum_{v \in A_X} \sum_{w \in W_X v} t^{|w|} \\ &= \sum_{v \in A_X} \sum_{u \in W_X} t^{|u|+|v|} \\ &= A_X(t) W_X(t). \qquad \square \end{aligned}$$

Recall that

$$\begin{aligned} A_X &= \{v \in W \mid v \text{ is } X\text{-minimal}\}, \\ B_X &= \{v \in W \mid v \text{ is } X\text{-minimal and is not } Y\text{-minimal}, \\ &\qquad\qquad \text{for any } Y \text{ with } X \subsetneq Y \subset S\}. \end{aligned}$$

LEMMA 2. *We have*

$$B_X(t) = \sum_{Y \supseteq X} (-1)^{|Y-X|} \frac{W_S(t)}{W_Y(t)}. \tag{6}$$

PROOF: Let $Y \supsetneq X$. If v is Y-minimal, then v is X-minimal. Therefore

$$B_X = A_X - \bigcup_{Y \supsetneq X} A_Y,$$

thus

$$A_X = \coprod_{Y \supseteq X} B_Y$$

and

$$A_X(t) = \sum_{Y \supseteq X} B_Y(t).$$

Let E be a finite set. We denote by $\mathcal{P}(E)$ the set of all parts of E. Let G be an abelian group and $f, g : \mathcal{P}(E) \to G$ be two maps. It is easy to prove (and well known) that the following two assertions are equivalent (see, for example [**1**, p. 152]):

$$f(X) = \sum_{Y \supseteq X} g(Y), \quad \forall X \in \mathcal{P}(E), \tag{M1}$$

$$g(X) = \sum_{Y \supseteq X} (-1)^{|Y-X|} f(Y), \quad \forall X \in \mathcal{P}(E). \tag{M2}$$

This equivalence is called the *Möbius inversion.*

We apply the Möbius inversion to

$$E = S,\ G = \mathbb{Z}[[t]],\ f(X) = A_X(t) \text{ and } g(X) = B_X(t)$$

and we obtain

$$B_X(t) = \sum_{Y \supseteq X} (-1)^{|Y-X|} A_Y(t).$$

Lemma 2 follows from the previous Corollary. □

LEMMA 3. *Let (W, S) be a Coxeter system and let $w_0 \in W$. The following two assertions are equivalent:*

1) $w_0 \in B_\emptyset$,

2) $\forall w \in W$, we have $|w| + |w^{-1}w_0| = |w_0|$.

PROOF: Observe that $A_\emptyset = W$, so that

$$B_\emptyset = W - (\bigcup_{s \in S} A_{\{s\}}).$$

Since $A_{\{s\}} = \{w \in W \mid |w| < |sw|\}$, the element $w \in W$ is in $B_\emptyset$ if and only if $|w| > |sw|$ for all $s \in S$.

2) $\Rightarrow$ 1): If (2) holds, then we have $|w_0| > |w|$ for all $w \in W - \{w_0\}$, and in particular $|w_0| > |sw_0|$ for all $s \in S$, so that $w_0 \in B_\emptyset$.

1) $\Rightarrow$ 2): Suppose $w_0 \in B_\emptyset$ and let us show, by induction on $r = |w|$, that

$$|w| + |w^{-1}w_0| = |w_0|.$$

If $r = |w| = 1$, then $w = s$ for some $s \in S$. The element w_0 is not $\{s\}$-minimal, thus $|sw_0| < |w_0|$. Furthermore $|sw_0| \geq |w_0| - 1$, therefore

$$|w_0| = 1 + |sw_0| = |s| + |sw_0|.$$

Suppose $r = |w| > 1$. Choose $s \in S$ and $w' \in W$ such that $w = sw'$ and $|w'| = r - 1$. By induction, there exists a reduced form

$$w_0 = s_1 \dots s_{r-1} t_1 \dots t_q \quad (s_i, t_j \in S),$$

such that $w' = s_1 \dots s_{r-1}$ and $w'^{-1}w_0 = t_1 \dots t_q$ are also reduced forms. Since w_0 is not $\{s\}$-minimal, we have $|sw_0| < |w_0|$. Furthermore, if $w_0 = s'_1 \dots s'_n$ is a reduced form, then $sw_0 \neq ss_1 \dots \hat{s}'_i \dots \hat{s}'_j \dots s'_n$ (otherwise $w_0 = s_1 \dots \hat{s}'_i \dots \hat{s}'_j \dots s'_n$). It follows, from the exchange lemma, that

$$\text{(a)} \qquad \text{either } sw_0 = s_1 \dots \hat{s}_i \dots s_{r-1} t_1 \dots t_q,$$

$$\text{(b)} \qquad \text{or } sw_0 = s_1 \dots s_{r-1} t_1 \dots \hat{t}_j \dots t_q.$$

If (a) holds, then $sw' = w = s_1 \dots \hat{s}_i \dots s_{r-1}$. This is not possible because $|w| = r$. Therefore (b) holds. It follows that

$$w_0 = \underbrace{ss_1 \dots s_{r-1}}_{w} \underbrace{t_1 \dots \hat{t}_j \dots t_q}_{w^{-1}w_0}$$

is a reduced form and

$$|w_0| = |w| + |w^{-1}w_0|. \quad \square$$

COROLLARY. *1) If W is infinite, then $B_\emptyset = \emptyset$.*

2) If W is finite, then there exists a unique element w_0 of maximal length in W, and $B_\emptyset = \{w_0\}$.

PROOF: Lemma 3 shows that $B_\emptyset$ is precisely the set of elements of maximal length, and that this set has at most one element. The Corollary follows. □

REFERENCES

1. M. Aigner, "Combinatorial theory," Springer-Verlag, New York, 1979.
2. H. Bass, The degree of polynomial growth of finitely generated nilpotent groups, *Proc. Lond. Math. Soc.* **25** (1972), 603–614.
3. M. Benson, Growth series of finite extensions of $\mathbb{Z}^n$ are rational, *Invent. Math.* **73** (1983), 251–269.
4. M. Benson, On the rational growth of virtually nilpotent groups, in "Combinatorial group theory and topology," Ann. Math. Studies, 111, Princeton University Press, Princeton, 1987, pp. 185–196.
5. M. R. Bott, An application of the Morse theory to the topology of Lie groups, *Bull. Soc. Math. France* **84** (1956), 251–281.
6. N. Bourbaki, "Groupes et algèbres de Lie," Chapitres 4, 5 et 6, Hermann, Paris, 1968.
7. J. W. Cannon, The growth of the closed surface groups and the compact hyperbolic Coxeter groups, 1980, (unpublished).
8. J. W. Cannon, The combinatorial structure of cocompact discrete hyperbolic groups, *Geom. Dedicata* **16** (1984), 123–148.
9. W. J. Floyd and S. P. Plotnick, Growth functions on Fuchsian groups and the Euler characteristic, *Invent. Math.* **88** (1987), 1–29.
10. E. Ghys and P. de la Harpe, La propriété de Markov pour les groupes hyperboliques, in "Sur les groupes hyperboliques d'après Mikhael Gromov," Birkhäuser, Basel, 1990, pp. 165–187.
11. M. A. Grayson, "Geometry and growth in three dimensions," Ph. D. Thesis, Princeton University, Princeton, 1983.
12. R. I. Grigorchuk, On Milnor's problem of group growth, *Soviet Math. Dokl.* **28** (1983), 23–26.
13. M. Gromov, Groups of polynomial growth and expanding maps, *Publ. Math. I.H.E.S.* **53** (1981), 53–73.
14. M. Gromov, Hyperbolic groups, in "Essays in group theory," M.S.R.I. 8, Springer, New York, 1987, pp. 75–263.
15. P. de la Harpe, Groupes de Coxeter infinis non affines, *Expo. Math.* **5** (1987), 91–96.
16. P. de la Harpe, An invitation to Coxeter groups, (this volume).
17. M. Iwahori and H. Matsumoto, On some Bruhat decomposition and the structure of Hecke rings of p-adic Chevalley groups, *Publ. Math. I.H.E.S.* **25** (1965), 5–48.

18. J. Lewin, The growth function of a graph group, *Comm. in Algebra* **17** (1989), 1187–1191.
19. H. Matsumoto, Générateurs et relations des groupes de Weyl généralisés, *C. R. Acad. Sc. Paris* **258** (1964), 3419–3422.
20. J. Milnor, A note on curvature and fundamental group, *J. Differential Geometry* **2** (1968), 1–7.
21. J. Milnor, Growth of finitely generated solvable groups, *J. Differential Geometry* **2** (1968), 447–449.
22. W. Parry, Counterexamples involving growth series and the Euler characteristics, *Proc. Amer. Math. Soc.* **102** (1980), 49–51.
23. J. P. Serre, Cohomologie des groupes discrets, in "Prospects in Mathematics," Ann. Math. Studies, 70, Priceton University Press, Princeton, 1971, pp. 77–169.
24. N. Smythe, Growth functions and Euler series, *Invent. Math.* **77** (1984), 517–531.
25. L. Solomon, The orders of the finite Chevalley groups, *J. of Algebra* **3** (1966), 376–393.
26. A. S. Svarc, A volume invariant of coverings, *Dokl. Akad. Nauk. SSSR* **105** (1955), 32–34.
27. J. Tits, Groupes à croissance polynomiale, in "Séminaire Bourbaki 1981," Lecture Notes in Math., 901, Springer, Berlin, 1981, pp. 176–188.
28. P. Wagreich, The growth function of a discrete group, in "Proc. Conference on algebraic varieties with group actions," Lecture Notes in Math., 956, Springer, Berlin, 1982, pp. 125–144.
29. J. A. Wolf, Growth of finitely generated solvable groups and curvatures of Riemannian manifolds, *J. Differential Geometry* **2** (1968), 421–446.

III. NON-POSITIVELY CURVED POLYHEDRA

Constructions of hyperbolic groups via hyperbolizations of polyhedra

Frédéric PAULIN

C.N.R.S. SDI 6178
Laboratoire de Mathématiques, Ecole Normale Supérieure de Lyon
46 allée d'Italie, 69364 LYON CEDEX 07, FRANCE

ABSTRACT

These are lecture notes for the 26 March-6 April worshop on Group Theory from a Geometrical Viewpoint held at the I.C.T.P. in Trieste, Italy. The definition, due to A.D. Aleksandrov, of negatively curved metric spaces, from which Gromov's hyperbolicity is a global non-local version, is given. After specialization to polyhedra, Gromov's criterion for a polyhedron to be negatively curved is stated. Then Gromov's hyperbolization process which turns any polyhedron into a negatively curved one is defined. Taking the fundamental group, this allows the construction of a lot of hyperbolic groups.

1 Introduction

Let us recall (see the introductory lectures in this volume) the definition of hyperbolic spaces and hyperbolic groups we are going to use.

Definition 1.1 • *A geodesic in a metric space (X,d) is an isometric map $f : I \to X$, where I is some closed interval of* $\mathbf{R}$, *endowed with the usual distance defined by the absolute value. Note that a geodesic $f : I \to X$ is uniquely determined, up to orientation and translation of the interval I, by its image $f(I)$. So we will often identify a geodesic and its image. If $I = [a,b]$, then f is said to be a geodesic between $f(a)$ and $f(b)$.*

- *A geodesic triangle in a metric space (X,d) is a triple (a,b,c) of three geodesics a,b,c respectively between B and C, C and A, A and B, where A,B,C are three points in X, called the opposite vertices of respectively a,b,c.*

- *A metric space (X,d) is geodesic if for every x,y in X, there exists a geodesic $f : [a,b] \to X$ with $f(a) = x$ and $f(b) = y$.*

- *A geodesic metric space is hyperbolic if there exists $\delta \geq 0$ such that for every geodesic triangle (a,b,c) in X,*

$$\sup_{x\in a} d(x, b\cup c) \leq 2\delta$$

- *A finitely generated group Γ, endowed with a finite generating subset S (stable by inverse and not containing the identity) is (word) hyperbolic if its Cayley graph $\mathcal{G}(\Gamma, S)$, endowed with the word metric, is hyperbolic.*

Also recall the following basic facts. The hyperbolicity of a metric space is a quasi-isometry invariant, and in particular the hyperbolicity of a finitely generated group does not depend on the choice of a generating subset. If Γ is a word hyperbolic group (this supposes that Γ is finitely generated), then Γ is finitely presented.

There are two main classes of examples of hyperbolic groups : the small cancellation groups and the negatively curved manifold groups.

(a) Groups admitting a $C'(\frac{1}{6})$ small cancellation presentation.

If $< X, R >$ is a finite presentation of a group, with R a finite subset of cyclically reduced words of the free group with basis the finite set X, a *piece* is a cyclically reduced word m such that there exist two different words r_1 and r_2 among the cyclic conjugates of elements of R and their inverses with $r_1 = mr_1'$ and $r_2 = mr_2'$ (the equalities hold as words).

This condition is very easy to check : you have only a finite number of facts to verify. A finite "machine", knowing only how to manipulate finite lists of symbols, is able to decide whether a presentation of a group has the small cancellation property or not.

However, this condition depends on the presentation, and is not geometric in nature. Furthermore, in some precise sense, this class of group is not very large.

(b) Fundamental groups of compact, connected, smooth manifolds M without boundary, admitting a Riemannian metric of negative sectional curvature.

In order to understand this class of examples, you need to have a certain amount of knowledge in differential geometry : differentiable manifolds, Riemannian metrics, sectional curvature ... Since we won't have to use any object in that class, we won't give the definitions. The interested readers may read the following list of examples in that class (which is not complete !), and the others may skip the following enumeration. Note that the proof (see for instance [GH] chapitre 3) that these groups are indeed hyperbolic relies on the CAT(χ)-inequality (see section 2).

1. If the dimension of M is 2, then $\Pi_1(M)$ is hyperbolic except when M is the torus $\mathbf{T}^2$ or the Klein bottle $\mathbf{K}^2$.

2. If the dimension of M is 3, then according to Thurston's hyperbolization conjecture, proven in the case where M is Haken (see [Thu1][Thu2]), one suspects that $\Pi_1(M)$ is hyperbolic except if $\Pi_1(M)$ contains a subgroup isomorphic to $\mathbf{Z} \times \mathbf{Z}$. A weaker result has been proved by M. Bestvina and M. Feign [BF].

3. The discrete, cocompact subgroups of the isometry group of RH^n, CH^n, QH^n and CaH^2, the irreducible non compact globally symmetric spaces of rank 1, are hyperbolic (see [Hel] for a definition and note that if a negatively curved, connected, simply connected, complete Riemannian manifold N, with a transitive isometric action of a Lie group, has a discrete cocompact group of isometries, then N is globally symmetric [Hei]).

   - Infinitely many cocompact arithmetic (hence discrete) subgroups of the above groups of isometries have been constructed by A. Borel [Bor] and A. Borel and Harish-Chandra [BHC].

     Examples of reflection groups (i.e. generated by orthogonal symmetries with respect to hyperplanes) in Isom(RH^n) that are cocompact and arithmetic have been constructed by V. Makarov [Mak] ($n = 3$), E. Vinberg [Vin1] [Vin2] [Vin3] [Vin6] ($n \leq 7$), and more recently by V.O. Bugaenko ($n = 8$). E. Vinberg proved [Vin4][Vin5] that a reflection group of isometries of RH^n does not exist in dimension greater than 29. See also P. de la Harpe lectures [Har] for more informations on reflection groups.

   - The first non arithmetic discrete cocompact subgroups have been constructed by :

     V. Makarov [Mak] and E. Vinberg [Vin1][Vin2] in Isom(RH^n) for $n \leq 5$; these examples are reflection groups.

     Mostow [Mos] and Deligne-Mostow [DM] in Isom(CH^n) for $n \leq 3$.

   - A major step has been achieved by M. Gromov and I. Piatetski-Shapiro [GPS] who constructed infinite families of non arithmetic examples in Isom(RH^n) in every dimension, by gluing two arithmetic examples.

4. It seems difficult to find groups which are fundamental groups of negatively curved compact connected manifold not of type 1, 2 or 3. The first examples are due to Mostow-Siu [MS]. M. Gromov and W. Thurston [GT] have constructed important examples. Let V be a C^∞ compact Riemannian manifold with constant sectional curvature -1. Let V' be a closed oriented totally geodesic submanifold of codimension 2 (for instance a closed geodesic in dimension 3), which is homologous to zero. Let $\tilde{V}_i$ be the cyclic covering of order i of V branched over V'. If $n \geq 4$, and i big enough, then M. Gromov and W. Thurston [GT] proved that $\tilde{V}_i$ has no Riemannian metric with constant curvature -1 (even if we change the differentiable structure). By smoothing the singularities of the induced metric, it can be proved (see [GT]) that $\tilde{V}_i$ has a Riemannian metric with negative sectional curvature. Furthermore the fundamental group of $\tilde{V}_i$ is not even isomorphic to a discrete cocompact subgroup of type 3, according to a theorem of Farell-Jones [FJ1][FJ2].

These last examples are quite similar to hyperbolized polyhedra, that we are going to describe.

Since small cancellation groups are sparse and the oeconstruction of negatively curved manifolds is not elementary, the theory of hyperbolic groups needs a vast class of easily constructed examples, in order to check conjectures and to have pictures in mind. This will be provided by the class of hyperbolized polyhedra (see section 4). In the first two sections, we give some criteria for a metric space to be hyperbolic. Indeed, the conditions for a geodesic metric space to be a hyperbolic space seem very difficult to check in general : it is a global metric condition, not a combinatorial one ; one has to check the condition of thinness over an infinite number of big triangles. That's why we need a criterion of hyperbolicity which is easy to verify, at least for some smaller class of topological spaces than the geodesic metric spaces.

The following lecture notes are based on the paper [Gro3] of M. Gromov, pages 114-124, with the help of the papers of W. Ballman (chapter 10 of [GH]), E. Ghys and A. Haefliger (chapter 12 of [GH]), S. Alexander and R. Bishop [AB], M. Davis and T. Januskiewicz [DJ], R. Charney and M. Davis [CD] and M. R. Bridson [Bri]. I would like to thank A. Haefliger for his kindness and numerous remarks on a preliminary version.

Contents

2 Negatively curved metric spaces

The notion of a hyperbolic metric space is not a local one : any compact geodesic metric space is hyperbolic. A.D. Aleksandrov (see [Ale1][Ale2][ABN][AZ]) has introduced a notion of bounds of curvature for general metric spaces, which turns out to be a local one.

2.1 *The CAT(χ)-inequality*

Definition 2.1 *The length, denoted by $\ell(f)$, of a continuous map $f : [a,b] \rightarrow X$ in a metric space X is the upper bound over all subdivisions $a = x_0 < x_1 < \cdots < x_n = b$ of the sums $\sum_{i=1}^{n} d(f(x_{i-1}), f(x_i))$. A length space (see [Gro1][Gro2] page 4) is a metric space such that the distance between two points is the lower bound of the length of all continuous paths between the two points.*

Note that the length of a path may be infinite (as $f : [0,1] \rightarrow \mathbf{R}^2$ with $f(x) = x \sin \frac{1}{x}$ if $x \neq 0$ and $f(0) = 0$).

Definition 2.2 *Let (X, d) be a metric space and Y a subset of X such that two points of Y may be connected by a path in Y of finite length. For every x, y in Y, let $\tilde{d}(x, y)$ be the lower bound of the length of all continuous paths in Y between x and y. Then $\tilde{d}$ is a metric on Y, called the induced length metric. Furthermore $(Y, \tilde{d})$ is a length space (see [Gro2] page 4).*

If two points of a metric space (X, d) are not connected by a path of finite length, we define $\tilde{d}(x, y) = +\infty$.

Proposition 2.3 *(Cohn-Vossen [CV], see [Gro2] page 4) If (X, d) is a locally compact, complete length space, then (X, d) is a geodesic space and the balls of finite radius are compact.*

Exercise 2.4 a) If $f : [a,b] \rightarrow X$ is a geodesic in a metric space, then its length is the minimal length of paths between $f(a)$ and $f(b)$ and $d(f(a), f(b)) = |a - b|$.

b) Prove the statements of the above definition.

c) The space $\mathbf{R}^2 - 0$ with the induced metric is a length space, but not a geodesic space. Prove the above proposition.

We will denote by $\mathbf{R}^n$ the Euclidean space of dimension n endowed with its usual metric

$$d(x,y) = \sqrt{(x_1 - y_1)^2 + \cdots + (x_n - y_n)^2}$$

if $x = (x_1, \cdots, x_n)$ and $y = (y_1, \cdots, y_n)$.

We will denote by $\mathbf{S}^n$ the spherical space of dimension n. It is the unit sphere

$$\{(x_1, \cdots, x_{n+1}) \in \mathbf{R}^{n+1} / x_1^2 + \cdots + x_{n+1}^2 = 1\}$$

in $\mathbf{R}^{n+1}$ endowed with the induced length metric.

Exercise 2.5 $\mathbf{S}^n$ is a geodesic metric space.

We will denote by $\mathbf{H}^n$ the hyperbolic space of dimension n. It is the open unit ball

$$\mathbf{B}^n = \{(x_1, \cdots, x_n) \in \mathbf{R}^n / x_1^2 + \cdots + x_n^2 < 1\}$$

endowed with the following distance : for every x, y in $\mathbf{B}^n$, $d(x,y)$ is the lower bound over all paths $\alpha : [0,1] \to \mathbf{B}^n$ of class C^1 of the integrals, where $|| \ ||$ is the Euclidean norm :

$$\int_0^1 \frac{2||\alpha'(t)||dt}{1 - ||\alpha(t)||^2}$$

Note that $\mathbf{R}^0$ (resp. $\mathbf{H}^0$) is a point, and $\mathbf{S}^0$ is a pair of points.

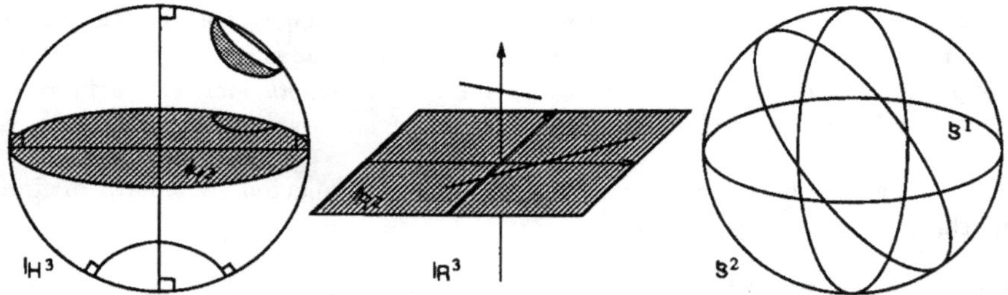

Figure 1 : Examples of geodesics and planes in the model spaces $\mathbf{H}^n$, $\mathbf{R}^n$, $\mathbf{S}^n$.

Exercise 2.6 • Characterize geometrically in $\mathbf{H}^n$, $\mathbf{R}^n$, $\mathbf{S}^n$ the k-planes (i.e. the subspaces isometric to $\mathbf{H}^k$, $\mathbf{R}^k$, $\mathbf{S}^k$ for $0 \leq k \leq n$). Note that the 1-planes are geodesics.

• A *flag* in $\mathbf{H}^n$, $\mathbf{R}^n$, $\mathbf{S}^n$ is a sequence $F_0, F_1, \cdots, F_n$ where F_k is a k-plane in $\mathbf{H}^n$, $\mathbf{R}^n$, $\mathbf{S}^n$ and $F_{k-1} \subset F_k$ for $k = 1 \cdots n$. Prove that the isometry group of $\mathbf{H}^n$, $\mathbf{R}^n$, $\mathbf{S}^n$ acts transitively on flags (i.e. given two flags, there exists an isometry of $\mathbf{H}^n$, $\mathbf{R}^n$, $\mathbf{S}^n$ sending one onto the other.).

• For every x, y, z in $[0, +\infty[$ satisfying the triangle inequalities (i.e. $x \leq y + z$ and cyclic permutations), there exists a geodesic triangle in $\mathbf{R}^n$ and $\mathbf{H}^n$ with side lengths x, y, z; if $x + y + z < 2\pi$, then there exists a geodesic triangle in $\mathbf{S}^n$ with side lengths x, y, z. In the three cases, such a triangle is unique up to isometry.

Exercise 2.7 • $\mathbf{H}^n$ is a geodesic metric space, hyperbolic with best possible constant

$$\delta = \log(1+\sqrt{2}).$$

• For every x in $\mathbf{R}^n$, $\mathbf{H}^n$ or $\mathbf{S}^n$, the visual sphere of center x in $\mathbf{R}^n$, $\mathbf{H}^n$ or $\mathbf{S}^n$ (which is the set of local geodesics $f : [0,+\infty[\to \mathbf{R}^n$, $\mathbf{H}^n$ or $\mathbf{S}^n$ with $f(0) = x$), endowed with the distance which is the angle at x between the two geodesics, is isometric to $\mathbf{S}^n$. The angle between two geodesics $f, g : [0,+\infty[\to \mathbf{H}^n$ (or $\mathbf{S}^n$) that originate from a same point x is the angle at x of the oriented straight lines of $\mathbf{R}^n$ (or $\mathbf{R}^{n+1}$) tangent at x to the geodesics.

• For every x in $\mathbf{R}^n$, $\mathbf{H}^n$ or $\mathbf{S}^n$, the sphere of center x in $\mathbf{R}^n$, $\mathbf{H}^n$ or $\mathbf{S}^n$ and radius ρ is isometric to $\mathbf{S}^{n-1}$, with $\rho = 1, -\log(\sqrt{2}-1), \frac{\pi}{2}$ respectively for $\mathbf{R}^n$, $\mathbf{H}^n$ or $\mathbf{S}^n$.

Notation 2.8 *If X is a metric space with metric d, and $\lambda < 0$, let λX be the set X with the metric λd.*

If χ is in $\mathbf{R}$, set

$$\begin{aligned} \mathbf{M}^n_\chi &= \tfrac{1}{\sqrt{\chi}}\mathbf{S}^n && \text{if } \chi > 0 \\ &= \mathbf{R}^n && \text{if } \chi = 0 \\ &= \tfrac{1}{\sqrt{-\chi}}\mathbf{H}^n && \text{if } \chi < 0 \end{aligned}$$

We use the convention that if χ is nonpositive, then $\frac{2\pi}{\sqrt{\chi}}$ and $\frac{\pi}{\sqrt{\chi}}$ are $+\infty$.

Let (X, d) be a metric space and (a, b, c) a geodesic triangle in X. The *perimeter* of (a, b, c) is the sum of the lengths of a, b and c. A geodesic triangle $(\overline{a}, \overline{b}, \overline{c})$ in $\mathbf{M}^n_\chi$ is a *comparison triangle* for (a, b, c) if the lengths of the corresponding sides are the same. According to exercise 2.6, a comparison triangle exists and is unique up to isometry if the perimeter of (a, b, c) is less than $\frac{2\pi}{\sqrt{\chi}}$. Let A (resp. $\overline{A}$) be the vertex opposite to a (resp. $\overline{a}$). If $p \in a$, let $\overline{p}$ be the unique point of $\overline{a}$ with $d(\overline{p}, \overline{A}) = d(p, A)$. Call $\overline{p}$ the point corresponding to p.

This last notation is dangerous, since the dependence on the geodesic triangle and on the side is forgotten.

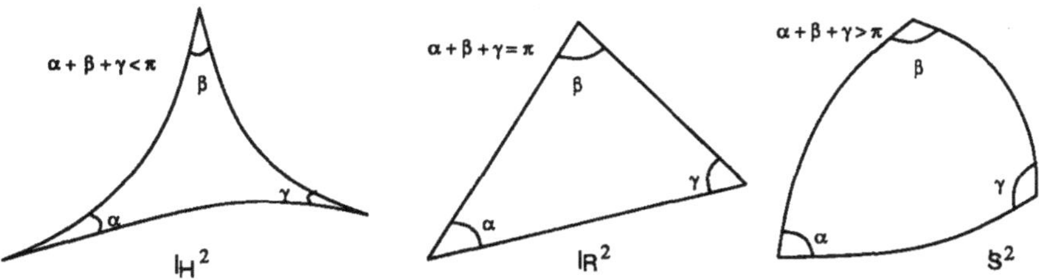

Figure 2 : Examples of geodesic triangles in the model spaces $\mathbf{H}^n$, $\mathbf{R}^n$, $\mathbf{S}^n$.

Definition 2.9 *If $\chi \in \mathbf{R}$, a geodesic triangle (a,b,c) of perimeter less than $\frac{2\pi}{\sqrt{\chi}}$ in a metric space (X,d) satisfies the CAT(χ)-inequality if for any comparison triangle $(\overline{a},\overline{b},\overline{c})$ in $\mathbf{M}^2_\chi$, for every two points p,q on two sides, if $\overline{p},\overline{q}$ are the corresponding points on the corresponding sides, then*

$$d(p,q) \leq d(\overline{p},\overline{q}).$$

A geodesic metric space is said to satisfy the CAT(χ)-inequality if every geodesic triangle of perimeter strictly less than $\frac{2\pi}{\sqrt{\chi}}$ satisfies the CAT(χ)-inequality.

A geodesic metric space is said to have curvature less than or equal to χ if any point x in X has a neighborhood V_x such that every geodesic triangle of perimeter strictly less than $\frac{2\pi}{\sqrt{\chi}}$ contained in V_x satisfies the CAT(χ)-inequality.

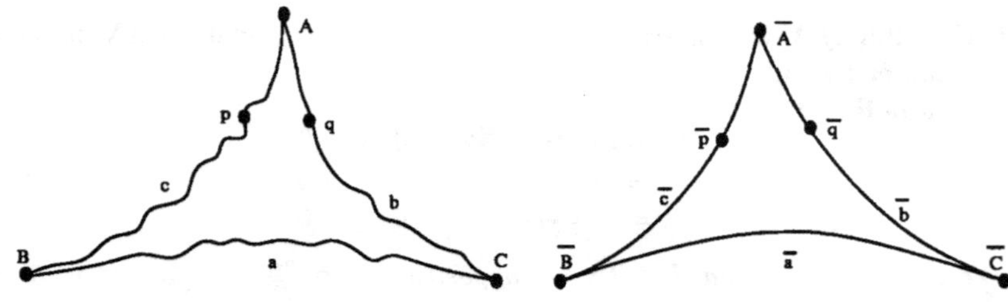

Figure 3 : The CAT(χ)-inequality.

This is a notion of an upper bound of curvature for a general metric space. By changing the inequalities, we could have defined a notion of a lower bound of curvature for a general metric space, but we won't need it.

Exercise 2.10 A space satisfying the CAT(χ)-inequality satisfies the CAT(χ')-inequality for every $\chi' \geq \chi$.

Exercise 2.11 (see for instance [GH] page 31) A geodesic space X has curvature $-\infty$ (i.e. satisfies the CAT(χ)-inequality for every $\chi \in \mathbf{R}$) if and only if it is an $\mathbf{R}$-tree (i.e. a metric space X such that between two points x, y there exists one and only one arc — in the sense of the image of an injective continuous map from $[0,1]$ to X sending 0 to x and 1 to y — and furthermore this arc is isometric to an interval in $\mathbf{R}$). See the lectures of P. Shalen [Sha] for more on $\mathbf{R}$-trees.

The following lemma is crucial. It will be used in several places in these notes. We state it here to show how to manipulate the above definition.

Lemma 2.12 *(M. Gromov - W. Ballmann) Let* (a_0, b, c_0) *and* (a_1, b, c_1) *be two geodesic triangles in a metric space* X*, of perimeter less than* $\frac{2\pi}{\sqrt{\chi}}$ *with common side* b *and common vertices* A *and* C *(see Figure 3 (a)). Suppose* $c_0 \cup c_1$ *is a geodesic from* B_0 *to* B_1 *and*

$$\ell(a_0) + \ell(a_1) + \ell(c_0) + \ell(c_1) < \tfrac{2\pi}{\sqrt{\chi}}.$$

If (a_0, b, c_0) *and* (a_1, b, c_1) *satisfy the CAT*(χ)*-inequality, then* $(a_0, a_1, c_0 \cup c_1)$ *also does.*

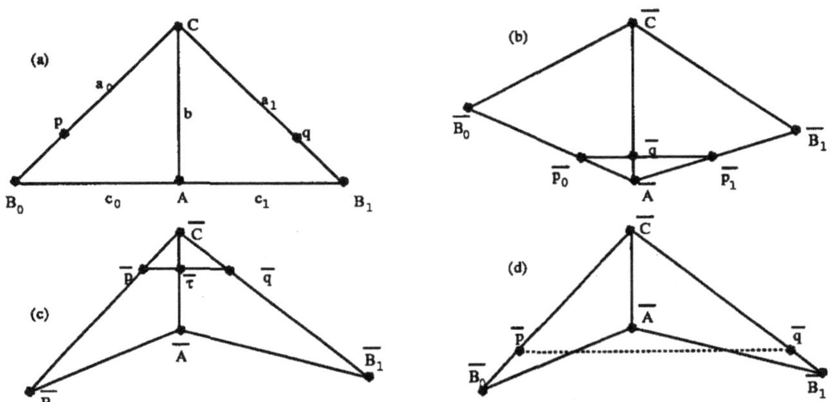

Figure 4 : The gluing lemma on CAT(χ) triangles.

Proof. (W. Ballman [GH] page 183-184) Note first that the comparison triangles $(\overline{a_0}, \overline{b}, \overline{c_0})$ and $(\overline{a_1}, \overline{b}, \overline{c_1})$ glued along $\overline{b}$ form a quadrilateral with interior angle bigger than or equal to π at the vertex $\overline{A}$.

Indeed, if the interior angle at $\overline{A}$ is strictly less than π (see Figure 4 (b)), then for p_0 on $c_0 - \{A\}$ and p_1 on $c_1 - \{A\}$ sufficiently close to A, the geodesic from $\overline{p_0}$ to $\overline{p_1}$ will meet $\overline{b}$ in $\overline{q} \neq A$. Then we have

$$d(p_0, q) \leq d(\overline{p_0}, \overline{q}) \text{ and } d(p_1, q) \leq d(\overline{p_1}, \overline{q})$$

But since $c_0 \cup c_1$ is a geodesic, we also have

$$\begin{aligned} d(p_0, p_1) &= d(p_0, A) + d(A, p_1) \\ &= d(\overline{p_0}, \overline{A}) + d(\overline{A}, \overline{p_1}) \\ &> d(\overline{p_0}, \overline{q}) + d(\overline{q}, \overline{p_1}) \end{aligned}$$

(since $\overline{A}$ does not lie on the geodesic between $\overline{p_0}$ and $\overline{p_1}$, and $\overline{q}$ does)

$$\begin{aligned} &\geq d(p_0, q) + d(q, p_1) \\ &\geq d(p_0, p_1). \end{aligned}$$

This is a contradiction.

Consider first the case $p \in a_0$ and $q \in a_1$, with the geodesic between $\overline{p}$ and $\overline{q}$ contained in the interior of the quadrilateral bounded by $\overline{a_0}, \overline{a_1}, \overline{c_0}, \overline{c_1}$ (see Figure 4 (c)). This geodesic cuts $\overline{b}$ in a point $\overline{r}$. By hypothesis, we have

$$\begin{aligned} d(p,q) &\leq d(p,r) + d(r,q) \\ &\leq d(\overline{p},\overline{r}) + d(\overline{r},\overline{q}) = d(\overline{p},\overline{q}). \end{aligned}$$

When looking at the comparison triangle for $(a_0, a_1, c_0 \cup c_1)$, obtained by imaginating that $\overline{a_0}, \overline{a_1}, \overline{c_0}, \overline{c_1}$ are rigid bars articulated at $\overline{A}, \overline{C}, \overline{B_0}, \overline{B_1}$, by holding the articulated system at the point $\overline{C}$ and by straightening the interior angle at $\overline{A}$ of the sector $(\overline{a_0}, \overline{a_1})$ so that $\overline{A}$ moves on the vertical line downwards, the distance between $\overline{p}$ and $\overline{q}$ is in fact increased. This proves the required inequality in this case. The cases $p \in a_0$ and $q \in c_1$ or $p \in c_0$ and $q \in a_1$, with the geodesic between $\overline{p}$ and $\overline{q}$ contained in the interior of the quadrilateral, or $p \in a_0$ and $q \in c_0$ or $p \in a_1$ and $q \in c_1$, are proved similarly.

The other cases are treated in the same way as the following one : $p \in a_0$ and $q \in a_1$, with the geodesic between $\overline{p}$ and $\overline{q}$ leaving the interior of the quadrilateral (see Figure 4 (d)).

In this case, we have

$$d(p,q) \leq d(p,A) + d(A,q) \leq d(\overline{p},\overline{A}) + d(\overline{A},\overline{q})$$

As above, by straightening $\overline{c_0} \cup \overline{c_1}$ to a straigth line to obtain the comparison triangle for $(a_0, a_1, c_0 \cup c_1)$, since the angles at $\overline{B_0}$ and $\overline{B_1}$ increase, the distances $d(\overline{p},\overline{A})$ and $d(\overline{A},\overline{q})$ increase. At some moment of the straightening process, the geodesic between $\overline{p}$ and $\overline{q}$ will meet $\{\overline{A}\}$, and the above argument applies. □

Let us derive two easy consequences of the CAT(χ)-inequality definition. First, since the geodesic triangles in a geodesic metric space satisfying the CAT(χ)-inequality are by definition thinner than those in $\mathbf{M}^2_\chi$, and according to exercise 2.7, we have

Proposition 2.13 *If a geodesic metric space X satisfies the CAT(χ)-inequality for some negative χ, then X is hyperbolic (with constant $\frac{1}{\sqrt{\chi}} \log(1+\sqrt{2})$).*

Note that the converse result is false, since for instance $\mathbf{S}^n$ is compact hence hyperbolic, but is CAT(χ) for no negative χ.

2.2 *Convexity and the Cartan-Hadamard theorem*

Definition 2.14 *If $f : [a, b] \to X$ is a geodesic in a metric space (X, d), then the map $[0,1] \to X$ defined by $t \mapsto f(a + (b - a)t)$ is called a geodesic parametrized proportionally to arclength. If f is constant, it is a constant map.*

A convex space is a geodesic metric space (X, d) such that for every pair of geodesics parametrized proportionally to arclength $f : [0,1] \to X$ and $g : [0,1] \to X$, the map $[0,1] \to \mathbf{R}$ defined by $x \mapsto d(f(x), g(x))$ is convex, i.e. for every x, y, t in $[0,1]$, we have

$$d(f(tx + (1-t)y), g(tx + (1-t)y)) \leq td(f(x), g(x)) + (1-t)d(f(y), g(y)).$$

Exercise 2.15 A geodesic metric space (X, d) is convex if and only if for every pair of geodesics $f : [a, b] \to X$ and $g : [c, d] \to X$,

$$d(f(\frac{a+b}{2}), g(\frac{c+d}{2})) \leq \frac{1}{2}(d(f(a), g(c)) + d(f(b), g(d))).$$

It is easy to see that $\mathbf{R}^2$ is convex. Moreover, we have

Proposition 2.16 *A geodesic metric space satisfying the CAT(χ)-inequality with $\chi \leq 0$ is convex.*

Proof. The space satisfies the CAT(0)-inequality according to exercise 2.10. If X is a geodesic metric space satisfying the CAT(0)-inequality, let $f : [0,1] \to X$ and $g : [0,1] \to X$ be geodesics parametrized proportionally to arclength, and $h : [0,1] \to X$ (resp. h', h'') be a geodesic parametrized proportionally to arclength between $f(0)$ and $g(1)$ (resp. $f(0)$ and $g(0)$, $f(1)$ and $g(1)$). Let p (resp. q, r) be the midpoint of f (resp. g, h). Then

$$d(p, q) \leq d(p, r) + d(r, q) \leq d(\overline{p}, \overline{r}) + d(\overline{r}, \overline{q})$$

where we apply the CAT(0)-comparison to the triangles (f, h, h'') and (h, g, h'). Since $\mathbf{R}^2$ is convex,

$$d(p, q) \leq \frac{1}{2}d(\overline{f(1)}, \overline{h(1)}) + \frac{1}{2}d(\overline{h(0)}, \overline{g(0)}) = \frac{1}{2}(d(f(0), g(0)) + d(f(1), g(1))).$$

We then apply the above exercise. □

Definition 2.17 *A metric space X is locally convex if for every x in X, there is a neighborhood V of x such that every two points of V are joined by a geodesic contained in V and for every pair of geodesics parametrized proportionally to arclength $f : [0,1] \to X$ and $g : [0,1] \to X$ with image contained in V, the map $x \mapsto d(f(x), g(x))$ is convex. Such a V will be called a neighborhood of convexity.*

In particular, a nonpositively curved geodesic metric space X is locally convex. Indeed, according to exercise 2.10, X is locally CAT(0). Let V_x be a neighborhood of x as in the definition 2.9, and let ϵ be such that the ball of radius ϵ and center x is contained in V_x. If V is the ball of center x and radius $\frac{\epsilon}{2}$, then any geodesic γ between two points y, y' of V will be contained in V_x. Indeed this geodesic has length less than $2\frac{\epsilon}{2}$, so every point of the geodesic is at distance less than $\frac{\epsilon}{2}$ of y or y', hence at distance less than ϵ from x. Using the geodesic triangle with sides γ and the geodesics between x and y, x and y', using the convexity of $\mathbf{R}^2$, this geodesic will be contained in V. The second statement follows from the proof of proposition 2.16.

Note that a convex space is *uniquely geodesic*, that is between two points, there exists one and only one geodesic. Indeed, a nonnegative convex map, with endpoints value 0 is identically 0.

Note that a convex space X is *contractible*, that is there exist a base point x_0 in X and a continuous map $h : [0,1] \times X \to X$ such that $h(0,x) = x$ and $h(1,x) = x_0$ for every x in X. Indeed, the map defined by $h(t,x) = f_x((1-t)d(x_0,x))$, where $f_x : [0, d(x_0,x)] \to X$ is the unique geodesic between x_0 and x, has the required properties, since

$$\begin{aligned} d(f_x((1-t)d(x_0,x)), f_{x'}((1-t')d(x_0,x'))) \\ &\leq d(f_x((1-t)d(x_0,x)), f_x((1-t')d(x_0,x))) + \\ &\qquad d(f_x((1-t')d(x_0,x)), f_{x'}((1-t')d(x_0,x'))) \\ &\leq |t-t'|d(x_0,x) + d(x,x'). \end{aligned}$$

Definition 2.18 • *A local geodesic is a map $f : I \to X$, where I is some closed interval of $\mathbf{R}$, such that for every t in I, there exists an $\epsilon > 0$ such that the restriction of f to $[x-\epsilon, x+\epsilon] \cap I$ is a geodesic.*

If $f : [a,b] \to X$ is a local geodesic, the map $[0,1] \to X$ defined by $t \mapsto f(a + (b-a)t)$ is called a local geodesic parametrized proportionally to arclength.

- *Let (X,d) be a geodesic metric space and x a point in X. The space $G(X)$ (resp. $G_x(X)$) of local geodesics parametrized proportionally to arclength $f : [0,1] \to X$ (resp. and with $f(0) = x$), endowed with the topology of the uniform convergence (i.e. with the distance $d(f,g) = sup\{d(f(t),g(t))/t \in [0,1]\}$) is called the space of geodesics of X (resp. the space of geodesics of X based at x).*

- *Let (X,d) be a geodesic metric space and x a point in X. The map $exp_x : G_x(X) \to X$ defined by $f \mapsto exp_x(f) = f(1)$ is called the exponential map*

Note that $G_x(X)$ is contractible (using $h(t,f) = g$ where $g(u) = f((1-t)u)$), and if X is a convex space, then the exponential map is an isometry. The uniform metric on $G_x(X)$ is complete if X is complete. Indeed a uniform limit of geodesics is a geodesic.

Theorem 2.19 *(M. Gromov) A complete, simply connected, locally convex, geodesic metric space X is convex.*

Proof. (S. Alexander-R. Bishop [AB]) **Step 1** : Let x be a base point in X. Let us first prove that the exponential map $exp_x : G_x(X) \to X$ is a local isometry.

Fix a local geodesic parametrized proportionally to arclength $f : [0,1] \to X$. By compactness of $f([0,1])$, there exists a positive ϵ such that for every u on $f([0,1])$, the closed ball $\overline{B}(u,\epsilon)$ of radius ϵ and center u is a neighborhood of convexity. (Indeed, cover $f([0,1])$ by a finite number of open balls $B(x_i,\epsilon_i)$ which are neighborhood of convexity centered on $f([0,1])$, and take for ϵ one third of the lower bound of the distance between two distinct points of the intersection of $f([0,1])$ and the spheres $S(x_i,\epsilon_i)$).

Let us prove by induction on n the following statement :

> for every $a < b$ in $[0,1]$ with $b - a \leq (\frac{3}{2})^n \epsilon$, if p, q are points in X with $d(p, f(a)) \leq \frac{\epsilon}{2}$ and $d(q, f(b)) \leq \frac{\epsilon}{2}$, then there exists a local geodesic g parametrized proportionally to arclength between p and q contained in the $\frac{\epsilon}{2}$-neighborhood of $f([a,b])$ and if g and h are local geodesics parametrized proportionally to arclength contained in the $\frac{\epsilon}{2}$-neighborhood of $f([a,b])$ then the map $t \mapsto d(g(t),h(t))$ is convex.

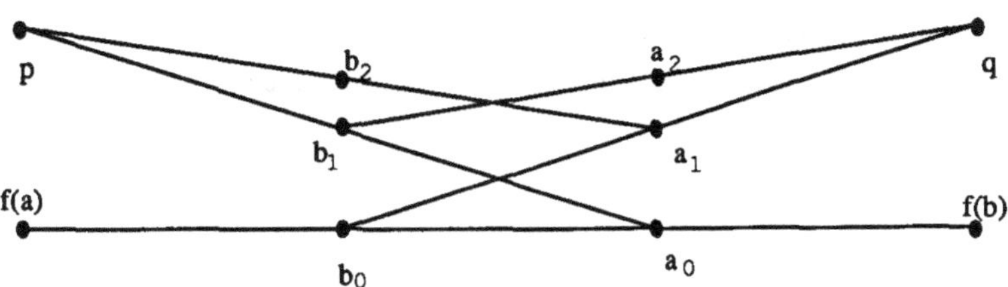

Figure 5 : The exponential map is locally isometric.

The case $n = 0$ follows by definition of ϵ, the closed ball of center the midpoint of a and b and radius ϵ being a neighborhood of convexity. Let $g_0(t) = f(a + \frac{2}{3}(b-a)t)$ and $h_0(t) = f(a + (b-a)(\frac{2}{3}t + \frac{1}{3}))$ for every t in $[0,1]$. Let $a_0 = h_0(\frac{1}{2})$ and $b_0 = g_0(\frac{1}{2})$. By another induction, using the induction hypothesis, there exist, for every nonnegative integer m, local geodesics parametrized proportionally to arclength g_{m+1} (resp. h_{m+1}) between p and a_m (resp. b_m and q) contained in the $\frac{\epsilon}{2}$-neighborhood of $g_0([0,1])$ (resp. $h_0([0,1])$) such that the map $t \mapsto d(g_{m+1}(t), g_m(t))$ (resp. $t \mapsto d(h_{m+1}(t), h_m(t))$) is convex. Let $a_{m+1} = h_{m+1}(\frac{1}{2})$ and $b_{m+1} = g_{m+1}(\frac{1}{2})$.

Since

$$d(g_{m+1}(t), g_m(t)) \leq d(g_{m+1}(1), g_m(1)) = d(h_m(\tfrac{1}{2}), h_{m-1}(\tfrac{1}{2})) \leq \tfrac{1}{2} d(h_m(1), h_{m-1}(1)),$$

it is easy to see by induction that

$$d(g_{m+1}(t), g_m(t)) \leq (\frac{1}{2})^m \frac{\epsilon}{2} \text{and} d(h_{m+1}(t), h_m(t)) \leq (\frac{1}{2})^m \frac{\epsilon}{2}.$$

So that $(g_m(t))_{m\in\mathbf{N}}$ and $(h_m(t))_{m\in\mathbf{N}}$ are uniformly Cauchy sequences. They converge to local geodesics parametrized proportionally to arclength $g(t)$ and $h(t)$, contained in the $\frac{\epsilon}{2}$-neighborhood of $f([a,b])$. It is easy to see that $g(t+\frac{1}{2}) = h(t)$ for $t \in [0, \frac{1}{2}]$. Hence the map $k(t) = g(\frac{3}{2}t)$ if $t \in [0, \frac{2}{3}]$ and $k(t) = h(\frac{3}{2}t - \frac{1}{2})$ if $t \in [\frac{2}{3}, 1]$ is a local geodesic parametrized proportionally to arclength. The required convexity property for two local geodesics parametrized proportionally to arclength contained in the $\frac{\epsilon}{2}$-neighborhood of $f([a,b])$ follows by subdividing and using the induction hypothesis, since the convexity of a function of a real variable is a local notion (if $f : [0,2] \to \mathbf{R}$ and $g : [1,3] \to \mathbf{R}$ are convex and coincide on [0,1], then $f \cup g : [0,3] \to \mathbf{R}$ is convex).

If n is big enough, the induction statement implies that exp_x is an isometry from the $\frac{\epsilon}{2}$-neighborhood of f onto the $\frac{\epsilon}{2}$-neighborhood of $exp_x(f)$.

Step 2 : Now let us prove that the exponential map $exp_x : G_x(X) \to X$ is a covering. Since $G_x(X)$ is contractible, hence connected and simply connected, and X is simply connected, exp_x will be an homeomorphism (see [Mas]), and hence injective. That is between two points, there is a unique local geodesic parametrized proportionally to arclength which is ipso facto a geodesic.

Lemma 2.20 *(S. Alexander-R. Bishop) If $Y, \tilde{Y}$ are complete length spaces, and Y is locally convex, then every surjective local isometry $\phi : \tilde{Y} \to Y$ is a covering.*

Proof. Let p be a point in Y and ϵ be a positive number such that the open ball $B(p,\epsilon)$ is a neighborhood of convexity of p. It is enough to show that $\phi^{-1}(B(p,\epsilon))$ is the disjoint union of the $B(\tilde{p},\epsilon)$ for all preimages $\tilde{p}$ of p , and ϕ is an homeomorphism from $B(\tilde{p},\epsilon)$ to $B(p,\epsilon)$.

Note that a local isometry between two length spaces is distance decreasing, since it preserves the length of paths. Hence $\phi(B(\tilde{p},\epsilon))$ is contained in $B(p,\epsilon)$. By a local isometry, since $\tilde{Y}$ is complete, every geodesic γ of Y may be lifted, starting from any given point of the preimage of the starting point of γ, to a unique geodesic in $\tilde{Y}$, with the same length. Hence $\phi(B(\tilde{p},\epsilon)) = B(p,\epsilon)$ since every two points of $B(p,\epsilon)$ are endpoints of a geodesic contained in $B(p,\epsilon)$.

Let $\tilde{p}, \tilde{p}'$ be two preimages of p. Suppose $\tilde{y}$ (resp. $\tilde{y}'$) lies in $B(\tilde{p},\epsilon)$ (resp. $B(\tilde{p}',\epsilon)$) and y, y' have the same image y under ϕ. Since $\tilde{Y}$ is a length space, there is a continuous curve $\tilde{\alpha}$ (resp. $\tilde{\alpha}'$) between $\tilde{y}$ and $\tilde{p}$ (resp. $\tilde{y}'$ and $\tilde{p}'$) lying in $B(\tilde{p},\epsilon)$ (resp. $B(\tilde{p}',\epsilon)$). Its image curve α (resp. α') in Y lies in $B(p,\epsilon)$ so consider the unique geodesic h_t between $\alpha(t)$ and $\alpha'(t)$. Lift h_t to a geodesic $\tilde{h}_t$ in $\tilde{Y}$ starting at $\tilde{\alpha}(t)$. The other endpoint of $\tilde{h}_t$ is easily seen to form a continuous curve $\tilde{\alpha}''$, lying over α'.

If $\tilde{p} = \tilde{p}'$, then since ϕ is a local isometry, $\tilde{\alpha}'' = \tilde{\alpha}'$, therefore $\tilde{y} = \tilde{y}'$. (Note that a local homeomorphism between topological spaces has not necessarily the property of unique path lifting, as may be seen with the standard example of the non Hausdorff bifurcating line.) This proves that the restriction of ϕ to $B(\tilde{p}, \epsilon)$ is injective, hence an homeomorphism onto its image.

Applying the same argument with the hypothesis $\tilde{y} = \tilde{y}'$ shows that $\tilde{p} = \tilde{p}'$, hence that $B(\tilde{p}, \epsilon)$ and $B(\tilde{p}', \epsilon)$ are disjoint if $\tilde{p} \neq \tilde{p}'$. □

The problem in applying the preceeding lemma is that $G_x(X)$ is not in general a length space (for instance, in the unit circle in $\mathbf{R}^2$, the upper half and the lower half of the circle (based at 1) are at distance π for the uniform metric. But the shorter path between these two has length 2π).

According to the above lemma, one has then to find a new metric d^* on $G_x(X)$, inducing the same topology as the uniform metric d, such that d^* is a complete length metric. Let d^* be the length metric induced by d, that is if f, g are points in $G_x(X)$, $d^*(f, g)$ is the lower bound of the length (using the distance d) of paths in $G_x(X)$ between f and g. Note that $d^*(f, g) \leq \ell(f) + \ell(g)$, using the canonical retraction of $G_x(X)$. Since $(G_x(X), d)$ is locally isometric to X, which is locally convex, the distances d and d^* coincide locally. Hence exp_x is a local isometry from $(G_x(X), d^*)$ to X. Since $d \leq d^*$, every Cauchy sequence for d^* converges for d, hence for d^*. By definition (2.2) of the induced length metric, d^* is a length metric.

Step 3 : Finally, let us prove that X is convex. Let $f : [0, a] \to X$ and $g : [0, b] \to X$ be two geodesics. According to exercise 2.15, we have to show that $d(f(\frac{a}{2}), g(\frac{b}{2}) \leq \frac{1}{2}(d(f(0), g(0)) + d(f(a), g(b)))$. Let $h : [0, c] \to X$ be the geodesic between $f(0)$ and $g(a)$. Since $d(f(\frac{a}{2}), g(\frac{b}{2})) \leq d(f(\frac{a}{2}), h(\frac{c}{2})) + d(h(\frac{c}{2}), g(\frac{b}{2}))$, and by applying a symmetric argument to the geodesic triangle with vertices $f(0), g(0), g(b)$, we have to show that $d(f(\frac{a}{2}), h(\frac{c}{2})) \leq \frac{1}{2}d(f(a), h(c))$.

Let $k : [0, d] \to X$ be the geodesic between $f(a)$ and $h(c)$, and for every t in $[0, d]$, let h_t be the unique (according to step 2) geodesic parametrized proportionally to arc length between $f(a)$ and $k(t)$. Let t_u be the upper bound of the non empty set of t in $[0, d]$ such that $d(f(\frac{a}{2}), h_t(\frac{1}{2})) \leq \frac{1}{2}d(f(a), h_t(1))$. According to step 2, the maps $t \mapsto h_t(\frac{1}{2})$ and $t \mapsto h_t(1)$ are continuous. Hence t_u is in fact a maximum. By the local results of step 1, we have $t_u = b$, and the result is proved. □

Corollary 2.21 *(Cartan-Hadamard-Aleksandrov-Gromov) Let X be a complete, simply connected, geodesic space with curvature less than or equal to χ, with $\chi \leq 0$. Then X satisfies the $CAT(\chi)$-inequality.*

Note that this result is false if $\chi > 0$. Indeed, let X be the subset of the unit sphere $\mathbf{S}^2$ in $\mathbf{R}^3$ contained in the half-space $\{(x, y, z) \in \mathbf{R}^3 / z \leq \frac{1}{2}\}$, endowed with the induced length metric. This metric space is compact, geodesic, simply connected, with curvature less than or equal to 1, and its boundary $\mathbf{S}^2 \cap \{(x,y,z) \in \mathbf{R}^3 / z = \frac{1}{2}\}$ is a local geodesic. But X does not satisfy the CAT(1)-inequality. This may be seen by looking at the two geodesics in the boundary between two opposite points.

Proof. According to the theorem 2.19, it is enough to prove that a convex space with curvature less than or equal to χ, with $\chi \leq 0$, satisfies the CAT(χ)-inequality.

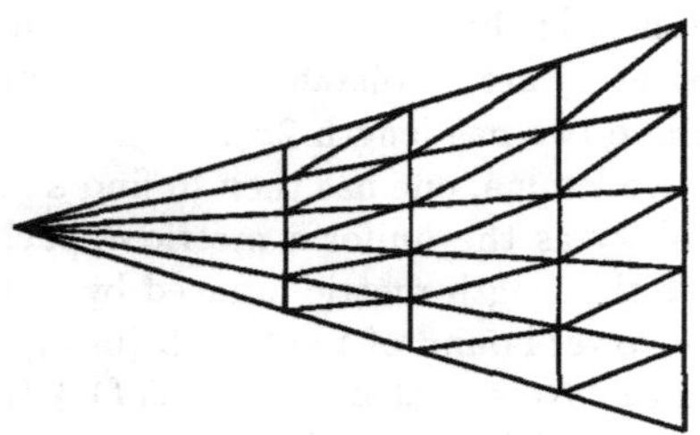

Figure 6 : Locally CAT(χ) and simply connected implies globally CAT(χ).

We can subdivide a geodesic triangle (a, b, c) as in Figure 6, such that every small triangle is contained in a neighborhood as in Definition 2.9. Indeed, the space T of points on the geodesic γ_s between the point $a(s)$ on the side a and its opposite vertex A is compact, since the map $[0,1] \times [0,1] \to X$ defined by $(s,t) \mapsto \gamma_s(t)$ is continuous (geodesics are parametrized proportionally to arclength). Hence T can be covered by finitely many neighborhoods as in Definition 2.9. Since geodesics vary continuously with their endpoints, we can subdivide such that every triangle is contained in one of these neighborhoods.

By applying inductively Lemma 2.12, the result follows. □

3 Negatively curved polyhedra

In order to have computable criteria of hyperbolicity, we have to restrict the class of spaces we are going to consider.

3.1 *Cells*

In this subsection, fix some $\chi \in \mathbf{R}$. The following definitions may be found (in the Euclidean case) in elementary textbooks on algebraic topology or piecewise-linear topology. We use [RS] as a general reference.

The standard simplex (resp. cube) of dimension n, or standard n-simplex (resp. standard n-cube) is the space

$$\Delta^n = \{(x_1, \cdots, x_{n+1}) \in \mathbf{R}^{n+1} / \sum_{i=1}^{n} x_i = 1 \text{ and } x_i \geq 0\}$$

(resp. $\square^n = [-1,1]^n$). The faces (of dimension k or k-faces) of Δ^n (resp. $\square^n$) are the subsets of the form $\{(x_1, \cdots, x_{n+1}) \in \Delta^n / x_{i_1} = \cdots = x_{i_{n-k}} = 0\}$ with $1 \leq i_1 < i_2 < \cdots < i_{n-k} \leq n+1$ (resp. $\{(x_1, \cdots, x_n) \in \square^n / \forall j \in \{i_1, \cdots i_{n-k}\}, x_j = -1 \text{ or } 1\}$ with $1 \leq i_1 < i_2 < \cdots < i_{n-k} \leq n$). The center of a standard n-simplex (resp. n-cube) is the point $(\frac{1}{n+1}, \cdots, \frac{1}{n+1})$ (resp. $(0, \cdots, 0)$).

For $0 \leq k \leq n$, define a k-plane in M^n_χ to be a subset E of M^n_χ isometric to M^k_χ for $0 \leq k \leq n$ and call k the dimension of E. A 1-plane (resp. $(n-1)$-plane) is also called a line (resp. hyperplane) of M^n_χ. A *closed halfspace* of M^n_χ is the closure of one of the two connected components of M^n_χ minus some hyperplane. A *flag* of M^n_χ is a sequence $F_0, F_1, \cdots, F_n$ where F_k is a k-plane in M^n_χ and $F_{k-1} \subset F_k$ for $k = 1 \cdots n$. Note that the isometry group of M^n_χ acts transitively on the flags of M^n_χ (see exercise 2.6).

A *convex subset* of a geodesic space is a subset S such that every geodesic between two points of S lies in S. The *convex hull* of a subset S of a geodesic space X is the intersection of all convex subsets of X containing S. Note that the convex hull of a subset of X is convex.

Recall that the *diameter* of a subset S of a metric space X is the smallest upper bound of the $d(x,y)$ for every x, y in S.

Definition 3.1 *A cell (or polytope) in M^n_χ is a non empty compact subset with diameter less than $\frac{\pi}{2\sqrt{\chi}}$ which is the intersection of finitely many closed halfspaces.*

The *dimension* k of a cell C is the smallest integer k such that C is contained in a k-plane. The *interior* of a cell C, denoted by $\overset{\circ}{C}$, is the interior of C in the k-plane containing C where k is the dimension of C. The *boundary* of C is defined by $\partial C = C - \overset{\circ}{C}$. The center of a cell C of M^n_χ is the center of the ball of M^n_χ of smallest radius containing C. Note that a ball of smallest radius containing a given cell is unique, since if two balls have the same radius strictly less than $\frac{\pi}{2\sqrt{\chi}}$, and are distinct, then their intersection is contained in a ball of strictly smaller radius.

Let C be a cell in M^n_χ, for $x \in \partial C$ define (x, C) to be the union of all lines L through x in M^n_χ such that $C \cap L$ is an arc with x in its interior. If there are no such line, define $(x, C) = x$ and call x a *vertex*. Define a *face* F of C to be a subset of M^n_χ of the form $F = (x, C) \cap C$. By convexity, (x, C) is some k-plane of M^n_χ and we say that F is a k-face or face of dimension k. A 1-face is also called an edge. A face of C of dimension strictly less than the dimension of C is called a proper face.

A simplex (resp. a cube) is a cell in some $\mathbb{M}^n_\chi$ homeomorphic to an Euclidean standard simplex (resp. cube) by an homeomorphism sending faces onto faces.

Exercise 3.2 • The intersection of two cells is a cell. The product of two Euclidean cells is a cell. A cell is the convex hull of its vertices. A subset of $\mathbb{M}^n_\chi$ of diameter less than or equal to $\frac{\pi}{2\sqrt{\chi}}$ is a cell if and only if it is the convex hull of finitely many points. A cell in $\mathbb{M}^n_\chi$ of dimension m is simplicial if and only if it is the convex hull of $m+1$ points in $\mathbb{M}^n_\chi$ in general position (i.e. m of these points are not contained in an $(m-1)$-dimensional subspace).

- A cell has finitely many faces. A face of a cell is a cell of lower or equal dimension. A face of a face of a cell C is a face of C. The intersection of two faces is a face. If C is a cell, then ∂C is the disjoint union of the $\mathring{F}$, where F is a face of C.

Definition 3.3 *Let C be a cell od dimension n in $\mathbb{M}^n_\chi$. A flag in C is a nested sequence $F_0 \subset F_1 \subset \cdots \subset F_{n-1}$ where each F_j is a j-face of C. The group of C, denoted by $G(C)$, is the group of all isometries of $\mathbb{M}^n_\chi$ under which C is invariant.*

The cell C in V is said to be regular if $G(C)$ acts transitively on the set $\mathcal{F}(C)$ of flags of C.

In particular two k-faces of a regular cell are isometric. For more informations, see P. de la Harpe lectures [Har].

Note that at the group theoretic level, there are no differences between the different χ, since the center of a cell is fixed by the group $G(C)$ and the subgroup of the isometry group of $\mathbb{M}^n_\chi$ fixing a point is isomorphic to $O(n)$.

Examples : (see P. de la Harpe lectures [Har])

(1) The standard simplex Δ^n is regular with group the symmetric group Σ_{n+1}.

(2) The standard cube $\square^n$ is regular, with group the semi-direct product of $(\mathbb{Z}/2\mathbb{Z})^n$ by Σ_n with the natural action of Σ_n on the direct product of n copies of the cyclic group of order 2.

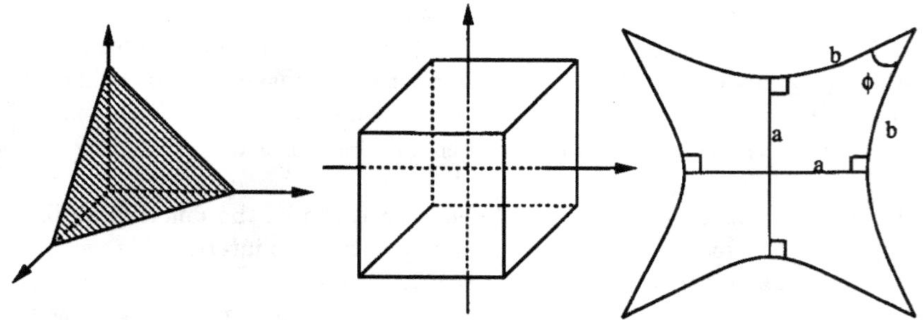

Figure 1 : Regular simplices and cubes.

We are going to construct a regular n-simplex Δ^n_χ (resp. n-cube $\square^n_\chi$) in $\mathbf{M}^n_\chi$ in order that Δ^n_χ (resp. $\square^n_\chi$) is isometric to any n-face of Δ^{n+1}_χ (resp. $\square^{n+1}_\chi$). These regular n-simplices and n-cubes depend on one parameter $\lambda_1 > 0$, and if $\chi > 0$, they are defined only for n less than N_{λ_1}. So this will force a upper bound on the dimension of multipolyhedra with $\chi > 0$ and regular cubes. Note that N_{λ_1} tends to $+\infty$ when λ_1 tends to 0. Unless stated otherwise, we will use these regular simplices and cubes.

Let x_0 be a point in $\mathbf{M}^n_\chi$ and fix n pairwise orthogonal oriented lines L_i through x_0, called the coordinate axes. For $i = 1, \cdots, n$, let e_i (resp. e_{i+n}) be the point on L_i at distance λ_n from the origin x_0 on the positive (resp. negative) side, where $\lambda_n \in [0, \frac{\pi}{4\sqrt{\chi}}[$ (where $\frac{\pi}{4\sqrt{\chi}}$ is $+\infty$ if $\chi \leq 0$). Define Δ^{n-1}_χ to be the convex hull of e_i for $i = 1, \cdots, n$. For $i = 1, \cdots, 2n$, let H_i be the hyperplane containing e_i orthogonal to L_i and H_i^+ be the closed halfplane defined by H_i containing x_0. Then define $\square^n_\chi$ to be the intersection of the H_i^+ for $i = 1, \cdots, 2n$.

The λ_n are choosen inductively such that $\square^n_\chi$ is isometric to a n-face of $\square^{n+1}_\chi$. We take $\lambda_1 < \frac{\pi}{4\sqrt{\chi}}$ and by induction

$$\begin{aligned}
\lambda_{n+1} &= \tfrac{1}{\sqrt{\chi}} \arcsin \tan(\sqrt{\chi}\lambda_n) && \text{if } \chi > 0 \\
&= \lambda_n && \text{if } \chi = 0 \\
&= \tfrac{1}{\sqrt{-\chi}} \operatorname{argsinh} \tanh(\sqrt{-\chi}\lambda_n) && \text{if } \chi < 0
\end{aligned}$$

This may be seen by looking at the formulae for the quadrilateral with three right angles (see [Bea] page 157 for instance). If the other angle has measure ϕ, and the sides adjacent to this angle have length b, the other two sides having length a, then if $\chi > 0$ (see figure 1)

$$\sinh^2 a = \cos\phi \quad \text{and} \quad \cosh a = \cosh b \sin\phi\ .$$

Note that if $\chi < 0$, then the sequence $(\lambda_i)_{i \in \mathbf{N}^*}$ decreases to 0 and if $\chi > 0$, the sequence $(\lambda_i)_{i \in \mathbf{N}^*}$ increases, and since the map $t \mapsto \arcsin \tan t$ is defined only for $t \leq \frac{\pi}{4}$ and has no fixed point other than 0, the sequence stops to be defined at some n.

Let C be a cell of dimension n in $\mathbf{M}^n_\chi$ and $x \in C$. The *link* (or direction space) of x in C, denoted by $\text{Link}(x, C)$, is the set of local geodesics $f : [0, +\infty[\to \mathbf{M}^n_\chi$ with $f(0) = x$ and $f(t) \in C$ for t small enough. Recall that local geodesics are parametrized by arclength (see definition 2.18). The distance between two geodesics is the angle they make at x. Note that $\text{Link}(x, C)$ is isometric to the subset of $\mathbf{S}^{n-1} = \mathbf{M}^{n-1}_1$, viewed as the intersection of the set of the above geodesics and of the sphere in $\mathbf{M}^n_\chi$ of center x and radius $\frac{\pi}{2\sqrt{\chi}}$, 1, $-\frac{\log(1+\sqrt{2})}{\sqrt{-\chi}}$ respectively for $\chi > 0$, $\chi = 0$, $\chi < 0$ (see exercise 2.7). In particular $\text{Link}(x, C)$ is a cell of dimension $(n-1)$ in the space $\mathbf{M}^{n-1}_1$ of constant curvature $+1$. If x is in the interior of a cell C of dimension n, then Link(x, C) is the whole sphere $\mathbf{S}^{n-1}$. The space Link(x, C) may be seen as the set of equivalence classes (called *germs of geodesics*) of geodesics $f : [0, \epsilon_f[\rightarrow \mathsf{M}^n_\chi$ with values in C, $f(0) = x$ and $\epsilon_f > 0$, two such geodesics f, g being identified if they coincide on $[0, \epsilon[$ for some $\epsilon > 0$ with $\epsilon < \min\{\epsilon_f, \epsilon_f\}$.

Let C be a cell in M^n_χ and F a face of C. The *link* (or orthogonal direction space) of F in C, denoted by Link(F, C), is the set of local geodesics $f : [0, +\infty[\rightarrow \mathsf{M}^n_\chi$ with $f(0) = x$, $f(t) \in C$ for t small enough, and f orthogonal to F at x, where x is the center of F (any other point of F would do as well).

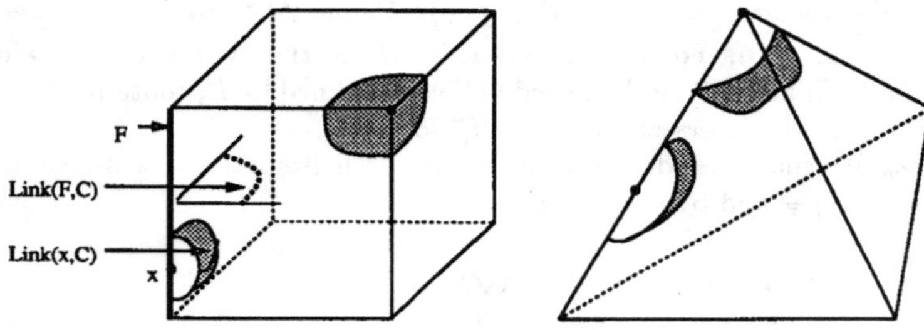

Figure 2 : Links.

3.2 *Polyhedra with piecewise constant curvature*

We are going to work in the category of multipolyhedra, which are cells isometrically glued together such that every cell is embedded, but two cells may be glued along more than one face.

Definition 3.4 *Let $\chi \in \mathbf{R}$. A (cell) multipolyhedron of piecewise constant curvature χ (for short M_χ-multipolyhedron, or multipolyhedron if χ is understood) is a disjoint set E of cells (not necessarily in the same M^n_χ) and an equivalence relation $\sim$ on $\cup E$ such that*

1. *for every C in E and x, y in C, $x \sim y$ if and only if $x = y$,*

2. *for every C, C' in E, $\mathcal{D}_{C',C} = \{x \in C / \exists y \in C',\ x \sim y\}$ is a (possibly empty) union of faces of C, and the restriction of $\sim$ to $C \times C'$ is the graph of an homeomorphism from $\mathcal{D}_{C',C}$ to $\mathcal{D}_{C,C'}$ which is an isometry on every face of $\mathcal{D}_{C',C}$.*

If, for every two cells C, C' of E, we have $\mathcal{D}_{C',C}$ is empty or one face of C, the multipolyhedron is called a *polyhedron*. With abuse, we will called multipolyhedron P the set of equivalence classes $\cup E/\sim$, and E and $\sim$ as above will be understood. By cell of a multipolyhedron P (resp. vertex, edge, etc), we mean the image in $\cup E/\sim$ of a face of a cell of E (resp. of a vertex of a cell of E, of an edge of a cell of E, etc). A cell will be endowed with its natural metric.

A multipolyhedron P is called *simplicial* (resp. *cubical*) if every cell of P is a simplex (resp. cube). The smallest upper bound of the dimensions of the cells of a multipolyhedron P is called the *dimension* of P. A *subpolyhedron* Q of P is a subset of P which is a union of cells of P. The *k-skeleton* of a multipolyhedron is the subpolyhedron of all cells of dimension less than or equal to k. A multipolyhedron P is said to be *finite* (resp. *countable*) if it has only a finite (resp. countable) number of cells. It is said to be *locally finite* if every cell of P meets only finitely many cells of P. An *isomorphism* between two multipolyhedra P and Q is a bijection $f : P \to Q$ such that, for every cell C in P, the restriction of f to C is an homeomorphism onto a cell of Q sending faces onto faces, and the same thing for f^{-1}.

Note that if two cells are glued along a face of dimension greater than or equal to 2, then they must have the same χ. There are no more $\mathbf{M}_\chi$-multipolyhedron than $\mathbf{M}_0$-multipolyhedron, since there are no more cells in $\mathbf{M}^n_\chi$ than in $\mathbf{R}^n$. That is, every $\mathbf{M}_\chi$-multipolyhedron is isomorphic to a $\mathbf{M}_0$-multipolyhedron and conversely. This may be seen by stereographic projection in the spherical case, and by taking the Klein model in the hyperbolic case. We won't define the Klein model here, see for instance [Thu1].

Examples. The circle $S^1 = \{(x,y) \in \mathbf{R}^2/x^2+y^2=1\}$ with two cells of dimension 1, $X_1 = \{(x,y) \in S^1/x \leq 0\}$ and $X_2 = \{(x,y) \in S^1/x \geq 0\}$, and the obvious gluings, is a multipolyhedron. If we look at the circle as the boundary of the standard 2-simplex, then we get a Euclidean polyhedron. The sphere $S^2 \in \mathbf{R}^3$, viewed as two standard 2-simplices glued along their boundary, is a Euclidean simplicial multipolyhedron. The sphere $S^2 \in \mathbf{R}^3$, viewed as the boundary of the standard 3-simplex, is a Euclidean simplicial polyhedron. Viewed as the boundary of the standard 3-cube, it is a Euclidean cubical polyhedron.

The *derived polyhedron* (also called *barycentric subdivision*) P' of a multipolyhedron P is the multipolyhedron defined inductively by : a cell of P' is the convex hull of the union of the center of some cell C of P and a cell of the barycentric subdivision of the boundary of C. In the case of the 1-simplex, we subdivide it into two segments by its midpoint. The second derived polyhedron (or second barycentric subdivision) is $P'' = (P')'$. Note that if Q is a subpolyhedron of P, then Q' is a subpolyhedron of P'.

If Q is a subpolyhedron of P, the *regular neighborhood* $N(Q,P)$ of Q in P is the subpolyhedron of P'' of the cells A in P'' such that there exist cells B in P''

and C in Q'' with A and C faces of B. We denote by $\partial N(Q,P)$ the union of cells of $N(Q,P)$ that do not meet Q'' and set $\overset{\circ}{N}(Q,P) = N(Q,P) - \partial N(Q,P)$.

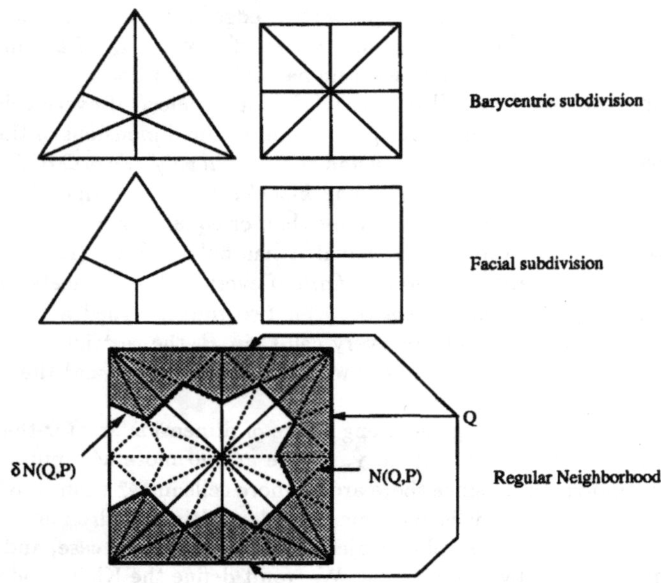

Figure 3 : Barycentric and facial subdivision ; regular neighborhood.

The derived polyhedron of a multipolyhedron is obviously a simplicial polyhedron. But, if we start with a cubical multipolyhedron, we may want to subdivide and still get a cubical multipolyhedron. Let P be a regular cubical multipolyhedron, the *facial subdivision* P^* of P is the obvious one, which subdivides every 1-simplex into two segments by its midpoint, and every n-cube $\square^n$ into 2^n regular cubes of side length the half of the side length of $\square^n$, which are the closures of the connected components of the complementary set in $\square^n$ of the n coordinate hyperplanes (see figure 3).

The *facial subdivision* P^* of a simplicial multipolyhedron P is defined inductively by : the vertices of P^* are the centers of the cells of P ; the edges are the segments between the center of a cell of P and the center of the faces of codimension one of this cell ; in general, a n-cell in P^* is the convex hull of the center of some cell C of P, the center of a codimension n proper face F of C, and the centers of all the cells of ∂C properly containing the face F. It is easy to see that the facial subdivision of a simplicial or cubical multipolyhedron is a cubical polyhedron. So that, up to subdivision, there are no more simplicial multipolyhedra than cubical multipolyhedra.

We will also use another kind of subdivision (called ***star subdivision***), whose result is to turn any point of a multipolyhedron into a vertex. If x is point in a cell C, then the star subdivion of C at x is the polyhedron whose cells are the convex hulls of x and the faces of C which do not contain x. If x is point in a multipolyhedron P, then the star subdivions of all cells of P containing x fit together to give the star subdivision of P at x. The star subdivision of a simplicial multipolyhedron at any point is a simplicial multipolyhedron.

If C is a cell of dimension m in $\mathbf{M}^n_\chi$, and $v \in \mathbf{M}^n_\chi$ is not in the m-plane containing C, then the *cone* with vertex v and base C is the convex hull of v and C. If P is a multipolyhedron, then a *multipolyhedral cone* with base P is a multipolyhedron whose cells are cones over the cells of P, and their faces, with the new gluings extending the old ones, such that the vertices of the cones over the cells of P are all glued together.

If X is a set, the abstract cone over X is the set ${}^{X\times[0,1]}/_{\sim}$, where $\sim$ is the equivalence relation defined by $(x,t) \sim (y,u)$ if and only if $(x,t) = (y,u)$ or $t = u = 1$. It is clear that the abstract cone over a regular simplicial multipolyhedron P has a natural structure of a multipolyhedral cone with base P. Moreover, it may be proved that the abstract cone over any multipolyhedron P has a natural structure of a multipolyhedral cone with base P. (Take some isomorphic piecewise constant zero curvature multipolyhedron. Subdivide it in order to have a polyhedron P, then embed it in $\mathbf{R}^{\{vertices\}}$. In the product with $\mathbf{R}^{\{vertices\}} \times \mathbf{R}$, take the cone over P with vertex having a non zero coordinate in the last $\mathbf{R}$. Then flatten every cell inductively on the dimension to go back to the original multipolyhedron, and return to the original curvature.)

Definition 3.5 *Let x be a point of a $\mathbf{M}_\chi$-multipolyhedron P. The link (or direction space) of x in P, denoted by Link(x, P), is the union of the Link(x, C) for all cells C of P, where for every common face F of cells C, C' of P, the cells Link(x, C), Link(x, C') are glued along Link(x, F).*

Let x be a point in a $\mathbf{M}_\chi$-multipolyhedron P. The (closed) star of x in P, denoted by Star(x, P), is the union of the cells C of the barycentric subdivision P' of P such that there is a cell D in P' with $x \in D$ and C a face of D.

Let F be a cell of a $\mathbf{M}_\chi$-multipolyhedron P. The link (or orthogonal direction space) of F in P, denoted by Link(F, P), is the union of the Link(F, C) for all cells C of P containing F, where for every common face F' of cells C, C' of P with $F \subset F'$, the cells Link(F, C), Link(F, C') are glued along Link(F, F').

Note that if C is a cell of P, then Link(C, P) is a not necessarily connected $\mathbf{M}_1$-multipolyhedron. If $x \in P$, possibly after subdividing P so that x is a vertex, Link(x, P) also has a structure of a $\mathbf{M}_1$-multipolyhedron. It is easy to see that the link of a cell of a cubical $\mathbf{M}_\chi$-multipolyhedron is a simplicial $\mathbf{M}_1$-multipolyhedron.

If x is a vertex of P, then the vertices y of Link(x, P) corresponds to the edges E of P containing x, and Link$(y, \text{Link}(x, P))$ is isomorphic, by an isomorphism which is an isometry on each cell, to Link(E, P). More generally, if A is a cell in P of dimension k, then the cells B in Link(A, P) of dimension ℓ correspond to the cells C in P of dimension $(k+\ell+1)$ having A as a face, and Link$(B, \text{Link}(A, P))$ is isomorphic, by an isomorphism which is an isometry on each cell, to Link(C, P).

If $C \subset \mathbf{S}^k$ is a spherical cell of dimension k and $\mathbf{S}^k, \mathbf{S}^\ell$ are embedded in $\mathbf{S}^{k+\ell+1}$ as the intersections with $\mathbf{S}^{k+\ell+1}$ of two orthogonal subspaces of $\mathbf{R}^{k+\ell+2}$ of dimension $k+1, \ell+1$ respectively. Define the *ℓ-fold suspension of* A, denoted by $\mathbf{S}^{\ell-1} \star C$ to be the convex hull in $\mathbf{S}^{k+\ell+1}$ of C and $\mathbf{S}^\ell$ (see [DJ] and R. Charney and M. Davis [CD], appendix). With the obvious subdivision, it is a spherical polyhedron. If Q is a $\mathbf{M}_1$-multipolyhedron, then the *ℓ-fold suspension of* Q, denoted by $\mathbf{S}^{\ell-1} \star Q$ is the union of $\mathbf{S}^{\ell-1} \star C$ for all cells C in Q, glued in the natural way : if two cells C, C' in Q are glued along a cell F, then $\mathbf{S}^{\ell-1} \star C$ and $\mathbf{S}^{\ell-1} \star C'$ are glued along $\mathbf{S}^{\ell-1} \star F$. We obviously have $\mathbf{S}^{n+1} = \mathbf{S}^0 \star \mathbf{S}^n$, and more generally $\mathbf{S}^0 \star (\mathbf{S}^{\ell-1} \star P)$ is isometric to $\mathbf{S}^\ell \star P$.

Note that if x is a vertex of P', then Star(x, P) is a simplicial $\mathbf{M}_\chi$-multipolyhedron, which is a multipolyhedral cone with vertex x and base a $\mathbf{M}_\chi$-multipolyhedron $\partial(\text{Star}(x, P))$. Furthermore, if x is a vertex of P, then $\partial(\text{Star}(x, P))$ is isomorphic to the barycentric subdivision of Link(x, P). We have to take the barycentric subdivision here to avoid the problems of multipolyhedra that are not polyhedra. Define the open star $\overset{\circ}{\text{Star}}\,(x, P)$ to be the union of the interior of the cells C of the barycentric subdivision of P with $x \in C$. The open star is then Star$(x, P) - \partial(\text{Star}(x, P))$.

Note that for every point x in a $\mathbf{M}_\chi$-multipolyhedron P, there is a well-defined *radial projection* r from Star$(x, P) - \{x\}$ into Link(x, P) which associates to a point y the germ of the unique geodesic from x to y in a cell C containing x and y (which does not depend on the cell).

In the 2-dimensional case, the space of directions Link(x, P) at a point x is homeomorphic (with the induced length topology) to a (maybe non locally finite) graph. If x is not on the 1-skeleton, then Link(x, P) is isometric to the circle $\mathbf{S}^1$. If x is on the 1-skeleton, but not a vertex, then Link(x, P) is isometric to the union of half circles of length π, glued along their pairs of endpoints, one for each 2-face containing x. In particular, every simple closed curve in Link(x, P) has in that case length greater than or equal to 2π. So what is important is the space Link(x, P) for the vertices x.

3.3 *Completeness and geodesics*

Let P be a $\mathbf{M}_\chi$-multipolyhedron. A map f from $[a, b]$ to P is called a *broken geodesic* if there is a subdivision $a = t_0 < t_1 < \cdots < t_{p+1} = b$ such that $f([t_i, t_{i+1}])$ is contained in some cell and the restriction of f to $[t_i, t_{i+1}]$ is a geodesic inside that cell, for $i = 1 \cdots p$. Then define the length $\ell(f)$ of the broken geodesic map f to be

$$\sum_{i=0}^{p} \ell(f_{|[t_i,t_{i+1}]}) = \sum_{i=0}^{p} d(f(t_i), f(t_{i+1})),$$

the length inside a cell being mesured with respect to the metric of the cells. Then define $\tilde{d}(x,y)$, for every two points x, y in P, to be the lower bound of the lengths of broken geodesics from x to y. This is clearly a symmetric map, satisfying the triangle inequality, called the *length pseudo-distance* (or intrinsic pseudo-metric as in [Bri]). If two points x, y of P are not joined by a broken geodesic, we say that P is not connected and define $\tilde{d}(x,y)$ to be $+\infty$. Note that subdividing does not change anything to the length pseudo-distance.

The weak topology on a multipolyhedron is the one for which a set is open if and only if its intersection with every cell is open (see for example [Spa]). Note that a multipolyhedron with the weak topology is metrizable if and only if it is locally finite (see [Spa]). The weak topology is in general weaker than the topology induced by the length pseudo-distance. They coincide in particular if the multipolyhedron is finite or locally finite.

Note that the length pseudo-distance on a multipolyhedron may not be a metric (i.e. may be non Hausdorff). One of the easiest examples (given by M. Bridson) is the one with 2 vertices joined by countably many edges, the n-th edge having length $\frac{1}{n}$. Note that even if the length pseudo-distance on a multipolyhedron is a metric, it may be non complete. For instance $]0,1]$, seen as the locally finite polyhedron, union of the closed intervals $[\frac{1}{n+1}, \frac{1}{n}]$, is non complete. Note that even if the length pseudo-distance on a connected multipolyhedron is a complete metric, it may be non geodesic. For instance, the multipolyhedron with 2 vertices joined by countably many edges, the n-th edge having length $1 + \frac{1}{n}$. Note that the cells may be non convex (two intervals of different lengths glued by their endpoints), and even the geodesic in a cell between a vertex and a point of that cell may not be a geodesic in the multipolyhedron.

If the length pseudo-distance on a M_χ-multipolyhedron P is Hausdorff, complete and geodesic, if x is a point in P, then the length pseudo-distance on the M_1-multipolyhedron Link(x, P) is Hausdorff, but not necessarily complete, nor geodesic on connected components if complete. The first fact follows from the lemma 3.8 below. For the second fact, take the cone over $]0,1]$, seen as the union of the closed intervals $[\frac{1}{(n+1)^3}, \frac{1}{n^3}]$, with the rays piecewise linearily scaled so that the ray joining the vertex of the cone to the point $\frac{1}{n^3}$ has length $\frac{1}{n}$. For the third fact, take the union of the angular sectors A_n of the closed unit disk in $\mathbf{R}^2$ with angle $\pi + \frac{1}{n}$, with $te^{i(\pi+\frac{1}{n})} \in A_n$ identified with $te^{i(\pi+\frac{1}{n+1})} \in A_{n+1}$ and $t \in A_n \cap \mathbf{R}$ identified with $t \in A_{n+1} \cap \mathbf{R}$ for $t \in [0,1]$.

If P is not locally finite, in some cases, the space Link(x, P) may not be considered as (up to a scaling of the metric) the subset of points at distance ϵ from x, for some ϵ small enough depending on x (see examples above).

The following remark is due to M. Davis [DJ][CD]. If P is a $\mathbf{M}_\chi$-multipolyhedron, C is a cell of dimension k in P and x is a point in the interior $\overset{\circ}{C}$ of C, then $\mathrm{Link}(x, P)$ is isometric (for the length pseudo-distances) to the k-fold suspension $\mathbf{S}^{k-1} \star \mathrm{Link}(C, P)$ of $\mathrm{Link}(C, P)$.

One of the most powerful theorem asserting that the length pseudo-distance is Hausdorff, complete and geodesic on every connected component is the following. It furthermore gives a characterization of the local geodesics. A broken geodesic $f : [a, b] \to P$ is *taut* if for every $t \in]a, b[$, the distance in $\mathrm{Link}(f(t), P)$ between the two germs of geodesics $f_{|]t-\epsilon,t]}$ and $f_{|[t,t+\epsilon[}$ is greater than or equal to π. (Our terminology differs from Bridson's [Bri]).

Theorem 3.6 *(M.R. Bridson [Bri]) If a multipolyhedron P has finitely many isometry types of cells, then the length pseudo-distance, on every connected component of P, is a complete geodesic metric. The local geodesics are the taut broken geodesics.*

One of the main applications of the above theorem is to Tit's buildings (see K. Brown's lectures [Bro] for an introduction), that may be non locally finite. The theorem is also valid (as may be seen in the proof) if for every N in $\mathbf{N}$ and some x in P, the ball $B_{\bar{d}}(x, N)$ of center x and radius N for the length pseudo-distance meets only finitely many cells of P. But the balls are difficult to compute in general. The length pseudo-distances on the links are also Hausdorff, complete and geodesic on every connected component, since they obviously satisfy the hypothesis of the theorem (in fact, it is proved inductively). Note that the hypothesis of finitely many isometry types of cells implies that the multipolyhedron P is finite dimensional.

The following example, due to M.R. Bridson, shows a counterexample in an infinite dimensional case. For every n in $\mathbf{N} - \{0\}$, we have $\Delta^n \subset \Delta^{n+1}$, induced by $(x_1, \cdots, x_{n+1}) \mapsto (x_1, \cdots, x_{n+1}, 0)$ from $\mathbf{R}^{n+1}$ to $\mathbf{R}^{n+2}$. Let Δ^∞ be the union of Δ^n for n in $\mathbf{N} - \{0\}$. Obviously, Δ^∞ is a regular simplicial Euclidean polyhedron. Take two copies $\Delta^\infty_+, \Delta^\infty_-$ of Δ^∞ and glue them along the opposite faces of the first vertex (that is along $\Delta^n_\pm \cap \{(x_1, \cdots, x_{n+1}) \in \mathbf{R}^{n+1} / x_1 = 0\}$ for every n in $\mathbf{N} - \{0\}$). Then there is no geodesic between the first vertices x_+, x_- of $\Delta^\infty_+, \Delta^\infty_-$. Indeed the height a_n of Δ^n (i.e. the distance between any vertex and the opposite face) decreases to the limit $\frac{1}{\sqrt{2}}$, and the length pseudo-distance between x_+, x_- is twice the limit, that is $\sqrt{2}$, which is not reached by any length of path between x_+, x_-. If b_n denotes the distance between the barycenter B_n of Δ^n and any of its vertices $V_1, \cdots, V_{n+1}$, then we have inductively $a_n = \sqrt{1 - b_{n-1}^2}$ and $B_n = \frac{1}{n+1}(V_1 + \cdots + V_{n+1})$, so that $b_n^2 = \frac{1}{(n+1)^2}(||V_1V_2 + \cdots + V_1V_{n+1}||^2) = \frac{n+n(n-1)\cos\frac{\pi}{3}}{(n+1)^2}$ hence $b_n = \sqrt{\frac{n}{2(n+1)}}$ and $a_n = \sqrt{\frac{n+1}{2n}}$.

A lot of people have made contributions to results related to the theorem 3.6. In the 2-dimensional case, see [AZ][GH]. In the locally finite case, see

[Bus][Ale1][Glu][Sto], and [Mou] whose arguments were adapted by Bridson. If the length pseudo-distance is Hausdorff and complete, the existence of geodesics on every connected component follows if the multipolyhedron is locally compact (see proposition 2.3). In particular, we have the following corollary of the first two steps of the proof of 3.6. The *height* of a cell C is the minimum of the distances $d(F,F')$ in C, where F,F' are disjoint faces (including vertices) of C.

Proposition 3.7 *(G. Moussong [Mou]) If P is a locally finite multipolyhedron, if there is a positive lower bound on the heights of all cells of P, then the length pseudo-distance on every connected component is Hausdorff, complete and geodesic. The local geodesics are the taut broken geodesics.*

Proof of 3.6 [Bri] : Since the paper [Bri] of M.R. Bridson is included in these proceedings, we only give here a sketch of the proof. The proof uses an induction on the dimension.

Step 0 : (G. Moussong) If x is a point of a multipolyhedron P, define $\delta(x)$ to be the infimum, over all cells C containing x, of the distance in C between x and $C - \text{Star}(x,P)$. If $\delta(x) > 0$ for every x in P, then the pseudo-distance $\tilde{d}(x,y)$, on every connected component of P, is indeed a metric. Moreover the distance between x and y is equal to the length of the geodesic g between x,y in any cell containing both x and y, thus g is a geodesic in P. If y,z are points at distance less than $\frac{\delta(x)}{2}$ from x, then any geodesic between y and z, if it exists, is contained in $\text{Star}(x,P)$. Note that $\delta(x) > 0$ if there is a positive lower bound on the heights of all cells of P containing x, since there is also one in P'. In particular $\delta(x) > 0$ if there are only finitely many isometry types of cells containing x, or if P is locally finite.

To prove that the above assertions are true, suppose x and y are at pseudo-distance less than $\delta(x)$. Then y has to belong to the closed star of x (any broken geodesic f which goes out of $\text{Star}(x,P)$ has length at least $\delta(x)$, as may be seen by induction on the number of cells crossed by f). But the length of any broken geodesic $f : [a,b] \to P$, whose image is inside $\text{Star}(x,P)$, between $x = f(a)$ and $z = f(b)$ is bigger than or equal to the length of the geodesic between x and z in any cell containing both x and z. Indeed, if $a = t_0 < t_1 < \cdots < t_{p+1} = b$ is a subdivision such that $f([t_i,t_{i+1}])$ is a geodesic in some cell, if B_p is any cell containing $x, f(t_p), f(t_{p+1})$ then, by induction on p,

$$\begin{aligned} d_{B_p}(x,f(t_{p+1})) &\leq d_{B_p}(x,f(t_p)) + d_{B_p}(f(t_p),f(t_{p+1})) \\ &\leq \ell(f_{|[a,t_p]}) + d_{B_p}(f(t_p),f(t_{p+1})) = \ell(f). \end{aligned}$$

Step 1 : Let δ be the greatest lower bound of the heights of all cells in P. Note that $\delta > 0$ if there are only finitely many isometry types of cells and that if $\delta > 0$, then one has $\delta(y) > 0$ for every y in P. Then the length pseudo-distance, which is a true distance by the step 0, is complete.

Indeed, let α be the greatest lower bound of the distances $d(y,z)$ where u is the center of a cell of P, y is a point of Star(u,P') and z is a point in $P-$Star(u,P). By hypothesis, $\alpha > 0$. Consider a Cauchy sequence $(x_i)_{i\in\mathbb{N}}$ and suppose $\tilde{d}(x_i,x_j) < \frac{\alpha}{2}$ for $i,j \geq N$. Since the closed sets Star(u,P') for all centers u of all cells of P cover P, let x be a center of a cell of P with $x_N \in$ Star(x,P'). Then $x_i \in$ Star(x,P) for every $i \geq N$. Then two cases are possible, either there is a subsequence of $(x_i)_{i\in\mathbb{N}}$ converging to x, and hence the sequence converges to x, or if r is the radial projection from Star$(x,P)-\{x\}$ into Link(x,P), the sequence $(r(x_i))_{i\in\mathbb{N}}$ is a Cauchy sequence. Indeed if an almost geodesic between x_n, x_{n+1} comes closer and closer to x, then x_n tends to x. Moreover, if C is a cell, v a vertex of C, y,z two points of C and the geodesic between y,z at distance from v in the interval $[\rho_1,\rho_2]$ with $\rho_1 > 0$, $\rho_2 < +\infty$, then

$$K_1 d_C(y,z) \leq d_{\mathrm{Link}(v,C)}(r(y),r(z)) \leq K_2 d_C(y,z)$$

where K_1, K_2 are constants depending only on χ, ρ_1, ρ_2. By induction on the dimension in the case of finitely many isometry types of cells (or since the links are finite in the locally compact case), $(r(x_i))_{i\in\mathbb{N}}$ converges. Since $(x_i)_{i\in\mathbb{N}}$ remains at bounded distance from x, $(x_i)_{i\in\mathbb{N}}$ has a subsequence which converges, hence converges.

Step 2 (M.R. Bridson) : If $f : [a,b] \rightarrow P$ is a broken geodesic, define the *combinatorial length* of f to be the minimum m in $\mathbb{N}$ such that there exists a subdivision $a = t_0 < t_1 < \cdots < t_m = b$ with $f([t_{i-1},t_i])$ contained in some cell of P. Amongst the broken geodesics between two given points x,y in P with combinatorial length less than or equal to a given $M \in \mathbb{N}$, there is one with a shorter length.

Indeed, since there is a finite number of isometry types of cells in P, the broken geodesics of combinatorial length less than or equal to M are, up to isometry, contained in a finite number of finite multipolyhedra, obtained by gluing M cells of P. Since those multipolyhedra are compact, since the length of a broken geodesic f of combinatorial length M is a continuous map of the finite number of "breaking " points $f(t_i)$, since a limit of a sequence of broken geodesics of combinatorial length less than or equal to M is also a broken geodesic of combinatorial length less than or equal to M, then the minimum is reached.

If, amongst the broken geodesics between two given points in P with combinatorial length less than a given $M \in \mathbb{N}$, $f : [a,b] \rightarrow P$ has a minimal length, then f is *semi-taut*, in the following sense : if $a = t_0 < t_1 < \cdots < t_m = b$ is a subdivision such that $f([t_{i-1},t_i])$ is contained in some cell C_i of P and m is minimum, then the restriction of f to $[t_{i-1},t_i]$ is geodesic and the distance in Link$(f(t_i), C_i \cup C_{i+1})$ between $f([t_{i-1},t_i])$ and $f([t_i,t_{i+1}])$ (which may be $+\infty$) is bigger than or equal to π. Indeed, by convexity we could otherwise shorten the broken geodesic, keeping the same combinatorial length (see Figure 4).

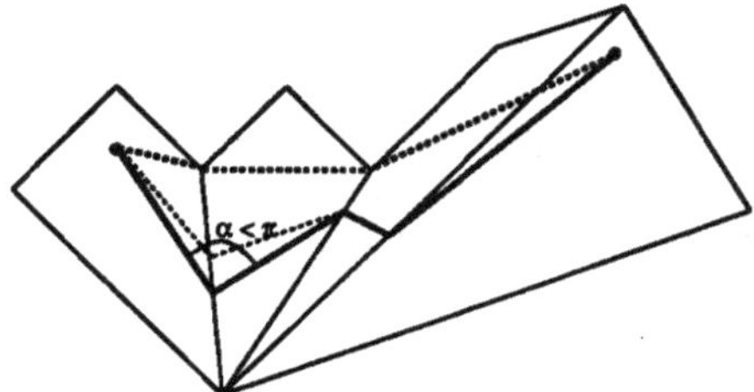

Figure 4 : Shortening of broken geodesics.

Step 3 First start with a lemma which is basic for the induction process.

Lemma 3.8 *(W. Ballman) Suppose x is a vertex in a multipolyhedron Q. Let r_x be the radial projection from Star$(x, Q) - \{x\}$ into Link(x, Q). The map $r_x \circ f$, for every geodesic (resp. semi-taut broken geodesic) $f : [a, b] \to Q$ with image contained in Star$(x, Q) - \{x\}$, is a geodesic (resp. semi-taut broken geodesic), after a suitable reparametrization, of length in Link(x, Q) strictly less than π.*

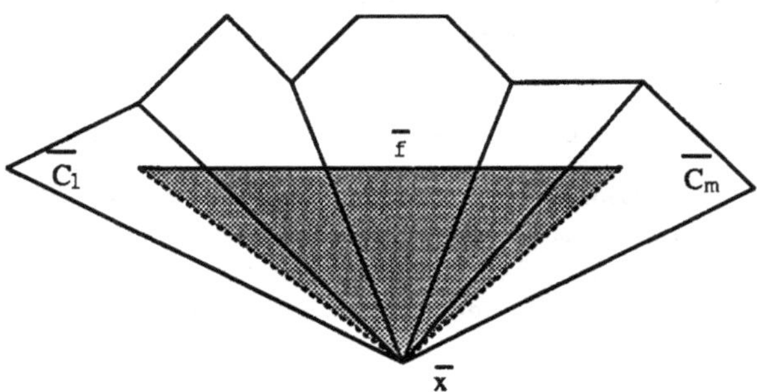

Figure 5 : Developing cone $\overline{D}_f$ of a geodesic f in a link.

Proof. (W. Ballman, [GH] page 197) If y is a point of Star$(x, Q) - \{x\}$, $\gamma_y : [0, 1] \to Q$ denotes the unique maximal geodesic parametrized proportionally to arclength in Star(x, Q) starting at x and containing y in any cell in Star$(x, Q) \subset Q'$ containing both x and y. There is a subdivision $a = t_0 < t_1 < \cdots < t_m = b$ such that $f([t_{i-1}, t_i])$ is contained in some cell for $i = 1 \cdots m$. If $r_x \circ f$ is not constant, then the map $\zeta : (s, t) \mapsto \gamma_{f(t)}(s)$ for $0 \leq s \leq 1$ and $t_{i-1} \leq t \leq t_i$ defines a 2-dimensional cell $\overline{C_i} = \{(s, t)\}$ (with the pull back metric) in $\mathbf{M}^2_\chi$ (in general $C_i = \zeta(\overline{C_i})$ does not belong to the triangulation). Glue the $\overline{C_i}$ in $\mathbf{M}^2_\chi$ for

$i = 1 \cdots m$ consecutively along their sides. We obtain a (not necessarily convex) polygon $\overline{D}_f$ as in figure 5. Denote by $\overline{x}$ the point of $\overline{D}_f$ corresponding to x and by $\overline{f}$ the corresponding broken geodesic on $\overline{D}_f$. The angle of $\overline{D}_f$ at $\overline{x}$ is exactly the length of $r_x \circ f$ in Link(x, Q). If f is a geodesic or a semi-taut broken geodesic, the distance from $\overline{f}(t)$ to $\overline{x}$ is a convex fonction, (otherwise we could shorten f keeping its combinatorial length). Here we use the fact that the diameter of the cells is less than $\frac{\pi}{2\sqrt{\chi}}$. Hence the angle at $\overline{x}$ has to be strictly less than π if $\chi \leq 0$ and less than or equal to π if $\chi < 0$. But the equality implies that $d(x, f(a)) = d(x, f(b)) = \frac{\pi}{2\sqrt{\chi}}$), which is impossible since the diameter of the cells of P' is strictly less than $\frac{\pi}{2\sqrt{\chi}}$.

Suppose for instance that f is geodesic and let us prove that $r_x \circ f$ is also geodesic (the following proof extends with minor changes to the case of semi-taut broken geodesics). Indeed suppose that $g : [c, d] \to \text{Link}(x, Q)$ is a broken geodesic between $r_x(f(a))$ and $r_x(f(b))$. Take a subdivision $c = u_0 < u_1 < \cdots < u_m = d$ such that $g([u_{i-1}, u_i])$ is contained in Link(x, D_i) for some cell D_i in Q for $i = 1, \cdots, m$. There is a cell F_i in $\mathbf{M}^2_\chi$ of dimension 2 which is isometric to the union of the maximal geodesics in D_i corresponding to $g(s)$ for $s \in [u_{i-1}, u_i]$. Glue F_i to F_{i+1} in $\mathbf{M}^2_\chi$ along the geodesic corresponding to $g(u_i)$ to obtain a polygon $\overline{\overline{D}}_g$ and denote by $\overline{\overline{x}}$ the point corresponding to x. There is a formula in a geodesic triangle (a, b, c) in $\mathbf{M}^2_\chi$ expressing the length of one side c in terms of a, b and the angle at the opposite vertex C. Since f is geodesic, any homothetic of f in Star(x, Q) closer to x has to be a geodesic, hence the angle of $\overline{D}_f$ at $\overline{x}$ has to be less than or equal to the angle of $\overline{\overline{D}}_g$ at $\overline{\overline{x}}$ □

A good understanding of the local geodesics is given by the following result.

Lemma 3.9 *(G. Moussong) Let Q be a multipolyhedron. The local geodesics in Q are taut broken geodesics. Conversely, if for every vertex x of Q, there is a positive lower bound on the heights of all cells containing x, and if Q is locally geodesic (that is for every point x of Q, for every y, z in Star(x, Q) there is a geodesic in Q between x and y), then the taut broken geodesics are local geodesics.*

Proof. It is obvious that if a path is a local geodesic, then it is a broken geodesic. Furthermore if a broken geodesic $f : [a, b] \to Q$ is a local geodesic then it is taut. This may be seen as above by using Ballman's tool of the "developing cone of a geodesic".

Indeed suppose that, for some $t \in]a, b[$, there is a broken geodesic $g : [c, d] \to \text{Link}(f(t), Q)$ between the two germs of geodesics $f_{|]t-\epsilon,t]}$ and $f_{|[t,t+\epsilon[}$ whose length is strictly less than π. Take a subdivision $c = t_0 < t_1 < \cdots < t_m = d$ such that $g([t_{i-1}, t_i])$ is contained in Link(x, D_i) for some cell D_i in Q for $i = 1, \cdots, m$. There is a cell F_i in $\mathbf{M}^2_\chi$ of dimension 2 which is isometric to the union of the maximal geodesics in D_i corresponding to $g(s)$ for $s \in [t_{i-1}, t_i]$. Glue F_i to F_{i+1} in $\mathbf{M}^2_\chi$ along the geodesic corresponding to $g(t_i)$ to obtain a polygon and denote by $\overline{\overline{x}}$ the point corresponding to $f(t)$.

Then (see Figure 5) we get an angular sector in M^2_χ in a neighborhood of $\overline{\overline{x}}$, with angle strictly less than π. Take the geodesic, contained in this angular sector, between two points x^+, x^-, close to but different from $\overline{\overline{x}}$, one on each side of the angular sector corresponding to $f_{|]t-\epsilon,t]}$ and $f_{|[t,t+\epsilon[}$. Then the corresponding broken geodesic in Q between the corresponding points $f(t^+), f(t^-)$ is a shorter broken geodesic than f between these two points.

To prove the converse, let $f : [a,b] \to Q$ be a taut broken geodesic. Recall from step 1 that if y, z are points in $\mathrm{Star}(f(t), Q)$ close enough to $f(t)$, then a geodesic between y, z, which exists by assumption, has to stay in $\mathrm{Star}(f(t), Q)$. Recall also that if y is close enough to $f(t)$, the geodesic between $f(t)$ and y in any cell of $\mathrm{Star}(f(t), Q)$ is a geodesic in P. What we have to prove is that if ϵ is small, then f is a geodesic between $f(t-\epsilon)$ and $f(t+\epsilon)$. Note that $f([t-\epsilon, t+\epsilon])$ is contained in $\mathrm{Star}(f(t), Q)$. Let g be any geodesic between $f(t-\epsilon)$ and $f(t+\epsilon)$. We may suppose that $f(t)$ is not on g otherwise $g = f$. By lemma 3.8, the length of the (broken) geodesic $r_x \circ g$ would be strictly less than π, contradicting the definition of a taut broken geodesic. □

Step 4 (M.R. Bridson) :

Lemma 3.10 *(M.R. Bridson) For every multipolyhedron P of finite dimension D having a positive lower bound δ' on the heights of the cells of P' and of the cells of the links of the cells of P' and so on, there is a constant $\tau > 0$ depending only on D, δ' satisfying the following property. If $f : [a,b] \to P$ is a semi-taut broken geodesic in P with combinatorial length $N(f)$, then the length of f is bigger than or equal to $\tau N - \frac{1}{\tau}$.*

Proof. (M.R. Bridson) We claim first that there is an m in $\mathbb{N} - \{0\}$ and a subdivision $a = t_0 < t_1 < \cdots < t_m = b$ such that for $i = 2, \cdots, m-1$,

- $f([t_{i-1}, t_i])$ is contained in $\mathrm{Star}(x_i, P)$ for some vertex x_i of P and the length $\ell(f_{|[t_{i-1},t_i]})$ is bigger than or equal to some constant $\tau_1 > 0$ depending only on δ'.

- $f([t_0, t_1]), f([t_{m-1}, t_m])$ are contained in the stars $\mathrm{Star}(x_1, P), \mathrm{Star}(x_m, P)$ of vertices of P respectively.

This is proved by induction on the combinatorial length of f. Indeed, the closed stars of the vertices of P cover P and the open stars are disjoint. What may happen if we take the closed stars crossed by f and the length is not bounded below is that, in the developing cone $\overline{D}_{f_{|\mathrm{Star}(x,P)}}$ of the geodesic in one of the chosen vertex x of P (see Figure 5), f comes close to a vertex of $\overline{D}_{f_{|\mathrm{Star}(x,P)}}$ which is not $\overline{x}$. This vertex corresponds to some cell C of P and since f is a semi-taut broken geodesic, either f stays in the link of every vertex of C, or f goes out of the link of some vertex of C, hence has in that link a length more than some τ_1 (the heights of the cells of P' are bounded below by δ' ; any path which starts near the vertex and goes out of the link has length at least $\frac{\delta'}{2}$).

Suppose that $g : [c,d] \to P$ is a taut broken geodesic, with image contained in Star(x,P). The combinatorial length $N(g)$ of g, which is the same as the combinatorial length of $r_x \circ g$, is by induction, since the length of $r_x \circ g$ is bounded above by π according to the lemma 3.8, bounded above by a constant $\tau_2 > 0$ depending only on D, δ'. Note that if g goes through x, then the combinatorial length of g is at most two and its length is at least $\delta' > 0$.

Hence we have $\ell(f) \geq (m-2)\tau_1$ and $N(f) = \sum_{i=1}^{m} N(f_{|[t_{i-1},t_i]}) \leq m\tau_2$. Hence $\ell(f) \geq \frac{1}{\tau_2}N(f) - 2\tau_1$, which proves the result. □

End of the proof of the theorem 3.6 : If x, y are two points in the same connected component of P, if we take a sequence $(f_i)_{i\in\mathbf{N}}$ of broken geodesics between them whose lengths tend to $d(x,y)$, then we may replace f_i by a shorter broken geodesic f_i' minimizing the length amongst the ones with combinatorial length less than or equal to $N(f_i)$, which are semi-taut broken geodesic, according to the step 2. But the step 4 says that the combinatorial length of semi-taut broken geodesics with bounded length is bounded. Hence there is a geodesic between x and y. □

Proposition 3.11 *For every χ in* $\mathbf{R}$, *every locally finite dimensional if $\chi \leq 0$ or finite dimensional if $\chi > 0$ simplicial or cubical multipolyhedron has a structure of a* M_χ*-multipolyhedron with a Hausdorff, complete, length pseudo-distance, geodesic on every connected component.*

Proof. We identify every k-simplex (resp. k-cube) of P with a fixed regular k-simplex (resp. k-cube) in M_χ^n, in a compatible gluing way, which is possible by the examples following the definition 3.3. Then apply Bridson's theorem 3.6. □

3.4 *Negatively curved polyhedra*

Definition 3.12 *Subsequently in the paper, by* M_χ*-multipolyhedron, we understand a* M_χ*-multipolyhedron whose length pseudo-distance is Hausdorff, complete, geodesic on every connected component and for which, for every cell C of P, the* M_1*-multipolyhedron Link(C,P) is complete, geodesic on every connected component. A negatively curved multipolyhedron is a* M_χ*-multipolyhedron having curvature less than or equal to χ, for some $\chi < 0$.*

Note that under the hypotheses of the definition, Link(C,P) is Hausdorff and Link(x,P) is Hausdorff, complete, geodesic on every connected component for every x in P (use M. Davis' remark above the theorem 3.6, and arguments of the proof of the theorem 3.6 : the step 1 for the completeness and the step 3 for the geodesics). Note that according to lemma 3.8, between two points of $\text{Link}(x, P)$ at pseudo-distance strictly less than π, there is a geodesic. Furthermore if there are only finitely many isometry types of cells, then, according to the theorem 3.6, the preliminary assumptions on a M_χ-multipolyhedron are automatically satisfied.

Question (E. Ghys) : Is every compact connected smooth Riemannian manifold with negative curvature homeomorphic (by a quasi-isometric homeomorphism ?) to a finite negatively curved multipolyhedron ?

This question is unsolved (the answer should be yes, according to M. Gromov), except in the case of constant negative curvature.

Question (M. Gromov) : Is every hyperbolic group isomorphic to the fundamental group of some finite negatively curved multipolyhedron ?

We are going to look for criteria for a M_χ-multipolyhedron P to be negatively curved. Define a *closed geodesic* of length ℓ in a metric space X to be an isometric embedding of the circle $S_\ell = \frac{\ell}{2\pi}\mathbf{S}^1$ with the geodesic metric of total length ℓ. The (first) *systole* of X is the greatest lower bound $\text{Sys}(X)$ of the positive lengths of closed geodesics in X (which may be 0). A *digon* of length ℓ is a pair (f, g) of two distinct geodesics $f, g : [a, b] \to P$ between two points x_0, y_0, meeting only at their endpoint, with length $b - a = \ell$. The *injectivity radius* of X is the greatest lower bound $\text{Injrad}(X)$ of the length of a digon in X. The CAT(χ)-*perimeter* of X is the greatest lower bound $\text{Peri}_{\text{CAT}(\chi)}$ of the perimeters of geodesic triangles in X that do not satisfy the CAT(χ)-inequality. (The three numbers defined above may be $+\infty$.)

Theorem 3.13 *(M. Gromov) A M_χ-multipolyhedron P, such that for every vertex v of P the heights of all cells of P containing v have a positive lower bound, has curvature less than or equal to χ if and only if for every vertex v of P, the space $\text{Link}(v, P)$ satisfies the CAT(1)-inequality.*

The mild assumption on the heights is satisfied for example if P has locally finitely many isometry types of cells or is locally finite. Furthermore, it is a sufficient condition for the length pseudo-distance to be Hausdorff, as seen in the step 0 of the proof of the theorem 3.6.

Proof. Step 1 (W. Ballman, [GH] Chapter 10) : We are going to prove first that P has curvature less than or equal to χ if for every point x of P, the space $\text{Link}(x, P)$ satisfies the CAT(1)-inequality. We have to find for every point x in P a neighborhood of x such that every triangle with perimeter less than $\frac{2\pi}{\sqrt{\chi}}$ contained in that neighborhood satisfies the CAT(χ)-inequality.

We may subdivide so that x is a vertex and $r_x : \text{Star}(x, P) - \{x\} \to \text{Link}(x, P)$ is the radial projection. Let V be the ball of radius $\frac{\delta'}{2}$ centered at x, with $\delta' > 0$ the greatest lower bound of the cells containing x in the multipolyhedron P' subdivided so that x is a vertex. If y a point of $\mathrm{Star}(x,P)-\{x\}$, let γ_y be the geodesic between x and y contained in any cell of $\mathrm{Star}(x,P)-\{x\}$ containing x,y. According to the steps 0,1 of the proof of the theorem 3.6, γ_y is a geodesic in P if y is in V.

Claim Let $f:[s,t]\rightarrow P$ be a geodesic of P whose image is contained in $V-\{x\}$. The geodesic triangle $(\gamma_{f(s)},\gamma_{f(t)},f)$ in P satisfies the CAT(χ)-inequality.

This follows from the gluing lemma 2.12 and since it is obtained by gluing the triangles $(\gamma_{f(t_{i-1})},\gamma_{f(t_i)},f_{|[t_{i-1},t_i]})$ consecutively along their sides and f is a geodesic, with the notations of the proof of the lemma 3.8.

We want to show that every geodesic triangle (a,b,c) contained in V (necessarily of perimeter less than $\frac{2\pi}{\sqrt{\chi}}$ since δ' is bounded by the diameter of the cells, hence by $\frac{\pi}{2\sqrt{\chi}}$) satisfies the CAT(χ)-inequality.

Case 1 : The point x lies on $a\cup b\cup c$.

If x lies on a, for instance, apply the gluing lemma 2.12 and the claim to the two triangles with vertices x,A,B and x,A,C, where A,B,C are the vertices opposite to a,b,c.

Case 2 : The point x does not lie on $a\cup b\cup c$ and

$$d(r_x(A),r_x(B))+d(r_x(B),r_x(C))+d(r_x(C),r_x(A))\geq 2\pi$$

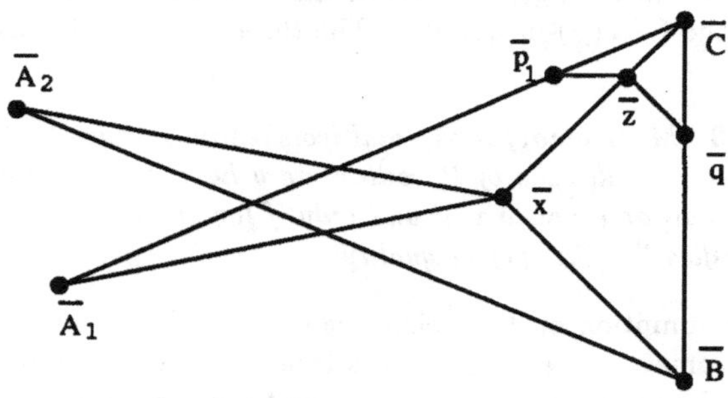

Figure 6 : Overlapping comparison triangles.

Let p be on b and q on a, for instance. Let $\overline{p},\overline{q}$ be the corresponding point on a comparison triangle for (a,b,c). We want to show that

$$d(p,q)\leq d(\overline{p},\overline{q})$$

We glue the comparison triangle $(\overline{\gamma_{A_1}},\overline{\gamma_B},\overline{c})$ for (γ_A,γ_B,c) with the comparison triangle $(\overline{\gamma_B},\overline{\gamma_C},\overline{a})$ for (γ_B,γ_C,a) along $\overline{\gamma_B}$, and $(\overline{\gamma_B},\overline{\gamma_C},\overline{a})$ with the comparison triangle $(\overline{\gamma_{A_2}}, \overline{\gamma_C}, \overline{b})$ for (γ_A, γ_C, b) along $\overline{\gamma_C}$ (see Figure 6). To obtain the comparison triangle for (a, b, c), we have to move $\overline{A}_1$ and $\overline{A}_2$ to the intersection $\overline{A}$ of the circles of radius $\ell(c)$ centered at $\overline{B}$ and $\ell(b)$ centered at $\overline{C}$, keeping $\overline{B}, \overline{C}, \overline{x}$ fixed. We denote by $\overline{p_1}$ the point correponding to p on the comparison triangle for (γ_A, γ_B, c). For every $\overline{y}$ on the geodesic between $\overline{x}$ and $\overline{B}$, and every $\overline{z}$ on the geodesic between $\overline{x}$ and $\overline{C}$, we have, since $\overline{A}_1$ is moved upwards,

$$d(\overline{p}_1, \overline{y}) \leq d(\overline{p}, \overline{y}) \text{ and } d(\overline{p}_1, \overline{z}) \leq d(\overline{p}, \overline{z}).$$

After straightening, the geodesic from $\overline{p}$ to $\overline{q}$ will meet either the segment from $\overline{x}$ to $\overline{B}$ in a point $\overline{y}$, or the segment from $\overline{x}$ to $\overline{C}$ in a point $\overline{z}$. In the second case (the first one is similar),

$$\begin{aligned} d(\overline{p}, \overline{q}) &= d(\overline{p}, \overline{z}) + d(\overline{z}, \overline{q}) \\ &\geq d(\overline{p}_1, \overline{z}) + d(\overline{z}, \overline{q}) \\ &\geq d(p, q). \end{aligned}$$

Case 3 : The point x does not lie on $a \cup b \cup c$ and

$$d(r_x(A), r_x(B)) + d(r_x(B), r_x(C)) + d(r_x(C), r_x(A)) < 2\pi$$

If we glue the comparison triangles for (γ_A, γ_B, c), (γ_B, γ_C, a) (γ_C, γ_A, b) along $\overline{\gamma_B}$, $\overline{\gamma_C}$, $\overline{\gamma_A}$, we obtain a polyhedral cone $\overline{C_x}$ in $\mathbf{M}^3_\chi$ (see the following picture). If the dimension of P is 2, according to the lemma 3.8, this proves that there is a closed local geodesic in Link(x, P) with length less than 2π. This would contradict the hypothesis.

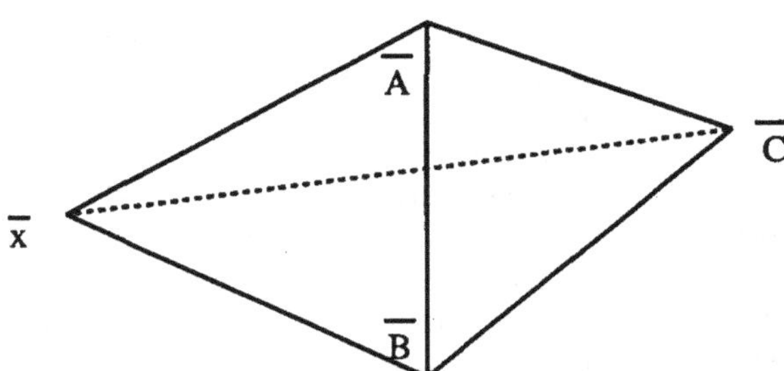

Figure 7 : Non-overlapping comparison triangles.

According to the lemma 3.8, $(r_x \circ a, r_x \circ b, r_x \circ c)$ is a geodesic triangle (after a suitable reparametrization) in Link(x, P). Let $r_{\overline{x}}$ be the radial projection from Star$(\overline{x}, \overline{C_x}) - \{\overline{x}\}$ to Link$(\overline{x}, \overline{C_x})$ (seen as lying in the ball of center $\overline{x}$ in $\mathbf{M}^3_\chi$ with the right radius). Then the image by $r_{\overline{x}}$ of the geodesic triangle $(\overline{A}, \overline{B}, \overline{C})$ is the comparison triangle for $(r_x \circ a, r_x \circ b, r_x \circ c)$. Since any triangle of perimeter strictly less than 2π satisfies the CAT(1)-inequality, we have

$$d(r_x(p), r_x(q)) \leq d(r_{\overline{x}}(\overline{p}), r_{\overline{x}}(\overline{q}))$$

for every p on a and q on c, say.

Let $f : [u, v] \to P$ be a geodesic between p and q. According to the lemma 3.8, $r_x \circ f$ is a geodesic (parametrized proportionally to arclength) whose length is $d(r_x(p), r_x(q))$. The angle $\tilde{\alpha}$ of the developing cone $\tilde{D}_f$ (see the proof of the lemma 3.8) at the point $\tilde{x}$ corresponding to x is $d(r_x(p), r_x(q))$. Let $\tilde{p}, \tilde{q}$ be the points of $\tilde{D}_f$ corresponding to p, q. Note that the length of the side c of a geodesic triangle (a, b, c) in M^2_χ is an increasing function of the opposite angle (at C) if the lengths of a and b are fixed. We have

$$d(\tilde{x}, \tilde{p}) = d(x, p) = d(\overline{x}, \overline{p}) \text{ and } d(\tilde{x}, \tilde{q}) = d(x, q) = d(\overline{x}, \overline{q}).$$

If $\overline{\alpha}$ is the angle at $\overline{x}$ in M^3_χ between the rays going through $\overline{p}, \overline{q}$, then

$$\tilde{\alpha} = d(r_x(p), r_x(q)) \leq d(r_{\overline{x}}(\overline{p}), r_{\overline{x}}(\overline{q})) = \overline{\alpha}.$$

Hence $d(p, q) = d(\tilde{p}, \tilde{q}) \leq d(\overline{p}, \overline{q})$.

Step 2 (W. Ballman, [GH] Chapter 10) : If there is a triangle in Link(x, P) which does not satisfies the CAT(1)-inequality, then the proof of the case 3 of the step 1 easily shows that P has, arbitrarily close to x, triangles that do not satisfy the CAT(χ)-inequality.

Step 3 : We have to prove that if for every vertex v of P, the space Link(v, P) satisfies the CAT(1)-inequality, then for every point x of P, the space Link(x, P) satisfies the CAT(1)-inequality, the converse being obvious. Indeed, let x be a point in P which is not a vertex, C be a cell of dimension k containing x in its interior and v be a vertex of C. According to the remarks above the theorem 3.6, Link(x, P) is isometric to a k-fold suspension of Link(C, P). If D is the cell of Link(x, P) corresponding to C, we have seen at the end of the section 3.2 that Link(C, P) is isometric to Link$(D, \text{Link}(v, P))$. If Link(v, P) satisfies the CAT(1)-inequality, then for every y in the interior of D, Link$(y, \text{Link}(v, P))$ also satisfies the CAT(1)-inequality by applying the step 2, and is isometric to a $(k-1)$-fold suspension of Link$(D, \text{Link}(v, P))$. So the result follows from the following fact.

Lemma 3.14 *(G. Moussong [Mou], R. Charney - M. Davis, appendix of [CD]) Let X be a M_1-multipolyhedron. Then X satisfies one of the following three conditions if and only if the suspension $\mathsf{S}^0 \star X$ does :*

- *the length pseudo-distance is Hausdorff, complete, geodesic,*
- *the CAT(1)-inequality is satisfied,*

- *the injectivity radius is greater than or equal to π,*
- *the systole is greater than or equal to 2π.*

Proof. The first assertion has already been checked after the definition 3.12 The others follow from the following remarks. The spaces $X, \mathbf{S}^0$ are naturally contained in $\mathbf{S}^0 \star X$ and there is a natural retraction, called the radial projection, of $(\mathbf{S}^0 \star X) - \mathbf{S}^0$ onto X. The space X separates $\mathbf{S}^0 \star X$ into two halves, and is isometric to the link of any one of the two points N, S of $\mathbf{S}^0$. If a geodesic goes from a point of X to one of the two points of $\mathbf{S}^0$, then its length is at least $\frac{\pi}{2}$. From the proof of the lemma 3.8, it follows that the radial projection from $(\mathbf{S}^0 \star X) - \mathbf{S}^0$ into X of a geodesic that does not go through N, S is a geodesic in X having length less than or equal to π, with equality if and only if the endpoints of the geodesic are in X, and at distance π in X.

If two points of X are at distance strictly less than π in $\mathbf{S}^0 \star X$, then every geodesic f between x and y is contained in X (this remark is due to M. Gromov [Gro3] page 122) ; indeed, if g is a geodesic of length strictly less than π, between two points of X, contained in the closure of the half space defined by N, the developing cone of g, that is the union of all geodesics of length $\frac{\pi}{2}$ between N and a point of X containing a point on f, is isometric to a filled geodesic triangle in $\mathbf{S}^2$ with two sides of length $\frac{\pi}{2}$ and one side of length strictly less than π.

Hence in particular, any geodesic of X of length less than or equal to π is a geodesic of $\mathbf{S}^0 \star X$. Furthermore, the radial projection of a digon of length strictly less than π (resp. a closed geodesic of length strictly less than 2π) contained in $\mathbf{S}^0 \star X$, that does not go through N nor S, is a digon in X of length strictly less than π or a closed geodesic in X of length strictly less than 2π. The last ingredient, for proving the CAT(1)-condition, is the gluing lemma 2.12 and the proof of the steps 1,2. □ □

Theorem 3.15 *(M. Gromov) A M_χ-multipolyhedron P, having finitely many isometry types of cell, or being locally finite, has curvature less than or equal to χ if and only if one of the following equivalent conditions is satisfied*

- *for every cell C of P, the injectivity radius of Link(C, P) is greater than or equal to π,*
- *for every cell C of P, the systole of Link(C, P) is greater than or equal to 2π*
- *for every vertex v of P, the space Link(v, P) satisfies the CAT(1)-inequality.*

Proof. It is an easy consequence of the theorem 3.13 and of the following theorem 3.16. In order to apply it to verify the assertions in the theorem 3.15, one uses an induction on the dimension (transfinite induction if the multipolyhedron is not locally finite dimensional) and the step 3 of the proof of the theorem 3.13. One has to apply the theorem 3.16 at each stage of the induction, that's why we need the first two conditions for every cell of P. The level 0 of the induction follows from the fact that a graph endowed with a geodesic metric has negative curvature, as it is locally a tree.

Theorem 3.16 *(M. Gromov) Let Q be a complete locally compact geodesic metric space, or a $\mathbf{M}_\chi$-multipolyhedron with finitely many isometry types of cell, having curvature less than or equal to χ, with $\chi > 0$. Then the following assertions are equivalent :*

- *the injectivity radius of Q is greater than or equal to $\frac{\pi}{\sqrt{\chi}}$,*
- *the systole of Q is greater than or equal to $\frac{2\pi}{\sqrt{\chi}}$,*
- *the space Q satisfies the $CAT(\chi)$-inequality.*

So the theorem holds for a locally finite $\mathbf{M}_\chi$-multipolyhedron with a positive lower bound on the height of the cells, according to the theorem 3.7. According to the corollary 2.21 and the remark following it, this theorem is interesting in positive curvature (locally CAT(1) may not imply globally CAT(1)).

Proof. (R. Charney - M. Davis [CD]) According to the theorem 3.6, the length pseudo-distance on Q is Hausdorff, complete, geodesic.

One obviously has that if Q satisfies the CAT(χ)-inequality, then there exists no isometric embedding of S_ℓ with $\ell < \frac{2\pi}{\sqrt{\chi}}$. Otherwise, we subdivide the image into three parts and the CAT(χ)-inequality, applied to the geodesic triangle defined that way, gives a contradiction.

Let us show now by absurd that if the systole is strictly less than $\frac{2\pi}{\sqrt{\chi}}$, then the injectivity radius is less than $\frac{\pi}{\sqrt{\chi}}$. So suppose there exists a digon of length strictly less than $\frac{\pi}{\sqrt{\chi}}$. Since Q has only finitely many isometry types of cells (or by Ascoli's theorem if Q is a locally compact complete geodesic space, since the balls are then compact (see the proposition 2.3)), there exists a digon of minimum length (see the proof of the theorem 3.6). To see that the two geodesics of a digon of minimal length strictly less than $\frac{\pi}{\sqrt{\chi}}$ are distinct, note that there exists $\epsilon > 0$ (depending only on Q) such that if (f, g) is a digon of length strictly less than $\frac{\pi}{\sqrt{\chi}}$, then the image of f is not contained in the ϵ-neighborhood of the image of g. Indeed, since there are only finitely many isometry types of cells (or by compactness), there exists $\epsilon > 0$ such that for every z in Q (or in some big ball), every geodesic triangle in the ball $B(z, 2\epsilon)$ satisfies the CAT(χ)-inequality. If the image of f is contained in the ϵ-neighborhood of the image of g, we may subdivide (as in the proof of corollary 2.21) the two geodesics f, g into a finite number of corresponding pairs of segments, each pair lying in a model neighborhood in which the CAT(χ)-property is satisfied. By joining the two subdivided geodesics by small geodesic triangles, and applying inductively the lemma 2.12 as in the proof of corollary 2.21, we have that $f = g$ by uniqueness of geodesics of length strictly less than $\frac{\pi}{\sqrt{\chi}}$ in spaces satisfying the CAT(χ)-property.

We claim that a digon $D_{\min}$ of minimum length strictly less than $\frac{\pi}{\sqrt{\chi}}$ is a closed geodesic. Indeed, a digon D is a closed geodesic if and only if any pair of opposite points of D are at distance precisely the length of D. So suppose that x, y are two opposite points on $D_{\min}$ at distance strictly less than the length of $D_{\min}$. The sides of $D_{\min}$ and a geodesic between x, y define two geodesic triangles T_1, T_2 such that the distance between two points of T_1 (resp. T_2) is strictly less than the injectivity radius of Q. If both T_1 and T_2 satisfy the CAT(χ)-inequality, then by the gluing lemma 2.12, D will satisfy the CAT(χ)-inequality, which is impossible. So the result follows from the following claim.

Claim : Let T be a geodesic triangle in Q such that any two points of T are at distance strictly less than $\text{Injrad}(Q) < \frac{\pi}{\sqrt{\chi}}$. Then T satisfies the CAT(χ)-inequality.

Proof. (If $\chi \leq 0$, this may be found between the lines in the proof of the theorem 2.19 and its corollary 2.21, where uniqueness was the difficult part to get. Of course the method will be the same in the case $\chi > 0$). Let us begin with the following lemma. Let f be a geodesic parametrized proportionally to arclength of length $< \text{Injrad}(Q)$. If ϵ is as above, we may also suppose that $\epsilon < \frac{\pi}{4\sqrt{\chi}}$ and $\epsilon \leq \ell(f) \leq \text{Injrad}(Q) - \epsilon < \frac{\pi}{\sqrt{\chi}} - \epsilon$.

Lemma 3.17 *There exists $A \geq 1$ (depending only on $\epsilon > 0$ and $\chi > 0$), such that if x, y are points in X with $d(x, f(0)) < \frac{\epsilon}{3A}$ and $d(y, f(1)) < \frac{\epsilon}{3A}$, then there exists a unique geodesic g parametrized proportionally to arclength between x and y, such that f (resp. g) is contained in the $\frac{\epsilon}{3}$-neighborhood of g (resp. f).*

Proof. (S. Alexander - R. Bishop [AB]) This is exactly the same proof as the first step of the proof of the theorem 2.19. We prove by induction on n the following statement :

> for every $a < b$ in $[0, 1]$ with $b - a \leq (\frac{3}{2})^n \frac{\epsilon}{2}$, if p, q are points in X with $d(p, f(a)) \leq \frac{\epsilon}{3A}$ and $d(q, f(b)) \leq \frac{\epsilon}{3A}$, then there exists a unique geodesic h parametrized proportionally to arclength between p and q with $h([0, 1])$ (resp. $f([a, b])$) contained in the $\frac{\epsilon}{3}$-neighborhood of $f([a, b])$ (resp. $h([0, 1])$).

The uniqueness statement follows immediately from the definition of ϵ, since the length of h hqs to be strictly less than $\frac{\pi}{\sqrt{\chi}}$.

But we will need a more quantitative version of this. Recall that for a geodesic triangle (a,b,c) in $\mathbf{S}^2$ with $\ell(b) = \ell(c)$, we have (see [Ber] for instance)

$$\sin\frac{\ell(a)}{2} = \sin\ell(b)\sin\frac{\alpha}{2} \quad \text{and} \quad \cos\alpha = \frac{\cos\ell(a) - \cos^2\ell(b)}{\sin^2\ell(b)}$$

where α is the angle at A opposite to a. Furthermore $\sin u \leq u$ if $u \geq 0$, and $\sin v \geq \frac{2v}{\pi}$ if $0 \leq v \leq \frac{\pi}{2}$. Hence if $0 < \eta \leq \ell(b) \leq \pi - \eta$, if we parametrize b, c so that $b(0) = c(0) = A, b(\ell(b)) = C$ and $c(\ell(c)) = B$, if $d(B,C) \leq \eta' \leq \eta$ then $d(b(t), c(t)) \leq \alpha \leq \frac{\pi}{2\sin\eta}\eta'$. Also note that the maximum for $d(b(t), c(t))$ is obtained when $t = \frac{\pi}{2}$ (extending the geodesics if necessary).

The case $n = 0$ follows by definition of ϵ. Let $g_0(t) = f(a + \frac{2}{3}(b-a)t)$ and $g_0'(t) = f(a + (b-a)(\frac{2}{3}t + \frac{1}{3}))$ for every t in $[0,1]$. Let $a_0 = g_0'(\frac{1}{2})$ and $b_0 = g_0(\frac{1}{2})$. By another induction, using the induction hypothesis, there exist, for every nonnegative integer m, geodesics g_{m+1} (resp. g'_{m+1}) parametrized proportionally to arclength between p and a_m (resp. b_m and q) contained in the $\frac{\epsilon}{3}$-neighborhood of $g_0([0,1])$ (resp. $g_0'([0,1])$). Let $a_{m+1} = g'_{m+1}(\frac{1}{2})$ and $b_{m+1} = g_{m+1}(\frac{1}{2})$.

Note that in $\mathbf{S}^2$, if two geodesics originating from a same fixed point have lengths strictly less than $\frac{2\pi}{3}$, then the distance between their midpoints is strictly less than the distance between their free endpoints. Making this quantitative as above, there is a constant λ with $0 < \lambda < 1$, depending only on ϵ and χ such that $d(a_{m+1}(t), a_m(t)) \leq (\lambda)^m \frac{\epsilon}{3A}$ and $d(b_{m+1}(t), b_m(t)) \leq (\lambda)^m \frac{\epsilon}{3A}$. So that $(g_m(t))_{m\in\mathbf{N}}$ and $(g'_m(t))_{m\in\mathbf{N}}$ are uniformly Cauchy sequences. They converge to local geodesics $g(t)$ and $g'(t)$ parametrized proportionally to arclength, contained in the $\frac{\epsilon}{3}$-neighborhood of $f([a,b])$. It is easy to see that $g(t+\frac{1}{2}) = g'(t)$ for $t \in [0, \frac{1}{2}]$. Hence the map $h(t) = g(\frac{3}{2}t)$ if $t \in [0, \frac{2}{3}]$ and $h(t) = g'(\frac{3}{2}t - \frac{1}{2})$ if $t \in [\frac{2}{3}, 1]$ is a local geodesic parametrized proportionally to arclength.

The map h is a geodesic. Indeed since we are in a space whose balls are compact or in a multipolyhedron with finitely many isometry types of cells, if h is not a geodesic, then there is a $t \in]0,1[$ such that $h_{|[0,t]}$ is a geodesic and there exists a distinct geodesic between $h(0)$ and $h(t)$. But in our case, this would contradict the fact that $\ell(h) < \mathrm{Injrad}(Q)$. □

The claim now follows by looking at a continuous 2-dimensional ruled surface obtained by taking geodesics from one vertex of T to the opposite side using the above lemma 3.17 (and the compactness of $[0,1]$ and the uniqueness of geodesics between two points of T), and then subdividing this surface into small geodesic triangles as in the corollary 2.21 and applying inductively the lemma 2.12 to show that T satisfies the CAT(χ)-property. □

Corollary 3.18 *(M. Gromov) Let P be a M_χ-multipolyhedron of dimension 2. The following propositions are equivalent :*

(a) P has curvature less than or equal to χ,

(b) every simple closed curve in the space of directions $Link(x, P)$ for every vertex x of P has length greater than or equal to 2π.

Proof. The link of a vertex in a 2-dimensional multipolyhedron is a 1-dimensional multipolyhedron, i.e. a graph. The corollary follows easily from the theorem 3.15. □

Let us give some applications of the corollary 3.18.

Proposition 3.19 *(M. Gromov) Let P be a finite 2-dimensional simplicial (resp. cubical) multipolyhedron such that for every vertex p, every simple closed curve in the link of p as at least seven (resp. five) edges. Then P has a structure of a negatively curved multipolyhedron, hence the fundamental group of every connected component of P is hyperbolic. If every simple closed curve in the link of p as at least six (resp. four) edges, then P has a structure of a non positively curved multipolyhedron, hence every connected component of P has a contractible universal covering.*

Proof. Identify every 2-cell of P with a geodesic triangle (resp. square) in $\mathbf{H}^2$ with sides of the same small length ϵ and same angles $\theta < \frac{\pi}{3}$ close to $\frac{\pi}{3}$ (resp. $\theta < \frac{\pi}{2}$ close to $\frac{\pi}{2}$). Endow P with the length pseudo-distance. Since $7 \times \frac{\pi}{3}$ (resp. $5 \times \frac{\pi}{2}$) is greater than 2π, every simple closed curve in the space of directions at a vertex p has length greater than 2π. Hence by the corollary 3.18, the multipolyhedron P is negatively curved. Its fundamental group is hyperbolic according to the corollary 2.21 and the proposition 2.13. In the same way, under the second set of hypotheses on the links, using standard Euclidean simplices or cubes, th result follows form the corollary 2.21, the proposition 2.18 and the remarks above the definition 2.16. □

This is the first purely combinatorial condition that yields a hyperbolic group. This condition is checkable by a "machine". Note that every finitely presented group is isomorphic to the fundamental group of a finite 2-dimensional polyhedron. But there exist non hyperbolic groups. So it is not possible to subdivide any given finite 2-dimensional multipolyhedron in order to get the above condition on the links. Note that the combinatorial condition of hyperbolicity is not invariant under subdivision.

Exercise 3.20 No subdivision of the boundary of the standard 3-simplex is such that for every vertex p, every simple closed curve in the link of p as at least seven edges.

The above proposition, in the cubical case, is a corollary of the following theorem. We don't know how to prove any analogous result in the case of a simplicial multipolyhedron.

Theorem 3.21 *(M. Gromov) Let P be a finite cubical polyhedron. Suppose P satisfies the "no Δ no $\square$ condition" :*

> *for every cube C of P, in the link Link(C,P) of C, every simple closed curve in the 1-skeleton consisting of three (resp. four) edges is the boundary of a 2-simplex of Link(C,P) (resp. the boundary of the union of two 2-simplices glued along one of their edges).*

Then P has a structure of a negatively curved polyhedron, and hence, if P is connected, its fundamental group is hyperbolic.

Proof. (M. Gromov [Gro3] page 122) **Step 1.** Identify every cube of P with a regular euclidean cube of the same dimension, in a compatible gluing way. Denote by d_0 the length pseudo-distance on P (and on every link), which is Hausdorff, complete and geodesic (Theorem 3.6). Let us prove that if for every cube C of P the space Link(C,P) satisfies the no Δ condition, then Sys(Link$(C,P)) \geq 2\pi$, hence P has non positive curvature according to theorem 3.15. Furthermore, if for every cube C of P the space Link(C,P) satisfies the no Δ no $\square$ condition, then Sys(Link$(C,P)) > 2\pi$.

We are going to prove that there exists in Link(C,P) a closed geodesic of minimum length contained in the 1-skeleton. Note that by the compactness of P and Ascoli's theorem, there does exist a geodesic of minimum length. So let f be a closed geodesic of minimum length such that the maximum m of the dimension of a cell D of Link(C,P) with f meeting the interior of D, is minimum over all closed geodesics of minimum length. Also suppose that the number of cells with dimension m whose interior meets f is minimum. The interior of any cell meets f along a finite number of subarcs. Suppose this number is minimum for some cell of dimension m that does meet f in its interior, over all f satisfying the previous requirements. By absurd, suppose $m \geq 2$. Let D be a cell of dimension m whose interior meets f in a non zero minimum number of subarcs, and choose one of these subarcs. At least one, call it v, of the vertices of D is not on that subarc, since f is geodesic in D. Consider the radial projection from $S(v) - \{v\}$ onto Link$(v, \text{Link}(C,P))$, where $S(v)$ is the union of all cells containing v. Recall that Link(C,P) is built of regular spherical simplices with $\chi = +1$ and side length $\frac{\pi}{2}$. We are going to use arguments of the proof of the lemmas 3.14 and 3.8. In particular Link$(v, \text{Link}(C,P))$ may be isometrically identified with the subset of points at distance $\frac{\pi}{2}$ from v and every geodesic of Link$(v, \text{Link}(C,P))$ of length less than or equal to π is a geodesic of Link(C,P). The only closed geodesics contained in $S(v)$ that do not contain v are contained in Link$(v, \text{Link}(C,P))$. Hence the image of f meets $S(v)$ in (at least one) segment $f([a,b])$ with $f(a), f(b)$ in Link$(v, \text{Link}(C,P))$ and the interior of the segment $f([a,b])$ meeting the interior of D. The radial projection g of $f([a,b])$ is a geodesic in Link$(v, \text{Link}(C,P))$ according to the lemma 3.8, hence a geodesic in Link(C,P) of the same length, less than or equal to π. The union of the images of f and g, minus $f([a,b])$, is a closed geodesic, otherwise f would not be a closed geodesic. But this contradicts the minimality assumptions on f.

So let f be a closed geodesic contained in the 1-skeleton of Link(C,P). Note that if f meets the interior of one edge, then that edge is contained in f, and that the edges have length $\frac{\pi}{2}$. Since P is a polyhedron, f contains at least 3 edges of Link(C,P). But by the no Δ condition, f contains at least 4 edges (the boundary of a regular spherical 2-simplex of side length $\frac{\pi}{2}$ is not geodesic). By the no $\square$ condition, f contains at least 5 edges (the boundary of two regular spherical 2-simplices of side length $\frac{\pi}{2}$ glued by one side is not geodesic). But $5\frac{\pi}{2} > 2\pi$.

Step 2. If we identify every cube of dimension m of P with a regular cube of the same dimension in $\mathbf{M}^m_\epsilon$, with $\epsilon < 0$, in a compatible gluing way, we get a length pseudo-distance d_ϵ on every Link(C,P), which is Hausdorff, complete and geodesic. Also note that Link(C,P) is made of regular spherical simplices of side length strictly less than $\frac{\pi}{2}$, but tending to $\frac{\pi}{2}$ as ϵ goes to 0. Also note that $d_\epsilon(x,y) \leq d_{\epsilon'}(x,y)$ if $\epsilon \geq \epsilon'$ (so the proposition page 124 in [Gro3] is useless). But let us prove that

$$\limsup_{\epsilon\to 0} \mathrm{Sys}(\mathrm{Link}(C,P), d_\epsilon) \geq \mathrm{Sys}(\mathrm{Link}(C,P), d_0).$$

Indeed, it is an easy corollary of Ascoli's theorem. Let X be the topological space Link(C,P) and $f_n : \mathbf{S}^1 \to (X, d_{\frac{1}{n}})$ be a closed geodesic parametrized by arclength, of minimal length ℓ_n. We may suppose that $\ell_n < +\infty$ and eventually after extracting a subsequence that $\ell_n > 0$, otherwise P has negative curvature, since P is finite and the second condition of the theorem 3.15 is obviously satisfied since empty. The maps $d_{\frac{1}{n}} : X \times X \to \mathbf{R}$ converge pointwise to $d_0 : X \times X \to \mathbf{R}$, since these spaces are geodesic and the geodesics are broken geodesics (see the theorem 3.6), and the length of a broken geodesic depends continuously on n. Furthermore, $X \times X$ is compact and $d_{\frac{1}{n}}(x,y) \leq d_{\frac{1}{m}}(x,y)$ if $n \leq m$. According to Dini's theorem, $d_{\frac{1}{n}}$ converges uniformly to d_0. Since $d_{\frac{1}{n}}(f_n(a), f_n(b)) = \frac{\ell_n}{2\pi} d(a,b)$ and ℓ_n is bounded by the diameter of $(X, d_{\frac{1}{n}})$, hence by the diameter of (X, d_0), it follows that the maps $f_n : \mathbf{S}^1 \to (X, d_0)$ are equicontinuous, hence have a converging subsequence. The limit is obviously a closed geodesic (parametrized proportionally to arclength) or a point. In order to exclude the last case, it is sufficient to prove that the ℓ_n have a positive lower bound. This follows by induction from the proof that Link(C,P) has curvature less than or equal to $+1$ with some uniformity argument (see the proof of the theorem 3.15). $\square$

If we were interested only in a purely combinatorial condition on presentations for a group to be hyperbolic, there is a statement, analogous to the proposition 3.19, due to S. Gersten [Ger] and S. Pride [Pri1] [Pri2].

Let $\mathcal{P} = < S; R >$ be a finite presentation of a group G, where every r in R is cyclically reduced. We suppose that R is stable by cyclic permutations and undertaking inverses. Let P^{st} be the link of the vertex in the Cayley 2-complex associated to the presentation. Combinatorically, it is the graph, called ***star graph***, whose set of vertices is $S \cup S^{-1}$ (disjoint union), set of oriented edges R, the initial vertex of $r \in R$ being the first symbol of r, the final vertex being the inverse of the last symbol of r. The edge inverse to $r \in R$ is r^{-1}, and the realizations of the edges r and r^{-1} are identified.

Theorem 3.22 *(S. Gersten - S. Pride) Suppose there exists a non-negative angle map $\theta : R \to \mathbf{R}^+$ such that*

1. *the sum of the angles of the edges of every non-empty reduced (i.e. with no edge followed by its inverse) closed path in P^{st} is greater than or equal to 2π,*

2. *for every $r = x_1 x_2 \cdots x_n$ in R with $x_i \in S \cup S^{-1}$,*

$$\epsilon_r = (n-2)\pi - \sum_{i=1}^{n} \theta(x_i \cdots x_m x_1 \cdots x_{i-1}) > 0$$

then with $\epsilon = \min\{\epsilon_r / r \in R\}$, the presentation P satisfies a linear isoperimetric inequality with constant $\frac{3\pi+\epsilon}{\epsilon}$. Hence G is hyperbolic.

Proof. (In a particular case) A complete combinatorial proof (see [Ger][Pri1]) applies the Gauβ-Bonnet formula to some Van-Kampen diagram. But we may use a method analogous to what we have already done, to obtain a weaker result, in the case where θ takes the same non zero value on two cyclic conjugates in R and the length of every element of R is bigger than or equal to 3 (this last hypothesis is very mild, since every finitely presented group has a finite presentation with relators of length at least 3).

Recall that the Cayley 2-complex of the presentation is the cell complex (which is not a multipolyhedron in our sense, since the closed edges are not embedded) obtained in the following way : take a wedge of oriented circles, one oriented circle for every element of S and one (regular Euclidean) 2-cell with n edges for every conjugacy class of the elements of R of length n ; label by elements of S and orient the edges of these cells so that the word we read cyclically is the corresponding element of R ; glue these cells to the wedge of circles preserving labels and orientations. So the angle function θ is a map from the angular sectors of vertices of these cells to $\mathbf{R}^+$.

As A. Haefliger told us, if $\theta_1, \cdots, \theta_n$ are positive and satisfy $\sum_{i=1}^{n} \theta_i < (n-2)\pi$, there may not exist a Riemannian metric on a polygon with n geodesic sides of length 1 such that the angles are the θ_i in cyclic order and the curvature is negative (even non constant). For instance in the triangle case, the three angles have to be less that or equal to $\frac{\pi}{3}$ using a comparaison with an equilateral triangle in $\mathbf{R}^2$. In the case where the θ_i for $i = 1 \cdots n$ are equal, we simply take the regular polygon with $n \geq 3$ sides in $\mathbf{H}^2$ with angles θ_1, and normalize the metric so that the length of the sides is one.

Identify every 2-cell in the Cayley 2-complex of the presentation with a negatively curved polygon having the prescribed angles. The first condition says that we may apply the corollary 3.18 to show that the fundamental group of the Cayley 2-complex is hyperbolic. Note that the corollary 3.18 still works even if the Cayley 2-complex neither is a multipolyhedron nor has piecewise constant curvature.

If the angle map θ takes a zero value, we may suitably approximate it by a positive map, satisfying the two conditions, and then use the reasonning of the step 2 of the proof of the theorem 3.21. □

Example 1 A. Douady's napkin (guaranteed un-ironable) : Consider the polyhedral surface obtained by gluing a countable number of equilateral Euclidean triangles, with exactly two triangles by edges and seven triangles by vertex. With the induced length metric, this space is quasi-isometric to $\mathbf{H}^2$.

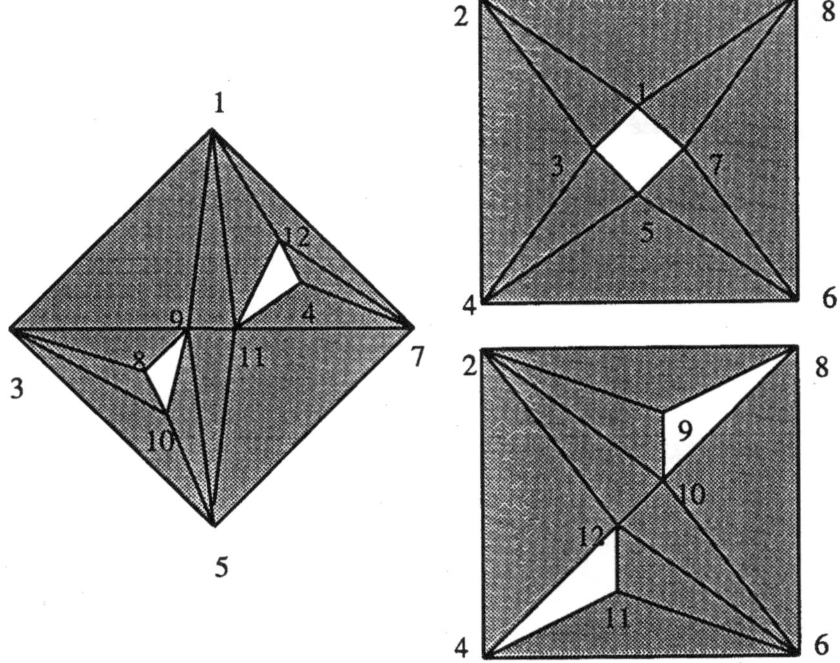

Figure 8 : A triangulated surface with valency 7.

Example 2 (C. Bavard-A. Marin) : Consider the surface obtained by gluing the triangulated surfaces of the figure 9, one annulus (drum) and two pairs of pants, so that the labeled vertices coincide. One obtain a triangulation of the oriented, compact, connected surface without boundary, of genus 2, such that at every vertex, there are 7 adjacent edges. Every oriented, compact, connected, surface S without boundary of genus greater than or equal to 2 may be obtained as a covering of this one, hence has such a triangulation.

If S_{-1} is a non orientable (compact, connected, without boundary) surface of Euler characteristic -1, then S_{-1} has no triangulation in the usual sense with at least 7 adjacent edges per vertex. Indeed, if there was such a triangulation, with number of vertices s, number of edges a and number of faces f, then we would have

$$f - a + s = -1$$

by the property of the Euler characteristic,

$$3f = 2a$$

since every face has exactly 3 edges, and every edge is an edge of exactly 2 faces,

$$7s \leq 2a$$

since every edge has exactly 2 vertices, and from every vertex start at least seven edges. Hence one has

$$s \leq 6.$$

This is a contradiction : there is at least 1 vertex x, and from this vertex 7 others, joined by an edge to x, which are pairwise distinct if we suppose the triangulation to be in the usual sense. Remark that S_{-1} has a smooth Riemannian metric with constant negative curvature -1 and that one can find a simplicial multipolyhedral structure with at least 7 edges per vertex. Note that the example of the figure 9 is one with the least possible number of vertices 12.

4 Hyperbolization of polyhedra

4.1 *Hyperbolization of 2-dimensional polyhedra*

We are going to define three hyperbolization processes for 2-dimensional polyhedra.

0) Let S be a triangulated, compact, connected, oriented surface of genus g bigger than or equal to 1, with boundary ∂S one circle, with induced triangulation having three vertices, such that

(i) at every vertex in $S - \partial S$, there are at least seven adjacent edges,

(ii) at every vertex in ∂S, there are at least three adjacent edges, not contained in ∂S.

For instance, take if $g \geq 2$ the surface of the example 2 in the previous section (figure 8), and remove one 2-face.

Let P be a finite 2-dimensional simplicial multipolyhedron. For every 2-simplex F of P, let S_F be a copy of S. Define a multipolyhedron $\mathrm{Hyp}_0(P)$ by removing from P the interior of every 2-simplex F, and by gluing S_F such that the edges of ∂F and those of S_F are matched isometrically. For instance, if one starts with the 2-sphere, triangulated as the boundary of the standard 3-simplex, one gets a compact, oriented, connected surface of genus $4g$.

We can also use hyperbolic surfaces (i.e. surfaces with a Riemannian metric of constant curvature -1) with a connected geodesic boundary, and use in the proof of the proposition 4.1 the gluing lemma 4.4 instead of the proposition 3.19.

1) Let P be a finite simplicial or cubical polyhedron of dimension 2. In particular, two 2-simplices, glued along two distinct edges are in fact the same. Let P^1 be the 1-skeleton of P. Define a finite cubical polyhedron $\mathrm{Hyp}_1(P)$ of dimension 2 as follows : take two copies P_1^1 and P_2^1 of the 1-skeleton P^1, and for every 2-face F of P, an annulus cubated as $\partial F \times [-1,+1]$. Then glue these spaces by identifying $\partial F \times \{-1\}$ in $\partial F \times [-1,+1]$ with ∂F in P_1^1 and $\partial F \times \{+1\}$ in $\partial F \times [-1,+1]$ with ∂F in P_2^1 isometrically on every edge.

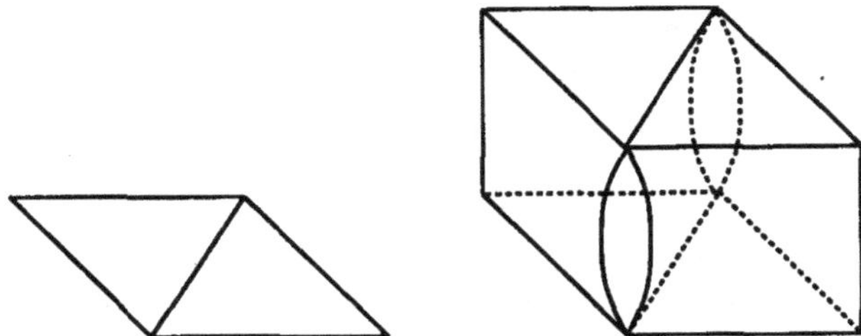

Figure 1 : Suspension hyperbolization.

For instance, if one starts with the 2-sphere, triangulated as the boundary of the standard 3-simplex Δ_3, one gets a cubated, compact, oriented, connected surface of genus three, which may be seen as the boundary of a regular neighborhood of the 1-skeleton of Δ_3 in $\mathbf{R}_3$ (seen as the hyperplane $x_1 + x_2 + x_3 + x_4 = 1$).

2) Let $\mathcal{M}_2$ be the Moebius strip, cubated (as a multipolyhedron, since two distinct squares share two distinct edges) in the following way :

$$\mathcal{M}_2 = {}^{(\partial\Box_2)\times[-1,+1]}/_{\sim}$$

where $\square_2 = [-1,+1] \times [-1,+1]$ is the standard 2-cube and $\sim$ is the equivalence relation defined by

$$(t_1, t_2, t_3) \sim (t'_1, t'_2, t'_3) \text{ if and only if } (t_1, t_2, t_3) = \pm(t'_1, t'_2, t'_3).$$

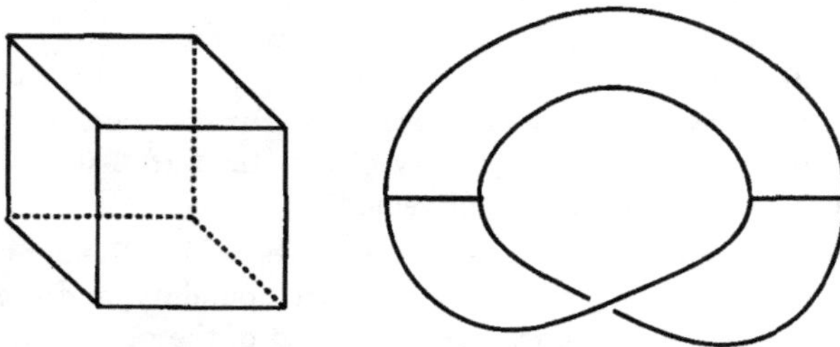

Figure 2 : Moebius strip.

Let P be a finite cubical multipolyhedron of dimension 2, and for every 2-cube C of P, let $\mathcal{M}_2(C)$ be a copy of $\mathcal{M}_2$. Note that the boundary of $\mathcal{M}_2$ is, in a canonical way, homeomorphic to $\partial\square_2$ (a circle subdivided into four arcs). Define a polyhedron $\mathrm{Hyp}_2(P)$ by removing from P the interior of every 2-cube C, and by gluing $\mathcal{M}_2(C)$ so that the edges of ∂C and those of $\partial\mathcal{M}_2(C)$ are matched isometrically. For instance, if one starts with the 2-sphere, cubated as the boundary of the standard 3-cube, one gets a compact connected non oriented surface which is the connected sum of six projective planes $\mathbf{P}_2$.

Examples 0) and 2) are due to M. Gromov [Gro3], and example 1) to M. Davis and T. Januszkiewicz [DJ][Jan], who also give a far more general way of constructing hyperbolization tools.

Note that if P is connected, then $\mathrm{Hyp}_i(P)$ is connected, for $i = 0, 1, 2$. If P is homeomorphic to a surface (possibly with boundary), so is $\mathrm{Hyp}_i(P)$. Indeed, the link of every vertex in $\mathrm{Hyp}_i(P)$ is homeomorphic to the link of the corresponding vertex in P (or in S for the case $i = 0$). Hence, if the link of every point in P is a circle, the link of every point in $\mathrm{Hyp}_i(P)$ is a circle and every point in $\mathrm{Hyp}_i(P)$ has a neighborhood which is a disk

Furthermore, in the hyperbolization process, any edge of the link of a vertex has been subdivided into at least 4 edges in case 0), into two edges in case 1) and 2).

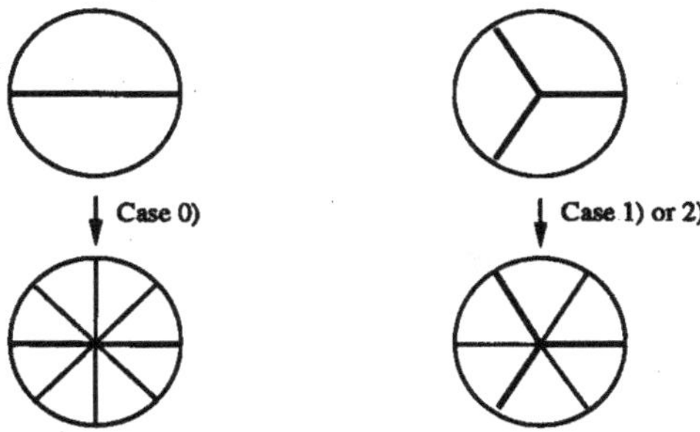

Figure 3 : Hyperbolization of links.

Proposition 4.1 *If P is a finite 2-dimensional simplicial multipolyhedron, then $Hyp_0(P)$ has a structure of a negatively curved simplicial polyhedron. If P is a finite 2-dimensional simplicial or cubical polyhedron, then $Hyp_1(P)$ has a structure of a negatively curved cubical polyhedron. If P is a finite 2-dimensional cubical polyhedron then $Hyp_2(P)$ has a structure of a negatively curved cubical polyhedron. Hence the fundamental group of every connected component of $Hyp_i(P)$ is hyperbolic for $i = 0, 1, 2$.*

Proof. It is easy to check, according to the above discussion, that every simple closed curve in the link of a vertex has at least seven edges in the case 0, and five edges in the cases 1,2, and we apply the proposition 3.19. □

For every spanning tree T (i.e. a subtree of the 1-skeleton of P, containing every vertex of P) for a simplicial or cubical multipolyhedron P, one has an associated presentation of $\Pi_1(P)$, if P is finite and connected, in the following way. One has to imagine that T is just one vertex (which may be realized by a homotopy equivalence). The set S of generators is the set of edges of P that are not in T. To every 2-cell F of P, associate a word r_F in $S \cup S^{-1}$ (disjoint union) in the following way. First orient arbitrarily every edge of P. Then r_F is the sequence of letters in $S \cup S^{-1}$ obtained by following the closed curve ∂F, starting from any vertex of F, in any direction, where we read only the edges not in T, with an inverse if the orientations do not match.

Let P be a finite simplicial multipolyhedron (resp. a finite cubical polyhedron) with a choice of spanning tree, and associated presentation $< S; R >$ with

$$S = \{x_i \,/ i = 1, \cdots, n\} \quad R = \{r_j \,/ j = 1, \cdots, m\}$$

Then $\mathrm{Hyp}_0(P)$ (resp. $\mathrm{Hyp}_2(P)$) has a presentation for its fundamental group of the form

$$S = \{x_i \ (i = 1 \cdots n), a_{1,j}, b_{1,j}, \cdots, a_{g,j}, b_{g,j} \ (j = 1 \cdots m)\}$$

$$R = \{r_j[a_{1,j}, b_{1,j}] \cdots [a_{g,j}, b_{g,j}] \ (j = 1 \cdots m)\}$$

(resp.

$$S = \{x_i \ (i = 1 \cdots n), y_j \ (j = 1 \cdots m)\}$$
$$R = \{r_j y_j^2 \ (j = 1 \cdots m)\})$$

Note that in fact the first presentation is easily seen to have the small cancellation property. It is easy to see that the hypotheses of the theorem 3.22 do apply here, since we can take $+1$ for the weight of an edge between vertices in $S \vee S^{-1}$ and every other weights $\frac{1}{4g}$.

4.2 *Asphericalization of polyhedra*

In this section, unless stated otherwise, the cells of a multipolyhedron are Euclidean ones.

An hyperbolization process is a process with turns a finite simplicial or cubical multipolyhedron into one admitting a structure of a negatively curved multipolyhedron. For the applications we are aiming to, we only have to look for asphericalization processes. These are processes turning finite simplicial or cubical multipolyhedra into aspherical ones, in the following sense, as a consequence of nonpositive curvature. But note that according to the theorem 3.21, the asphericalization processes we are going to define are indeed hyperbolization processes.

Definition 4.2 *A finite connected multipolyhedron is aspherical if its universal covering is contractible.*

Let us define two constructions, which we will prove to be asphericalization processes. They both proceed by induction on the dimension n of a cubical multipolyhedron, by defining the asphericalization of the standard cube of dimension n, and then the asphericalization of a cubical multipolyhedron of dimension n. The first one generalizes the Hyp_1 construction, the second the Hyp_2 construction, of the previous section.

Construction 1 (M. Davis - T. Januszkiewicz)

Let us define by induction on the dimension n a functor Asp_1 from the category $\mathcal{C}^n$ whose objects are the finite Euclidean standard cubical multipolyhedron P^n of dimension n and morphisms the one-to-one maps $f^n : P^n \to Q^n$ such that the restriction of f^n to every cell of P^n is an isometry onto a cell of Q^n.

If P^1 has dimension 1, let $\mathrm{Asp}_1(P) = P$ and for any morphism $f : P^1 \to Q^1$ let $\mathrm{Asp}_1(f) = f$. Suppose by induction that $\mathrm{Asp}_1(P^k)$ and $\mathrm{Asp}_1(f^k)$ have been defined for every object P^k and morphism f^k of dimension k less than or equal to n, with $n \geq 1$.

Let P^{n+1} be a finite Euclidean standard cubical multipolyhedron of dimension $n+1$. Let P^n be the n-skeleton of P^{n+1} and $C_1, \cdots, C_p$ be the $(n+1)$-cubes of P^{n+1}, with $f_i : C_i \to P^{n+1}$ being the natural morphisms. Denote by $\partial f_i : \partial C_i \to P^n$ the induced morphisms. Set

$$\mathrm{Asp}_1(\square_{n+1}) = \mathrm{Asp}_1(\partial\square_{n+1}) \times [-1,1].$$

By induction, any marking isomorphism $\square_{n+1} \to C_i$ induces a marking isomorphism $\mathrm{Asp}_1(\square_{n+1}) \to \mathrm{Asp}_1(C_i)$. The multipolyhedron $\mathrm{Asp}_1(P^{n+1})$ is obtained by gluing the $\mathrm{Asp}_1(C_i)$ for $i = 1 \cdots p$ and two copies Q_1, Q_2 of $\mathrm{Asp}_1(P^n)$ in the following way. Identify $\mathrm{Asp}_1(\partial C_i)\times\{+1\}$ in $\mathrm{Asp}_1(C_i)$ with the isomorphic subpolyhedron $\mathrm{Asp}_1(\partial f_i)(\mathrm{Asp}_1(\partial C_i))$ in the first copy Q_1 and $\mathrm{Asp}_1(\partial C_i) \times \{-1\}$ in $\mathrm{Asp}_1(C_i)$ with the isomorphic subpolyhedron $\mathrm{Asp}_1(\partial f_i)(\mathrm{Asp}_1(\partial C_i))$ in the second copy Q_2, in a face marking preserving way for every $i = 1 \cdots p$. Note that in $\mathrm{Asp}_1(P^{n+1})$ there are two disjoint subpolyhedra canonically isomorphic to $\mathrm{Asp}_1(P^n)$, hence by induction, 2^k (resp. 2^n) disjoint subpolyhedra canonically isomorphic to $\mathrm{Asp}_1(P^{n+1-k})$ for $k = 0, \cdots, n$ (resp. $k = n+1$).

If $f^{n+1} : P^{n+1} \to Q^{n+1}$ is a morphism, then $\mathrm{Asp}_1(f^{n+1})$ is defined in an obvious way, by sending the two copies of $\mathrm{Asp}_1(P^n)$ in $\mathrm{Asp}_1(P^{n+1})$ to the two copies of $\mathrm{Asp}_1(Q^n)$ in $\mathrm{Asp}_1(Q^{n+1})$ by $\mathrm{Asp}_1(f^n)$, where f^n is the restriction of f^{n+1} to the n-skeleton, and then sending the asphericalized $(n+1)$-cubes of P^{n+1} to the asphericalized image $(n+1)$-cubes of Q^{n+1}. The facts that have to be checked (in particular the functorial properties $\mathrm{Asp}_1(id_{P_n}) = id_{\mathrm{Asp}_1(P_n)}$ and $\mathrm{Asp}_1(f^n \circ g^n) = \mathrm{Asp}_1(f^n) \circ \mathrm{Asp}_1(g^n)$) are obvious by induction.

Construction 2 (M. Gromov)

By a similar induction process, let us define a functor Asp_2 from the category $\mathcal{C}^n_*$ whose objects are the finite Euclidean standard cubical multipolyhedron P^k of dimension k less than or equal to n and morphisms the one-to-one maps $f : P^k \to Q^{k'}$ with $k \leq k' \leq n$ such that the restriction of f to every cell of P^k is an isometry onto a cell of $Q^{k'}$.

If P^1 is a object of dimension 1 and $f^1 : P^1 \to Q^1$ a morphism of dimension 1

$$\mathrm{Asp}_2(P^1) = P^1 \text{ and } \mathrm{Asp}_2(f^1) = f^1.$$

Let $s_{n+1} : \square_{n+1} \to \square_{n+1}$ be the morphism of dimension $n+1$ defined by $s(x) = -x$. Let $\partial s_{n+1} : \partial\square_{n+1} \to \partial\square_{n+1}$ be the induced map $x \mapsto -x$ which is a fixed point free involution, and $\mathrm{Asp}_2(s) : \mathrm{Asp}_2(\partial\square_{n+1}) \to \mathrm{Asp}_2(\partial\square_{n+1})$ be the fixed point free involution obtained by induction and the functorial properties. Define

$$\mathrm{Asp}_2(\square_{n+1}) = {}^{\mathrm{Asp}_2(\partial\square_{n+1})\times[-1+1]}/_{(\mathbf{Z}/2\mathbf{Z})}$$

where $1 \in \mathbf{Z}/2\mathbf{Z}$ acts on $\mathrm{Asp}_2(\partial\square_{n+1}) \times [-1,+1]$ by the fixed point free involution $(\mathrm{Asp}_2(s), x \mapsto -x)$. Note that by the assertion (iii) of the inductively proved theorem 4.3, $\mathrm{Asp}_2(\Box_{n+1})$ is a (non orientable) topological manifold with boundary, which is a fiber bundle over the manifold $\mathrm{Asp}_2(\partial\Box_{n+1})$ with fiber $[-1,+1]$, and whose boundary is canonically homeomorphic to $\mathrm{Asp}_2(\partial\Box_{n+1})$.

It is then easy to define $\mathrm{Asp}_2(P^{n+1})$ and $\mathrm{Asp}_2(f)$, for every object P^{n+1} of dimension $n+1$, and morphism $f : P^k \to Q^{k'}$ with $k \leq k' \leq n+1$ by identifying, for every $(n+1)$-cube C of P^{n+1}, the boundary of $\mathrm{Asp}_2(C)$ with the image under $\mathrm{Asp}_2(\partial g)$ of $\mathrm{Asp}_2(\partial C)$ in $\mathrm{Asp}_2(P_n)$, where $\partial g : \partial C \to P^n$ is the injection in the n-skeleton.

The facts that have to be proved are obvious.

Note that even if $\mathrm{Asp}_2(\Box_2)$ is by definition a quotient of $\mathrm{Asp}_1(\Box_2)$ by $\mathbb{Z}/2\mathbb{Z}$, in general $\mathrm{Asp}_2(P)$ can not be seen as a natural quotient of $\mathrm{Asp}_1(P)$. For example, if P is the union of two copies $\Box_2^1, \Box_2^2$ of $\Box_2$ glued along one edge, then the actions of $\mathbb{Z}/2\mathbb{Z}$ on $\mathrm{Asp}_1(\Box_2^1), \mathrm{Asp}_1(\Box_2^2)$ can not be glued continuously to yield $\mathrm{Asp}_2(P)$. Nevertheless, the link of a vertex in $\mathrm{Asp}_2(P)$ is isomorphic to the link of any of the $2^{\dim(P)-1}$ corresponding vertices in $\mathrm{Asp}_1(P)$ (see the proof of the theorem 4.3).

Theorem 4.3 *(M. Gromov) Let P be a finite connected Euclidean standard cubical polyhedron. Then for $i = 1, 2$, $Asp_i(P)$ is a finite Euclidean standard connected multipolyhedron such that*

(i) $Asp_i(P)$ has a word hyperbolic fundamental group,

(ii) $Asp_i(P)$ is aspherical,

(iii) if P is homeomorphic to a manifold (resp. manifold with boundary), then so is $Asp_i(P)$; furthermore, $\partial Asp_2(P)$ is naturally isomorphic to $Asp_2(\partial P)$, and $\partial Asp_1(P)$ is naturally isomorphic to the disjoint union of two copies of $Asp_1(\partial P)$,

(iv) $Asp_i(P)$ has a structure of a negatively curved multipolyhedron.

For other properties of the functors Asp_i, see [Gro3] [DJ] [Jan]. The basic lemma in the proof of the theorem 4.3 is the following.

Lemma 4.4 *(M. Gromov) Suppose either that*

a) X is the disjoint union of two complete, locally compact geodesic spaces X_1 and X_2, and $Y_i \subset X_i$ is for $i = 1, 2$ a non empty locally convex closed subspace,

or

b) X is a complete locally compact geodesic metric space and Y_1, Y_2 are two disjoint locally convex closed subspaces.

Let $f : Y_1 \to Y_2$ be an isometry of the induced metrics, and $\hat{X}$ be the space built from X by identifying Y_1 and Y_2 via f. Then $\hat{X}$, with the induced length metric is a geodesic space. If X has curvature less than or equal to χ, with $\chi \leq 0$, then so does $\hat{X}$.

By definition, a locally convex subspace Y in a geodesic metric space X is a subspace Y such that for every z in Y, there is a neighborhood U_z of z in X such that for every x, y in U_z, every geodesic in X between x and y lies in fact in Y.

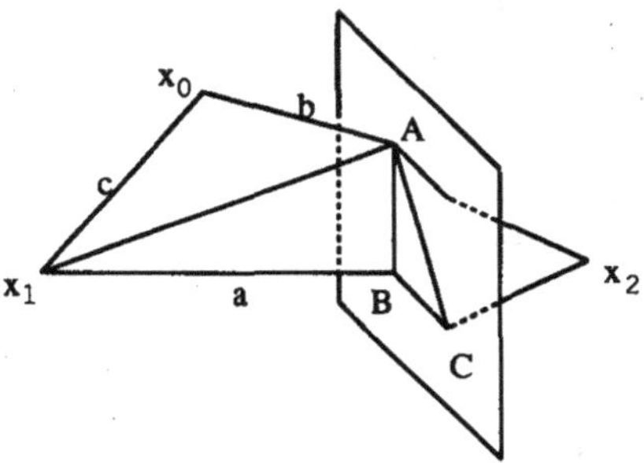

Figure 4 : Gluing lemma.

Proof. (M. Davis, T. Januszkiewicz [DJ]) Let $\hat{Y}$ be the image of Y_1 and Y_2 in $\hat{X}$. In case b), we set $X_1 = X_2 = X$. If $\hat{x} \in \hat{X}$ is the image of $x \in X - (Y_1 \cup Y_2)$, then since $Y_1 \cup Y_2$ is closed, $\hat{x}$ has a compact neighborhood $\hat{V}(\hat{x})$, image of a neighborhood $V(x)$ of x that does not meet $Y_1 \cup Y_2$. We may suppose that the diameter of $V(x)$ is strictly less than the distance between $V(x)$ and $Y_1 \cup Y_2$. A point $\hat{x}$ in $\hat{Y}$, having $x_1 \in Y_1$ and $x_2 \in Y_2$ as preimages, has a neighborhood $\hat{V}(\hat{x})$ in $\hat{X}$ made of a compact neighborhood V_1 of x_1 in X and a disjoint compact neighborhood V_2 of x_2 glued along the compact isometric subsets $V_1 \cap Y_1$ and $V_2 \cap Y_2$. We may suppose that V_1, V_2 are neighborhoods as in the definition of a locally convex subspace, and we denote by $\hat{V}_1, \hat{V}_2$ their images in $\hat{X}$. Hence $\hat{X}$ is locally compact.

Consider the pseudo-distance defined to be the greatest lower bound of the lengths of continuous curves in $\hat{X}$ between two points. Since $\hat{Y}$ is closed in $\hat{X}$, every continuous curve meets $\hat{Y}$ into a disjoint family of closed intervals, and $\hat{X} - \hat{Y}$ into a disjoint countable family of open intervals. The length of a curve id defined as the sum of the length of the pieces, which is well defined in $\hat{Y}$, since Y_1 and Y_2 are isometric. Between two points there obviously is a curve of finite length. Note that there are locally no shortcuts in the following sense. If $\hat{x}$ comes from a point in $X - (Y_1 \cup Y_2)$, then the pseudo-distance of two points in $\hat{V}(\hat{x})$ is the length of any geodesic between the two corresponding points in $V(x)$. If $\hat{x} \in \hat{Y}$, the distance between two points $\hat{y}, \hat{z}$ of $\hat{V}(\hat{x})$ is, by the local connectedness, the length of any geodesic between the two corresponding points in $V(x)$ if $\hat{y}, \hat{z}$ are in the same $\hat{V}_i$ or if $\hat{y} \in \hat{V}_1, \hat{z} \in \hat{V}_2$ say, the minimum (attained by compactness) of the sum of the length of a geodesic in $\hat{V}_1$ between $\hat{y}$ and a point $\hat{w}$ in $\hat{V}(\hat{x}) \cap \hat{Y}$ and the length of a geodesic in $\hat{V}_2$ between $\hat{w}$ and $\hat{z}$. Hence the pseudo-distance is a complete metric and according to the proposition 2.3, $\hat{X}$ is a geodesic metric space. We must show that every point p in $\hat{X}$ has a neighborhood V_p as in the definition 2.17. We may suppose p is in $\hat{Y}$.

Let x_0, x_1, x_2 be three points in $V_p = \hat{V}(p)$, and suppose for instance x_0, x_1 in $\hat{V}_1$ and x_2 in $\hat{V}_2$. Consider three geodesics a, b, c respectively between x_1 and x_2, x_0 and x_2, x_0 and x_1. Let A be the first point in $\hat{Y}$ on b starting from x_0, B the first point in $\hat{Y}$ on a from x_1, and C the first point in $\hat{Y}$ on a starting from x_2 (see the figure 4). Fix a geodesic between x_1 and A, A and B, A and C. Since $\hat{Y}$ is locally convex, the geodesics between A and B, B and C, C and A are in $\hat{Y}$. By applying the lemma 2.12, to the triangles (x_1, A, B) and (A, B, C), and once more to (x_1, A, C), (A, C, x_2), we obtain that (x_1, A, x_2) satisfies the CAT(χ)-property. Applying the lemma 2.12 again to the triangles (x_0, x_1, A) and (x_1, A, x_2), one has that (x_0, x_1, x_2) satisfies the CAT(χ)-property. □

Proof. (of 4.3.) Of course the assertions (i), (ii) follows from (iv). But we give some simpler arguments. The assertion (i) follows immediately from the proposition 4.1 in the case of Asp_2, since the fundamental group is carried by the 2-skeleton and the 2-skeleton of $\mathrm{Asp}_2(P)$ is $\mathrm{Asp}_2(P^2) = \mathrm{Hyp}_2(P^2)$, where P^2 is the 2-skeleton of P. But in the case of Asp_1, this does no longer hold. We do have to use the assertion (iv).

To prove assertion (ii), remind that every n-cube of $\mathrm{Asp}_i(P)$ is identified with the standard Euclidean n-cube. First, it is immediate to verify that $\mathrm{Asp}_i(\square_n)$ has a locally convex boundary. Indeed, if X is a geodesic metric space, then any continuous path in $X \times [-1, +1]$ between two points in $X \times \{t\}$ has length bigger than its vertical projection on $X \times \{t\}$, and even strictly bigger if the path is not contained in $X \times \{t\}$. In fact the geodesics are the preimages by the vertical projection of geodesics of X, that have constant slopes.

By induction, the link of any vertex in $\mathrm{Asp}_2(P)$ is isomorphic to the link of any of the $2^{\dim(P)-1}$ corresponding vertex in $\mathrm{Asp}_1(P)$. Indeed, this is clear by induction at the level of the hyperbolized cubes, and then follows since the combinatoric of the gluings of the $\mathrm{Asp}_i(C_k)$ along the subpolyhedra $\partial\mathrm{Asp}_i(C_k)$ where the C_k are the dim(P)-cubes of is the same as the one for the C_k along the ∂C_k. According to the theorem 3.15, we hence may prove the assertions (ii), (iii), (iv) only for Asp_2.

Furthermore, the link of a cell of $\mathrm{Asp}_2(\square_n)$ is isomorphic (hence isometric) either to the link of a cell in $\mathrm{Asp}_2(\partial\square_n)$ or to a spherical cone (half of a suspension) over a cell in $\mathrm{Asp}_2(\partial\square_n)$. Hence by induction, the link of every vertex in $\mathrm{Asp}_2(P)$ is isomorphic (but not isometric) to the barycentric subdivision of the link of the corresponding vertex in P. Indeed, any cone over the barycentric subdivision of the boundary of the standard m-simplex is (abstractly) isomorphic to the $(m+1)$-simplex. Moreover this proves that any simple closed curve in the 1-skeleton consisting of 3 or 4 edges bounds a 2-simplex or the union of two 2-simplices glued by an edge.

Next, if C is a n-cube of a polyhedron P of dimension n, then $\mathrm{Asp}_i(\partial C)$ is locally convex in $\mathrm{Asp}_i(P^{n-1})$. This comes from the fact that in a spherical cone over a piecewise spherical polyhedron, the base is locally convex (see the lemma 3.14). By induction, using the above lemma 4.4, it follows that $\mathrm{Asp}_i(P)$ has a complete geodesic metric with non positive curvature. We then apply the theorem 2.19 to the universal covering of $\mathrm{Asp}_i(P)$, and the remark just before the definition 2.18.

The assertion (iii) easily follows from the fact that the link of every point of $\mathrm{Asp}_2(P)$ is homeomorphic to the link of the corresponding point in P, as seen above, and the fact that a cubical multipolyhedron is homeomorphic to a manifold if and only if the link of every point is homeomorphic to a sphere (of the right dimension), since the cone over a sphere is homeomorphic to a ball, and every point of a cubical multipolyhedron has a neighborhood homeomorphic to the cone over its link.

Let us prove (iv). The cubical multipolyhedron $\mathrm{Asp}_i(P)$ does admit a structure of a negatively curved polyhedron. To see this, we use the theorem 3.21, whose hypothesis was checked above. Note that the idea was to give to every cube of dimension n of $\mathrm{Asp}_i(P)$ the structure of a regular n-cube in $\mathbf{M}^n_\chi$, with χ negative and close to zero. But we can not use the lemma 4.4 now, since the boundary of $\mathrm{Asp}_i(\Box_2)$ is no longer locally convex. □

Note that there is a natural map $f_i(P) : \mathrm{Asp}_i(P) \to P$ satisfying nice homological properties (see [Gro3] [DJ] [Jan]). In the case $i = 2$, the construction by induction on the dimension is the following one. If P^1 has dimension 1, then $f_2(P^1)$ is the identity. The map $f_2(\Box_2) : \mathrm{Asp}_i(\Box_2) = \mathcal{M}_2 \to \Box_2$ sends the boundary of the Moebius band to the boundary of the disk canonically isomorphically and the soul of the Moebius band (that is the image of $\partial\Box_2 \times \{0\}$) to the center of the disk $\Box_2$, and extends radially using the fibration by intervals of the Moebius band. So if P^2 is a 2-dimensional polyhedron, then $f_2(P^2)$ is the identity on the 1-skeleton of P^2, and on every 2-cube is the above map. By induction, $f_2(\Box_{n+1})$ is the map $[(x,t)] \mapsto |t|\, f_2(\partial\Box_{n+1})$ if $x \in \mathrm{Asp}_2(\partial\Box_{n+1})$ and $t \in [-1,+1]$ and if P is a polyhedron of dimension $n+1$ and n-skeleton P^n, then $f_2(P)$ induces $f_2(P^n)$ on $\mathrm{Asp}_2(P^n)$ and the above maps on the $(n+1)$-cubes.

4.3 *Relative asphericalization of polyhedra*

We are going to give Gromov's transformation process on polyhedra, related to the ones above, which leaves a given subpolyhedron almost unchanged.

Let P be a finite Euclidean regular cubical multipolyhedron, and P_0 a subpolyhedron. We are going to define $\mathrm{Asp}_2(P, P_0)$, a finite (Euclidean regular) cubical multipolyhedron, called the asphericalization of P relative to P_0.

Recall that if R is a cubical (resp. simplicial) multipolyhedron then the barycentric subdivision R' (resp. the facial subdivision R^*) is a simplicial (resp. cubical) polyhedron (see the section 3.2). Also recall (see the section 3.2) that the cone $\mathrm{Cone}(R)$ over a finite simplicial multipolyhedron R has a natural structure of a simplicial multipolyhedron.

Let $\mathrm{Cone}(\partial N(P_0, P))$ be the cone over the boundary of the regular neighborhood (see the section 3.2) of P_0 in P, and glue it to $P'' - N(P_0, P)$ by identifying the base $B = \partial N(P_0, P) \times \{0\}$ and $\partial N(P_0, P)$. Finally let

$$\hat{K} = [(P'' - N(P_0, P)) \cup \mathrm{Cone}(\partial N(P_0, P))]^*$$

Let x_0 be the image in the cubical polyhedron $\hat{K}$ of the cone vertex (corresponding to $\partial N(P_0, P) \times \{1\}$). Note that the link (space of directions) of x_0 in $\hat{K}$ is isomorphic precisely to $\partial N(P_0, P)$, since in the facial subdivision, the links of vertices are not subdivided.

Let $\hat{x}_0$ be the point in $\mathrm{Asp}_2(\hat{K})$ corresponding to x_0. As we have seen in the proof of the theorem 4.3, the link of $\hat{x}_0$ is isomorphic to the barycentric subdivision of the link of x_0 in $\hat{K}$.

Define $\mathrm{Asp}_2(P, P_0)$ to be $\mathrm{Asp}_2(\hat{K})$ minus the star of $\hat{x}_0$ (defined without taking the baricentric subdivision of $\mathrm{Asp}_2(\hat{K})$), glued with $(N(P_0, P))'$ along their isomorphic boundary $(\partial N(P_0, P))'$.

Definition 4.5 *A cobordism W between two (resp. cubated) topological manifolds M_0 and M_1 without boundary is a (resp. cubated) topological manifold W with boundary equal to the disjoint union of M_0 and M_1.*

Theorem 4.6 *(M. Gromov) Let P_0 be a finite cubical polyhedron, homeomorphic to a topological manifold without boundary. Let $P = P_0 \times [0, 1]$. Then $\mathrm{Asp}_2(P, P_0 \times \{0\})$ is a cobordism between $P_0 \times \{0\}$ and $\mathrm{Asp}_2(P_0 \times \{1\})$.*

Note that the functor $\mathrm{Asp}_2(P, P_0)$ satisfies other nice properties (see [Gro3] [DJ]). For a statement in purely combinatorial terms, see [BP].

Proof. The proof is clear, using the theorem 4.3. □

4.4 *Topological applications*

Theorem 4.7 *(M. Gromov) Any compact triangulable manifold without boundary is cobordant to an aspherical compact manifold without boundary.*

Proof. This follows from theorem 4.6 and theorem 4.3. □

Let us give without proof the following very nice results of M. Davis and T. Januzkiewicz that use the same constructions.

Theorem 4.8 *[DJ] For each $n \geq 5$, there exists a nonpositively curved piecewise flat polyhedron P with the following properties.*

(i) P is homeomorphic to a compact topological manifold without boundary of dimension n

(ii) P is not homeomorphic to $\mathbf{R}^n$.

Theorem 4.9 *[DJ] There exists a finite polyhedron P, which is homeomorphic to a smooth manifold of dimension 5, and which has a structure of a negatively curved polyhedron for which the hyperbolic boundary of its universal covering is not homeomorphic to a manifold.*

See [DJ][Jan] for other results.

References

[AB] S.B. Alexander, R.L. Bishop, *"The Hadamard-Cartan theorem in locally convex spaces"*, preprint University of Illinois 1989, to appear in L'Ens. Math..

[ABN] A.D. Aleksandrov, V.N. Berestovskii, and I.G. Nikolaev, *"Generalized Riemannian spaces"*, Russian Math. Surveys **41:3** (1986), 1-54.

[Ale1] A.D. Aleksandrov, *"A theorem on triangles in a metric space and some applications"*, Trudy Math. Inst. Steklov. **38** (1951), 5-23 (in Russian, translation by Stallings, University of California at Berkeley).

[Ale2] A.D. Aleksandrov, *"Uber eine Verallgemeinnerung der Riemannschen Geometrie"*, Schr. Forschunginst. Math. **1** (1957), 33-84.

[AZ] A.D. Aleksandrov, V.A. Zalgaller, *"Intrinsic geometry of surfaces"*, Transl. Math. Monographs **15**, Amer. Math. Soc., 1967.

[Bea] A. Beardon, *"The geometry of discrete groups"*, Springer-Verlag 1983.

[Ber] M. Berger, *"Géométrie 5 : la sphère pour elle-même, géométrie hyperbolique, l'espace des sphères*, Cedic-Fernand Nathan, Paris, 1977.

[BF] M. Bestvina, M. Feighn, *"A combination theorem for negatively curved groups"*, preprint University of California at Los Angeles 1990.

[BHC] A. Borel, Harish-Chandra, *"Arithmetic subgroups of algebraic groups"*, Ann. of Math. **75** (1962), 485-535.

[Bor] A. Borel, *"Compact Clifford-Klein forms of symmetric spaces"*, Topology **2** (1963), 111-122.

[BP] W. A. Bogley, S. J. Pride, *"Aspherical relative presentations"*, preprint, University of Glasgow, 1990.

[Bri] M.R. Bridson, *"Geodesics and geometry in metric simplicial complexes"*, these proceedings.

[Bro] K. Brown, *"Buildings"*, these proceedings.

[Bus] H. Busemann, *"Spaces with non-positive curvature"*, Acta Math. **80** (1948), 259-310.

[CD] R. Charney, M. Davis *"Singular metrics of nonpositive curvature on branched covers of Riemannian manifolds"*, preprint Ohio State Univ., 1990.

[CE] J. Cheeger-D.G.Ebin, *"Comparison theorems in Riemaniann geometry"*, North Holland, 1975.

[CV] S. Cohn-Vossen, *"Existenz Kurzester Wege"*, Doklady SSSR **8** (1935) 339-342.

[DJ] M.W. Davis, T. Januszkiewicz, *"Hyperbolization of polyhedra"*, preprint, Ohio State University 1989, to appear in J. of Diff. Geom..

[DM] P. Deligne, G.D. Mostow, *"Monodromy of hypergeometric functions and non-lattice integral monodromy"*, Pub. Math. I.H.E.S. **63** (1986), 5-90.

[FJ1] F.T. Farrel, L.E. Jones, *"A topological analogue of Mostow's rigidity theorem"*, J. Amer. Math. Soc. **2** (1989), 257-369.

[FJ2] F.T. Farrel, L.E. Jones, *"Rigidity and other topological aspects of compact nonpositively curved manifolds*, Research anouncement, Bull. Amer. Math. Soc. **22**(1990) 59-64.

[Ger] S. Gersten, *"Reducible diagrams and equations over groups"*, Essays in group theory, M. S. R. I. Pub. **8**, Springer-Verlag, 1987.

[GH] E. Ghys, P. de la Harpe (ed.), *"Sur les groupes hyperboliques d'après M. Gromov"*, Progress in Math. **83**, Birkhaüser (1990).

[Glu] H.R. Gluck, *"Piecewise linear methods in Riemannian geometry"*, University of Pennsylvania lecture notes, 1972.

[GPS] M. Gromov, I. Piatetski-Shapiro, *"Non-arithmetic groups in Lobachevsky spaces"*, Pub. Math. I.H.E.S. **66** (1988), 93-103.

[Gro1] M. Gromov, *"Hyperbolic manifolds, groups and actions*, in Riemann Surfaces and Related Topics, ed. I. Kra and B. Maskit, Ann. of Math. Studies **97**, Princeton Univ. Press 1981.

[Gro2] M. Gromov, *"Structures métriques pour les variétés riemanniennes"*, written with J. Lafontaine and P. Pansu, Cedic/Fernand Nathan, Paris, 1981.

[Gro3] M. Gromov, *"Hyperbolic groups"*, In : Essays in group theory, ed. : S.M. Gersten, M.S.R.I. Pub. **8**, 75-263, Springer-Verlag (1987).

[GT] M. Gromov, W. Thurston, *"Pinching constants for hyperbolic manifolds"*, Inv. Math. **89** (1987), 1-12.

[Har] P. de la Harpe, *"An invitation to Coxeter groups"*, these proceedings.

[Hei] E. Heintze, *"Compact quotients of homogeneous negatively curved Riemannian manifolds"*, Math. Zeit. **140** (1974), 79-80.

[Hel] S. Helgason, *"Differential geometry, Lie groups and symmetric spaces"*, Academic Press, 1978.

[Jan] T. Januszkiewicz, *"Hyperbolizations"*, these proceedings.

[Mak] V.S. Makarov, *"A class of partitions of a Lobachevskii space"*, Soviet Math. Dokl. **6** (1965), 400-401.

[Mas] W.S. Massey, *"Algebraic topology : an introduction"*. Graduate texts in Mathematics **56** (1967), Springer-Verlag.

[Mos] G. D. Mostow, *"Strong rigidity of locally symmetric spaces"*, Ann. Math. Studies **78**, Princ. Univ. Press (1973).

[Mou] G. Moussong, *"Hyperbolic Coxeter group"*, Doctoral Dissertation, Ohio State University, 1988.

[MS] G.D. Mostow, Y.T. Siu, *"A compact kähler surface of negative curvature not covered by the ball"*, Ann. Math. **112** (1980), 321-360.

[Pri1] S. J. Pride, *"Star complexes and the dependance problems for hyperbolic 2-complexes"*, Glasgow Math. J. **30** (1988), 155-170.

[Pri2] S. J. Pride, *"Involutary presentations, with applications to Coxeter groups, NEC-groups and groups of Kaneskii"*, J. Algebra **120** (1989), 200-223.

[RS] C.P. Rourke, B.J. Sanderson, *"Introduction to piecewise-linear topology,* Springer-Verlag, 1982.

[Sha] P.B. Shalen, *"Dendrology and its applications"*, these proceedings.

[Spa] E.H. Spanier, *"Algebraic topology"*, Tata McGraw Hill (1966).

[Sto] D.A. Stone, *"Geodesics in piecewise linear manifolds"*, Trans. Amer. Math. Soc. **215** (1976), 1-44.

[Thu1] W.P. Thurston, *"Geometry and topology of* 3*-manifolds"*, Lecture notes, Princeton University, 1976-1979.

[Thu2] W.P. Thurston, *"Hyperbolic structures on* 3*-manifolds I : Deformation of acylindrical manifolds"*, Ann. of Math. **124** (1986), 203-246.

[Vin1] E.B. Vinberg, *"Discrete groups generated by reflections in Lobachevskii spaces"*, Math. U.S.S.R. Sbornik **1** (1967), 429-444.

[Vin2] E.B. Vinberg, *"Some examples of crystallographic groups in Lobachevskii spaces"*, Math. U.S.S.R. Sbornik **7** (1969), 617-622.

[Vin3] E.B. Vinberg, *"Some arithmetical discrete groups in Lobachevskii spaces"*, Discrete subgroups of Lie groups, Oxford Univ. Press (1975), 323-348.

[Vin4] E.B. Vinberg, *"Absence of crystallographic groups of reflections in Lobachevskii spaces of large dimension"*, Funct. Ana. and its App. **15** (1981), 128-130.

[Vin5] E.B. Vinberg, *"The absence of crystallographic groups of reflections in Lobachevskii spaces of large dimension"*, Trans Moscow Math. Soc. **47** (1985), 75-112.

[Vin6] E.B. Vinberg, *"Hyperbolic reflection groups"*, Russian Math. Surveys **40** (1985), 29-66.

Geodesics and Curvature in Metric Simplicial Complexes

MARTIN R. BRIDSON[1]
Cornell University

ABSTRACT: We prove that if a metric simplicial complex has only finitely many isometry types of cells then it is a complete geodesic metric space. We then analyse the geometry of such complexes, and establish the equivalence of a variety of characterisations of non-positive curvature (both local and global), characterisations which are due to Gromov, Alexandrov, Bruhat and Tits, and others. These results provide new techniques for studying simplicial group actions on complexes which are not necessarily locally finite. Specific examples are given to illustrate how one can establish the existence, or non-existence, of a metric of non-positive curvature on a given complex. In particular, we prove that the Culler-Vogtmann complex does not support an $Out(F_n)$-equivariant metric of non-positive curvature for $n \geq 3$.

Introduction

Groups which act cocompactly on simplicial trees were completely classified by the work of Bass and Serre [24]. The elegance of this theory is such that the prospect of extending it to higher dimensions is an extremely enticing one. As a first step one must identify a suitable higher-dimensional analogue of a tree. A strong candidate for this role is provided by the class of *non-positively curved piecewise Euclidean complexes* (which are defined below). The study of such complexes is not new, but has been brought to the fore in recent years by Gromov, who made extensive use of these spaces in his remarkable work on hyperbolic groups.

In his seminal article [17] Gromov (pp. 119–120) states several theorems which are important to the understanding of the geometry of these complexes — theorems which concern the existence of geodesics, and the relationship between local and global definitions of non-positive curvature in simply connected spaces. However, until now the validity of these results had only been established for locally finite complexes. In the study of groups acting on trees the spaces under consideration are not required to be locally finite, and such a restriction would be very limiting in the higher-dimensional case. Thus we are presented with a serious technical difficulty.

In Sections 1–3 of this paper we remove this difficulty by proving these theorems for a large class of spaces. This class includes any piecewise Euclidean complex which admits a cocompact action by a group of isometries. We also relate Gromov's ideas to earlier work of others, notably Bruhat and Tits [10], and Alexandrov [2].

These results allow us to analyse the structure of groups which act cocompactly on a simplicial complex by using curvature to convert local (combinatorial) information about the complex into global information which relates to the group action. In

[1] Partially supported by an Alfred P. Sloan Doctoral Dissertation Fellowship.

Sections 4 and 5 we give a number of examples to show how, in the presence of sufficient local information about the complex, one can establish the existence, or non-existence, of a metric of non-positive curvature.

In order to state our results we need the following definitions. A *geodesic metric space* is a metric space in which every pair of points can be joined by a *geodesic segment* — a topological arc which, with the induced metric, is isometric to a closed interval of the real line. A large class of examples is provided by *piecewise Euclidean complexes:* Given a simplicial complex K one can metrize the cells of the geometric realisation of K so that each cell is isometric to a Euclidean simplex, and if two cells intersect then the induced metrics on their common face agree. There is a well defined notion of piecewise linear (PL) paths in K and a consistent way of measuring their length. The *intrinsic pseudometric* on K is defined by setting the distance between two points equal to the greatest lower bound on the length of PL paths joining them.

If K is connected and locally finite then this pseudometric is actually a metric, and if K is finite then this construction yields a geodesic metric space. Moreover, this construction works equally well if one replaces Euclidean simplices by hyperbolic or spherical ones, and we use the term *metric simplicial complex* to refer to a complex which is piecewise-Euclidean, -spherical, or -hyperbolic.

If the complex K is not locally finite then in general the intrinsic pseudometric is not a metric. A simple example is the 1–complex with two vertices joined by countably many edges (subdivided to make the complex simplicial), the n–th of which has length $1/n$. To avoid the type of limiting behaviour present in this example we restrict our attention to metric simplicial complexes with only finitely many isometry types of cells.

Theorem 1.1: *If K is a metric simplicial complex with only finitely many isometry types of cells then K is a complete geodesic metric space.*

We say that K is of *type B* if it satisfies the hypothesis of Theorem 1.1. We do *not* require spaces of type B to be locally compact. We shall also consider (locally finite) metric simplicial complexes all of whose closed bounded subsets are compact. Such complexes are said to be of *type A*.

Any finite-dimensional simplicial complex can be metrized as a piecewise Euclidean complex of type B by metrizing each cell as a regular simplex of unit edge length. It then satisfies the hypothesis of our Theorem 1.1 and hence is a complete geodesic metric space. Any simplicial action on the space becomes an action by isometries, and the potential exists for studying groups which act in this way via the geometry of geodesics in the complex. This technique works particularly well in the presence of *non-positive curvature*.

Our Main Theorem, which is stated below, establishes the equivalence of various characterisations of non-positive curvature for $M(\kappa)$–complexes of type A or B, where $\kappa \leq 0$ denotes the sectional curvature of the simplices in the given complex.

For clarity of exposition, we state and prove our results for piecewise Euclidean complexes (Section 2) before generalising to the case $\kappa \leq 0$ (Section 3).

In Section 2.1 we study complexes for which there is a unique shortest path between every pair of points, and obtain the following convexity result.

Theorem 2.1: *If K is a piecewise Euclidean complex of type A or B which has unique geodesic segments then any geodesic segments α_0 and α_1 with $\alpha_0(0) = \alpha_1(0)$ satisfy*

$$d(\alpha_0(t), \alpha_1(t)) \leq t \cdot d(\alpha_0(1), \alpha_1(1)) \quad \forall t \in [0,1].$$

We then show, in Section 2.2, that K has unique geodesic segments if and only if it has non-positive curvature as defined by the CAT(0) inequality of Gromov [17], the CN inequality of Bruhat-Tits [10], and Alexandrov's condition on the excess of geodesic triangles:

Theorem 2.7: *If K is a piecewise Euclidean complex of type A or B then the following global characterisations of non-positive curvature are equivalent:*

I) *K has unique geodesic segments.*
II) *K satisfies CAT(0).*
III) *K satisfies CN.*
IV) *Every geodesic triangle in K has non-positive excess.*

This leads to a fixed point theorem in the manner of Bruhat-Tits. In dimension 2 this fixed point theorem is due to Gersten [15] and plays a central role in his work with Stallings on triangles of groups [26].

Fixed Point Theorem: *If K satisfies any of the conditions* I *to* IV *given in Theorem 2.7, and a group Γ acts on K by isometries, such that there is a bounded orbit, then the fixed point set of Γ is non-empty and contractible.*

In Section 2.3 we describe the relationship between local definitions of curvature and prove:

Theorem 2.8: *If K is a piecewise Euclidean complex of type A or B then the following local characterisations of non-positive curvature are equivalent:*

I) *K has unique geodesic segments locally.*
II) *The metric on K is convex locally.*
III) *K satisfies CAT(0) locally.*
IV) *K satisfies CN locally.*
V) *Every point of K has a neighbourhood such that any geodesic triangle in that neighbourhood has non-positive excess.*
VI) *K satisfies the link condition.*

Further, if K is of type B then each of the above conditions is equivalent to

VII) *There exists $\epsilon_0 > 0$ such that for all $x \in K$ the ball $B_{\epsilon_0}(x)$ is geodesically convex and has unique geodesic segments.*

In Section 2.4 we prove the following theorem, which provides the vital link between local and global definitions of non-positive curvature in a simply connected space.

Theorem 2.12: *If K is a piecewise Euclidean complex of type A or B which has unique geodesic segments locally then for every pair of points $x, y \in K$ there is a unique shortest path in each homotopy class of paths from x to y in K.*

This result, for a restricted class of locally-finite complexes, is due to Stone [27]. The strategy of our argument is modelled on the proof of a similar result for smooth manifolds, which can be found in Milnor's book on Morse theory [22].

In Section 3 we generalise the preceding results to $M(\kappa)$−simplicial complexes and obtain:

Main Theorem: *If K is a simply connected $M(\kappa)$−simplicial complex of type A or B and $\kappa \leq 0$ then the following 13 conditions are equivalent:*

Global conditions:

I) *K has unique geodesic segments.*
II) *K satisfies $CAT(\kappa)$ globally.*
III) *K satisfies $CAT(\chi)$ globally, for some χ.*
IV) *K satisfies CN globally.*
V) *The metric on K is convex.*
VI) *Every geodesic triangle in K has non-positive excess.*

Local conditions:

VII) *K has unique geodesic segments locally.*
VIII) *K satisfies $CAT(\kappa)$ locally.*
IX) *K satisfies $CAT(\chi)$ locally, for some χ.*
X) *K satisfies CN locally.*
XI) *The metric on K is convex locally.*
XII) *Every point of K has a neighbourhood such that any geodesic triangle in that neighbourhood has non-positive excess.*
XIII) *K satisfies the link condition.*

The *link condition* listed as condition XIII) is essentially a condition on the combinatorics of the links of vertices in K. Thus, if we have sufficient local (combinatorial) information about the space K then we can deduce global (topological and algebraic) information. (e.g. If K is a non-positively curved complex of type A or B then its universal cover is contractible.) In Section 4 we give a number of examples where one can verify the link condition directly. In particular we discuss the work of Gersten and Stallings on triangles of groups. We also work through a specific example to show how the Fixed Point Theorem can lead to a classification of finite subgroups in a group which acts by isometries on a complex of non-positive curvature.

In Section 5 we describe a technique for deciding whether or not a complex can be given a metric of non-positive curvature which is equivariant with respect to a given group action. Using this technique we prove the following result.

Theorem 5.6: *If $n \geq 3$ then there does not exist an $Out(F_n)$−equivariant piecewise Euclidean (or piecewise hyperbolic) structure of non-positive curvature on the Culler-Vogtmann complex K_n.*

The Culler-Vogtmann complex is closely analogous to the Teichmüller space of a Riemann surface, and Theorem 5.6 provides an interesting analogue of known results about the curvature of Teichmüller space (see [21]).

The study of curvature in locally finite complexes, particularly PL manifolds, is not new. Work in this area was done by Banchoff and Stone in the sixties, and more recently by Gromov and Thurston [18]. Aitchison and Rubinstein [1] have shown that this theory has powerful applications in the geometry of 3–manifolds. We should also note that in the case of locally finite complexes Ballman ([16], Chapter 10) has given a lucid account of the theorems of Gromov to which we referred at the beginning.

The role which non-positive curvature plays in the study of orbihedra is highlighted by results of Gromov ([17], pp. 127–130), which have been explained in detail by Haefliger ([16], Chapter 11). Recently Haefliger [19] has generalised these results using the techniques which we introduce in this paper.

Contents:

1. The Existence of Geodesics in Metric Simplicial Complexes

In this section we prove the following theorem:

Theorem 1.1: *If K is a metric simplicial complex with only finitely many isometry types of cells then K is a complete geodesic metric space.*

Notice that we do not require the underlying simplicial complex K to be locally finite. In fact, if K is locally finite then we can weaken the condition that K has only finitely many isometry types of cells, and instead require only that in the intrinsic metric every closed bounded subset of K is compact. Then one can prove that K is a geodesic metric space by using the Arzela-Ascoli Theorem to show that the Birkhoff curve-shortening process converges (cf. [16] Chapter 10). Alternatively, one can view the existence of geodesics in this case as a formal consequence of the case where K has finitely many isometry types of cells; since for any two points $x, y \in K$ with $d(x, y) < N$ we need only consider paths between them which lie in the minimal subcomplex of K containing the ball of radius N about x, and this is a finite complex.

This section is organised as follows: In Section 1.1 we give a precise definition of a metric simplicial complex K, and define the intrinsic pseudometric d in terms of *m–chains* (which provide a useful method for describing piecewise-geodesic paths combinatorially). In Sections 1.2–1.5 we establish the existence of geodesic segments in K, and in Section 1.6 we prove that the metric d is complete. For 2–dimensional complexes our proof simplifies considerably, and in Section 1.7 we give the details in this restricted setting to illustrate the ideas involved in the proof of the general case.

1.1 Definitions

A *geodesic segment* in a metric space (X, d) is a topological arc which is isometric to a closed interval of the real line. (So in particular the length of a geodesic segment is equal to the distance between its endpoints.) (X, d) is a *geodesic metric space* if every pair of points in X can be connected by a geodesic segment.

An n*−simplex* in Euclidean n−space $\mathbf{E^n}$ or hyperbolic n−space $\mathbf{H^n}$ is the convex hull of $(n+1)$ points in general position. An n*−simplex* in spherical n−space $\mathbf{S^n}$ is the intersection of $\mathbf{S^n}$ with the positive cone spanned by $n+1$ linearly independent vectors in $\mathbf{R^{n+1}}$.

An **E** (respectively **S**, **H**) *simplicial complex* consists of the following information:

I) An (abstract) simplicial complex K.

II) A set $Shapes(K)$ of Euclidean (respectively spherical, hyperbolic) simplices $\sigma_i \subset \mathbf{E}^{\mathbf{n}_i}$ (respectively $\mathbf{S}^{\mathbf{n}_i}$, $\mathbf{H}^{\mathbf{n}_i}$).

III) For every closed simplex $B \subset K$ a simplicial isomorphism $f_B : B \to \sigma(B)$ where $\sigma(B) \in Shapes(K)$ and

$$f_B \circ \left(f_C|_{(B\cap C)}\right)^{-1}$$

is an isometry for all simplices $B, C \subset K$.

Similarly, for any value of κ, we can define an $M(\kappa)$*−simplicial complex*, by requiring that $Shapes(K)$ consist of simplices of constant curvature κ. A *metric simplicial complex* is an $M(\kappa)$−simplicial complex, for some κ. We say that K is *connected* if it is connected in the weak topology.

Remark: It is purely for convenience that we work with simplicial complexes rather than more general complexes in which the cells are metrized as convex polyhedra, and since any such polyhedral complex can be made simplicial by subdivision this involves no loss of generality.

A *line segment* in K is the inverse image of a geodesic arc in $f_B(B)$ for some simplex B. By a *PL path* in K we mean a path which is the concatenation of finitely many such line segments. Notice that the maps f_B induce metrics d_B on the individual simplices of K. These metrics agree on faces of intersection and give a well defined notion of length for PL paths in K. To describe these paths combinatorially we use the following definition:

An *m-chain from x to y* is an $(m+1)$−tuple $C = (x_0, x_1, \ldots, x_m)$ of points in K, such that $x = x_0$, $y = x_m$, and for each i there exists a simplex $B(i)$ containing x_i and x_{i+1}. We call m the *size* of C, and define the *length* of C to be:

$$\lambda(C) = \sum_{i=0}^{m-1} d_{B(i)}(x_i, x_{i+1}).$$

Every m-chain determines a PL path in K, given by the concatenation of the line segments $[x_i, x_{i+1}]$. We denote this path $p(C)$.

If K is connected we can define a distance function:

$$d(x,y) = inf\,\{\lambda(C) : C \text{ a chain from } x \text{ to } y\}.$$

We call d the *intrinsic pseudometric* on K. The example given in the introduction shows that in general d does not define a metric. However, in Section 2 we show that if K is any connected metric simplicial complex with $Shapes(K)$, the set of isometry classes of cells of K, *finite* then this formula does define a metric on K. We then refer to d as the *intrinsic metric* on K.

For the remainder of Section 1 we fix an arbitrary κ and write M in place of $M(\kappa)$. We always assume that K is a connected M−simplicial complex, and that $Shapes(K)$ is a finite set. The letter d always denotes the intrinsic pseudometric on K.

1.2 The Intrinsic Metric

We begin by proving that $d(x, y)$ defines a metric on K. The fact that it is a pseudometric is immediate, the only difficulty is in showing that the distance between distinct points is nonzero.

Notation: For $x \in K$ we denote the open star of x (i.e., the union of the interiors of the cells containing x) by $st(x)$, and denote the closed star of x by $St(x)$.

Given $x \in K$ let $\epsilon(x) = inf\{d_B(x, B - st(x)) : B \text{ a closed simplex in } K, x \in B\}$.

Lemma 1.2: *$\epsilon(x) > 0$ for all $x \in K$.*

Proof: Fix x and let B_0 denote the unique simplex of K which contains x in its interior. Define an equivalence relation on the closed simplices of K that contain x by $B \sim B'$ if and only if there is an isometry from B to B' which restricts to the identity on B_0. Notice that $d_B(x, B - st(x))$ is well-defined on equivalence classes.

Because $Shapes(K)$ is finite, there are only finitely many equivalence classes under $\sim$. Choose representatives $\{B_1, \ldots, B_m\}$ for these classes. Then

$$\begin{aligned}\epsilon(x) &= inf\,\{d_B(x, B - st(x)) : B \text{ a closed simplex in } K,\ x \in B\}\\ &= min\,\{d_{B_i}(x, B_i - st(x)) : i = 1, \ldots, m\}\end{aligned}$$

and hence $\epsilon(x) > 0$. □

The positivity of the metric now follows from Lemma 1.3, which also shows that for every simplex B the metric d agrees with d_B locally. However, it should be noted that in general the metric which d induces on B is *not* equal to the metric d_B, but instead we have the inequality $d_B \geq d$. But this difficulty (which is illustrated by the following example) is a minor one, which can be rectified by taking a suitable barycentric subdivision of the model simplices in $Shapes(K)$ and giving K the induced simplicial structure.

Example: Consider the 1–complex K with vertices $\{x\}, \{y\}, \{z\}$ and edges $\{x, y\}, \{x, z\}, \{y, z\}$ of length 1, 1 and 4 respectively. If we let $B = \{y, z\}$ then $d_B(a, b) > d(a, b)$ whenever $d_B(a, b) > 3$. However, for every simplex B' in the first barycentric subdivision of K we have $d_{B'}(a, b) = d(a, b)$, for all $a, b \in B'$.

Lemma 1.3: *If $x \in K$ and $d(x, y) < \epsilon(x)$ then there exists a simplex B containing both x and y such that $d_B(x, y) = d(x, y)$.*

Proof: It is enough to show that there exists a simplex B such that $x, y \in B$ and $d_B(x, y) \le d(x, y)$. This follows immediately from the following assertion, which we prove by induction on m, the size of the given chain:

If $C = (x_0, \ldots, x_m)$ is an m–chain in K and $\lambda(C) < \epsilon(x_0)$ then $x_i \in st(x_0)$ for all i, and for some (and hence any) simplex $B(m)$ containing x_0 and x_m the inequality $d_{B(m)}(x_0, x_m) \le \lambda(C)$ holds.

The case $m = 1$ follows immediately from the definition of $\epsilon(x_0)$. For $m > 1$: Because C is a chain there is a simplex $B(m)$ containing both x_{m-1} and x_m. Applying our inductive hypothesis to $C' = (x_0, \ldots, x_{m-1})$, we may assume that $x_{m-1} \in st(x_0)$. Hence $x_0 \in B(m)$, and

$$\begin{aligned} d_{B(m)}(x_0, x_m) &\le d_{B(m)}(x_0, x_{m-1}) + d_{B(m)}(x_{m-1}, x_m) \\ &\le \lambda(C') + d_{B(m)}(x_{m-1}, x_m) \\ &= \lambda(C) \\ &< \epsilon(x_0). \end{aligned}$$

It follows from the definition of $\epsilon(x_0)$ that $x_m \in st(x_0)$. This completes the induction. □

Note: The metric topology which we have constructed on K coincides with the topology given by barycentric coordinates, and hence is strictly smaller than the weak topology except when K is locally finite. However, the identity map from K equipped with the weak topology to K equipped with the metric topology is a homotopy equivalence [14]. *This latter observation is important because it implies that if K is simply connected and has non-positive curvature then it is contractible in both the metric topology and the weak topology.*

1.3 The Existence of Shortest m-chains

Suppose that $C = (x_0, x_1, \ldots, x_m)$ is an m–chain from x to y in K satisfying $\lambda(C) = d(x, y)$. Then the PL path obtained by concatenation of the line segments

$[x_i, x_{i+1}]$ is a geodesic segment. Therefore, to prove that K is a geodesic metric space it is enough to show that for all $x, y \in K$ the infimum in the definition of $d(x, y)$ is attained. This we do in two stages. Firstly, in this section, we show that if for a fixed integer m there exists some m−chain from x to y in K then there exists an m-chain of minimal length from x to y in K. Then, in Section 1.5, we show that there exists a linear function f such that for every pair of points $x, y \in K$

$$\begin{aligned} &inf\{\lambda(C) : C \text{ an } m\text{−chain from } x \text{ to } y,\ m > 0\} \\ =&inf\{\lambda(C) : C \text{ an } m\text{−chain from } x \text{ to } y,\ m \leq f(d(x,y))\}. \end{aligned}$$

It follows that the infimum in the definition of $d(x, y)$ is attained, and hence K is a geodesic metric space.

Lemma 1.4: *(Moussong) If L is a finite M-simplicial complex, and two points x and y can be joined by an m-chain in L, where m is a fixed integer, then there is a shortest m-chain from x to y in L.*

Proof: Let $X \subset L^{m+1}$ denote the set of m-chains from x to y in L. We show that X is closed and hence compact. The length function λ on m-chains is continuous on X, and therefore attains a minimum if X is closed.

Notice that for any $w, z \in L$, the set $st(w) \cap st(z)$ is empty if and only if there is no closed cell of L containing both w and z. Thus, if $\mathbf{z} = (z_0, z_1, \ldots, z_m)$ is *not* an m-chain then for some i between 0 and $m-1$ we have $st(z_i) \cap st(z_{i+1}) = \emptyset$. Hence $\mathbf{z}$ has a neighbourhood disjoint from X, namely $(L \times \ldots L \times st(z_i) \times st(z_{i+1}) \times L \times \ldots \times L)$. So X is closed. □

Given K with $Shapes(K)$ finite, and a positive integer m, one can build a finite set of *"models"*. That is, connected complexes obtained by taking at most m (not necessarily distinct) simplices from $Shapes(K)$ and identifiying faces by isometries. Any subcomplex K_0 of K which can be expressed as the union of at most m closed cells must be isometrically isomorphic to one of these models. The existence of this finite set of models allows us to pass from the case of compact complexes (Lemma 1.4) to the case of interest, complexes with $Shapes(K)$ finite (Lemma 1.5).

Notice that K_0 with the induced metric from K is not in general isometric to the model with its intrinsic metric. However, since the length of a chain is defined in terms of the local metrics d_B, a given m-chain in K_0 and the corresponding m-chain in the model have the same length. This is the key to the following lemma.

Lemma 1.5: *If K is an M-simplicial complex with $Shapes(K)$ finite, and two points x and y can be joined by an m-chain in K, where m is a fixed integer, then there is a shortest m-chain from x to y in K.*

Proof: For any fixed pair of elements x and y there are only finitely many bipointed models, $(K'; x', y')$, for $(K_0; x, y)$ as K_0 runs over all subcomplexes of K which contain both x and y and can be expressed as the union of at most m closed cells. Thus, any m-chain from x to y in K corresponds to an m-chain of the same length from x' to y' in one of the finitely many models under consideration, and vice versa. The present lemma now follows by application of Lemma 1.4 to each of these models. □

Remark: Suppose that one were to weaken the condition that $Shapes(K)$ is finite, and require instead that there be uniform bounds on how small and thin the simplices in $Shapes(K)$ may be. Then a straightforward adaptation of the arguments given in Section 1.2 shows that this weaker assumption is sufficient to ensure that the intrinsic pseudometric on K is indeed a metric. However, under this weaker hypothesis Lemma 1.5 is no longer true, and in general if K is locally inifinite then it is *not* a geodesic metric space, as the following example shows.

Example: We construct a Euclidean 2–complex L as follows: L has a subcomplex L' consisting of three vertices $\{x\}, \{y\}$ and $\{z\}$, and two edges $e_y = \{x, y\}$ and $e_z = \{x, z\}$, both of which are metrized to have unit length. For every integer $n \geq 2$ we metrize a barycentrically subdivided 2–simplex $\sigma(n)$ as an isosceles triangle with two sides of unit length meeting at an angle $(\frac{\pi}{3} + \frac{\pi}{n})$. We then attach each $\sigma(n)$ to L' by identifying one of its edges of unit length with e_y, and the other with e_z. The result is a Euclidean simplicial complex L which is not a geodesic metric space. To see this notice that $d(y, z) = 1$ whereas every m–chain from y to z has length strictly larger than 1. In fact, there exist 2–chains C from y to z with $\lambda(C)$ arbitrarily close to $1 = d(y, z)$, so the conclusion of Lemma 1.5 fails to hold in this case.

1.4 Taut Chains and Local Geodesics

In this subsection we introduce the notion of a *taut chain*. Intuitively speaking, this is an $m-$chain whose length cannot be shortened by perturbation unless one allows the integer m to increase. In Lemma 1.6 we show that the minimising chains yielded by Lemma 1.5 are taut. Then in Section 1.5 we formalise the idea that the size of a taut chain is directly related to its length, and this leads to the desired bound on the size of chains which one must consider when seeking a geodesic segment between a given pair of points in K.

In the present subsection we also define what it means for an $m-$chain to be a local geodesic. Local geodesics do not play an essential role in the proof of Theorem 1.1, but they do (as the name suggests) describe the local behaviour of geodesic segments, and the understanding which this description provides proves to be important in later sections. In fact, in Section 2 we show that in simply-connected complexes of non-positive curvature there is a one-to-one correspondence between geodesic segments and taut local geodesics. This observation provides a link between the results presented here and the work of Gersten and Stallings on non-positively curved 2–complexes [26], where they proved the existence of geodesics using a local criterion.

Before defining what it means for an m-chain to be taut, we must make some observations about small subcomplexes of K. Suppose that B and B' are closed simplices in K which have non-empty intersection, and consider $L = B \cup B'$ equipped with its intrinsic metric. (Notice that in general the intrinsic metric on L is not equal to restriction of the intrinsic metric on K.) Suppose that x and y lie in the same simplex of L. Then it follows immediately from the triangle inequalities for d_B and $d_{B'}$ that the line segment $[x, y]$ is a geodesic segment in L. (Here we are using the fact that K is simplicial, and thus if $x, y \in B \cap B'$ then the line segments joining x to y in B and B' coincide.) From this it follows that if $x \in B$ and $y \notin B$ then $d(x,y) = inf\{d_B(x,z)+d_{B'}(z,y) : z \in B \cap B'\}$. And since $z \mapsto d_B(x,z)+d_{B'}(z,y)$ is a continuous function on the compact set $B \cap B'$ it attains a minimum. Therefore L is a geodesic metric space, and the minimal m-chain associated to any geodesic segment has size at most 3.

Definition: An m-chain $C = (x_0, x_1, \ldots, x_m)$ in K is *taut* if it satisfies the following two conditions for $i = 1, \ldots, m-1$: Firstly, there is no simplex containing $\{x_{i-1}, x_i, x_{i+1}\}$. Secondly, if $x_{i-1}, x_i \in B(i)$ and $x_i, x_{i+1} \in B(i+1)$ then the concatenation of the line segments $[x_{i-1}, x_i]$ and $[x_i, x_{i+1}]$ is a geodesic segment in $L = B(i) \cup B(i+1)$.

Notice that if a chain is taut then only its first and last entries can lie in the interior of a top dimensional simplex of K.

Lemma 1.6: *For a fixed integer m, if C is an m-chain from x to y in K of minimal length then $p(C)$, the path in K determined by C, is the path determined by some taut n-chain with $n \leq m$.*

Proof: Let $C = (x_0, x_1, \ldots, x_m)$. Suppose that for some i there exists a simplex B containing x_{i-1}, x_i and x_{i+1}, and let C^i denote the $(m-1)$-chain obtained from C by deleting the entry x_i. The triangle inequality for d_B gives $d_B(x_{i-1}, x_{i+1}) \leq d_B(x_{i-1}, x_i) + d_B(x_i, x_{i+1})$, with equality if and only if x_i lies on the line segment $[x_{i-1}, x_{i+1}]$. Thus $\lambda(C^i) \leq \lambda(C)$. But $\lambda(C)$ is minimal, so in fact $\lambda(C^i) = \lambda(C)$, and hence x_i must lie on the line segment $[x_{i-1}, x_{i+1}]$. This implies that C^i determines the same path as C. We can repeat this procedure until no simplex of K contains three successive entries of the resulting chain — the first condition for tautness.

We now show that C satisfies the second condition for tautness. Let B and B' be any cells containing $\{x_{i-1}, x_i\}$ and $\{x_i, x_{i+1}\}$ respectively. Every geodesic segment in the complex $L = B \cup B'$ can be expressed as the concatenation of at most two line segments. So if the concatenation of the line segments $[x_{i-1}, x_i]$ and $[x_i, x_{i+1}]$ were not a geodesic segment in $L = B(i) \cup B(i+1)$, then we could replace x_i by some $x_i' \in B \cap B'$ to obtain an m-chain from x_0 to x_m which would be strictly shorter than C, contradicting the minimality of $\lambda(C)$. □

As an immediate consequence of Lemma 1.6 we obtain:

Corollary 1.7: *$d(x,y) = inf\{\lambda(C) : C$ a taut chain from x to $y\}$.*

We now turn our attention to local geodesics. In order to describe the local behaviour of PL paths we need a precise notion of angle between line segments in K. For this we introduce a spherical metric on the link of each point in K.

Let B be an $n-$simplex in M^n (the unique simply-connected, complete $n-$manifold of constant sectional curvature κ). Given $x \in B$, the tangent cone of B at x is defined to be $T_xB = \{v \in T_xM^n : exp_x(\epsilon v) \in B, some\ \epsilon > 0\}$. We identify $\mathbf{S}^{n-1}$ with the unit sphere in T_xM^n, and define $LK(x,B)$ to be $T_xB \cap \mathbf{S}^{n-1}$. This is a simplex in $\mathbf{S}^{n-1}$ if x is a vertex of B, the whole of $\mathbf{S}^{n-1}$ if x is an interior point of B, and otherwise a closed convex subset of a hemisphere in $\mathbf{S}^{n-1}$.

Now suppose $x \in B \subseteq K$. We define T_xB to be $T_{f_B(x)}f_B(B)$ and $LK(x,B)$ to be $LK(f_B(x), f_B(B))$.

Definition: The *geometric link* $LK(x,K)$ of x in K is defined to be the disjoint union of the cells $\{LK(x,B) : x \in B \subseteq K\}$, modulo the natural identifications. Specifically, $u \in LK(x,B)$ is identified with $v \in LK(x,B')$ if and only if the differential of $f_{B'}f_B^{-1}$ maps u to v. T_xK is defined similarly.

$LK(x,K)$ is a spherical complex, which can be made simplicial by subdivision. Further, since $Shapes(K)$ is finite, so too is $Shapes(LK(x,K))$. In particular we have an intrinsic metric on connected components of $LK(x,K)$.

Definition: An m-chain $C = (x_0, x_1, \ldots, x_m)$ in K is a *local geodesic* if $x_i \neq x_{i+1}$ for all i, and for $i = 1, \ldots, m-1$ the distance between the points of $LK(x_i, K)$ determined by the line segments $[x_{i-1}, x_i]$ and $[x_i, x_{i+1}]$ is at least π. (If these points lie in different path components of $LK(x,K)$ then we say that they are an infinite distance apart.)

We now prove that if the path associated to an $m-$chain is a geodesic segment then the chain is a local geodesic. Intuitively this is clear, since the definition of a local geodesic is just a precise statement of the fact that the path determined by the given chain has no sharp corners. If such a sharp corner did exist then one could shorten the path by "cutting the corner". The following proof, which is due to Moussong [23], is simply a restatement of this fact in more precise language.

Lemma 1.8: *Let $C = (x_0, x_1, \ldots, x_m)$ be an $m-$chain in K, and suppose that $x_i \neq x_{i+1}$ for all i. If $p(C)$, the path determined by C, is a geodesic segment then C is a local geodesic.*

Proof: Let u and v denote the points of the spherical complex $LK(x_i, K)$ which are determined by the line segments $[x_{i-1}, x_i]$ and $[x_i, x_{i+1}]$ respectively.

If $d(u,v) < \pi$ in $LK(x_i, K)$ then there exists a sector S of the unit disc in $\mathbf{E}^2$, and a local isometry $\phi : S \to T_{x_i}L$ which maps the circular arc component of the boundary of S, which we call σ, to a piecewise geodesic path of length $< \pi$ joining u to v in $LK(x_i, K)$.

Consider the line in S which joins the endpoints of σ. This has length $2 - 3\delta$ for some positive δ. The map ϕ sends this line to a PL path from u to v in $T_{x_i}K$. The endpoints of the line segments in this path determine an m–chain $(a_0, a_1, \ldots, a_n)$ in $T_{x_i}L$.

On each cell of $T_{x_i}K$ we have the exponential map to the corresponding cell of K, and where two cells meet this is well defined on their common face. So for sufficiently small ϵ we have the following chain joining x_0 to x_m in K: $C' = (x_0, \ldots, x_i, exp_x(\epsilon a_0), \ldots, exp_x(\epsilon a_n), x_{i+1}, \ldots, x_m)$. The map ϕ is a local isometry, and we can choose ϵ so that the exponential map restricted to the ϵ–neighbourhood of the origin increases distances by at most a factor of $1/(1-\delta)$. Hence $\lambda(C') < \lambda(C) - \delta$, contradicting the fact that $p(C)$ is a geodesic segment. □

Remark: Suppose that $C = (x_0, \ldots, x_m)$ is taut. Lemma 1.6, together with the second condition for tautness, implies that for $i = 1, \ldots, m-1$ the distance between the points of $LK(x_i, B(i) \cup B(i+1))$ determined by the line segments $[x_{i-1}, x_i]$ and $[x_i, x_{i+1}]$ is at least π. In fact, if x_i is not a vertex of $B(i) \cap B(i+1)$ then this distance is exactly π, and it follows that the image of the concatenation of $[x_{i-1}, x_i]$ and $[x_i, x_{i+1}]$ under any local isometry into $M(\kappa)^n$ is a geodesic arc.

Example: Figure 1.1 illustrates the fact that a taut chain is not necessarily a local geodesic. Here K is a planar 2–complex with three 2–simplices, and the chain (a, b, c) is taut but not a local geodesic.

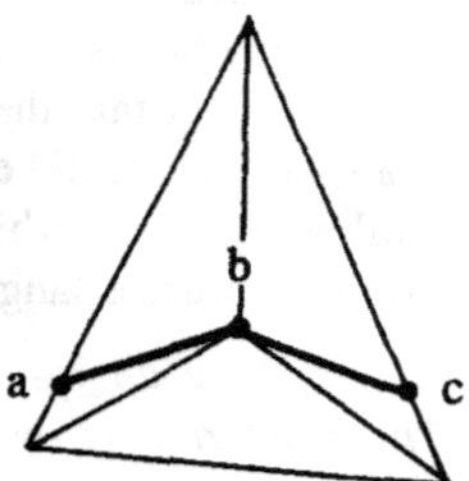

Figure 1.1: A taut chain which is not a local geodesic.

1.5 The Growth of Taut Chains

In this section we prove that for large m the length of taut $m-$chains grows at least linearly with respect to m (Theorem 1.11). Combining this with Corollary 1.7, we see that the infimum in the definition of $d(x, y)$ need only be taken over $m-$chains with m less than a certain linear function of $d(x, y)$. This, as we explained at the beginning of Section 1.3, completes the proof that K is a geodesic metric space.

We begin by noting a simple fact about spherical geometry. We shall use the term "spherical" to describe complexes whose cells are modelled on a sphere of some fixed radius, which may not be 1. The usage should be clear from the context.

Claim: If P is the vertex of a (2–dimensional) geodesic triangle $\triangle$ in $\mathbf{S^n}$ and S is the boundary of an $\epsilon-$neighbourhood of P in the usual (arclength) metric on $\mathbf{S^n}$, where ϵ is suitably small, then $\triangle \cap S$ is a geodesic arc on S.

Proof: The idea of the proof is to reduce to the case $n = 2$ where the result is clear. Consider $\mathbf{S^n}$ as the unit sphere in $\mathbf{R^{n+1}}$ centred at the origin O. Then S is an $(n-1)$-sphere of Euclidean radius $sin(\epsilon)$ whose centre, which we denote by Q, lies on the line OP. Let V be the three dimensional subspace of $\mathbf{R^{n+1}}$ determined by the vertices of $\triangle$. The intersection of $\mathbf{S^n}$ with V is a 2–sphere of unit radius centred at O. The line OP lies in V, so in particular Q does, and $S \cap V$ consists of those points on the unit 2–sphere in V which are a Euclidean distance $sin(\epsilon)$ from Q. Thus $S \cap V$ is a great circle on S and since $\triangle \cap S$ is an arc of this circle we are done. □

Note that if in the above Claim we replace $\mathbf{S^n}$ by $M(\kappa)^n$ (the unique complete simply-connected $n-$manifold of constant curvature κ) then the conclusion still holds. In the hyperbolic case this can be seen most easily by thinking of P as the centre of the Poincaré disc model.

Lemma 1.9: *Let K be an M-simplicial complex with $Shapes(K)$ finite. Then there exists a constant ϵ, depending only on $Shapes(K)$, such that for every vertex P of K the set $S(P) = \{x \in K : d(x, P) = \epsilon\}$ is an* **S**-*simplicial complex, whose dimension is one less than that of K.*

Proof: Define
$\epsilon = min\{d_\sigma(v, \tau) : \sigma \in Shapes(K), v$ a vertex of σ, and τ a face of σ with $v \notin \tau\}$.

With this choice of ϵ it follows from Lemma 1.3 that for every simplex $B \subseteq K$ the metric d_B agrees with d, the intrinsic metric on K, in the $\epsilon-$neighbourhood of each vertex of B. So $S(P)$ is the union of the sets $\{x \in B : d_B(P, x) = \epsilon, P \in B\}$. And by the preceding claim these are spherical simplices. □

Remark: Notice that there is a natural identification of $LK(P, K)$ with $S(P)$.

To complete the inductive step in our proof of Theorem 1.11 we shall need the following notion of radial projection from a vertex in K. The crucial property of this map is that it takes taut chains to taut chains (Lemma 1.10).

Suppose that x is a vertex of a simplex σ in M^n, and that $\epsilon < d(x, \sigma - st(x))$. Then there is a well defined notion of radial projection from x, taking $(\sigma - x)$ onto the intersection of σ with the boundary of the ϵ–ball about x.

Let P and ϵ be as in Lemma 1.9. For every simplex $B \ni P$ we let $f_B(B)$ and $f_B(P)$ play the roles of σ and x in the previous paragraph. We can then pull back the radial projection map, by means of the map f_B, to obtain a map from B onto $S(P) \cap B$. The pull-backs obtained in this way agree on common faces, and so combine to give a *radial projection map* from $St(P)$ to $S(P)$. We denote this map by ρ.

Lemma 1.10: *Let K and P be as in Lemma 1.9. If C is a taut chain in $st(P)$ which does not pass through P then the image of C under radial projection ρ from P to $S(P)$ is a taut chain.*

Proof: Let $C = (x_0, \dots, x_m)$, and let $C' = (\rho(x_0), \dots, \rho(x_m))$.

For every $k > 0$ the map ρ gives a 1–1 correspondence between the k–simplices of $S(P)$ and those $(k+1)$–simplices in $St(P)$ which meet P. If three sucessive entries $\rho(x_{i-1}), \rho(x_i), \rho(x_{i+1})$ of C' were to lie in some simplex of $S(P)$ then the set $\{x_{i-1}, x_i, x_{i+1}\}$ would be contained in the corresponding simplex of $St(P)$, contradicting the fact that C is taut. Thus C' satisfies the first condition for tautness.

To see that C' satisfies the second condition, consider simplices $\rho(B(i))$ and $\rho(B(i+1))$ which contain the line segments $[\rho(x_{i-1}), \rho(x_i)]$ and $[\rho(x_i), \rho(x_{i+1})]$ respectively. Then the simplices $B(i)$ and $B(i+1)$ contain the line segments $[x_{i-1}, x_i]$ and $[x_i, x_{i+1}]$ respectively. We can extend the line segments $[P, x_{i-1}]$ and $[P, x_i]$ until they meet $(B(i) - st(P))$, and the resulting line segments form two sides of a unique geodesic triangle in $B(i)$. (Here we mean a 2–dimensional triangle which is totally geodesic with respect to the metric $d_{B(i)}$.) We denote this triangle by Δ_i. In the same way, the line segments $[P, x_i]$ and $[P, x_{i+1}]$ determine a unique geodesic triangle Δ_{i+1} in $B(i+1)$. These triangles have a common edge, and we can map them into $M(\kappa)^2$ by a local isometry ϕ to obtain the planar 2–complex $k(x_i)$ shown in Figure 1.2.

Because C is taut, the path in $L = B(i) \cup B(i+1)$ determined by the 3–chain $C_i = (x_{i-1}, x_i, x_{i+1})$ is a geodesic segment, so its image under ϕ is a geodesic arc in $M(\kappa)^2$. Moreover, the angle θ_{x_i} which this arc subtends at $\phi(P)$ must have length less than π, because otherwise the path in $k(x_i)$ determined by the 3–chain $(\phi(x_{i-1}, \phi(P), \phi(x_{i+1}))$ would pull back to a path from x_{i-1} to x_{i+1} in L of length less than $\lambda(C_i)$, contradicting the fact that the path determined by C_i is a geodesic segment.

To prove that C satisfies the second condition for tautness we must show that the path which the 3–chain $C'_i = (\rho(x_{i-1}), \rho(x_i), \rho(x_{i+1}))$ determines in the spherical complex $L' = \rho(B(i)) \cup \rho(B(i+1))$ is a geodesic segment. The map ϕ, which we defined in the previous paragraph, sends this path isometrically onto an arc of the circle of radius ϵ about $\phi(P)$, allowing us to express $\lambda(C'_i)$ as a monotone function of the angle θ_{x_i}. If C_i' were not a geodesic segment in L' then there would be a shorter 3–chain $(\rho(x_{i-1}), \rho(y), \rho(x_{i+1}))$ from $\rho(x_{i-1})$ to $\rho(x_{i+1})$ in L'. But then the vertex angle θ_y in the corresponding planar 2–complex $k(y)$ would be strictly smaller than θ_{x_i}, which in turn would imply that the geodesic segment joining the images of x_{i-1} and x_{i+1} in $k(y)$ would pull back to a path from x_{i-1} to x_{i+1} in L of length less than $\lambda(C_i)$. The existence of such a path would contradict the fact that C_i is a geodesic segment in L. Hence C_i' must be a geodesic segment in L'. □

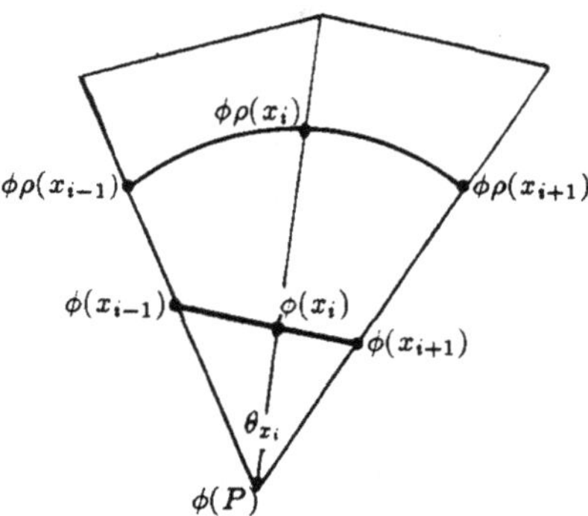

Figure 1.2: The complex $k(x_i)$.

Remark: Using simliar methods one can show that the property of being a local geodesic is also preserved under the radial projection map ρ.

We are now in a position to prove Theorem 1.11. The idea of the proof is to use induction to show that a large taut chain cannot be contained in the neighbourhood of any point. We first use a colouring argument to localise the problem to the star of a single vertex, then we radially project from the vertex to obtain a taut chain in a complex of lower dimension, and use our inductive hypothesis to complete the proof.

Theorem 1.11: *Let K be an M-simplicial complex with $Shapes(K)$ finite. Then there exist constants N and α, depending only on $Shapes(K)$, such that every taut chain of size at least N has length at least α.*

Proof: We proceed by induction on the dimension of K. It is important to the induction that the constants N and α depend only on $Shapes(K)$ and not on the global structure of K itself. For 1–dimensional complexes the result is clear, and the two dimensional case is included as an example in Section 1.7. Assume that the result holds for dimension $n-1$, and that K has dimension n.

In Lemma 1.9 we defined a constant ϵ, depending only on $Shapes(K)$, such that the ϵ–neighbourhood of any vertex $P \in K$ is contained in $st(P)$. We let $\eta_0 = \epsilon/3$

and paint the η_0–neighbourhood of each vertex of K with colour γ_0. Then, for $i = 1, 2, \ldots, n-2$ we paint the η_i–neighbourhood of each i–simplex in K with colour γ_i (except that we do *not* paint over points already coloured with a colour of lower index). The constants η_i (which we define below) depend only on $Shapes(K)$.

We define the constants η_i inductively. At each stage we require that η_i is small enough to ensure that the γ_i–regions corresponding to distinct i–cells are disjoint. It then follows that each point $x \in K$ which is painted with colour γ_i lies in the η_i–neighbourhood of a *unique* i–simplex, and we make the additional requirement that the distance from x to the link of the barycentre of this unique i–simplex must be at least η_i. In the previous paragraph we defined η_0 so that these conditions hold in the case $i = 0$. Suppose now that we have defined η_{i-1} with the desired properties. Consider a simplex $\sigma \in Shapes(K)$ of dimension at least $i+1$, and let τ^i be an i–dimensional face of σ with barycentre $b_0(\tau^i)$. The complement in τ^i of the open η_{i-1}–neighbourhood of its $(i-1)$–skeleton is compact, and we have chosen η_{i-1} so that this set is non-empty. Let η_{τ^i} denote the distance from this set to the link of $b_0(\tau^i)$ in σ. Then $\eta_i = \frac{1}{3} min\{\eta_{\tau^i} : \tau^i \subseteq \sigma, \sigma \in Shapes(K)\}$ has the desired properties.

Notice that $\eta_{i+1} \leq \eta_i$ for all i. Thus if $x \in K$ is painted with colour γ_i, and b_0 is the barycentre of the unique i–simplex responsible for this colouring, then η_{n-2} is a lower bound on the distance from x to $(St(b_0) - st(b_0))$, the link of b_0 in K.

After we have painted a neighbourhood of the $(n-2)$–skeleton of K in this way, we paint the remainder of K white, which we call colour γ_∞.

Given an m–chain $C = (x_0, x_1, \ldots, x_m)$ we shall refer to the open line segments (x_i, x_{i+1}) as the intervals of C. If $i \neq 0$ or $(m-1)$ then we call (x_i, x_{i+1}) an interior interval. The colouring on K induces a colouring on the intervals of any chain in K by the rule: paint each interval with the colour γ_j where $j = min\{k$: some point of $[x_i, x_{i+1}]$ is coloured $\gamma_k\}$.

Every interval of C has the same length as a geodesic arc in some model simplex $\sigma \in Shapes(K)$. If C is taut then only its initial and terminal entries can lie in the open star of an adjacent entry, so for an interior interval the corresponding geodesic segment in σ must have its endpoints in distinct faces of $\partial\sigma$. Moreover, if the interval is painted white then the corresponding geodesic segment must be contained in the complement of the η_{n-2}–neighbourhood of the $(n-2)$–skeleton of σ. There is a lower bound on the length of such geodesic segments, and taking the minimum of these lower bounds as σ ranges over $Shapes(K)$ we obtain a lower bound, which we call ℓ, on the length of interior intervals of C which are painted white.

If a taut m-chain $(m \geq 3)$ is not entirely contained in the open star of the barycentre of any simplex in K, then either it must contain an interior interval which is painted white, or else it must contain a coloured interval and a point which is not in the star of the barycentre of the cell responsible for that colouring. Thus if we set $\alpha = min\{\ell, \eta_{n-2}, |2\pi\kappa|\}$ then Theorem 1.11 reduces to the following Claim.

Claim: Let κ denote the curvature of the model simplices for K. Suppose that b_0 is the barycentre of a simplex in K, and that C is a taut m–chain contained in $st(b_0)$. If $\kappa \leq 0$, or $\kappa > 0$ and $\lambda(C) < 2\pi\kappa$, then $m \leq N$, where N is a constant depending only on $Shapes(K)$.

Proof of claim: We wish to reduce to the case where b_0 is a vertex of K. If b_0 is not a vertex then we change the underlying simplicial structure of K by adding a vertex at b_0 and forming the simplicial join of b_0 with its link. Then for each simplex $B \subseteq st(b_0)$ we must add to $Shapes(K)$ the simplices resulting from the corresponding subdivision of $\sigma(B)$. (However, we do not delete $\sigma(B)$ from $Shapes(K)$, since in general there will exist simplices $B' \subseteq (K - st(B))$ with $\sigma(B) = \sigma(B')$, and these simplices have not been subdivided.) Let $K'(b_0)$ denote the resulting metric simplicial complex. Notice that as b_0 varies there are only finitely many possibilities for $Shapes(K'(b_0))$, and the set of these possibilities, which we denote $\{S_1, \ldots, S_r\}$, depends only on $Shapes(K)$.

Fix a barycentre b_0, and suppose that $Shapes(K'(b_0)) = S_j$. Let $C = (x_0, x_1, \ldots, x_m)$ be a taut m–chain in K satisfying the hypotheses of the Claim. Each of the line segments $[x_i, x_{i+1}] \subseteq K$ will be a PL path in $K'(b_0)$. Let $C^i = (x_i, x_i^1, \ldots, x_i^{n_i}, x_{i+1})$ denote the unique chain of minimal length representing this path, and let $\tilde{C} = (x_0, x_0^1, \ldots, x_0^{n_0}, x_1, \ldots, x_{m-1}^{n_{m-1}}, x_m)$. This a taut m'–chain in $K'(b_0)$ with $\lambda(\tilde{C}) = \lambda(C)$ and $m' \geq m$. Suppose that the Claim were true in the case where the barycentre in the statement is a vertex. Then there would exist a constant N_j, depending only on $S_j = Shapes(K'(b_0))$, such that $m' \leq N_j$. We would then have a constant $N = max\{N_1, \ldots, N_r\}$ depending only on $Shapes(K)$ such that $m \leq m' \leq N_j \leq N$ as required. So it is enough to consider the case where b_0 is a vertex of K.

Assume that this is so, and let C be as in the statement of the Claim. If $p(C)$, the path determined by C, were to pass through b_0 then because b_0 is a vertex it would have to occur as an entry in C. The first condition for tautness would then imply that C had size at most two. We are only interested in large chains, so we assume $C = (x_0, x_1, \ldots, x_m)$ with $m > 2$ and hence $p(C)$ does not pass through b_0. If we radially project from b_0 onto the spherical complex $S(b_0)$ (which we defined in Lemma 1.8) then, according to Lemma 1.9, the image of C under this map is a taut m-chain. We denote this m–chain in $S(b_0)$ by C'.

Let $B(i)$ be a simplex containing $[x_i, x_{i+1}]$. We can extend the line segments $[b_0, x_i]$ and $[b_0, x_{i+1}]$ until they meet $(B(i) - st(b_0))$, and the resulting line segments form two sides of a unique geodesic triangle in $B(i)$. (Here we mean a 2–dimensional triangle which is totally geodesic with respect to the metric $d_{B(i)}$.) We denote this triangle by Δ_i. For each i the triangles Δ_i and Δ_{i+1} have a common edge and we can map these triangles isometrically into $M(\kappa)^2$ to obtain the complex shown in Figure 1.3. Notice that the in general the triangles Δ_i are not 2–simplices in the simplicial structure on K, and we do not have a bound on the vertex angles.

Because the chain C is taut the image of $p(C)$ under this map is a geodesic arc in $M(\kappa)^2$ of length $\lambda(C)$. Further, this map takes C' isometrically onto an arc of the circle of radius ϵ about the image of b_0. We are assuming that if $\kappa > 0$ then $\lambda(C) < 2\pi\kappa$, so the image of $p(C)$ in $M(\kappa)^2$ cannot be a closed geodesic, and hence it subtends an angle $< 2\pi$ at the image of b_0. Thus the image of C' in $M(\kappa)^2$, and hence C' itself, has length less than $L_\kappa(S_\epsilon)$, the length of a circle of radius ϵ in the plane of constant curvature κ. By induction on $n = dim(K)$ there exists an integer $N(b_0)$, depending only on $Shapes(S(b_0))$, such that every taut chain in the $(n-1)$–dimensional complex $S(b_0)$ of size at least $N(b_0)$ has length at least $L_\kappa(S_\epsilon)$. Hence $size(C) = size(C') < N(b_0)$.

There are only finitely many possibilities for $Shapes(S(b_0))$, and the set of these possibilities depends only on $Shapes(K)$. So setting N equal to the minimum of the corresponding integers $N(b_0)$ finishes the proof. □

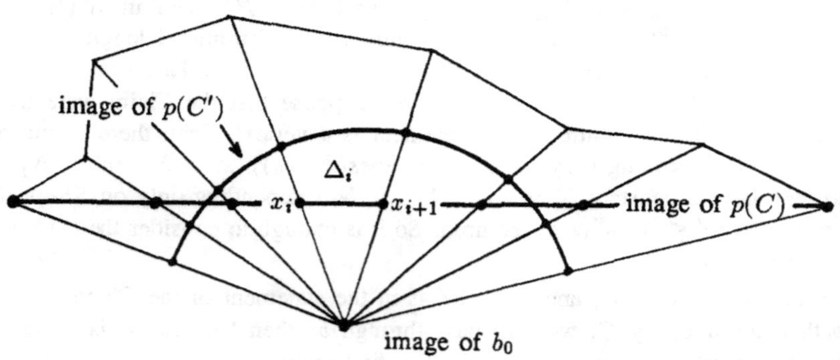

Figure 1.3.

1.6 The Completeness of the Metric

Theorem 1.12: *If $Shapes(K)$ is finite then d, the intrinsic metric on K, is complete.*

Proof: Let (x_n) be a Cauchy sequence in (K, d), and $\epsilon > 0$. Fix an integer R such that $d(x_n, x_m) < \epsilon$ for all $m, n \geq R$.

By Theorem 1.11, there is an upper bound, N say, on the size of taut chains from x_R to $x_m (m > R)$. It follows that any geodesic segment joining x_R to x_m in K lies is some subcomplex $K_0(m)$ which can be expressed as the union of at most N closed simplices.

Because $Shapes(K)$ is finite there are only finitely many pointed models for $(K_0(m), x_R)$. Let (K'_0, x'_R) be a model which occurs for infinitely many m, and consider the image in K'_0 of the sequence (x_m). Because the model is compact this has a convergent subsequence, which we denote (x'_{n_i}), with limit x'. We shall prove that (x_{n_i}), the corresponding subsequence in K, converges to a point in the preimage of x'.

We first describe the preimage of x' in K. In general K will have infinitely many bipointed subcomplexes $(K_\lambda; x_R, x_\lambda)$ giving rise to the bipointed model $(K'_0; x'_R, x')$. However, the pair (B_0, x_λ), where B_0 is the unique simplex of K which contains x_λ in its interior, is well defined up to isometry type. Suppose $B \in Shapes(K)$ has a face isometric to B_0. For each choice ϕ of isometry from B_0 to a face of B we can measure the distance from $\phi(x_\lambda)$ to $(B - st(\phi(x_\lambda)))$. (So if B_0 is a top dimensional cell this is equal to $d(x_\lambda, \partial B_0)$.) Let a_1 be the minimum of these distances over all choices of B and ϕ, and let a_2 be the minimum distance between distinct points in the orbit of x_λ under the action of the isometry group of B_0.

If we set $\epsilon = \frac{1}{3}min\{a_1, a_2\}$ then the ϵ-balls about the x_λ are disjoint. In fact the distance between distinct balls is at least ϵ. Further, for every λ the closed ball $B_\epsilon(x_\lambda)$ is contained in $st(x_\lambda)$, so by Lemma 1.3 the line segment joining x_λ to any point in $B_\epsilon(x_\lambda)$ is a geodesic segment in K. The definition of x' implies that the sequence (x_{n_i}) must lie in the union of the $B_\epsilon(x_\lambda)$ after some finite stage, and since this sequence is Cauchy it must eventually be contained in a single ball. Let x denote the x_λ corresponding to this ball. For large n we have:

$$\begin{aligned} d(x, x_{n_i}) &= length\ of\ the\ line\ segment\ [x, x_{n_i}] \\ &= length\ of\ the\ line\ segment\ [x', x'_{n_i}] \\ &\to 0 \quad as \quad n_i \to \infty. \end{aligned}$$

Hence the subsequence (x_{n_i}) converges to x, and since the sequence (x_m) is Cauchy it too converges to x. □

1.7 The Two Dimensional Case

To illustrate the ideas involved in the proof of Theorem 1.11 we now restrict our attention to the 2–dimensional case where many of the details are considerably easier. For simplicity we shall only consider Euclidean complexes — the general case requires only that one consider pictures drawn in $M(\kappa)^2$ rather than $\mathbf{E}^2$.

Throughout this section K denotes a connected 2–dimensional **E**-simplicial complex with $Shapes(K)$ finite and d denotes its intrinsic metric.

We say that an m–chain $C = (x_0, \ldots, x_m)$ in K has *property* τ if it satisfies the following conditions: Firstly, there is no simplex in K containing three successive entries of C. Secondly, if B and B' are 2–simplices in K which contain the line segments $[x_{i-1}, x_i]$ and $[x_i, x_{i+1}]$ respectively, and x_i lies in the interior of an edge common to B and B', then the images of x_{i-1}, x_i and x_{i+1} under any local isometry from $B \cup B'$ into $\mathbf{E}^2$ are colinear.

Property τ is weaker than the property of being taut (which we definied in Section 1.4).

In Section 1.3 we showed that for a fixed integer m there exists a shortest m-chain joining any two points in K. Let C be such a chain. We can delete entries of C until no three sucessive entries lie in the same simplex. The resulting chain must then satisfy property τ, since otherwise we could perform simple moves to produce a shorter chain of the same size and with the same endpoints, as illustrated below.

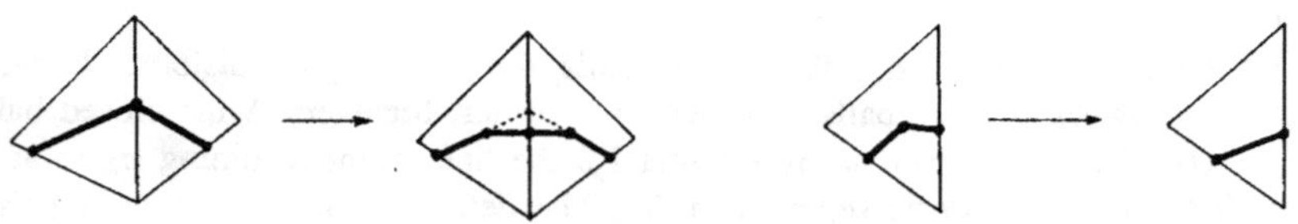

Figure 1.4: Shortening a chain which does not have property τ.

The existence of geodesic segments in the 2–dimensional case now follows from the following lemma.

Lemma 1.13: *There exist constants N and α such that every chain in K which has property τ and size greater than N has length at least α.*

Proof: Choose ϵ small enough to ensure that the ϵ–neighbourhoods of distinct vertices of K are disjoint. Colour these neighbourhoods black and the remainder of K white. Note that there is a lower bound ℓ_1 on the distance from any black region to the link of the corresponding vertex, and a lower bound ℓ_2 on the length of line segments which lie entirely in the white region of some 2–simplex, and have endpoints in the boundary of that simplex.

Let $C = (x_0, \ldots, x_m)$ be an m–chain in K with propery τ. Suppose $m > 3$ and that $p(C)$, the PL path determined by C, is not contained in the open star of any vertex of K. Then either there exists some $i \in \{1, \ldots, m-2\}$ such that the line segment $[x_i, x_{i+1}]$ lies entirely in the white region of K, or else $p(C)$ must contain a point in a black region and a point in the link of the corresponding vertex. In either case C must have length at least $\alpha = min\{\ell_1, \ell_2\}$. So it is enough to show that if m is sufficiently large then $p(C)$ cannot be contained in the open star of any vertex.

For this one simply observes that if $p(C)$ were contained in $st(b_0)$ then we could join each line segment $[x_i, x_{i+1}]$ to b_0 and map the resulting triangles isometrically into $\mathbf{E}^2$ to obtain a complex of the type shown in Figure 1.3. Because $Shapes(K)$ is finite there is a lower bound, θ say, on the angles subtended at the vertex, so taking $N = \pi/\theta + 2$ we are done. □

2. Non-Positive Curvature in Piecewise Euclidean Complexes

In this section we prove the following special case of the Main Theorem:

Theorem : *If K is a simply connected piecewise Euclidean complex of type A or B then the following 11 conditions are equivalent:*

Global conditions:

I) *K has unique geodesic segments.*
II) *K satisfies CAT(0) globally.*
III) *K satisfies CN globally.*
IV) *The metric on K is convex.*
V) *Every geodesic triangle in K has non-positive excess.*

Local conditions:

VI) *K has unique geodesic segments locally.*
VII) *K satisfies CAT(0) locally.*
VIII) *K satisfies CN locally.*
IX) *The metric on K is convex locally.*
X) *Every point of K has a neighbourhood such that any geodesic triangle contained in that neighbourhood has non-positive excess.*
XI) *K satisfies the link condition.*

The proof of this theorem involves a number of auxillary results about the geometry of piecewise Euclidean complexes, which are of interest in their own right. These include not only the theorems stated in the introduction, but some of the supporting lemmas aswell.

The most difficult step in the proof is in passing from the local to the global situation. This we do in Section 2.4, modelling our argument on that used by Milnor in his book on Morse theory [22] to prove the Cartan-Hadamard Theorem, and subsequently by Stone [27] to prove the corresponding result for PL manifolds. Although the same strategy of proof works here, the details of our proof have little in common with those of the original, due to the absence of both local compactness and a smooth structure.

As a postscript we show in Section 2.5 that geodesic segments in a piecewise Euclidean complex of non-positive curvature can be extended indefinitely, provided that during the extension the endpoint of the extended geodesic does not have a contractible link.

Terminology

We recall the regularity conditions A and B which we defined in the introduction:

A) *Every subset of K which is closed and bounded in the intrinsic pseudometric is compact.*

B) *$Shapes(K)$ is finite.*

We do not require complexes of type A to be finite dimensional, nor do we require complexes of type B to be locally compact. (In the literature spaces satisfying condition A are sometimes called proper.)

As we explained in the introduction, our main interest is in complexes of type B, but some of our proofs require knowledge of the corresponding result for spaces of type A, and including spaces of type A into our development requires only a minimal amount of extra work.

For the remainder of Section 2 the letter K, without further qualification, shall denote a piecewise Euclidean complex of type A or B.

In Section 1 we descibed paths combinatorially using $m-$chains. In this section we adopt a more analytic approach, and to do so we use the following definitions.

Definition: A *geodesic segment* joining x to y in K is a PL path $\alpha : [0,1] \to K$ with $\alpha(0) = x, \alpha(1) = y$ which is parameterised proportional to arc length, and has length $L(\alpha) = d(x,y)$.

Definition: A PL path $\beta : [0,1] \to K$ is a *local geodesic* if it is parameterised proportional to arc length and every $t \in [0,1]$ has a neighbourhood $[s,s']$ such that $d_K(\beta(t'),\beta(t'')) = L(\beta)|t' - t''|$ for all $t',t'' \in [s,s']$.

Our insistence that all paths under consideration are defined on $[0,1]$ and parameterised proportional to arc length can prove inconvenient, but it allows us to describe convexity properties of the metric by means of inequalities involving geodesic segments.

Any PL path $\alpha : [0,1] \to K$ has a unique minimal expression as the concatenation of line segments in K, and the endpoints of these line segments form an $m-$chain. It is easy to check that α is a local geodesic if and only if this $m-$chain is a local geodesic (as defined in Section 1.3). If α is a geodesic segment then this chain is also taut.

2.1 The Convexity of the Metric in Spaces with Unique Geodesic Segments.

In this subsection we prove that if K has unique geodesic segments (i.e. every pair of points can be joined by a unique geodesic segment in K) then every local geodesic in K is a geodesic segment, and the intrinsic metric is convex in the following sense: If α and β are geodesic segments in K then the function $f_{\alpha,\beta} : [0,1] \to K$ given by $f(t) = d(\alpha(t),\beta(t))$ is convex. The latter assertion follows easily from the following result, the proof of which occupies the remainder of this subsection.

Theorem 2.1: *If K has unique geodesic segments, and α_0 and α_1 are geodesic segments in K with $\alpha_0(0) = \alpha_1(0)$ then*

$$d(\alpha_0(t), \alpha_1(t)) \leq t \cdot d(\alpha_0(1), \alpha_1(1)) \quad \forall t \in [0,1].$$

The first step in the proof of Theorem 2.1 is to show that geodesic segments in K vary continuously with their endpoints (this statement will be made precise in Lemma 2.3). To prove this fact for complexes of type B we introduce the notion of a *corridor*.

Definition: An $M(\kappa)$*−corridor* is an $M(\kappa)$−simplicial complex which can be expressed as the union of finitely many closed simplices $\{\sigma_1, \ldots, \sigma_m\}$ such that the following conditions hold for $i = 1, \ldots, m-1$: Firstly, $\tau_i = \sigma_i \cap \sigma_{i+1}$ is a non-empty proper subface of both σ_i and σ_{i+1}. Secondly $\tau_i \cap \tau_{i+1}$ (which may be empty) is properly contained in both τ_i and τ_{i+1}. Thirdly, $\sigma_i \cap \sigma_{i+r} = \bigcap_{j=0}^{r-1} \tau_{i+j}$ for all $r > 0$. We will denote such a corridor by $\Gamma = (\sigma_1, \ldots, \sigma_m)$.

This third condition ensures that any path in Γ which has one of its endpoints in each of σ_1 and σ_m must intersect all of the closed simplices σ_i. Notice that Γ is compact and hence a geodesic metric space of type A.

We shall be concerned with corridors which arise in the following way: Suppose that $C = (x_0, \ldots, x_m)$ is an m−chain in an $M(\kappa)$−simplicial complex K, and that no three sucessive entries of C lie in any closed simplex of K (we shall call an m−chain with this property *minimal*). We let $B(i)$ denote the unique closed simplex of smallest dimension which contains the line segment $[x_{i-1}, x_i]$, and consider the disjoint union of the m (not necessarily distinct) simplices $\sigma(B(i)) \in Shapes(K)$, modulo the equivalence relation generated by: $f_{B(i)}(x) \sim f_{B(i+1)}(x)$ whenever $x \in B(i) \cap B(i+1)$. (The maps f_B are those defined in the definition of a metric simplicial complex in Section 1.1.) It is easy to check that because C is minimal this is an $M(\kappa)$−corridor. We denote this corridor Γ_C. If C is the unique taut chain associated to a geodesic segment α then we denote this corridor by $\Gamma(\alpha)$.

Definition: An $M(\kappa)$−corridor Γ is said to *occur in* K if $\Gamma = \Gamma_C$ for some minimal m−chain C in K.

Remark: In general the map from Γ_C into K induced by $f_{B(i)}^{-1} : \sigma(B(i)) \to B(i)$ is *not* an injection.

If K is of type B then for every integer m there are only finitely many isometry types of corridors of length m which can occur in K. As in Section 1.3, we represent these isometry types by *model* corridors, and according to Theorem 1.11 any geodesic segment in K whose length is bounded by a certain linear function of m is the image of a geodesic segment in one of these models.

If $\kappa \leq 0$ then the corridor Γ has unique geodesic segments. This follows by application of the following lemma to Γ and its sub-corridors.

Lemma 2.2: *Suppose $\Gamma = (\sigma_1, \dots \sigma_m)$ is an $M(\kappa)$–corridor where $\kappa \leq 0$. If $x \in \sigma_1$ and $y \in \sigma_m$ then there is a unique geodesic segment from x to y in Γ.*

Proof: In Theorem 1.1 we proved that there exists a geodesic segment from x to y in Γ, so the only question is that of uniqueness. Suppose that there are two geodesic segments from x to y, given by distinct m-chains $C_1 = (a_0, \dots, a_m)$ and $C_2 = (b_0, \dots, b_m)$, where $a_0 = b_0 = x, a_m = b_m = y$ and $a_i, b_i \in \sigma_i \cap \sigma_{i+1}$. Let c_i be the midpoint of the line segment $[a_i, b_i]$. The simplex σ_i is isometric to a convex subset of $M(\kappa)^n$ for some n. Hence

$$d_{\sigma_i}(c_i, c_{i+1}) \leq \frac{1}{2}(d_{\sigma_i}(a_i, a_{i+1}) + d_{\sigma_i}(b_i, b_{i+1}))$$

with strict inequality in the case $i = min\{j : a_j \neq b_j\}$. Summing over i shows that $C = (c_0, \dots, c_m)$ is an m–chain from x to y in Γ with

$$\lambda(C) < \frac{1}{2}(\lambda(C_1) + \lambda(C_2)) = d_\Gamma(x, y)$$

which is a contradiction. □

We now prove that if K has unique geodesic segments then they vary continuously with their endpoints. For paths $\alpha, \beta : [0,1] \to K$ we use the notation $\|\alpha - \beta\| = sup\{d(\alpha(t), \beta(t)) : t \in [0,1]\}$.

Lemma 2.3: *Suppose that K has unique geodesic segments. Given $x, y \in K$, and sequences of points $x_i \to x$ and $y_i \to y$, let α denote the geodesic segment from x to y and let α_i denote the geodesic segment from x_i to y_i. Then $\|\alpha - \alpha_i\| \to 0$ as $i \to \infty$.*

Proof: First we prove the weaker statement that $\alpha_i(t) \to \alpha(t)$ as $i \to \infty$ for all t, an easy compactness argument then shows that the convergence is uniform.

Case 1: Suppose that K is of type A. Then the closed ball of radius $2d(x,y)$ about x is compact. After some finite stage all the α_i must be contained in this set. Suppose, for contradiction, that for some t it is not true that $d(\alpha_i(t), \alpha(t)) \to 0$ as $i \to \infty$. Then by compactness $(\alpha_i(t))$ has a convergent subsequence $(\alpha_j(t))$ with limit $z \neq \alpha(t)$. So we have

$$\begin{aligned} &d(x_j, z) - d(x_j, \alpha_j(t)) \to 0 \\ &d(y_j, z) - d(y_j, \alpha_j(t)) \to 0 \qquad as\ j \to \infty. \end{aligned}$$

But on the other hand,

$$\begin{aligned} d(x_j, \alpha_j(t)) &= t\, d(x_j, y_j) \to t\, d(x, y) \\ d(y_j, \alpha_j(t)) &= (1-t)\ d(x_j, y_j) \to (1-t)\ d(x, y) \qquad as\ j \to \infty. \end{aligned}$$

So z satisfies

$$\begin{aligned} d(x, z) &= \lim_{j \to \infty} d(x_j, \alpha_j(t)) = t\, d(x, y) \quad \text{and} \\ d(y, z) &= \lim_{j \to \infty} d(y_j, \alpha_j(t)) = (1-t)\ d(x, y). \end{aligned}$$

But $\alpha(t)$ is the unique point of K which satisfies these equations. Hence $z = \alpha(t)$, contrary to hypothesis.

Case 2: Now suppose that K is of type B. After some finite stage $x_i \in st(x)$ and $y_i \in st(y)$. Hence there is an upper bound on the distances $d(x_i, y_i)$, and so by Theorem 1.11 there exists an integer N such that each α_i is the path determined by a taut chain of size at most N. Thus α_i is the image of a geodesic segment $\tilde{\alpha}_i$ in at least one of the finitely many bipointed model corridors $(\tilde{x}, \tilde{y}; \sigma_0, \ldots, \sigma_N)$, where $\sigma_i \in Shapes(K)$ and $\tilde{w}$ denotes the image of a point $w \in K$ under the identification of a corridor in K with the corresponding model corridor.

Suppose that μ is a model which contains infinitely many of the model geodesic segments $\tilde{\alpha}_i$, and let (α_j) denote corresponding subsequence of (α_i). We proved in Lemma 2.2 that μ satisfies the hypotheses of Case 1, so the $\tilde{\alpha}_j$ converge pointwise to the unique geodesic segment $\tilde{\alpha}^\mu$ from $\tilde{x}$ to $\tilde{y}$ in μ. Hence

$$\begin{aligned} d_K(x_j, y_j) &= L(\alpha_j) = L(\tilde{\alpha}_j) = d_\mu(\tilde{x}_j, \tilde{y}_j) \\ &\to d_\mu(\tilde{x}, \tilde{y}) = L(\tilde{\alpha}^\mu) \quad as\ j \to \infty. \end{aligned}$$

But $d_K(x_j, y_j) \to d_K(x, y)$ as $j \to \infty$, so $d_\mu(\tilde{x}, \tilde{y}) = d_K(x, y)$ and hence $\tilde{\alpha}^\mu$ models the unique geodesic segment α from x to y in K. In particular, because α is unique, any occurence of the bipointed model μ in K must contain α. Also notice that the image in K of the geodesic segment from $\tilde{\alpha}_j(t)$ to $\tilde{\alpha}^\mu(t)$ in μ is a path of the same length from $\alpha_j(t)$ to $\alpha(t)$ in K. Hence $d_\mu(\tilde{\alpha}^\mu(t), \tilde{\alpha}_j(t)) \geq d_K(\alpha(t), \alpha_j(t))$ for all t.

There are only finitely many models μ, so for sufficiently large integers R we can decompose $(\alpha_i)_{i>R}$ into finitely many infinite subsequences, each of which consists of geodesic segments which can be modelled in some fixed μ. For each such subsequence (α_j^μ) and for every $t \in [0, 1]$

$$d_K\left(\alpha(t), \alpha_j^\mu(t)\right) \leq d_\mu\left(\tilde{\alpha}^\mu(t), \tilde{\alpha}_j^\mu(t)\right) \to 0 \quad as\ j \to \infty.$$

Hence the sequence (α_i) converges pointwise to α.

Uniform Convergence: Assume that i is sufficiently large so that $L(\alpha_i) < L(\alpha)+1$. Given ϵ and $t \in [0, 1]$, we have proved the existence of an integer $N(t)$ such that $d(\alpha(t), \alpha_i(t)) < \epsilon$ whenever $i > N(t)$. Hence

$$\begin{aligned} d(\alpha(t+\delta), \alpha_i(t+\delta)) &\leq d(\alpha(t+\delta), \alpha(t)) + d(\alpha(t), \alpha_i(t)) + d(\alpha_i(t), \alpha_i(t+\delta)) \\ &< |\delta|\ L(\alpha) + \epsilon + |\delta|\ L(\alpha_i) \\ &< 2\epsilon \qquad\qquad whenever\ \ |\delta| < \frac{\epsilon}{2L(\alpha)+1} \end{aligned}$$

The result now follows from the compactness of $[0, 1]$. □

Remark: More generally, the conclusion of Lemma 2.3 holds in any open subset of K with the property that any two points in that subset can be joined by a unique geodesic segment, and that this geodesic segment lies entirely within the given subset. The same remark applies to Lemma 2.4. This importance of this observation is that it allows us to appeal to Lemmas 2.3 and 2.4 when dealing with complexes which have unique geodesic segments locally. (See Section 2.3.)

We shall use Lemma 2.3 to reduce the proof of Theorem 2.1 to the case where the geodesic segments under consideration are uniformly close. We now give the proof in this restricted setting.

Lemma 2.4: *If K has unique geodesic segments and α is a geodesic segment in K then there exists $\epsilon > 0$ such that any geodesic segment β with $\beta(0) = \alpha(0)$ and $\|\alpha - \beta\| < \epsilon$ satisfies*

$$d(\alpha(t), \beta(t)) \leq t \cdot d(\alpha(1), \beta(1)) \quad \forall t \in [0,1].$$

Proof: Represent α and β by the taut chains $(a_0, \ldots, a_m)$ and $(b_0, \ldots, b_r)$ respectively. Choose η sufficiently small so that $B_{4\eta}(a_i) \subset st(a_i)$ and the open balls $B_{4\eta}(a_i)$ are disjoint. Then choose $\epsilon < \eta$ so that for every i the ϵ–neighbourhood of each compact arc $[a_i, a_{i+1}] - (B_{4\eta}(a_i) \cup B_{4\eta}(a_{i+1}))$ is contained in $st(a_i) \cap st(a_{i+1})$. Decompose $\{b_0, \ldots, b_r\}$ into sets

$$S_0 = \{b_0, b_1, \ldots, b_{n_0}\}, S_1 = \{b_{n_0}, b_{n_0+1}, \ldots, b_{n_1}\}, \ldots, S_m = \{b_{n_{m-1}}, \ldots, b_r\}$$

such that $S_i \subseteq st(a_i)$. (With our choice of η and ϵ this can be done for any chain which determines a path in K uniformly ϵ–close to α.)

Let B_j be a closed simplex of K which contains the line segment $[b_j, b_{j+1}]$. If $b_j \in st(a_i)$ then $B_j \cap st(b_j) \subseteq B_j \cap st(a_i)$ is non-empty. In particular if $j = n_i$ then there is a unique geodesic triangle (with respect to the metric d_{B_j}) in B_j with vertices $\{a_i, a_{i+1}, b_{n_i}\}$. We can map this (possibly degenerate) triangle isometrically into $\mathbf{E}^2$ to obtain the geodesic triangle Δ_i shown in Figure 2.1. Similarly, for every $b_j \in S_i$ we have a unique geodesic triangle $\Delta(b_j, b_{j+1}, a_i)$ in B_j. We map each of these triangles isometrically into $\mathbf{E}^2$, as indicated in Figure 2.1. Here x' denotes the image in $\mathbf{E}^2$ of $x \in K$.

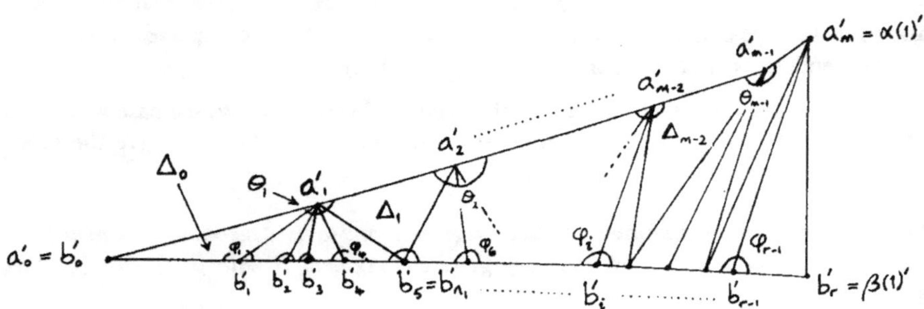

Figure 2.1

By barycentrically subdividing the simplices in $Shapes(K)$ and giving K the induced simplicial structure, we may assume that the restriction of the intrinsic metric on K to each simplex B agrees with the local metric d_B. Hence a line segment joining x' to y' in $\mathbf{E}^2$ lifts to a geodesic segment in K if x' and y' lie in the same triangle, while in general the line segment joining x' to y' in $\mathbf{E}^2$ lifts to a path from x to y in K of length $\geq d_K(x,y)$. Thus

$$\begin{aligned} &d_{E^2}\left(\alpha(1)', \beta(1)'\right) = d_K\left(\alpha(1), \beta(1)\right), \quad \text{and} \\ &d_{E^2}\left(\alpha(t)', \beta(t)'\right) \geq d_K\left(\alpha(t), \beta(t)\right) \;\; \forall t \in [0,1]. \end{aligned} \tag{1}$$

The angles θ_i and ϕ_i shown in Figure 2.1 are all $\geq \pi$, because α and β are geodesic segments. Therefore the derivative (which is defined almost everywhere) of the continuous piecewise linear function $\psi : t \mapsto d_{E^2}(\alpha(t)', \beta(t)')$ is non-decreasing. It is also non-negative, so ψ is convex and since $\psi(0) = 0$ we have the following inequality

$$d_{E^2}\left(\alpha(t)', \beta(t)'\right) \leq t \cdot d_{E^2}\left(\alpha(1)', \beta(1)'\right) \;\; \forall t \in [0,1]. \tag{2}$$

Combining (1) and (2) completes the proof. □

Examining the proof of Lemma 2.4 we see that the essential use of the fact that α and β are geodesic segments is to infer that $a_{i-1} \notin st(a_i)$ for $i \neq 1$, and that the angles θ_i and ϕ_i are all $\geq \pi$. To ensure that these conditions hold, one need only assume that α and β are local geodesics. Thus we obtain the following strengthening of Lemma 2.4.

Lemma 2.4*: *If K has unique geodesic segments and α is a local geodesic in K then there exists $\epsilon > 0$ such that any local geodesic β with $\|\alpha - \beta\| < \epsilon$ and $\beta(0) = \alpha(0)$ satisfies*

$$d(\alpha(t), \beta(t)) \leq t\, d(\alpha(1), \beta(1)) \qquad \forall t \in [0,1].$$

An important consequence of this generalisation is the following result, which is needed in the proof of Theorem 2.7.

Corollary : *If K has unique geodesic segments then every local geodesic in K is a geodesic segment. Hence there is a unique local geodesic joining any two points in K.*

Proof: Consider a local geodesic $\gamma : [0,1] \to K$ and let ϵ be as in Lemma 2.4. We will prove that $S = \{s : d(\gamma(0), \gamma(s)) = \text{length of } \gamma|_{[0,s]}\}$ is the whole of $[0,1]$. The set S is defined by a closed condition and contains 0, so it suffices to prove that it is open. Fix $s \in S$. Lemma 2.3 implies that for sufficiently small δ the geodesic segment from $\gamma(0)$ to $\gamma(s+\delta)$ is uniformly ϵ–close to the local geodesic $\gamma'(t) = \gamma(t(s+\delta))$. So by Lemma 2.4* these paths must coincide. □

We now turn to the proof of Theorem 2.1, and then conclude this section by giving two easy corollaries.

Theorem 2.1: *If K has unique geodesic segments, and α_0 and α_1 are geodesic segments in K with $\alpha_0(0) = \alpha_1(0)$ then*

$$d(\alpha_0(t), \alpha_1(t)) \leq t \cdot d(\alpha_0(1), \alpha_1(1)) \quad \forall t \in [0,1].$$

*Proof:*Let σ denote the unique geodesic segment from $\alpha_0(1)$ to $\alpha_1(1)$ and let α_s denote the geodesic segment from $\alpha_0(0)$ to $\sigma(s)$. We will prove that the set $\Sigma = \{s : d(\alpha_0(t), \alpha_s(t)) \leq td(\alpha_0(1), \alpha_s(1)) \forall t \in [0,1]\}$ is the whole of $[0,1]$.

Σ contains 0 and is closed by Lemma 2.3, because if for some t the inequality $d(\alpha_0(t), \alpha_s(t)) > td(\alpha_0(1), \alpha_s(1))$ holds at s then it holds in a neighbourhood of s. To see that Σ is open: Fix $s \in \Sigma$. Then by Lemma 2.4 there exists $\epsilon > 0$ such that if $\|\alpha_s - \alpha_{s+\delta}\| \leq \epsilon$ then

$$d(\alpha_s(t), \alpha_{s+\delta}(t)) \leq t \cdot d(\alpha_s(1), \alpha_{s+\delta}(1)) \quad \forall t \in [0,1].$$

Lemma 2.3 implies that $\|\alpha_s - \alpha_{s+\delta}\| \leq \epsilon$ for sufficiently small δ. For such δ we have the following inequality, which shows that $(s+\delta) \in \Sigma$.

$$\begin{aligned} d(\alpha_0(t), \alpha_{s+\delta}(t)) &\leq d(\alpha_0(t), \alpha_s(t)) + d(\alpha_s(t), \alpha_{s+\delta}(t)) \\ &\leq t \cdot d(\alpha_0(1), \alpha_s(1)) + t \cdot d(\alpha_s(1), \alpha_{s+\delta}(1)) \\ &= t \cdot d(\alpha_0(1), \alpha_{s+\delta}(1)) \qquad \forall t \in [0,1]. \end{aligned}$$

Hence Σ is open. □

Corollary : *If K has unique geodesic segments then for any geodesic segments α and β in K the function $f_{\alpha,\beta}(t) = d(\alpha(t), \beta(t))$ is convex.*

Proof: We show that for any geodesic segments α and β in K

$$d(\alpha(t), \beta(t)) \leq t\, d(\alpha(1), \beta(1)) + (1-t)\, d(\alpha(0), \beta(0)) \quad \forall t \in [0,1].$$

The result then follows by applying this inequality to the initial segments of α and β. Let γ denote the geodesic segment from $\alpha(0)$ to $\beta(1)$. Applying Theorem 2.1 twice, once using α, γ and once using γ^*, β^* (where the star denotes reverse orientation), yields

$$\begin{aligned} d(\alpha(t), \beta(t)) &\leq d(\alpha(t), \gamma(t)) + d(\gamma(t), \beta(t)) \\ &\leq t\, d(\alpha(1), \gamma(1)) + (1-t)\, d(\gamma(0), \beta(0)) \\ &= t\, d(\alpha(1), \beta(1)) + (1-t)\, d(\alpha(0), \beta(0)) \quad \forall t \in [0,1], \end{aligned}$$

as required. □

Another immediate consequence of Theorem 2.1 is the following.

Corollary : *If K has unique geodesic segments and α_x denotes the geodesic segment from $x \in K$ to a fixed point $x_o \in K$, then $F : K \times I \to K$ given by $F(x,t) = \alpha_x(t)$ is a Lipschitz contraction of K to x_0.*

Proof: We have $d(F(x,t), F(y,t)) = d(\alpha_x(t), \alpha_y(t)) \leq t d(x,y)$ for all $t \in [0,1]$, hence result with Lipschitz constant 1. □

2.2 Global Characterisations of Non-Positive Curvature, and the Fixed Point Theorem.

In this section we define three global criteria for non-positive curvature and show that if K is a piecewise Euclidean complex of type A or B then each of these criteria is equivalent to requiring that K has unique geodesic segments. One of these criteria is the CN-inequality of Bruhat and Tits, and this leads to the Fixed Point Theorem stated in the introduction. The other descriptions of non-positive curvature which we consider in this section are the $CAT(0)$ inequality, as defined by Gromov, and Alexandrov's condition on the excess of geodesic triangles. We begin by describing this last condition and proving that it is equivalent to the uniqueness of geodesic segments in K.

Definition: Given geodesic segments α and β in K with common initial point x, the *angle* between α and β at x is the distance between the points determined by α and β in the spherical complex $LK(x, K)$.

Definition: A *geodesic triangle* T in K consists of three points in K (the vertices of T) and a choice of geodesic segment between each pair of vertices (the edges of T).

We use the notation $T = \Delta(x_0, x_1, x_2)$, where x_0, x_1, x_2 are the vertices of T. But it should be noted that T is not uniquely determined by its vertices unless geodesic segments are unique in K.

Definition: Let $T = \Delta(x_0, x_1, x_2)$ be a geodesic triangle in K. A triangle $T' = \Delta(x'_0, x'_1, x'_2)$ in $\mathbf{E}^2$ is called a *comparison triangle* for T if $d_{E^2}(x'_i, x'_j) = d_K(x_i, x_j)$ for $i, j \in \{0, 1, 2\}$.

Alexandrov [2] defined curvature via the notion of *the excess of a triangle*, relating the sum of the *sup angles* of a geodesic triangle in the space under consideration to the sum of the angles in the comparison triangle. If K is a complex of type A or B and α and β are geodesic segments in K with a common initial point x then the sup angle between α and β at x is the lesser of π and the angle between α and β as defined above. Given a geodesic triangle Δ in K, let $|\Delta|$ denote the sum of the sup angles at the vertices.

Definition: The *excess* of a geodesic triangle Δ in K is $|\Delta| - \pi$.

Alexandrov defined non-positive curvature in geodesic metric spaces by the condition that every geodesic triangle in the given space has non-positive excess. In order to prove that K satisfies this condition if and only if it has unique geodesic segments we must first introduce some notation.

Suppose that K has unique geodesic segments and consider geodesic segments α and β in K with $\alpha(0) = \beta(0) = x, \alpha(1) = y$, and $\beta(1) = z$. Let Θ denote the angle between α and β at x, and suppose that $\Theta \leq \pi$. For sufficiently small t the geodesic segment from $\alpha(t)$ to $\beta(t)$ is contained in $st(x)$. We can form the join of this geodesic segment with x and map the resulting 2–complex isometrically into $\mathbf{E}^2$, as shown in Figure 2.2. Let x', y', z' be points in $\mathbf{E}^2$ with the property that $d_K(x, y) = d_{E^2}(x', y'), d_K(x, z) = d_{E^2}(x', z')$, and the angle between the line segments $[x', y']$ and $[x', z']$ (which we call α' and β' respectively) is Θ.

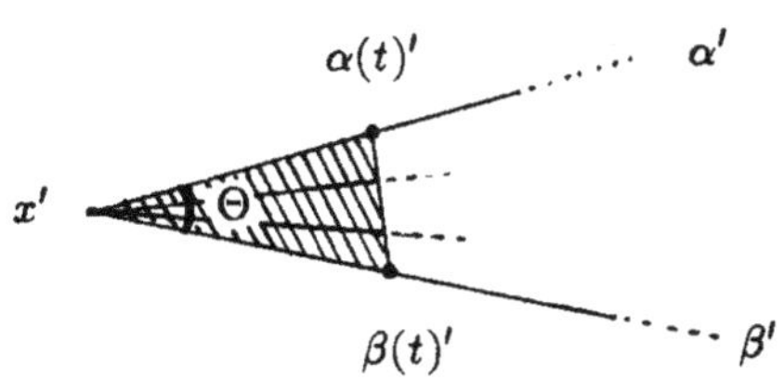

Figure 2.2

Lemma 2.5: *If y, z, y' and z' are as above then $d_K(y,z) \geq d_{E^2}(y',z')$.*

Proof: If ϵ is sufficiently small then $d_K(\alpha(\epsilon),\beta(\epsilon)) = d_{E^2}(\alpha'(\epsilon),\beta'(\epsilon))$, because Figure 2.2 is isometric to its preimage in K. By elementary Euclidean geometry

$$d_{E^2}\left(\alpha'(\epsilon),\beta'(\epsilon)\right) = \epsilon\, d_{E^2}\left(\alpha'(1),\beta'(1)\right) = \epsilon\, d_{E^2}\left(y',z'\right).$$

And by Theorem 2.1

$$d_K\left(\alpha(\epsilon),\beta(\epsilon)\right) \leq \epsilon\, d_K\left(\alpha(1),\beta(1)\right) = \epsilon\, d_K\left(y,z\right).$$

Combining these inequalities finishes the proof. □

Remark: It is not difficult to show that Lemma 2.5 is equivalent to Theorem 2.1.

Gersten calls Lemma 2.5 the Topogonov inequality (cf. Theorem 3.1), and in [15] he derives the CN-inequality directly from this result. He was concerned only with the 2–dimensional case and used a local definition of geodesic, but with the benefit of the corollary to Lemma 2.4* his argument would work equally well in the present setting.

Proposition 2.6: *K has unique geodesic segments if and only every geodesic triangle in K has non-positive excess.*

Proof: Suppose that every geodesic triangle in K has non-positive excess, and fix $x, y \in K$. Consider a geodesic triangle formed by taking two geodesic segments α and β from x to y and adding a third vertex z at the midpoint of α. Since α is a geodesic segment, the angle at z is $\geq \pi$. Hence the angles at x and y are zero and consequently α and β coincide near their endpoints. The same argument shows that at no point can α and β diverge, and hence they coincide.

Conversely, suppose that K has unique geodesic segments and that T is a geodesic triangle in K. If one of the edges of T is the concatenation of the other two sides then the sup angles of T are $\pi, 0$ and 0. If not then, because every local geodesic in K is a geodesic segment, each vertex angle must be less than π. It then follows from Lemma 2.5 that each vertex angle is no greater than the corresponding angle in a comparison triangle for T, and hence the sum of these angles is no greater than π. □

Gromov [17] defines a geodesic metric space X to be non-positively curved if it satisfies $CAT(0)$, the comparison axiom of Alexandrov-Topogonov.

CAT(0): Let $T = \Delta(x_0, x_1, x_2)$ be a geodesic triangle in X, and let y a point on the side of T which has endpoints x_1 and x_2. Choose a comparison triangle $T' = \Delta(x_0', x_1', x_2')$ in $\mathbf{E}^2$ and let y' denote the unique point on the line segment $[x_1', x_2']$ such that $d_{E^2}(x_i', y') = d_K(x_i, y)$ for $i = 1, 2$. Then $d_{E^2}(x_0', y') \geq d_K(x_0, y)$.

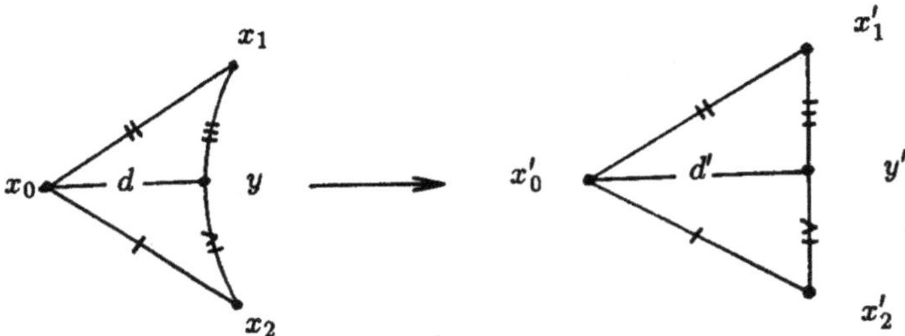

Figure 2.3: *CAT(0)*

CN: A metric space X satisfies the CN-condition of Bruhat-Tits [10] if for every pair of elements $x_1, x_2 \in X$ there exists a point m with $d(x_1, m) = d(x_2, m)$ such that for all $x_0 \in X$

$$d(x_1, x_0)^2 + d(x_2, x_0)^2 \geq 2d(m, x_0)^2 + \frac{1}{2}d(x_1, x_2)^2 . \tag{CN}$$

The main result of this subsection is the following:

Theorem 2.7: *If K is a piecewise Euclidean complex of type A or B then the following global characterisations of non-positive curvature are equivalent:*

I) *K has unique geodesic segments.*
II) *K satisfies CAT(0).*
III) *K satisfies CN.*
IV) *Every geodesic triangle in K has non-positive excess.*

Proof: The equivalence of I and IV was proved in Proposition 2.6. We will now show II $\Rightarrow$III $\Rightarrow$I $\Rightarrow$II. Most of the work is in the final implication.

II $\Rightarrow$III: Suppose that K satisfies $CAT(0)$. Fix $T = \Delta(x_0, x_1, x_2)$ with comparison triangle $T' = \Delta(x'_0, x'_1, x'_2)$. Let m denote the midpoint of the edge of T which has endpoints x_1 and x_2, and let m' denote its comparison point. It is easy to check that $\mathbf{E}^2$ satisfies the CN condition (in fact, by the parallelogram law, one gets equality in all cases) so the inequality (CN) holds for x'_0, x'_1, x'_2, m'. Removing the primes the value of each term in (CN) stays the same with the exception of the first term on the right hand side, for which by $CAT(0)$ we have $d_{E^2}(x'_0, m') \geq d_K(x_0, m)$.

III $\Rightarrow$I: Suppose that there are two geodesic segments from x_1 to x_2 in K. Shortening them if necessary, we may assume that their midpoints, which we denote by m and x_0, are distinct. Then $d(x_1,x_0)^2 + d(x_2,x_0)^2 = \frac{1}{2}d(x_1,x_2)^2$, so (CN) implies that $d(m,x_0) = 0$, contrary to hypothesis.

I $\Rightarrow$II: Let $T = \Delta(x_0,x_1,x_2) \subset K$, and let y be some point on the unique geodesic segment from x_1 to x_2 (for which we adopt the notation $[x_1,x_2]$). Let θ denote the angle between $[y,x_1]$ and $[y,x_0]$ at y, and let ϕ denote the angle between $[y,x_2]$ and $[y,x_0]$ at y.

First suppose that $\theta \geq \pi$, then it follows that the concatenation of $[x_1,y]$ and $[y,x_0]$ is a local geodesic from x_1 to x_0. As a corollary to Lemma 2.4* we showed that every local geodesic in K is a geodesic segment. Hence

$$d_K(x_0,y) = d_K(x_0,x_1) - d_K(x_1,y).$$

The triangle inequality applied to the comparison triangle $T' = \Delta(x'_0,x'_1,x'_2)$ in $\mathbf{E}^2$ yields

$$\begin{aligned} d_{E^2}(x'_0,y') &\geq d_{E^2}(x'_0,x'_1) - d_{E^2}(x'_1,y') \\ &= d_K(x_0,x_1) - d_K(x_1,y). \end{aligned}$$

Hence $d_K(x_0,y) \leq d_{E^2}(x'_0,y')$, so we are done if $\theta \geq \pi$, and similarly if $\phi \geq \pi$.

So we may assume that θ and ϕ are both less than π. This enables us to construct the planar 1–complex shown in Figure 2.4. Here $d_K(x_1,y) = d_{E^2}(x''_1,y''), d_K(x_2,y) = d_{E^2}(x''_2,y''), d_K(x_0,y) = d_{E^2}(x''_0,y'')$ and the angles θ' and ϕ' are equal to θ and ϕ respectively. Notice that since $[x,y]$ is a geodesic segment $\phi' + \theta' \geq \pi$.

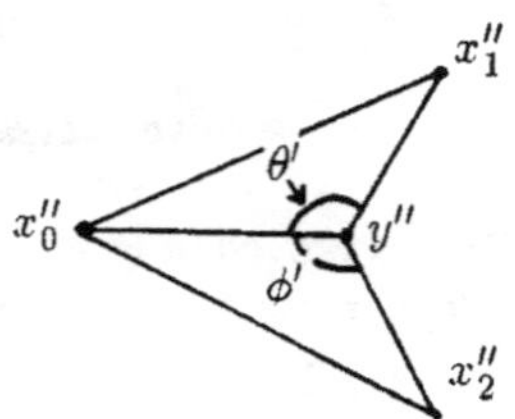

Figure 2.4

We now transform Figure 2.4 into a comparison triangle for T by motions which do not decrease $d_{E^2}(x''_0,y'') = d_K(x_0,y)$, thus completing the proof of the theorem.

By Lemma 2.5, $d_K(x_0,x_1) \geq d_{E^2}(x''_0,x''_1)$ and $d_K(x_0,x_2) \geq d_{E^2}(x''_0,x''_2)$. Also, since $\theta < \pi$ and geodesic segments are unique in K, we have $d_K(x_0,y)+d_K(y,x_1) >$

$d_K(x_0, x_1)$. So without changing the length of the edges incident at y'' we can increase the angles θ' and ϕ' until the distances $d_{E^2}(x_0'', x_1'')$ and $d_{E^2}(x_0'', x_2'')$ reach $d_K(x_0, x_1)$ and $d_K(x_0, x_2)$ respectively. This brings us to the situation shown in Figure 2.5. We now increase the angle ψ until it is equal to π. While doing so we keep all edge lengths constant, except that $d_{E^2}(x_0'', y'')$ is allowed to increase. This completes the construction of the comparison triangle, and with it the proof of Theorem 2.7. □

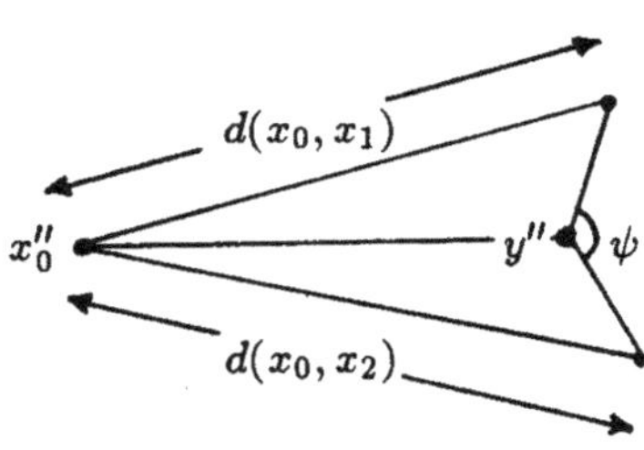

Figure 2.5

Bruhat and Tits showed that if a metric space X is complete and satisfies the CN condition then any group G which acts on X by isometries and has a non-empty bounded orbit stabilises some point of X. An elegant proof of this, due to Serre, can be found in [8] pp.157–158. Moreover, if X is a *geodesic* metric space then there is a unique geodesic segment connecting any two points which are fixed by G. Since G acts by isometries it must fix this geodesic segment pointwise. It follows that the fixed point set of G is contractible. So we have the following consequence of Theorem 2.7.

Fixed Point Theorem: *If K satisfies any of the conditions* I *to* IV *given in Theorem 2.7, and a group G acts on K by isometries, such that there is a bounded orbit, then the fixed point set of G is non-empty and contractible.*

2.3 Local Characterisations of Non-Positive Curvature.

One would like to define what it means for a Euclidean simplicial complex to have non-positive curvature locally. The obvious way to do this is to select one of the global characterisations of non-positive curvature given in Sections 2.1 and 2.2 and require that every point of the given complex have a neighbourhood in which this condition holds. Theorem 2.8 (which is stated below) shows that this definition does not depend on which global charactersiation we choose.

Theorem 2.8 also relates local curvature to the structure of links in the given complex by means of the link condition. It is this relationship which enables us to determine whether or not certain simplicial complexes, whose local (combinatorial) structure is sufficiently well understood, support a structure of non-positive curvature (see Proposition 4.4 and Theorem 5.6).

We maintain the convention that the letter K, without further qualification, denotes a piecewise Euclidean complex of type A or B.

Definition: K satisfies the *link condition* if for every $x \in K$ two points y', z' in the spherical simplicial complex $LK(x, K)$ can be joined by a *unique* geodesic segment in $LK(x, K)$ whenever $d(y', z') < \pi$.

Definition: An open set U in K is said to be *geodesically convex* if for all $x, y \in U$ *every* geodesic segment joining x to y in K is contained in U.

Definition: K is said to satisfy a property *locally* if K can be covered by geodesically convex open sets each of which satisfies the given property.

Theorem 2.8: *For piecewise Euclidean complexes K of type A or B the following local characterisations of non-positive curvature are equivalent:*

I) *K has unique geodesic segments locally.*
II) *The metric on K is convex locally.*
III) *K satisfies $CAT(0)$ locally.*
IV) *K satisfies CN locally.*
V) *Every point of K has a neighbourhood such that any geodesic triangle contained in that neighbourhood has non-positive excess.*
VI) *K satisfies the link condition.*

Further, if K is of type B then each of the above conditions is equivalent to

VII) *There exists $\epsilon_0 > 0$ such that for all $x \in K$ the ball $B_{\epsilon_0}(x)$ is geodesically convex and has unique geodesic segments.*

If K is of type A, then condition VII is strictly stronger than the other conditions, as the following example shows.

Example: Let L be the 1–complex which has vertices v_n, and which for every integer n has three 1–cells $\{v_{2n-1}, v_{2n}\}, \{v_{2n-1}, v_{2n+1}\}, \{v_{2n}, v_{2n+1}\}$, each of length $1/n$. This is a Euclidean simplicial complex of type A which satisfies conditions I to VI of Theorem 2.8, but does not satisfy condition VII.

Proof of Theorem 2.8: We begin by showing that conditions I to IV are equivalent. Let U be a geodesically convex open subset of K, and suppose that geodesic segments are unique in U. The Remark following the proof of Lemma 2.3 implies that the conclusions of Lemmas 2.3 and 2.4 are valid in U. Theorem 2.1 follows from these results in an entirely formal way, hence the metric on U is convex. Then, a formal translation of the arguments given in Section 2.2 proves that U satisfies $CAT(0)$, CN and Alexandrov's condition on the excess of a geodesic triangle. The argument given in the proof of Proposition 2.6 shows that this last condition implies that geodesic segments are unique in U. Hence conditions I to V are equivalent.

It is clear that VII⇒I, so it suffices to show that I and VI are equivalent and that if K is of type B then VI⇒VII. This follows from the following three results.

First we show that *any* **E**-simplicial complex of type A or B can be covered with geodesically convex open sets. Here, as in Lemma 1.2, $\epsilon(x)$ denotes the distance from x to its link in K.

Lemma 2.9: *Given $x \in K$, the open ball $B_\epsilon(x)$ is geodesically convex whenever $\epsilon < \frac{1}{2}\epsilon(x)$.*

Proof: Fix $y, z \in B_\epsilon(x)$ and notice that if α is a geodesic segment joining y to z in K then its image is contained in $B_{\epsilon(x)} \subseteq st(x)$. If we form the join of α with x and map the resulting 2–complex isometrically into $\mathbf{E}^2$ then we obtain the situation illustrated in Figure 2.6. It follows that

$$d(\alpha(t), x) \leq \max\{d(x,y), d(x,z)\} < \epsilon$$

as required. □

Lemma 2.10: *If K is of type B then there is a constant $\epsilon_0 > 0$ such that for every $x \in K$ there exists y with $B_{\epsilon_0}(x) \subseteq B_{\frac{1}{2}\epsilon(y)}(y)$.*

Proof: The idea of the proof is to construct a constant ϵ_0 such that if $x \in K^{(n)}$ lies in the $2\epsilon_0$–neighbourhood of the $(n-1)$–skeleton of K then there exists $y \in K^{(n-1)}$ such that $B_{\epsilon_0}(x) \subseteq B_{\frac{1}{2}\epsilon(y)}(y)$.

Let Σ denote the disjoint union of $\{\sigma : \sigma \in Shapes(K)\}$, and define $\eta : \Sigma \to (0, \infty)$ by

$$\eta(s) = \min\{d(s, F) : \sigma \in Shapes(K), F \text{ a face of } \sigma,\ s \in (\sigma - F)\}.$$

Note that η is continuous on the interior of simplices.

Let η_0 be $\frac{1}{4}$ of the minimum value attained by η on $\Sigma^{(0)}$, the 0–skeleton of Σ. Then inductively for $n \leq D = dim(K)$ we define η_n to be $\frac{1}{4}$ of the minimum value attained by η on the compact set obtained by deleting the η_{n-1} neighbourhood of $\Sigma^{(n-1)}$ from $\Sigma^{(n)}$. Notice that $4\eta_n \leq \eta_{n-1}$. Set $\epsilon_0 = \eta_D$.

Arguing by induction on n it is easy to see that if $x \in K^{(n)} - K^{(n-1)}$ then either $2\eta_n < \epsilon(x)$ or else there exists $y \in K^{(n-1)}$ with $B_{2\eta_n}(x) \subseteq B_{\frac{1}{2}\epsilon(y)}(y)$. □

Proposition 2.11: *K satisfies the link condition if and only if for every $x \in K$ geodesic segments are unique in $B_\epsilon(x)$ whenever $\epsilon < \epsilon(x)/2$.*

Proof: Fix $x \in K$ and let $\epsilon < \epsilon(x)/2$. The exponential map allows us to identify $LK(x, K)$ with the boundary of the closed ball $B_\epsilon(x)$, by a map which scales the metric by a factor of ϵ. Given $y, z \in B_\epsilon(x)$ let y', z' denote the points in $LK(x, K)$ determined by the line segments $[x, y], [x, z]$. Lemma 1.2 shows that if $y' = z'$ then $[y, z]$ is the unique geodesic segment joining y to z in K. So we may assume that $y' \neq z'$.

Suppose that there exists a geodesic segment α joining y to z in $(B_\epsilon(x) - x)$. In Lemma 1.10 we showed that α', the image of α under radial projection, is represented by a taut chain in $LK(x, K)$, and $L(\alpha') < \pi$ (see Figure 1.3). On the other hand, any piecewise geodesic path α' of length $< \pi$ which joins y' to z' in $LK(x, K)$ is the image of a path α from y to z in $(B_\epsilon(x) - x)$, as illustrated in Figure 2.6.

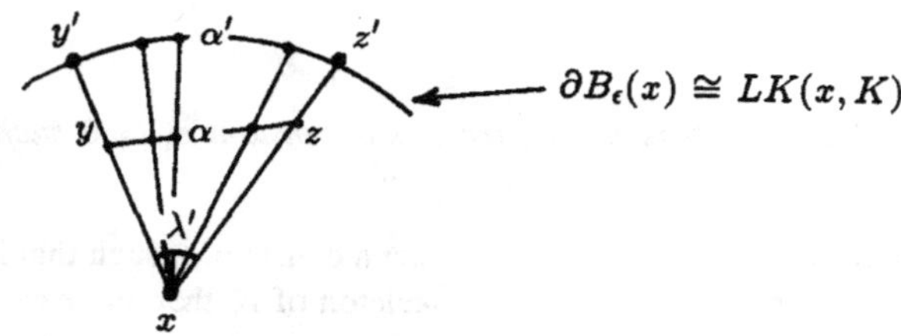

Figure 2.6

Let $\lambda = L(\alpha)$ and $\lambda' = L(\alpha') < \pi$. By the cosine rule we have

$$\lambda^2 = d(x, y)^2 + d(x, z)^2 - 2d(x, y)\,d(x, z)\cos\lambda'.$$

Hence minimising λ is equivalent to minimising λ'. So if $d(y', z') < \pi$ then there is a unique gedoesic segment joining y' to z' in $LK(x, K)$ if and only if there is a unique geodesic segment joining y to z in $(B_\epsilon(x) - x)$.

To complete the proof we must show that if $d(y', z') \geq \pi$ then there is a unique geodesic segment from y to z in $B_\epsilon(x)$. But this is clear, because by the arguments given above there is no local geodesic from y to z in $(B_\epsilon(x) - x)$, and the line segment $[x, w]$ is the unique geodesic segment from x to w for all $w \in B_\epsilon(x)$. Hence the concatenation of the line segments $[y, x]$ and $[x, z]$ is the unique geodesic segment from y to z in $B_\epsilon(x)$. □

We now return to the proof of Theorem 2.8. If condition I holds, then Lemma 2.9 implies that for every $x \in K$ there exists $\delta(x) > 0$ such that $B_\epsilon(x)$ is geodesically convex and has unique geodesic segments whenever $\epsilon < \delta(x)$. It then follows from the "if" implication of Proposition 2.11 that K satisfies the link condition. Conversely, the "only if" implication of Proposition 2.11 shows that VI$\Rightarrow$I.

In fact, these arguments shows that if condition I holds then geodesic segments are unique in $B_{\epsilon(y)/2}(y)$ for every $y \in K$. Moreover, the uniqueness of geodesic segments implies that the restriction of the metric to each such ball is convex. It follows that if $B_{\epsilon_0}(x) \subseteq B_{\epsilon(y)/2}(y)$ then $B_{\epsilon_0}(x)$ is geodesically convex and has unique geodesic segments. The equivalence of conditions I and VII for complexes of type B now follows from Lemma 2.10. □

2.4 Passing From the Local to the Global Situation.

We now turn to the the most difficult step in the proof of the Main Theorem. Namely, that of relating local characterisations of non-positive curvature to global characterisations.

Theorem 2.12: *If K is a piecewise Euclidean complex of type A or B which has unique geodesic segments locally then for all $x, y \in K$ there is a unique shortest path in each homotopy class of paths from x to y in K.*

Corollary : *If K is simply connected and satisfies any of the local characterisations of non-positive curvature given in Theorem 2.8 then it satisfies all of the global characterisations of non-positive curvature given in Theorems 2.1 and 2.7.*

Thus Theorem 2.12 completes the proof of the special case of the Main Theorem which we stated at the beginning of Section 2, and shows that for simply connected simplicial complexes the existence of a piecewise Euclidean metric of non-positive curvature is, as one would expect, a local condition.

Throughout Section 2.4 we assume that K satisfies the hypotheses of Theorem 2.12, and that $p, q \in K$ are fixed.

Let Ω denote the set of PL paths $\alpha : [0,1] \to K$ such that $\alpha(0) = p$ and $\alpha(1) = q$, equipped with the metric topology given by $||\alpha - \beta|| = sup\{d(\alpha(t), \beta(t)) : t \in [0,1]\}$. We continue to denote the length of α by $L(\alpha)$, and denote the length of $\alpha|_{[0,t]}$ by $L(\alpha, t)$. Notice that $L(\alpha, t)$ is a piecewise linear function of t and hence is differentiable almost everywhere.

Definition: If $\alpha \in \Omega$ then the *energy* of α is $E(\alpha) = \int_o^1 (\frac{dL}{dt}(\alpha, t))^2 dt$.

The Cauchy-Schwarz inequality implies that $E(\alpha) \geq L(\alpha)^2$, with equality if and only if α is parameterised proportional to arc length.

Although E is not continuous on Ω, it is continuous on the the following subspace of "broken geodesics".

$$\Omega(n) = \left\{\alpha \in \Omega : t \mapsto \alpha\left(\frac{i+t}{n}\right) \text{ is a geodesic segment for } i = 0, \dots, n-1\right\}$$

On this subspace E is given by the formula

$$E(\alpha) = \sum_{i=0}^{n-1} \int_{t_i}^{t_{i+1}} \left(\frac{d}{dt} L(\alpha, t)\right)^2 dt = \sum_{i=0}^{n-1} \frac{d(\alpha(t_{i+1}), \alpha(t_i))^2}{t_{i+1} - t_i}$$
$$= n \sum_{i=0}^{n-1} d(\alpha(t_{i+1}), \alpha(t_i))^2$$

where $t_i = i/n$. To prove Theorem 2.12 we analyse the convexity properties of E on the subspace

$$\Omega(n, e) = \{\alpha \in \Omega(n) : E(\alpha) \leq e\}.$$

Proof of Theorem 2.12

The strategy of our argument is as follows (cf. [22] Theorem 19.2, and [27] Theorem 1). Fix a homotopy class of paths from p to q in K and let α_0 and α_1 be two shortest paths in this class. In Lemma 2.13 we show that there exist constants n and e such that α_0 and α_1 lie in the same path component of $\Omega(n, e)$. Moreover, we can choose e so that every $\alpha \in \Omega(n, e)$ is uniquely determined by the sequence of points $(\alpha(t_i))_{i=1}^{n-1}$. This allows us to embed $\Omega(n, e)$ in K^{n-1}. We then extend E to a continuous function on the whole of K^{n-1} and prove that it has strong convexity properties (Proposition 2.17). This leads to the following result (see Lemmas 2.19 and 2.20):

The set of local minima of $E|_{\Omega(n,e)}$ is discrete, and there is a strong deformation retraction of $\Omega(n, e)$ onto this set.

This will complete the proof of Theorem 2.12, because the fact that α_0 is a shortest path in its homotopy class implies that E restricted to the corresponding component of $\Omega(n, e)$ attains its minimum value at α_0. The same is true of α_1. But E has a unique local minimum in each component of $\Omega(n, e)$, hence α_0 and α_1 must coincide.

Lemma 2.13: *Suppose that K is of type B, and let ϵ_1 be such that for every $x \in K$ the ball $B_{4\epsilon_1}(x)$ is geodesically convex and has unique geodesic segments. If α_0 and α_1 are two shortest paths in the same path component of Ω then there exists a positive integer n and a constant $e > 0$ such that α_0 and α_1 lie in the same path component of $\Omega(n, e)$. Moreover, if we let $t_i = i/n$ then $d(\alpha(t_i), \alpha(t_{i+1})) < \epsilon_1$ for every $\alpha \in \Omega(n, e)$ and $i \in \{1, \dots, n-1\}$.*

Proof: Let $D : I \times I \to K$ be a homotopy between the maps α_0 and α_1, and denote $D(s,\tau)$ by $\alpha_s(\tau)$. Because D is uniformly continuous, we can find an integer m such that if $\tau_i = i/m$ then $d(\alpha_s(\tau_i), \alpha_s(\tau_{i+1})) < \epsilon_1$ for all $s \in I$ and $i \in \{0, \ldots, m-1\}$. We can then replace each α_s by the path α'_s which is defined to be the concatenation of the *unique* geodesic segments joining $\alpha_s(\tau_i)$ to $\alpha_s(\tau_{i+1})$ in K.

More precisely, we reparameterise the geodesic segment joining $\alpha_s(\tau_i)$ to $\alpha_s(\tau_{i+1})$ to obtain a path $\sigma_s^i : [\tau_i, \tau_{i+1}] \to K$ which is parameterised proportional to arc length, and define $\alpha'_s : [0,1] \to K$ by $\alpha'_s|_{[\tau_i,\tau_{i+1}]} = \sigma_s^i$. Notice that $\alpha_0 = \alpha'_0$ and $\alpha_1 = \alpha'_1$ because α_0 and α_1 are shortest paths in their homotopy class and $B_{\epsilon_1}(x)$ is simply connected (indeed contractible) for all x.

Geodesics vary continuously with their endpoints in $B_{\epsilon_1}(x)$ for every $x \in K$, and hence $s \mapsto \alpha'_s$ is a continuous path from α_0 to α_1 in $\Omega(m)$. The energy function E is continuous along this path and hence attains a maximum value, which we denote by e.

For every α'_s we have the following inequality

$$E(\alpha'_s) = \sum_{i=0}^{m-1} \int_{\tau_i}^{\tau_{i+1}} \left(\frac{d}{dt} L(\alpha'_s, \tau)\right)^2 d\tau = m \sum_{i=0}^{m-1} d(\alpha'_s(\tau_{i+1}), \alpha'_s(\tau_i))^2 \le m^2 \epsilon_1^2,$$

which implies that $e \le m^2 \epsilon_1^2$.

Thus if we let $n = m^2$ and $t_i = i/n$ then for every $\alpha \in \Omega(n,e)$ and $i \in \{0, \ldots, m-1\}$ we have

$$\begin{aligned} d(\alpha(t_i), \alpha(t_{i+1}))^2 &= \left(\int_{t_i}^{t_{i+1}} \frac{d}{dt} L(\alpha,t)\, dt\right)^2 \\ &\le (t_{i+1} - t_i) \int_0^1 \left(\frac{d}{dt} L(\alpha,t)\right)^2 dt \\ &\le \frac{1}{n} e \le \epsilon_1^2. \end{aligned}$$

Where the second line comes from the Cauchy-Schwarz inequality in $L^2[0,1]$ applied to $\frac{d}{dt}L(\alpha,t)$ and the characteristic function of $[t_i, t_{i+1}]$. □

Before proceeding we must dispense with an irksome technicality. Namely, if K is of type A then in general the fact that K has unique geodesic segments locally does not imply that there exists a constant ϵ_0 such that geodesic segments are unique in $B_{\epsilon_0}(x)$ for every $x \in K$. However, one can apply the method of Proposition 2.11 to show that for every integer R there exists $\epsilon(R) > 0$ such that if $d(x,p) < R$ then $B_{\epsilon(R)}(x)$ is geodesically convex and has unique geodesic segments. This allows us to modify the proof of Lemma 2.13 to obtain an analogous result if K is of type A, as we now indicate.

For the purposes of this discussion we retain the notation which we introduced in the proof of Lemma 2.13. Fix an integer R so that the image of the homotopy disc D is contained in the ball of radius R about p. We can choose m so that $d(\alpha_s(\tau_i), \alpha_s(\tau_{i+1})) < \epsilon(R)$ for all $s \in [0,1]$ and $\tau = i/m$ where $i \in \{0, \dots, m-1\}$. As in the proof of Lemma 2.13, we replace each α_s by α'_s, the concatenation of the geodesic segments joining $\alpha_s(\tau_i)$ to $\alpha_s(\tau_{i+1})$. This gives a continuous path from α_0 to α_1 in $\Omega(m)$. Let e denote the maximum value which E attains on this path.

If $\alpha \in \Omega$ and $E(\alpha) < e$ then the image of α is contained in the ball of radius $\sqrt{e}$ about p. Fix an integer R' so that $R' \gg \sqrt{e}$. The final inequality in Lemma 2.13 shows that if we take n sufficiently large then $d(\alpha(t_i), \alpha(t_{i+1})) < \epsilon(R')$ for every $\alpha \in \Omega(n,e)$ and $i \in \{0, \dots, n-1\}$, where $t_i = i/n$.

Thus, if we write ϵ_1 in place of $\epsilon(R')/4$ then the essential conclusion of Lemma 2.13 remains valid if K is of type A. Namely, a path $\alpha \in \Omega(n,e)$ is determined by the sequence of points $(\alpha(t_i))$, and for every $x \in K$ which we shall need to consider (i.e. those which lies in a small neighbourhood of the set $\{\alpha(t) : \alpha \in \Omega(n,e)\}$) the ball $B_{\epsilon_1}(x)$ is geodesically convex and has unique geodesic segments.

For the remainder of this section we assume that n and e are as in Lemma 2.13 or as in the preceding discussion, according to whether K is of type B or of type A. We also retain the notation t_i for i/n.

Lemma 2.14: *The map $\Phi : \Omega(n,e) \to K^{n-1}$ given by $\alpha \mapsto \langle \alpha(t_1), \dots, \alpha(t_{n-1}) \rangle$ is an injection.*

Proof: Elements of $\Omega(n,e)$ are geodesic on the subintervals $[t_i, t_{i+1}]$, and we chose ϵ_1 so that if $d(\alpha(t_i), \alpha(t_{i+1})) < \epsilon_1$ then there is a unique geodesic segment from $\alpha(t_i)$ to $\alpha(t_{i+1})$ in K. □

To complete the proof of Theorem 2.12 we study the image of $\Omega(n,e)$ under the map Φ, which we defined in Lemma 2.14. We begin by describing the metric structure of K^{n-1}, whose elements we denote by $X = \langle x_1, \dots, x_m \rangle$. Each closed cell $\Sigma \subset K^{n-1}$ is the $(n-1)$–fold Cartesian product of simplices in K. The partial metric on $\Sigma = \sigma_1 \times \dots \times \sigma_{n-1}$ is given by $d(X,Y)^2 = \sum_{i=1}^{n-1} d(x_i, y_i)^2$; and K^{n-1} is a piecewise Euclidean space of type A or B according to the type of K.

Lemma 2.15: $\alpha(t) = \langle \alpha_1(t), \dots, \alpha_{n-1}(t) \rangle$ *is a geodesic segment in K^{n-1} if and only if α_i is a geodesic segment in K for every i.*

Proof: According to the Cauchy-Schwarz inequality, a path α is a geodesic segment if and only if it has minimal energy among all paths which have the same endpoints as α. Therefore it is enough to consider paths which are parameterised proportional arc length, and show that $E(\alpha) = \Sigma_{i=1}^{n-1} E(\alpha_i)$.

Suppose that α is represented by the $m-$chain $(\alpha(\tau_0), \ldots, \alpha(\tau_m))$. Then α_i is represented by the $m-$chain $(\alpha_i(\tau_0), \ldots, \alpha_i(\tau_m))$, and is parameterised proportional to arc length on each subinterval $[\tau_i, \tau_{i+1}]$. Hence

$$E(\alpha) = \sum_{i=0}^{m-1} \int_{\tau_i}^{\tau_{i+1}} \left(\frac{d}{dt} L(\alpha, t)\right)^2 dt = \sum_{i=0}^{m-1} \frac{d(\alpha(\tau_{i+1}), \alpha(\tau_i))^2}{\tau_{i+1} - \tau_i}$$

$$\sum_{j=1}^{n-1} E(\alpha_j) = \sum_{j=1}^{n-1} \int_0^1 \left(\frac{d}{dt} L(\alpha_j, t)\right)^2 dt = \sum_{j=1}^{n-1} \sum_{i=0}^{m-1} \frac{d(\alpha_j(\tau_{i+1}), \alpha_j(\tau_i))^2}{\tau_{i+1} - \tau_i}$$

The last terms in each row are equal because of the definition of the metric on the individual cells of K^{n-1}. □

We have proved that the map $\Phi : \alpha \mapsto \langle \alpha(t_1), \ldots, \alpha(t_m) \rangle$ is injective, and it is easy to see that it is continuous $(d(\Phi(\alpha), \Phi(\beta)) \leq (n-1)^{\frac{1}{2}} \|\alpha - \beta\|)$. In fact Φ is a homeomorphism of $\Omega(n, e)$ onto its image, because in the region of K under consideration geodesic segments vary continuously with their endpoints in balls of radius $4\epsilon_1$, and $L(\alpha|_{[t_i, t_{i+1}]}) < \epsilon_1$ for all $\alpha \in \Omega(n, e)$.

We denote the image of Φ in K^{n-1} by P_e. We extend the map $(E \circ \Phi^{-1}) : P_e \to [0, \infty)$ to a continuous function (which we shall also call E) defined on the whole of K^{n-1}, by the formula

$$E(X) = n \left(d(p, x_1)^2 + \sum_{i=1}^{n-2} d(x_i, x_{i+1})^2 + d(x_{n-1}, q)^2 \right).$$

We shall prove (Proposition 2.17) that this extended function has a strong convexity property which enables us to strong deformation retract P_e (and hence $\Omega(n, e)$) onto a discrete set of local minima of E, thus completing the proof of Theorem 2.12.

We shall need the following consequence of Theorem 2.8.

Lemma 2.16: *Let $x \in K$, and suppose that $B_\delta(x)$ is geodesically convex and has unique geodesic segments. If $\alpha : [0, 1] \to B_\delta(x)$ is a geodesic segment in $B_\delta(x)$ such that $\alpha(0) \neq \alpha(1)$ and $D = d(x, \alpha(0)) = d(x, \alpha(1)) > 0$ then $d(x, \alpha(t)) < D$ for all $t \in (0, 1)$.*

Proof: By Theorem 2.8 $CAT(0)$ holds in $B_\delta(x)$, so it is enough to consider the case $K = \mathbf{E}^2$, where the result is clear. □

The proof of the following result is essentially due to Stone [27].

Proposition 2.17: *If $X, Y \in K^{n-1}$ are distinct points in the ϵ_1-neighbourhood of P_e and $d(X,Y) < \epsilon_1$ then there is a unique geodesic segment from X to Y in K^{n-1}, and E is strictly convex along this geodesic segment.*

Proof: Notice that $d(X,Y) < \epsilon_1$ implies that $d(x_i, y_i) < \epsilon_1$ for all i. Hence there exists a unique geodesic segment from x_i to y_i in K, and we denote this by β_i. The uniqueness of the β_i, together with Lemma 2.15, implies that $\beta(t) = \langle \beta_1(t), \ldots, \beta_{n-1}(t) \rangle$ is the unique geodesic segment from X to Y in K^{n-1}.

It remains to show that E is strictly convex along β. We are assuming that there exists $Z \in P_e$ such that $d(Z, X) < \epsilon_1$. Hence

$$d(x_i, x_{i+1}) \leq d(x_i, z_i) + d(z_i, z_{i+1}) + d(z_{i+1}, x_{i+1}) < 3\epsilon_1.$$

It follows that β_i and β_{i+1} are both contained in $B_{4\epsilon_1}(x_i)$, which is geodesically convex and has unique geodesic segments. Further, by the convexity of the metric on $B_{4\epsilon_1}(x_i)$ we have

$$\begin{aligned} d(p, \beta_1(t)) &\leq (1-t)\, d(p, x_1) + t\, d(p, y_1) \\ d(\beta_{n-1}(t), q) &\leq (1-t)\, d(x_{n-1}, q) + t\, d(y_{n-1}, q) \\ d(\beta_i(t), \beta_{i+1}(t)) &\leq (1-t)\, d(x_i, x_{i+1}) + t\, d(y_i, y_{i+1}). \end{aligned}$$

By Lemma 2.15 the first of these inequalities is strict for $t \in (0,1)$ unless $d(p, x_1) \neq d(p, y_1)$ or $x_1 = y_1$. And if $x_1 = y_1$ then by the same argument the bottom inequality is strict for $i = 1$ and $t \in (0,1)$ unless $d(x_1, x_2) \neq d(y_1, y_2)$ or $x_2 = y_2$. The points $X, Y \in K^{n-1}$ are distinct, so proceeding in this way we see that one of the inequalities given above is strict for all $t \in (0,1)$, or else there is some i for which $d(x_i, x_{i+1}) \neq d(y_i, y_{i+1})$. (Here we have adopted the convention that $z_0 = p$ and $z_n = q$ for all $Z \in K^{n-1}$.)

Using the convexity of the function $f(a) = a^2$ one can show that for any non-negative numbers a, b, c if $a \leq (1-t)b + tc$ then $a^2 \leq (1-t)b^2 + tc^2$, with equality only if the original inequality was actually an equality, and either $t \in \{0, 1\}$, or $b = c = a$. Applying this observation to the inequalities given above, and adding the resulting inequalities we obtain

$$E(\beta(t)) \leq (1-t)\, E(X) + t\, E(Y)$$

with strict inequality for all $t \in (0,1)$. □

By subdividing the simplices of K we may assume that every closed cell in K^{n-1} which meets P_e has diameter bounded above by ϵ_1. Let L denote the union of these closed cells, and notice that if two points in L are $4\epsilon_1$-close then they can be joined by a unique geodesic segment in L. It follows from Lemma 2.4* that if these points are elements of the same closed cell in L then the line segment joining them in L

(which is *a priori* only a local geodesic) is a geodesic segement. In particular, by Proposition 2.17, E is strictly convex along this line segment. Hence $E|_\Sigma$ attains its minimum value at a unique point $X_\Sigma \in \Sigma$ for every closed cell Σ in L. Inductively, in order of increasing dimension, we star each cell with this point to obtain a (simplicial) subdivision of L.

More precisely, if Σ is a 1–cell and $\partial\Sigma = \{X_0, X_1\}$ then we introduce new edges $\{X_0, X_\Sigma\}$ and $\{X_\Sigma, X_1\}$ into the underlying cell structure of L. Then inductively we assume that we have constructed the desired simplicial subdivision on the $(i-1)-$skeleton of L and consider an $i-$cell Σ. For every $j-$dimensional face τ^j of $\partial\Sigma$ that does not contain X_Σ we introduce the union of the line segments $\{[X_\Sigma, Y] : Y \in \tau^j\}$ as a new $(j+1)-$simplex in the cell structure on L. Notice that we have not changed the metric structure on L, only the underlying combinatorial structure.

Let J denote the union of the closed cells in this subdivision which meet P_e. Notice that if C is a closed cell in J then E is strictly convex along all of the line segments in C and $E|_C$ attains its unique minimum at a vertex.

Proposition 2.18: *The set $\{E(v) : v$ is a vertex of $J\}$ is finite.*

Proof: If K is of type A then the result is trivial because J is compact. So we may assume that K is of type B. The vertices of J are of two types, those which were vertices of L, and those which we introduced in the construction of J. First we show that $\{E(X) : X$ is a vertex of $L\}$ is a finite set.

Every vertex of K^{n-1} is of the form $X = \langle v_1, \ldots, v_{n-1}\rangle$ where each of the coordinates v_i is a vertex of K. If $X \in L$ then (as we showed in the proof of Proposition 2.17) $d(v_i, v_{i+1}) < 3\epsilon_1$ for all i. By Theorem 1.11, there exists an integer N such that any geodesic segment in K of length less than $3\epsilon_1$ can be modelled in one of the finitely many corridors which occur in K and have length less than N. Let Λ denote the set of such corridors. Then for $i \in \{1, \ldots, n-2\}$ we have $d(v_i, v_{i+1}) \in \{d(u, v) : u, v$ vertices of some $\mu \in \Lambda\}$. This last set is finite, so $d(v_i, v_{i+1})$ takes on only finitely many values as X varies over the vertices of L. A similar argument shows that $d(p, v_i)$ and $d(q, v_{i+1})$ can only take on finitely many values, and since $E(X)$ depends only on these quantities the set $\{E(X) : X$ is a vertex of $L\}$ is finite.

It remains to show that the set of values which E can take at vertices which were introduced in the construction of J is finite. This set can be written as $\{inf E|_\Sigma : \Sigma$ a closed cell in L and $\Sigma \cap P_e \neq \emptyset\}$, and we will use this description to show that the set is finite. The idea of the proof is to show that $inf E|_\Sigma$ depends only on the metric structure of a set of complexes which we obtain by concatenating $n-$tuples of corridors in Λ. Because Λ is a finite set there are only finitely many possible isometric models for such complexes.

Let $\Sigma = (B_1 \times \ldots \times B_{n-1})$ be a closed cell in L which meets P_e, and fix $X \in \Sigma$. Because $d(x_i, x_{i+1}) < 3\epsilon_1$ and geodesic segments are unique in $B_{4\epsilon_1}(x)$ for all $x \in K$, there is a unique geodesic segment from x_i to x_{i+1} in K for every $i \in \{0, \ldots, n-1\}$ (recall the convention that $x_0 = p, x_n = q$). We parameterise this geodesic segment proportional to arc length to obtain $\gamma_X^i : [t_i, t_{i+1}] \to K$, and denote the concatenation of the γ_X^i by $\gamma_X : [0,1] \to K$. Notice that $\gamma_X \in \Omega(n)$ and $E(X) = E(\gamma_X)$ (which is greater than e if $X \notin P_e$).

Let Γ_i denote the model corridor $\Gamma(\gamma_i^X)$ (here we are using the notation which we introduced with the definition of a corridor at the beginning of Section 2.1), and let $\tilde{\gamma}_X^i$ denote the preimage of γ_X^i in Γ_i. We concatenate the corridors Γ_i by identifying the unique face of Γ_{i+1} which contains $\tilde{\gamma}_X^{i+1}(t_{i+1})$ in its interior with the unique face of Γ_i which contains $\tilde{\gamma}_X^i(t_{i+1})$ in its interior. We denote the resulting complex by $C(X) = \Gamma_0 * \ldots * \Gamma_{n-1}$, and identify Γ_i with its image in $C(X)$. Notice that the defining maps $\Gamma(\gamma_i^X) \to K$ induce a map $\phi : C(X) \to K$ which is length preserving.

The concatenation of the paths $\tilde{\gamma}_X^i$ form a path from $\tilde{\gamma}_X^0(0) = \tilde{p}$ to $\tilde{\gamma}_X^{n-1}(1) = \tilde{q}$ in $C(X)$, and if we denote this path by $\tilde{\gamma}_X$ then $\phi \circ \tilde{\gamma}_X = \gamma_X$. In particular, because ϕ is length preserving $E(\gamma_X) = E(\tilde{\gamma}_X)$.

On the other hand, if $\tilde{\gamma}$ is any path in $C(X)$ with the property $\tilde{\gamma}(0) = \tilde{p}$, $\tilde{\gamma}(1) = \tilde{q}$ and for every i the image of $\tilde{\gamma}|_{[t_i, t_{i+1}]}$ is contained in Γ_i then $E(\tilde{\gamma}) \geq E(Y)$ for some $Y \in \Sigma$. To see this notice that our construction of $C(X)$ was such that $\Gamma_i \cap \Gamma_{i+1}$ is the unique simplex of $C(X)$ which contains $\tilde{\gamma}_X(t_i)$ in its interior. Hence for every i the point $\phi \circ \tilde{\gamma}(t_i)$ lies in the unique simplex of K which contains x_i in its interior, and this implies that $\langle \phi \circ \tilde{\gamma}(t_1), \ldots, \phi \circ \tilde{\gamma}(t_{n-1}) \rangle$ is an element of Σ. Let Y denote this element. If $\phi \circ \tilde{\gamma}|_{[t_i, t_{i+1}]} = \gamma_Y^i$ for every i then $E(\tilde{\gamma}) = E(\phi \circ \tilde{\gamma}) = E(\gamma_Y)$, if not then $E(\tilde{\gamma}) = E(\phi \circ \tilde{\gamma}) > E(\gamma_Y)$.

Thus if we let $e(C(X); \tilde{p}, \tilde{q})$ denote the infimum of $E(\tilde{\gamma})$ taken over all such paths $\tilde{\gamma}$, then the minimum value which E attains on Σ is

$$inf \{ e(C(X); \tilde{p}, \tilde{q}) : X \in \Sigma \}.$$

The value of this expression depends only on the metric structure of the bipointed complex $(C(X); \tilde{p}, \tilde{q})$, and since $C(X)$ was obtained by concatenating corridors from the finite set Λ, there are only finitely many possible isometric models for $(C(X); \tilde{p}, \tilde{q})$. □

We regard $V = \{E(v) : v \in P_e \text{ is a vertex of } J\}$ as the set of *critical values* of $E|_J$. Proposition 2.18 implies that this set is finite, and we write $V = \{e_1, \ldots, e_m\}$ where $0 < e_1 < \ldots < e_m \leq e$. Notice that J was constructed so that V contains all the local minima of $E|_{P_e}$.

We consider the following subcomplexes of J:

$$H_i = \bigcup\{\sigma \subset P_e : E(W) \leq e_i \ \forall\, vertices\ W \in \sigma\}, \quad 1 \leq i \leq m.$$

The strict convexity of $E|_J$ along geodesic segments implies that H_i is a *full* subcomplex of J and that $H_m \subseteq \{X \in J : E(X) \leq e\} = P_e$. It also implies that no two adjacent vertices (i.e., vertices that cobound a 1–cell in J) correspond to the same e_i. In particular, H_1 is a discrete set consisting of global minima for $E|_{P_e}$.

The following two lemmas complete the proof of Theorem 2.12.

Lemma 2.19: *P_e strong deformation retracts onto H_m.*

Proof: We constructed J as a simplicial subdivision of a neighbourhood of P_e in L. So any closed simplex σ of J which is not contained in P_e can be written uniquely as a join $(\sigma' \star \sigma'')$ where $\sigma' \subset H_m$ and σ'' is disjoint from P_e. The straight line retraction of $(\sigma - \sigma'')$ onto σ' takes $(P_e \cap \sigma)$ into itself because E is strictly decreasing along each of the line segments from σ'' to σ'. These straight line retractions of individual simplices agree on common faces, and hence combine to give the desired retraction of P_e onto H_m. □

Lemma 2.20: *H_i strong deformation retracts onto the union of H_{i-1} with a discrete set of vertices at which E has a local minimum.*

Proof: Consider a vertex W with $E(W) = e_i$. If W is an isolated point of H_i then the value of E at all of the adjacent vertices of J is greater than e_i and hence E has a local minimum at W.

H_i is a *full* subcomplex of J, so if W is not an isolated point of H_i then $St(W, J) \cap H_i$ is the cone $W \star (link(W, J) \cap H_{i-1})$, i.e. it is the union of the line segments $\{[X, W] : X \in link(W, J)$ with $E(X) < e_i\}$. We use this cone structure to define a strong deformation retraction D of $St(W, J) \cap H_i$ onto $link(W, J) \cap H_{i-1}$ as follows:

Fix a point $X_0 \in link(W, J) \cap H_{i-1}$ and let $\alpha : [0,1] \to St(W, J)$ denote the line segment $[W, X_0]$. We define D by requiring that at time t the map $Y \mapsto D(Y, t)$ sends the line segment $[X, W]$ linearly onto the unique geodesic segment from X to $\alpha(t)$ in J, for every $X \in (link(W; J) \cap H_{i-1})$ and $t \in [0, 1]$. Lemma 2.3 ensures that the homotopy is continuous. And the convexity of $E|_J$ implies that $[X, \alpha(t)]$ is contained in H_i for all $t \in [0, 1]$, and is contained in H_{i-1} for $t = 1$. (In general the image of D may not be contained in $St(W)$, but this is not important.) □

2.5 Extending Geodesics

In Section 1 we showed that if K is a complex of type A or B then its intrinsic metric is complete. One would like to deduce from this that if K is simply connected and has non-positive curvature then any isometric embedding of an interval of the real line can be extended to an isometric embedding of the whole line (i.e., geodesic segments in K can be extended indefinitely). This is not true as stated, but to make it true we need only make an exclusion analogous to that which one would make for boundary points in a complete Riemannian manifold of non-positive sectional curvature.

Proposition 2.21: *Suppose that K is a simply connected piecewise Euclidean complex of type A or B which satisfies any of the characterisations of non-positive curvature given in Theorems 2.7 and 2.8. Then any isometric map from an interval of the real line into K can be extended to an isometric embedding of the whole real line, or to an embedding of a closed interval the image of whose endpoints lie in the set*

$$\Sigma = \{x \in K : \ LK(x, K) \ \ is \ \ contractible\} .$$

Proof: K is complete, so any isometric map of an interval of the real line into K can be extended to the closure of that interval. So it is enough to prove that we can extend any isometry $\alpha : I \to K$ where I has a finite endpoint a with $x = \alpha(a) \notin \Sigma$.

We are assuming that K satisfies the link condition, so there must exist $Q \in LK(x, K)$ with $d(P, Q) \geq \pi$. For if this were not the case then there would be a unique geodesic segment from P to every point of $LK(x, K)$, implying that $LK(x, K)$ is contractible, contrary to hypothesis.

Let q be the image of Q under the natural identification of $LK(x, K)$ with the sphere $S_{\epsilon(x)/2}(x)$. The angle between the geodesic segments α and $[x, q]$ at x is $d(P, Q)$, and hence is at least π. So extending α by $[x, q]$ we get a local geodesic. But any local geodesic in K is a geodesic segment. □

3. M(κ)-Complexes of Non-positive Curvature

All of the results which we proved for piecewise Euclidean complexes in Section 2 have analogues in the world of piecewise hyperbolic complexes. Indeed, it was only for clarity of exposition that we did not give a unified treatment of these results in the previous section. This we do now, by considering $M(\kappa)-$simplicial complexes with $\kappa \leq 0$.

3.1 The Proof of The Main Theorem

The following definitions are needed for the statement of the Main Theorem.

If $\chi > 0$ then we say that a geodesic triangle is $\chi-$*small* if the sum of the lengths of its sides is no greater than $2\pi/\sqrt{\chi}$. (If $\chi \leq 0$ then the condition of being $\chi-$*small* is vacuous.)

Following Gromov [17] we say that a geodesic metric space X has curvature $\leq \chi$ if it satisfies the following condition locally.

CAT(χ): Let $T = \Delta(x_0, x_1, x_2)$ be a $\chi-$small geodesic triangle in X, and let y be a point on the side of T which has endpoints x_1 and x_2. Choose a comparison triangle $T' = \Delta(x'_0, x'_1, x'_2)$ in $M(\kappa)^2$ (the plane of constant curvature κ) and let y' denote the unique point on the geodesic segment $[x'_1, x'_2]$ such that $d_{M(\kappa)^2}(x'_i, y') = d_K(x_i, y)$ for $i = 1, 2$. Then $d_{M(\kappa)^2}(x'_0, y') \geq d_K(x_0, y)$.

The Main Theorem: *If K is a simply connected $M(\kappa)-$simplicial complex of type A or B and $\kappa \leq 0$ then the following 13 conditions are equivalent:*

Global conditions:

I) *K has unique geodesic segments.*
II) *K satisfies $CAT(\kappa)$ globally.*
III) *K satisfies $CAT(\chi)$ globally, for some χ.*
IV) *K satisfies CN globally.*
V) *The metric on K is convex.*
VI) *Every geodesic triangle in K has non-positive excess.*

Local conditions:

VII) *K has unique geodesic segments locally.*
VIII) *K satisfies $CAT(\kappa)$ locally.*
IX) *K satisfies $CAT(\chi)$ locally, for some χ.*
X) *K satisfies CN locally.*

XI) *The metric on K is convex locally.*

XII) *Every point of K has a neighbourhood such that any geodesic triangle contained in that neighbourhood has non-positive excess.*

XIII) *K satisfies the link condition.*

In Section 2 we proved a special case of the Main Theorem, and for the most part the proofs which we gave there are equally valid (*mutatis mutandis*) in the present setting. When this is the case we do not repeat the details of the argument, or the accompanying motivation.

Remark: At first glance it may seem strange that in this class of spaces $CAT(\kappa)$ is equivalent to $CAT(\chi)$, where χ is arbitrary. However this merely reflects the fact that if we metrize the simplices of a given complex to have constant negative curvature, then in doing so we redistribute the natural curvature of the complex, and concentrate it in the skeleton of codimension 2. In particular, if the given complex does not support a structure of negative curvature (e.g. if it is not aspherical) then forcing the simplices to be negatively curved concentrates an infinite amount of positive curvature at certain points in the complex, and thus any condition which gives a bound on the local curvature of the space, such as $CAT(\chi)$, fails in the neighbourhood of such a point.

For the remainder of Section 3 the letter K, without further qualification denotes an $M(\kappa)$–simplicial complex of type A or B.

Intuitively speaking, a geodesic metric space X has curvature $\leq \kappa$ if geodesic segments in X diverge more quickly than geodesics in $M(\kappa)^2$. To make this notion precise we introduce the following "divergence function".

Given $\theta \in [0, \pi]$ and $a, b \in [0, \infty)$ we can choose geodesic segments $\alpha, \beta : [0,1] \to M(\kappa)^2$ with lengths $L(\alpha) = a$ and $L(\beta) = b$, which have a common initial point and meet at an angle θ. We then define

$$D(\theta, a, b) = d(\alpha(1), \beta(1)).$$

The homogeneity of $M(\kappa)^2$ ensures that this definition is independent of the choices of α and β.

In the proof of Theorem 3.1 we shall need the fact that for fixed a and b the function $\theta \mapsto D(\theta, a, b)$ is monotone increasing. This follows from the cosine rule in $M(\kappa)^2$. We shall also need the fact that for all $a, b, c > 0$, and all $\theta, \theta', \theta'' \in [0, \pi]$ such that $\theta = \theta' + \theta''$,

$$D(\theta, a, c) \leq D(\theta', a, b) + D(\theta'', b, c).$$

This is simply a restatement of the triangle inequality in $M(\kappa)^2$.

The following comparison theorem is closely analoguous to the Topogonov inequality for manifolds of non-positive curvature [5].

Theorem 3.1: *Suppose that* K *has unique geodesic segments, and let* α_0 *and* α_1 *be geodesic segments in* K *with common initial point* $\alpha_0(0) = \alpha_1(0) = x$. *If the angle between* α_0 *and* α_1 *at* x *(which we denote by* $\theta_{0,1}$*) is less than* π *then*

$$D(\theta_{0,1}, L(\alpha_0), L(\alpha_1)) \leq d(\alpha_0(1), \alpha_1(1)).$$

Our proof of Theorem 3.1 relies on the following lemma, which can be proved using the construction given in Lemma 2.4. (See Figure 2.1.)

Lemma 3.2: *If* K *has unique geodesic segments then for every* $\eta > 0$ *and every local geodesic* α *in* K, *there exists a constant* $\epsilon > 0$ *with the following property: If* β *is a local geodesic in* K *for which* $\beta(0) = \alpha(0)$ *and* $\|\alpha - \beta\| < \epsilon$ *then* Θ, *the angle between* α *and* β *at* $\alpha(0)$, *is less than* η *and*

$$D(\Theta, L(\alpha), L(\beta)) \leq d(\alpha(1), \beta(1)).$$

Corollary : *If* K *has unique geodesic segments then every local geodesic in* K *is a geodesic segment. Hence there is a unique local geodesic joining any two points in* K.

Proof of Theorem 3.1: Let σ denote the unique geodesic segment from $\alpha_0(1)$ to $\alpha_1(1)$. We denote the geodesic segment from $\alpha_0(0)$ to $\sigma(s)$ by α_s, and the angle between α_s and α_t at x by $\theta_{s,t}$. It follows from the preceding corollary that $\theta_{s,t} < \pi$ for all $s, t \in [0,1)$.

We shall prove that the set

$$\Sigma = \{s : D(\theta_{0,s}, L(\alpha_0), L(\alpha_s)) \leq d_K(\alpha_0(1), \alpha_s(1))\}$$

is the whole of $[0,1]$.

Lemma 3.2 implies that $\theta_{0,s}$ and $L(\alpha_s)$ vary continuously with s, and hence Σ is closed. To see that it is open, fix $s < 1$ with $s \in \Sigma$. In Lemma 2.3 we proved that geodesic segements vary continuously with their endpoints, so it follows from Lemma 3.2 that if $\delta > 0$ is small enough then $\theta_{s,s+\delta} < \pi - \theta_{0,s}$ and

$$D(\theta_{s,s+\delta}, L(\alpha_s), L(\alpha_{s+\delta})) \leq d(\alpha_s(1), \alpha_{s+\delta}(1)).$$

The following inequality shows that $s + \delta \in \Sigma$.

$$\begin{aligned} d(\alpha_0(1), \alpha_{s+\delta}(1)) &= d(\alpha_0(1), \alpha_s(1)) + d(\alpha_s(1), \alpha_{s+\delta}(1)) \\ &\geq D(\theta_{0,s}, L(\alpha_0), L(\alpha_s)) + D(\theta_{s,s+\delta}, L(\alpha_s), L(\alpha_{s+\delta})) \\ &\geq D(\theta_{0,s} + \theta_{s,s+\delta}, L(\alpha_0), L(\alpha_{s+\delta})) \\ &\geq D(\theta_{0,s+\delta}, L(\alpha_0), L(\alpha_{s+\delta})). \end{aligned}$$

Here we have used the triangle inequality in $LK(x, K)$ to deduce that $\theta_{0,s+\delta} \leq \theta_{0,s} + \theta_{s,s+\delta}$. □

Lemma 3.3: *If a geodesic metric space X satifies $CAT(0)$ then the metric on X is convex.*

Proof: Fix $t \in [0,1]$. Given geodesic segments α and β in X with $\alpha(0) = \beta(0)$, and $s \in [0,1]$, we consider the geodesic triangles $\Delta_s \subseteq X$ which have vertices $\{\alpha(0), \alpha(1), \beta(s)\}$. Let $\Delta_s' \subseteq \mathbf{E}^2$ be a comparison triangle for Δ_s. We denote the image of $z \in \Delta_s$ in Δ_s' by z_s', and the vertex angle at $\alpha(0)_s'$ by ϕ_s.

$CAT(0)$ applied to Δ_1 yields $d(\beta(t), \alpha(1)) \leq d(\beta(t)_1', \alpha(1)_1')$. And since $d(\beta(t)_t', \alpha(1)_t') = d(\beta(t), \alpha(1))$, this implies that $\phi_t \leq \phi_1$. Hence

$$d\left(\alpha(t)_t', \beta(t)_t'\right) \leq d\left(\alpha(t)_1', \beta(t)_1'\right).$$

$CAT(0)$ applied to Δ_t yields $d(\alpha(t), \beta(t)) \leq d(\beta(t)_t', \alpha(t)_t')$. Hence

$$d(\alpha(t), \beta(t)) \leq d\left(\alpha(t)_1', \beta(t)_1'\right).$$

Finally, from elementary Euclidean geometry, we have

$$d\left(\alpha(t)_1', \beta(t)_1'\right) = t\ d\left(\alpha(1)_1', \beta(1)_1'\right) = t\ d(\alpha(1), \beta(1)).$$

Hence

$$d(\alpha(t), \beta(t)) \leq t\ d(\alpha(1), \beta(1)).$$

The fact that the function $t \mapsto d(\alpha(t), \beta(t))$ is a convex function for any geodesics segments α and β in K follows easily from the case $\alpha(0) = \beta(0)$, as we showed in Section 2.1. □

Proof of The Main Theorem:

I⇒VI : Theorem 3.1 implies that if geodesic segments are unique in K then each of the vertex angles of a geodesic triangle in K is no larger than the corresponding angle of a comparison triangle in $M(\kappa)^2$.

I⇒II : In the proof of Theorem 2.7 we used a special case of the Topogonov inequality (Lemma 2.5) to show that a piecewise Euclidean complex with unique geodesic segments satisfies $CAT(0)$. A direct translation of that argument, employing Theorem 3.2 in place of Lemma 2.5, proves the present implication.

II⇒IV : If X is any geodesic metric space which satisfies $CAT(\kappa)$ then it satisfies $CAT(\chi)$ for all $\chi \geq \kappa$. And in the proof of Theorem 2.5 we proved that if X satisfies $CAT(0)$ then it satisfies CN.

II⇒V: This is the content of Lemma 3.3.

IV, V, VI⇒I: These implications follow easily from the definition of the given conditions, as we verified in Section 2.3.

Local versions of the preceding arguments prove the equivalence of conditions VII, VIII, XI, X and XII.

XIII⇔VII: For the case $\kappa = 0$ this is the content of Proposition 2.11. In the proof of that proposition the only point at which we used the hypothesis that K was piecewise Euclidean was to employ the cosine rule when comparing the length of a chain in $st(x, K)$ to the length of the projected chain in $LK(x, K)$. The cosine rule in hyperbolic geometry serves the desired purpose equally well. Hence the conclusion of Proposition 2.11 is valid for piecewise hyperbolic complexes.

II⇒III⇒IX: Trivial.

IX⇒VII: Immediate from the definition.

As in the case $\kappa = 0$ we pass from the local to the global situation by proving VII⇒I (which completes the proof of the theorem). To do so we observe that the results presented in Section 2.4 depend only on the local convexity properties of the metric on K, and hence are equally valid for piecewise hyperbolic complexes. One might be concerned by the fact that in the hyperbolic case the $n-$fold Cartesian product K^n is not itself piecewise hyperbolic, but this is of no consequence since our results concerning the geometry of K^n relied only on the fact that a path in K^n is a geodesic segment if and only if its projection onto each coordinate is a geodesic segment. And this observation remains valid in the piecewise hyperbolic case. □

3.2 Alternative Forms of the Link Condition

The *link condition*, which we defined in Section 2.3, is a condition which bounds the curvature of the spherical simplicial complexes which occur as links in some given complex K. It seems natural to ask whether one can prove a result, analogous to the Main Theorem, which would relate this condition to other characterisations of bounded curvature in spherical simplicial complexes. Such a result might then yield alternative ways of verifying the link condition for a given complex.

In fact the Main Theorem does not have a direct analogue in the world of spherical simplicial complexes, but in this subsection we outline the proof of a partial analogue (Lemma 3.4) which gives two alternative characterisations of bounded curvature in the link complexes of an $M(\kappa)-$simplicial complex of type A or B. Each of these conditions is equivalent to the link condition, and in Proposition 4.5 we use one of these characterisations to show that a certain class of 3–complexes are non-positively curved.

Using the techniques of Section 2 one can show that if L is a spherical simplicial complex of type A or B then it satisfies Gromov's $CAT(1)$ condition if and only if there is a *unique* geodesic segment from x to y in L whenever $d(x, y) < \pi$. Similarly, one can show that L has unique geodesics segments locally if and only if it satisfies

$CAT(1)$ locally. However, if L is simply connected and satisfies $CAT(1)$ locally then it does *not* follow that L satisfies $CAT(1)$ globally, as the following example shows.

Example: Consider a geodesic triangulation of the standard 2–sphere cut open along a 1–simplex of length less than π. Let L denote the spherical 2–complex obtained by identifying two copies of this space along the boundary loop γ. L is (topologically) a 2–sphere, and satisfies $CAT(1)$ locally. However, it does not satisfy $CAT(1)$ globally, because the image of γ in L is a geodesic circle of length strictly less than π.

This example highlights the difficulty which arises when trying to relate local descriptions of bounded curvature in spherical simplicial complexes to global descriptions. Namely, that one must take account of the possible occurence of short geodesic circles (i.e., isometric embeddings of a circle of length less than 2π).

This motivates the following definition. The *systole* of L, which we denote $sys(L)$, is the infimum of the lengths of geodesic circles in L. We wish to relate $sys(L)$ to the *injectivity radius* of L, which is defined by $inj(L) = sup\{r$: there is a unique geodesic segment from x to y in L whenever $d(x, y) \leq r\}$.

The argument given by Charney and Davis in Lemma 1.3 of [11], shows that if L is a finite spherical simplicial complex with $inj(L) > 0$ then it fails to satisfy $CAT(1)$ if and only if $\frac{1}{2}sys(L) = inj(L) < \pi$. The hypothesis that the complex is finite is only used to ensure that L has the following property: If $inj(L) \in (0, \infty)$ then there is a (non-degenerate) geodesic bigon in L whose sides have length exactly $inj(L)$.

By employing the method of finite models (which we used in Lemmas 1.5 and 2.3) it is easy to show that complexes of type B also have this property. Moreover, if L is of type B then Lemma 2.10 implies that $inj(L) > 0$ if and only if L has unique geodesics locally. Hence we obtain the following result (cf. [17], p.120).

Lemma 3.4: *If K is an $M(\kappa)$–simplicial complex of type A or B (where κ is arbitrary) then the following are equivalent:*

I) *K satisfies the link condition. (i.e., $inj(LK(x, K)) \geq \pi$ for every $x \in K$).*

II) *For every $x \in K$ the complex $LK(x, K)$ has unique geodesic segments locally and $sys(LK(x, K)) \geq 2\pi$.*

III) *For every $x \in K$ the complex $LK(x, K)$ satisfies $CAT(1)$ globally.*

4. Examples

In this section we describe some examples of complexes of non-positive curvature. In particular we are interested in complexes which are not locally finite. Interesting examples of this type arise in the work of Gersten and Stallings on triangles of groups, and in Section 3.1 we describe their results in the context of the work presented here. Then in Section 3.2 we discuss non-positively curved 3–complexes which have planar links. A particular example of such a complex arises in the study of the group $GL_n(Z[\omega])$, where ω is a primitive sixth root of unity. We use this example to illustrate how the Fixed Point Theorem proved in Section 2.2 can be used to classify the finite subgroups of a group which acts on a complex of non-positive curvature. Before proceeding to Section 2.1 we briefly mention two other classes of examples.

A particularly rich source of non-positively curved complexes is provided by Euclidean buildings. The theory of buildings is extensive and well-documented (see for example [8], [9], [10], [28], and [29]). So too is the role of non-positive curvature in understanding the geometry of buildings of Euclidean type. We do not attempt to develop anything of this theory here, but merely note that Euclidean buildings provide interesting examples of piecewise Euclidean complexes of type B.

Another class of examples which is well-understood is that of metric simplicial trees. Any metric simplicial tree satisfies $CAT(\kappa)$ for all κ, and hence "a non-positively curved simply connected 1–dimensional Euclidean simplicial complex of type B" is just another name for a metric simplicial tree in which the set of edge lengths is finite. Groups which act on such complexes were completely classified by the work of Bass and Serre [24].

4.1 Simplices of Groups.

The simplest example of a group which acts cocompactly (but not freely) on a tree is an amalgamated free product. So it seems reasonable that as a first step towards generalising Bass-Serre theory to higher dimensions one should look for a 2–dimensional analogue of an amalgamated free product. This is the starting point for the work of Gersten and Stallings on triangles of groups.

Definition: An *n-simplex* of groups is a contravariant functor from the poset of faces of an $n-$simplex ordered by inclusion, into the category of groups and monomorphisms.

We think of such a functor T as a diagram of groups, and refer to the image under T of a vertex as a vertex group, the image of a 1–simplex as an edge group, and so on. We denote the direct limit (or generalised pushout) of this diagram in the category of groups by $\Gamma(T)$.

Definition: An $n-$simplex of groups T is said to be *realisable* if the canonical map from each vertex group of T into $\Gamma(T)$ is an injection.

There is an obvious paradigm in this situation. Namely, if a group acts without inversions on a simplicial complex, and the quotient is a single $n-$simplex then the diagram of stabilisers and inclusions in a fundamental domain is a realisable $n-$simplex of groups. (A group is said to act without inversions if the fixed point set of any element is a simplicial subcomplex.)

Definition: We say that an $n-$simplex of groups is *geometric* if it arises as the diagram of stabilisers and inclusions for the fundamental domain of the action of a group of isometries on a piecewise Euclidean complex of non-positive curvature. (Since the quotient is compact the complex will necessarily be of type B.)

Remark: This use of the term "geometric" is not a standard one.

If T is a 1–simplex of groups then $\Gamma(T)$ is the amalgamated free product of the vertex groups over the edge group. It is well-known that every 1–simplex of groups is realisable and geometric [24]. This is far from true in dimension 2, but a sufficient condition for a triangle of groups to be both realisable and geometric has been given by Gersten and Stallings, using the idea of the angle between subgroups of an arbitrary group.

Consider a group G and subgroups A, B, C with $C \subseteq A \cap B$. Fix a set of coset representatives $\{g_i\}_{i \in I}$ for G/C. Let $L_G(A, B; C)$ be the (unoriented) graph with vertex set $G/A \coprod G/B$ and 1–simplices $\{\{g_iA, g_iB\} \mid i \in I\}$.

Definition: The *angle in G between A and B as measured over C* is $2\pi/n$, where n is the length of the shortest reduced circuit in $L_G(A, B; C)$.

Definition: Given a triangle of groups T one associates to each vertex the angle between the incident edge groups as measured over the 2–cell group. The triangle is said to be *non-spherical* if the sum of the vertex angles is no greater than π.

The following theorem is proved in [26]. (An alternative proof is given in [7].)

Theorem 4.1: *(Gersten-Stallings) Every non-spherical triangle of groups is realisable.*

Gersten and Stallings also proved that a non-spherical triangle of groups is geometric. We shall now outline a proof of this fact. The proof which we give here is somewhat different to that which was originally given by Gersten and Stallings.

Fix a non-spherical triangle of groups T, and let Γ denote the direct limit of the corresponding diagram of groups. We assume that each of the vertex angles is non-zero. (If one of the vertex angles is zero, then the triangle degenerates to an amalgamated free product.) We denote the edge groups by E_1, E_2, E_3 and the vertex group with incident edge groups E_i and E_j by $V_{i,j}$. The 2–cell group shall be denoted C.

We define an (abstract) simplicial 2–complex $K(T)$ as follows: (For convenience we write all indices mod 3.)

$$K(T)^0 = \coprod_{i<j} \Gamma/V_{i,j}$$

$$K(T)^1 = \coprod_{i=1}^{3} \{\{e^i_\lambda V_{i,i+1}, e^i_\lambda V_{i,i-1}\} \mid e^i_\lambda \text{ coset reps for } \Gamma/E_i\}$$

$$K(T)^2 = \{\{c_\eta V_{1,2}, c_\eta V_{2,3}, c_\eta V_{3,1}\} \mid c_\eta \text{ coset reps for } \Gamma/C\}.$$

There is a natural action of $\Gamma(T)$ on K, given by left multiplication of cosets. The quotient space for this action is a single 2–simplex, and the cosets containing the identity element form a fundamental domain. Moreover, the pattern of stabilisers in this fundamental domain is precisely the original triangle of groups T. One can also show that the geometric realisation of $K(T)$ is simply connected [7].

Notice that $K(T)$ has a natural labelling, for example a vertex corresponding to a coset of $V_{1,2}$ in Γ is thought of as being labelled $\{1,2\}$, a 1–simplex corresponding to a coset of E_3 in Γ is thought of as having label $\{3\}$, and so on. If we fix a (hyperbolic or Euclidean) triangle Δ with vertices indexed 1 to 3, and each vertex angle equal to the group theoretic angle at the corresponding vertex of T, then this labelling of $K(T)$ induces a simplicial isomorphism from each 2–simplex in $K(T)$ to Δ. The collection of these maps satisfies the axioms given in Section 1.1, and hence we obtain a metric simplicial complex of type B.

The labelling on $K(T)$ also gives a graph isomorphism from the link of any vertex v which is labelled $\{i,j\}$ to the graph $L_{V_{i,j}}(E_i, E_j; C)$. The metric on the link of a vertex in a 2–dimensional complex is given by the angular measure at the vertex. Thus each edge of the metric graph $LK(v, K(T))$ has length π/n. But n was defined to be the length of the shortest reduced circuit in $L_{V_{i,j}}(E_i, E_j; C)$. Thus for every vertex $v \in K$ the graph $LK(v, K(T))$ contains no geodesic circles of length less than 2π, and hence $K(T)$ satisfies the link condition. Thus we have proved:

Theorem 4.2: *(Gersten-Stallings) Every non-spherical triangle of groups is geometric.*

The following result now follows immediately from the Fixed Point Theorem, which we proved at the end of Section 2.2.

Theorem 4.3: *(Gersten-Stallings) If T is a non-spherical triangle of groups then every bounded (e.g. finite) subgroup of $\Gamma(T)$ is conjugate to a subgroup of one of the vertex groups.*

Recently, Haefliger [19] has shown that any non-positively curved orbihedron which has only finitely many isometry types of cells arises as the quotient of a non-positively curved piecewise Euclidean complex by a group of isometries. This generalises Theorems 4.1 and 4.2.

4.2 3–Complexes With Planar Links.

The link of a vertex in a 2–dimensional metric simplicial complex is a graph, with edge lengths given by the angular measure at the vertex. Thus to verify the link condition it is enough to calculate the lengths of reduced circuits in each such graph. In higher dimensions things are much more delicate, because it is difficult to identify geodesics in the link complexes. However, if one has sufficiently explicit knowledge about the structure of the links then it may still be possible to decide whether a given complex satisfies the link condition. (See for example [23], [11] and Section 5 below.) One can also prove more general results for classes of complexes in which the structure of the links is sufficiently simple. For example:

Proposition 4.4: *Suppose that K is an (abstract) simplicial 3–complex such that the link of each vertex is simplicially isomorphic to a triangulation of the plane in which every vertex has valence at least 6. Fix a regular tetrahedron T in $M(\kappa)^3$ (the unique simply connected 3–manifold of constant sectional curvature $\kappa \leq 0$). If we metrize each 3–simplex in K by means of a simplicial isomorphism to T then K satisfies the link condition.*

Our proof of Proposition 4.4 requires the following simple fact from spherical geometry: Suppose that the paths $\alpha, \beta : [0,1] \to S^2$ satisfy $\alpha(0) = \beta(0)$ and $\alpha(1) = \beta(1)$, and that they cobound an embedded disc $D \subset S^2$. We say that D is *good* if β is an arc of a great circle, and α is a piecewise gedoesic path of length strictly less than π with the property that the angle between its sucessive geodesic subarcs, as measured in D, is at least π.

Lemma 4.5: *Every good disc in S^2 contains an equilateral spherical triangle of side $\pi/2$.*

Proof: We may assume that $\alpha(0) = \beta(0)$ is the north pole, that the initial segment of β follows the line of longitude $0°W$, and that the longitudinal coordinate of $\alpha(t)$ is $a(t)°W$, where a is an increasing function of t.

Notice that $a(t) \geq 180$ for some $t \in (0,1]$. Hence α does not meet the equator between $0°W$ and $90°W$. For if $\alpha(t)$ did lie on this arc then the paths $\alpha|_{[0,t]}$ and $\alpha|_{[t,1]}$ would both have length at least $\pi/2$, contradicting the fact that the length of α is strictly less than π. Therefore the south-west octant of the sphere is contained in D. □

Corollary : *If Γ is a 2–dimensional* **S**-*corridor whose cells are equilateral spherical triangles of side $l < \pi/2$, then there is a unique local geodesic from x to y for all $x, y \in \Gamma$.*

Proof: By Theorem 1.1 there is a geodesic segment α from x to y in Γ. Any geodesic segment is a local gedoesic, so the only issue is that of uniqueness. Suppose that γ is another local geodesic from x to y. Restricting to subpaths if necessary we may assume that α and γ cobound an embedded disc $D \subset \Gamma$. If we orient D then for every $x \in D$ we have a well-defined notion of the clockwise and anticlockwise directions in the spherical complex $LK(x,\Gamma)$ (which is a circle or an arc of a circle).

Because α and γ are local geodesics, the paths in $LK(\alpha(t),\Gamma)$ and $LK(\gamma(t),\Gamma)$ determined by D each have length at least π, for every $t \in [0,1]$. In fact, we may assume that the path which D determines in $LK(\gamma(t),\Gamma)$ has length exactly π. For if this is not the case then we can replace γ by the unique local geodesic β which has the same initial segment and for which the distance in $LK(\beta(t),\Gamma)$ from the point determined by the forward direction of β to that determined by the backward direction is exactly π, when measured in the anticlockwise direction. We can then replace α by an initial segment so that the disc D' cobounded by α and β is non-singular.

However, if such a disc $D' \subset \Gamma$ were to exist then it would be isometric to a good disc in S^2, and Lemma 4.5 would then give an isometry from an equilateral spherical triangle of side $\pi/2$ into Γ. No such map exists. □

Proof of Proposition 4.4: Fix $x \in K$, and write $L = LK(x,K)$. The 2–cells of L are equilateral spherical triangles with vertex angles strictly between $\pi/3$ and $\pi/2$, so the link of a point $p \in L$ is a circle of length $> 2\pi$ if p is a vertex of L, and of length 2π if p is not a vertex. Thus L satisfies the link condition and has unique geodesic segments locally. So by Lemma 3.4 it is enough to prove that there are no geodesic circles of length less than 2π in L.

We use the phrase *a tesselation line in* L to describe a locally isometric embedding of the real line into L whose image lies in the 1–skeleton. Notice that every edge can be extended to a tesselation line, and that if a geodesic circle in L crosses a tesselation line then it must cross it at least twice.

Suppose that $\sigma \subset L$ is a geodesic circle of length less than 2π. Because L is homeomorphic to the plane, σ bounds an open disc in L. Consider a tesselation line which meets this disc in an arc γ which is outermost among all tesselation lines which intersect the disc. Let α denote the (short) arc of σ which has the same endpoints as γ. Because γ is outermost, there is a corridor in L which contains both α and γ. But these paths are both local geodesics, and the length of α is strictly less than π. This contradicts the preceding corollary. □

Remark: The 3–dimensional nature of Proposition 4.4 is something of an illusion, because any complex K which satisfies the given hypotheses strong deformation retracts onto the following 2–dimensional subcomplex of its first barycentric subdivision.

$$L = K' - \bigcup \{st\,(v,K') \ : v \text{ a vertex of } K\}$$

L can be metrized as a Euclidean simplicial complex of non-positive curvature in such a way that any simplicial isomorphism of L is an isometry. To do this one first observes that for every vertex $v \in K$ the complex $link(v, K')$ is simplicially isomorphic to the first barycentric subdivision of a tesselation of the plane in which every vertex has valence at least 6. Let T_v denote this tesselation. We metrize each 2–simplex in T_v as a Euclidean equilateral triangle of side 1, and give $link(v, K')$ the induced metric. This defines a Euclidean simplicial structure on L which is of both type A and type B, and it is easy to check that it also satisfies the link condition.

Notice that if G is a group which acts on K by simplicial isomorphisms then the retraction of K onto L can be done $G-$equivariantly, and the induced action is by isometries. In particular, if the vertex stabilisers for this induced action are finite then G is (equi-)semihyperbolic in the sense of [3].

A complex for $\mathbf{GL_2(Z[\omega])}$

An example of a simplicial complex which satisfies the hypotheses of Proposition 4.4 is the following complex which was studied by Roger Alperin [4].

Let R be a ring, and consider the set Λ of free direct summands for R^2. We say that $L_1, L_2 \in \Lambda$ are independent if $L_1 + L_2 = L_1 \oplus L_2 = R^2$. Let $K(R)$ be the (abstract) simplicial complex whose vertices are the elements of Λ and whose $q-$simplices are those subsets of Λ of the form $\{L_0, \ldots, L_q\}$ for which L_i, L_j are independent for $0 \leq i \neq j \leq q$. The action of $GL_2(R)$ on Λ preserves the relation of independence, and hence gives an induced action on $K(R)$.

Let ω be a primitive sixth root of unity. In the case $R = GL_2(Z[\omega])$ one can show that $K(R)$ is a simply connected 3–complex, and that the link of every vertex is simplicially isomorphic to the standard tesselation of the Euclidean plane by equilateral triangles. It follows from Proposition 4.4 that $K(R)$ can be metrized as a piecewise hyperbolic complex of negative curvature. In particular $K(R)$ is contractible. In [4] Alperin showed that $K(R)$ was contractible by studying a filtration of the space by subcomplexes. He then calculated the Euler charcteristic and homology of $SL_2(Z[\omega])$ by studying its action on the 2–dimensional retract of $K(R)$ which we described in the preceding remark.

According to our Fixed Point Theorem any finite subgroup of $GL_2(Z[\omega])$ stabilises a point of $K(R)$. We shall show that this leads to a classification of the finite subgroups in $GL_2(Z[\omega])$.

The centre of $GL_2(Z[\omega])$ acts trivially on $K(R)$, so we get an induced action of $PGL_2(Z[\omega])$. It is this action which we study. For convenience we write Γ in place of $PGL_2(Z[\omega])$.

If a finite subgroup $H \subset GL_2(Z[\omega])$ fixes a vertex in $K(R)$ then it acts by simplicial isomorphisms on the link of that vertex. If we metrize this link as a Euclidean plane in the standard way then H acts by isometries, and hence has a fixed point. Thus any finite subgroup of Γ stabilises the barycentre $b_0(\sigma)$ of some simplex $\sigma \subset K(R)$ of dimension at least 1. The complex $link(b_0, K)$ is finite, and hence its full group of symmetries, which we denote $sym(b_0)$, is also finite.

It is easy to see that no non-trivial element of Γ fixes a 2–simplex in $K(R)$ pointwise. It follows that if σ is a simplex of dimension at least 1 and $b_0(\sigma)$ is its barycentre then $stab_\Gamma(b_0(\sigma))$ injects into $sym(b_0(\sigma))$.

Γ acts transitively on the set of simplices in $K(R)$ in each dimension. Thus if we choose simplices $\sigma_1, \sigma_2, \sigma_3$ of dimension 1, 2 and 3 respectively, then every finite subgroup of Γ is conjugate to a subgroup of $stab_\Gamma(b_0(\sigma_1)), stab_\Gamma(b_0(\sigma_2))$, or $stab_\Gamma(b_0(\sigma_3))$. We now give explicit descriptions of these stabilisers.

The complex $link(b_0(\sigma_2), K(R))$ is the suspension of the triangle $\partial\sigma_2$. Thus $sym(b_0(\sigma_2))$ is isomorphic to $S_3 \times \mathbf{Z}_2$. Because no non-trivial element of Γ fixes σ_2 pointwise, $stab_\Gamma(b_0(\sigma_2))$ must be contained in the S_3 factor. In fact it is isomorphic to S_3. To see this notice that if we let σ_2 be the 2–simplex with vertices $(1,0), (0,1)$ and $(1,1)$ then $stab_\Gamma(b_0(\sigma_2))$ contains a 3–cycle a and a transposition b given by the following matrices.

$$A = \begin{pmatrix} 0 & 1 \\ -1 & 1 \end{pmatrix} \qquad B = \begin{pmatrix} 0 & 1 \\ 1 & 0 \end{pmatrix}$$

The complex $link(b_0(\sigma_1), K(R))$ is the suspension of a hexagonal 1–complex, and hence $sym(b_0(\sigma_1))$ is isomorphic to $D_{12} \times \mathbf{Z}_2$. Because no non-trivial element of Γ fixes a 2–simplex in $K(R)$ pointwise, $stab_\Gamma(b_0(\sigma_1))$ does not contain the unique symmetry of the link which interchanges the suspension points and leaves the other vertices fixed. Hence $stab_\Gamma(b_0(\sigma_1))$ is isomorphic to a subgroup of D_{12}, the dihedral group of order 12.

Finally, the complex $link(b_0(\sigma_3), K(R))$ is the boundary of a tetrahedron, and hence $sym(b_0(\sigma_3))$ is isomorphic to S_4. One can show that no $\gamma \in \Gamma$ acts transitively on the vertices of a 3–simplex in $K(R)$, so $stab_\Gamma(b_0(\sigma_3))$ contains no 4–cycles. Hence it is isomorphic to a subgroup of A_4, the alternating group on four letters.

At this stage we have shown that $stab_\Gamma(b_0(\sigma_2)) \cong S_3$, $stab_\Gamma(b_0(\sigma_3)) \hookrightarrow A_4$ and $stab_\Gamma(b_0(\sigma_1)) \hookrightarrow D_{12}$. Thus if we can exhibit subgroups of Γ which are isomorphic to D_{12} and A_4 then these groups must occur as the stabilisers of a 1–simplex and a 3–simplex in $K(R)$ respectively.

Let c and d denote the elements of Γ determined by the matrices C and D (which are given below). These elements generate the subgroup $G_{c,d} = \langle c, d \mid c^6 = b^2 =$

$(cd)^2 = 1\rangle$, which is a dihedral group of order 12.

$$C = \begin{pmatrix} \omega & 0 \\ 0 & 1 \end{pmatrix} \qquad D = \begin{pmatrix} 0 & 1 \\ 1 & 0 \end{pmatrix}$$

On the other hand, the elements $d, e \in \Gamma$ determined by the matrices D and E generate the subgroup $G_{e,f} = \langle e, f \mid e^3 = f^2 = (e^{-1}f)^3 = 1\rangle$, which is isomorphic to A_4.

$$E = \begin{pmatrix} 0 & 1 \\ -1 & 1 \end{pmatrix} \qquad F = \begin{pmatrix} 0 & \omega \\ \omega^2 & 0 \end{pmatrix}$$

Thus any finite subgroup of $PGL_2(Z[\omega])$ is conjugate to a subgroup of one of the groups $G_{a,b}, G_{c,d}$ or $G_{e,f}$. It follows that any finite subgroup of $GL_2(Z[\omega])$ is conjugate to a subgroup in the preimage of one of these three groups.

5. The Curvature of the Culler-Vogtmann Complex

In this section we exemplify a method for deciding if a given complex can be given a piecewise Euclidean structure of non-positive curvature. The idea is to develop a good understanding of the local (combinatorial) structure of the space, and use this to decide whether or not the given complex can satisfy the link condition.

More specifically: let G be a group which acts on a simplicial complex K with finite quotient, and suppose that the quotient space, the local structure of K, and the $G-$stabilisers of the vertices of K can be described explicitly. One would like to know whether or not the space K can be given a $G-$equivariant piecewise Euclidean structure of non-positive curvature. We assume (for the sake of argument) that K/G has been metrized as a piecewise Euclidean complex such that the induced structure on K has non-positive curvature. In Section 1 we described the induced spherical simplicial structure on the links of points in K. By analysing the symmetries of these spherical complexes we can identify "conjugate points", i.e., pairs of points which cannot be joined by a unique geodesic segment in the link complex. We are supposing that K satisfies the link condition, so any path joining a pair of conjugate points must have length at least π. Thus from each each choice of such a path we obtain an inequality involving the angles of the cells in the quotient space. The aim is to gather sufficient information to determine these angles, or else to obtain contradictory bounds and hence deduce that K does not support a $G-$equivariant piecewise Euclidean structure of non-positive curvature. The same method can of course be applied when dealing with piecewise hyperbolic structures.

In Section 5.1 we shall describe the action of $Out(F_n)$, the group of outer automorphisms of the free group of rank n, on the Culler-Vogtmann complex K_n. It has been proved ([12], [13], [20]) that K_n is contractible and that the fixed point set of any finite subgroup of $Out(F_n)$ is non-empty and contractible. We saw in Section 2 that these properties are indicative of non-positive curvature, so it seems reasonable to ask whether or not K_n supports an $Out(F_n)-$equivariant piecewise Euclidean structure of non-positive curvature. The answer to this question is yes in the case $n = 2$, where it is known that $Out(F_2)$ is isomorphic to $SL_2(\mathbf{Z})$, and that its action on K_2 is the usual action of $SL_2(\mathbf{Z})$ on the Serre tree (see [13]). For $n \geq 3$ we shall prove that K_n does *not* support an $Out(F_n)-$equivariant piecewise Euclidean (or piecewise hyperbolic) structure of non-positive curvature, by using the technique described above.

5.1 The Culler-Vogtmann Complex K_n

In this subsection we describe the Culler-Vogtmann complex K_n as the geometric realisation of a certain poset consisting of marked graphs. We then describe the natural action of $Out(F_n)$ on this poset. This action is order preserving and hence induces a simplicial action of $Out(F_n)$ on K_n.

Definitions

A *graph* G is a connected 1–dimensional CW-complex. G is said to be *admissible* if it is not homotopy equivalent to any proper subgraph, all of its vertices have valence at least three, and the complement of every edge is connected. Henceforth all graphs are required to be admissible. A graph R is called a *rose* if it has a single vertex, which we denote $v(R)$.

Fix a rose R_0 and identify F_n with $\pi_1(R_0, v(R_0))$. We call this the *standard rose.* A *marking* is a homotopy equivalence $g : R_0 \to G$, where G is an admissible graph. Two markings $g_1 : R_0 \to G_1$ and $g_2 : R_0 \to G_2$ are said to be equivalent if there is a graph isomorphism $i : G_1 \to G_2$ such that the following diagram commutes up to free homotopy.

$$\begin{array}{ccc} R_0 & \xrightarrow{g_1} & G_1 \\ \| & & \downarrow i \\ R_0 & \xrightarrow{g_2} & G_2 \end{array}$$

A *marked graph* is an equivalence class of markings, and the class containing $g : R_0 \to G$ is denoted (g, G).

Let e be an edge of the graph G. We say that (g', G') is obtained from (g, G) by *blowing down the edge* e if there is a cellular homotopy equivalence $d : G \to G'$ which collapses e, is one-to-one on the complement of e, and satisfies $d \circ g \simeq g'$.

Description of K_n

We define a partial ordering on the set of marked graphs by $(g', G') \preceq (g, G)$ if and only if (g', G') is obtained from (g, G) by blowing down finitely many edges. K_n is defined to be the geometric realisation of this partially ordered set. It is easy to prove that an admissible graph whose fundamenal group is free of rank n has at most $(3n-3)$ edges, hence K_n has dimension $(2n-3)$. An example of a maximal dimensional cell in the case $n = 3$ is shown in Figure 5.2.

Labelled graphs

We wish to represent a given vertex (g, G) of K_n pictorially, in such a way that both the graph G and the marking g are determined by the picture. To do this we choose a maximal tree T in G and a homotopy inverse to g which sends T to the vertex $v(R_0)$. This map sends each oriented edge of $(G - T)$ to a loop in R_0 based at $v(R_0)$, and we label the edge with the corresponding word in $F_n = \pi_1(R_0, v(R_0))$.

It is important to notice that the representation of a given (g, G) by a labelled graph is *not* unique. Different choices of maximal tree and homotopy inverse will give rise to different markings; in particular, altering the labels on a given labelled graph by the action of an inner automorpism of F_n produces a labelled graph representing the same marked graph. Figure 5.1 shows four different labelled graphs representing the same point in K_n.

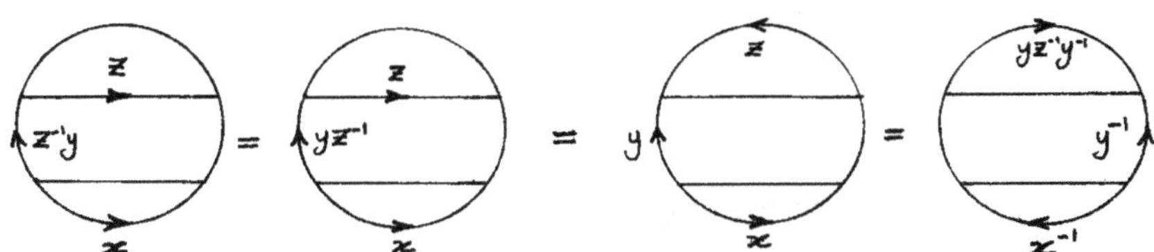

Figure 5.1

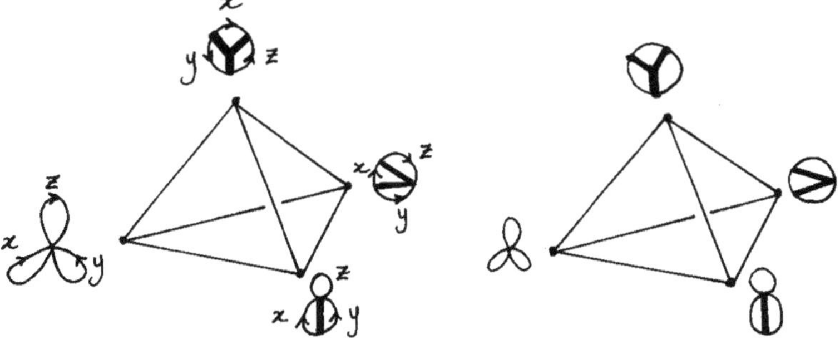

Figure 5.2

Figure 5.3

The action of $Out(F_n)$

There is a natural left action of $Out(F_n)$ on the complex K_n, which for our purposes is best described in terms of marked graphs. Given $\phi \in Aut(F_n)$ and a vertex $(g, G) \in K_n$ determined by a labelled graph with edge labels $\{w_1, \ldots, w_n\}$ we define $\phi \cdot (g, G)$ to be the vertex represented by the same graph with edge labels $\{\phi(w_1), \ldots, \phi(w_n)\}$.

It can be shown that this action is well-defined and preserves the partial ordering $\preceq$, hence it is simplicial. Further, the penultimate sentence of the paragraph on labelled graphs implies that the action of $Inn(F_n)$ is trivial, so we get an induced action of $Out(F_n)$ on K_n.

The $Out(F_n)$–stabiliser of a vertex in K_n is isomorphic to its group of graph automorphisms. This follows from the definition of equivalence for marked graphs, as is proved in [25]. For example, in the case $n = 3$ the isotropy group of a rose is isomorphic to Ω_3, the semi-direct product of $\mathbf{Z}_2\times\mathbf{Z}_2\times\mathbf{Z}_2$ with S_3, which occurs naturally as the full group of isometries of a cube in Euclidean 3–space.

We shall represent cells in the quotient complex $K_n/Out(F_n)$ by omitting the edge labels but continuing to draw the edges which are to be blown down in boldface. For example the image of the cell shown in Figure 5.2 will be drawn as in Figure 5.3.

5.2 The Link of a Rose in K_3

In order to carry out the program of proof outlined in the introduction to this section it is essential that we have explicit knowledge of the links of vertices in K_3. The most complicated link which we need to consider is that of a rose. The Gertrude Stein lemma of [13] shows that the links of any two roses are isomorphic via the action of some element of $Out(F_n)$. Moreover, since we are assuming that $Out(F_n)$ acts by isometries, it follows that they are isometric as spherical simplicial complexes. For notational convenience we shall study the link of the standard rose $\rho_0 = (id, R_0)$ and represent all vertices of the link by graphs with labels $\{a, b, c\}$, where $\{a, b, c\}$ is the standard basis for $\pi_1(R_0, v(R_0))$.

Note: In this subsection we are only concerned with the *combinatorial* structure of the link, not the metric structure.

As we remarked earlier, the stabiliser of a rose is Ω_3 the full group of symmetries of a cube in Euclidean 3–space. Our explicit description of $lk(R_0)$ distinguishes a triangulated 2–sphere embedded in the link. This can be viewed as a cube in a natural way, and the action of $\Omega_3 = stab(\rho_0)$ on $lk(\rho_0)$ is entirely determined by its usual action on this cube. This description of the action of $stab(\rho_0)$ exhibits the symmetries of $lk(\rho_0)$ with sufficient clarity for us to identify conjugate points, i.e. pairs of points in $LK(\rho_0)$ which cannot be joined by a unique geodesic segment. If K_n satisfies the link condition then any path joining a pair of conjugate points must have length at least π. We use this fact to obtain the inequalities given in Section 5.3.

Notation: The ordered triple (a', b', c') denotes the automorphism of F_3 given by $(a, b, c) \mapsto (a', b', c')$, and $[a', b', c']$ denotes the corresponding outer automorphism class.

A vertex of K_3 lies in $st(\rho_0)$ if and only if it can be represented by a labelled graph with labels $\{a, b, c\}$. To list all such vertices one could first list those which have six (the maximal number) of edges and then blow down edges to obtain all the other vertices. Alternatively, one could start at ρ_0 and "blow up" edges (see [13] for details). In any case, the process of assembling and validating such a list is an exhausting one, and for the sake of brevity is omitted. Instead, we simply describe all the 2–cells in $lk(\rho_0)$ (of which there are 408), grouped into convenient subcomplexes, and assemble $lk(\rho_0)$ from these subcomplexes in a suggestive way.

Figure 5.4 shows 32 two-cells which fit together to form a subcomplex of $lk(\rho_0)$ which we call *"the face* $A(+)$". Changing the labels on all the vertex graphs of $A(+)$ by the action of the automorphims $(a^{-1}, b, c), (b, a, c), (b^{-1}, a, c), (c, b, a,)$ and (a^{-1}, b, c) respectively, we obtain 5 more "face" subcomplexes $A(-), B(+), B(-), C(+)$ and $C(-)$, each consisiting of 32 two-cells. Making all possible identifications between these faces we obtain the subcomplex shown in Figure 5.5. Here we see the cube referred to earlier, and the action of $stab(\rho_0)$ in permuting the labels $\{a^{\pm 1}, b^{\pm 1}, c^{\pm 1}\}$ on the graphs corresponds to the usual action of Ω_3 on the cube.

Figure 5.6 shows a further subcomplex consisting of 36 two-cells, which we call *"the hexagon* $H(c)$". The letter c in this notation denotes the fact that this is the label on the edge that is "against the flow" on the central graph. Permuting the labels on all graphs by the automorphisms $(c, b, a), (a, c, b)$ and (a, b, c^{-1}) respectively we obtain "hexagonal" subcomplexes $H(a), H(b)$ and $H(0)$ (in $H(0)$ the arrows associated to the edges labelled a, b, c on the central graph are all confluent). Notice that there are no edge identifications between the hexagons, and that their boundary edges correspond to all marked graphs in $lk(\rho_0)$ which are of the type shown in Figure 5.7. These edges also occur as cells in the 1–skeleton of the cube shown in Figure 5.5, and making the necessary identifications defines gluing maps from the boundary of each hexagon into the cube. This fits the hexagons into our previous picture beautifully, as shown in Figure 5.8.

The remaining 72 two-cells in $lk(\rho_0)$ are shown in Figure 5.9, where they have been grouped to form six discs corresponding to the six possible pairings of distinct hexagons from $\{H(c), H(a), H(b), H(0)\}$. Figure 5.10 illustrates how these fit into the subcomplex of $lk(\rho_0)$ which we have so far constructed. This completes the description of the link of a rose in K_3.

Remark: Our analysis of the link differs from that of Culler and Vogtmann, who described it as "a torus with ten discs attached, having the homotopy type of a wedge of eleven 2–spheres" (see [6] for details). In our construction we obtained $lk(\rho_0)$ (topologically) by taking a 2–sphere and attaching ten 2–discs by injective maps on their boundaries, so the above description of the homotopy type is clear. The torus desribed in [13] is shown in Figure 5.11.

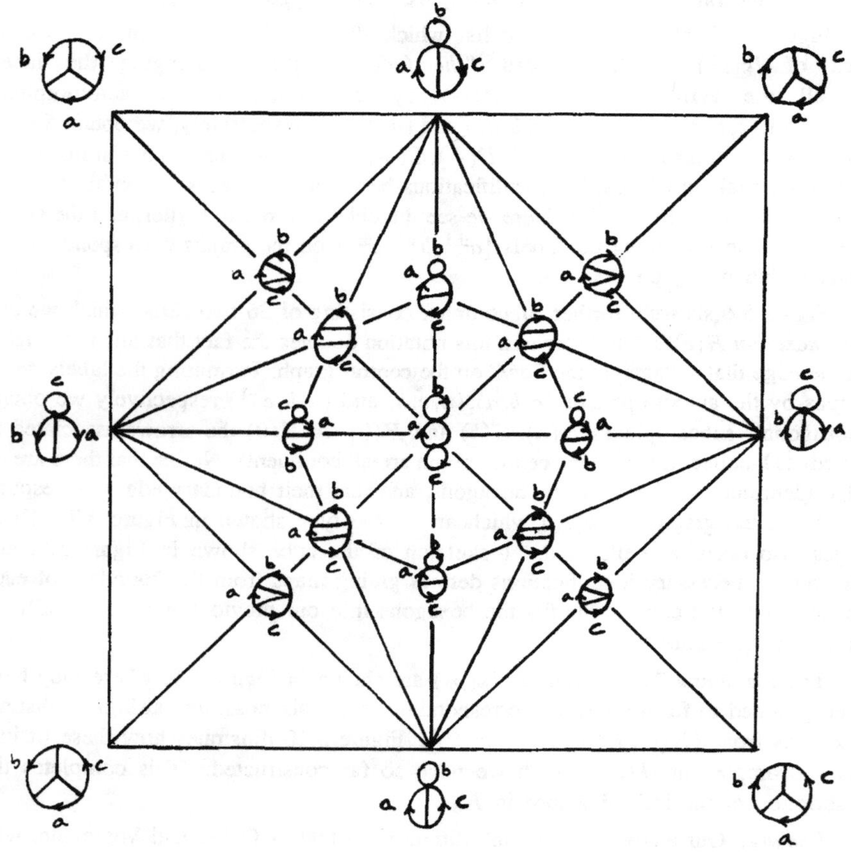

Figure 5.4: The face A(+)

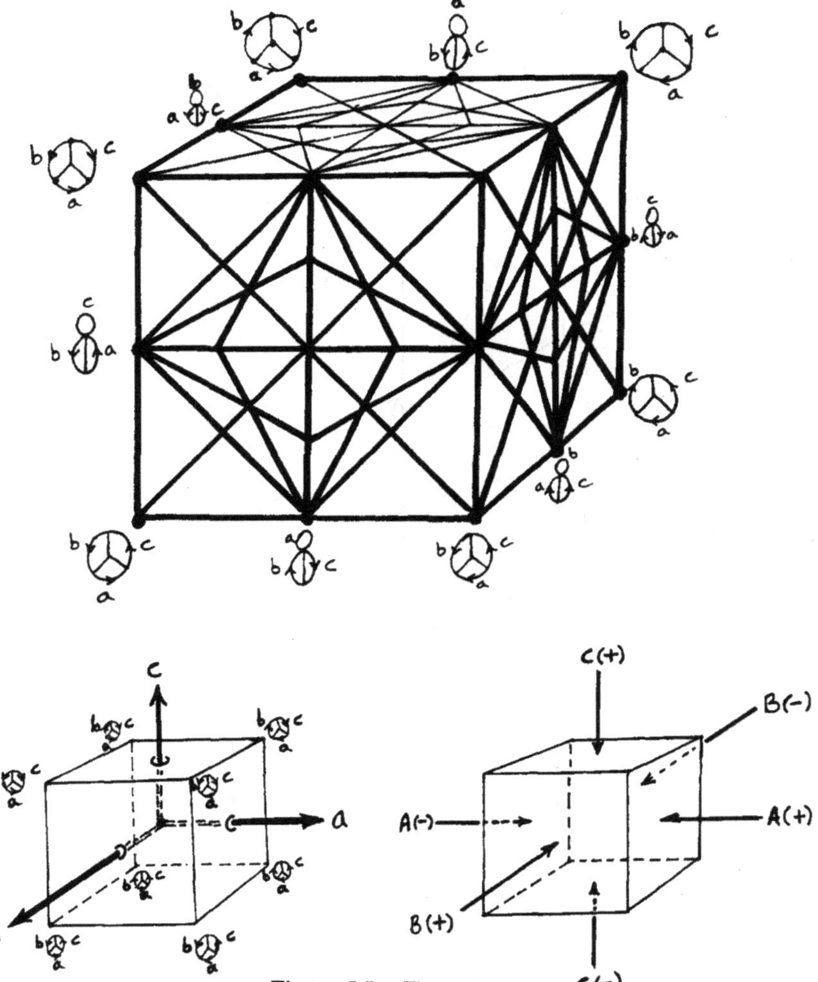

Figure 5.5: The cube

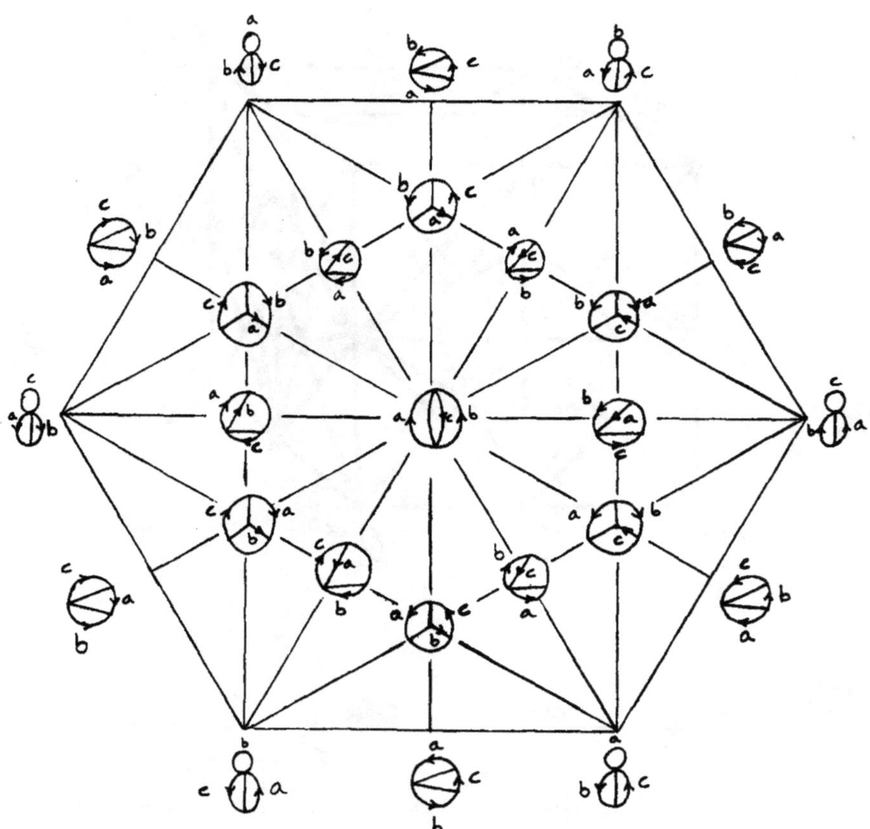

Figure 5.6: The hexagon $\underset{.}{H}(c)$

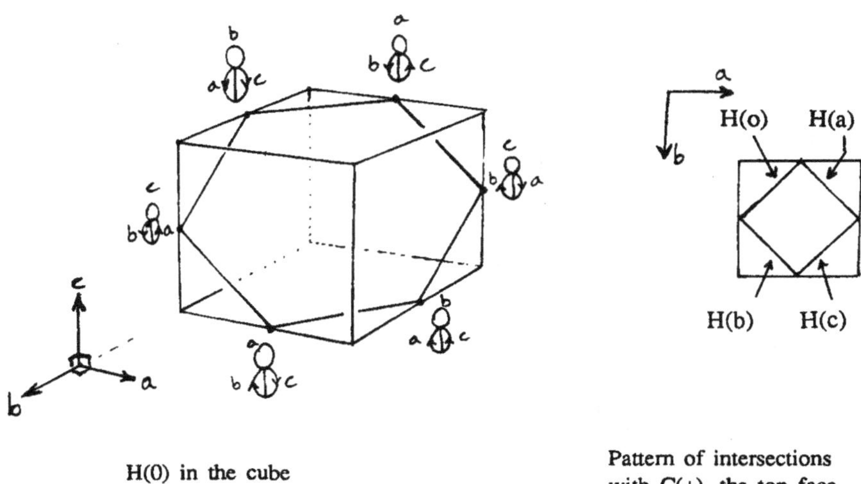

H(0) in the cube

Pattern of intersections with C(+), the top face

Figure 5.7: Fitting the hexagons into the cube

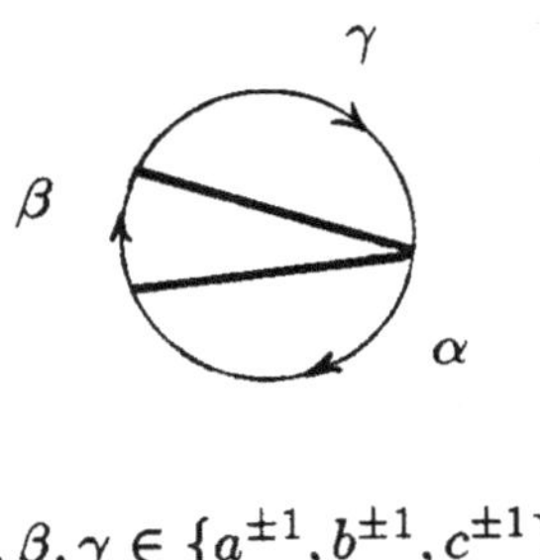

$$\alpha, \beta, \gamma \in \{a^{\pm 1}, b^{\pm 1}, c^{\pm 1}\}$$

Figure 5.8: The graphs in 1–to-1 correspondence with the hexagons

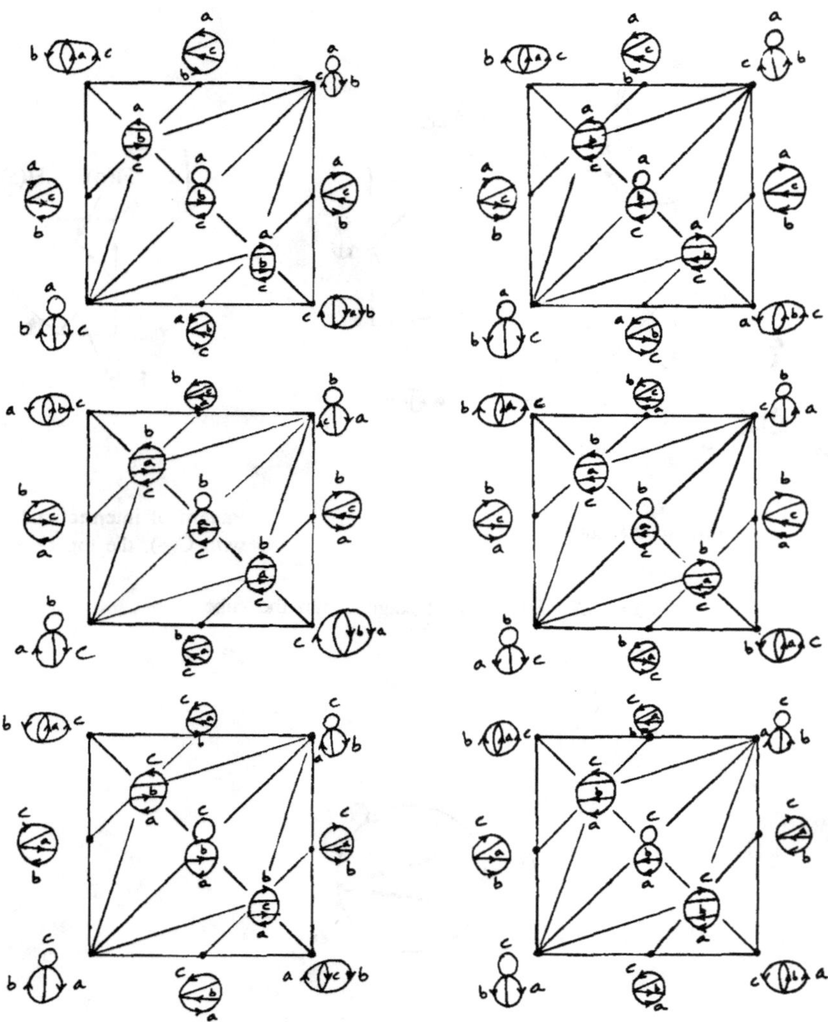

Figure 5.9: The remaining seventy two 2–cells

S(b,c) S(o,a)

S(a,c) S(o,b)

S(b,a) S(o,c)

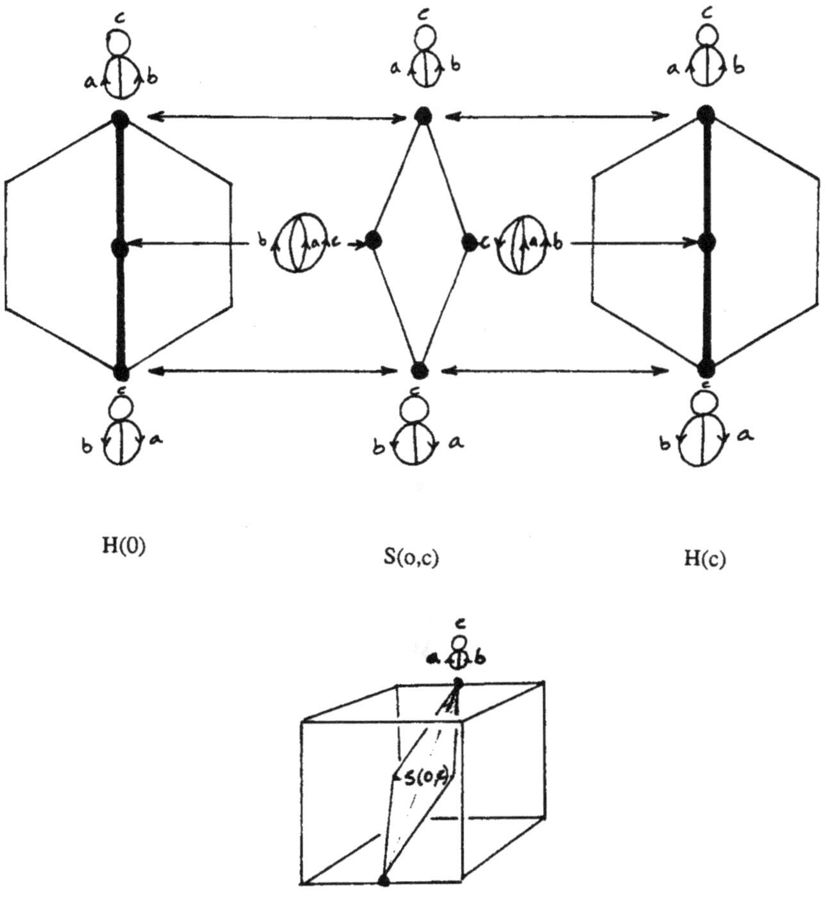

Figure 5.10: Attaching the disc S(o,c)

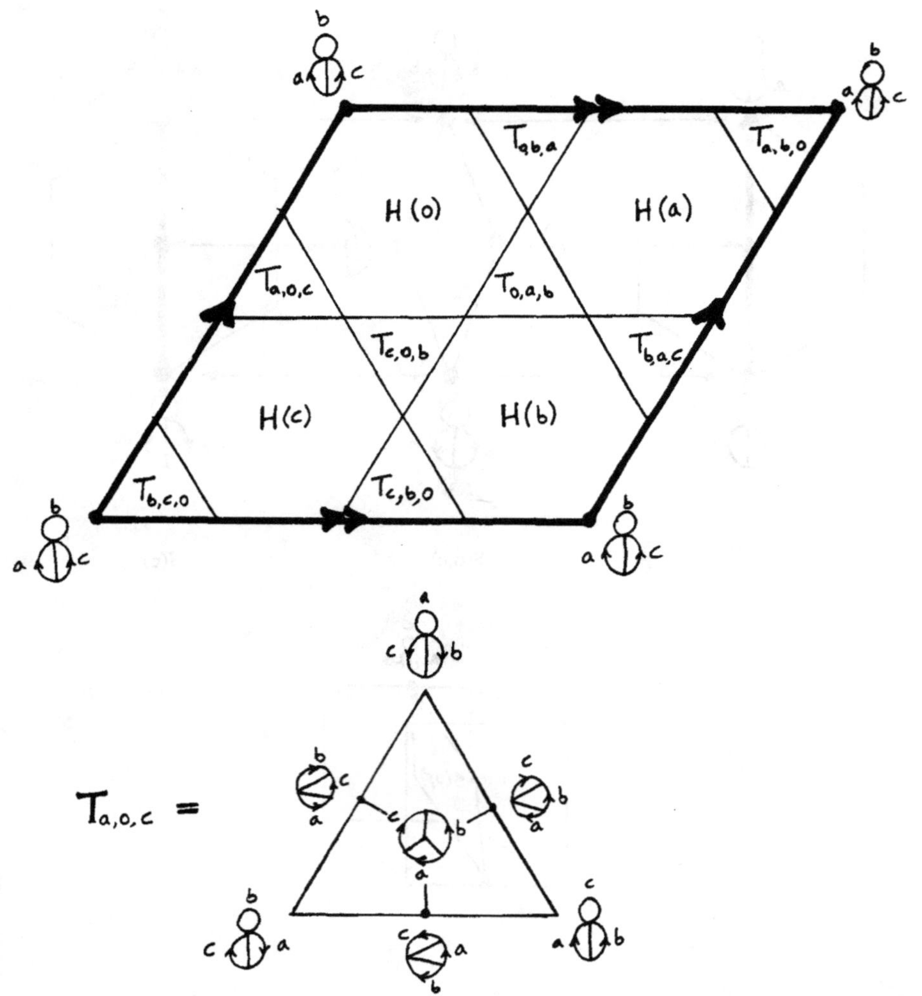

Figure 5.11: The torus

5.3 The Curvature of K_3

Theorem 5.1: *There does not exist an $Out(F_3)$–equivariant piecewise Euclidean (or piecewise hyperbolic) structure of non-positive curvature on K_3.*

The strategy of the proof was described in the introduction to this section. *We assume that the cells of $K_3/Out(F_3)$ have been metrized as Euclidean simplices so that the induced piecewise Euclidean structure on K_3 has non-positive curvature.* Then, by studying the links of vertices we obtain contradictory bounds on the angles of the cells in $K_3/Out(F_3)$. Our arguments centre on three 2–cells in $K_3/Out(F_3)$, which are shown in Figure 5.12. We shall prove the following lemmas under the hypothesis that K_3 is non-positively curved (the greek letters denote the angles defined in Figure 5.12).

Lemma 5.2:

$$\frac{\beta}{2} + \gamma + \delta \geq \frac{\pi}{2} \, .$$

Lemma 5.3:

$$\psi \geq \frac{\pi}{2} \, .$$

Lemma 5.4:

$$\lambda + \mu + \eta \geq \pi \, .$$

Lemma 5.5:

$$\theta + \phi \geq \pi \, .$$

Summing these inequalities we get

$$(\beta + \gamma + \delta) + \psi + (\lambda + \eta + \mu) + (\theta + \phi) \gneqq 3\pi$$

contradicting the fact that these are the angles of three Euclidean triangles. Thus Lemmas 5.2–5.5 together constitute a proof of Theorem 5.1.

Since all our arguments involve bounding the length of paths in links of vertices, it is worth recalling that the lengths of the 1–simplices in each of the spherical simplicial complexes $LK((g, G))$ are given by the angles in the 2–simplices of the quotient space $K_3/Out(F_3)$. It is by bounding the lengths of paths in the link complexes $LK((g, G))$ that we obtain bounds on the angles of the 2–simplices in the quotient space.

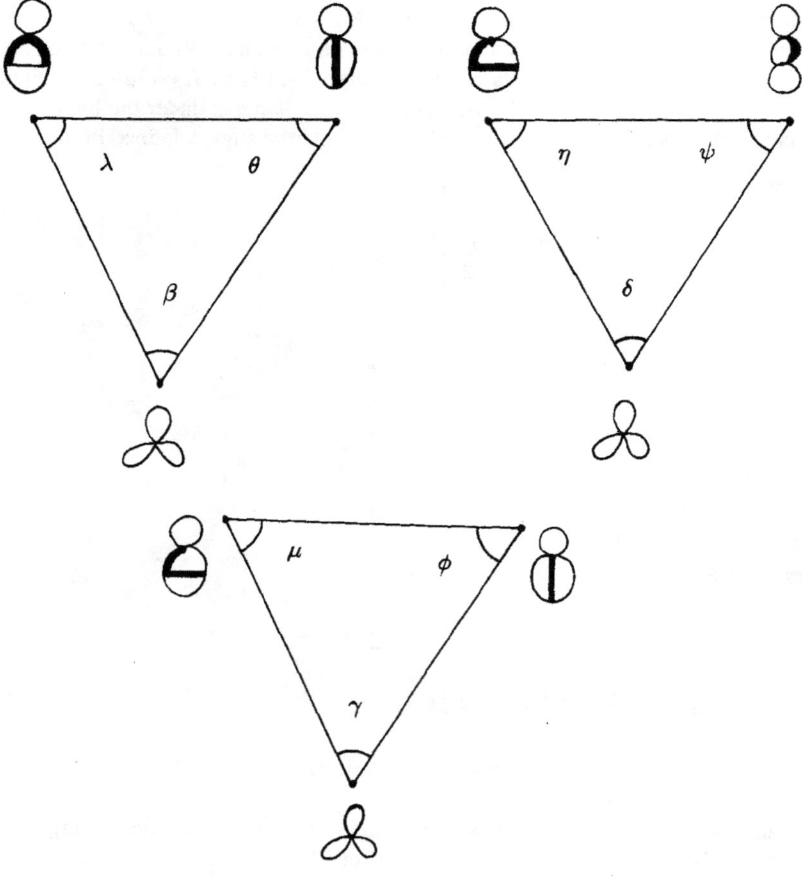

Figure 5.12

Before proceeding to the proof of Lemma 5.2 we make a simple (yet crucial) observation about the action of $Out(F_n)$. Adjacent vertices in K_n have underlying graphs which are not homeomorphic. So if an element of $Out(F_n)$ fixes a point in the interior of a simplex in K_n then it fixes the whole simplex pointwise. It follows that if w is a vertex in $lk(v)$, and $\Gamma \subseteq stab(v)$ fixes w but none of the adjacent vertices in $lk(v)$, then Γ acts freely on $lk(w) \cap lk(v)$. In particular there is no Γ–invariant topolgical arc in $lk(v)$ which has an endpoint at w.

Lemma 5.2:

$$\frac{\beta}{2} + \gamma + \delta \geq \frac{\pi}{2}\,.$$

Proof: The proof is based on the structure of $lk(\rho_0)$. Let v_1, v_2 be the vertices of the spherical simplicial complex $LK(\rho_0)$ represented by the graphs shown in Figure 5.13. Each is fixed by the action of both $[a^{-1}, b, c], [a^{-1}, b^{-1}, c^{-1}] \in stab(\rho_0)$. (The action of $[a^{-1}, b, c]$ on the cube shown in Figure 5.5 is by reflection in the (b, c)–plane. The action of $[a^{-1}, b^{-1}, c^{-1}]$ is by rotation through π about the axis (v_1, v_3)). The link of v_2 in $LK(\rho_0)$ has four vertices, represented by the graphs v_4, v_5, v_6, v_7 in Figure 5.13. None of these vertices is fixed by the action of $[a^{-1}, c^{-1}, b^{-1}]$, so neither is any topological arc in $LK(\rho_0)$ with an endpoint at v_2. In fact v_1 and v_2 are the only points of $LK(\rho_0)$ fixed by the action of both $[a^{-1}, b, c]$ and $[a^{-1}, b^{-1}, c^{-1}]$.

If there were a unique geodesic segment from v_1 to v_2 in $LK(\rho_0)$ it would be a topological arc fixed by any automorphism fixing both v_1 and v_2. Therefore such a segment does not exist and, since we are assuming that K_3 satisfies the link axiom, it follows that $d_{LK(\rho_0)}(v_1, v_2) \geq \pi$. Consider P_1, the path shown in Figure 5.14. This joins v_1 to v_2 in $LK(\rho_0)$, and hence must have length at least π. But its actual length is $(\beta + 2\gamma + 2\delta)$. $\square$

Lemma 5.3:

$$\psi \geq \frac{\pi}{2}\,.$$

Proof: Let (g_0, G_0) be as in Figure 5.15. To prove the lemma we analyse $lk(g_0, G_0)$. (Notice that (g_0, G_0) is the central vertex of the face $B(+)$ of the cube shown in Figure 5.5.)

Given a vertex u in the geometric realisation of any poset, its link can be described as the join of its link in the subcomplex spanned by vertices greater than u (its upper link) and its link in the subcomplex spanned by vertices less than u (its lower link). This makes the calculation of $lk((g_0, G_0))$ a simple matter, since its upper link is a

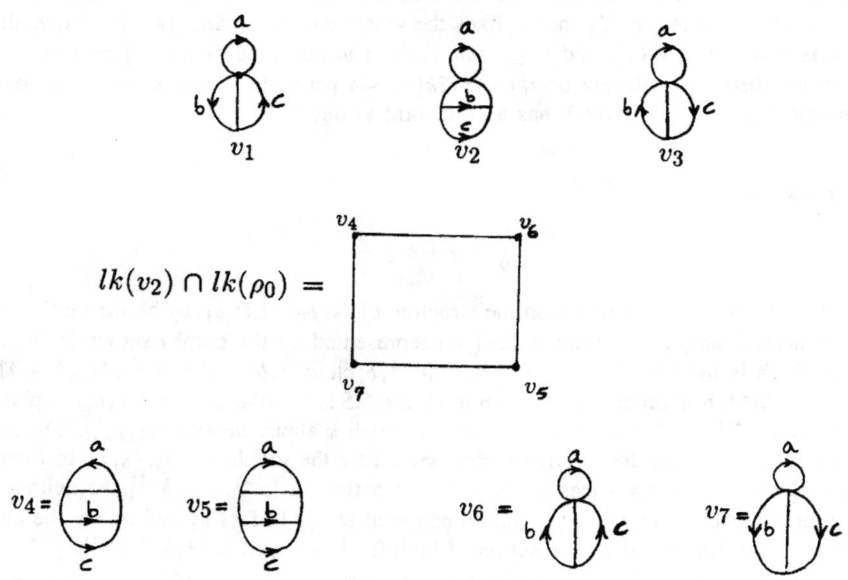

Figure 5.13

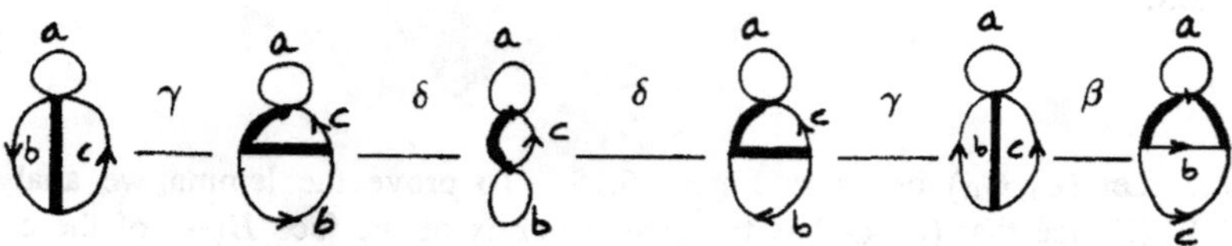

Figure 5.14: The path P_1

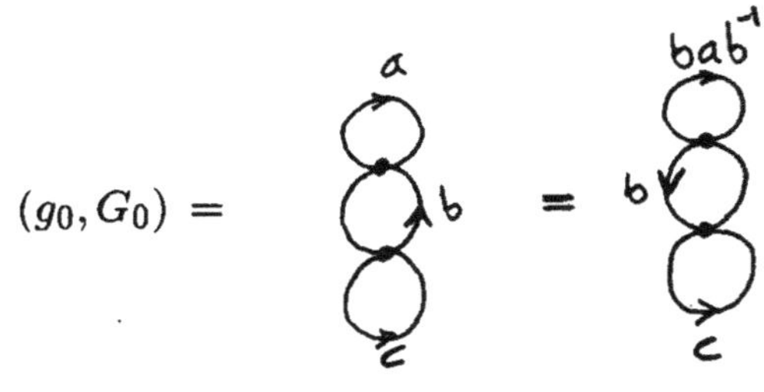

$(g_0, G_0) =$

Figure 5.15

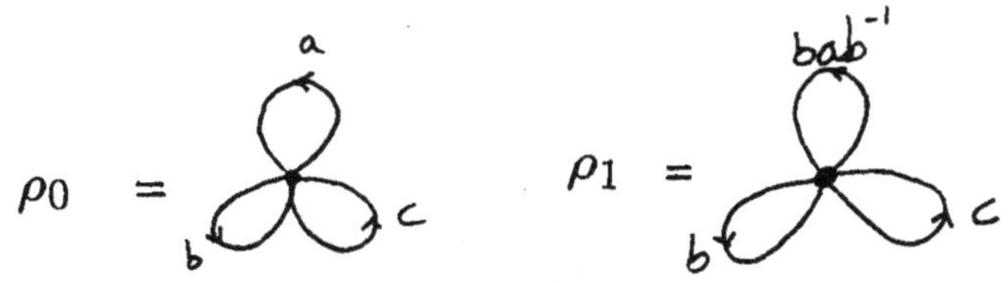

$\rho_0 =$ $\rho_1 =$

Figure 5.16

ρ_0 ρ_1

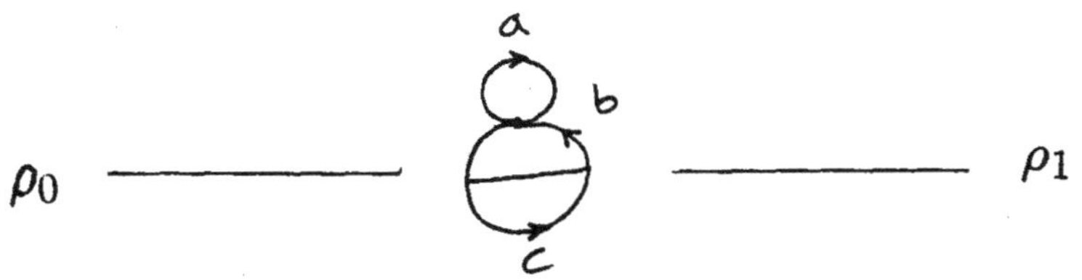

Figure 5.17: The Path P_2 of Lemma 5.3

subcomplex of $lk(\rho_0)$, which we have already caclulated, and its lower link consists only of the roses ρ_0 and ρ_1 shown in Figure 5.16. We shall prove that there does not exist a unique geodesic segment from ρ_0 to ρ_1 in $Lk((g_0, G_0))$.

The group $stab(\rho_0) \cap stab(\rho_1) \cap stab((g_0, G_0))$ is isomorphic to the dihedral group D_8, which is the stabiliser of the face $B(+)$ of the cube in Figure 5.5 under the action of $\Omega_3 = stab(\rho_0)$. No point in the upper link of (g_0, G_0) $(= lk(\rho_0) \cap lk((g_0, G_0)))$ is fixed by this action. Therefore, as in Lemma 5.2, we deduce that $d_{LK((g_0,G_0))}(\rho_0, \rho_1) \geq \pi$ and hence the path P_2, which is shown in Figure 5.17, is no shorter than π. The actual length of P_2 in $LK((g_0, G_0))$ is 2ψ. □

Remark: The dihedral group referred to in the proof of Lemma 5.3 is:

$$\{[a^{\pm 1}, b, c], [a^{\pm 1}, b, c^{-1}], [c^{\pm 1}, b^{-1}, a], [c^{\pm 1}, b^{-1}, a^{-1}]\}.$$

Lemma 5.4:

$$\lambda + \mu + \eta \geq \pi \, .$$

Proof: For this lemma we need to understand $lk((g_1, G_1))$, which is described in Figure 5.18. We prove that there is not a unique geodesic segment from the vertex ρ_0 to the vertex v' in $LK((g_1, G_1))$.

The automorpism $[a^{-1}, b, c] \in stab((g_1, G_1))$ interchanges the two elements of the upper link of (g_1, G_1) and fixes the lower link pointwise, so if there were a unique geodesic segment from ρ_0 to v' in $LK((g_1, G_1))$ then it would have to be contained in the lower link. But, every path from ρ_0 to v' in the lower link has length at least $\lambda + \mu + \eta$, and there are two paths of precisely this length. Hence the required inequality. □

Lemma 5.5:

$$\theta + \phi \geq \pi \, .$$

Proof: Let (g_2, G_2) be as in Figure 5.19. In Section 5.2 we saw this graph as the midpoint of an edge of the cube in Figure 5.6, and calculated its upper link. Its lower link consists of the roses ρ_0, ρ_2, and ρ_3 shown in Figure 5.19. The action of $[a^{-1}, b, c] \in stab((g_2, G_2))$ fixes each of these roses, aswell as three vertices of the upper link of (g_2, G_2). Figure 5.20 shows the 1–dimensional subcomplex of $LK((g_2, G_2))$ fixed by $[a^{-1}, b, c]$. We call this complex L. Each of the edges which is drawn as a solid line has length ϕ and each of the edges which is drawn as a broken line has length θ.

$(g_1, G_1) \quad = \quad \quad = $

Upper link of $(g_1, G_1) \quad = \quad \{ \quad , \quad \}$

Lower link of (g_1, G_1):

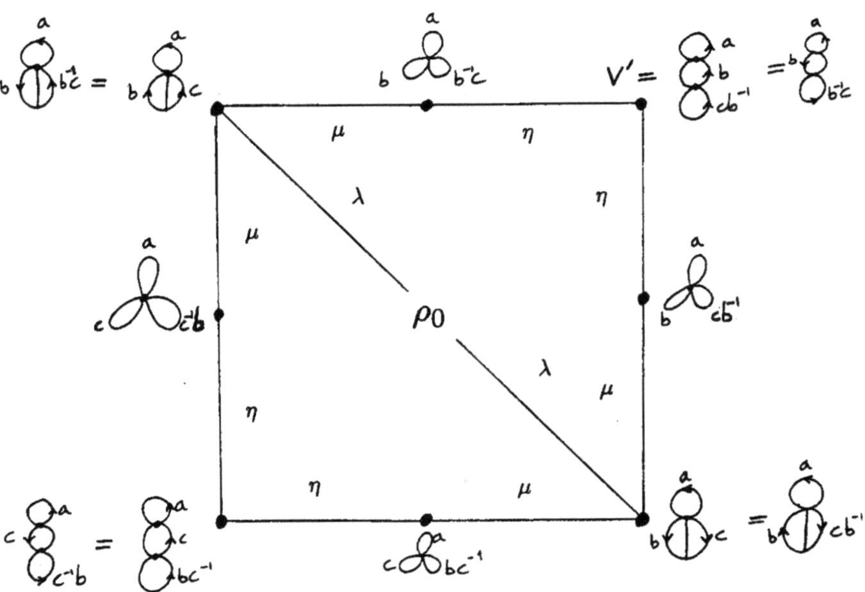

Figure 5.18

$(g_2, G_2) \quad =$

$\rho_0 \quad = \qquad \rho_2 \quad = \qquad \rho_3 \quad =$

$u_1 \quad = \qquad u_2 \quad = \qquad u_3 \quad =$

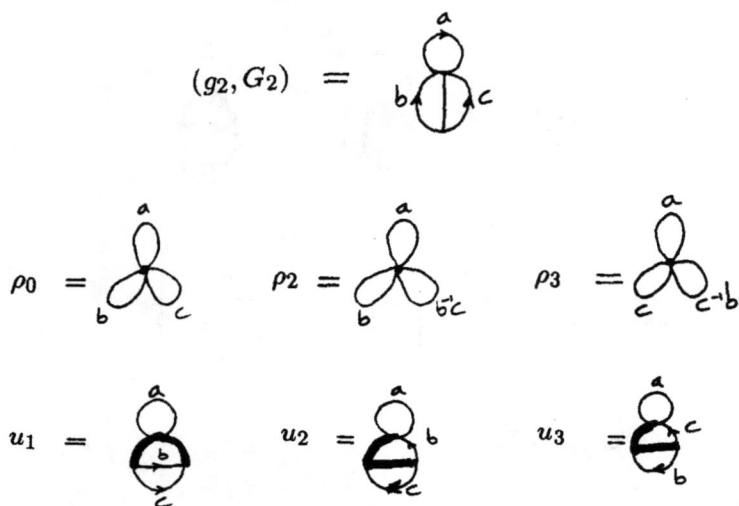

Figure 5.19

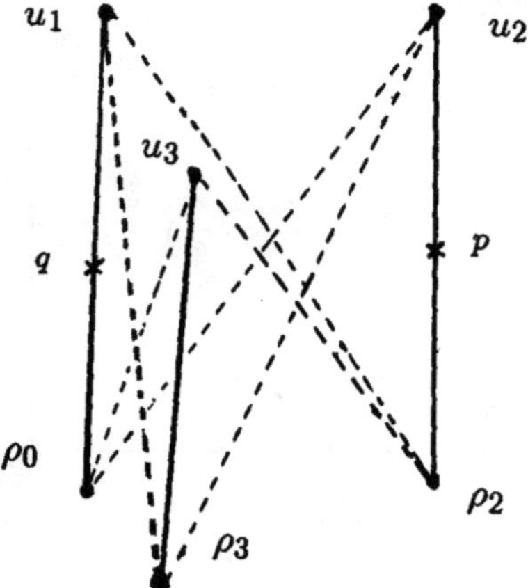

Figure 5.20: The 1–complex L

Let p and q be the midpoints of distinct solid edges. If there were a unique geodesic segment from p to q in $LK((g_2, G_2))$ then it would be contained in L. But, all paths from p to q in L have length at least $(\theta + \phi)$ and there are two paths joining them which are of precisely this length. Hence there is not a unique geodesic segment from p to q in $LK((g_2, G_2))$, which imples that $(\theta + \phi) \geq \pi$. □

5.4 The Curvature of K_n

The complexity of the links of vertices in K_3 (particularly that of a rose) makes the prospect of explicitly calculating the links of vertices in K_n for $n > 3$ seem daunting, if not impossible. However, we circumvent this difficulty by restricting our attention to the fixed point set of Γ_n, a particular finite subgroup of $Out(F_n)$. The links of vertices in this set are 2–dimensional, and admit an easy description in terms of the links of vertices in K_3. We use this description to obtain a contradiction in the manner of Section 5.3.

Theorem 5.6: *If $n \geq 3$ then there does not exist an $Out(F_n)$–equivariant piecewise Euclidean (or piecewise hyperbolic) structure of non-positive curvature on K_n.*

Notation: Fix a basis $\{a_1, \ldots, a_{n-2}, b, c\}$ *for the free group of rank n.*

Let Γ_n denote the subgroup of $Out(F_n)$ generated by the $n-2$ involutions $[a_1^{-1}, a_2, \ldots, a_{n-2}, b, c], \ldots, [a_1, a_2, \ldots, a_{n-2}^{-1}, b, c]$ together with the outer automorphism classes of those automorphisms which fix b and c and act by permutations on the set $\{a_1, \ldots, a_{n-2}\}$. Thus Γ_n is a semidirect product of $\mathbf{Z}_2 \times \ldots \times \mathbf{Z}_2$ ($n-2$ factors) with the symmetric group on $n-2$ letters.

Lemma 5.7 describes the fixed point set of Γ_n. Before stating this lemma we need to make some observations about the geometry of embedded circles in a marked graph and their relationship to cyclic words in the free group.

In the paragraph on labelled graphs we noted that there were certain ambiguities in the choice of labels. However, it is important to notice that the marking g assigns unique labels to reduced circuits in G in the following sense: If we join the endpoints of a labelled edge by the unique topological arc between them in the chosen maximal tree then we obtain an embedded circle in G. This oriented circle corresponds, via the marking g, to a unique conjugacy class of words in F_n (i.e., a cyclic word). Moreover, this conjugacy class, which contains the given edge label, does not depend on the choice of maximal tree. Because the graph under consideration is assumed to have no separating edges it is the union of the circles obtained from the labelled edges by this construction. Further, any two of these circles meet in a (possibly empty) subarc of our chosen maximal tree.

If Φ denotes the isometry of G induced by $\phi \in stab((g, G))$ then the image under Φ of an embedded oriented circle which corresponds to a given cyclic word w is the embedded oriented circle corresponding to the cyclic word $\phi(w)$. In particular, if $\phi(w) = w$ then the embedded circle is mapped to itself by an orientation preserving homeomorphism, and if $\phi(w) = w^{-1}$ then it is mapped to itself by an orientation reversing homeomorphism.

Lemma 5.7: *Suppose that the vertex $(g, G) \in K_n$ can be represented by a labelled graph with the following properties: it has edge labels $\{a_1, \ldots, a_{n-2}, \omega_1, \omega_2\}$ for some words ω_1 and ω_2 involving only the letters b and c, and contains a bouquet of $n-2$ circles with labels $\{a_1, \ldots, a_{n-2}\}$ based at some vertex. Then Γ_n stabilises a vertex $(h, H) \in lk((g, G))$ if and only if (h, H) can be represented by a labelled graph with the same properties.*

Proof: Sufficiency is immediate from the definition of the action of $Out(F_n)$ on K_n. To prove that the given condition on (h, H) is necessary we first observe that *any* vertex in the lower link of (g, G) can be represented by a labelled graph with the desired properties. To see this, consider blowing down an edge of (G, g): The labelling of G by $\{a_1, \ldots, a_{n-2}, \omega_1, \omega_2\}$ arises from choosing a particular maximal tree. If we wish to blow down an edge which is not in this tree, then we must rechoose the tree. Since the new tree must lie in the complement of the loops labelled $\{a_1, \ldots, a_{n-2}\}$ these labels remain unchanged, while the labels ω_1 and ω_2 may be multiplied by words in b and c (see Figure 5.1).

On the other hand, any vertex (h, H) in the upper link of (g, G) can be represented by a labelled graph with edge labels $\{a_1, \ldots, a_{n-2}, \omega_1, \omega_2\}$. Suppose that the automorphism $[a_1^{-1}, a_2, \ldots, a_{n-2}, b, c]$ (which for brevity we call ϕ) stabilises (h, H). If the edge labelled a_1 were not a loop, then the embedded circle obtained by joining its endpoints by the unique arc between them in the chosen maximal tree would intersect the circle determined by some other labelled edge in a non-trivial arc. But this is impossible, because the isometry Φ sends the circle determined by the edge labelled a_1 to itself by an orientation reversing homeomorphism, and sends the circle determined by any other labelled edge to itself by an orientation preserving homeomorphism.

Repeating this argument with a_i in place of a_1 shows that if the graph (h, H) is fixed by the action of Γ_n then the edge labelled a_i is a loop, for all $i \in \{2, \ldots, n-2\}$. The action of the symmetric group S_{n-2}, which occurs in the semi-direct product structure of Γ_n, permutes these loops, hence they must all be based at the same vertex. □

Definition: Suppose $(g, G) \in K_3$ can be represented by a labelled graph with edge labels $\{a, b, c\}$, such that the edge labelled a is a loop. We define $(g, G)(n)$ to be the vertex of K_n represented by the labelled graph obtained from (g, G) by replacing the edge labelled a with a bouquet of circles based at the same vertex, and labelled

$a_1, \ldots, a_{n-2}$. (It is easy to check that $(g,G)(n)$ depends only on (g,G) and not on the particular labelling chosen.)

The following is an immediate consequence of Lemma 5.7.

Corollary : *The map* $(h,H) \mapsto (h,H)(n)$ *is a simplicial isomorphism from the fixed point set of* $[a^{-1},b,c]$ *acting on* $lk((g,G))$ *to the fixed point set of* Γ_n *acting on* $lk((g,G)(n))$.

Proof of Theorem 5.6: The proof centres on the three 2–cells of $K_n/Out(F_n)$ which correspond to the three 2–cells shown in Figure 5.12. More precisely, for each of the 2–cells in Figure 5.12 we choose a 2–cell in K_3 which lies in its preimage and for which each of the vertex graphs has a loop labelled a. We then take the image in K_n of this 2–cell under the map $(H,h) \mapsto (H,h)(n)$, and project this down into $K_n/Out(F_n)$. We retain the names used for the corresponding angles in the rank 3 case.

Suppose that $K_n/Out(F_n)$ can be metrized so that K_n, with the induced structure, is non-positively curved. If this were the case then the sum of the angles shown in Figure 5.12 would be 3π. However, by virtue of the preceding corollary, the arguments given in Section 5.3 apply *(mutatis mutandis)* in the present setting to yield

$$(\beta+\gamma+\delta)+\psi+(\lambda+\eta+\mu)+(\theta+\phi) \gneqq 3\pi.$$

We rephrase Lemmas 5.1–5.4 to clarify this remark.

If $K_n/Out(F_n)$ can be metrized so that K_n, with the induced structure, is non-positively curved then the following assertions hold:

Lemma 5.2 (The case n=3): *The only points of* $lk(\rho_0)$ *fixed by both* $[a^{-1},b,c]$ *and* $[a^{-1},b^{-1},c^{-1}]$ *are the vertices* v_1 *and* v_2 *shown in figure 5.13. Hence*

$$\frac{\beta}{2}+\gamma+\delta \geq \frac{\pi}{2}\,.$$

Lemma 5.2′ (The general case): *The only points of* $lk(\rho_0)$ *fixed by the action of* $\Gamma_n \times \{1,[a_1,\ldots,a_{n-2},b^{-1},c^{-1}]\}$ *are the vertices* $v_1(n)$ *and* $v_2(n)$. *Where* v_1 *and* v_2 *are shown in figure 5.13. Hence*

$$\frac{\beta}{2}+\gamma+\delta \geq \frac{\pi}{2}\,.$$

Lemma 5.3 (The case n=3): *The only points of $LK((g_0, G_0))$ fixed by both $[a^{-1}, b, c]$ and $[a, b^{-1}, c]$ are the roses ρ_0 and ρ_1 shown in figure 5.16. Hence*

$$\psi \geq \frac{\pi}{2} .$$

Lemma 5.3′ (The general case): *The only points of $LK((g_0, G_0)(n))$ fixed by the action of $\Gamma_n \times \{1, [a_1, \ldots, a_{n-2}, b^{-1}, c^{-1}]\}$ are the roses $\rho_0(n)$ and $\rho_1(n)$. Where ρ_0 and ρ_1 are as shown in figure 5.16. Hence*

$$\psi \geq \frac{\pi}{2} .$$

Lemma 5.4 (The case n=3): *The automorpism $[a^{-1}, b, c] \in stab((g_1, G_1))$ interchanges the two elements of the upper link of (g_1, G_1) and fixes the lower link pointwise. Let v' be as shown in figure 5.18. Every path from ρ_0 to v' in the lower link of (g_1, G_1) has length at least $\lambda + \mu + \eta$, and there are two paths of precisely this length. Hence*

$$\lambda + \mu + \eta \geq \pi .$$

Lemma 5.4′ (The general case): *The group $\Gamma_n \subset stab((g_1, G_1))$ has no fixed points in the upper link of $(g_1, G_1)(n)$, and fixes the lower link pointwise. Let v' be as shown in figure 5.18. Every path from $\rho_0(n)$ to $v'(n)$ in the lower link of $(g_1, G_1)(n)$ has length at least $\lambda + \mu + \eta$, and there are two paths of precisely this length. Hence*

$$\lambda + \mu + \eta \geq \pi .$$

Lemma 5.5 (The case n=3): *The fixed point set for the action of $[a^{-1}, b, c]$ on $LK((g_2, G_2))$ is isomorphic to the graph L shown in figure 5.20, with vertices $\rho_0, \rho_2, \rho_3, u_1, u_2, u_3$ as defined in figure 5.19. If L is metrized so that the solid edges have length ϕ and the broken edges have length θ, then any path from p to q in L has length at least $\theta + \phi$ and there are two paths of precisely this length. Hence*

$$\theta + \phi \geq \pi .$$

Lemma 5.5′ (The general case): *The fixed point set for the action of Γ_n on $LK((g_2, G_2)(n))$ is isomorphic to the graph L shown in figure 5.20, with vertices $\rho_0(n), \rho_2(n), \rho_3(n), u_1(n), u_2(n), u_3(n)$, where $\rho_0, \rho_2, \rho_3, u_1, u_2, u_3$ are as shown in figure 5.19. If L is metrized so that the solid edges have length ϕ and the broken edges have length θ, then any path from p to q in L has length at least $\theta + \phi$ and there are two paths of precisely this length. Hence*

$$\theta + \phi \geq \pi .$$

This concludes the proof of Theorem 5.6. □

Acknowledgements: I would like to thank Karen Vogtmann and Marshall Cohen for their many helpful comments regarding the content and presentation of this material.

References

1. I. R. Aitchison and J. H. Rubinstein, *An introduction to polyhedral metrics of non-positive curvature on 3–manifolds*, preprint, University of Melbourne, 1989.

2. A. D. Alexandrov, *A theorem on triangles in a metric space and some applications*, (In Russian), Trudy Mat. Inst. Steks. **38** (1951), 5–23.

3. J. M. Alonso and M. R. Bridson, *Semihyperbolic Groups*, preprint, Cornell University, 1990.

4. R. Alperin, *Homology of* $SL_2(Z[\omega])$, Comm. Math. Helv. **55** (1980), 364–377.

5. W. Ballmann, M. Gromov, and V. Schroeder, *Manifolds of nonpositive curvature*, Progress in mathematics. vol. 61, Birkhäuser, Boston, 1985.

6. T. Brady, *The integral cohomology of* $Out_+(F_3)$, Ph.D Thesis, Cornell University, 1988.

7. M. R. Bridson, *Normal forms in triangles of groups*, (in prepartion).

8. K. S. Brown, *Buildings*, Springer-Verlag, New York, 1989.

9. K. S. Brown, *Five lectures on buildings*, this volume.

10. F. Bruhat and J. Tits, *Groupes réductifs sur un corps local I. Données radicielles valuées*, I.H.E.S. Sci. Publ. Math. 41 (1972), 5–251.

11. R. Charney and M. Davis, *Singular metrics of non-positive curvature on branched Riemannian manifolds*, preprint, The Ohio State University, 1990.

12. M. Culler, *Finite groups of outer automorphisms of a free group*, Contributions to group theory, pp. 197–207, Contemp. Math., 33, Amer. Math. Soc., Providence, R.I., 1984.

13. M. Culler and K. Vogtmann, *Moduli of graphs and automorphisms of free groups*, Invent. Math **84** (1986), 91–119.

14. C. H. Dowker, *The topology of metric complexes*, Amer. J. Math. **74** (1952), 555–577.

15. S. M. Gersten, *Convexity properties of non positive curvature*, preprint MSRI, 1989.

16. E. Ghys and P. de la Harpe (Eds.), *Sur les groupes hyperboliques d'après Mikhael Gromov*, Progress in Mathematics, vol. 83, Birkhäuser, Boston 1990.

17. M. Gromov, *Hyperbolic groups*, "Essays in group theory" (S. M. Gersten, Ed.) pp. 75–264, Springer-Verlag, 1987.

18. M. Gromov and W. P. Thurston, *Pinching constants for hyperbolic manifolds*, Invent. Math. **89** (1987), 1–12.

19. A. Haefliger, *Complexes of group and orbihedra*, this volume.

20. S. Krstic, K. Vogtmann, (in prepartation).

21. H. Masur, *On a class of geodesics in Teichmüller space*, Ann. of Math. (2) **90** (1969), 1–8.

22. J. Milnor, *Morse theory*, Ann. of Math. Studies, no. 51, Princeton Univ. Press, 1963.

23. G. Moussong, *Hyperbolic coxeter groups*, Ph.D. thesis, The Ohio State University, 1988.

24. J-P. Serre, *Trees*, Springer-Verlag, Berlin, 1980, translation of "Arbres, Amalgames, SL_2", Astérique **46**, 1977.

25. J. Smillie and K. Vogtmann, *A generating function for the Euler charcteristic of $Out(F_n)$*, J. Pure Appl. Algebra **44** (1987), 329–348.

26. J. R. Stallings, *Non-positively curved triangles of groups*, this volume.

27. D. A. Stone, *Geodesics in piecewise linear manifolds*, Trans. Amer. Math. Soc. **215** (1976), 1–44.

28. J. Tits, *On buildings and their applications to group theory*, in "Proceedings of the International Congress of Mathematics", Vancouver 1974, vol. 1, Canad. Math. Congress, Montreal, 1975, pp. 209–220.

29. J. Tits, *Immeubles de type affine*, in "Buildings and the Geometry of Diagrams" (L. A. Rosati, Ed.), Proc. CIME Como 1984, Lecture Notes in Mathematics 1181, Springer-Verlag, Berlin, 1986, pp. 159–190.u

DEPARTMENT OF MATHEMATICS, WHITE HALL, CORNELL UNIVERSITY, ITHACA, NY 14853.

E-mail: bridson@mssun7.msi.cornell.edu

HYPERBOLIZATIONS

TADEUSZ JANUSZKIEWICZ*

Instytut Matematyczny, Uniwersytet Wrocławski
pl. Grunwaldzki 2/4, 50–384 Wrocław, Poland

INTRODUCTION

The idea of hyperbolization is related to the notion of curvature of piecewise flat spaces. A triangulation of a (say locally finite) polyhedron can be used to produce a metric on that polyhedron, by declaring all the simplices in the triangulation isometric to the standard simplex of an appropriate dimension. This gives a collection of compatible metrics on pieces of the polyhedron, and we can extend it in an obvious way to a genuine metric. Moreover it is a geodesic space, that is, there is a path minimizing the distance between any two points; this is how the metric is constructed. Even though most of the polyhedron is flat, there is an interesting information about the "curvature" encoded in the geometry of links, which carry natural piecewise constant (+1) curvature metric. Gromov isolated a precise condition (which we call "large links") on the link which forces the geodesics in the metric space to behave the way geodesics in a nonpositively curved Riemannian manifold do, i.e. to diverge.

Hyperbolization is a procedure (invented by Gromov), which at the expense of modifying topology, changes the geometry so that all the links are large. Thus it is a way of converting a triangulation of a polyhedron into a nonpositively curved space. Since this can be done in many ways, it is suggestive to consider the two dimensional case first. To simplify the discussion we will hyperbolize only two dimensional manifolds, but this is in no way essential, and the methods, with slight modifications, work for polyhedra.

One of the possible hyperbolizations is to take a ramified covering of a triangulated two-manifold. Let us just recall in an elementary way the construction of two-fold ramified cover. In a two-manifold M we pick k pairs of (distinct) points (p_i, q_i) and a system of k *cuts*, i.e. nonintersecting embedded arcs joining p_i and q_i. Each cut has the left and the right side. Now we take two copies of M,

*The author was partially supported by Polish Scientific Grant RP.I.10. He also enjoyed hospitality of CPT Marseille Luminy while parts of this paper were being written.

slit it open along the system of cuts and glue the right side of a cut in one copy of M to the left side of the corresponding cut in the other. This can be done many ways: even if the set of all $\{p_i, q_i\}$ (which we call the branching locus) is given we can vary the cut system.

Now we hyperbolize M with a triangulation on it by taking *a* two-fold ramified covering, which branches at the vertices of the triangulation. There is a problem with actually doing so if the number of vertices is odd, but then we can do the following: take two vertices, and form a two-fold covering branched at those points. Then in the covering space the preimage of the vertices of the original triangulation consists of an even number of points, and we can form a ramified covering with that set as the branching locus. It is clear that the lifted triangulation has the property that each vertex is adjacent to at least six triangles. This is the "large links" condition in dimension two.

Another hyperbolization removes the interiors of all the two simplices, and glues back in surfaces (which are not discs) whose boundaries are circles. Thus for each two simplex in the triangulation we have a surface with boundary, and the boundary of the surface is identified with the boundary of the corresponding two simplex. We then identify the boundary points of the surfaces following the identification pattern in the triangulation. The resulting two complex carries a triangulation with large links, since any surface (different from the disc) admits a triangulation such that each vertex on the boundary is adjacent to at least three, and each vertex in the interior is adjacent to at least six triangles.

In both cases we can produce "smaller" hyperbolizations by ramifying only at the points where less than six triangles meet, or by substituting a triangle with a surface only the triangles adjacent to the vertices where less than six triangles meet.

We are concerned with generalizations of these two constructions to higher dimensions. The plan of the paper is as follows.

In Section 1 we discuss basic aspects of the geometry of spaces of piecewise constant curvature.

Section 2 contains the fiber product construction of hyperbolization.

Section 3 contains a description of topological applications in [5] and a simple combinatorial application.

Section 4 discusses ramified coverings and describes some applications to differential and algebraic geometry, which partly predates hyperbolization technique.

Section 5 is a discussion of some open problems, though many questions are scattered in other Sections.

Finally in the Appendix we provide for the readers' convenience, a method for constructing contractible manifolds, taken from a beautiful paper [9].

The present paper is a significantly expanded version of the talk I gave at

Trieste. I am grateful to the organizers of Trieste Workshop, especially to Etienne Ghys, for providing an opportunity to present these ideas. Many of them I either learned from, developed jointly with or discussed at some point with Mike Davis.

1. Geometry of nonpositively curved spaces

Since the basic geometric concepts related to nonpositively curved spaces have been discussed extensively elsewhere in these Proceedings we will be brief. The exposition here follows very closely [5].

A *geodesic segment* in a metric space X is an isometric map from an interval into X. A *triangle* in X consists of three points (vertices), together with three geodesic segments (edges) connecting them. A metric space X is a *geodesic space* if any two points can be joined by a geodesic segment. A subset Y in a geodesic space X is a *totally geodesic* (sometimes also called *convex*) subspace if, locally every geodesic segment in X with endpoints in Y is actually contained in Y.

For each real number ε, let $M^2(\varepsilon)$ be the simply connected space of constant curvature ε. If T is a triangle in X, then a comparison triangle T' in $M^2(\varepsilon)$ is a triangle with the same edge lengths as T. The comparison triangle is unique up to an isometry. Comparison triangles always exist for $\varepsilon \leq 0$; for $\varepsilon > 0$, a comparison triangle exists if T has edge lengths $\leq 2\pi/\sqrt{\varepsilon}$. Suppose that T is a triangle in X with vertices x_0, x_1, x_2, and y is a point on the geodesic segment $[x_1, x_2]$. Let T' be a comparison triangle in $M^2(\varepsilon)$ with corresponding vertices x'_0, x'_1, x'_2, and y' be the point on $[x'_1, x'_2]$ corresponding to y. The pair (T, y) *"satisfies $CAT(\varepsilon)$"* if $d(x_0, y) \leq d(x'_0, y')$. The space X *"satisfies $CAT(\varepsilon)$"* if every pair (T, y) satisfies $CAT(\varepsilon)$.

Definition. [11] A geodesic space X has curvature $\leq \varepsilon$ if it satisfies $CAT(\varepsilon)$ locally.

According to [11] a simply connected geodesic space of nonpositive curvature satisfies $CAT(0)$ globally; this implies (convexity of the distance function and thus) contractibility. The following Gluing Lemma (see Gromov [11, p. 124], [5, Lemma 2.4]) is useful in constructing examples of nonpositively curved spaces.

Gluing Lemma 1.1. *Suppose either that*

(a) *X is the disjoint union of two geodesic spaces X_1 and X_2 and $Y_i \subset X_i$ is a totally geodesic closed subspace, or*

(b) *X is a geodesic space and Y_1 and Y_2 are disjoint totally geodesic closed subspaces.*

Let $f : Y_1 \longrightarrow Y_2$ be an isometry and let $\hat{X}$ be the space formed from X by identifying Y_1 with Y_2 via f. Then $\hat{X}$, with the obvious metric, is a geodesic space. If the curvature of both components is $\leq \varepsilon \leq 0$, then the same is true for $\hat{X}$.

These gluings are geometric versions of well known constructions of combinatorial group theory: free products with amalgamation and HNN extensions. In fact they induce them on fundamental groups.

The ideal boundary. For a point x in a metric space X, denote by $B_x(r)$ ($\bar{B}_x(r)$, $S_x(r)$) respectively the open metric ball (closed ball or a sphere) of radius r about x. Suppose that X is a $CAT(0)$ space. It follows that any two points can be joined by a unique geodesic. Define a map $c_r : X - B_x(r) \longrightarrow S_x(r)$, called geodesic contraction, by sending a point y to the point on the geodesic joining x and y belonging to $S_x(r)$. It is straightforward to check that c_r is a continuous deformation retraction.

Definition. The *ideal boundary* (or the *visual sphere*) at x is the set of geodesic rays emanating from x. We denote it by $S_x(\infty)$.

Since every geodesic ray emanating from x intersects $S_x(r)$ in a unique point, we have an identification $S_x(\infty) = \varprojlim S_x(r)$. This identification equips $S_x(\infty)$ with the topology. It is proved in [5, Theorem 2.6] that $S_x(\infty)$ is homeomorphic to the so called horospherical infinity in some interesting cases. On the other hand it should be stressed that usually it is *not homeomorphic* to the Floyd boundary defined elsewhere in these Proceedings, unless we are dealing with spaces of strictly negative curvature.

Subsection on piecewise constant curvature spaces. Let $M^n(\varepsilon)$ denote the complete simply connected space of constant curvature ε, that is as $\varepsilon = +1, 0, -1$, $M^n(\varepsilon)$ is the n-sphere, Euclidean n-space and hyperbolic n-space. Recall that a (geodesic) k-simplex in $M^n(\varepsilon)$ is the convex hull of (independent) $(k+1)$ points.

Let K be a n-dimensional, locally finite simplicial complex. Suppose that each simplex of K is identified with a simplex in $M^n(\varepsilon)$, and the obvious compatibility condition is satisfied: if two simplices intersect along a k-dimensional simplex σ, then the two identifications of σ with a k-simplex in $M^n(\varepsilon)$ are isometric. This allows one to define the length of a curve as the sum of the lengths of its intersections with (interiors of) each simplex. We assume here that the curve is rectifiable in the sense that its intersection with any simplex is rectifiable. Thus we have a metric on K: the distance between points x and y is the infimum of lengths of rectifiable curves joining x and y. The complex with such a metric is called a (triangulated) space of piecewise constant ($= \varepsilon$) curvature.

Suppose that σ is an n-simplex in $M^n(\varepsilon)$ and that v is a vertex of σ. The set of all unit tangent vectors to a geodesic ray which emanates from v and enters σ constitutes a spherical $(n-1)$-simplex, denoted $link(v, \sigma)$. More generally if u is a k-face of σ, and x is in the relative interior of u, the unit tangent vectors at x, which point into σ and are normal to u form a spherical $(n-k-1)$-simplex *link*(u,σ) (or to be more precise, *link*(x,u,σ); it is in fact (up to an isometry) independent of x). The set of *all* vectors at x pointing into σ is a $(k-1)$-fold suspension of *link*(u,σ) where k is the codimension of the face containing x in its relative interior. Recall that the k-fold suspension of a simplex σ is the orthogonal join of σ and S^{k-1}, i.e. the union of all geodesic segments in S^{n+k} from a point in S^{k-1} to a point in σ, where we embedded S^n and S^{k-1} into S^{n+k} so that the corresponding linear spaces are orthogonal. In particular the n-fold suspension of an empty set is S^{n-1} and the n-fold suspension of a point is an n-hemisphere.

Let K be a triangulated space of piecewise constant curvature and u be a simplex in K. As x runs over the set of all simplices containing u, the simplices *link*(u,x) form a piecewise spherical triangulated space which we denote by *link*(u,K). We can also define *link*(x,K) for any point in a triangulated space of piecewise constant curvature, using appropriate suspensions of *link*(u,K). Up to an isometry *link*(x,K) is independent of the triangulation, in the sense that it depends on the metric only.

Definition. A piecewise spherical complex L is *large* if any two points x and y in K with $d(x,y)<\pi$ can be joined by a unique geodesic segment.

For example if L is one dimensional, L is large if any closed loop has length $\geq 2\pi$.

It follows that if L is large then for any $v\in L$ the open ball $B_\pi(v)$ of radius π around v is contractible. Since a spherical complex is large if and only if its suspension is large, it make sense to say that K has large links if *link*(x,K) is large for any x. The following Lemma of Gromov ([11, p. 120]) relates large links to $CAT(0)$.

Lemma 1.2. *Suppose that K is a complex of piecewise constant curvature $\leq\varepsilon$. Then the curvature of K is $\leq\varepsilon$ if and only if K has large links.*

From now on by "$K\leq 0$ space" we mean a piecewise constant nonpositive curvature complex with large links.

Shadows. We now want to discuss the concept of the infinitesimal shadows; later it will be useful in studying large spheres in simply connected $K\leq 0$ spaces. Suppose that K is a piecewise constant curvature space, and that we have a geodesic segment with endpoint $x\in K$. If the link of x is small, then it may happen that we cannot extend the geodesic past x. On the other hand if the link is large, then unlike the case of smooth Riemannian manifolds, the local extension may be nonunique. In other words a point casts a "shadow". The nonuniqueness of the geodesic continuation is measured by a subset of *link*(x,K) which we call the infinitesimal shadow of x with respect to the geodesic segment.

Suppose that we have a geodesic through x; it defines two points in the link of x: the incoming and outgoing directions. For a given point v in $link(x, K)$ define $shad(x, v)$ to be the set of all points w in $link(x, K)$ such that there is a geodesic through x, for which v is the incoming, and w is outgoing direction.

Lemma 1.3 (2.22 of [5]). *Supose that K is a piecewise constant curvature space, $x \in K$ and $v \in link(x, K)$. Then $shad(x, v)$ is the complement of the open ball $B_v(\pi)$.*

This Lemma is important, since it is the basic ingredient in analyzing the topology of simply connected $K \leq 0$ spaces. There are two theorems about such spaces which play a role in our story. One is an extension of Cartan-Hadamard Theorem to PL manifolds. The other is a theorem which says that there is a limit to possible extensions of Cartan-Hadamard theorem from Riemannian manifolds to more general manifolds of nonpositive curvature. We need some definitions first.

An $(n-1)$-dimensional piecewise spherical polyhedron L is a *PL-sphere* if it is PL-homeomorphic to S^{n-1}. Suppose K is a piecewise constant curvature polyhedron, and a homology manifold (recall that it means that the relative homology groups $H_*(K, K - \{x\})$ are the same as $H_*(\mathbb{R}^n, \mathbb{R}^n - \{0\})$) a point is *metrically nonsingular* if its link is isometric to the standard sphere. It is a *PL-nonsingular* if its link is a PL-sphere. K is a PL-manifold if the set of PL-singular points is empty.

Theorem 1.1 (3.2 of [5]). *Suppose that K is a simply connected PL-manifold. Then*

(i) *For each $x \in K$, $\bar{B}_x(r)$ is homeomorphic to the standard ball,*
(ii) *K is homeomorphic to $\mathbb{R}^n$,*
(iii) *The visual sphere $S_x(\infty)$ is homeomorphic to S^{n-1}.*

This Theorem (except for the claim about the visual sphere) has been first proved by D. Stone [22]. It can be significantly generalized (say to appropriately defined piecewise variable nonpositive curvature manifolds). The proof presented in [5] uses some heavy topological machinery: Approximation Theorem for cell-like maps. The fact that the geodesic contraction map is cell-like follows from the properties of shadows: the fact that for PL manifolds they are cell-like.

To discuss the non-PL case we need more definitions. A polyhedral homology n-manifold is a *generalized homology sphere* if it has the same homology as S^n. An inverse system of groups $\{f_i : G_i \to G_{i-1}\}$ has the *Mittag-Leffler property* if for each i the descending chain $G_i \supset f_{i+1}(G_{i+1}) \supset f_{i+2}(G_{i+2}) \cdots$ eventually stabilizes.

Suppose that X is a countable increasing union of compact spaces $X = \bigcup C_i$. We say that X is *one ended* if the inverse limit of the inverse system of sets

$\pi_0(X - C_i)$ consists of one element. Suppose that X is one ended and choose the C_i so that each $X - C_i$ is path connected. The space X is *semistable at infinity* if the inverse system $\{\pi_1(X - C_i)\}$ is Mittag-Leffler.

Theorem 1.2 (3.5 of [5]). *Suppose that K is a simply connected $K \leq 0$ space and a homology manifold. Then for each $x \in K$*

(i) *$\bar{B}_x(r)$ is a contractible homology n-manifold,*
(ii) *$S_x(r)$ is a generalized homology sphere,*
(iii) *If $s > r$, then the geodesic contraction $S_x(s) \to S_x(r)$ is a degree one map and thus induces a surjection on fundamental groups,*
(iv) *K is semistable at infinity,*
(v) *The fundamental group at infinity of K is $\pi_1^\infty = \varprojlim \pi_1(S_x(r))$,*
(vi) *If for some r and x, $\pi_1(S_x(r)) \neq 0$ then K is not simply connected at infinity.*

Usually we can be more specific about metric spheres. There is a general method for getting a good hold on the spheres, which works in a most transparent way in the proof of the following Theorem.

Theorem 1.3 (Proposition 3.7 of [5]). *Suppose that K is a $K \leq 0$ homology manifold and that the set of PL-singular points is discrete. Let $\{s_1, \ldots, s_k\}$ denote the set of singular points in $B_x(r)$. Then $S_x(r)$ is homeomorphic to the connected sum*

$$S_x(r) = link(s_1, K)\# \cdots \#link(s_k, K)$$

The detailed proof is presented in [5]. Since we will use the method of the proof in many arguments let us instead of repeating it here, give the intuition. We are growing the spheres starting from x. A very small sphere is essentially the same as the link of x. The proof of our version of Cartan-Hadamard for PL-manifolds tells us that nothing happens to the topology of the sphere as long as our sphere grows through the region free of singular points. Now as the sphere crosses the singularity, it acquires the link of the singular point as a connected sum summand. "Infinitesimally" it is obvious from the study of shadows and a little argument is needed that it is really the case.

2. Fiber product hyperbolization

In this Section we will describe an approach of [5], based on the concept of fiber product, to hyperbolization procedure invented by Gromov [11]. Actually this approach was used by Williams [24] quite some time ago, for a similar purpose.

A *space over σ^n* is a pair (X, π) where X is a topological space and $\pi : X \longrightarrow \sigma^n$ is a continuous map. A *simplicial complex over σ^n* is a pair (L, f) where L is a simplicial complex, and $f : L \longrightarrow \sigma^n$ is a nondegenerate (that is, injective on each simplex) simplicial map. Such a map may fail to exist, but in the case of pseudomanifolds, if it exists it is essentially (up to the action of the symmetric group in the range) unique. The letter f stands for "folding". This is indeed how the map is constructed: by defining it arbitrarily on one of top dimensional simplices, and then extending it by folding adjacent simplices. The first barycentric subdivision of any complex admits a canonical map into the (top dimensional) simplex.

The construction. Suppose that (X, π) is a space over σ^n and (L, f) is a simplicial complex over σ^n. The fiber product of f and π will be denoted by $X \Delta L$. In other words $X \Delta L$ is the subset of $X \times L$ consisting of all pairs (x, l) with $f(l) = \pi(x)$. The natural projections of the product onto factors X, L restricted to $X \Delta L$ are denoted by π_X and f_L respectively. If J is any subcomplex of σ^n, we put $X_J = \pi^{-1}(J)$. In particular if J is a closed face of σ^n, X_J is called a *face* of X.

The intuitive content to this construction is that we replace each simplex of L with a copy of X, which has faces X_J, and glue them according to the gluing pattern of the triangulation.

Now we would like to list some conditions on π, and discuss their consequences.

C_{conn} X is path connected and for each codimension one face J of σ^n, X_J is nonempty.

C_{trans} X is a compact n-dimensional PL-manifold with boundary and for each k-dimensional face J of σ^n, X_J is a k-dimensional PL-submanifold of ∂X and $\partial(X_J) = X_{\partial J}$. The map π is required to be piecewise linear. It should be remarked that it follows from C_{trans} that π is transverse to each face of σ^n in the PL sense.

C^{∞}_{trans} (A smooth version of C_{trans}) X is a compact smooth manifold with corners, and for each k-dimensional face J of σ^n, X_J is a union of k-dimensional strata. Moreover the map π is required to be smooth and transverse to each proper face of σ^n.

C_{deg} X is oriented, satisfies C_{trans} (or C^{∞}_{trans}) and $\pi : (X, \partial X) \longrightarrow (\sigma^n, \partial\sigma^n)$ is a map of degree one.

C^2_{deg} X satisfies C_{trans} (or C^{∞}_{trans}) and in addition $\pi : (X, \partial X) \longrightarrow (\sigma^n, \partial\sigma^n)$ is a map of degree one mod 2.

C_{tan} X satisfies C_{trans} or C^{∞}_{trans} and τ_X is trivial. This condition deals with the tangential properties of π. Here X is a smooth or PL-manifold, and τ_X is its stable tangent bundle. In the PL case we understand that τ_X is the "stable tangent block bundle in the sense of [23].

C_{asph} X is an aspherical complex and if J is any subcomplex of σ^n, then each component of X_J is aspherical and the inclusion $X_J \to X$ induces a monomorphism of fundamental groups.

C_{conv} X is a nonpositively curved complex and if J is any subcomplex of σ^n then each component of X_J is convex (totally geodesic). Sometimes it is also useful to have a hyperbolized simplex which is negatively curved.

These conditions have the following effect on $X\Delta L$:

1. Suppose π satisfies C_{conn}; suppose L is connected. Then $X\Delta L$ is path connected, and the map π_L induces a surjection $\pi_1(X\Delta L) \longrightarrow \pi_1(L)$,
2. (Williams [24]). If (X, π) satisfies C^2_{deg} then the following map $H_*(X\Delta L, \mathbb{Z}_2) \longrightarrow H_*(L, \mathbb{Z}_2)$ induced by π_L is onto. Moreover if (X, π) satisfies C_{deg} then the map $H_*(X\Delta L, \mathcal{A}) \longrightarrow H_*(L, \mathcal{A})$ induced by π_L is onto, where $\mathcal{A}$ is any local coefficient system on L,
3. Conditions C^{∞}_{trans} have consequences for the local structure of $X\Delta L$. Namely, they guarantee that the preimage of a simplex in L by π_L is a manifold. Moreover the transversality assumption on the map π implies that links of preimages of simplices in L are PL isomorphic to links in L. Thus if L is a manifold or a homology manifold, the same holds for $X\Delta L$,
4. The transversality condition also implies that the normal bundle of $X\Delta L$ in $X \times L$ is trivial; but then C_{tan} implies that the map $\pi_L : X\Delta L \longrightarrow L$ is covered by the isomorphism of stable tangent bundles. This is very useful, since the (stable) characteristic classes of $X\Delta L$ are the images of the characteristic classes of L. In particular, characteristic numbers are the same for hyperbolizations satisfying C_{tan} and C_{deg},
5. It can be shown that C_{conv} implies C_{asph} and this in turn that $X\Delta L$ is aspherical. Moreover, the Gluing Lemma implies that $X\Delta L$ is nonpositively curved if X satisfies C_{conv}.

Definition. By a hyperbolization we mean a fiber product construction satisfying C_{conn}, C^{∞}_{trans} and C_{conv}.

Actually this definition is a deliberate lie committed for the sake of being definite. We will present several constructions, which are legitimate hyperbolizations, but do not satisfy some of these conditions. One should also note that our hyperbolization produces spaces which are only semihyperbolic in the sense of Gromov, unless the hyperbolized simplex is negatively curved.

Relative Construction. Let J be a subcomplex of K. Let $R(J, K)$ be the standard derived neighborhood of J in K, that is the sum of all the simplices in the first barycentric subdivision of K which meet J. Let $\hat{K}$ denote the simplicial complex formed by deleting the interior of $R(J, K)$ from K and attaching the cone over $\partial R(J, K)$. Let c_0 be the cone point. The complex $\hat{K}$ is in a natural way a complex over σ^n, and c_0 maps to (say) 0. Consider points v_i in $X\Delta\hat{K}$ which maps to c_0 (there may be many of them). The condition C_{trans} guarantees that each v_i has a neighborhood of the form $Cone(\partial R(J,K))$. Remove interiors of these neighborhoods and paste in $R(J,K)$. The result, denoted by $X\Delta(K,J)$, is the relative fiber product construction on (K,J).

This relative construction has a beautiful application (cf. [11, 3.4.C p. 117]). Applying it to cobordisms yields the following facts:

1. Any manifold is cobordant to an aspherical manifold,
2. An aspherical manifold M bounds iff it bounds an aspherical manifold,
3. Two aspherical manifolds are cobordant iff they are cobordant through an aspherical cobordism.

Construction of hyperbolized simplices. The constructions we described so far are pretty dry, so now we want to present some examples of hyperbolized simplices. In fact, the reader is challenged to find her own examples. The first example is mainly pedagogical, but it is also useful. For example it is an ingredient in a proof that fundamental groups of polyhedra hyperbolized according to Gromov's Möbius band hyperbolization of cubical polyhedra ([11, 3.4]) (to which it is closely linked) are K-amenable in the sense of Cuntz. Anyway I cannot resist presenting it here because it is so simple. Its definition is inductive over the dimension: $(\sigma^0)_h = \sigma^0$ (nothing happens to points), and $\sigma^n_h = (\partial\sigma^n)_h \times [0,1]$. That is, we hyperbolize the boundary of σ^n using $(n-1)$-dimensional procedure and then we cross it with the interval. The map $\sigma^n_h \longrightarrow \sigma^n$ is defined as follows: $(\partial\sigma^n)_h$ is equipped with the map to $\partial\sigma^n$ by $(n-1)$-dimensional hyperbolization. We extend that map from the boundary of $(\partial\sigma)_h \times [0,1]$ (two copies of $(\partial\sigma)_h$) to the cone over $\partial\sigma^n$, which happens to be a simplex, so that the extension is a product map on the collar neighborhood of the boundary. This construction produces a hyperbolic simplex, which obviously satisfies all the conditions we listed, except C^2_{deg}.

For topological applications of the next Section we need a hyperbolized simplex which satisfies C_{tan} and C_{deg}, so that the M_h has the same characteristic numbers as M. The idea here (again provided by Gromov) is as follows. We again do it inductively and again $(\sigma_0)_h = \sigma_0$. To get σ^n_h, we hyperbolize the barycentric subdivision of $\partial\sigma^n$, and we notice that the symmetries of σ^n still act on $(\partial\sigma^n)_h$. In particular any reflection, which used to divide $\partial\sigma^n$ into two halves acts on $\partial\sigma^n_h$, and divides it into two halves. Now we take the product $(\partial\sigma^n)_h \times S^1$ and slit it open along a half H of $(\partial\sigma^n)_h$. Alternatively, we take the product $\partial\sigma^n_h \times [-1,1]$ and divide it by the relation

$$(a,\epsilon) \sim (a^*,\epsilon^*) \text{ if and only if } a,a^* \in H \text{ and } \epsilon,\epsilon^* = \pm 1.$$

The boundary of the space thus obtained is clearly simplicially isomorphic to $(\partial\sigma^n)_h$, the map into σ^n obtained as before by coning the map $(\partial\sigma^n)_h \longrightarrow \partial\sigma^n$

is clearly degree one and it is shown in [5] using a little argument that C_{tan} is satisfied.

Some remarks. At the end of this Section we want to make two remarks of general nature.

1. We have seen that a map $M \longrightarrow \sigma^n$ is a very convenient piece of data for doing hyperbolization. One may wonder if it is really necessary, and indeed sometimes it is not. This happens when the hyperbolized simplex is symmetric, that is if σ_h^n carries an action of the symmetric group Σ_{n+1} so that the map $\sigma_h^n \longrightarrow \sigma^n$ is equivariant (with respect to the standard action of Σ_{n+1} on σ^n). Intuitively, all the faces look the same and one does not have to keep track of which face gets glued to which. To perform the construction it is again convenient to use the fiber product:

First notice that the quotient σ^n/Σ_{n+1} can be identified with the standard simplex σ^n, and the quotient map $\sigma^n \longrightarrow \sigma^n/\Sigma_{n+1} \simeq \sigma^n$ with the *folding map* of the first barycentric subdivision of σ^n. In the case when there is an equivariant map $\sigma_h^n \longrightarrow \sigma^n$ the above identification gives the map $\pi_\Sigma : \sigma_h^n/\Sigma_{n+1} \longrightarrow \sigma^n$.

Now we can take any triangulation, form its barycentric subdivision, take the folding map $f_{bar} : (M, \Delta_{bar}) \longrightarrow \sigma^n$, and then take the fiber product of f_{bar} with π_Σ, again denoted by M_h. Both C_{trans} and C_{conv} may fail here but this is not important, since their consequences hold anyway.

When we apply this construction to the simplex σ^n we get σ_h^n. Thus it follows that M_h has indeed large links: every simplex in (M, Δ) has been replaced by σ_h^n and we can apply the Gluing Lemma. Moreover, the transversality holds near the skeleta of the original (before the subdivision) triangulation, hence M_h is a (homology) manifold if M and σ_h^n were (homology) manifolds.

2. Is there something like *the* hyperbolization? We certainly have both simple and complicated hyperbolizations. For example we can iterate any hyperbolization twice, and the result will be a (very) complicated hyperbolization. This suggests that there is a notion of partial ordering of the hyperbolizations. Within the framework of the fiber product construction we propose the following definition:

The hyperbolization h_1 *dominates* h_2 if there is a (distance decreasing, perhaps) map $f : \sigma_{h_1} \longrightarrow \sigma_{h_2}$ such that the diagram

$$\begin{array}{ccc} \sigma_{h_1} & \xrightarrow{f} & \sigma_{h_2} \\ \pi_1 \downarrow & & \pi_2 \downarrow \\ \sigma^n & \xrightarrow{id} & \sigma^n \end{array}$$

commute. Clearly such a map produces the map $M_{h_1} \longrightarrow M_{h_2}$ for any M and one can think that M_{h_1} is more complicated than M_{h_2}. Any two hyperbolizations clearly have a common refinement: the hyperbolized simplex of the refinement is the fiber product of two hyperbolized simplices. As apparently there is no pushout in this category, the maximal element—the simplest hyperbolization—does not exist. But certainly we have maximal elements in the set of hyperbolizations. This is as close to *the* hyperbolization as we can get.

3. Applications

In this Section we present examples of some aspherical manifolds form [5], as well as a simple fact about combinatorics of triangulations of four-manifolds. One should remember to count as an application of (relative) hyperbolization Gromov's results on cobordism of aspherical manifolds mentioned in Section 2.

Theorem 3.1 (Universal covering not homeomorphic to $\mathbb{R}^n$). *For any $n \geq 4$ there is a closed topological manifold N such that its universal covering is not homeomorphic to $\mathbb{R}^n$.*

Proof. We start with Σ a non simply connected PL homology sphere which bounds a PL-manifold M, and also bounds a contractible topological manifold C. A rich supply of such manifolds can be found in the Appendix. We attach the cone over the boundary to M and get a (triangulated) homology manifold $\overline{M}$. Hyperbolize it with any hyperbolization you like. The Theorem 2.3 tells you that the universal cover of $\overline{M}_h$ is not simply connected at infinity. Now we cut off the singular point and glue in C. Since C is contractible, the homotopy type of $\overline{M}_h$ does not change, but now the modified $\overline{M}$ is a topological manifold. Now an easy argument shows that the modified $\overline{M}_h$ is also non simply connected at infinity. This argument works in every dimension in which exotic contractible manifolds exist, i.e. for $n \geq 4$. Taking Σ that bounds a smooth (or PL) contractible manifolds one gets $\overline{M}_h$ which are smooth (or PL) and its universal covers is not simply connected at infinity. □

Remark. The first examples of that kind were produced by Mike Davis [4] using reflection groups. We will discuss his method in Section 4.

Theorem 3.2 (Nontriangulable manifold). *There exists a closed topological four-manifold N^4 with the following properties:*

(i) *N^4 is not homotopy equivalent to a PL-manifold,*
(ii) *N^4 is not homeomorphic to a simplicial complex.*

Proof. The proof here uses rather serious four dimensional topology. Let E_8 be the manifold with boundary obtained by plumbing eight copies of the unit tangent disc bundles of the two sphere according to the E_8 Dynkin diagram (see [2, pp. 116–126]). The boundary is a homology sphere, so attaching the cone over it we get what is called the E_8 homology manifold. Take any triangulation on it and hyperbolize using the hyperbolized simplex satisfying in addition C_{tan} and C_{deg}. The result is an aspherical homology manifold $\hat{N}$ with an isolated PL singular point. Now it is a hard theorem of M. Freedman [8], that the boundary of E_8 plumbing bounds a contractible topological manifold F^4. Cutting off the cone point of $\hat{N}$ and gluing in the Freedman contractible manifold we obtain a topological manifold homotopy equivalent to the hyperbolization of the E_8 homology manifold. It is a spin manifold in the sense that the second Stiefel-Whitney class is zero. Its signature is eight (by homotopy invariance of the signature it is the same as the signature of the E_8 homology manifold). Thus Rohlin's Theorem (the signature of a PL spin manifold is divisible by 16) tells us that it is not homotopy equivalent to a PL-manifold, and moreover a recent theorem of Casson that the Rohlin invariant of a homotopy sphere is zero implies that N cannot be triangulated.

This argument is identical to the proof that the E_8 manifold obtained by gluing the Freedman contractible manifold to the boundary of the E_8 plumbing cannot be triangulated (see [17]). The only difference is that the manifold we are dealing with is aspherical. □

Using Farrell's Fibering Theorem we can show that $N^4 \times T^{n-4}$ is a closed aspherical manifold not homotopy equivalent to a PL-manifold.

Both examples presented so far had exotic universal covers, but due to chopping off singular points and regluing in a rather mysterious (from the geometric point of view) contractible manifold, they are not nonpositively curved (or at least we don't know if they are). We can produce examples with exotic universal covers *and* nonpositive curvature, using non PL triangulations. This answers a question of Gromov ([10]), raised before Davis' examples were known.

Theorem 3.3 (Exotic universal covering of $K \leq 0$ manifold). *For every $n \geq 5$ there is $K \leq 0$ complex Q which is a closed topological manifold such that its universal cover is not simply connected at infinity.*

Proof. Let A^{n-1} be a compact acyclic smooth manifold (with boundary) such that

1. the map $\pi_1(\partial A^{n-1}) \longrightarrow \pi_1 A^{n-1}$ is onto,
2. the double of A (a homology $(n-1)$-sphere) is not simply connected.

It is easy to construct such A: let H^{n-2} be a nonsimply connected homology sphere, and C^{n-2} is the complement of an open ball. Then we can put $A = C \times [0,1]$. It is straightforward to check that both 1 and 2 are satisfied. Choose a triangulation on A and form a simplicial complex Z by attaching to A the cone over its boundary. Let K be the suspension of Z.

In fact K is miraculously a manifold. This follows from the Manifold Characterization Theorem (see [6, p. 119]), a consequence of which says that a triangulated homology manifold is a topological manifold if and only if the links of vertices are simply connected. Hence, as it is easy to check that K is a simply connected generalized homology sphere, the Poincare Conjecture implies that it is a sphere.

Now we hyperbolize the (non-PL) triangulation on K obtained by suspending a triangulation on Z. Arguments similar to the proof of Theorem 2.3 show that the hyperbolization has the required properties. □

Exotic topology of the universal covering implies that manifolds constructed above do not carry smooth or PL metrics of nonpositive curvature. Studying the sphere at infinity in more detail, we can exhibit examples of topological $K \leq 0$ manifolds covered by $\mathbb{R}^n$, which carry no smooth or PL metric of $K \leq 0$. We thus have the following theorem (which is strictly speaking, a corollary of Theorem 3.3):

Theorem 3.4. (Top $K \leq 0$ manifold with no PL $K \leq 0$ metric). *There is a 5-dimensional smooth manifold carrying a (non-PL) triangulation with $K < 0$, which carries no PL metric with $K \leq 0$, and no $K \leq 0$ Riemannian metric.*

Proof. We again use non-PL triangulations. Let Σ be an nonsimply connected homology sphere. Take its double suspension M. Manifold Characterization Theorem tells us that it is homeomorphic to a sphere. The natural triangulation on M obtained by suspending a triangulation on Σ is not a PL triangulation. Now hyperbolize such a triangulation. Using again the method of the proof of Theorem 2.3, it can be shown that the universal cover is simply connected at infinity, thus by Stallings Theorem it is homeomorphic to $\mathbb{R}^n$, but the ideal boundary is *not homeomorphic* to a sphere. If we use a hyperbolization which produces polyhedra of strictly negative curvature, we can take advantage of the fact that the ideal boundary of the universal covering has a very strong invariance property, in particular, one can read off the fundamental group of the hyperbolized polyhedron.

Thus the smooth part of the statement follows from the (classical) Cartan-Hadamard theorem and the PL part follows from our version of it. □

One might ask if the phenomena exhibited in the last two theorems happen in dimension 4.

Subsection on Combinatorics. Our aim in this subsection is to apply a formula of Cheeger, Müller and Schrader [3] to hyperbolized polyhedra to prove an inequality for the Euler characteristic, analogous to the inequality for smooth manifolds of nonpositive curvature predicted by the conjecture of Hopf. For a discussion of the Conjecture we refer reader to Section 5. A consequence of our inequality is a bound on the number of cells in a triangulation (or a more general cell decomposition) of a four-manifold. It seems to us that the possibility of. *conjecturing* such a bound, based on general notions, is quite an attractive principle.

We should start with notation for the angle measure. Let $\sigma^k \subset \sigma^l$ be simplices. By (σ^k, σ^l), or simply by (k, l), we will denote the measure of the $link(\sigma^k, \sigma^l)$, normalized so that the measure of the whole sphere is one. We will call it the *interior angle*. We will also need the *exterior angle* $(\sigma^k, \sigma^l)^*$: it is the normalized measure of the *dual* spherical simplex, consisting of all the unit vectors making an angle $\geq \pi/2$ with every vector of $link(\sigma^k, \sigma^l)$.

The formula for the Euler characteristic of piecewise flat spaces proved in [3], specified to manifolds reads as follows: we have to sum over all vertices local contributions depending on the geometry of the link. These local contributions look as follows:

$$P_\chi(\sigma^0) = 1 + \sum_{flags} (-1)^l (2i_0, 2i_1)(2i_1, 2i_2) \cdots (2i_{l-1}, 2i_l).$$

Here the sum is taken over all flags $\sigma^0 \subset \sigma^{2i_0} \subset^{2i_1} \subset \cdots \subset \sigma^{2i_l}$ of *even dimensional* simplices containing σ^0. It will be important to us to note that, as the consequence of the *proof* given in [3], this formula is also true for decompositions of piecewise flat spaces into cubes rather than simplices (with all the necessary grammatical changes in the statements and definitions).

Let us say a few words about the proof of the formula above. It starts with the obvious fact that the sum of the *exterior* angles of a simplex of arbitrary dimension is one. Then it uses an argument which is essentially the Möbius inversion formula, to express the interior angles in terms of exterior ones, starting with the formula for the volume of the join $N * N^{dual}$. The result is a formula similar to the one we quoted, but involving all the flags, and not even dimensional ones only. The final formula requires an additional argument and in a sense it is the hardest part of the proof. It uses a formula of Schläfli.

The reason why this formula is true for cubes is that the starting point—the formula for the sum of the exterior angles holds true for cubes. The remaining part of the argument does not involve the specific shape of the cell at all.

It is an easy consequence of the formula that the local contribution vanishes if the link of a point is a join with a standard sphere, that is if the geometry of the space is that of a local product by a flat Euclidean space.

Now let us try to think what this formula tells us in the case of a hyperbolization, say cross-with-the-interval hyperbolization.

The only points where the local contribution is nonzero are preimages of the original triangulation or cubical decomposition. The angles are all standard, and it remains to count the number of cells of appropriate dimension. But this is elementary, let us do it in dimension four. The formula becomes

$$P_\chi(\sigma_0) = 1 - (0,2)f_2 - (0,4)f_4 + (0,2)(2,4)f_{2,4}$$

Here f_2, f_4 and $f_{2,4}$ denotes respectively the number of two-cells, four-cells and flags of two-cell in a four-cell, containing the vertex σ_0. It is clear that $f_{2,4} = 6f_4$. Recall that each cell of the hyperbolization is a cube of appropriate dimension. As for the angles after the hyperbolization, they are the same for simplicial and cubical cell decompositions: $(0,2) = (2,4) = \frac{1}{4}$, $(0,4) = \frac{1}{16}$.

We can express f_2 and f_4 in terms of number of cells at the vertex σ_0 of the original cubization, using a fact that *the link of a vertex in the new cubization is combinatorially isomorphic to the first barycentric subdivision of the original link.* Explicitly, with k_i denoting the number of i-cells in the original decomposition containing σ_0, $f_4 = 24k_4$ and $f_2 = 2k_2 + 6k_3 + 14k_4$; moreover $k_3 = 2k_4$. Thus the final result is

$$P_\chi(\sigma_0) = 1 - \frac{1}{2}k_2 + k_4$$

Now it is fairly clear that the local contribution is positive. If there is a 2-cell containing σ_0 which belongs to exactly two 4-cells, then the link is necessarily a double of the simplex and $P_\chi(\sigma_0) = 0$. Otherwise every 2-cell belongs to at least three 4-cells and $P_\chi(\sigma_0)$ is nonnegative. Thus we have established

Theorem 3.5. *The Euler characteristic of a hyperbolized with cross-with-the-interval hyperbolization triangulated four-manifold is nonnegative.*

It is an easy task to compute the Euler characteristic of hyperbolizations directly from the combinatorics of the cell decomposition and Euler characteristics of hyperbolized cell. The nonnegativity result proved above results in the following theorem

Theorem 3.6. *Let c_i (s_i) denote the number of i-cells in a cubization (triangulation) of any four-manifold. Then $c_0 - c_1 + 2c_3 \geq 0$ ($s_0 - s_1 + s_3 \geq 0$)*

We should make two additional remarks here. First, that the inequality is true for a cell decomposition such that *each cell is embedded*. In this class they are optimal: the hyperbolized double of a cell has the Euler characteristic zero. Second, what is probably more important, our proof works for pseudomanifolds which are *Eulerian complexes*, that is, for which the Euler characteristic of the link of any simplex of codimension i is $1 + (-1)^i$. It would be interesting to apply other hyperbolizations to triangulated manifolds to get better (or different) inequalities.

4. Ramified coverings.

In this section we present a method of constructing examples using ramified coverings. We do not have a general streamlined definition of a ramified covering which would provide us with anything really new. We think that the approach to constructing interesting examples through ramified is in a way dual to (perhaps relative) hyperbolization. Namely a folding map of a triangulation, and the (induced by folding) map $M_h \to \sigma_h^n$ should be considered as ramified covering (in a very general sense). From this example one sees that there is a tremendous variation of species of ramified covering. Another example of that duality is discussed in the part dealing with Davis' method of reflection groups.

Ramified coverings have an advantage of preserving close relationship between objects on M and M_h. So far this has been the source of their success in differential geometry. They also appear very naturally in algebraic and analytic complex geometry. An elementary example of this apperance is the fact that any holomorphic map between Riemann surfaces is a ramified covering. We refer to the book of Namba [21] for a discussion of analytic aspects of ramified coverings.

So what we mean here by a ramified covering is a map $f : X \to Y$ which has some specific local model (see below). We adopt the following terminology:
the ramification locus is the set of points x such that f is not injective on some neighborhood of x; the branching locus is the image by f of the ramification locus.

A standard example is as follows. Let M^n be a manifold, and N^{n-2} be a codimension two embedded submanifold. Suppose that there exists a two-fold covering of $M - N$ which, when restricted to a fiber of a normal sphere bundle of N, is nontrivial. The covering of $M - N$ can be compactified by adding N to it, and the result, call it $\hat{M}$, is a manifold. If we had taken the trivial covering of $M - N$, the resulting space would not be a manifold. It is straightforward to check that such a nontrivial covering exists iff the fundamental homology class of N is zero in $H_{n-2}(M, \mathbb{Z}_2)$. Alternatively one may define such a ramified covering by saying that it is modelled on the map $f : \mathbb{C} \times \mathbb{R}^k \longrightarrow \mathbb{C} \times \mathbb{R}^k$ given by $f(z, v) = (z^2, v)$. That is, the map is generically a covering, and at the point of the domain where the jacobian is zero it looks locally like the model map.

Now we want to discuss the geometry of $\hat{M}$. We assume that M is a $K \leq 0$ space, and that N is piecewise totally geodesic, so that the induced metric on N converts it into $K \leq 0$ space. We pull back the metric from M to $\hat{M}$. Then we have the following fact.

Proposition 4.1. *$\hat{M}$ is $K \leq 0$ space if and only if N is totally geodesic.*

Proof. Suppose it is not. It means it is not totally geodesic at some point n, that is, there exists a point $n \in N$, and two points $x, y \in link(n, M)$, such that the distance (in the link) between them is less than π and the (unique since the link of n in M is large) geodesic between them is not contained in $link(n, N)$, moreover we can assume that the interior of the geodesic is *not* in $link(n, N)$. Let $\hat{n}$ be a point of $\hat{M}$ hanging over n. The $link(\hat{n}, \hat{M})$ is itself a ramified covering of $link(n, M)$. Take the preimage in $link(\hat{n}, \hat{M})$ of the geodesic from x to y. It consists of many branches, if we take just two of them, we get a *geodesic* and its length is less than 2π. Thus $link(\hat{n}, \hat{M})$ is not large. On the other hand it is fairly clear that if N is totally geodesic $\hat{M}$ is $K \leq 0$, since a short closed geodesic in the link would cross the part of the link corresponding to N and this would give a contradiction. □

Already this simple construction provides examples interesting from the differential geometric point of view. Namely in [13] Gromov and Thurston observe that for manifolds $M = H^n/\Gamma$, where Γ is arithmetic, there is plenty of totally geodesic submanifolds N^{n-2}, such that $N = 0$ in $H_{n-2}(M, \mathbb{Z})$. Then there are ramified coverings $\hat{M}_p$ of arbitrary degree, nontrivially branched at N. By constructing explicit smoothings of the singular metric they show that $\hat{M}_p$ are smooth Riemannian manifolds of negative curvature. Then using an ingenious combination of Mostow Rigidity Theorem and Wang Finiteness Theorem, they conclude that all but finitely many of $\hat{M}_p$'s are *not* homeomorphic to constant curvature manifolds. A stronger version of the finiteness theorem allows us in fact to conclude that almost all $\hat{M}_p$'s cannot be pinched, that is they are not homeomorphic to smooth manifolds with sectional curvatures K satisfying $-k^2 \leq K \leq -1$ for any given k.

First examples of not locally symmetric manifolds of negative curvature were obtained by Mostow and Siu in [18] also using ramified coverings. Their example is much more complicated than [13] and the method of constructing the metric is by smoothing, but it has additional nice features: it is an algebraic surface with a Kähler metric.

Another application is the following example of Fornari and Schroeder [7]. Start with N_1, N_2 two surfaces, with C_1, C_2 two closed geodesics such that the tubular neighborhood of C_i in N_i is metrically a product $C_i \times [-r, r]$, and C_1 separates N_1. Then

(i) For any k there exist a k-fold covering N of $N_1 \times N_2$,
(ii) N carries a smooth $K \leq 0$ Riemannian metric,
(iii) N cannot carry an analytic $K \leq 0$ metric.

The existence of the covering is elementary and the smoothing is constructed by writing down the metric explicitly. The reason for the last claim is that any element of $\pi_1(\hat{M})$ carried by the ramification locus T^2, has a centralizer which is not a fundamental group of a compact $K \leq 0$ manifold. Thus existence of an analytic $K \leq 0$ metric would contradict a theorem of Lawson and Yau [16].

It happens quite often that if the ramification locus is not a submanifold, $\hat{M}$ is a $K \leq 0$ space, even though N is not totally geodesic. The simplest example of such a situation is as follows. Consider first *the local model* of our ramified cover; here it will be the map $\mathbb{C}^n \to \mathbb{C}^n$ given by $f(z_1 \ldots z_n) = (z_1^{\alpha_1}, \ldots z_n^{\alpha_n})$. It is a ramified covering and if the α_is are integers ≥ 2 the branching locus consist of coordinate hyperplanes. Thus it is definitely not convex. Moreover it is easy to prove (say using induction on n) that the metric induced by f on the source $\mathbb{C}^n$ (that is pulled back) is $K \leq 0$. Now we can globalize the local model as follows.

Proposition 4.2. *Suppose that M is a $K \leq 0$ PL-manifold and $N_1, \dots N_k$ are totally geodesic submanifolds of codimension 2, which intersect transversally, and moreover at any point of intersection the angle between N_i and N_j is at least $\frac{\pi}{2}$. Assign to N_i an integer $\alpha_i \geq 2$. Then the ramified covering associated to this data (if it exists) is a $K \leq 0$ PL-manifold.*

Proof. It comes down to disentangling the notion of the angle between N_i and N_j at the point of intersection. Since $N_i \cap N_j$ is a submanifold of codimension 4, at each point of the intersection, $link(N_i \cap N_j, M)$ is S^3, in which we see two S^1's corresponding to N_i and N_j. The angle condition says that the spherical distance between the two S^1's is at least $\frac{\pi}{2}$, for any point of $N_i \cap N_j$. Then the Proposition follows by an inductive argument as in the case of the local model. □

One of the nicest applications of ramified coverings with the local models as above is in the work of Hirzebruch [15], where he constructed many beautiful examples of compact complex surfaces carrying metrics of constant holomorphic curvature. The construction starts with a branching locus, which is taken to be a configuration of lines in $\mathbb{CP}^2$. If the configuration of lines has points of triple intersection, the ramified covering space is singular, and one has to resolve it. Fortunately, such a resolution can be done in a very elementary and explicit fashion: one is either taking the normalization or blows up all the triple points, and takes the ramified covering of the blown up space along the divisor which is a preimage of the original configuration of lines. As this divisor has only normal crossings, the covering space is nonsingular.

One needs to know the behaviour of the Euler characteristic and the signature under the ramified covering (and blow-ups), and finally one has to prove some positivity statement about characteristic forms, usually phrased as the fact that the ramified covering is a surface of general type.

For some specific configurations of lines one obtains surfaces, such that the ratio of the Euler number and the signature is the same as for the complex plane (or by Hirzebruch proportionality for a complex hyperbolic surface) that is 3. Then a difficult theorem of Miyaoka-Yau tells us that indeed they are complex hyperbolic, that is they carry metrics of constant negative holomorphic curvature. The final product—the constant curvature metric—is very different from the singular pullback metric and its smoothings we have considered so far.

Another application of such ramified coverings, due to Gromov, is a construction of intriguing aspherical manifolds with no $K \leq 0$ metrics. To describe them, we need to generalize the discussion to odd dimensional manifolds. Our local models will be a product $f \times id : \mathbb{C}^n \times \mathbb{R} \to \mathbb{C}^n \times \mathbb{R}$, where f is a local model as before. In this context the proposition is true essentially with the same proof. Now Gromov [11, pp. 125-126], notes the following. If we take T^{2k+1} for M, and a family of orthogonal subtori for $N_1, \ldots, N_k$, so that the fundamental classes of N_i span $H_{2n-1}(M, \mathbb{Z})$, the resulting ramified covering is hyperbolic (in Gromov's sense). A generic homomorphism $T^{2k+1} \to T^1$ induces a fibration, $\hat{M} \to T^1$, whose fiber F is a ramified coverings of T^{2k} branched along a family of codimension 2 tori. *The monodromy of this fibration is highly nontrivial.* As a consequence, the fibers are not totally geodesic submanifolds of the total space. Since both $\hat{M}$ and T^1 are aspherical, the fibers are also aspherical. They do not carry $K < 0$ metric since this would contradict the theorem 5.4.A of [11] that the outer automorphism group of a hyperbolic group is finite (in fact the homological version of 5.4.A proved in [11] is sufficient here). The metric induced on F by the ramified covering of T^{2k} is not $K \leq 0$ generically, since the tori intersect at angles $\leq \frac{\pi}{2}$. I do not have a geometric explanation of the asphericity of F not mentioning the fibration $\hat{M} \to T^1$. These examples are reminiscent of Atiyah's [1] examples (constructed using ramified coverings) of fibrations $M^2 \to V^4 \to N^2$ for which the signature of V is not zero. They are related to the the topology of the moduli spaces of Riemann surfaces. A similar moduli space should appear in the present situation.

Next we want to mention ramified coverings with more complicated local models. Quite generally one can take for f a quotient map $f : \mathbb{R}^n \to \mathbb{R}^n/G$ where G is any finite group acting linearly on $\mathbb{R}^n$. This is especially useful when $\mathbb{R}^n/G$ is homeomorphic to $\mathbb{R}^n$. Our previous examples are particular cases of this construction: one takes for G a product of cyclic groups acting diagonally on $\mathbb{C}^n$ (or $\mathbb{C}^n \times \mathbb{R}$). There are two particularly interesting possibilities for G. One is the case when G is a *complex* reflection group, that is a finite subgroup of $GL(n, \mathbb{C})$ generated by complex reflections, i.e. elements that are diagonalizable with eigenvalue 1 of multiplicity $n-1$ (the other eigenvalue is a root of unity).

The fact that $\mathbb{C}^n/G$ is homeomorphic, or in fact biholomorphic to $\mathbb{C}^n$ is called Chevalley Theorem. Usually it is stated as follows. A finite group G acting linearly on $\mathbb{C}^n$ is generated by pseudoreflections if and only if its ring of invariants is a polynomial ring $\mathbb{C}[x_1, \ldots, x_n]$.

One should be aware that the coordinate system on $\mathbb{C}^n/G$ given by invariant polynomials is not well adapted to metric features of the ramified covering. Let us look at the case when G is the symmetric group on $n+1$ letters, Σ_{n+1}, acting on $\mathbb{C}^n$ by the standard reflection representation. That is $\mathbb{C}^n$ is the subspace of $\mathbb{C}^{n+1}$, $\mathbb{C}^n = (z_1, \ldots, z_{n+1} : \sum z_i = 0)$, and we restrict the permutation action of Σ_{n+1} on $\mathbb{C}^{n+1}$ to $\mathbb{C}^n$. The map $(z_1, \ldots, z_{n+1}) \to (\sum z_i^2, \sum z_i^3, \ldots, \sum z_i^{n+1})$ is the quotient map (remember $\sum z_i = 0$). One recognizes in this setup Chevalley theorem as the fundamental theorem on symmetric functions.

The ramification locus consists of planes $z_i - z_j = 0$. To get the equation for the branching locus, one has to express $\prod(z_i - z_j)^2$ in terms of symmetric functions $\sum z_i^k$. One sees that the triangulation we need to construct in order to check that f is a ramified covering has to be curvilinear. Ramified coverings with such local models have as yet seen little success, despite the beautiful geometry involved.

Another interesting family of examples is related to real reflection groups. Namely, we take a reflection group G acting on $\mathbb{R}^n$, and then we take the subgroup G_0 of orientation preserving elements. The quotient $\mathbb{R}^n/G$ is the infinite cone over a simplex (this is one of the most fundamental features of reflection groups), thus the quotient $\mathbb{R}^n/G_0$ is the infinite cone over the double of the simplex, hence it is homeomorphic to $\mathbb{R}^n$. Let us look again at the symmetric group case. The reflection action of Σ_{n+1} on $\mathbb{R}^n$ is defined as before. The ramification locus for G_0 consists of $\{(x_1 \dots x_n) : x_i - x_j = x_j - x_k = 0\}$ for some i, j, k (the equation is $\prod(x_i - x_j)^2 + (x_j - x_k)^2 = 0$). The branching locus is the cone over the condimension 2 skeleton of the simplex. There is a problem with giving the quotient a canonical smooth structure. This is related to the fact that invariant polynomials do not form a polynomial ring, or equivalently, that the quotient $\mathbb{C}^n/G_0$ is not a manifold.

When thinking of examples of this kind it is very natural to ask the following **Question**: what is a neccesary and sufficient condition on an orientation preserving linear action of a finite group G on $\mathbb{R}^n$, that guarantees that $\mathbb{R}^n/G$ is homeomorphic to $\mathbb{R}^n$?

My current guess is that it is the following: G is generated by "codimension 2" reflections: elements that have eigenvalue 1 of multiplicity $n-2$. This would be a rather nice extension of Chevalley theorem. It is true in small dimensions, and it is a fairly straightforward consequence of a theorem of A. Haefliger and Quach Ngoc Du [14] that it is a necessary condition.

Subsection on Davis' reflection groups method. Now we would like to discuss Davis' method of reflection groups. It is very closely related to the ramified coverings associated to a real reflection group we just discussed, and in fact it may be considered to be its special case.

The construction goes as follows. Take a triangulated manifold K. Associate with it a *Coxeter group* Γ, generators of which are indexed by the vertices of the triangulation, and which has the property that the subgroup spanned by generators $\gamma_1, \dots, \gamma_k$ is finite iff the corresponding vertices span a simplex. It is unclear for which triangulations it can be done. Suppose however that the triangulation is sufficiently fine, that is it has the property that vertices $\gamma_1, \dots, \gamma_k$ span a simplex iff any two are joined by an edge; barycentric subdivision of any triangulation has that property. Then such a Coxeter group can be given by relations $(\gamma_i\gamma_j)^2 = 1$ iff γ_i and γ_j are joined by an edge.

Now consider the cell decomposition dual to the triangulation. Top dimensional cells of the decomposition of K are called mirrors (or walls) of the cone over K and correspond to the vertices of the original triangulation, hence to the generators of Γ.

With each point p in K we associate a group Γ_p generated by all γ such that p belongs to a (closed) cell corresponding to γ. Now consider the quotient space $K(\Gamma) = cone(K) \times \Gamma / \sim$ where the relation $\sim$ is defined as follows

$$(p, \gamma) \sim (q, \eta) \text{ iff } p = q \text{ and } \gamma\eta^{-1} \in \Gamma_p$$

Then Γ acts properly on $K(\Gamma)$ and Γ', a torsion free subgroups of finite index (which Γ does contain being a linear group) acts freely, and quotients by such actions are the examples we were after.

Well, not quite. In reality there are several adjustments to be made. First, the fact we take the cone over K is irrelevant: any filling of K which has a collared boundary is as good. Second, if we want an aspherical example we should better take a contractible filling of K, thus (if we want to get a manifold) K has to be a homology sphere. Davis proves in [4], using combinatorics of Coxeter groups that the examples are indeed aspherical, and moreover if we take for K a not simply connected homology sphere, their universal covers are not simply connected at infinity.

Actually, with another adjustment, this result is a corollary of Theorem 1.3. Namely one has to extend the discussion from manifolds to complexes. It has been indicated by Gromov in [12] and proved by Gabor Moussong [19] that any Coxeter group acts on a piecewise flat complex of nonpositive curvature (the complex is very natural and has been described by Davis). The space $K(\Gamma)$ we described above is an example of Davis complex. Thus the asphericity follows from general features of nonpositively curved spaces and the claim about the fundamental group at infinity is the Theorem 1.3.

Now this construction is related to ramified coverings as follows. Suppose that we take the double of the cone over K (i.e. glue along K two copies of the $cone(K)$). We declare codimension 1 (in K) skeleton of the dual (to the original triangulation) cell decomposition to be the ramification locus. Locally (in the double) it looks like the cone over the codimension one skeleton of the boundary of the simplex, embedded in the simplex, thus we recognize one of our local models. In the case we described the finite reflection groups are products of $\mathbb{Z}_2$. We cannot at this point be much more specific about the covering, since in the construction we declared Γ' to be a quite arbitrary torsion free subgroup. However suppose that we restrict ourselves to the case of the barycentric subdivision of a triangulation (or to a triangulation having a folding map onto σ^{n-1}).

Then we have a canonical choice of the torsion free subgroup Γ'. To show this, notice that since each mirror is marked with the color of the corresponding vertex (the vertex of σ^n to which is its image by the folding map), the folding map induces a homomorphism $\Gamma \to \mathbb{Z}_2^n$. We take the kernel of this map to be Γ'. It is fairly easy to show that Γ' is torsion free. Now coming back to the ramified covering we can declare that the multiplicity of the covering over each stratum of the branching locus is exacly the same as for local models, $f : \mathbb{R}^n \to \mathbb{R}^n/\mathbb{Z}_2^{n-1}$. Then the ramified covering of the double gives us $K(\Gamma)/\Gamma'$.

In the restricted context of the barycentric subdivision of a triangulation we can also express $K(\Gamma)/\Gamma'$ as a fiber product construction, to make it look just like a hyperbolization. Namely we have the folding map from the double (along K) of the cone over K $dcone(K)$ to the double (along σ^{n-1}) of the cone over σ^{n-1}, $dcone(\sigma^{n-1})$. Over $dcone(\sigma^{n-1})$ hangs the space $dcone_h(\sigma^{n-1})$, which is the cube, and the map $dcone_h(\sigma^{n-1}) \to dcone(\sigma^{n-1})$ is the quotient map by the $\mathbb{Z}_2^{n-1}$ action of orientation preserving elements in the $\mathbb{Z}_2^n$ acting diagonally by reflections in coordinate hyperplanes. $K(\Gamma)/\Gamma'$ is the fiber product over $dcone(\sigma^{n-1})$ of these two maps. If we want to perform Davis' construction with a filling of K different from the cone, the construction above can be modified by the use of relative hyperbolization.

The reflection groups method is conceptually as simple as fiber product hyperbolization, it is adequate for producing examples exhibited in Theorems 3.1 and 3.3, however it is less flexible. Manifolds obtained from it are stably parallellizable. Moreover in dimensions higher than 3 it is really difficult to use it to produce strictly hyperbolic examples, and in fact it follows from the method of the proof of Vinberg's theorem on the nonexistence of cocompact crystallographic groups on hyperbolic spaces that it cannot be done at all in dimensions ≥ 29 (probably the bound is much lower). Thus it is hard to produce examples similar to those of Theorems 3.2 and 3.4.

5. Questions

The first two questions I would like to mention are well known and illustrate the amount of our ignorance about aspherical manifolds. They are two conjectures, mutually exclusive, about the homological structure of aspherical manifolds. As far as I know both were formulated by Thurston; the second conjecture is mentioned in Kirby's 1977 collection of problems from Stanford.

Conjecture 1. *For any closed manifold different from S^2 or $\mathbb{RP}^2$ there exists a closed aspherical manifold with isomorphic cohomology groups. A stronger version of the Conjecture states that in fact cohomology rings should be isomorphic.*

Conjecture 2. *The Euler characteristic of an aspherical manifold of dimension* $2n$ *has the same sign as the Euler characteristic of the product of surfaces, that is* $(-1)^n\chi(M^{2n}) \geq 0$.

The evidence for the first conjecture is a theorem of Kan-Thurston, later improved by Maunder, that homology of any complex is isomorphic to the homology of a $K(\pi, 1)$ for some group π. Moreover if the complex we started with is finite, then $K(\pi, 1)$ has the homotopy type of a finite complex.

The basis for the second Conjecture is the conjecture of Hopf, stating that the conclusion of the Conjecture holds in the restricted context of smooth Riemannian manifolds of nonpositive curvature. So far this has been proved only in dimension 4 by Chern (Abh. Math. Sem. Hamburg 20(1955)). Thus it is perhaps reasonable to restrict the conjecture to dimension 4. On the other hand the discussion of Section 3 suggests that perhaps the conjecture should be extended from manifolds to aspherical pseudomanifolds which are Eulerian complexes. Another possibility would be to ask the same question for *piecewise flat nonpositively curved* manifolds or pseudomanifolds which are Eulerian complexes. Using hyperbolization, the inequality on the Euler characteristic would result in a restriction on the possible numbers of k-simplices in the triangulation of a manifold as in Section 3.

Clearly these two conjectures are mutually exclusive. A counterexample to the second conjecture would be provided by an aspherical six dimensional homology sphere. It is interesting to note, that an example of a 5-dimensional aspherical homology sphere is known by work of Mumford: in [20] he constructed using methods of characteristic p algebraic geometry an example of an algebraic surface of general type, with homology of $\mathbb{CP}^2$; in fact this manifold is aspherical, as it carries a metric of constant negative holomorphic curvature. The total space of an S^1 bundle with the first Chern class dual to a generator of H_2 is a homology sphere which is clearly aspherical.

Periodic $\mathbb{R}^4$? The successful application of hyperbolization in constructions of exotic universal coverings leads to the following question: can one exhibit (preferably using techniques similar to those described here) an example of a smooth aspherical four manifold whose universal covering is diffeomorphic to an exotic $\mathbb{R}^4$. Of course one can be much more ambitious and try to construct **every** exotic $\mathbb{R}^4$ as the universal covering of some compact smooth aspherical four manifold. Notice that since we know uncountably many exotic structures on $\mathbb{R}^4$ such a construction would imply another dramatic example: a compact topological manifold with uncountably many inequivalent smooth structures.

Smoothing. Another major question about hyperbolization relates it to differential geometry. We have seen that special ramified coverings are smoothable in the sense that they carry smooth metrics of nonpositive curvature "originating" from the singular ones. On the other hand exotic topology of the boundary of the universal covering prevent the existence of the smooth metric of nonpositive curvature on the hyperbolization of the double suspension of a homology sphere.

It seems important to understand the interplay between smooth and singular metrics, in particular to prove that a "generic" example obtained by a hyperbolization is unsmoothable. Exercise number one in Ballman-Gromov-Schroeder's book "Manifolds of nonpositive curvature" provides an explicit candidate.

6. Appendix

In this appendix we present, following [9], a way of constructing four dimensional contractible manifolds. Actually we start with homology spheres. Let us recall that an n-dimensional $\mathcal{A}$-homology sphere is an n-dimensional manifold whose homology with coefficients in $\mathcal{A}$ is the same as that of the n-sphere. The method of [9] constructs new homology spheres from old ones and *knots* in them. So let M^3 be a homology sphere, K be a knot (i.e. smoothly embedded S^1) in M, and X be the knot complement, i.e. the complement of the open tubular neighborhood of K. Its boundary is T^2.

A particular feature of knots in homology spheres, is a canonical (up to isotopy) identification of ∂X with the standard T^2. It is picked up as follows. The tubular neighborhood of K is a trivial D^2 bundle over S^1. We need a trivialization, or what amounts to the same thing, a nonvanishing section. Now given a section we can isotop S^1 in the direction of the section. The resulting homology class in $H_1(X, \mathbb{Z})$ depends on the isotopy class of the section only. Since M is a homology sphere, $H_1(X, \mathbb{Z}) \simeq \mathbb{Z}$ (the isomorphism is given by the linking number). Thus we associated an integer to a section, the canonical trivialization is picked so that the integer is zero. Before describing the construction, let us recall a classical fact that invertible matrices with integer entries acts on T^2 and every diffeomorphism of T^2 is isotopic to a linear one.

Now the construction is as follows: for $i = 1, 2$ let M_i be a homology sphere, K_i be a knot (i.e. a smoothly embedded S^1) in M_i, and X_i be the knot complement, that is the complement of the open tubular neighborhood of K_i. Let

$$A = \begin{pmatrix} \alpha & \beta \\ \gamma & \delta \end{pmatrix}$$

such that $det(A) = -1$. We glue boundaries of X_i by A. The result is a closed orientable three manifold M. Gordon points out that if $\gamma = \pm 1$, M is a homology sphere. This is a corollary of the Mayer-Vietoris sequence, once we realize how A acts on the homologies in sight.

The same method can be used to produce contractible four-manifolds. Recall that a (classical) knot cobordism between K_0 and K_1 is a pair $(S^3 \times I, S^1 \times I)$, such that $(S^3 \times 0, S^1 \times 0) \simeq K_0$ and $(S^3 \times 1, S^1 \times 1) \simeq K_1$.

Consider two such cobordisms. We can now glue the exteriors of tubular neighborhoods of $S^1 \times I$, whose boundaries are $S^1 \times \partial D^2 \times I$ using $A \times Id$. In this situation Gordon proves that if $\gamma = \pm 1$, one of the knots is slice (i.e. cobordant to an unknot) and if in addition either the other knot is slice or $\alpha = 0$, then the result is a simply connected homology cobordism between S^3 and a homology sphere. If we attach a four ball to S^3, we obtain a contractible four manifold.

This construction can be generalized to knots in other homology spheres, and higher dimensions. To obtain a specific four dimensional example of a nonstandard contractible manifold the reader can experiment with some ribbon knots (which are slices).

References

1. M.F. Atiyah, *The signature of fiber bundles*, Global Analysis. Papers in Honor of K. Kodaira, Princeton Univ. Press, 1969.
2. W. Browder, *Surgery on simply connected manifolds*, Springer-Verlag, New York and Berlin, 1972.
3. J. Cheeger, W. Müller and R. Schrader, *On the curvature of piecewise flat spaces*, Commun. Math. Phys. **92** (1984), 405–454.
4. M. Davis, *Groups generated by reflections and aspherical manifolds not covered by Euclidean space*, Ann. of Math. **117** (1983), 293–325.
5. M. Davis and T. Januszkiewicz, *Hyperbolization of polyhedra*, Journal of Differential Geometry (to appear).
6. R. D. Edwards, *The topology of manifolds and cell-like maps*, Proc. of ICM Helsinki (1978), 111–127.
7. Susana Fornari and V. Schroeder, *Ramified coverings with nonpositive curvature*, preprint (1989).
8. M. Freedman, *The topology of four-dimensional manifolds*, Journal of Differential Geometry **17** (1982), 357–453.
9. C. McA. Gordon, *Knots, homology spheres, and contractible 4-manifolds*, Topology **14(2)** (1975), 151–172.
10. M. Gromov, *Hyperbolic manifolds, groups and actions*, Ann of Math. Studies, vol 97, Princeton Univ. Press, 1981, pp. 183–215.
11. ———, *Hyperbolic groups*, Essays in group theory (S. Gersten, ed.), Springer-Verlag, New York and Berlin, 1988, pp. 75–264.
12. ———, *Infinite groups as geometric objects*, Proceedings of the International Congress of Mathematicians, Warszawa, 1982, Vol 1, pp. 385–391.
13. M. Gromov and W. Thurston, *Pinching constants for hyperbolic manifolds*, Invent. Math. **89** (1987), 1–12.
14. A. Haefliger and Quach Ngoc Du, *Une presentation du groupe fondamental d'une orbifold*, Asterisque **116** (1984), 98–107.

15. F. Hirzebruch, *Arrangements of lines and algebraic surfaces*, Arithmetic and Geometry, vol II, Geometry. Papers dedicated to I.R. Shafarevich (M. Artin, J. Tate, eds.), Birkhäuser, Boston-Basel-Stuttgart, 1983, pp. 113–140.
16. B. Lawson and S.T. Yau, *Compact manifolds of nonpositive curvature*, Journal of Differential Geometry **7** (1972), 211–228.
17. A. Marin, *Un nouvel invariant pour les spheres d'homologie (d'apres Casson)*, Asterisque **161–162** (1988), 151–164.
18. G.D. Mostow and Y.T. Siu, *A compact Kähler surface of negative curvature not covered by the ball*, Ann. of Math. **112** (1980), 321–360.
19. G. Moussong, *Hyperbolic Coxeter groups*, Ph.D. thesis, The Ohio State University, 1988.
20. D. Mumford, *An algebraic surface with K ample,* $K^2 = 9$ *and* $p_g = q = 0$, Am. J. of Math. **109** (1979), 233–244.
21. M. Namba, *Branched coverings and algebraic functions*, Pitman Research Notes in Math., vol 161, Longman Sci and Tech., 1987.
22. D. Stone, *Geodesics in piecewise linear manifolds*, Trans. Amer. Math. Soc. **215** (1976), 1–44.
23. C. Rourke and B. Sanderson, *Block bundles I*, Ann. of Math. **87** (1968), 1–28.
24. R.F. Williams, *A useful functor in topology*, Trans. Amer. Math. Soc **106** (1963), 319–329.

NON-POSITIVELY CURVED TRIANGLES OF GROUPS

JOHN R. STALLINGS

ABSTRACT. The "angle" of a triad of groups is defined. This enables one to define a non-positively curved triangle of groups. The colimit of such a triangle contains the vertex groups; it acts on a piecewise-Euclidean complex of non-positive curvature, with quotient a single triangle (there are exceptional cases to consider when some angle is zero). A finite subgroup of the colimit (or, more generally, a bounded subgroup) must be conjugate to some subgroup of a vertex group.

§0. INTRODUCTION

This is a sketch of joint work with S. M. Gersten, University of Utah.

The theory of group-actions on trees, which was described by Bass and Serre, led to new understanding of purely combinatorial concepts such as amalgamated free products of groups. The ideas to be described in this paper are intended to be a start on the theory of higher-dimensional cases; we are concerned with a special kind of two-dimensional picture, a "triangle of groups." Examples of these triangles have shown up before in certain special cases, in hyperbolic geometry and algebraic number theory. In analogy with the geometric case, we can define angles and discuss curvature; in the "positively curved" case, we have no results of interest, but in the non-positively curved case we can prove several results by using geometric constructions and techniques.

§1. DEFINITIONS AND STATEMENTS OF RESULTS

1.1. The angle between a pair of subgroups. Given two subgroups A and B of a group G, how should one define the "angle" between A and B in G? We want to do this so that, in the dihedral group of order $2n$, given by generators a, b and relations $a^2 = b^2 = 1$, $(ab)^n = 1$, the "angle" between the cyclic groups generated by a and b is π/n. The definition which works well is this:

The inclusions $A \to G$ and $B \to G$ determine a homomorphism $\phi: A * B \to G$, where $A*B$ denotes the free product. If ϕ is injective, we say that the angle between A and B is 0. Otherwise, we look for the minimum length of a non-trivial element

1980 *Mathematics Subject Classification* (1985 *Revision*). 20E05, 20F32.

Key words and phrases. group actions, piecewise-Euclidean, buildings.

This work is based on research partly supported by the National Science Foundation under Grants No. DMS-8600320 and DMS-8905777.

in the kernel of ϕ; this minimum length is even, say $2n$; and we define the angle to be π/n.

Another way to describe this is combinatorially: Consider the sets of left cosets, and the set maps:

$$G/A \leftarrow G \rightarrow G/B.$$

We imagine the elements of G to be "edges"; the "initial vertex" of g is the coset gA, and the "terminal vertex" of g is the coset gB. Thus, the above diagram of sets describes a directed graph $\Gamma(A, B)$; this graph may have a non-trivial reduced closed path in it; the shortest such has number of edges equal to $2n$, and we define the angle between A and B to be π/n. If $\Gamma(A, B)$ is a forest, then the angle is 0. Thus, if we give to each edge of $\Gamma(A, B)$ the length equal to the radian measure of the "angle between A and B", then any closed curve γ in $\Gamma(A, B)$ having its length less than 2π is a contractible curve, and so, combinatorially speaking, it must contain a backtracking.

1.2. The angle of a triad of subgroups. We can generalize the above to a more general situation, where A and B are subgroups of G, and there is also given a subgroup $C \subset A \cap B$. Then we talk about "angle (A,B;C)." We have a homomorphism of the amalgamated free product $\phi: A *_C B \rightarrow G$, and look for the smallest length of nontrivial elements of the kernel; this length is $2n$ and this angle is π/n. Or we look at the directed graph determined by

$$G/A \leftarrow G/C \rightarrow G/B.$$

The smallest number of edges of a non-trivial closed reduced path in this graph is $2n$, and the angle is π/n.

For example, if $A \cap B$ is strictly larger than C, the angle is π.

It is generally quite difficult to determine the angle, even if we are looking at finite groups. But in specific cases, such as the dihedral group case, we can figure it out. An example to be wary of, in the infinite case, is this:

Fix some integer n. Let $\mathcal{P} = \langle a,\ b \mid (ab)^n b = 1 \rangle$. Let G be the group of this presentation, and let A and B be the infinite cyclic groups generated by a and b. Then the angle between A and B in G is not π/n, as one would guess at first; it is π, because in this group $(a^n)(b^{n+1}) = 1$.

1.3. A triangle of groups. Let us now consider a triangle of groups. That is to say, a commutative diagram of groups and injective homomorphisms:

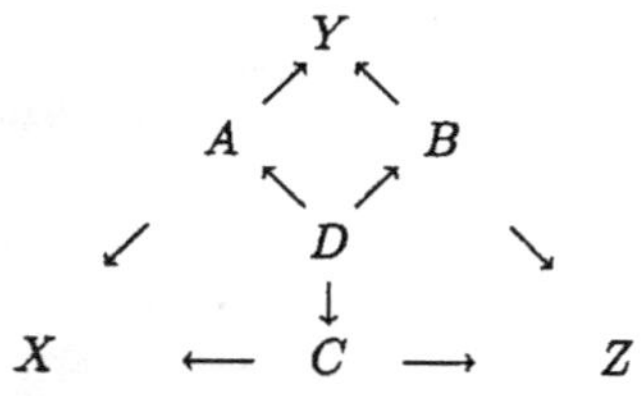

We can consider the colimit (or "generalized pushout") of this diagram, which we call G. In this picture, we call the groups X, Y, and Z the "vertex groups," the groups A, B, and C the "edge groups," and the group in the middle, D, the "2-cell group." We think of A as the intersection of X and Y, and so on. The presence of D in the picture signifies that we imagine all the triple intersections, such as $(X \cap Y) \cap (Y \cap Z)$, to be the same. At each vertex, there is an angle of subgroups; at X, there is angle$(A, C; D)$; at Y, there is angle$(A, B; D)$; and at Z, there is angle$(B, C; D)$. We call the triangle of groups "non-spherical" when the sum of these three angles is less than or equal to π. If the sum of the angles were greater than π, we would call it "spherical"; if this sum were strictly less than π, we would call it "hyperbolic." Our major theorems are for the non-spherical case.

1.4. A spherical example. But here is a spherical example to think about: $D = \{1\}$; A, B, and C are infinite cyclic, generated respectively by a, b, and c. The vertex groups are given by:

$$X = \langle\, c,\ a \mid cac^{-1} = a^2 \,\rangle$$

$$Y = \langle\, a,\ b \mid aba^{-1} = b^2 \,\rangle$$

$$Z = \langle\, b,\ c \mid bcb^{-1} = c^2 \,\rangle.$$

The colimit group G is then given by all three generators and all three relations; this is a standard example, from which we deduce that G is the trivial group. The angles are each equal to $\pi/2$, and thus add up to more than π. In this case, for example, the natural homomorphism $X \to G$ is not injective.

1.5. Embedding of vertex groups. However, in the non-spherical case, we have a general theorem which says that these homomorphisms are indeed injective:

Theorem 1. *Consider any non-spherical triangle of groups, as above. Then the natural maps $X \to G$, $Y \to G$, and $Z \to G$ are injective.*

Thus, we can consider, in the non-spherical case, that the vertex groups are subgroups of the colimit group G. We give, in Section **3.2**, a topological proof of this, based upon an analysis of mappings into a space modeled on the triangle of groups. Chermak [Ch] has a slightly different way of looking at this, based on group-actions on trees.

1.6. Bounded subgroups. Now, given any triangle of groups, we can consider an element s of the colimit G. For each such s, we can write s as a product of elements of X, Y, and Z; in other words, there is a natural homomorphism of the free product $X * Y * Z$ onto G. The minimal length, in terms of this free product, of an element representing s will be called the "length of s." If S is a subgroup of G, we shall say that S is *bounded* when there is an integer N such that every $s \in S$ has length $\leq N$. Our other major theorem is:

Theorem 2. *Suppose that G is the colimit of a non-spherical triangle of groups as above. Let S be any bounded subgroup of G. Then S is conjugate in G to a subgroup of a vertex group.*

In particular, all the finite subgroups of G are conjugate to subgroups of the vertex groups. The proof of this uses the construction of a 2-complex L on which the group G acts; this 2-complex has properties of non-positive curvature; it has a unique geodesic between any pair of points; this enables it to be given the structure of a complete metric space on which G acts by isometries; and furthermore, every pair of points has a unique midpoint. Thus we can apply a theorem of Bruhat and Tits [BT] to derive the fact that every bounded subgroup has a fixed point and thus deduce the Theorem. In this paper we include the details of the Bruhat-Tits fixed point theorem in Section **3.8**, as explained by J.-P. Serre in a talk at MSRI in 1987.

§2. Geometric lemmas

2.1. Generalities. The proofs of these theorems involve geometry. There are theorems about planar graphs, basically ways of stating the fact that the Euler characteristic of the 2-sphere is 2; such results are at the heart of classical small cancellation theory. There are some more purely geometrical theorems also, concerning complexes made up of Euclidean triangles, geodesics, geometric inequalities, and some ideas due to Bruhat-Tits, etc. A considerable amount of the geometric detail has been done by the Russian school centered around A. D. Alexandrov; but the connection between Alexandrov's type of geometry and the small cancellation theory has been slow in coming.

The impetus for all this comes from a paper by M. Gromov [Gr]. Steve Pride [Pr] also has some results that predated and led the way for our definition of the angle between subgroups.

For Gromov, a hyperbolic group is a combinatorial generalization of the notion of the fundamental group of a compact hyperbolic manifold without boundary; a free group is a particular kind of hyperbolic group also.

Now, a certain generalization of the notion of a free group is the notion of a group acting on a tree, such as an amalgamated free product. The next higher-dimensional case is represented explicitly by our colimit G of a non-spherical triangle of groups; such a G does act on a certain 2-dimensional simplicial complex L, which we call its "building." We set this up so that L/G is actually a triangle Δ. In the amalgamated free product analogy, L would be a tree and L/G would consist of a single closed interval. Our groups G are not contained in the class of Gromov's hyperbolic groups, unless the vertex groups are finite and the sum of the angles is strictly less than π; nevertheless, they are clearly in the same spirit as Gromov's idea. The possibility of having infinite vertex groups means that the building L may not be a locally finite complex; this definitely complicates some of the details of the subject, such as proving the existence of geodesics in L.

If we try to generalize our picture beyond simple-looking triangles, the analogy between the theory of group-actions on trees and group-actions on higher-dimensional "non-spherical" or "hyperbolic" complexes becomes less clear. In particular, if L/G is more complicated than a triangle (for example, it might be a triangulated 2-manifold), then we cannot recover G from the pattern of the stabilizers of the simplexes representing those of the quotient; there is at the very least an extension phenomenon which has to be encoded into the picture; the possibility that L/G is a triangulated "dunce-hat" also defies easy understanding even though the dunce-hat is a contractible complex. Haefliger [Ha] has analyzed some of these problems.

Another sort of technical complication seems to occur when L/G is not compact. We could go on and on with the ways in which 2-dimensional complexes can act more strangely than 1-dimensional ones. The even more mysterious question of generalizing to complexes of dimension greater than 2 has hardly been attacked; but since there are interesting examples of hyperbolic manifolds of dimension greater than 2, this kind of theory can surely be extended into that world.

2.2. Planar graphs, weights, lengths, valences. One of the kinds of geometry concerns planar graphs. We consider a finite graph Γ embedded in the plane R^2 or in the sphere S^2. The complementary regions are the connected components of $S^2 \setminus \Gamma$; if Γ is connected, then each complementary region in S^2 is simply connected. On the other hand, in R^2 the infinite complementary region is not simply connected unless Γ is empty. At each vertex of Γ, there are edges radiating outward arranged in a cyclic order. The angular space between two adjacent edges at a vertex is called a "corner." Thus, to each vertex we have certain corners associated, and to the boundary of each simply connected region of $S^2 \setminus \Gamma$ we have certain corners. In Figure 1, we associate α, β, γ to the vertex v; and we associate α, β, δ, ϵ, η, θ, and ζ to the region Σ.

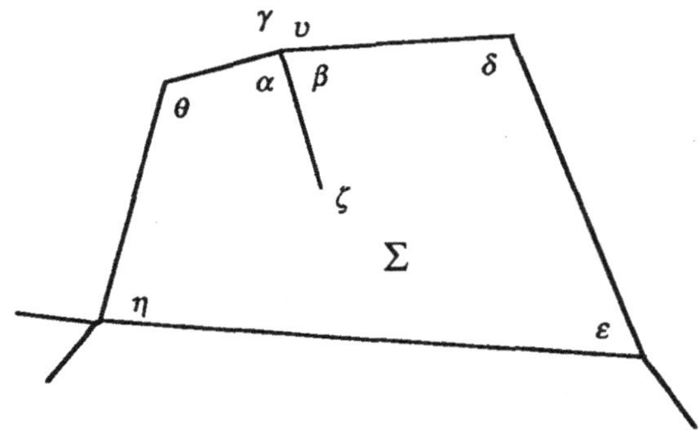

Figure 1

Two things can be noticed here: The bounded region Σ has the two corners α

and β associated to it. And a vertex of valence one exists, which has just one corner ζ associated to it.

Now, we shall consider an assignment of a real number $w(\alpha)$ to each corner α. We call w a "weight." If we insist that for each α, the weight $w(\alpha) \geq 0$, we call w a "non-negative" weight. Now, to each vertex v, we can associate the sum of the weights of the corners at v; we call this $w(v)$. Each component R of $S^2 \setminus \Gamma$ has several boundary components; the number of boundary components is 0 only when $\Gamma = \emptyset$; otherwise, it is 1 for simply-connected R and greater than 1 for non-simply-connected R. For each boundary component C of R, we can add up the weights of the corners which are involved with C; we call this $w(C)$.

A few more definitions: For v a vertex, we let $n(v)$ denote the number of edges incident to v (graph-theorists sometimes call $n(v)$ the "valence" of the vertex v in the graph); for example, the vertex in the picture corresponding to the corner ζ has $n(v) = 1$; if there is an edge making a loop, joining v to itself, this contributes 2 to $n(v)$. If C is a boundary component of a region R, we run along C and count the number of edges there, and call this the length of C or $\ell(C)$; we may have to count an edge twice if R lies on both sides of that edge. Another way of saying this is that $n(v)$ is the number of corners around v and $\ell(C)$ is the number of corners around C. The terminology $\operatorname{Bd} R$ means the boundary of R, and is particularly nice to use when R has a single boundary component C (in which case, $C = \operatorname{Bd} R$).

2.3. Statement of combinatorial lemmas. There are now three combinatorial lemmas, basically very easy to prove, based on the fact that the Euler characteristic of the 2-sphere is 2.

Theorem 3. *Let Γ be a finite graph embedded in R^2. Let w be an assignment of weights to each corner. Suppose*

(1) w is non-negative.

(2) For each vertex v of Γ, $w(v) \leq n(v) - 2$.

(3) For each simply-connected component Σ of $R^2 \setminus \Gamma$, if Σ is not all of R^2, then $w(Bd\,\Sigma) \geq 2$.

Then we conclude: Γ is empty.

Theorem 4. *Let Γ be a finite graph embedded in S^2. Let w be an assignment of weights to each corner. Suppose*

(1) For each vertex v of Γ, we have $w(v) \geq 2$.

(2) For each boundary component C of each component of $S^2 \setminus \Gamma$, we have $w(C) \leq \ell(C) - 2$.

(Note that here we do not assume that w is non-negative.)

Then: Γ is empty.

Theorem 5. *Suppose we consider a cell structure on a 2-cell D. Imagine the graph Γ to be the 1-skeleton of this cell structure. Imagine that we have two designated vertices a and b which are on the boundary of D. Suppose there is a weight associated to each corner within D, such that:*

(1) The weight w is non-negative.

(2) For the boundary C of each 2-cell in the cell structure, we have $w(C) \leq \ell(C) - 2$.

(3) For each vertex v in the interior of D, we have $w(v) \geq 2$.

(4) For each vertex v' on the boundary of D, except for the designated vertices a and b, we have the sum of the weights of the corners inside D, which we still call $w(v')$, is at least 1.

Then: $w(a) = w(b) = 0$.

3. Sketches of proofs

3.1. The colimit as π_1 of a functorial construction. In the proof of Theorem 1, we construct, out of our triangle of groups, a space which has as its fundamental group the appropriate colimit, as follows: First, we think of the Eilenberg-Mac Lane functor $K(\Pi, 1)$ as a functor from groups to CW-complexes, which takes injective homomorphisms to injective maps of complexes. We take the disjoint union of $K(X,1)$, $K(Y,1)$, and $K(Z,1)$; one for each vertex group. For each edge group A, B, and C, we construct its $K(\Pi,1)$ and take the product with the unit interval $I = [0,1]$, getting $K(A,1)\times I$, etc. Using the inclusion $A \to X$, we identify $K(A,1)\times 0$ with a subcomplex of $K(X,1)$; using the inclusion $A \to Y$, identify $K(A,1)\times 1$ with a subcomplex of $K(Y,1)$; and so on around the boundary of the triangle. What we have at this point is a space whose fundamental group is an HNN-extension of $X *_A Y *_B Z$ which contains the two copies of C in X and Z made conjugate to each other by a stable letter t. Now, we consider the group in the middle, D, and form $K(D,1)\times\Delta$, where Δ is a triangle; the inclusions we have, and the commutative diagrams they make, allow us to identify $K(D,1)\times \operatorname{Bd}\Delta$ with a subcomplex of the previously defined complex. Putting this all together, we get the final result, the complex K with $\pi_1(K) = G$ the colimit of the diagram. (This result for fundamental groups could have been attained if we had ignored the 2- and higher-dimensional cells of the edge complexes $K(A,1)$ etc., and the 1- and higher-dimensional cells of the 2-cell complex $K(D,1)$. But the asphericity of the picture is important to us, and this is why it is important to include in the picture the entire $K(\Pi,1)$ complexes.) In case $D = \{1\}$, what we have is the circular complex with the HNN-extension of the amalgamated free product as its fundamental group, with a triangular 2-cell added whose effect is to kill the stable letter t.

Now, there is a natural map of K onto the triangle Δ, $r: K \to \Delta$. Consider a point p in the middle of Δ, and draw three perpendicular lines out to the edges; call the resulting triod T. To each corner of Δ, in paragraph **1.2** we have associated an angle; there is a corner of the triod T facing the given corner of Δ, and to this corner we associate a weight, which is the group-theoretic "angle", divided by π. — Of course, we should probably never have said "π" at all.

3.2. Proof of Theorem 1. Recall that we are assuming that G is the colimit of a non-spherical triangle of groups, and X is one of the vertex groups; we construct K out of $K(\Pi,1)$ spaces, as in Section 3.1. To show that $X \to G$ is injective, we consider a curve in $K(X,1)$ which is contractible in K, and want to show that it is contractible in $K(X,1)$. Thus, we have a continuous map $f: D^2 \to K$, mapping $\mathrm{Bd}\, D^2$ into $K(X,1)$. Smooth it out, and look at $(rf)^{-1}(T) = \Gamma$; this is a graph in the interior of D^2; we first get rid of the parts of the graph which are simple closed curves, and then every vertex has valence 3; each corner around a vertex maps to a corner around p in T, and we can pull back the weights to get a weighting of Γ in D^2. The weight of each vertex is, because the triangle of groups is assumed to be non-spherical, at most 1, which is, of course, $n(v) - 2$. If the length $\ell(C)$ of a boundary component C of a simply connected region of $D^2 \setminus \Gamma$ should happen to be less than 2, then $f(C)$ will represent a word in one of the amalgamated products, say in $A *_D B$, that maps to 1 in Y, and this word will be shorter than the length of any non-trivial element of this kernel; thus $f(C) = 1$ in $A *_D B$. What this implies, after a little thought, is that we can change f by a homotopy so as to reduce by 2 the number of vertices of $\Gamma = (rf)^{-1}(T)$. At this point, we may have introduced new simple-closed curves into Γ, which we must get rid of, also by a homotopy. The end result will have Γ satisfying the assumptions in Theorem 3, and so Γ will be empty, and the result (Theorem 1) is proved.

3.3. The functorial space is aspherical. Now, we look at the universal cover of K, that is, at the covering projection $\phi: \tilde{K} \to K$. We have $r\phi: \tilde{K} \to \Delta$; identify each connected component of the inverse image of each point of Δ to a single point; this yields the "monotone-light" factorization of $r\phi$, giving a complex L and a map $\psi: L \to \Delta$. This complex L is of dimension two, a union of triangles, each of which is mapped by ψ homeomorphically to Δ; it can be shown to have several interesting geometric properties relating to the original triangle of groups in paragraph **1.3**; we call L the "building" of this triangle of groups.

The group G which was the fundamental group of K, acts freely on $\tilde{K}$, and thus L inherits an action of G (no longer free) on it. L is a 2-dimensional complex, and L/G can be identified with the triangle Δ. If we pick a triangle in L mapping homeomorphically onto Δ, then the stabilizers of the vertices, edges, and the triangle itself form a diagram of groups isomorphic to the diagram which we started with in paragraph **1.3**. If we look at a vertex of L, say over the "Y" vertex of Δ, then the link of that vertex in L is isomorphic to the directed graph described in terms of cosets by the diagram:

$$Y/A \leftarrow Y/D \rightarrow Y/B.$$

We can now imagine a 2-sphere being mapped into L. It is possible to use an analysis of this map by the geometric lemma, Theorem 3, to show that this map is null-homotopic; or, alternatively, one can use a slightly different, dual construction, using Theorem 4, to prove this fact.

Now, L is simply-connected, and a geometric argument proves that $\pi_2(L) = 0$, and L is 2-dimensional. This implies that L is contractible. The map from $\tilde{K}$ to L has the property that the inverse image of each point is contractible (each such is a copy of the universal cover of some $K(\Pi, 1)$). Thus, it follows that $\tilde{K}$ is itself contractible, and hence that K is a $K(G,1)$ space.

3.4. The piecewise-Euclidean structure of the building L. In what follows, we will assume, for simplicity, that the group-theoretic angles of the original triangle of groups are non-zero. The cases that arise when one or more angles are zero can be handled separately by the classical theories of amalgamated free products; it is also possible to modify the following geometrical discussion to work in this case.

We are assuming that the sum of the angles of the triangle is less than or equal to π. We can then find a Euclidean triangle whose three angles are perhaps a little bit larger than the group-theoretic angles; this triangle, with its Euclidean metric, is what we take for the model triangle Δ in the preceding discussion. Now we have the map $\psi: L \to \Delta$, which takes each triangle in L onto Δ; we metrize each triangle of L so as to make ψ into an isometry on that triangle. Notice that L itself has not been made into a metric space, only that L is a cell-complex, in which each 2-cell happens to be a triangle, and each 2-cell is metrized to be a particular shape of Euclidean triangle. These metrics on the 2-cells are compatible in that they agree on common faces where they overlap.

One can, in a sort of obvious way, abstract this situation into what one might call a "piecewise-Euclidean" or "PE" complex. Thus, we now have L described as a PE 2-complex, on which the group G acts.

The link of a vertex v in L is a 1-dimensional complex $\mathrm{Lk}(v, L)$, in which a vertex is given by a 1-cell e radiating out of v, and in which an edge is given by a corner of a 2-cell having v at one corner. That corner of a 2-cell is a Euclidean angle, having a certain radian measure, and we put a metric structure on $\mathrm{Lk}(v, L)$ so that its 1-cells have length equal to the radian measure of the corresponding angle. Because of the relation of the group-theoretic angles to the metric structure of $\mathrm{Lk}(v, L)$, we can see that any closed path in $\mathrm{Lk}(v, L)$ of length less than 2π is contractible in $\mathrm{Lk}(v, L)$. We call this rather interesting fact the "link condition" on the PE 2-complex L. — The whole thing could be phrased in terms of weights, in which case one divides out the "π."

3.5. Geodesics in the building L. In a PE complex, such as our building L, we can define a "geodesic." The definition is local: A geodesic is a path in L, made up of a finite number of pieces; each piece is a straight line segment in one of the Euclidean simplexes of L; the pieces fit together at their endpoints. If two adjacent pieces belong to the same simplex, then they fit together in a straight line. If they belong to different triangles, and the endpoint in common is a point interior to a 1-simplex, then we imagine mapping the two triangles isometrically into the Euclidean plane, and the two segments must then form a straight line. If the common endpoint is a 0-simplex v, then the two pieces define two points in $\mathrm{Lk}(v, L)$, which has its metric defined in terms of angles; the condition is that the smallest path in $\mathrm{Lk}(v, L)$ between these two points has length at least π.

Theorem 6. *In the building L of the non-spherical triangle of groups, for every pair of points a, $b \in L$, there exists one and only one geodesic joining a and b.*

The existence of a geodesic is a subtle matter; this is discussed in some detail by Bridson [Br]; the classical case of this theorem, due to the school of A. D. Alexandrov and others (see [Ba]), assumes that the building is a locally finite complex; this is not the case here, but Bridson has shown that the local finiteness condition can be replaced by the assumption that there are only a finite number of isometry types of triangles; in our case, there is only one isometry type. The uniqueness is a geometric fact involving Theorem 5; suppose we had two geodesics from a to b; these are homotopic since L is simply-connected; thus there is a map from D^2 into L describing this homotopy. This map can be simplified, and the angular weights pulled back to a cell-structure on D^2, so as to satisfy the hypothesis of Theorem 5 (the basic point is that L satisfies the link condition); this lemma will imply that the two geodesics agree near a and b and thus, eventually, they agree everywhere.

It is worth noting that if you start at a point p, there is a direction to go away from p, namely, any particular point in $\mathrm{Lk}\,(p, L)$. This starts a geodesic away from p in L. When we hit the edge of a triangle, there may be many other triangles with that same edge, and we have a choice of which triangle to continue the geodesic in; if we hit the corner of a triangle, the set of choices is even greater. With this warning in mind, however, the general geometric picture of L resembles greatly the way things are in a Riemannian manifold whose sectional curvature is everywhere less than or equal to zero. In our building L, the negative curvature is concentrated, so to speak, at the vertices, where the negativeness of curvature has to do with the link condition, that every closed path in the link of length less than 2π is contractible; or that there are not any closed geodesics in the link of length less than 2π.

Now we can define a metric on the building L. The distance between two points is the length of the unique geodesic between them. It turns out that L is then a complete metric space, and that the geodesic between two points is the same as the piecewise-linear path of smallest length joining them.

3.6. The Toponogov inequality. Suppose we now have three points p, x, y in L and construct the geodesic triangle in L, consisting of the geodesics $[p, x]$, $[p, y]$, and $[x, y]$, with lengths respectively $d(p, x)$, $d(p, y)$, and $d(x, y)$. The geodesics $[p, x]$ and $[p, y]$ start off at p at two points in $\mathrm{Lk}\,(p, L)$; their distance in this link can then be regarded as the angle between these geodesics; call this angle θ. If we were to draw lines in the Euclidean plane with these lengths and angle, we could compute a formula for the Euclidean distance between the points corresponding to x and y. The next result, the "Toponogov inequality," says that the distance in L is at least that big. This is a numerical version of the concept of non-positive curvature which is codified in the link condition on L.

Theorem 7. *In the above situation,*

$$d(x, y)^2 \geq d(p, x)^2 + d(p, y)^2 - 2\, d(p, x)\, d(p, y) \cos\theta.$$

3.7. The midpoint inequality. Again suppose we have three points x, y, and z chosen in the building L. Draw the geodesics. Find the midpoint p of the geodesic $[y, z]$. Let α be the angle at p between $[p, x]$ and $[p, y]$, and let β be the angle between $[p, x]$ and $[p, z]$. The fact that $[y, z]$ is a geodesic and contains p implies that $\alpha + \beta \geq \pi$. For the following result, the case that α or β is greater than π can be handled particularly; otherwise, we have $\cos\alpha + \cos\beta \leq 0$, and Theorem 7 applied to these two cases yields the result. The picture is in Figure 2.

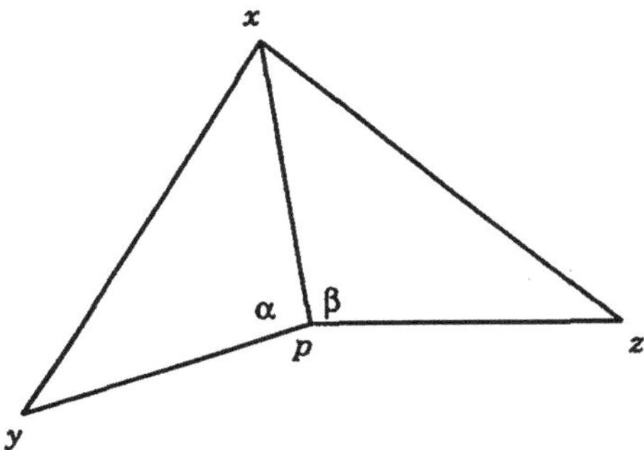

Figure 2

Theorem 8. *If x, y, z are any points of the building L, and p is the midpoint of the geodesic $[y, z]$, then:*

$$d(x,y)^2 + d(x,z)^2 \geq 2\,d(p,x)^2 + 2\,d(p,y)^2.$$

Note that the second term on the right could also be written as one-half of $d(y,z)^2$.

3.8. The Bruhat-Tits fixed point theorem. The significance of the above midpoint result (another numerical version of non-positive curvature), is that it allows us to use a fixed-point result of Bruhat and Tits. The explanation which follows is from Stallings' notes on Serre's talk at MSRI in 1987:

Suppose that X is a metric space with metric d. Let A be a bounded subset of X. Thus, for any point $p \in X$, the set $\{\, d(p,a) \mid a \in A \,\}$ is a bounded set of real numbers. We define:

$$\operatorname{supdist}(p,A) = \sup\{\, d(p,a) \mid a \in A \,\}.$$

Given A, if there exists a point $p \in X$, such that for all $q \in X$, $\operatorname{supdist}(p,A)$ is less than or equal to $\operatorname{supdist}(q,A)$, we call p a "center" of A in X. In other words, a center of A is a point of X such that A can be included in a ball of smallest radius about the center.

Theorem 9. *Suppose that X is a complete metric space, and suppose that every two-point set in X has a center. In addition, suppose that the "midpoint inequality" is satisfied: For every pair of points y, $z \in X$, there exists a center p of the two-point set $\{y, z\}$, such that for all $x \in X$,*

$$d(x,y)^2 + d(x,z)^2 \geq 2\,d(p,x)^2 + \frac{1}{2}\,d(y,z)^2.$$

Then for every bounded subset A of X, there is a unique center of A.

The proof consists of considering any sequence $\{p_n\}$ of elements of X, such that supdist (p_n, A) converges to the infimum of $\{\text{supdist}\,(q, A)\}$. This sequence turns out to be a Cauchy sequence in X; its limit is a center of A; and the limit is unique, because we can alternate a sequence converging to one center and another, and this is Cauchy and so has a unique limit.

Thus the crux of the proof is to show that $\{p_n\}$ is a Cauchy sequence. This will follow from the following fact:

Let $E = \inf\{\text{supdist}\,(q, A) \mid q \in X\}$. Suppose $\delta > 0$ and $y, z \in X$, such that

$$\text{supdist}\,(y, A) \leq E + \delta, \qquad \text{supdist}\,(z, A) \leq E + \delta.$$

Then $d(y,z)^2 \leq 8E\delta + 4\delta^2$.

To prove this, let p be the center of $\{y, z\}$ which satisfies the hypothesis in Theorem 9. Then, for any $\epsilon > 0$, there exists $a \in A$, such that $d(p, a) \geq E - \epsilon$. We have $d(y, a) \leq E+\delta$ and $d(z, a) \leq E+\delta$. The midpoint inequality in the hypothesis of Theorem 9 for a, y, z, yields

$$\tfrac{1}{2}d(y, z)^2 \leq d(a, y)^2 + d(a, z)^2 - 2d(p, a)^2,$$

$$\text{and this is } \leq 2(E + \delta)^2 - 2(E - \epsilon)^2.$$

Letting $\epsilon \to 0$, we get the claimed result. This implies the claim that $\{p_n\}$ is a Cauchy sequence, by substituting p_n and p_m, for large n and m, into this lemma for y and z.

3.9. Proof of Theorem 2. We are now closing in on Theorem 2. We have our triangle of groups, whose angles satisfy the non-spherical condition. By a topological construction, this yields a contractible 2-complex L on which the colimit group G acts. We throw in a little geometry by making L into a piecewise-Euclidean complex, modeled on a triangle whose angles are greater than or equal to the group-theoretic angles. In L we can define geodesics, and prove that any two points are joined by a unique geodesic, and thus define a metric on L. This metric is complete, and furthermore, using the Toponogov inequality, we can prove a midpoint inequality. Now, in fact the midpoint on the geodesic between two points is the center of that two-point set.

Now, the hypothesis in Theorem 2, that S is a bounded subgroup of the colimit group G, translates into the statement that if we restrict the action of G on the building L to S, then each orbit is a bounded set. Let A be one such orbit. By Theorem 9, there is a unique center of A in L. Since S acts by isometries and the center is defined in terms of the metric alone, it follows that this center is left fixed by every element of S. I.e., if S has a bounded orbit, then it has a 1-point orbit. It therefore follows that S is contained in the stabilizer of some vertex of L. These stabilizers, considering the G action again, are exactly the same as the conjugates of the vertex groups X, Y, or Z, in the colimit group G.

REFERENCES

[Ba] W. Ballman, Sur les groupes hyperboliques d'après Mikhael Gromov, E. Ghys and P. de la Harpe, editors, Birkhäuser, 1990.

[Br] M. Bridson, *Geodesics and curvature in metric simplicial complexes*, preprint, Cornell University, 1990.

[BT] F. Bruhat and J. Tits, *Groupes réductifs sur un corps local, I. Données radicielles valuées*, I. H. É. S. Publ. Math., vol. 41, 1972.

[Ch] A. Chermak, *R-trees, small cancellation, and convergence*, preprint, Kansas State University.

[Gr] M. Gromov, *Hyperbolic groups*, Essays in group theory, S. Gersten, editor, vol. 8, Springer-Verlag, MSRI series, 1987.

[Ha] A. Haefliger, *Complexes of groups and orbihedron*, this volume.

[Pr] S. Pride, *The diagrammatic asphericity of groups given by presentations in which each defining relator involves exactly two types of generators*, Arch. Math. **50** (1988), 570–574.

MATHEMATICS DEPARTMENT, UNIVERSITY OF CALIFORNIA, BERKELEY CA 94720

E-mail: stall@cartan.berkeley.edu

Complexes of groups and orbihedra

ANDRE HAEFLIGER
Section de Mathématiques
C.P. 240
CH-1211 Genève 24

ABSTRACT

We extend the main features of the classical theory of graph of groups, due to Bass and Serre, to the higher dimensional case of complex of groups. We also view a complex of groups as an orbihedron. A complex of groups is called developable if it is associated to a simplicial action of a group on a simply connected simplicial cell complex. Conditions for developability are given.

Given a simplicial action of a group G on a simply connected simplicial complex $\widetilde{X}$, how can one reconstruct this action from data on the quotient $X = G\backslash\widetilde{X}$? When $\widetilde{X}$ is 1-dimensional, there is a complete answer in terms of Bass-Serre theory of graph of groups (cf. Serre [13]). In the higher dimensional case, one is led to introduce the notion of complex of groups on X. In case X is 1-dimensional, one recovers the theory of graph of groups, and if X is a 2-simplex, the notion of a triangle of groups studied by Gersten and Stallings [14].

We work in the category of simplicial cell complexes, and simplicial actions without inversions so that the quotient is again a simplicial cell complex (for the precise definitions, see § 1). A complex of groups $G(X)$ on a simplicial cell complex X associates to the barycenter of each cell σ of X a group G_σ and to each edge a of the barycentric subdivision of X, with its natural orientation, a monomorphism ψ_a of the group attached to its initial point in the group attached to its terminal point. Moreover, for each pair of composable edges a, b of the barycentric subdivision of X, an element $g_{a,b}$ of the group associated to the terminal point of a is given, such that two conditions are satisfied (cf. § 2). A simplicial action without inversion of a group G on a simplicial cell complex $\widetilde{X}$ gives on the quotient $X = G\backslash\widetilde{X}$ a complex of group $G(X)$ unique up to isomorphism. A complex of groups obtained in this way is called developable.

Given a complex of groups $G(X)$ on X and a base point σ_0, one can construct (§ 3) an abstract group $G = \pi_1(G(X), \sigma_0)$ and an action of this group on a simply connected complex $\widetilde{X}$ such that $X = G\backslash\widetilde{X}$ (cf. § 4). But in dimension bigger than one, the complex of groups associated to this action is in general only a quotient of $G(X)$. It will be isomorphic to $G(X)$ if and only if the natural homomorphisms of the G_σ in G defined up to conjugation are injections.

Behind the notion of complex of groups is an equivalent more geometric one, namely the notion of an orbihedron structure on X (§ 5). Such a structure associates to the star $St\sigma$ in the barycentric subdivision of each cell σ of X an action of a group G_σ on the star $St\widetilde{\sigma}$ of a cell $\widetilde{\sigma}$ in a simplicial cell complex so that $St\sigma$ is the quotient of $St\widetilde{\sigma}$ by the action of G_σ. The orbispace structure on X is given by a topological groupoid Γ with space of units the disjoint union of the $St\widetilde{\sigma}$'s so that its restriction to each $St\widetilde{\sigma}$ is given by the action of G_σ. The notion of fundamental group can be naturally defined for topological groupoids as well as the notion of covering or universal covering (this is recalled in the appendix), and leads to the corresponding notions for complex of groups. Although the theory of complex of groups can be developed without those considerations, we feel that the concepts of covering, fundamental group, and so on, are more natural in this framework; the corresponding concepts for a complex of groups are just an algebraic translation of those notions.

Suppose that each simplex σ of X is modelled on a geodesic simplex Δ_σ in the hyperbolic space or the Euclidean space, in a consistent way. Suppose also that the set of isometry types of the simplices Δ_σ is finite. Then following a recent work of Bridson [3], one can define naturally a metric on each $St\widetilde{\sigma}$ invariant by the action of G_σ, and more generally under the action of the groupoid Γ. If this metric is non-positively curved in the sense of Alexandrov (see Gromov [7] and Ballman in [5]), then the given complex of groups $G(X)$ on X is developable, namely there exists a group G acting by isometries on a non-positively curved simply connected simplicial cell complex $\widetilde{X}$ such that $X = G\backslash\widetilde{X}$ and that the complex of groups associated is isomorphic to $G(X)$; moreover $\widetilde{X}$ is contractible.

This main theorem is stated in § 6 without a detailed proof. It has been proved previously in two particular cases, namely by Gersten and Stallings [14] when X is a triangle and by Haefliger ([5], Chapitre 11), following a statement of Gromov, when all the groups G_σ are finite and X is a finite complex. The proof in the general case applies essentially the techniques of Bridson [3] to extend the validity of the methods of Ballman and Haefliger in [5], Chapter 10 and 11. Details of the proof should appear in the thesis of B. Spieler.

Our study of orbispaces originated from the pages 128-129 in Gromov [7] (cf. Chapitre 11 in GH [5]). I thank Barry Spieler who mentioned to me, in connection with orbispaces, the existence of the preprint [14] of Gersten and Stallings; after reading this paper, I realized that the concept of orbihedron could be coded in terms of complex of groups. I also thank John Stallings who pointed out to me the paper of Bridson [5]. In his thesis [4], J.M. Corson has developed independently a similar theory for 2-complexes of groups;his notion of labelled complex of groups corresponds to what we call here a complex of groups.

1. The category of simplicial cell complexes.

This is a natural generalization of the concept of graph.

1.1. Simplicial cell complex. A simplicial cell complex X is a cell complex such that each (open)-n-cell σ has an affine structure. More precisely, for each n-cell σ, there is a set of $(n+1)!$ continuous maps (called orderings of σ) of the standard n-simplex Δ^n in X such that

i) the restriction of an ordering to the interior of Δ^n is a homeomorphism on the cell σ

ii) two orderings differs by an affine isomorphism of Δ^n

iii) the composition of an affine injection of Δ^k on a k-face of Δ^n with an ordering of σ is an ordering for a k-cell of X.

1.2. A good example to keep in mind is the dunce hat, the simplicial cell complex obtained by identifying in the 2-simplex Δ^2 with vertices v_0, v_1, v_2 the edge (v_0, v_1) with (v_0, v_2) and with (v_1, v_2) with the indicated orientation. This 2-dimensional simplicial cell complex has one cell in each dimension 0,1 and 2 and is contractible as a topological space.

1.3. A simplicial map f of a simplicial cell complex X in another one Y is a continuous map such that the composition of an ordering of an n-cell of X with f is an ordering of an n-cell of Y. In particular the restriction of f to a cell is a homeomorphism on its image.

An **action without inversion of a group** G **on** X is an action of G by simplicial homeomorphisms such that if an element of G maps a cell σ onto itself, then its restriction to σ is the identity. It follows that $G\backslash X$ is naturally a simplicial cell complex so that the projection of X on $G\backslash X$ is a simplicial map.

1.4. The **barycentric subdivision** of a simplicial cell complex X is well defined and is again a simplicial cell complex such that each cell has a natural ordering. Its vertices are the barycenters of the cells (and will be labelled with the name of the corresponding cell). An edge a of the barycentric subdivision of X contained in an n-cell σ is the image by an ordering $f : \Delta^n \to X$ of an edge of the barycentric subdivision of Δ^n issued from its barycenter (there are $2^{n+1} - 2$ such edges); the origin (or initial point) of a is the barycenter of σ, in notation $i(a) = \sigma$, and its extremity (or terminal point of a) is the barycenter of a k-cell τ (with $k < n$), in notation $t(a) = \tau$.

1.5. The category $C(X)$. The set $E(X)$ of edges of the barycentric subdivision of X together with the set $V(X)$ of cells of X is the set of morphisms of a category

$C(X)$ whose set of objects is $V(X)$. Two edges a and b are composable if $i(a) = t(b)$ and the composition $c = ab$ is the edge c with $i(c) = i(b)$ and $t(c) = t(a)$ such that a, b and c form the boundary of a 2-simplex of the barycentric subdivision. The set of objects is naturally graded by the dimension of cells.

For instance, in the dunce hat X, $E(X)$ has 8 elements a, a', b, b', b'', c, c', c'' as indicated in the figure 1 (it is understood that the edges with the same label must be identified) with the relations

$$ab = a'b' = c''$$

$$a'b'' = a'b = c'$$

$$ab' = ab'' = c$$

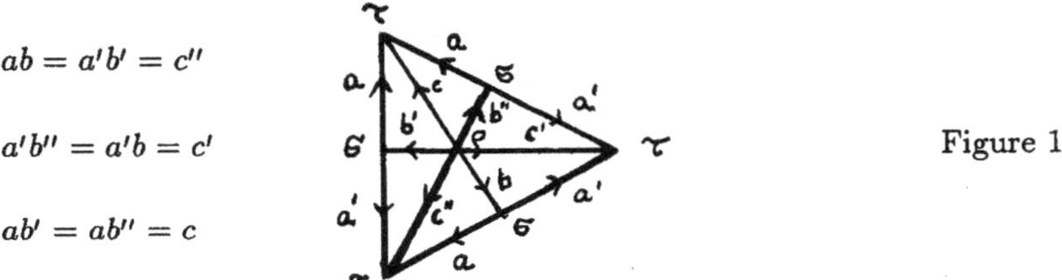

Figure 1

A simplicial map f of X in Y induces a homomorphism (or functor) of $C(X)$ in $C(Y)$ still denoted by f.

One can reconstruct the barycentric subdivision of X from the category $C(X)$: the vertices are the objects $V(X)$ of $C(X)$, and for $k > 0$, the k- simplices of the barycentric subdivision of X are the non-degenerate k-simplices of the geometric realization (Milnor [8]) of the nerve (cf. Segal [12]) of the category $C(X)$, namely the sequence of k composable elements of $E(X)$. For an object σ of $C(X)$, let n be the maximum length of a sequence of composable elements with initial object σ; the union of the vertex σ and the simplices of the nerve of $C(X)$ corresponding to sequences of composable element with initial object σ will be the n-cell of X with barycenter σ.

1.6. Generalization.

All the constructions in the following sections can be generalized by replacing the category $C(X)$ by a small category C without loop in the following sense.

The objects of the category C form a set $V(C)$ called the set of vertices of C and the morphisms which are not identities form a set $E(C)$ called the set of edges of C. For an element a of $E(C)$ which is a morphism of the object σ on the object τ, we denote σ by $i(a)$ and τ by $t(a)$. To C is associated a directed graph whose set of vertices is $V(C)$ and whose set of oriented edges is the set $E(C)$, the initial point of an edge a being $i(a)$ and its terminal point $t(a)$.

We say that C is without loop if, for each sequence $a_1, \dots, a_n$ of composable elements of $E(C)$, then $t(a_1)$ is distinct from $i(a_n)$. This means that in the associated graph there is no closed edge path made up of edges with their given orientation. Nevertheless there might be several edges with the same initial and terminal vertices.

The nerve of each small category without loop has a geometric realization which is a simplicial cell complex with set of vertices $V(C)$ and set of $k-$simplices, with $k > 0$, represented by the sequences of k composable elements of $E(C)$. Each k-simplex has a natural ordering and its closure is homeomorphic to a standard closed k-simplex, but a simplex is not determined by the sequence of its vertices. The 1-skeleton is isomorphic to the directed graph associated to C. This complex has a natural "polyhedral decomposition", namely for each vertex σ is associated the union of those simplices with initial vertex σ.

For instance, if we start from a polyhedron of dimension 2 whose faces are polygons with s sides, then X is the geometric realization of a small category without loop. Its objects are the barycenters of the faces, sides and vertices. The edges join the barycenters of faces to barycenters of sides and to vertices, or barycenters of sides to vertices. The barycenter of a face is the initial point of $2s$ edges. Each (open) face is the union of its barycenter, $2s$ simplices of dimension 1 and $2s$ simplices of dimension 2.

The generalization in this context of the notion of simplicial map will be a functor f from a small category without loop C in another one C' such that, for each $\sigma \in v(C)$, the restriction of f to the set $E(C)_\sigma$ of edges in C with initial point σ is injective. An action without inversion of a group G on a small category C without loop is an action of G on C associating to each $g \in G$ an isomorphism of C which is injective on each $E(C)_\sigma$ and the identity on $E(C)_\sigma$ if $g\sigma = \sigma$.

2. Complexes of groups.

Let X be a simplicial cell complex as defined above. The notions introduced in 2.1 and in 2.2 are motivated by the example 2.3.

2.1. Definition of a complex of groups $G(X)$ on X.

A complex of groups $G(X) = (X, G_\sigma, \psi_a, g_{a,b})$ on X is given by

1) a group G_σ for each n-cell σ of X,

2) an injective homomorphism $\psi_a : G_{i(a)} \to G_{t(a)}$ for each edge $a \in E(X)$ of the barycentric subdivision of X (cf. 1.4),

3) for two composable edges a and b in $E(X)$, an element $g_{a,b}$ of $G_{t(a)}$ is given such that

2.1. *i)* $$Ad(g_{a,b})\psi_{ab} = \psi_a \psi_b$$

where $Ad(g_{a,b})$ denotes the conjugation of $G_{t(a)}$ by $g_{a,b}$, and such that the following cocycle condition holds for a triple a, b, c of composable elements of $E(X)$.

2.1 ii) $$\psi_a(g_{b,c})g_{a,bc} = g_{a,b}g_{ab,c}.$$

This cocycle condition is non empty only if the dimension of X is at least 3.

When $dim\, X = 1$, our notion is equivalent to the notion of a graph of groups [13].

If one gives for each element a of $E(X)$ an element g_a of $G_{t(a)}$, one gets a complex $G'(X)$ of groups on X where $G'_\sigma = G_\sigma$,

$$\psi'_a = Ad(g_a)\psi_a$$
$$g'_{a,b} = g_a \psi_a(g_b) g_{a,b} g_{ab}^{-1} \ .$$

This complex of groups is said to be deduced from $G(X)$ by the coboundary of $\{g_a\}$.

2.2. Homomorphisms of complexes of groups. Let f be a simplicial map of X in a simplicial cell complex X' and let $G(X') = (X',\ G'_{\sigma'},\ \psi'_{a'},\ g'_{a',b'})$ be a complex of groups on X'.

A homomorphism Φ of $G(X)$ in $G(X')$ over $f : X \to X'$ is given by

1) a homomorphism $\varphi_\sigma : G_\sigma \to G'_{f(\sigma)}$ for each n-cell σ of X

2) an element g'_a of $G'_{tf(a)}$ for each element a of $E(X)$ *such that*

2.2. *i)* $$Ad(g'_a)\psi'_{f(a)}\varphi_{i(a)} = \varphi_{t(a)}\psi_a \ ,$$

and such that for composable elements a, b of $E(X)$

2.2. *ii)* $$\varphi_{t(a)}(g_{a,b})g'_{ab} = g'_a \psi'_{f(a)}(g'_b) g'_{f(a),f(b)}.$$

Suppose that an element $\overline{g}_\sigma \in G'_{f(a)}$ is given for each cell $\sigma \in V(X)$. We can construct from Φ an homomorphism $\overline{\Phi}$ of $G(X)$ in $G(X')$ given by

$$\overline{\varphi}_\sigma = Ad(\overline{g}_\sigma)\varphi_\sigma \qquad \text{and}$$
$$\overline{g}_a = \overline{g}_{t(a)} g'_a \psi'_{f(a)}(\overline{g}_{i(a)}^{-1}).$$

We say that $\overline{\Phi}$ is deduced from Φ by the coboundary of the cochain $\overline{g}_\sigma$ and $\overline{\Phi}$ will be called equivalent to Φ.

If $X' = X$ and if $G'(X)$ is deduced from $G(X)$ by the coboundary of $\{g_a\}$, then the collections φ_σ = identity of G_σ and $g'_a = g_a^{-1}$ give an isomorphism of $G(X)$ on $G'(X)$ over the identity of X.

2.3. Developable complex of groups.

Let G be a group acting simplicially without inversion on a simplicial cell complex $\widetilde{X}$ and let p be the natural projection of $\widetilde{X}$ on $G\backslash\widetilde{X} = X$. We shall associate to this action a complex of groups $G(X)$ on X, unique up to isomorphism.

For each cell σ of X, choose a cell $\widetilde{\sigma}$ of $\widetilde{X}$ such that $p(\widetilde{\sigma}) = \sigma$. Define G_σ as the stability subgroup $G_{\widetilde{\sigma}}$ of $\widetilde{\sigma}$. For each a of $E(X)$ with $i(a) = \sigma$, let $\widetilde{a}$ be the edge in $\widetilde{\sigma}$ whose projection by p is a. Choose an element h_a of G such that $t(h_a(\widetilde{a})) = \widetilde{\tau}$, where $\tau = t(a)$; if $t(\widetilde{a}) = \widetilde{\tau}$, we choose h_a to be the identity element of G. We define $\psi_a : G_{i(a)} \to G_{t(a)}$ by $\psi_a(g) = h_a g h_a^{-1}$. For two composable edges a and b of $E(X)$, we define $g_{a,b} = h_a h_b h_{ab}^{-1}$ (figure 2). Then the conditions 2.1. i) and ii) are satisfied.

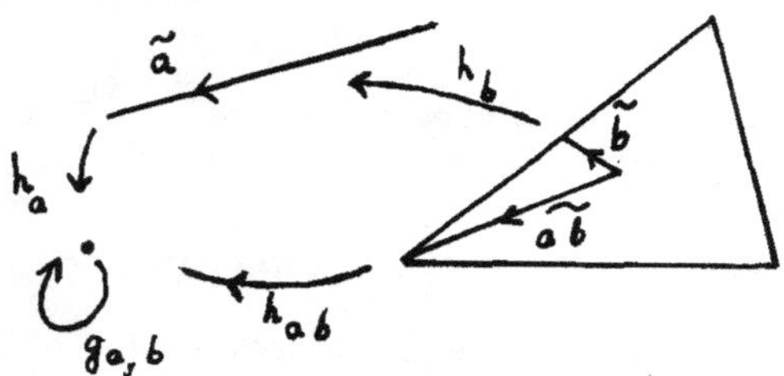

Figure 2

Another choice for the representatives in $\widetilde{X}$ of cells of X and for the $h'_a s$ will give a complex of groups $G'(X)$ over X with an isomorphism of $G(X)$ on $G'(X)$ over the identity of X.

A complex of groups is called *developable* if it is isomorphic to a complex of groups associated to an action of a group G on a simplicial cell complex $\widetilde{X}$ as above.

Let G' be a group acting simplicially without inversion on a simplicial cell complex $\widetilde{X}'$. Let p' be the projection of $\widetilde{X}'$ on $X' = G'\backslash\widetilde{X}'$. Let $G'(X')$ be the complex of groups associated to this action after the choice of representatives $\widetilde{\sigma}'$ and elements h'_a as above. Let φ be a homomorphism of G in G' and let $\widetilde{f}$ be a

φ-equivariant simplicial map of $\widetilde{X}$ in $\widetilde{X}'$. Then $\widetilde{f}$ induces a homomorphism Φ of $G(X)$ in $G'(X')$ over the map f of X in X'.

To see this, choose for each cell σ of X an element k_σ in G' such that $k_\sigma \widetilde{f}(\widetilde{\sigma})$ is the representative of $f(\sigma)$. Define $\varphi_\sigma : G_\sigma \to G'_{f(\sigma)}$ as the homomorphism mapping g on $k_\sigma \varphi(g) k_\sigma^{-1}$ and g'_a by

$$g'_a h'_{f(a)} k_{i(a)} = k_{t(a)} \varphi(h_a) .$$

Conditions 2.2.i) and ii) are satisfied. Another choice for k_σ leads to a homomorphism equivalent to Φ.

2.4. Examples. Suppose that X is the standard simplex Δ^n of dimension n. Any complex of groups $G(\Delta^n)$ on Δ^n is isomorphic to a complex of groups for which the elements $g_{a,b}$ are all the identity; such a complex of groups is given by attaching to each barycenter of the simplices of Δ^n a group together with injective homomorphisms along the edges of the barycentric subdivision to obtain a commutative diagram, a so called simplex of groups (cf. Stallings [14]).

To see it, let C_k be the subcategory of $C(X)$ whose set $E(C_k)$ of edges is the set of elements a of $E(X)$ such that $i(a)$ is the barycenter of a face of dimension $\geq k$. We argue by descending induction on k; assume that $g_{a,b}$ is the identity for all $a, b \in E(C_k)$. Let $a \in E(X)$ be an edge such that $i(a)$ is the barycenter of a face of dimension $k-1$ of Δ^n. Let d be the unique edge such that $i(d)$ is the barycenter of Δ^n and $t(d) = i(a)$; define $g_a \in G_{t(a)}$ by $g_a = g_{a,d}^{-1}$.

Modification of the complex of groups by the coboundary of the chain associating g_a to $a \in E(C_{k-1}) - E(C_k)$ and the identity to the other edges yields a complex of groups $G'(X)$ such that $g'_{e,f} = 1$ for all $e, f \in E(C_{k-1})$.

Indeed this is true for $e, f \in E(C_k)$, so we have to check it only for $e = a \in E(C_{k-1}) - E(C_k)$. If a and b are composable, there is an edge c such that $i(c)$ is the barycenter of Δ_n and $bc = d$. By the cocycle condition, we have

$$\phi'_a(g'_{b,c}) g'_{a,bc} = g'_{a,b} g'_{ab,c} ,$$

but $g'_{b,c} = 1$, $g'_{ab,c} = 1$, $g'_{a,bc} = 1$, hence $g'_{a,b} = 1$.

Other examples are given in 3.4 and 6.6.

2.5. The category associated to a complex of groups.

Let $G(X) = (X,\ G_\sigma,\ \psi_a,\ g_{a,b})$ be a complex of groups on X. Recall that $C(X)$ is the small category whose set of morphisms is the union of the set $V(X)$ of cells of X (or vertices of the barycentric subdivision of X) and the set $E(X)$ of edges of the barycentric subdivision of X. We denote by i and t the maps of $C(X)$ on $V(X)$ associating to an element of $C(X)$ its right and left unit (hence on $V(X)$, i and t are the identity).

For α, β composable elements of $C(X)$, define $g_{\alpha,\beta}$ as before if α and β belongs to $E(X)$ and as 1 if α or β belongs to $V(X)$.

We define a small category $CG(X)$ whose elements are the pairs (g, α) with $\alpha \in C(X)$ and $g \in G_{t(\alpha)}$. The composition $(g,\ \alpha)(h,\ \beta)$ is defined if $t(\beta) = i(\alpha)$ and is equal to $(g\psi_\alpha(h)g_{\alpha,\beta},\ \alpha\beta)$. Taking account of 2.i), the condition 2,ii) is equivalent to the associativity property of this law of composition. Note that the set of objects of $CG(X)$ is in bijection with the set of objects $V(X)$ of $C(X)$.

A homomorphism Φ of $G(X)$ in $G(X')$ (given by $\{\varphi_\sigma\}$ and $\{g'_a\}$ in the notations of 2.2) determines a homomorphism (or functor) $C(\Phi)$ of $CG(X)$ in $CG'(X')$ defined by

$$C(\Phi)(g, \sigma) = (\varphi_\sigma(g), f(\sigma))$$
$$C(\Phi)(g, a) = (\varphi_{t(a)}(g)g'_a, f(a)).$$

Conversely, any functor of $CG(X)$ in $CG(X')$ projecting on the functor of $C(C)$ in $C(X')$ determined by f comes from a homomorphism of $G(X)$ in $G(X')$.

If $\overline{\Phi}$ is deduced from Φ by the coboundary of $\{\overline{g}_\sigma\}$ (cf.2.2), then $\{\overline{g}_\sigma\}$ defines a natural transformation of $C(\Phi)$ on $C(\overline{\Phi})$, and conversely.

2.6. The classifying space $BG(X)$ of $G(X)$.

It is the geometric realization $BG(X)$ of the nerve of the category $CG(X)$. For instance, if $G(X)$ is the trivial complex of groups on X, then $BG(X)$ is X itself with its barycentric subdivision. If X is a point σ, then $BG(X)$ is the Eilenberg-Mac Lane complex $K(G_\sigma, 1)$. The natural projection of $CG(X)$ on $C(X)$ induces a simplicial map of $BG(X)$ on X. The inverse image of a cell σ of X is an Eilenberg-MacLane complex $K(G_\sigma, 1)$.

We mention without proof the following facts. If X is a graph (resp. if X is a triangle), then BG(X) has the homotopy type of the space constructed in Scott-Wall [11], p. 156, (resp. in Stallings [14], 3.1). For X connected, the fundamental group of $G(X)$ as constructed in the next section will be isomorphic to the fundamental group of $BG(X)$ (see 3.1, a)).

Any homomorphism Φ of $G(X)$ in $G(X')$ over f induces a simplicial map of $BG(X)$ in $BG(X')$ over f.

Let $G(X)$ be a complex of groups associated to a simplicial action without inversion of a group G on a simplicial cell complex $\widetilde{X}$; then $BG(X)$ has the homotopy type of the bundle with fiber $\widetilde{X}$ associated to the universal bundle EG for G. So its cohomology is the equivariant cohomology of the G-space $\widetilde{X}$.

3. The fundamental group of a complex of groups.

3.1. Definition of the fundamental group.

Let X be a simplicial cell complex. We denote by $E^{\pm}(X)$ the set of symbols a^+ and a^-, where a belongs to the set $E(X)$ of edges of the barycentric subdivision of X. The elements e of $E^{\pm}(X)$ should be considered as the edges of the barycentric subdivision of X together with an orientation. We define $i(a^+) = t(a^-) = i(a)$ and $t(a^+) = i(a^-) = t(a)$. An edge path is a sequence $e_1, \ldots, e_n$ of elements of $E^{\pm}(X)$ such that $t(e_k) = i(e_{k+1})$ for $1 \leq k \leq n-1$. The element $i(e_1)$ (resp. $t(e_n)$) is the origin (respectively the extremity) of the edge path.

Let $G(X) = (X, G_\sigma, \psi_a, g_{a,b})$ be a complex of groups on X. We first define (compare with Serre [13]) the group $FG(X)$ as the group generated by

$$\begin{cases} \text{the elements of } G_\sigma, \text{ where } \sigma \text{ belongs to the set } V(X) \text{ of cells of } X, \\ \text{the elements of } E^{\pm}(X) \end{cases}$$

subjected to the relations

$$\begin{cases} \text{the relations in } G_\sigma \\ (a^+)^{-1} = a^- \quad \text{and} \quad (a^-)^{-1} = a^+ \\ \psi_a(g) = a^- g a^+ \quad \text{for} \quad g \in G_{i(a)} \\ (ab)^+ = b^+ a^+ g_{a,b}. \end{cases}$$

The image of G_σ in $FG(X)$ will be denoted by $\overline{G}_\sigma$.

Following Serre [13], we give two definitions of the fundamental group of $G(X)$.

3.1.a) First definition.

Given σ_0 and σ in $V(X)$, a $G(X)$- path c from σ_0 to σ is a sequence $g_0,\ e_1,\ g_1, \ldots,\ e_n,\ g_n$, where $e_1, \ldots,\ e_n$ is an edge path in the barycentric subdivision of X with $i(e_1) = \sigma_0$ and $t(e_n) = \sigma$, and where g_k belongs to $G_{t(e_k)} = G_{i(e_{k+1})}$. The $G(X)$-path c represents the element $g_0 e_1 g_1 \ldots e_n g_n$ of $FG(X)$. Two $G(X)$-paths from σ_0 to σ are homotopic if they represent the same element of $FG(X)$. We define $\pi_1(G(X), \sigma_0, \sigma)$ as the subset of $FG(X)$ made up of elements represented by $G(X)$-paths from σ_0 to σ. When $\sigma_0 = \sigma$, this is a subgroup denoted by $\pi_1(G(X), \sigma_0)$ and called *the fundamental group of* $G(X)$ with respect to the base point σ_0. Of course, $\pi_1(G(X), \sigma_0, \sigma)$ is a $\pi_1(G(X), \sigma_0)$-coset in $FG(X)$ (i.e. an orbit for the action of $\pi_1(G(X), \sigma_0)$ on $FG(X)$ by left translation).

Moreover $\overline{G}_\sigma$ acts by composition on the right on $\pi_1(G(X), \sigma_0, \sigma)$.

Note that the conjugation in $FG(X)$ by an element of $\pi_1(G(X), \sigma_0, \sigma)$ maps isomorphically $\pi_1(G(X), \sigma)$ on $\pi_1(G(X), \sigma_0)$ and $\overline{G}_\sigma$ on a subgroup of $\pi_1(G(X), \sigma_0)$.

Any homomorphism Φ of a complex of groups $G(X)$ in a complex of groups $G'(X')$ over a simplicial map f of X in X' induces a homomorphism φ of $FG(X)$ in $FG'(X')$ giving by restriction a homomorphism of $\pi_1(G(X), \sigma_0)$ in $\pi_1(G'(X')), f(\sigma_0))$. In the notations of 2.2, φ maps the generator $g \in G_\sigma$ on the generator $\varphi_\sigma(g) \in G_{f(\sigma)}$ and a^+ on $f(a)^+ g_a'^{-1}$.

If all the groups G_σ are trivial, then we get a presentation of the fundamental group of X, using the barycentric subdivision of X. Hence we always have a surjective homomorphism of $\pi_1(G(X), \sigma_0)$ on $\pi_1(X, \sigma_0)$.

A $G(X)$-loop determines an edge loop in $BG(X)$ (cf. 2.6) and it is easy to see that this induces an isomorphism between the fundamental groups of $G(X)$ and $BG(X)$.

3.1.b) Second definition.

Let T be a maximal tree in the 1-skeleton of the barycentric subdivision of X. Then the fundamental group $\pi_1(G(X), T)$ of $G(X)$ at T is the quotient of the group $FG(X)$ by the normal subgroup generated by all the elements a^+, where a is an element of $E(X)$ contained in T. In other words, this group is obtained by adding to the presentation of $FG(X)$ the relations

$$a^+ = 1, \quad \text{for} \quad a \in T .$$

3.2. Proposition. *If X is connected, $\pi_1(G(X), T)$ is isomorphic to $\pi_1(G(X), \sigma_0)$.*

Indeed the composition of the inclusion of $\pi_1(G(X), \sigma_0)$ in $FG(X)$ with the projection on the quotient $\pi_1(G(X), T)$ is an isomorphism (see Serre [13], p. 44-45).

When X is connected, the image of G_σ in $\pi_1(G(X), T)$ will still be denoted by $\overline{G}_\sigma$, because the image of G_σ in $FG(X)$ is mapped isomorphically on its image by the projection of $FG(X)$ on $\pi_1(G(X), T)$.

3.3. Corollary. *If all the groups G_σ are finitely presented and if X is a finite complex, then the fundamental group of $G(X)$ is finitely presented.*

Note that $\pi_1(G(X), T)$ is generated by the groups G_σ associated to the 0-cells σ of X and by the elements a^+ corresponding to the edges of the barycentric subdivision of X contained in the 1-skeleton of X.

3.4. Examples.

a) For a triangle of groups, the fundamental group is simply the colimit of the triangle of groups.

b) Suppose that X is the dunce hat. We use the notations of 1.5. We have three groups G_τ, G_σ, G_ϱ attached to the cells τ, σ, ϱ of dimension 0, 1, 2 and eight monomorphisms corresponding to the eight edges of the barycentric subdivision. After modification with a coboundary, we can assume that only $g_{a,b''}$ and $g_{a',b''}$ might be non trivial. Let T be the maximal tree in the barycentric subdivision of X which is the union of a' and b''. Then $\pi_1(G(X), T)$ is the group generated by the elements of G_τ and the element a, submitted to the relations in G_τ and the two relations

$$g_{a',b''}a = ag_{a,b''}\ a^{-1}$$
$$\psi_{a'}(g) = a\psi_a(g)a^{-1} \qquad \text{for all} \quad g \in G_\sigma \quad .$$

As a specific example, assume that $G_\tau = Z/2Z$, $G_\sigma = G_\varrho = 1$, and $g_{a',b''} = g_{a,b''}^{-1}$ is a generator s of G_τ. Then $\pi_1(G(X),\tau)$ has the presentation $< s, a;\ s^2 = 1, sa^2s = a >$, hence it is isomorphic to the group of permutations of three objects. The group G_τ injects in $\pi_1(G(X),T)$. We shall see in paragraph 4 that this implies that $G(X)$ is developable.

c) Assume that X is connected and that all the groups G_σ are isomorphic to the same group and that all the monomorphisms ψ_a are isomorphisms. Then one has an exact sequence

$$\pi_2(X,\sigma) \longrightarrow G_\sigma \longrightarrow \pi_1(G(X),\sigma) \longrightarrow \pi_1(X,\sigma) \longrightarrow 1$$

where the image of $\pi_2(X,\sigma)$ is in the center of G_σ. The group $\pi_1(G(X),\sigma)$ acts by automorphisms on the sequence, so that $G_\sigma \longrightarrow \pi_1(G(X),\sigma)$ is a crossed module.

If X is simply connected and G_σ abelian, then the isomorphism class of the complex of groups is completely characterized by the cohomology class in $H^2(X,G_\sigma)$ of the 2-cocycle $g_{a,b}$.

This example arises in the following geometric situation. Let $p : E \to B$ be a fiber space whose fiber is an Eilenberg-Mac Lane complex $K(G,1)$. For each σ, let s_σ be a continuous section of E over the open star of σ. Denote by σ^b the barycenter of σ and let G_σ be the group $\pi_1(p^{-1}(\sigma^b), s_\sigma(\sigma^b))$. If $a \in E(X), \sigma = i(a)$ and $\tau = t(a)$, then $s_\sigma(a)$ is a path in E from $s_\sigma(\sigma^b)$ to $s_\tau(\tau^b)$; choose a path ℓ_a in $p^{-1}(\tau^b)$ joining the terminal point of $s_\sigma(a)$ to $s_\tau(\tau^b)$. This choice defines an isomorphism ψ_a of G_σ on G_τ. If $b \in E(X)$ with $i(b) = \varrho$ and $t(b) = \sigma$, consider the loop $\ell_a^{-1}s_\sigma(a)^{-1}\ell_b^{-1}s_\varrho(b)^{-1}s_\varrho(ab)\ \ell_{ab}$. It is homotopic in the inverse image by p of the star of τ to a loop in $p^{-1}(\tau^b)$ whose homotopy class in G_τ is denoted by $g_{a,b}$.

Those data define a complex of groups $G(X)$. The exact sequence above is the homotopy exact sequence of the bundle $p : E \to B$. Conversely, given a complex of groups on X such that all ψ_a are isomorphisms, then $BG(X)$ (cf. 2.6) has the homotopy type of a bundle over X with fiber an Eilenberg-Mac Lane complex $K(G_\sigma, 1)$.

3.5. The canonical quotient $\overline{G}(X)$ of $G(X)$. Let $G(X) = (X, G_\sigma, \psi_a, g_{a,b})$ be a complex of groups on the simplicial cell complex X. We can associate to it a quotient complex of groups $\overline{G}(X) = (X, \overline{G}_\sigma, \overline{\psi}_a, \overline{g}_{a,b})$ on X as follows. The group $\overline{G}_\sigma$ is the image of G_σ in $FG(X)$ (see 2.3), $\overline{\psi}_a$ is obtained from ψ_a by passing to the quotient and $\overline{g}_{a,b}$ is the image of $g_{a,b}$ by the map of $G_{t(a)}$ on $\overline{G}_t(a)$. Note that $\overline{\psi}_a$ is injective because it is the restriction to $\overline{G}_{i(a)}$ of the conjugation in $FG(X)$ by a^-. The projection of $G(X)$ on $\overline{G}(X)$ is an isomorphism if and only if the natural map of G_σ in $FG(X)$ is injective for each cell σ in X.

Note that $FG(X)$ is canonically isomorphic to $F\overline{G}(X)$. Hence $\pi_1(G(X), \sigma_0)$ (resp. $\pi_1(G(X), T)$) is canonically isomorphic to $\pi_1(\overline{G}(X), \sigma_0)$ (resp. $\pi_1(\overline{G}(X), T)$). Note also that if $G(X)$ is developable, i.e. is associated to a simplicial action of a group G on a simplicial cell complex $\widetilde{X}$ with $X = G\backslash\widetilde{X}$, then $G(X) = \overline{G}(X)$. The aim of the next section is to prove the converse.

4. The universal covering of a complex of group.

4.1. Theorem. *Let $G(X)$ be a complex of groups on a connected simplicial cell complex X. Assume that for each cell σ of X the natural homomorphism of G_σ in $FG(X)$ is injective. Then $G(X)$ is developable.*

More precisely, let T be a maximal tree in the 1-skeleton of the barycentric subdivision of X. Then there is canonically a simply connected simplicial cell complex $\widetilde{X}$ and an action of $G = \pi_1(G(X), T)$ without inversion on $\widetilde{X}$, a simplicial map p of $\widetilde{X}$ on X inducing an isomorphism of $G\backslash\widetilde{X}$ on X, a natural lifting $\widetilde{T}$ of T in $\widetilde{X}$ so that $G(X)$ is naturally isomorphic to the complex of groups on X associated to this action.

If $G(X)$ is the complex of groups associated to an action without inversion of a group $\overline{G}$ on a connected simplicial cell complex $\overline{X}$ with $X = \overline{G}\backslash\overline{X}$ and with respect to a lifting $\overline{T}$ of T, then there is a surjective homomorphism φ of G on $\overline{G}$ and a φ-equivariant covering map f of $\widetilde{X}$ on $\overline{X}$ mapping $\widetilde{T}$ on $\overline{T}$ and projecting on the identity of X. Hence the kernel of φ is the Galois group of the covering f and so the fundamental group of $\overline{X}$.

4.2. Proof.

As by hypothesis the map of G_σ in $\pi_1(G(X),T)$ is injective, we shall denote its image by G_σ. The set $V(\widetilde{X})$ of cells of $\widetilde{X}$ will be the set of pairs (gG_σ,σ), where σ runs over the set $V(X)$ of cells of X and gG_σ runs over the set of right G_σ-cosets in $\pi_1(G(X),T)$. The projection p maps (gG_σ,σ) on σ. The set $E(\widetilde{X})$ of edges of the barycentric subdivision of $\widetilde{X}$ will be the set of pairs (gG_σ,a), where $a\in E(X)$ with $i(a)=\sigma$. We define $p(gG_\sigma,a)=a$, $i(gG_\sigma,a)=(gG_\sigma,i(a))$ and $t(gG_\sigma,a)=(ga^+G_{t(a)},t(a))$; this is well defined because, for $h\in G_{i(a)}$, we have $gha^+=ga^+\psi_a(h)$ in $FG(X)$. The elements $(gG_{i(a)},a)$, and $(hG_{i(b)},b)$ of $E(\widetilde{X})$ are composable if $t(b)=i(a)$ and $g^{-1}hb^+\in G_{i(a)}$ and their composition is the pair $(hG_{i(b)},ab)$. Those data define $C(\widetilde{X})$; its geometric realization is $\widetilde{X}$ and $p:\widetilde{X}\to X$ is a well defined simplicial map.

The action of $G=\pi_1(G(X),T)$ on $\widetilde{X}$ is the obvious one : for instance the element γ of $\pi_1(G(X),T)$ maps the edge $(gG_{i(a)},a)$ on the edge $(\gamma gG_\sigma,a)$. This action is without inversion and the map p induces an isomorphism of $G\backslash\widetilde{X}$ on X. We have a canonical lifting $\widetilde{T}$ of T associating to the vertex σ the vertex $\widetilde{\sigma}=(G_\sigma,\sigma)$ and to an edge a in T the edge $\widetilde{a}=(G_{i(a)},a)$ of $E(\widetilde{X})$. For $a\in E(X)$, we define $\widetilde{a}$ as the edge $(G_{i(a)},a)$ and $h_a=a^-$; note that $h_a(t(\widetilde{a}))=\widetilde{\tau}$, where $\tau=t(a)$. The stability subgroup of the cell $\widetilde{\sigma}$ is the subgroup G_σ of G; the homomorphism of $G_{i(a)}$ in $G_{t(a)}$ mapping h on $h_ahh_a^{-1}=a^-ha^+=\psi_a(h)$ is precisely ψ_a; moreover $h_ah_bh_{ab}^{-1}=a^-b^-(ab)^-=g_{a,b}$. Hence $G(X)$ is the complex of groups associated to the action of G on $\widetilde{X}$.

Let $\overline{G}$ be a group acting without inversion on a simplicial cell complex $\overline{X}$ such that $X=\overline{G}\backslash\overline{X}$. Suppose that $\overline{T}$ is a lifting of T with respect to the projection $\overline{p}$ of $\overline{X}$ on X; this determines a lifting $\overline{\sigma}$ in $\overline{T}$ of each $\sigma\in V(X)$ and a lifting $\overline{a}$ of each $a\in E(X)$ so that if $i(a)=\sigma$ and $t(a)=\tau$, then $i(\overline{a})=\overline{\sigma}$. Let $\overline{h}_a$ be elements of $\overline{G}$ such that $\overline{h}_a(t(\overline{a}))=\overline{\tau}$; we choose $\overline{h}_a$ to be the identity if $\overline{a}$ is contained in $\overline{T}$. We assume that the complex of groups associated to those choices like in 2.3 is equal to $G(X)$. In other words, G_σ is the subgroup of $\overline{G}$ fixing $\overline{\sigma}$, ψ_a is the homomorphism of $G_{i(a)}$ in $G_{t(a)}$ mapping h on $\overline{h}_ah\overline{h}_a^{-1}$ and $g_{a,b}=\overline{h}_a\overline{h}_b\overline{h}_{ab}^{-1}$.

Let φ be the map identifying G_σ to the subgroup of $\overline{G}$ leaving fixed $\overline{\sigma}$ and mapping a^+ onto $\overline{h}_a^{-1}$. This defines a homomorphism φ of $G=\pi_1(G(X,T)$ in $\overline{G}$ because this is compatible with the relations in G. Let f be the map of $C(\widetilde{X})$ in $C(\overline{X})$ mapping $(gG_\sigma,\sigma)\in V(\widetilde{X})$ on $\varphi(g)\overline{\sigma}\in V(\overline{X})$ and $(gG_{i(a)},a)\in E(\widetilde{X})$ on $\varphi(g)\overline{a}\in E(\overline{X})$. This is a morphism of category inducing a simplicial cell map f of $\widetilde{X}$ in $\overline{X}$, which is clearly φ -equivariant. It remains to show the following.

Lemma. *The maps φ and f are surjective. The kernel $Ker\varphi$ of φ acts freely on $\overline{X}$ and f induces an isomorphism of $Ker\varphi\backslash\widetilde{X}$ on $\overline{X}$. The space $\widetilde{X}$ is connected and simply connected.*

To prove this, we first introduce some notations. For $e \in E^{\pm}(X)$ contained in the cell σ, define $\widetilde{e} \in E(\widetilde{X})^{\pm}$ (resp. $\overline{e} \in E(\overline{X})^{\pm}$) as the unique oriented edge contained in the cell $\widetilde{\sigma}$ (resp. $\overline{\sigma}$) projecting by p (resp. $\overline{p}$) on e. Note that $f(\widetilde{e}) = \overline{e}$.

We also define the elements ${}^ih_e$ and ${}^th_e$ of G (see figure 3) by

$$ {}^ih_e = 1 \quad \text{and} \quad {}^th_e = h_a \quad \text{if} \quad e = a^+ $$

$$ {}^ih_e = h_a \quad \text{and} \quad {}^th_e = 1 \quad \text{if} \quad e = a^-. $$

Note that $e = {}^ih_e({}^th_e)^{-1}$ in $G = \pi_1(G(X), T)$.

We define similarly ${}^i\overline{h}_e$ and ${}^t\overline{h}_e$ by replacing in the definition above h_a by $\overline{h}_a$. They are the images by φ of ${}^ih_e$ and ${}^th_e$.

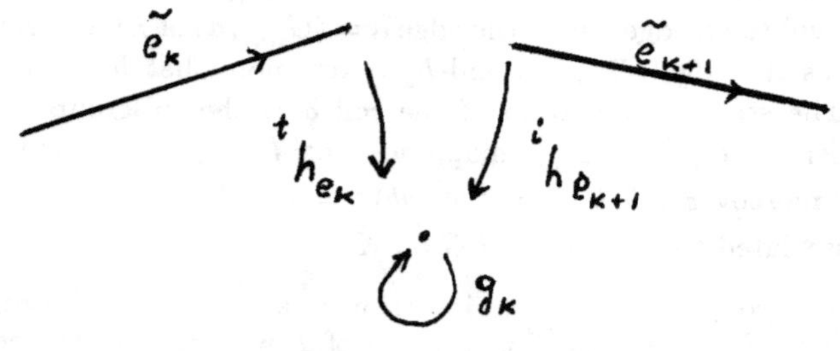

Figure 3

Let $c = (g_0, e_1, \cdots, e_n, g_n)$ be a $G(X)$-path from σ_0 to σ (cf. 3.1). To c we can associate the element $\lambda(c) = g_0 e_1 \cdots e_n g_n$ of G and the edge path $\Lambda(c) = \widetilde{g}_0\widetilde{e}_1, \widetilde{g}_0\widetilde{g}_1\widetilde{e}_2, \cdots, \widetilde{g}_0 \cdots \widetilde{g}_n\widetilde{e}_n$ in $\widetilde{X}$, where $\widetilde{g}_0 = g_0{}^ih_{e_1}, \widetilde{g}_n = ({}^th_{e_n})^{-1}g_n$, and $\widetilde{g}_k = ({}^th_{e_k})^{-1}g_k{}^ih_{e_{k+1}}$ for $0 < k < n$.

The origin of $\Lambda(c)$ is $\widetilde{\sigma}_0$ and its extremity is $\lambda(c)\widetilde{\sigma}$. As for each element g of G and each cell σ of X there is a $G(X)$-loop c at σ with $\lambda(c) = g$ (cf. 3.2), we see that $\Lambda(c)$ is an edge path in $\widetilde{X}$ from $\widetilde{\sigma}$ to $g\widetilde{\sigma}$, hence $\widetilde{X}$ is connected.

Let us prove now the surjectivity of f and φ. Any cell of $\overline{X}$ is of the form $\overline{g}\,\overline{\sigma}$ where $\overline{g} \in \overline{G}$ and $\sigma \in V(X)$. As $\overline{X}$ is connected, there is an edge path $f_1, \cdots, f_n$ in the barycentric subdivision of $\overline{X}$ from $\overline{\sigma}$ to $\overline{g}\,\overline{\sigma}$. Its projection by $\overline{p}$ is an edge path $e_1, \dots, e_n$ in X. For each $k, 1 \le k \le n$, there is an element $h_k \in \overline{G}$ such that $h_k(\overline{e}_k) = f_k$. Define (see figure 4)

$$g_0 = h_1({}^i\overline{h}_{e_1})^{-1}, g_k = {}^t\overline{h}_{e_k} h_k^{-1} h_{k+1}({}^i\overline{h}_{e_{k+1}})^{-1} \quad \text{and} \quad g_n = {}^t\overline{h}_{e_n} h_n^{-1}.$$

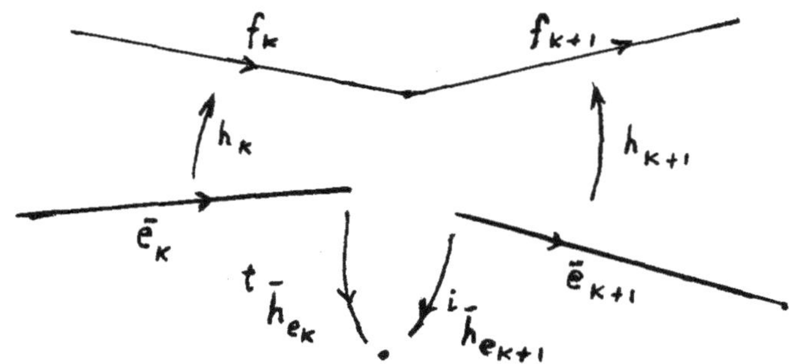

Figure 4

The sequence $c = g_0, e_1, \dots, g_{n-1}, e_n, g_n$ is a $G(X)$-loop at σ, because, if $\tau = t(e_k) = i(e_{k+1})$, then g_k belongs to the stability subgroup of $\overline{G}_{\overline{\tau}}$ which is equal to G_τ. The edge path $\wedge(c)$ in $\widetilde{X}$ is $\widetilde{g}_0 e_1, \dots, \widetilde{g}_0 \cdots \widetilde{g}_n \widetilde{e}_n$, where $\widetilde{g}_k$ is as defined above. Its image by f is the edge path $\overline{g}_0 \overline{e}_1, \dots, \overline{g}_0 \cdots \overline{g}_{n-1} \overline{e}_n$ in $\overline{X}$, where $\overline{g}_0 = g_0 {}^i\overline{h}_{e_1}, \overline{g}_k = {}^t\overline{h}_{e_k}^{-1} g_k {}^i\overline{h}_{e_{k+1}}$ because $\varphi(\widetilde{g}_k) = \overline{g}_k$, and this path is precisely $f_1, \dots, f_n$. Hence f is surjective. Also $\varphi(\lambda(c))\overline{\sigma} = \overline{g}\overline{\sigma}$, hence $\lambda(c)^{-1}\overline{g} = h \in \overline{G}_{\overline{\sigma}}$. It follows that $\overline{g} = \varphi(\lambda(c)h)$, hence φ is surjective.

Clearly Kerφ acts freely on $\widetilde{X}$ because it intersects trivially the stability subgroups of the cells of $\overline{X}$. If two cells $(g_1 G_\sigma, \sigma)$ and $(g_2 G_\sigma, \sigma)$ of $\widetilde{X}$ have the same image by f, then $\varphi(g_1)\overline{\sigma} = \varphi(g_2)\overline{\sigma}$. Hence there is an element h of $G_\sigma = \overline{G}_{\overline{\sigma}}$ such that $\varphi(g_1) = \varphi(g_2)h = \varphi(g_2 h)$, so $g_2 h g_1^{-1} \in Ker\varphi$. This implies that Ker$\varphi$ acts transitively on each fiber of f, and that $\overline{X}$ is the quotient of $\widetilde{X}$ by Kerφ.

It remains to check that $\widetilde{X}$ is simply connected. This follows from the fact that we can replace in the preceding considerations $\overline{X}$ by its universal covering using the construction in the paragraph 3 of the Appendix.

4.3. Example. Let (W, S) be a Coxeter system, namely W is the group generated by a finite set $S = \{s_0, \dots, s_n\}$, with the relations $(s_i s_j)^{m_{i,j}} = 1$, where $m_{i,j} = m_{j,i}$ is an integer > 1 or ∞ if $i \neq j$ and $m_{i,i} = 1$ (cf. Brown [2] or his lectures in this volume).

Let X be the standard n-simplex with vertices $0, 1, \dots, n$. We define a complex of groups $G(X)$ as follows. To the k-simplex σ in X with set of vertices $\{i_o, \dots, i_k\}$, we associate the subgroup G_σ of W generated by $S_\sigma = S - \{i_o, \dots, i_k\}$. It is known that (G_σ, s_σ) is the Coxeter system with Coxeter matrix obtained from the Coxeter matrix $(m_{i,j})$ of (W, X) by suppressing rows and columns indexed by the vertices of σ. If $a \in E(X)$, then ψ_a is the natural inclusion of $G_{i(a)}$ in $G_{t(a)}$. All the elements $g_{a,b}$ are trivial.

This complex of groups is developable. For $n > 1$, the fundamental group of $G(X)$ is naturally isomorphic to W and the universal covering $\widetilde{X}$ is the Coxeter complex of (W, S). For $n = 1$, the fundamental group is the infinite dihedral group and the universal covering of $G(X)$ is the line considered as the universal covering of the Coxeter complex of (W, S).

4.4. Construction of the universal covering in the general case.

When the complex of groups $G(X)$ is developable, then its universal covering is the simply connected simplicial cell complex $\widetilde{X}$ constructed above on which the fundamental group $G = \pi_1(G(X), T)$ acts.

In general, we can construct the simplicial cell complex $\widetilde{X}$ with the action of the group $G = \pi_1(G(X), T)$ as above, but using the quotient complex of groups $\overline{G}(X)$ as defined in 3.5 (remember that G is canonically isomorphic to $\pi_1(\overline{G}(X), T)$). In particular the cells of $\widetilde{X}$ are pairs $(g\overline{G}_\sigma, \sigma)$, where $\sigma \in V(X), g \in G$ and $\overline{G}_\sigma$ denotes the image of G_σ in G. For $h \in G_\sigma$, we denote by $\overline{h}$ its image in $\overline{G}_\sigma \subset G$.

The universal covering of the complex of groups X will be a complex of groups $G(\widetilde{X})$ on $\widetilde{X}$ defined as follows.

For each cell $\sigma' = (g\overline{G}_\sigma, \sigma)$ of $\widetilde{X}$ projecting by p on σ, choose a representative $g_{\sigma'}$ in the coset $g\overline{G}_{\sigma'}$ of $\widetilde{X}$. For $a' = (g\overline{G}_{i(a)}, a) \in E(\widetilde{X})$, choose $h_{a'} \in G_{t(a)}$ such that $g_{i(a')}a^+\overline{h}_{a'}^{-1} = g_{t(a')}$.

The complex $G(\widetilde{X})$ is defined as follows. For $\sigma' \in p^{-1}(\sigma)$, the group $G_{\sigma'}$ is the kernel of the natural homomorphism of G_σ in G and for $a' \in p^{-1}(a)$, then $\psi_{a'} = Adh_{a'}\psi_a$. Moreover for $b' \in p^{-1}(b)$, then $g_{a',b'} \in G_{t(a')}$ is defined by

$$h'_a\psi_a(h_{b'})g_{a,b} = g_{a',b'}h_{a'b'}.$$

We have a homomorphism Φ of $G(\widetilde{X})$ on $G(X)$ over p defined in the notations of 2.2. by

1) $\varphi_{\sigma'}$ is the inclusion of $G_{\sigma'}$ in G_σ,
2) $g_{a'} = h_{a'}$.

It is easy to check that the conditions 2.2,i) and ii) are satisfied.

The fundamental group G of $G(X)$ acts on $G(\widetilde{X})$ through isomorphisms in the following sense.

Let γ be an element of G. For each cell σ' in $\widetilde{X}$ choose $\ell^{\gamma}_{\sigma'} \in G_{\sigma'}$ such that $\overline{\ell}^{\gamma}_{\sigma'} = g^{-1}_{\gamma\sigma'}\gamma g_{\sigma'}$.

An automorphism Φ^{γ} of $G(\widetilde{X})$ is defined (in the notations of 2.2) by

$$\varphi^{\gamma}_{\sigma'} = Ad(\ell^{\gamma}_{\sigma'}) : G_{\sigma'} \to G_{\gamma\sigma'} = G_{\sigma'}$$

and

$$g^{\gamma}_{a'} = (\ell^{\gamma}_{t(a')}\psi_{a'}(\ell^{\gamma}_{i(a')})h_{a'}h^{-1}_{\gamma a'}$$

for each $\sigma \in V(\widetilde{X})$ and $a' \in E(\widetilde{X})$.

The composition $\Phi^{\gamma}\cdot\Phi^{\gamma'}$ is equal to $\Phi^{\gamma\gamma'}$ after modification by a coboundary (see 2.2).

4.5. Example. Suppose that $G(X)$ is a complex of groups on a triangle X with 0-cells τ, τ', τ'', 1-cells $\sigma, \sigma', \sigma''$ and one 2-cell ϱ. Assume that G_{ϱ} is trivial, that all the groups associated to the other cells are cyclic of order 2 except the group G_{τ} which is the dihedral group D_n of order $2n$ generated by two elements s' and s'' of order 2 which are the images of the generators of $G_{\sigma'}$ and $G_{\sigma''}$ by the monomorphisms associated to the edges with terminal point τ. The fundamental group of $G(X)$ is cyclic of order 2. The complex of groups $G(\widetilde{X})$ which is the universal covering of $G(X)$ is a complex of groups on the complex $\widetilde{X}$ which is the union of two triangles identified along their boundary. With the notations of the example 6.6.1, the group associated to each cell is trivial, except G_{τ} which is cyclic of order n; moreover, all the elements associated to two composable edges are trivial, except $g_{a',b'}$ which is a generator of G_{τ}. In the terminology of the next section, the orbihedron corresponding to the complex of groups $G(\widetilde{X})$ is the simply connected orbifold called by Thurston the tear drop.

5. Orbihedron.

5.1. Links and stars. Let X be a simplicial cell complex. The **star of** σ is the open set $St\sigma$ of X union of the cells of the barycentric subdivision of X whose closure contains a cell of the barycentric subdivision of σ. For $a \in E(X)$, the star of a is the union Sta of the cells of the barycentric subdivision whose closure contains a. Note that if σ and τ are cells such that $\dim\sigma < \dim\tau$, then $St\sigma \cap St\tau$ is the disjoint union of the Sta's with $i(a) = \tau$ and $t(a) = \sigma$.

The link of a n-cell σ of X is the simplicial cell complex $Lk\sigma$ whose k-cells are in bijection with the edges a of the barycentric subdivision of X such that $t(a) = \sigma$ and $i(a)$ is a $n+k+1$-cell. The set $E(Lk\sigma)$ of edges of the barycentric subdivision of $Lk\sigma$ are the pairs (a,b) of composable elements of $E(X)$ such that $t(a) = \sigma$. We have $i(a,b) = ab$ and $t(a,b) = a$. The edges (a,b) and (a',b') in $E(Lk\sigma)$are composable if $a' = ab$ and their composition is (a, bb').

For instance in the dunce hat X with the notations of 1.5, the link of the vertex has two o-cells and three 1-cells (cf.Figure 5).

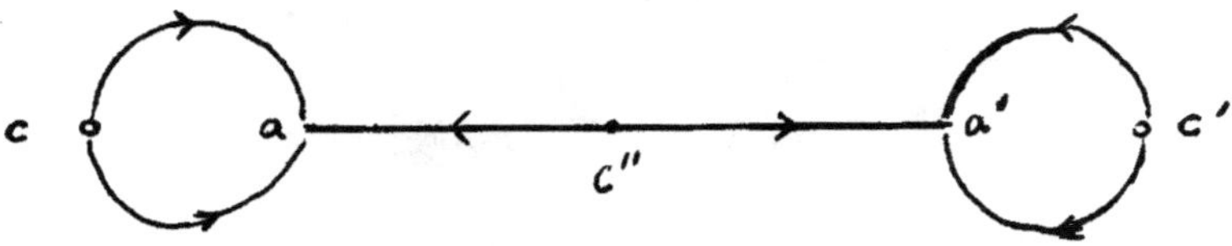

Figure 5

If X and Y are simplicial cell complexes, their join $X * Y$ is the simplicial cell complex, whose n cells are the join of a k-cell of X and a $(n-k-1)$-cell of Y, or a n-cell of X or Y. If we denote by $\overline{\sigma}$ the subcomplex of X whose underlying topological space is the closure of σ, then Stσ is naturally isomorphic to the star of $\sigma = \sigma * \phi$ in the join of $\overline{\sigma}$ and $Lk\sigma$.

5.2. Definition of an orbihedron structure.

Let X be a simplicial cell complex. An orbihedron structure on X is given by the following data.

1) For each cell σ of X, a simplicial cell complex $Lk\widetilde{\sigma}$ is given, as well as a simplicial action without inversion of a group G_σ on $Lk\widetilde{\sigma}$ and a simplicial projection $Lk(p_\sigma)$ of $Lk\widetilde{\sigma}$ on $Lk\sigma$ inducing an isomorphism of $G_\sigma \backslash Lk\widetilde{\sigma}$ on $Lk\sigma$.

On the join of the closure $\overline{\sigma}$ of σ with $Lk\widetilde{\sigma}$, we consider the simplicial action without inversion of G_σ which is the join of the identity on $\overline{\sigma}$ and the given action on $Lk\overline{\sigma}$. We denote by $\widetilde{\sigma}$ the simplex corresponding to σ in this join. Hence we get an action of G_σ on the star $St\widetilde{\sigma}$ of $\widetilde{\sigma}$ in this join, and a projection p_σ of $St\widetilde{\sigma}$ on $St\sigma$ deduced from $Lk(p_\sigma)$, and inducing an isomorphism of $G_\sigma \backslash St\widetilde{\sigma}$ on $St\sigma$.

2) Let a be an edge of the barycentric subdivision of X with $\tau = i(a)$ and $\sigma = t(a)$. Let us denote by $\widetilde{a}$ the edge of the barycentric subdivision of $\widetilde{\tau}$ projecting by p_τ on a. Clearly $St\widetilde{a} = p_{\widetilde{\tau}}^{-1}\ (Sta)$.

An injective homomorphism $\psi_a : G_\tau \to G_\sigma$ is given as well as a ψ_a-equivariant simplicial homeomorphism f_a of $St\tilde{a}$ on an open set of $p_{\tilde{\sigma}}^{-1}(Sta)$ projecting on the inclusion of Sta in $St\sigma$. Moreover the restriction of ψ_a to the subgroup of G_τ fixing a point x of $St\tilde{\tau}$ is an isomorphism on the subgroup of G_σ fixing $f_a(x)$.

3) For composable elements a, b of $E(X)$, an element $g_{a,b}$ of $G_{t(b)}$ is given such that $f_a f_b = g_{a,b} f_{ab}$ and that

i) $Ad(g_{a,b})\psi_{ab} = \psi_a \psi_b$

ii) $\psi_a(g_{b,c})g_{a,bc} = g_{a,b}g_{ab,c}$

for each triple a, b, c of composable elements of $E(X)$.

Clearly, if we retain only the groups G_σ, the homomorphisms ψ_a and the elements $g_{a,b}$, we get an associated complex of groups on X.

The orbihedron structure on X is called rigid if, for each cell σ, the action of G_σ on $St\tilde{\sigma}$ is rigid. This means that if the restriction of an element g of G_σ to an open set of $St\tilde{\sigma}$ is the identity, then g is the identity element of G_σ. In that case, $g_{a,b}$ is uniquely determined by the condition $f_a f_b = g_{a,b} f_{a,b}$ and the cocycle condition 3), ii) is automatically satisfied.

If, for each 0-cell σ, the group G_σ is finite and $St\tilde{\sigma}$ is a manifold, then we get an orbifold structure on X.

5.3. Example. We consider the example 2.3 with the same notations. The associated orbihedron structure on X is given by the action of G_σ on $St\tilde{\sigma}$, the homeomorphisms f_a being the restrictions to $St\tilde{a}$ of the elements h_a of G. An orbispace structure on X obtained in this way is called developable.

5.4. The groupoid associated to the orbihedron structure.

To the data giving an orbihedron structure on X is associated uniquely a topological groupoid $\Gamma(X) = \Gamma$ étale on its space of units U which is the disjoint union of the $St\tilde{\sigma}$'s (cf. Appendix). By definition, the restriction Γ_σ of Γ to $St\tilde{\sigma}$ is the topological groupoid $G_\sigma \times St\tilde{\sigma}$ associated to the action of G_σ on $St\tilde{\sigma}$ (see appendix, 2). For an edge $a \in E(X)$ with $i(a) = \sigma$ and $t(a) = \tau$, the space Γ_a of elements γ of Γ with $i(\gamma) \in St\tilde{a}$ and $t(\gamma) \in St\tilde{\tau}$ is the product $G_\tau \times St\tilde{a}$. Such an element γ will be represented by a triple (g, f_a, x) with $i(\gamma) = x, g \in G_\tau$ and $t(\gamma) = gf_a(x) = y$. The space of elements γ of Γ with $i(\gamma) \in St\tilde{\sigma}$ and $t(\gamma) \in St\tilde{\tau}$ is the union of the Γ_a's where $a \in E(X)$ with $i(a) = \sigma$ and $t(a) = \tau$.

The composition $(h, y)(g, f_a, x)$, where $(h, y) \in \Gamma_\tau$, is equal to (hg, f_a, x) and the composition $(g, f_a, x)(k, z)$, where $k \in G_{i(a)}$ and $z = k^{-1}x$, is equal to $(g\psi_a(k), f_a, z)$. If a and b are composable with $i(b) = \varrho$, and if (g', f_b, x') is such that $g' \in G_{i(a)}$ and $x' \in St\tilde{\varrho}$, then the composition $(g, f_a(x)(g', f_b, x')$ is defined if $x = g'f_b(x')$ and is equal to $(gg_{a,b}\psi_a(g'), f_{ab}, x')$. The conditions 3, i),

ii) assure that we have associativity. Finally, we introduce inverses $(g, f_a, x)^{-1}$ for the elements (g, f_a, x) and we extend consistently the law of composition to get the topological groupoid Γ whose restriction $St\widetilde{\sigma}$ is Γ_σ. The space of orbits $\Gamma\backslash U$ is naturally isomorphic to X.

In the example 5.3, the associated groupoid Γ is equivalent in the sense of the Appendix to the groupoid associated to the action of G on $\widetilde{X}$.

One can show that the space $BG(X)$ constructed in 2.6 has the same homotopy type as the classifying space of the topological groupoid Γ (cf. Haefliger [8]).

5.5. The orbihedron associated to a graph of groups.

The main result of the paragraph 5 will be that the concept of orbihedron is equivalent to the concept of complex of groups. Before stating in 5.6 the general result, we first look at the simple example where $G(X)$ is a complex of groups on a graph X. We want to associate to it an orbihedron structure on X.

For each 0-cell τ, we define $Lk\widetilde{\tau}$ as the 0-dimensional complex which is the disjoint union of the sets $G_\tau\backslash\psi_a(G_{i(a)})$, where a runs over the set of edges with $t(a) = \tau$. We form the cone $\widetilde{\tau} * Lk\widetilde{\tau}$ with vertex a point $\widetilde{\tau}$ and basis the discrete set $Lk\widetilde{\tau}$, and consider the star $St\widetilde{\tau}$ of $\widetilde{\tau} = \widetilde{\tau} * \phi$ in the join $\widetilde{\tau} * \widetilde{Lk}$. The group G_τ acts naturally on $Lk\widetilde{\tau}$, hence on $St\widetilde{\tau}$, and the quotient $G_\tau\backslash St\widetilde{\tau}$ is naturally isomorphic to $St\tau$.

For a 1-cell σ, then $St\widetilde{\sigma}$ will be isomorphic to $St\sigma$ with the trivial action of G_σ. If $a \in E(X)$ with $i(a) = \sigma$ and $t(a) = \tau$, let $\widetilde{a}$ be the open edge in $St\widetilde{\sigma}$ corresponding to a. Then f_a will be the homeomorphism of $\widetilde{a}$ on the edge of $St\widetilde{\tau}$ with terminal point $\widetilde{\tau}$ and initial point the barycenter of the 1-cell which is the join of $\widetilde{\tau}$ with the point $(a, \psi_a(G_\sigma))$ of $Lk\widetilde{\tau}$ (cf. figure 6).

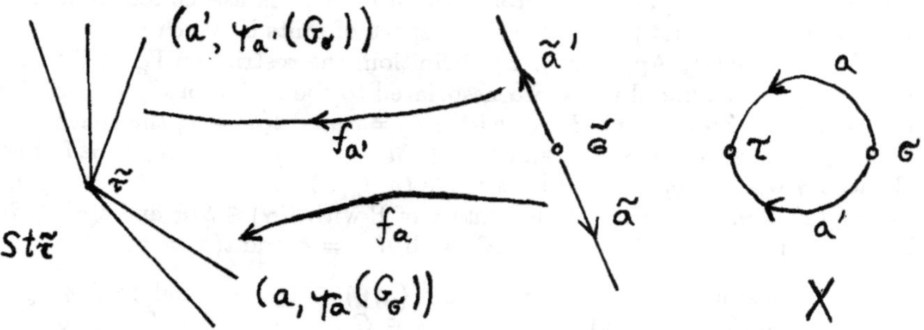

Figure 6

We now look at the general case.

5.6. The orbihedron associated to a complex of groups

Theorem. *Let $G(X)$ be a complex of groups on X. Then there is an orbihedron structure on X, unique up to isomorphism, such that the associated complex of groups is isomorphic to $G(X)$.*

For each cell σ, the set $V(Lk\widetilde{\sigma})$ of cells of $Lk\widetilde{\sigma}$ is the set of pairs $(g\psi_a(G_{i(a)}), a)$, where $g \in G_\sigma$ and $a \in E(X)$ with $t(a) = \sigma$.

The set $E(Lk\widetilde{\sigma})$ of edges of the barycentric subdivision of $Lk\widetilde{\sigma}$ are the pairs $(g\psi_{ab}(G_{i(b)}), a, b)$, where $g \in G_\sigma$, a and b are composable edges in $E(X)$ with $t(a) = \sigma$. The initial cell and terminal cell are defined by

$$i(g\psi_{ab}(G_{i(b)}), a, b) = (g\psi_{ab}(G_{i(b)}), ab)$$

$$t(g\psi_{ab}(G_{i(b)}), a, b) = (gg_{a,b}^{-1}\psi_a(G_{i(a)}), a)\ .$$

The composition of two edges $(g\psi_{ab}(G_{i(b)}), a, b)(h\psi_{a'c}(G_{i(c)}, a', c)$ is defined if $a' = ab$, $hg_{ab,c}^{-1}\psi_{ab}(G_{i(b)}) = g\psi_{ab}(G_{i(b)})$, and is equal to $(g\psi_{ab}(G_{i(b)}), a, bc)$.

The action of G_σ is given by left translations on the first element of the pairs.

Proof. For each cell σ of X, we first describe the complex Dσ, called the dual of σ, which is the cone with vertex the barycenter σ^b of σ and basis the barycentric subdivision of $Lk\sigma$. One can view Dσ as the geometric realization of the nerve of a category without loop : the set $V(D\sigma)$ of vertices is the union of σ^b with $V(Lk(\sigma))$. The set $E(D\sigma)$ of edges is the union of the set $E(Lk\sigma)$ with the set of pairs (σ^b, a), where $a \in V(Lk(\sigma))$, the initial vertex of (σ^b, a) being a and its terminal vertex σ^b. The composition of (σ^b, a) with (a, b) is (σ^b, ab).

We have a simplicial map of $D\sigma$ in the barycentric subdivision of X mapping the edge (σ^b, a) (resp. (a, b)) of $E(D\sigma)$ on the edge a (resp. b) of $E(X)$. This map is injective on the union of simplices of $D\sigma$ whose closure contains σ^b; such a k-simplex is represented by a sequence $a_1, \ldots, a_k$ of composable elements of $E(X)$ with $t(a_1) = \sigma$.

We consider the following complex of groups $G(D\sigma)$ on the category without loop associated to $D\sigma$. To the vertex $\sigma^b \in V(D_\sigma)$ is associated the group G_σ and to $a \in V(\text{Lk}\sigma) \subset V(D\sigma)$ is associated the group $G_{i(a)}$. The homomorphism associated to the edge (σ^b, a) is ψ_a and the one associated to the edge $(a, b) \in E(Lk\sigma) \subset E(D\sigma)$ is ψ_b. For two composable edges of the form (a, b) and (a', b'), we define $g_{(a,b),(a',b')} = g_{b,b'}$ and for two composables edges (σ^b, a) and (a, b), we define $g_{(\sigma^b,a),(a,b)} = g_{a,b}$. The conditions 2.1, i) and ii) are satisfied.

Let T be the maximal tree in $D\sigma$ which is the union of the edges (σ^b, a).

The fundamental group $\pi_1(G(D\sigma), T)$ is naturally isomorphic to G_σ. The isomorphism maps a generator $g \in G_\sigma$ onto itself, $g \in G_{i(a)}$ on $\psi_a(g)$, the generator $(\sigma, a)^+$ on 1 and $(a, b)^+$ on $g_{a,b}^{-1}$.

The analogue of the construction of 4.1 gives a category without loop; the geometric realization $D\widetilde{\sigma}$ of its nerve will be the cone over the barycentric subdivision of the simplicial cell complex Lk$\widetilde{\sigma}$ described in the statement of the theorem and $\widetilde{\sigma}^b$ as the vertex of the cone. We have a simplicial projection p_σ of $D\widetilde{\sigma}$ on $D\sigma$ in the barycentric subdivision of X mapping the edge $(g\psi_a(G_{i(a)}), \sigma^b, a)$ on a and the edge $(g\psi_{ab}(G_{i(b)}), a, b) \in E(Lk\widetilde{\sigma})$ on the edge $b \in E(X)$.

Finally $St\widetilde{\sigma}$ will be the star of $\widetilde{\sigma} = \sigma * \phi$ in the first barycentric subdivision of $\overline{\sigma} * Lk\widetilde{\sigma}$.There is one vertex which is the barycenter $\widetilde{\sigma}^b$ of $\widetilde{\sigma}$; the (open) simplices of $St\widetilde{\sigma}$ of dimension $n > 0$ correspond to the sequences $a_1, \dots, a_k, \widetilde{a}_{k+1}, \dots, \widetilde{a}_n$, where $a_1, \dots, a_k$ is a sequence of k composable edges of $E(X)$ with $i(a_k) = \sigma$ and $\widetilde{a}_{k+1}, \dots, \widetilde{a}_n$ a sequence of $n-k$ composable edges in $E(D\widetilde{\sigma})$ with $t(\widetilde{a}_{k+1}) = \widetilde{\sigma}^b$. Such a simplex can be interpreted as the join of the k-simplex of the barycentric subdivision of the cell σ corresponding to the sequence $a_1, \dots a_k$ with the $(n-k-1)$-simplex of the barycentric subdivision of $Lk\widetilde{\sigma}$ represented by the sequence $\widetilde{a}_{k+2}, \dots, \widetilde{a}_n$. An element g of the group G_σ acts on $St\widetilde{\sigma}$ by sending the n-simplex corresponding to the above sequence on the simplex represented by the sequence $a_1, \dots, a_k, g\widetilde{a}_{k+1}, \dots, g\widetilde{a}_n$. We have a natural projection p_σ of $St\widetilde{\sigma}$ on $St\sigma$ which maps the simplex corresponding to the sequence $a_1, \dots, a_k, \widetilde{a}_{k+1}, \dots, \widetilde{a}_n$ on the simplex of $St\sigma$ corresponding to the sequence $a_1, \dots, a_k, p_\sigma(\widetilde{a}_{k+1}), \dots, p_\sigma(\widetilde{a}_n)$. This projection induces an isomorphism of $G\backslash St\widetilde{\sigma}$ on $St\sigma$.

Let $a \in E(X)$ be such that $t(a) = \sigma$ and $i(a) = \varrho$. We have a simplicial map j_a of $Lk\varrho$ in $Lk\sigma$ mapping the edge $(b, c) \in E(Lk\varrho)$ on the edge $(ab, c) \in E(Lk\sigma)$. This map lifts as a ψ_a-equivariant simplicial map $\widetilde{j}_a$ of $Lk\widetilde{\varrho}$ in $Lk\widetilde{\sigma}$ associating to the cell $(g\psi_b(G_{i(b)}), b) \in V(LK\widetilde{\varrho})$ the cell $(\psi_a(g)g_{a,b}\psi_{ab}(G_{i(b)}), ab)$, and to the edge $g\psi_{ab}(G_{i(c)}), b, c) \in E(Lk\widetilde{\varrho})$ the edge $(\psi_a(g)g_{a,bc}\psi_{abc}(G_{i(b)}), ab, c) \in E(Lk\widetilde{\sigma})$. The conic extension of $\widetilde{j}_a$ to $D\widetilde{\varrho}$ is a map of $D\widetilde{\varrho}$ in $D\widetilde{\sigma}$ still denoted by $\widetilde{j}_a$. The reader should check that if a and b are two composable edges in $E(X)$, then $\widetilde{j}_a\widetilde{j}_b = g_{a,b}\widetilde{j}_{ab}$.

It remains to construct for an edge a, with $i(a) = \varrho$ and $t(a) = \sigma$, the injective ψ_a-equivariant map f_a of $p_\varrho^{-1}(Sta)$ in $p_\sigma^{-1}(Sta)$.

Denote by $St_\varrho a$ the union of simplices of the barycentric subdivision of the cell ϱ whose closures meet a; such a k-simplex is represented by a sequence $a_1, \dots, a_k$ of k composable edges of $E(X)$ such that their composition $a_1 \dots a_k$ is a. There is a natural lifting s_a of $St_\varrho a$ in $St\widetilde{\sigma}$ with respect to p_σ mapping a on the edge $(\sigma, \psi_a(G_\varrho), a)$ of $D\widetilde{\sigma}$. The image of s_a is left fixed by the subgroup $\psi_a(G_\varrho)$ of G_σ.

The open set $p_\varrho^{-1}(Sta)$ of $St\widetilde{\varrho}$ is the union of simplices represented by sequences $a_1, \dots, a_k, \widetilde{a}_{k+1}, \dots, \widetilde{a}_n$, where

$\widetilde{a}_{k+1}, \dots, \widetilde{a}_n$ is a sequence of composable edges in $D\widetilde{\varrho}$ (if $k < n$) with $t(\widetilde{a}_{k+1}) = a$, and

$a_1, \dots, a_k$ is a sequence of composable edges in $E(X)$ such that there is a integer $j \leq k$ with $a_j \dots a_k = a$.

The map f_a will map the n-simplex represented by such a sequence on the simplex of $St\widetilde{\sigma}$ represented by the sequence $a_1, \dots, a_{j-1}, s_a(a_j, \dots, a_k)$, $\widetilde{j}_a(\widetilde{a}_{k+1}), \dots, \widetilde{j}_a(a_n)$, where $s_a(a_j, \dots, a_k)$ denotes the sequence of composable edges of $E(D\widetilde{\sigma})$ representing the image by s_a of the simplex of $St_\varrho a$ represented by the sequence $a_j, \dots, a_k$. For two composable edges a and b of $E(X)$, we have $f_a f_b = g_{a,b} f_{ab}$.

So we have defined on X an orbihedron structure whose associated complex of groups is $G(X)$.

6. Non-positively curved orbihedron.

6.1. Non-positively curved simplicial complex.

Let us first recall some definitions (see Gromov [7], GH [5], Paulin [10], Bridson [3]). A metric space is geodesic if, for any two points x and y of X, there is a curve joining x to y whose length is equal to the distance $d(x,y)$ of x to y. Such a curve is called a geodesic from x to y and will be denoted by $[x,y]$.

Let M^n_χ be a complete simply connected manifold of constant curvature χ; so this is a sphere of dimension n if $\chi > 0$, the n-dimensional euclidean space if $\chi = 0$ and an hyperbolic space of dimension n if $\chi < 0$.

Given three points x, y, z in the geodesic metric space X and a geodesic triangle $\Delta = [x,y] \cup [y,z] \cup [z,x]$, a comparison triangle is a geodesic triangle $\overline{\Delta}$ in M^2_χ with vertices $\overline{x}, \overline{y}, \overline{z}$ and $d(x,y) = d(\overline{x},\overline{y}), d(y,z) = d(\overline{y},\overline{z}), d(z,x) = d(\overline{z},\overline{x})$. Such a comparison triangle always exists when $\chi \leq 0$ or when $\chi > 0$ and the perimeter of Δ is less than $2\pi/\sqrt{\chi}$. For a point p on the side $[y,z]$, the comparison point $\overline{p}$ is the point on the side $[\overline{x},\overline{z}]$ with $d(\overline{y},\overline{p}) = d(y,p)$. The space X satisfies the **Cat(χ)-inequality** if for any geodesic triangle Δ as above and any $p \in [y,z]$, then $d(x,p) \leq d(\overline{x},\overline{p})$. We say that X is **non-positively curved** (resp. **negatively curved**) if each point of X has a neighbourhood which satisfies the $Cat(0)$-inequality (resp. the $Cat(\chi)$-inequality for some $\chi < 0$).

An M_χ-simplicial cell complex is a simplicial cell complex X such that for each n-cell σ there is a continuous map f_σ of a geodesic n-simplex Δ_σ of M^n_χ in X whose restriction to the interior of σ is a homeomorphism on σ, and such that, if τ is a face of σ of dimension k, then the map f_τ of Δ_τ in X associated to τ is the composition of f_σ with an isometry of Δ_τ on a face of Δ_σ.

Following Bridson [3], we denote by Shapes (X) the set of isometry types of the simplices Δ_σ. We shall assume throughout this section that this set is finite. In such a complex, the length of a curve in X joining two points is well defined and the distance $d(x,y)$ between two points x and y of X will be the infimum of the length of curves joining x to y. This is a metric on X (assuming that Shapes (X) is finite, see Bridson [3]). For x in X, we denote by D_x the set of radial geodesics issued from x and contained in simplices whose closure contain x (see Balmann, Chapter 10 in GH [5]). This set is naturally a M_1- simplicial cell complex (in other words it has a natural spherical structure).

6.2. Theorem (Bridson). *Let X be a connected M_χ-simplicial cell complex such that the set of isometry types of the simplices is finite. Then*

a) X is a complete geodesic space,

b) if, for each vertex σ of X, the M_1- simplicial cell complex D_σ satisfies the $Cat(1)$-inequality and if $\chi \leq 0$, then X is non-positively curved and its universal covering satisfies the $Cat(\chi)$-inequality.

This theorem is proved in Bridson [3]. This is an analogue of a theorem stated by Gromov [7]. For a proof in the locally compact case, see Ballman in GH [5], chapter 10.

6.3. Definition. Let X be a M_χ-simplicial cell complex such that the set of isometry types of its cells is finite and let $G(X)$ be a complex of groups on X. Then for the associated orbihedron structure, for each 0-cell σ of X, the link $Lk\tilde{\sigma}$ has a natural M_1-structure. We say that a complex of groups $G(X)$ on X is **non-positively curved (resp. negatively curved)** if $\chi \leq 0$ (resp. $\chi < 0$) and the link $Lk\tilde{\sigma}$ satisfies the $Cat(1)$-inequality for each 0-cell σ of X.

This condition is always satisfied if $dimX = 1$. When $dimX = 2$, it means that for each 0-cell σ of X, in $Lk\tilde{\sigma}$ with its angular metric induced from the M_χ-structure on X, the length of any simple closed curve is $\geq 2\pi$ (see Gromov [7], Ballman [5], Bridson [3]).

6.4. Main theorem. *Let X be a connected M_χ-simplicial cell-complex with $\chi \leq 0$ and such that the set of isometry types of the cells is finite. Let $G(X)$ be a non-positively curved complex of groups on X. Then $G(X)$ is developable : there is a group G acting on a simply connected M_χ-simplicial cell complex $\widetilde{X}$ without inversion such that $G(X)$ is the complex of groups associated to this action. The group G is the fundamental group $\pi_1(G(X),T)$ of $G(X)$ and, for each vertex σ of X, the natural map of G_σ in G is injective.*

The space $\widetilde{X}$ is a complete geodesic space which satisfies the $Cat(\chi)$-inequality (in particular the $Cat(0)$-inequality, hence it is contractible).

This theorem is proved by Gersten-Stallings [14] (see also Bridson [3]) in the particular case of a triangle of groups. The proof given by Haefliger (see [5], chapitre 11) in the case where the groups G_σ are finite and X is finite, can be extended using the techniques of Bridson.

It follows from the last remark of 2.6 that, in the non-positively curved case, the cohomology of the group is the cohomology of the classifying space $BG(X)$.

In this context, the generalization of the theorem 2 in Stallings [14] is the following. Let G be the fundamental group $\pi_1(G(X), \sigma_0)$ of $G(X)$ with respect to a cell σ_0. A subgroup H of G is said to be **bounded** if there is an integer N such that each element of H is represented by a $G(X)$-loop (cf. 3.1) of length bounded by N.

6.5. Theorem. *Let $G(X)$ be a non-positively curved complex of groups on the M_χ-simplicial cell complex X. Then every bounded subgroup of G is conjugated to a subgroup of some G_σ.*

This follows from the fact that the universal covering $\widetilde{X}$ verifies the $Cat(0)$-inequality, that the hypothesis on H means that the H-orbit of $\widetilde{\sigma}_0$ is bounded, hence H has a fixed point, so is conjugated to a subgroup of G of the form G_σ (cf. Brown [2], p. 160-161, or Stallings [14]).

6.6. Examples of non positively curved complex of groups.

6.6.1. Assume that X is obtained by glueing two congruent triangles along their boundary. We have three 0-cells τ, τ', τ'', three 1-cells $\sigma, \sigma', \sigma''$ and two 2-cells ϱ, ϱ'. The edges of the barycentric subdivision of X are labelled as in the figure 7 (it is understood that edges with the same label must be identified).

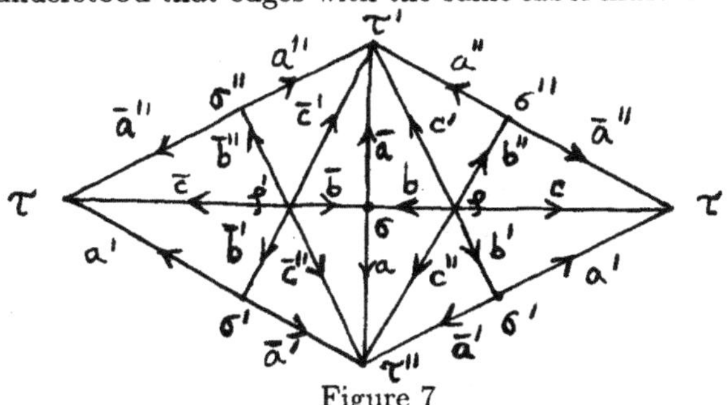

Figure 7

Let $G(X)$ be a complex of groups on X. We can assume that the elements associated to pairs of composable edges are all trivial except possibly the three elements $g_{a',b'} = h^{-1} \in G_\tau, g_{a'',b''} = h'^{-1} \in G_{\tau'}$ and $g_{a,b} = h''^{-1} \in G_{\tau''}$. Let

T be the maximal tree in the 1-skeleton of the barycentric subdivision of X which is the union of the edges $\bar{a}, \bar{a}', \bar{a}'', b, b', b'', \bar{b}$. Then the fundamental group $G = \pi_1(G(X), T)$ has the following presentation. It is generated by the groups $G_\tau, G_{\tau'}$ and $G_{\tau''}$ with the relations in those groups and the relations

$$\begin{aligned}
\psi_{\bar{a}}(g) &= h''\psi_a(g)h''^{-1} && \text{for all} \quad g \in G_\sigma \\
\psi_{\bar{a}'}(g') &= h\psi_{a'}(g')h^{-1} && \text{for all} \quad g' \in G_{\sigma'} \\
\psi_{\bar{a}''}(g'') &= h'\psi_{a''}(g'')h'^{-1} && \text{for all} \quad g'' \in G_{\sigma''} \\
hh'h'' &= 1.
\end{aligned}$$

In a way analogous to Stallings [14], we would like to define an angle α for the triangles of X at the vertex τ. Identify the groups $G_\tau, G_{\bar{\varrho}}, G_{\sigma'}, G_{\sigma''}$ with their images in G_τ by the monomorphisms $\psi_a, \psi_{\bar{a}}, \psi_{a'}, \psi_{a''}$. For the orbihedron associated to X, the link $Lk\tilde{\tau}$ (more precisely its barycentric subdivision) is the directed graph associated to the diagram

$$\begin{array}{ccc}
G_\tau/G_\varrho & \longrightarrow & G_\tau/G_{\sigma''} \\
\downarrow & & \downarrow \\
G_\tau/G_{\sigma'} & \longrightarrow & G_\tau/G_{\bar{\varrho}}
\end{array}$$

where all maps are induced by inclusions, except the left vertical one which maps gG_ϱ on $ghG_{\sigma'}$. Let $2n$ be the even integer which is the minimal number of edges in a simple closed path in $Lk\tilde{\tau}$. Define $\alpha = 2\pi/n$ if $n > 0$ and $\alpha = 0$ otherwise.

We define similarly α' and α'' for the other vertices τ' and τ''. Assume that $\alpha + \alpha' + \alpha'' < \pi$ (resp. $\alpha > 0, \alpha' > 0, \alpha'' > 0$ and $\alpha + \alpha' + \alpha'' = \pi$). Then in that case we can introduce on X a M_χ- structure such that the complex of groups $G(X)$ is negatively curved.

Indeed we can consider that X is obtained by glueing along their boundary two geodesic triangles in the hyperbolic plane (resp. the Euclidean plane) with positive angles bigger or equal to α, α' and α''. Then $G(X)$ is negatively (resp. non positively) curved, because the length of any simple closed curve in the links $Lk\tau, Lk\tau'$ and $Lk\tau''$ is $\geq 2\pi$ with respect to the angular measure.

In case some of the angles are zero and their sum is π, the result is still true, but its proof requires the special argument alluded to in Stallings [14].

A simple particular case is when all the groups are trivial, besides $G_\tau, G_{\tau'}$ and $G_{\tau''}$. Then $\alpha = \pi/n, \alpha' = \pi/n'$ and $\alpha'' = \pi/n''$, where n, n' and n'' are the orders of h, h' and h''. We are in the non-positively curved case when $1/n+1/n'+1/n'' \leq 1$.

6.6.2. Assume that X is obtained by identifying in a triangle two sides so that X has two 0-cells τ and $\overline{\tau}$, two 1-cells σ and $\overline{\sigma}$ and one 2-cell ϱ. The ten edges are labelled as in figure 8.

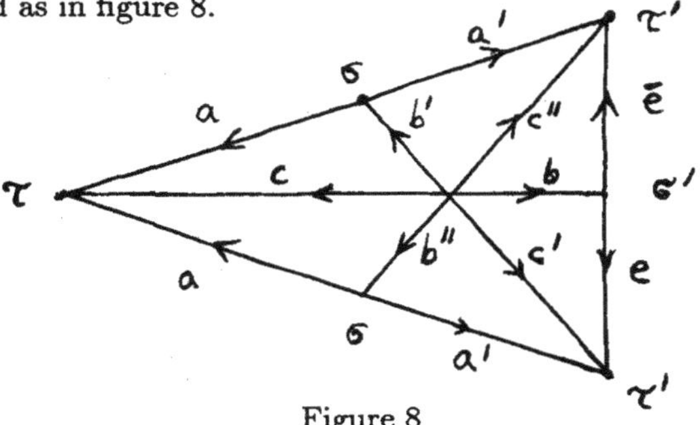

Figure 8

Let $G(X)$ be a graph of groups on X; we can assume that only $g_{a,b''} = h^{-1} \in G_\sigma$ might be non trivial. The fundamental group of $G(X)$ is the group generated by the elements of the groups G_τ and $G_{\tau'}$, submitted to the relations in these groups and the relations

$$\psi_a(g) = \psi_{a'}(g) \qquad \text{for all} \quad g \in G_\sigma$$
$$\psi_{\overline{e}}(g') = h\psi_e(g')h^{-1} \qquad \text{for all} \quad g' \in G_{\sigma'}.$$

If we identify G_ϱ and G_σ with the subgroups $\psi_c(G_\varrho)$ and $\psi_a(G_\sigma)$ of G_τ, according to 5.6, $Lk\widetilde{\tau}$ is the directed graph associated to the diagram of maps

where one of the maps is induced by the inclusion of G_ϱ in G_σ and the other one maps gG_ϱ on ghG_σ. Let $2n$ be the smallest number of edges of a simple closed edge path in this graph, and let α be $2\pi/n$ if $n > 0$ and zero otherwise.

Identify now $G_\varrho, G_\sigma, G_{\sigma'}$ and $\overline{G}_{\sigma'}$ to their images in $G_{\tau'}$ under the injections $\psi_{c'} = \psi_{c''}, \psi_{a'}, \psi_e$ and $\psi_{\overline{e}}$. Then $Lk\widetilde{\tau}'$ is the directed graph associated to the diagram

$$G_{\tau'}/\overline{G}_{\sigma'} \longleftarrow G_{\tau'}/G_\varrho \longrightarrow G_{\tau'}/G_\sigma \longleftarrow G_{\tau'}/G_\varrho \longrightarrow G_{\tau'}/G_{\sigma'}$$

where all the maps are induced by inclusions. Let $2m$ be the smallest number of edges of a simple closed edge path in this graph, and let β be π/m if $m > 0$ and zero otherwise.

If $\alpha + 2\beta < \pi$ (resp. if α and β are positive and $\alpha + 2\beta = \pi$), then one can introduce on X a M_χ-structure such that $G(X)$ is negatively curved (resp. non positively curved), hence developable.

Indeed, X can be obtained by identifying two sides of an hyperbolic triangle or Euclidean triangle with angles α, β and β, if α and β are both positive (or positive angles α', β' and β' with $\alpha' > \alpha, \beta' > \beta$). In the links $Lk\widetilde{\tau}$ and $Lk\widetilde{\tau}'$, the angular length of any simple closed curve is $\geq 2\pi$, hence the $Cat(1)$-condition is satisfied, and we can apply the main theorem.

6.6.3. The dunce hat. We keep the notations of 1.5. Let us identify G_σ and G_ϱ with their images by ψ_a and ψ_c in G_τ. We also define the subgroups $G'_\sigma = \psi_{a'}(G_\sigma), G'_\varrho = \psi_{c'}(G_\varrho), G''_\varrho = \psi_{c''}(G_\varrho)$ of G_τ. Note that G''_ϱ is a subgroup of G_σ and G'_σ. Let $h = g^{-1}_{a,b''}$ and $h' = g^{-1}_{a',b''}$. Note that G_ϱ and $h^{-1}G_\varrho h$ are subgroups of G_σ and that G'_ϱ and $h'^{-1}G'_\varrho h'$ are subgroups of G'_σ.

Then $Lk\widetilde{\tau}$ is the directed graph associated to the diagram

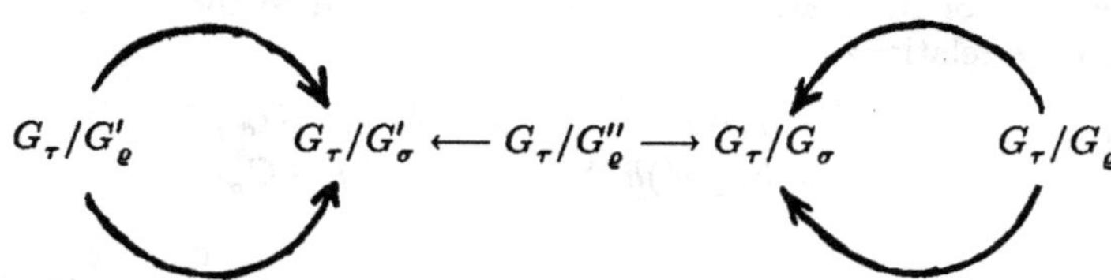

where all the maps are induced by inclusions except the upper left one which maps gG'_ϱ on $gh'G'_\sigma$ and the upper right one which maps gG_ϱ on ghG_σ.

Let $2n$ be the minimal number of edges in a simple closed edge path in this graph and let $\alpha = 2\pi/n$. If $\alpha \leq \pi/3$, then one can define on X a M_χ-structure such that the complex of groups $G(X)$ is non-positively curved, hence it is developable. Indeed we can build X by identifying the sides of an equilateral triangle in hyperbolic space or Euclidean space with angles α, and we can apply the main theorem.

6.6.4. An example of Gromov. In [6], 4 C", p. 389, Gromov claims that, for fixed integers $k \geq 6$ and $\ell \geq 2$, there exists a unique simply connected 2-dimensional polyhedron $\widetilde{X}$ such that

a) each 2-cell is a plane regular k-gon, the intersection of two k-gons (if non empty) being a common edge or a vertex,

b) every edge in $\widetilde{X}$ has ℓ adjacent k-gons,

c) every vertex in $\widetilde{X}$ has $\ell + 1$ adjacent edges (and hence $\ell(\ell+1)/2$ adjacent k-gons).

Understanding this claim was for us the motivation to study the concept of orbihedron. In fact for each $k \geq 3$ and $\ell \geq 2$, one can construct an orbihedron structure over a triangle X (considered as the quotient of a k-gon by the dihedral group D_k) with vertices τ, τ' and τ'' such that

$\widetilde{a}$) $St\widetilde{\tau}$ is the interior of a regular k-gon (with its barycentric subdivision),

$\widetilde{b}$) $St\widetilde{\tau}'$ is the union of ℓ triangles along a common edge,

$\widetilde{c}$) $St\widetilde{\tau}''$ is the cone over the barycentric subdivision of the 1-skeleton of a ℓ-simplex.

More precisely, the orbihedron is associated to the following triangle of groups. Let $Aut(0,1,\ldots,\ell)$ be the permutation group of the $\ell+1$ elements $0,1,\ldots,\ell$. Let D_n be the dihedral group of order $2n$. The triangle of groups is

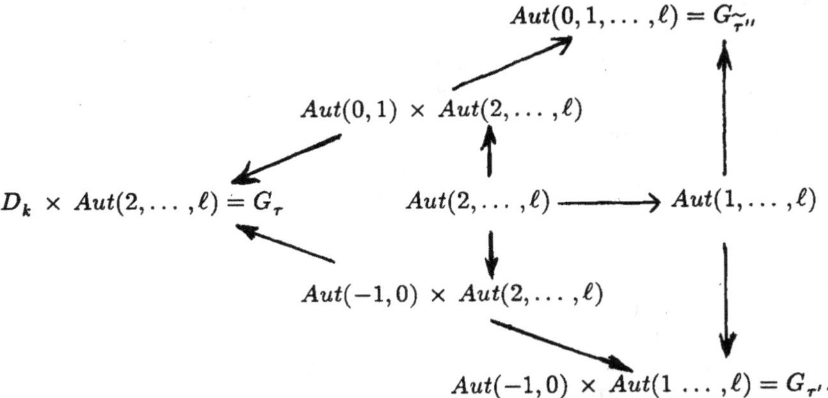

The inclusions are the obvious one, the generators of $Aut(0,1)$ and $Aut(-1,0)$ being mapped on the two generators of D_k.

One can consider X as a Euclidean triangle (resp. a hyperbolic triangle) with angles π/k at the vertex τ, and $\pi/2$ at τ'. If $k \geq 6$ (resp. $k \geq 7$), then this orbihedron is non positively curved (resp. negatively curved), hence developable. This shows the existence of $\widetilde{X}$ claimed by Gromov.

For $k \geq 3$ and $\ell = 2$, the orbihedron is also developable; its universal covering $\widetilde{X}$ is a tiling of the 2-sphere for $k \leq 5$, of the Euclidean plane for $k = 6$ and of the hyperboic plane for $k \geq 7$. It is also developable for $k = 3$ and $\ell \geq 2$ (in that case $\widetilde{X}$ is the 2-skeleton of the $(\ell+1)$-simplex) and for $k = 4$ and $\ell \geq 2$ (in that case $\widetilde{X}$ is the 2-skeleton of the cube of dimension $\ell+1$).

I have been informed by Alexis Marin that a student of his, Frederic Haglund, has proved that this triangle of groups is developable for all $k \geq 3$, $\ell \geq 2$.

Appendix. Covering theory for topological groupoids.

1. Topological groupoids étale on their space of units.

By definition, a groupoid Γ acting on a set U is a small category whose morphisms are all invertible and with set of objects $\Gamma^{(0)} = U$ (also called the set of units). We denote by i (resp. t) the map of Γ on U associating to $g \in \Gamma$ its left (resp. right) unit. We denote by $\Gamma \times_U \Gamma$ (or $\Gamma^{(2)}$) the set of pairs of composable elements of Γ, namely pairs (γ, γ') with $i(\gamma) = t(\gamma')$, and by $\gamma\gamma'$ the composition of γ and γ'.

Assume that Γ has a topology such that composition and taking inverses give continuous maps of $\Gamma \times_U \Gamma$ on Γ and of Γ on Γ (on $\Gamma \times_U \Gamma$, the topology is induced by the product topology). On the space of units U we consider the induced topology; then the maps i and t are continuous. With those assumptions we say that Γ is a topological groupoid acting on the space U. In what follows we shall assume that the maps i and t are local homeomorphisms (étale maps).

The space of orbits $\Gamma \backslash U$ is the quotient of U by the equivalence relation identifying x and y if there is an element γ of Γ with $i(\gamma) = x$ and $t(\gamma) = y$. For $x \in U$, we denote by Γ_x the group formed by the elements γ of Γ such that $i(\gamma) = t(\gamma) = x$.

2. Example. Let U be a topological space and G be a group acting by homeomorphisms on U (in general not effectively). The associated topological groupoid Γ is the product $G \times U$, where G has the discrete topology, i and t mapping (g, x) on x and gx respectively. The composition $(g, x)(g', x')$ is defined whenever $x = gx'$ and is equal to (gg', x).

3. Homomorphisms and equivalences. If Γ and Γ' are topological groupoids acting on U and U', a homomorphism Φ of Γ in Γ' is a continuous map such that $\Phi(\gamma\gamma') = \Phi(\gamma)\Phi(\gamma')$ and $\Phi(\gamma^{-1}) = \Phi(\gamma)^{-1}$. Assume that Γ and Γ' are étale on their space of units, that Φ is an étale map inducing a bijection of $\Gamma \backslash U$ on $\Gamma' \backslash U'$ and giving a bijection of Γ_x on $\Gamma'_{\Phi(x)}$ for each $x \in U$. Then we say that Φ is an **equivalence** of Γ on Γ', and that Γ is equivalent to Γ'. This generates an equivalence relation among groupoids étale on their space of units (for more details, see Haefliger [8]).

The condition in 5.2, 2) means that the map $(g, x) \longmapsto (\psi_a(g), f_a(x))$ of $G_{i(a)} \times p_{i(a)}^{-1}(Sta)$ on $G_{t(a)} \times p_{t(a)}^{-1}(Sta)$ is an equivalence.

Assume that $\Gamma = G \times U$ is the topological groupoid associated to an action of Γ on the topological space U (as in example 2). Assume that U is connected, locally connected and locally simply connected so that it has a universal covering $p : \widetilde{U} \to U$, where the points of $\widetilde{U}$ are homotopy classes $[c]$ relative to their extremities of continuous paths $c : [0,1] \to U$ with origin a base point x_0. Then we can construct a group $\widetilde{G}$ whose elements are the pairs $([c], g)$, where c is a path with $c(0) = x_0$ and $c(1) = gx_0$. The composition is defined by $([c],g)([c'],g') = ([c.gc'],gg')$ and $([c],g)^{-1} = ([g^{-1}c^{-1}],g^{-1})$. Obviously we have an exact sequence

$$1 \to \pi_1(U, x_0) \to \widetilde{G} \to G \to 1 .$$

$\widetilde{G}$ acts on $\widetilde{U}$ by $([c],g)[c'] = [c.gc']$. Let $\widetilde{\Gamma} = \widetilde{G} \times \widetilde{U}$ be the associated groupoid. The map Φ of $\widetilde{\Gamma}$ on Γ associating to $([c],g),[c'])$ the couple $(g, c'(1))$ is a homomorphism which is an equivalence.

If a complex of groups $G(X)$ is associated to an action of a group G on a simplicial cell complex $\widetilde{X}$, then the groupoid associated to the corresponding orbihedron as in 5.4 is equivalent to the groupoid $G \times \widetilde{X}$.

4. Γ-paths and fundamental group. Let Γ be a topological groupoid étale on its space of units U.

A Γ-path $c = (\gamma_0, c_1, \gamma_1, c_2, \gamma_2, \dots, c_n, \gamma_n)$ is a sequence, where $c_k : [t_{k-1}, t_k] \to U$ is a continuous path, γ_k an element of Γ with $t(\gamma_k) = c_k(t_k)$ for $k > 0, i(\gamma_k) = c_{k+1}(t_k)$ for $k < n$ and $0 = t_0 \leq t_1 \leq \dots \leq t_n = 1$. The origin $i(c)$ of c is $t(\gamma_0)$ and its extremity $t(c)$ is $i(\gamma_n)$. A Γ-loop based at x_0 is a Γ-path with origin and extremity x_0 (compare with 3.1a)).

Consider the following two operations on c ;

i) add a new subdivision point $s \in [t_{k-1}, t_k]$ and in c replace c_k by the sequence c'_k, γ', c''_k, where c'_k and c''_k are the restrictions of c_k to the intervals $[t_{k-1}, s]$ and $[s, t_k]$ resp. and γ' is the unit $c_k(s)$;

ii) consider a continuous map f of $[t_{k-1}, t_k]$ in Γ such that its composition with i is c_k, and replace c_k by the composition tf, γ_{k-1} by $\gamma_{k-1}f(t_{k-1})^{-1}$ and γ_k by $f(t_k)\gamma_k$.

Those two operations generate an equivalence relation among Γ-paths with fixed origin and extremity.

We consider now a third operation on c. Let s be a parameter in $[0,1]$ and let c^s_k be a continuous family of paths in X with $c^0_k = c_k$, and let γ^s_k be a path in Γ with $\gamma^0_k = \gamma_k$ so that the sequence $c^s = (\gamma^s_0, c^s_1, \dots, c^s_n, \gamma^s_n)$ is a Γ-path with the same origin and extremity as c. The third operation is

iii) replace c by the Γ-path c^1.

We say that two Γ-paths c and c' are homotopic if they are equivalent under the equivalence relation generated by i), ii) and iii).

If $c = (\gamma_0, c_1, \dots, c_n, \gamma_n)$ and $c' = (\gamma'_0, c'_1, \dots, c'_n, \gamma'_n)$ are two Γ-paths such that $t(c) = i(c')$, then their composition $c.c'$ is the Γ-path

$$(\gamma_0, \widetilde{c}_1, \dots \widetilde{c}_n, \gamma_n\gamma'_0, \dots, \widetilde{c}'_n, \gamma'_n)$$

, where $\widetilde{c}_k$ is the composition of c_k with the map $t \to t/2$ and $\widetilde{c}'_k$ the composition of c'_k with the map $t \to t/2 + 1/2$.

The homotopy classes of Γ-loops based at x_0 form a group $\pi_1(\Gamma, x_0)$ called the **fundamental group** of Γ based at x_0.

If X is arcwise Γ-connected, i.e. if any two points x_0 and x_1 are the origin and extremity of a Γ- path, then the fundamental groups of Γ based at two points are isomorphic. It is immediate to check that the fundamental group of Γ as defined above is isomorphic to the fundamental group of the classifying space $B\Gamma$ (cf. [8]).

As an example, if Γ is the groupoid associated to an action of G on a connected space U like in 2, then $\pi_1(\Gamma, x_0)$ is the group $\widetilde{G}$ constructed in 3.

5. Covering space theory. We assume that Γ is a topological groupoid, étale on its space of units U.

Let p be a continuous map of a topological space $\widetilde{U}$ in U. Let $\widetilde{\Gamma}$ be the subspace $\Gamma \times_U \widetilde{U}$ of $\Gamma \times \widetilde{U}$ formed by the pairs $(\gamma, \widetilde{x})$ with $i(\gamma) = p(\widetilde{x})$. An action of Γ on $\widetilde{U}$ over p is a continuous map of $\widetilde{\Gamma}$ on $\widetilde{U}$, associating to $(\gamma, \widetilde{x})$ an element of $\widetilde{U}$ denoted by $\gamma\widetilde{x}$, such that

$$p(\gamma\widetilde{x}) = t(\gamma), \quad \gamma'(\gamma\widetilde{x}) = (\gamma'\gamma)\widetilde{x} \quad \text{and} \quad \gamma\widetilde{x} = \widetilde{x} \quad \text{if } \gamma \text{ is the unit } p(\widetilde{x}).$$

Then $\widetilde{\Gamma}$ is itself a topological groupoid with space of units $\widetilde{U}$, the projections i and t being defined by $i(\gamma, \widetilde{x}) = \widetilde{x}$, $t(\gamma, \widetilde{x}) = \gamma\widetilde{x}$ and the composition $(\gamma', \widetilde{x}')(\gamma, \widetilde{x})$ by $(\gamma'\gamma, \widetilde{x})$. The projection P of $\widetilde{\Gamma}$ in Γ mapping $(\gamma, \widetilde{x})$ on γ is a homomorphism of topological groupoids.

We say that $\widetilde{\Gamma}$ is a covering of Γ with projection P if moreover $\widetilde{U}$ is a covering of U with projection p.

If $P : \widetilde{\Gamma} \to \Gamma$ and $P' : \widetilde{\Gamma}' \to \Gamma$ are two coverings of Γ, a covering homomorphism of P in P' is a homomorphism of $\widetilde{\Gamma}$ in $\widetilde{\Gamma}'$ commuting with the projections on Γ. It induces a covering map of $\widetilde{U}$ in $\widetilde{U}'$.

Let $\widetilde{x}_0$ be a base point in $\widetilde{U}$ projecting by p on x_0. Assume that p is a covering map. Then any Γ-path $c = (\gamma_0, c_1, \dots, c_n, \gamma_n)$ in U with origin x_0 lifts uniquely as a $\widetilde{\Gamma}$-path $\widetilde{c} = (\widetilde{\gamma}_0, \widetilde{c}_1, \dots, \widetilde{c}_n, \widetilde{\gamma}_n)$ with origin $\widetilde{x}_0$, where $\widetilde{\gamma}_k$ and $\widetilde{c}_k$ are defined by induction as follows. First $\widetilde{\gamma}_0 = (\gamma_0, \gamma_0^{-1}\widetilde{x}_0)$; if $\widetilde{\gamma}_{k-1}$ has been defined, then $\widetilde{c}_k$ is the unique lifting of c_k in $\widetilde{X}$ with origin $i(\widetilde{\gamma}_{k-1})$ and $\widetilde{\gamma}_k = (\gamma_k, \gamma_k^{-1}(\widetilde{c}_k(t_k)))$. If c'

is homotopic to c, then the lifting $\widetilde{c'}$ of c' with origin $\widetilde{x}_0$ has the same extremity as $\widetilde{c}$.

If $F = p^{-1}(x_0)$, then we get an action of $\pi_1(\Gamma, x_0)$ on F : given an element represented by a Γ- loop c at x_0, its action on $\widetilde{x} \in F$ is the extremity of the lifting of c^{-1} with origin $\widetilde{x}$.

Assume now that U in Γ-connected, locally arcwise connected and locally simply connected. Then the preceding correspondance gives a bijection between isomorphisms classes of coverings of Γ and isomorphism classes of actions of $\pi_1(\Gamma, x_0)$ on sets.

To see this, we first construct the **universal covering $\widetilde{\Gamma}$ of Γ** as follows. Let $\widetilde{U}$ be the set of homotopy classes of Γ-paths in U with origin x_0, the projection p of $\widetilde{U}$ on U associating to the homotopy class $[c]$ of c its extremity. The group $\pi_1(\Gamma, x_0)$ acts on $\widetilde{U}$ by composition of loops with paths. A fundamental system of neighbourhoods of $[c]$ is obtained by taking the homotopy classes of the composition of c with paths contained in a simply-connected neighbourhood of the extremity of c. Then $p : \widetilde{U} \to U$ is a Galois covering of U with Galois group $\pi_1(\Gamma, x_0)$. The universal covering $\widetilde{\Gamma} = \Gamma \times_U \widetilde{U}$ of Γ is obtained by considering the action of Γ on $\widetilde{U}$ given by $\gamma[c] = [c\gamma^{-1}]$, where $c\gamma^{-1}$ denotes the composition of c with the Γ-path $(\gamma'_0, c'_1, \gamma'_1)$ where $\gamma'_0 = \gamma^{-1}, c'_1$ the constant path at $t(\gamma)$ and γ'_1 the unit $t(\gamma)$. It is easy to check that $\widetilde{U}$ is $\widetilde{\Gamma}$-connected and that the fundamental group of Γ is trivial.

An action of $\pi_1(\Gamma, x_0)$ through automorphisms of $\widetilde{\Gamma}$ is given as follows : the action of the element $[c'] \in \pi_1(\Gamma, x_0)$ represented by the Γ-loop c' on $(\gamma, [c])$ is $(\gamma, [c'c])$.

Given an action of $\pi_1(\Gamma, x_0)$ on a set F, we construct the corresponding covering $\overline{\Gamma}$ of Γ. The covering $\overline{X}$ of X is the covering $\widetilde{U} \times_G F$ of U with fiber F associated to $\widetilde{U}$ by the action of $G = \pi_1(\Gamma, x_0)$ on F; the action of the element γ of Γ on the class of $([c], y)$ is the class of $([c\gamma^{-1}], \gamma y)$.

As an example, let $\Gamma = G \times U$ be the topological groupoid associated to an action of a group G on a topological space U. Assume that U is simply connected. The universal covering $\widetilde{\Gamma}$ of Γ given by the preceding construction is the groupoid $G \times (G \times U)$ associated to the action of G on $G \times U$ given by $h(g, u) = (gh^{-1}, h.u)$. But the inclusion $u \mapsto (1, 1, u)$ of U in $\widetilde{\Gamma}$ is an equivalence of U, considered as the groupoid associated to the action of the trivial group on U, with $\widetilde{\Gamma}$. We see in this example that the general construction above does not lead in general to the simplest representative of the universal covering in its equivalence class.

6. Orbispaces.

An orbispace structure on a topological space X is given by

i) an open covering $\{V_i\}_{i\in I}$ of X

ii) for each $i \in I$, a group G_i acting by homeomorphisms on a topological space U_i and a continuous map p_i of U_i on V_i inducing a homeomorphism of $G\backslash U_i$ on V_i,

iii) a topological groupoid Γ whose space of units is the disjoint union U of the U_i's and whose restriction Γ_i to each U_i is the groupoid $G_i \times U_i$ associated to the action of G_i on U_i.

Two such datas define the same orbispace structure on X if the corresponding topological groupoids are equivalent over the identity of X.

The orbispace structure on X is called developable if the groupoid is equivalent to the groupoid associated to an action of a group G on a topological space $\widetilde{X}$.

It is called rigid if the action of G_i on U_i is rigid for each i. This means that an element of G_i whose restriction to an open set of U_i is the identity must be the identity element of G_i.

An orbihedron is a particular case of an orbispace.

Let us assume that the V_i's are locally connected and simply connected. Let V_{ij}^a be a connected component of $V_i \cap V_j$ such that $U_{ij}^a = p_j^{-1}(V_{ij}^a)$ is connected and simply connected. Then the condition iii) implies that there is an injective homomorphism ψ_a of G_j in G_i and a ψ_a-equivariant homeomorphism f_a of V_{ij}^a on a connected component of $p_i^{-1}(V_{ij})$ projecting on the identity of V_{ij}^a . Moreover $\psi_a(G_j)$ is the subgroup of G_j leaving $f_a(V_{ij}^a)$ invariant (compare with 5.2.,2)).

Let us explicit the general construction of the universal covering in the particular case of orbispaces. We assume that each U_i is arcwise connected and simply connected locally and globally.

Choose a base point x_0 in U. Let G be the fundamental group $\pi_1(\Gamma, x_0)$. For each $i \in I$, choose a Γ-path c_i from x_0 to a point x_i in U_i and for each $x \in U_i$, let c_x be the composition of c_i with a path in U_i from x_i from x_i to x. (For an orbihedron structure on a complex X as described in 5.2, such a choice can be determined using a maximal tree T and the maps f_a).

The space of homotopy classes of Γ- paths from x_0 to the points of U_i is naturally identified to $G \times U_i$, the point (g, x) corresponding to the homotopy class of the composition of a Γ- loop representing g with the Γ-path c_x.

We have a continuous homomorphism λ of Γ on G (with the discrete topology) mapping $\gamma \in \Gamma$ on the homotopy class of the Γ-loop $(c_{t(\gamma)}, \gamma^{-1}, c_{i(\gamma)}^{-1})$. Its restriction to $G_i \times U_i$ gives a homomorphism λ_i of G_i in G with image $\bar{G}_i = \lambda_i(G_i)$.

The groupoid $\widetilde{\Gamma}$ universal covering of Γ, given by the general construction above ,is the product $\widetilde{\Gamma} = G \times \Gamma$. Its space of units is the disjoint union $\widetilde{U}$ of the $G \times U_i$'s. The projections i and t are defined by $i(g,\gamma) = (g, i(\gamma))$ and $t(g,\gamma) = g\lambda(\gamma^{-1})$. The composition $(g,\gamma)(g',\gamma')$ is defined if $i(\gamma) = t(\gamma')$ and $g = g'\lambda(\gamma'^{-1})$, and is then equal to $(g, \gamma'\gamma)$. The restriction of $\widetilde{\Gamma}$ to $G \times U_i$ is the groupoid associated to the action of G_i on GXU_i defined by $h(g,x) = (g\lambda_i(h)^{-1}, hx)$. An action of G through isomorphisms of $\widetilde{\Gamma}$ is given by left translations on the first factor of $\widetilde{\Gamma} = G \times \Gamma$.

The natural projection of $\widetilde{U}$ on U gives a continuous map p of $\widetilde{\Gamma}\backslash\widetilde{U} = \widetilde{X}$ on $\Gamma\backslash U = X$. The inverse image $p^{-1}(V_i)$ is homeomorphic to the product $G/\overline{G}_i \times \widetilde{G}_i\backslash U_i$ where $\widetilde{G}_i$ is the kernel of $\lambda_i : G_i \to G$.

Let $\widetilde{I}$ be the set of pairs (gG_i, i), where $i \in I$ and $gG_i \in G/\overline{G}_i$. For each $\tilde{\imath} = (gG_i, i)$, choose a representative $g_{\tilde{\imath}}$ in the class gG_i and let $U_{\tilde{\imath}} = \{g_{\tilde{\imath}}\} \times U_i$ (compare with 4.4). The restriction $\widetilde{\Gamma}'$ of $\widetilde{\Gamma}$ to the disjoint union of the $U_{\tilde{\imath}}$'s is equivalent to $\widetilde{\Gamma}$. In case Γ is the groupoid associated to an orbihedron corresponding to the complex of groups $G(X)$, then $\widetilde{\Gamma}'$ is the groupoid associated to the orbihedron corresponding to the universal covering $G(\widetilde{X})$ of $G(X)$ as described in 4.4.

References.

[1.] K.S. Brown, Presentations for groups acting on simply connected complexes, *J. Pure App. Algebra* **32**, (1984), p. 1-10.

[2.] K.S. Brown, *Buildings*, Springer Verlag, (1989).

[3.] M. Bridson, *Geodesics and Curvature in Metric Simplicial Complexes*, this volume.

[4.] J.M. Corson,*Thesis in combinatorial group theory*, The University of Michigan, (1990).

[5.] E. Ghys et P. de la Harpe, *Sur les groups hyperboliques d'après Michael Gromov*, Birkhauser, (1990).

[6.] M. Gromov, *Infinite groups as geometrical objects*, Proc. ICM Warzava, **Vol. 1**, (1984), p. 385-391.

[7.] M. Gromov, *Hyperbolic groups, Essays in group theory*, S. Gersten, editor, **Vol. 8**, Springer Verlag, MSRI series, (1987).

[8.] A. Haefliger, *Groupoides d'holonomie et classifiants, Structure transverse des feuilletages, Toulouse 1982*, Astérisque **No 116**, (1984), p. 70-97.

[9.] J. Milnor, *The geometric realization of a semi-simplicial complex*, Ann. of Math., **65**, (1957), p. 357-362.

[10.] F. Paulin, *Constructions of hyperbolic groups via hyperbolization of polyhedra*, this volume.

[11.] G.P. Scott and C.T.C. Wall, *Topological Methods in Group Theory, Homological Group Theory, LMS Lect.Notes* **36**, Cambridge Univ. Press, (1979), p. 137-203.

[12.] G. Segal, *Classifying spaces and spectral sequences*, Publications Mathématiques IHES, **No 134, 1968**, p. 105-112.

[13.] J.-P. Serre, *Trees*, Springer Verlag, Berlin, 1980. Translation of "Arbres, Amalgames, $S1_2$", Astérisque **46**, 1977.

[14.] J.R. Stallings, *Non-positively curved triangles of groups*, this volume.

IV. ACTIONS ON TREES

DENDROLOGY AND ITS APPLICATIONS

PETER B. SHALEN*

University of Illinois at Chicago

PREFACE

The organizers of the ICTP workshop on geometric group theory kindly invited me to give a series of lectures about group actions on real trees in the spirit of my earlier expository article [Sh]. In the five hours that were allotted, I was able to cover a number of aspects of the subject that I had not covered in [Sh], and some developments that had taken place since [Sh] was written. In the present article I have extended the scope much further still, and have treated in some depth a number of topics that were barely mentioned in my lectures. I have tried to show what a broad and rich subject this is and how much other mathematics it interacts with.

Section 1 includes Gupta and Sidki's construction of finitely generated infinite p-groups using simplicial trees, and Brown's interpretation of the Bieri-Neumann invariant in terms of real trees. In Section 2 I describe Bestvina and Handel's work on outer automorphisms of free groups and its connection with Culler-Vogtmann and Gersten's outer space and with exotic free actions of free groups on trees; one example is worked out in considerable detail. Section 2 also includes a sketch of Skora's proof that every small action of a surface groups on a real tree is dual to a measured foliation.

In Section 3 I give an account of the Bruhat-Tits tree for SL_2 of a valued field, and its application by Lubotzky, Phillips and Sarnak in their work on Ramanujan graphs. I also discuss in some depth my work with Culler on trees associated to ideal points of curves in the character variety of a group and their applications in 3-manifold theory, including the proof of the Cyclic Surgery Theorem by Culler, Gordon, Luecke and myself. In Section 4 I explain the connection between trees and hyperbolic geometry in much greater depth than in [Sh], and include accounts of both the original approach used in my work with Morgan and based on the Bruhat-Tits tree, and the approach of Bestvina and Paulin based on Gromovian notions of convergence of metric spaces. There is also a

*Partially supported by an NSF grant.

brief discussion of Paulin's work on finiteness of outer automorphism groups. The section concludes with some idle speculations.

In Section 5 the emphasis is on the aspects of my work with Gillet on rank-2 trees that were left out of [Sh], especially the notion of strong convergence. This section includes a brand-new, and therefore somewhat tentative, conjecture about how to extend our theory to arbitrary rank. I also discuss some surprising connections between the notion of strong convergence on the one hand, and both Bestvina-Handel theory and the contractibility of outer space on the other.

In order to make this article self-contained, or at least coherent, it has been necessary to allow some slight overlap with [Sh]. I have tried to minimize it by taking a different point of view from that of [Sh] wherever possible. In the few cases where I needed to repeat something that I had said in [Sh], I have tried to be very brief.

I have tried to maintain the same informal style as in [Sh], which is intended to approximate the tone of a lecture rather than a journal article. But in print one is at a disadvantage in that one cannot use one's hands—for example, to point to the board where some important theorem had been written before it was erased. To compensate for this, I have divided the article into lots of subsections and included lots of cross-references, which I hope will keep non-expert readers from getting lost. If you find cross-references irksome, I can only ask you to ignore them.

In writing this article I have had to come to grips with a number of aspects of the subject of which I was only dimly conscious before. I am indebted to Marc Culler for the many hours that he has spent listening to my thoughts on the material and helping me unscramble them.

The tentative conjecture stated in 5.5.6 was formulated with the help of Marc Culler, Henri Gillet and Richard Skora. However, I will take the responsibility if it falls flat.

I am grateful to Mladen Bestvina for working out and lucidly explaining the example given in 2.2.

I don't usually enjoy conferences much, and I confess I didn't look forward to stepping off a trans-Atlantic flight and giving five lectures. As it turned out, ICTP is such a delightful place, and the organizers—Alberto Verjovsky, André Haefliger and Etienne Ghys—gave me such a nice welcome, that it was better than being on vacation.

* * * * * *

The preceding remarks, and the rest of this article, were completed in October of 1990. In December I visited the Hebrew University and had the pleasure of meeting Eliahu Rips. Rips informed me that he has now proved Conjecture

2.5.1 below, characterizing the finitely generated groups that act freely on $\mathbf{R}$-trees. From the outline of his proof that he gave me, it appears that there are fascinating parallels with the methods of [GiS1] and with those of [M2] and [MSk]. Furthermore, his methods seem remarkably well-adapted to some of the other conjectures that I discuss in Sections 2 and 5, and it seems likely to me that they can be used to prove some version of the tentative conjecture of 5.5.6.

SECTION 1. GENERALITIES (AND DIGRESSIONS)

1.1. Simplicial trees. A *simplicial tree* is a connected, simply connected simplicial 1-complex T. Combinatorially, simple connectivity means that T contains no circuits, or equivalently that every edge separates T into two pieces.

1.1.1. A substantial part of classical combinatorial group theory can be interpreted as the study of (simplicial) actions of groups on simplicial trees. For example, a group is free if and only if it acts on a simplicial tree in such a way that no non-trivial element of the group leaves any simplex invariant. An immediate consequence of this fact is the Nielsen-Schreier theorem, which asserts that any subgroup of a free group is free.

1.1.2. It is usually better to think of a simplicial tree not as a set of simplices, but as a set of vertices with the structure given, say, by the adjacency relation. So when one says that a group acts *freely* on a simplicial tree one means that no non-trivial element fixes any vertex. For example, the tree with two vertices and one edge admits a free $\mathbf{Z}/2\mathbf{Z}$-action. Following Serre [Se], one says that a group acts *without inversions* if no element of the group leaves an edge invariant but interchange its endpoints. We can reformulate the assertion of 1.1.1 by saying that a group is free if and only if it acts freely and without inversions on some simplicial tree.

1.1.3. The assumption that a group acts without inversions is not a serious restriction, because any (simplicial) action of Γ on T induces an action without inversions on the first barycentric subdivision of T.

1.1.4. It is easy to show that every free group acts freely and without inversions on some simplicial tree: one need only take the tree to be the Cayley graph with respect to some free generating set. Topologists often prove the converse by saying that if Γ acts freely and without inversions on T then T/Γ is a graph with universal covering space T and deck transformation group Γ, so that Γ is isomorphic to the fundamental group of a graph and is therefore free. There is a good deal of machinery hidden in this argument, and it is instructive to give a direct combinatorial proof.

This can be done by constructing a fundamental domain for the action, i.e. a subtree K of the first barycentric subdivision of T whose endpoints are midpoints of edges of T, and such that (i) $T = \bigcup_{\gamma\in\Gamma} \gamma \cdot K$ and (ii) int $K \cap \gamma \cdot K = \emptyset$ for every $\gamma \neq 1$. Now for each endpoint x of K there is a unique endpoint $x' \neq x$ of K that lies in the same Γ-orbit as x, and $\tau\colon x \mapsto x'$ is an involution of the set of endpoints of K. Let S be a complete system of orbit representatives for τ, and for each $s \in S$ let x_s denote the unique element of Γ that maps s to $\tau(s)$. Then it is straightforward to show that $(x_s)_{s\in S}$ is a system of free generators for Γ. For example, to show that a non-trivial reduced word $\prod_{i=1}^{n} x_{s_i}^{\epsilon_i}$ (where $\epsilon = \pm 1$) cannot represent the identity, we consider the sets $K_m = \prod_{i=1}^{m} x_{s_i}^{\epsilon_i}(K)$ for $m = 0, \ldots n$. It follows from the definitions that whenever $1 \leq m < n$, the sets K_{m-1} and K_{m+1} meet K_m in distinct endpoints of K_m, and hence lie in distinct components of $T -$ int K_m. It follows easily that $K_0, \ldots K_n$ are all distinct. In particular $K_n \neq K_0$, and the assertion follows.

1.1.5. In any event, a free actions without inversions of a group Γ on a simplicial tree T determines a quotient graph T/Γ, and the classification of such actions up to equivariant simplicial isomorphism is equivalent to the classification of connected graphs. This picture was generalized by Bass and Serre in [Se]. They showed that the classification of arbitrary actions without inversions on simplicial trees is equivalent to the classification of what are called graphs of groups.

To define a *graph of groups* one must specify (i) a graph $\mathcal{G}$; (ii) for each cell (vertex or edge) c of $\mathcal{G}$, a group A_c; and (iii) for each oriented edge[1] e of $\mathcal{G}$ with terminal vertex v, an injective homomorphism $J_e\colon A_e \to A_v$, defined modulo inner automorphisms of A_v. An action without inversions of a group Γ on a simplicial tree T defines a graph of groups in a natural way. We set $\mathcal{G} = T/\Gamma$. A cell e of $\mathcal{G}$ is an orbit of simplices of T, and we define A_e to be the stabilizer of a simplex in this orbit; thus A_e is a subgroup of Γ defined up to conjugation. Any oriented edge e of $\mathcal{G}$ with terminal vertex v is represented by an oriented edge $\tilde{e}$ in T whose terminal vertex $\tilde{v}$ represents v. We define J_e to be the inclusion homomorphism from the stabilizer of $\tilde{e}$ to the stabilizer of $\tilde{v}$.

Bass and Serre showed that this construction gives a bijective correspondence between equivariant simplicial isometry classes of group actions without inversions on simplicial trees, on the one hand, and isomorphism classes of graphs of groups on the other. Furthermore, the inverse construction can be described explicitly. Thus the inverse correspondence assigns to each graphs of groups a group Γ, called the *fundamental group* of the give graph of groups, and an action of Γ on a tree T, called the *universal covering* tree. (These are not to be confused with the fundamental group and the universal covering tree of the underlying graphs.)

---

[1] Each edge of a graph corresponds to two oriented edges. Each oriented edge has a well-defined terminal vertex. If the edge is a loop, both orientations define the same terminal vertex.

Given the bijectivity, one can of course *define* the fundamental group of a graph of groups to be the unique group which acts without inversions on a simplicial tree so that the given graph of groups appears as the quotient. However, Bass and Serre give an explicit and purely group-theoretical description of the fundamental group.

For example, if the given graph has one edge and two vertices v_1 and v_2, then there are two oriented edges e_1 and e_2, where e_i has terminal vertex v_i. Let us set $C = A_{e_1} = A_{e_2}$, and $A_i = A_{v_i}$ for $i = 1,2$. Then $J_i = J_{e_i}$ is an injective homomorphism from C to A_i (defined modulo inner automorphisms). In this case, the fundamental group of the graph of groups is obtained from the free product $A_1 * A_2$ by adjoining the relations $J_1(\gamma) = J_2(\gamma)$ for all elements $\gamma \in C$. By definition this is the *free product of A_1 and A_2 with amalgamated subgroup C*, denoted $A_1 *_C A_2$. (The injective homomorphisms J_1 and J_2 are suppressed from the notation.)

The existence of the universal covering tree includes a number of classical facts about an arbitrary free product with amalgamation $\Gamma = A_1 *_C A_2$: in particular, the natural homomorphisms $A_i \to \Gamma$ are injective, so that the A_i can be identified with subgroups of Γ; and we have $A_1 \cap A_2 = C$. (This last statement is just the translation of the fact that the stabilizer of an edge in the universal covering tree is the intersection of the stabilizers of its endpoints.)

Another especially important example is a graph with one edge and one vertex v. In this case we have two oriented edges e_1 and e_2, two groups $A = A_v$ and $C = A_{e_1} = A_{e_2}$, and for $i = 1,2$ a homomorphism $J_i = J_{e_i} : C \to A$. In this case the fundamental group is obtained from the free product of A with an infinite cyclic group $\langle t \rangle$ by adjoining the relations $tJ_1(\gamma)t^{-1} = J_2(\gamma)$ for all elements $\gamma \in C$. This is called an *HNN* (Higman-Neumann-Neumann) *extension with base group A and associated subgroup C* and is denoted $A*_C$.

In general, the fundamental group of a finite graph of groups can be calculated by successively using the two special constructions, free products with amalgamation and HNN extensions, that I have described above. Furthermore, the fundamental group of an infinite graph of groups can be obtained as the direct limit of the fundamental groups of its finite subgraphs.

If Γ is the fundamental group of a graph of groups, for each vertex (or edge) c the groups A_c has a natural identification—modulo inner automorphisms—with a subgroup of Γ. This again follows from the existence of the universal covering tree. A conjugate of A_c in Γ is called a *vertex group* (or edge group) of Γ.

The proofs given by Bass and Serre are combinatorial, rather like the proof I gave in 1.1.4 that a group acting freely is free. An alternative approach to their theory, using topology, was developed by Scott and Wall [ScW]. I discussed this topological approach in [Sh].

1.2. Metric trees. One useful way to think about a simplicial tree is to regard the set of vertices as forming a metric space T in which the distance function takes integer values: any two vertices x and y are joined by a unique simplicial arc, and the distance $\mathrm{dist}(x, y)$ is the number of 1-simplices that make up this arc. This provides an equivalence between simplicial trees and integer metric spaces satisfying certain conditions. When one writes the conditions down, they are seen to make sense not only for integer metric spaces, but for real metric spaces and—what is curious but important—even for metric spaces in a more general sense. This leads to the following definition, which was first given in [MSh1].

1.2.1. Let Λ be an ordered abelian group, denoted additively. A Λ-*metric space* is a set X equipped with a "distance function" $d\colon X \times X \to \Lambda$ which satisfies the usual formal axioms for a metric space. A *segment* in a Λ-metric space X is a subset which is isometric to a closed interval in Λ.

1.2.2. A Λ-*tree* is a Λ-metric space T with the following properties:

(i) Any two points $x, y \in T$ are the endpoints of a unique segment $[x, y]$;
(ii) For any $x, y, z \in T$ we have $[x, y] \cap [x, z] = [x, w]$ for some $w \in T$; and
(iii) If $[x, y] \cap [x, z] = x$ then $[x, y] \cup [x, z] = [y, z]$.

1.2.3. A **Z**-tree is just the 0-skeleton of a simplicial tree with the natural **Z**-metric that I described above. In this article I will use the terms "simplicial tree" and "**Z**-tree" interchangeably.

1.2.4. On the other hand, **R**-trees are more exotic objects. It was proved in [MSh1] that an **R**-metric space is an **R**-tree if and only if (i) any two points are the endpoints of a unique topological arc (i.e. a subspace homeomorphic to a closed interval) and (ii) every topological arc is isometric to a closed interval. (One could also replace (ii) by the condition that the space is a path space, i.e. that the distance between any two points is the infimum of the lengths of the paths joining them.) I will be mentioning some examples very soon.

1.2.5. When I talk about an action of a group Γ on a Λ- tree, I will always mean an action by isometries. This is the natural generalization of the convention that a group acting on a simplicial tree is understood to act simplicially. There is also a natural generalization of the "no inversions" condition. A group Γ is said to act *without inversions* on a Λ-tree T if whenever an element γ of Γ leaves a segment $[x, y]$ invariant, either (i) γ fixes x and y, or (ii) $\mathrm{dist}(x, y)$ is divisible by 2 in L. (In the latter case, $[x, y]$ has a midpoint z, and γ fixes z.) An action of a group on an **R**-tree is automatically without inversions since every real number is divisible by 2.

1.2.6. There is a natural generalization of the observation (1.1.4) that every free group acts freely and without inversions on its Cayley tree. Any free product

Γ of copies of an arbitrary ordered abelian group Λ acts freely and without inversions on a Λ-tree. To see this, we recall that any element γ of Γ can be written uniquely as a reduced word; we define the ***length*** of γ, a non-negative element of Λ, to be the sum of the absolute values of the letters in the word. Then we can make Γ into a Λ-metric space by defining $\mathrm{dist}(x, y)$ to be the length of $x - y$. It's not hard to show that Γ is a Λ-tree and that the left regular action of Γ on itself is a free action by isometries having no inversions.

1.2.7. These remarks suggest one reason, not to be scoffed at, for introducing the general notion of Λ-tree: it provides a natural formal setting for the study of trees. Many arguments about simplicial trees go through equally well for Λ-trees, and the logic of an argument often becomes clearer if it is phrased in terms of Λ-trees.

Let me give another elementary illustration of this. It is a classical fact that any finite subgroup of a free product with amalagamation $A_1 *_C A_2$ is contained in a conjugate of one of the factors A_i. According to the Bass-Serre theory (1.1.5), this is a special case of the following fact about a $\mathbf{Z}$-tree T: *any action without inversions of a finite group G on T has a fixed point.* Now this is not hard to prove, not only for a $\mathbf{Z}$-tree, but for a Λ-tree, where Λ is any ordered abelian group.

Indeed, if x is any point of T, then there is a smallest subtree containing the orbit $G \cdot x$. Since $G \cdot x$ is finite, this subtree is finite in the sense that it is a finite union of segments. Any finite Λ-tree T_0 has a well-defined diameter $D = \max_{x,y \in T_0} \mathrm{dist}\,(x, y)$ and a well-defined *barycenter* m. We may define m to be the midpoint of any segment of length D in T_0; it is an exercise in using the tree axioms to show that any two such segments have the same midpoint. The barycenter of the smallest subtree containing $G \cdot x$ is obviously a fixed point for the action of G.

In [Sh] you will find other elementary arguments like the one above, for which Λ-trees seem to provide the natural formal context.

1.2.8. The construction that I described in 1.2.6 gives a free action of the group $\mathbf{R} * \mathbf{R}$ on an $\mathbf{R}$-tree. As I said in [Sh], it is good to try to visualize this tree. This example shows how very infinite $\mathbf{R}$-trees can be. On the other hand, it is really a completely tame example: it is simply the real analogue of the Cayley graph of $\mathbf{Z} * \mathbf{Z}$.

1.2.9. If $\Gamma = *_i \Gamma_i$ is a free product of subgroups of $\mathbf{R}$, we get a free action of Γ on an $\mathbf{R}$-tree by restricting the action described in 1.2.6.

It is not obvious whether these are the only groups that act freely on $\mathbf{R}$-trees. I shall return to this question in Section 2.

1.2.10. An interesting case of the actions I just described in 1.2.9 is the one in which all the subgroups are infinite cyclic—but with arbitrary, possibly incommensurable, real generators. The actions that one obtains in this case are examples of what I call ***polyhedral*** actions. By definition, an action on an **R**-tree is polyhedral if it is topologically (but not necessarily metrically) equivalent to a simplicial action.

If you assign lengths to the edges of a graph $\mathcal{G}$, the universal cover $\tilde{\mathcal{G}}$ inherits an **R**-tree metric, and the action of $\pi_1(\mathcal{G})$ on $\tilde{\mathcal{G}}$ is polyhedral. All free polyhedral actions without inversions arise in this way. So there is nothing mysterious about such actions. And yet as I will be explaining in Section 2, the starting point for the fundamental work of Culler and Vogtmann on the outer automorphism group of a free group is to consider such actions. This is another good example of how the generalized notion of tree can be useful in a purely conceptual way.

1.2.11. A second reason for studying Λ-trees is that there are examples of actions of groups on Λ-trees which simplicial experience would not predict. For example, in the simplicial case it is only the free groups that act freely and without inversions. From the analogy in 1.2.6 one might guess that the only groups that act freely on **R**-trees are free products of subgroups of **R**. But this is false, as we shall see in 2.3. Similarly, the obvious free actions of a free group on **R**-trees are the polyhedral actions–and those derived from them in a trivial way[2]. But as we shall see in 2.2, there are free actions of free groups that are not of this type.

So an action of a group on a Λ-tree defines a genuinely new kind of structure on the group. And any new kind of structure provides an opportunity to recover a bit of order from the seemingly chaotic world of infinite groups.

1.2.12. A third reason for studying actions on Λ-trees is that they arise in applications, and not only as a formal device: exotic actions of the kind I mentioned in 1.2.11 come up in applications. In particular, the theory first introduced in [MSh1], and further developed in [MSh2], [M1], [MO], [Be], [Pau2], [Bab] and [Ch], provides a connection between hyperbolic geometry and the study of Λ-trees. This allows one to reduce certain important questions about hyperbolic manifolds to questions about group actions on Λ-trees for suitable Λ. In this way it has been possible to re-prove and extend some fundamental results of Thurston's by proving appropriate theorems about Λ-trees; and certain conjectures about Λ-trees that I will be discussing in 2.5 and in Section 5 would allow one to extend these results much further.

I mentioned in 1.2.10 that the theory of Culler and Vogtmann begins by considering polyhedral actions on **R**-trees. But there are now important interactions, which I will discuss in 2.2, between this theory and the study of certain exotic actions. More generally, according to work of Paulin [Pau3] which I will

[2] I am referring here to actions for which the induced action on the minimal invariant subtree is polyhedral. See 1.5.

briefly explain in 4.3.5, for a group Γ which is hyperbolic in the sense of the Cannon-Gromov theory, the study of the outer automorphism group $\mathrm{Out}(\Gamma)$ is closely related to the study of actions of Γ on trees.

These applications of dendrology are all aspects of a general theory that I will be discussing in Section 4.

1.3. Foundations and conventions. The remainder of this section was originally planned as a quick trip through some foundational material and conventions that I will be referring to later in the article. However, I will be stopping on the way to point out a couple of three-star views.

1.3.1. The geometric realization of a simplicial tree becomes an $\mathbf{R}$-tree if we give each edge the linear metric of length 1. Restricting the $\mathbf{R}$-tree metric to the 0-skeleton gives the $\mathbf{Z}$-tree metric.

More generally, if $\Lambda_0 \subset \Lambda$ are ordered abelian groups then any Λ_0-tree T_0 has an embedding i in a Λ-tree T. The embedding i is an isometry of Λ_0-metric spaces, and every point of T lies in a segment with endpoints in $i(T_0)$. These properties are expressed by saying that T (or more precisely the pair (T, i)) is a Λ-*completion* of T_0. The completion is unique up to isometry: if (T', i') is another completion then there is a unique isometry $j : T \to T')$ such that $j \circ i = i'$. The existence and uniqueness of the Λ-completion were established in [AlpB], after some partial results in [MSh1]. We may write ΛT for the Λ-completion of T.

It follows from the uniqueness of the completion that any action of a group Γ on a Λ_0-tree T has a unique extension to ΛT. This makes it possible to think of actions on Λ_0-trees as being essentially a special case of actions on Λ-trees.

1.3.2. There is a natural generalization of the first barycentric subdivision of a simplicial tree. If Λ is any ordered abelian group, the order on Λ extends to an order on $\frac{1}{2}\Lambda = \Lambda \otimes_{\mathbf{Z}} (\frac{1}{2}\mathbf{Z}) \supset \Lambda$. If T is a Λ-tree then $\frac{1}{2}\Lambda T$ is a generalized first barycentric subdivision of T. Generalizing 1.1.3, one can show that the $\frac{1}{2}\Lambda$ completion of any group action on T is an action without inversions on $\frac{1}{2}\Lambda T$.

1.3.3. I am in the habit of saying that an action of a group Γ on a Λ-tree T is *trivial* if it has a fixed point, i.e. if there is a point of T which is fixed by all of Γ. This terminology is motivated by Bass-Serre theory. An action without inversions of a group Γ on a simplicial tree defines a *splitting* of Γ, i.e. an isomorphism between Γ and the fundamental group of a graph of groups. A splitting of a group is said to be *trivial* if some vertex group in the graph of groups corresponds to the entire group Γ under the isomorphism. In this sense, exhibiting Γ as a free product with amalgamation $A_1 \star_C A_2$ gives a splitting which is non-trivial if and only if the A_i are proper subgroups of Γ; this is true if and only if C is a proper subgroup of each A_i. On the other hand, exhibiting a group Γ as an HNN extension $A{*}_C$ always gives a non-trivial splitting of Γ. An action is trivial if and only if it corresponds to a trivial splitting.

Using Bass and Serre's group-theoretical definition of the fundamental group of a graph of groups, one can show that a group Γ admits a non-trivial splitting if and only if either (i) Γ admits a homomorphism onto $\mathbf{Z}$, or (ii) Γ is an amalgamated free product of two proper subgroups.

1.4. Trivial actions and deep mathematics. The use of the term *trivial* to describe an action with a fixed point should not obscure the fact that such actions can be extremely interesting. I would like to illustrate this by describing some beautiful work due to Gupta and Sidki [GuS].

A *torsion group* is a group whose elements are all of finite order. The classical Burnside problem asked whether a finitely generated torsion group must be finite. Golod [Go] gave a strong negative solution by showing that for every prime p there is a finitely generated infinite group which is a p-group in the sense that the order of every element is a power of p. Gupta and Sidki gave a marvelously simple and elementary proof of this for the case where p is odd by producing a two-generator subgroup of the automorphism group of a simplicial tree which is an infinite p-group.

For concreteness let us take $p = 3$. Consider the set $T = \bigcup_{n\geq 0}(\mathbf{Z}/3\mathbf{Z})^n$ of all finite sequences of elements of $\mathbf{Z}/3\mathbf{Z}$. We give T the structure of a simplicial tree by joining two vertices if and only if they are of the form $(r_1, \ldots, r_n)$ and $(r_1, \ldots, r_{n+1})$ for some $n \geq 0$ and some $r_1, \ldots, r_{n+1} \in \mathbf{Z}/3\mathbf{Z}$. (So T is the family tree of descendants of a Martian amoeba that reproduces by three-fold fission. See figure 1.4.1.)

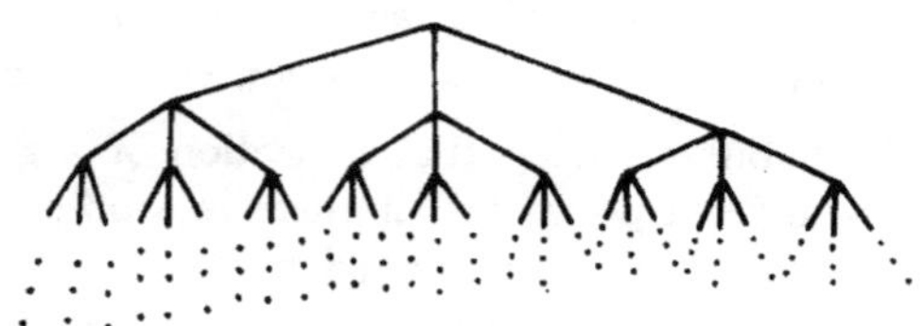

FIGURE 1.4.1

There is an obvious order-3 automorphism t of T defined by $t(r_1, r_2, \ldots, r_n) = (r_1 + 1, r_2, \ldots, r_n)$. In the above picture, t just cyclically permutes the subtrees lying below the points (1), (2) and (3). Now we define a second order-3 automorphism t as follows. Consider any vertex $v = (r_i)_{1 \le i \le n}$ of T. If $r_i = 0$ for $i = 1, \ldots, n-1$ we set $a(v) = v$. Otherwise let $k \le n-1$ be the smallest index such that $r_k \neq 0$, and set $a(v) = (r'_i)_{1 \le i \le n}$, where $r'_i = r_i$ for $i \neq k+1$, and $r'_{k+1} = r_{k+1} + r_k$.

When we draw the automorphism t in the above picture, something surprising happens. It fixes the top vertex $\emptyset$ and each of the depth-one vertices (1), (2) and (3). Each depth-one vertex (r) is the top point of a sub-tree T_r which looks like the full picture of the tree T. For $r = \pm 1$ the automorphism t acts on T_r exactly as a^r acts on T. But the action of t on T_0 looks exactly like the action of t *itself* on the whole tree T!

This recurrent or self-referential property of t is the key to understanding the properties of the group $\Gamma \subset \mathrm{Aut}(T)$ generated by a and t. In particular it is easy to write down an isomorphism J between an index-3 subgroup of Γ and a subdirect product of three isomorophic copies of Γ. The existence of such an isomorphism J immediately implies that Γ is infinite. A simple but ingenious combinatorial argument, also using the isomorphism J, shows that Γ is a 3-group. (However, the orders of elements of Γ are not bounded: all powers of 3 occur as orders of elements.)

1.5. Foundations and conventions continued: length functions. Let us now turn our attention to non-trivial actions. It was shown in [CuM] (for $\Lambda \le \mathbf{R}$) and independently in [AlpB] (for the general case) that if Γ acts non-trivially on a Λ-tree T then T has a unique minimal Γ-invariant Λ-subtree. Abstractly speaking, the basic problem in the subject is to classify minimal non-trivial actions up to equivariant isometry.

1.5.1. If a group Γ acts without inversions on an Λ-tree T, then

$$l(\gamma) = \min_{x \in T} \mathrm{dist}(x, \gamma \cdot x)$$

exists for every $\gamma \in \Gamma$. This follows from an elementary argument given in [MSh1] and [AlpB], and discussed in [Sh]. The non-negative-valued function $l\colon \Gamma \to \Lambda$ is called the *length function* associated to the given action. It is obvious that l is constant on each conjugacy class, and that $l(\gamma) = 0$ if and only if γ has a fixed point in T.

1.5.2. The simplest length function on any group Γ is the one that is identically zero; this is realized by the action of Γ on the Λ-tree consisting of a single point. The next simplest kind of length function is of the form $l(\gamma) = |h(\gamma)|$, where h is any homomorphism of a group Γ into Λ. We may realize this as a length function by regarding Λ itself as being a Λ-tree, and letting Γ act on Λ by $\gamma \cdot \lambda = \lambda + h(\gamma)$. Length functions of this form are often called *abelian*.

1.5.3. Let Γ be a finitely generated group. It was proved in [MSh1] and [AlpB], and discussed in [Sh], that the length function associated to a given action of Γ on a Λ-tree T is identically zero if and only if the action is trivial. For a non-trivial action, the length function is determined by the restriction of Γ to its minimal invariant subtree.

1.5.4. Conversely, it was proved in [CuM] (for $\Lambda \leq \mathbf{R}$) and in [AlpB] (for the general case), and discussed in [Sh] that if two non-trivial actions determine the same length function then either (i) their minimal invariant subtrees are equivariantly isometric, or (ii) the length function in question is abelian.

1.6. Abelian length functions and the Bieri-Neumann Strebel invariant. In 1.4 I pointed out that trivial actions—those that define the zero length function—can be extremely interesting. Actions that define abelian length functions can likewise be very interesting. In [Bro], Brown relates such actions to the theory of the Bieri-Neumann-Strebel invariant [BiNS], which is a powerful tool in combinatorial group theory.

Geometrically, to say that a non-trivial action of a group Γ on an $\mathbf{R}$-tree T defines an abelian length function means that it has an invariant end. An *end* of T is by definition an equivalence class of rays in T (i.e. subtrees isometric to $[0,\infty) \subset \mathbf{R}$), where two rays are defined to be equivalent if their intersection is a ray. Thus an action defines an abelian length function if and only if there is a ray $r \subset T$ such that $\gamma \cdot r$ is equivalent to r for every $\gamma \in \Gamma$.

In the obvious case where the minimal invariant subtree is a line, there are clearly two invariant ends for the action. When the minimal invariant subtree is not a line, the invariant end is unique. In the latter case I shall say that the action is an *exceptional abelian action.*

1.6.1. Examples of exceptional abelian actions are provided by *ascending HNN extensions.* Let A be a group and let J be an isomorphism of A onto a proper subgroup C of A. Consider a graph of groups with one vertex and one edge, both labelled with the graph A; the homomorphisms $A \to A$ corresponding to the two orientations of the edge are the identity and J. The fundamental group Γ of this graph of groups is the HNN extension (1.1.5) obtained from a free product $A \star \langle t \rangle$ by adjoining the relations $t\gamma t^{-1} = J(\gamma)$ for all $\gamma \in C$. The Bass-Serre theory gives an action of Γ on a $\mathbf{Z}$-tree, which is an exceptional abelian action as one can check. A simple example is obtained by taking $A = C = \mathbf{Z}$ and $J(n) = 2n$; this gives the group $\langle x, y | yxy^{-1} = x^2 \rangle$.

1.6.2. By definition an abelian length function l on Γ has the form $\gamma \mapsto |h(\gamma)|$ for some homomorphism $h\colon \Gamma \to \mathbf{R}$. I'll write $l = |h|$. Given a non-zero abelian length function l, there are two different homomorphisms with absolute value l. However, an exceptional abelian action of Γ on T gives rise in a canonical way to a homomorphism whose absolute value is the length function defined by the action. In fact, if e is the unique invariant end, then for any $\gamma \in \Gamma$ there exist a number $\epsilon = \pm 1$ and a ray r representing the end e such that $\gamma^{\epsilon}(r) \supset r$. Furthermore, $\epsilon = \epsilon(\gamma)$ is uniquely determined by γ (and the given action of Γ on T); and we have dist $(x, \gamma \cdot x) = l(x)$, where x is the endpoint of r and l is the length function defined by the action. The homomorphism h canonically associated with the action is defined by $h(\gamma) = \epsilon(\gamma) l(\gamma)$.

1.6.3. Let Γ be a finitely generated group. The set Hom $(\Gamma, \mathbf{R})$ of all homomorphisms from Γ to $\mathbf{R}$ is a vector space over $\mathbf{R}$, whose dimension is the number n of infinite cyclic summands in the commutator quotient of Γ. The multiplicative group $\mathbf{R}^+$ of positive reals acts on Hom $(\Gamma, \mathbf{R}) - \{0\}$ by homotheties, and the quotient space Hom $(\Gamma, \mathbf{R}) - \{0\}$ is a sphere S of dimension $n-1$. Let $p\colon$ Hom $(\Gamma, \mathbf{R}) - \{0\} \to S$ denote the projection map. We have a dense set $p(\text{Hom }(\Gamma, \mathbf{Q}) - \{0\}) = p(\text{Hom }(\Gamma, \mathbf{Z}) - \{0\})$ of *rational* points in S.

Let us say that a non-zero homomorphism $h\colon \Gamma \to \mathbf{R}$ is *exceptional* if it is the homomorphism canonically associated to some exceptional abelian action of Γ on an $\mathbf{R}$- tree. The exceptional homomorphisms form a subset of Hom $(\Gamma, \mathbf{R}) - \{0\}$ which is invariant under the action of $\mathbf{R}$. The image of this set under p is a closed subset of S whose complement in S is denoted $\Sigma = \Sigma(\Gamma)$. The open set $\Sigma \subset S$ is the *Bieri-Neumann-Strebel invariant* of the group Γ. If the isomorphism type of Γ is specified then Σ is a subset of the $(n-1)$-sphere $\mathbf{R}^n/\mathbf{R}^+$ which is well-defined modulo the natural action of $\mathrm{GL}_n(\mathbf{Z})$.

1.6.4. To illustrate what can be done with this invariant, let me outline a proof of the following striking and unexpected result, which is proved in [BiNS].

(1.6.4.1) *Let Γ be a finitely presented group whose commutator quotient has at least two infinite cyclic summands. Then either Γ has a finitely generated normal subgroup N with $\Gamma/N \cong \mathbf{Z}$, or Γ has a free subgroup of rank* 2.

The logic of the proof is as follows. First one proves:

(1.6.4.2) Let Γ be a finitely generated group and let $h\colon \Gamma \to \mathbf{ZR}$ be a non-zero homomorphism such that $p(h) \in \Sigma \cap (-\Sigma)$. (Here $-\Sigma$ of course denotes the image of Σ under the antipodal map.) Then $\ker h$ is finitely generated.

Second, one proves:

(1.6.4.3) If Γ is a finitely presented group which contains no free subgroup of rank 2, then $\Sigma(\Gamma) \cup (-\Sigma(\Gamma)) = S$.

Given (1.6.4.2) and (1.6.4.3), one can quickly prove (1.6.4.1) as follows. If the commutator quotient of the finitely presented group Γ has at $n \geq 2$ infinite cyclic summands, then $\Sigma = \Sigma(\Gamma)$ is a subset of a sphere S of dimension $n-1 \geq 1$. In particular S is connected. If Γ does not contain a free subgroup of rank 2, then by (1.6.4.3), S is the union of the two open sets Σ and $-\Sigma$. By connectedness we have $\Sigma \cap (-\Sigma) \neq \emptyset$. Since $\Sigma \cap (-\Sigma)$ is open and the set of rational points $p((\mathrm{Hom}\,(\Gamma, \mathbf{Z}) - \{0\})$ is dense in S, we have $p(h) \in \Sigma \cap (-\Sigma)$ for some non-zero homomorphism $h\colon \Gamma \to \mathbf{Z}$. By (1.6.4.2), $N = \ker h$ is finitely generated, and of course we have $\Gamma/N \cong \mathbf{Z}$.

In [Bro], Brown proves (1.6.4.2) from the dendrological point of view. We can assume that h is surjective. Let's fix an element $t \in \Gamma$ with $h(t) = 1$. Since h belongs to Σ, there is in particular no exceptional abelian action of Γ on a $\mathbf{Z}$-tree whose associated homomorphism is h. Translating this statement via Bass-Serre theory one concludes that Γ cannot be expressed as a properly ascending HNN extension with associated homomorphism h; that is, there is no proper subgroup B of $N = \ker h$ such that $tBt^{-1} \supset B$ and $N = \bigcup_{n \geq 0} t^n B t^{-n}$. Using this fact–and the hypothesis that Γ is finitely generated–it is elementary to conclude that N has a finitely generated subgroup C such that $N = \bigcup_{n \geq 0} t^{-n} C t^n$. This exhibits Γ as an ascending HNN extension with associated homomorphism $-h$; since $-h \in \Sigma$, this ascending HNN cannot be proper. Hence $C = N$, so that N is finitely generated.

A proof of (1.6.4.3) in the language of trees does not appear to have been written down, but one can translate the argument given by Bieri, Neumann and Strebel in roughly the following way. Since Γ is finitely presented, it can be realized as the fundamental group of a compact manifold M. Given any $h \in \mathrm{Hom}\,(\Gamma, \mathbf{R}) - \{0\}$, one can construct a piecewise-linear map $\tilde{f}$ from the universal cover $\tilde{M}$ of M to $\mathbf{R}$ such that $\tilde{f}(\gamma \cdot x) = \tilde{f}(x) + h(\gamma)$ for every $\gamma \in \Gamma$. Let N denote the universal abelian cover of M, i.e. the covering space corresponding to the commutator subgroup Γ' of $\Gamma = \pi_1(M)$. Then f induces a map $g\colon N \to \mathbf{R}$. Let $z \in \mathbf{R}$ be a regular value of g, so that $X_+ = g^{-1}[z, \infty)$ and $X_- = g^{-1}(-\infty, z]$ are manifolds-with-boundary of the same dimension as N, and $Y = g^{-1}(\{z\}) = \partial X_+ = \partial X_-$. For the moment let us assume that Y is connected. Then X_+ and X_- are connected, and $\Gamma' = \pi_1(N)$ is the free product of the groups $A_+ = \mathrm{im}(\pi_1(X_+) \to \pi_1(N)$ and $A_- = \mathrm{im}(\pi_1(X_-) \to \pi_1(N)$ amalgamated over the subgroup $C = \mathrm{im}(\pi_1(Y) \to \pi_1(N)$.

If h is exceptional then A_+ is a proper subgroup of Γ'. To prove this we consider an exceptional abelian action of Γ on an $\mathbf{R}$-tree T. We can take this action to be minimal. It is not hard to factor the map $\tilde{f}$ as $\alpha \circ \tilde{f}'$, where $\tilde{f}'\colon \tilde{M} \to T$ is a Γ-equivariant map and $\alpha\colon T' \to T$ satisfies $\alpha(\gamma \cdot y) = \alpha(y) + h(\gamma)$ for every $y \in T'$. Using that the action of Γ on T is exceptional, one can show that $P = \alpha^{-1}([z, \infty))$ is disconnected; this implies, using minimality, that $(\tilde{f}')^{-1}(P) = \tilde{f}^{-1}(Z)$ is disconnected. Thus X_+ has disconnected pre-image in the universal cover $\tilde{M}$ of N and therefore does not carry the fundamental group of N. This means that $A_+ \neq \Gamma'$, as asserted.

The same argument shows that if $-h$ is exceptional then A_- is a proper subgroup of Γ'. So if h and $-h$ are both exceptional, that is, if $p(h) \notin \Sigma \cup (-\Sigma)$, then Γ' is an amalgamated free product of two proper subgroups. By pushing this argument a bit further one can arrange that the amalgamated subgroup has infinite index in both factors. This implies that Γ' has a free subgroup of rank 2, contradicting the hypothesis. So we must have $h \in \Sigma \cup (-\Sigma)$.

I have sketched a proof that $h \in \Sigma \cup (-\Sigma)$ under the assumption that the set C, and hence the sets A_+ and A_-, are connected. In general this need not be the case. However, using the compactness of M (which reflects the hypothesis that Γ is finitely presented) one can show that there is always a unique component of A_+ whose image under g is an unbounded subset of $[0, \infty)$. The analogous assertion holds for A_-. Using this it is not hard to modify the argument so that it applies to an arbitrary $h \in \mathrm{Hom}\,(\Gamma, \mathbf{R}) - \{0\}$. This shows that $h \in \Sigma \cup (-\Sigma) = S$, as asserted in 1.6.4.3.

1.7. Foundations and conventions, concluded: coordinates for actions, and morphisms between trees. Apart from the fascinating ambiguity arising from abelian length functions, 1.5.4 says that minimal non-trivial actions are determined by their length functions. One can think of the length function associated with an action as defining coordinates for the action. This is a particularly useful point of view for the case $\Lambda = \mathbf{R}$.

1.7.1. To formalize it, we let $\mathcal{C}(\Gamma)$ denote the set of conjugacy classes in Γ, and we consider the Cartesian power $[0, \infty)^{\mathcal{C}(\Gamma)}$, where $[0, \infty)$ of course denotes the non-negative reals. We give $[0, \infty)^{\mathcal{C}(\Gamma)}$ the product topology and the subset $[0, \infty)^{\mathcal{C}(\Gamma)} - \{0\}$ the subspace topology. Any length function defined by a non-trivial action of Γ on an $\mathbf{R}$-tree may be regarded as a function on $\mathcal{C}(\Gamma)$ and may therefore be identified with a point of $[0, \infty)^{\mathcal{C}(\Gamma)} - \{0\}$. The set $\mathcal{L}(\Gamma)$ of all length functions for non-trivial actions is therefore identified with a subset of $[0, \infty)^{\mathcal{C}(\Gamma)} - \{0\}$. It turns out that this is a compact set; as I explained in [Sh], this was first proved in [CuM], and a more direct proof was given in [Par]. (I will mention yet another proof in 4.3.2.)

Life becomes even more pleasant if we work in the "projective space" $\mathcal{P}^{\Gamma}$ defined as the quotient of $[0, \infty)^{\Gamma} - \{0\}$ by the homothetic action

$$r \cdot (x_c)_{c \in \mathcal{C}(\Gamma)} = (r x_c)_{c \in \mathcal{C}(\Gamma)}$$

of the multiplicative group of positive reals. The space $\mathcal{P}^{\Gamma}$ is given the quotient topology, and the image $\mathcal{PL}(\Gamma)$ of $\mathrm{L}(\Gamma)$ in $\mathcal{P}^{\Gamma}$ is then a compact set. The points of $\mathcal{PL}(\Gamma)$ are called *projectivized length functions*.

1.7.2. Completing an action does not change its length function. More precisely, suppose that $\Lambda_0 \subset \Lambda$ are ordered abelian groups. If a group Γ acts without inversions on a Λ_0-tree T_0, then the completed action of Γ on $T = \Lambda T_0$ has no inversions, and it defines the same length function as the action of Γ on T. In particular the length function defined by the completed action takes values in L_0. Conversely, if Γ acts on a Λ-tree T and the length function defined by the action takes values in Λ_0, then the action is the completion of some action of Γ on a Λ_0-tree. These facts are proved in [AlpB].

1.7.3. It is often useful to think of Λ-trees, for any fixed Λ, as forming a category. Let T and T' be Λ-trees. A map of sets $f\colon T \to T'$ is called a *morphism* if each segment in T can be written as a finite union of subsegments, each of which is mapped isometrically into T' by f. (Thus a morphism crumples up any segment in at most a finite way.) An injective morphism is an isometry onto its image. If a morphism $f\colon T \to T'$ fails to be injective, it has a *fold*: that is, there are two segments in T with a common endpoint which are mapped isometrically by f onto the same segment in T'.

SECTION 2. OUTER SPACE, LIMITS AND EXOTIC FREE ACTIONS

2.1. Outer space and its boundary; small actions. As an example of how the coordinatization of actions described in 1.7 can be used, consider the free group F_n of rank $n < \infty$, and let Y_n denote the subset of $\mathcal{PL}(F_n)$ consisting of all projectivized length functions defined by free actions of F_n on $\mathbf{R}$-trees which are polyhedral in the sense of 1.2.10.

2.1.1. There is a natural action of the outer automorphism group $\mathrm{Out}(F_n)$ on Y_n given by $\alpha \cdot [l] = [l \circ \alpha^{-1}]$. Culler and Vogtmann [CuV1,2 and V] have used this action to study the group $\mathrm{Out}(F_n)$ through properties of the space Y_n. For example, they have shown that Y_n is a contractible triangulable space and have computed its dimension. (The contractibility of Y_n was also proved independently by S. Gersten. I will be discussing a proof of the contractibility later, in 5.4.) The group $\mathrm{Out}(F_n)$ acts properly discontinuously on this space; in particular, the stabilizer of any point in Y_n is a finite subgroup of $\mathrm{Out}(F_n)$. (This is because if a point of Y_n corresponds to an action of F_n on a polyhedral tree T, the stabilizer of the point is isomorphic to a group of automorphisms of the quotient graph T/F_n, and the automorphism group of a finite graph is finite.)

Using the action of $\mathrm{Out}(F_n)$ on Y_n, Culler and Vogtmann have succeeded in calculating the cohomological dimension of $\mathrm{Out}(F_n)$. The same formalism can be used to calculate, for example, the virtual Euler characteristic of $\mathrm{Out}(F_n)$. This was done by Smillie and Vogtmann in [SmV].

I am pleased to note that my own term *outer space* for the space Y_n, reflecting its close connection with the group $Out(F_n)$, has been gaining currency.

The compactness result stated in 1.7.1 implies that the closure $\hat{Y}_n$ of Y_n in $\mathcal{PL}(F_n)$ is a compact space which is again invariant under $\mathrm{Out}(F_n)$. There is considerable evidence by now that deeper properties of outer automorphisms of F_n can be understood by studying their extended action on $\hat{Y}_n$.

2.1.2. The space $\hat{Y}_n$ does not consist entirely of length functions defined by free actions, because the length functions defined by free actions do not form a closed subset of $\mathcal{PL}(F_n)$. However, there is a somewhat weaker property than being free which is closed. Let us say that a group is *small* if it has no rank-2 free subgroup. An action of a group Γ on a Λ-tree is termed *small* if the stabilizer of every non-degenerate segment is a small subgroup of Γ. A (projectivized) length function is *small* if it is defined by some small action. In [CuM], Culler and Morgan proved that for any finitely generated group Γ which is not small, the small projectivized length functions on Γ form a closed (and hence compact) subset $\mathcal{SPL}(\Gamma)$ of $\mathcal{PL}(\Gamma)$. In particular we have $\hat{Y}_n \subset \mathcal{SPL}(\Gamma)$.

The only small subgroups of a free group are cyclic subgroups. So $\hat{Y}_n$ consists of length functions defined by actions with cyclic segment-stabilizers.

2.2. Exotic actions of free groups. The space $\hat{Y}_2$ has been completely described by Culler and Vogtmann in their paper [CuV2]. In particular, the small actions that define the points of $\hat{Y}_2 - Y_2$ are all well understood. It turns out that none of these actions is free. In fact, a result due to Harrison [H] and re-proved geometrically by Morgan [M2] implies that the only free actions of F_2 on $\mathbf{R}$-trees are the free polyhedral actions, i.e. the actions whose length functions lie in Y_2.

By contrast, Bestvina and Handel have shown that for every $n \geq 3$ the set $\hat{Y}_n - Y_n$ does contain free actions. This is very striking because it shows that there exist free actions of F_3 on $\mathbf{R}$-trees which are exotic in the sense that they are not polyhedral. (I understand that G. Levitt has also constructed exotic free actions of F_n for $n \geq 3$ from a different point of view.) This is one way in which the theory of actions on $\mathbf{R}$-trees is genuinely different from the theory of actions on $\mathbf{Z}$-trees.

2.2.1. The Bestvina-Handel construction is very natural because it uses the action of $\mathrm{Out}(F_n)$. For any $[l] \in Y_n$ and any $\alpha \in \mathrm{Out}(F_n)$, we can consider the sequence $(\alpha \cdot [l])$. For many choices of $\alpha \neq 1$, Bestvina and Handel show that—for an arbitrary point $[l] \in Y_n$—this sequence converges and its limit is a point of $[l_\infty] \in \hat{Y}_n$ which is defined by a free action. The point l_∞ is clearly fixed by α. It follows that l_∞ cannot lie in Y_n. Indeed, if we had $l_\infty \in Y_n$ then l_∞ would have a finite stabilizer by 2.1.1, so that α would have finite order. But in this case the sequence $(\alpha \cdot [l])$ would not converge.

2.2.2. Let me illustrate this by a concrete example, which was kindly provided to me by Mladen Bestvina. Consider the free group $F = F_3$ on the generators x, y and z. The Cayley graph T_0 of F with respect to the generators x, y and z

is a simplicial tree, and its length function l_0 assigns to each element $\gamma \in F$ the length of a cyclically reduced word W in the conjugacy class of γ.

Let α be the automorphism of F defined by $\alpha(x) = xy$, $\alpha(y) = yz$ and $\alpha(z) = zxy$. For any cyclically reduced word W in F, we can of course calculate $\alpha(W)$ as a cyclically reduced word by replacing each generator in W by its image under α and making cyclic cancellations. (By a cyclic cancellation I mean either an ordinary cancellation, i.e. removing a subword of the form $u\bar{u}$ or $\bar{u}u$ where u is a generator and u is its inverse, or removing u from one end of the word and $\bar{u}$ from the other.) I'll say that a word W is ***legal*** if it is reduced and, in addition, it does not contain either $x\bar{z}$ or $z\bar{x}$ as a subword. If W is cyclically reduced and legal, and does not begin with $\bar{x}$ and ends with z, or begin with $\bar{z}$ and end with x, I'll say that W is ***cyclically legal.*** It is straightforward to check that if W is cyclically legal then no cyclic cancellations occur in calculating $\alpha(W)$, and what is more, that $\alpha(W)$ is itself a cyclically legal word.

For any cyclically legal word W, this makes it easy to study the behavior of $l_0(\alpha^n(W)$ as n increases. To do so let us associate to W the vector $v_W = (\xi, \eta, \zeta) \in \mathbf{R}^3$, where ξ denote the sum of the absolute values of the exponents of the generator x in W, and η and ζ are defined similarly in terms of the generators y and z. Then $l_0(W)$ is the sum of the coordinates of v_W. Since there are no cyclic cancellations involved in computing $\alpha(W)$, we have $v_{\alpha(W)} = A \cdot v_W$, where

$$A = \begin{pmatrix} 1 & 0 & 1 \\ 1 & 1 & 1 \\ 0 & 1 & 1 \end{pmatrix}.$$

Since, moreover, $\alpha(W)$ is cyclically legal, we can iterate this observation and conclude that $v_{\alpha^n(W)} = A^n \cdot v_W$ for every $n \geq 0$. So when W is cyclically legal, $l_0(\alpha^n(W)$ is simply the sum of the coordinates of $A^n \cdot v_W$.

Since A has non-negative entries, and A^2 has strictly positive entries, the Perron-Frobenius theory of positive matrices guarantees that A has a positive eigenvalue, and that if λ denotes its largest positive eigenvalue then, for any vector v in the open positive octant $(0, \infty)^3$, the sequence $(\lambda^{-n} A^n(v))_{n \leq 0}$ has a limit in $(0, \infty)^3$. It follows that $\lim_{n\to\infty} \lambda^{-n} l_0(\alpha^n(W)$ exists and is strictly positive for every non-trivial cyclically legal word W.

2.2.3. Remarkably enough, for every non-trivial element γ of F, there is a positive power α^k of α such that $\alpha^k(W)$ is conjugate to a cyclically legal word. This implies that $\lim_{n\to\infty} \lambda^{-n} l_0(\alpha^n(W)$ exists and is strictly positive for every non-trivial $\gamma \in F$. So the sequence $([l_0 \circ \alpha^n])_{\alpha \geq 0}$ has a limit $l_\infty \in \mathcal{SPL}(F)$, and furthermore we have $l_\infty(\gamma) \neq 0$ for every $\gamma \neq 1$. This means that l_∞ is the length function for a free action of F on an $\mathbf{R}$-tree. As I pointed out in 2.2.1, we have $l_\infty \notin Y_n$; that is, the free action defining l_∞ cannot be polyhedral.

2.2.4. The assertion of 2.2.3, that every $\gamma \neq 1$ is mapped to a conjugate of a cyclically legal word by some positive power of α, is proved by an elementary combinatorial argument. The key step is to show that if U and V are legal words then the element $\alpha^2(UV)$ is always represented by a legal word. In proving this, one can assume, after possibly replacing U and V by subwords, that the word UV is reduced. If UV is legal, then of course so are $\alpha(UV)$ and $\alpha^2(UV)$. If UV is reduced but illegal, we may assume that U ends with x and that V begins with $\bar{z}$. Let's write $U = U'x$ and $V = \bar{z}V'$. Then $\alpha(UV) = \alpha(U')\bar{z}\alpha(V')$. Since U' and V' are legal, $\alpha(U')$ and $\alpha(V')$ are legal, and moreover $\alpha(U')$ does not end with x. Hence if $\alpha(U')\bar{z}\alpha(V')$ is reduced then it is legal, so that the element $\alpha(UV)$ is represented by a legal word, and therefore so is the element $\alpha^2(UV)$.

So we may assume that $\alpha(U')\bar{z}\alpha(V')$ is not reduced. Now since V is reduced, V' does not begin with z, and hence $\alpha(V')$ does not begin with z either. So $\alpha(U')\bar{z}\alpha(V')$ can fail to be reduced only if the word $\alpha(U')$ ends with z; this means that U' must end with y, i.e. U must end with yx. By continuing this sort of analysis we see that U must in fact end with y^2x, and that V must begin with either $\bar{z}\bar{x}$ or $\bar{z}^2$. In the first case we can write $U = U''y^2x$ and $V = \bar{z}\bar{x}V''$. This gives $\alpha^2(UV) = \alpha^2(U'')yz^2\alpha^2(V'')$. Here, since $\alpha^2(U)$ is legal and in particular reduced, $\alpha^2(U'')$ is legal and does not end with $\bar{y}$. On the other hand, since $\alpha^2(V'')$ is the image of legal path under α, it does not begin with $\bar{z}^2$. Hence the word $\alpha^2(U'')yz^2\alpha^2(V'')$ representing the element $\alpha^2(UV)$ is legal, as required. The case in which V begins with $\bar{z}^2$ is handled similarly.

2.2.5. To prove the assertion of 2.2.3 we consider an element $\gamma \neq 1$ of F. Let n denote the smallest integer such that W is a product of n legal words. If $n > 1$, i.e. if γ is not represented by a legal word, it follows from 2.2.4 that α^2 is a product of fewer than n legal words. By induction it follows that $\alpha^{2k}(\gamma)$ is represented by a legal word. It is then easy to adapt the argument of 2.2.4 to conclude that the conjugacy class of $\alpha^{2k+2}(\gamma)$ is represented by a cyclically legal word.

2.2.6. Bestvina and Handel's general theory applies to any outer automorphism α of F_n which is *irreducible* in the sense that (say) no power of α leaves the conjugacy class of any free factor of F_n invariant. (Their definition of irreducibility is actually somewhat weaker than this.) For an irreducible α they prove that there exist a finite graph $\mathcal{G}$, an isomorphic identification of $\pi_1(\mathcal{G})$ with F_n and a map $f\colon \mathcal{G} \to \mathcal{G}$ inducing α, such that for any edge e of $\mathcal{G}$ and any $n \geq 0$, the map $\alpha^n|e\colon e \to \mathcal{G}$ is a reduced path, i.e. a locally 1-1 map of e into $\mathcal{G}$. Such a graph $\mathcal{G}$ is said to define a *train-track structure* for α. For example, for the automorphism α defined in 2.2.2 we can take $\mathcal{G}$ to be a three-leaf clover graph with one vertex and three edges. More generally, if an automorphism α is *positive* in the sense that it maps each generator to a positive word, then we can take $\mathcal{G}$ to be an n-leaf clover graph. Using the train-track structure and an analysis which is qualitatively similar to the one that I sketched in 2.2.2–2.2.5, Bestvina and Handel show that if for any irreducible outer automorphism α of F_n and any point $l \in Y_n$, the sequence $(\alpha \cdot [l])$ converges to a point $l_\infty \in \hat{Y}_n - Y_n$. Furthermore, they show that if no positive power of α fixes any conjugacy class in F_n then l_∞ is defined by a free action of F_n on an $\mathbf{R}$-tree.

The existence of a train-track structure for an irreducible automorphism has a number of other important consequences. In particular, Bestvina and Handel used it to prove a conjecture of G. P. Scott's, that if α is an *arbitrary* automorphism of F_n then the fixed elements of α form a subgroup of rank at most n.

2.2.7. The non-polyhedral free actions that I have described are by construction *limits* of free polyhedral actions, in the sense that the points of $\mathcal{SPL}(\Gamma)$ that they define are limits of sequences of points defined by free polyhedral actions. But it is easy to see that every free polyhedral action is a limit of simplicial actions: one need only approximate the lengths of the edges of the quotient graph by rational numbers. (This gives a $\frac{1}{m}\mathbf{Z}$-valued length function approximating the given polyhedral length function. But any $\frac{1}{m}\mathbf{Z}$-valued length function defines the same point of $\mathcal{SPL}(\Gamma)$ as some $\mathbf{Z}$-valued length function.) So the actions described in 2.2.1–2.2.6 are limits of simplicial actions.

2.3. Free actions of surface groups. I mentioned in 1.2.9 that the only groups that obviously admit free actions on $\mathbf{R}$-trees are free products of subgroups of $\mathbf{R}$. In 2.2 we saw that free groups, which of course belong to this class, admit surprising free actions. There are also examples of groups that are not free products of subgroups of $\mathbf{R}$ and which admit free actions. The first such examples, given by Alperin and Moss in [AlpM], are infinitely generated. In [MSh3], Morgan and I showed, by applying fundamental work of Thurston's, that the fundamental groups of most closed surfaces admit free actions on $\mathbf{R}$-trees.

2.3.1. As I pointed out in 2.2.7, the surprising free actions that I described for a free group F_n can be obtained as limits (in $\mathcal{SPL}(F_n)$) of free simplicial actions. General principles guaranteed that these limit actions would be small, and in fact they were often free. In the case of a surface group Γ, one can obtain surprising actions, including free actions, as limits of small simplicial actions. In this case the simplicial actions cannot be free, since the group is not free. Nevertheless, the limit actions are often free.

2.3.2. So I will begin by talking about some examples of small actions on $\mathbf{Z}$-trees. (Since I also described these examples in [Sh], I will be brief.) Let Σ be a closed surface and let $C \subset \Sigma$ be a disjoint union of two-sided simple closed curves. Then the universal cover $\tilde{\Sigma}$ of Σ is a simply-connected surface, and the pre-image $\tilde{C}$ of C in $\tilde{\Sigma}$ under the covering map $p\colon \tilde{\Sigma} \to \Sigma$ is a properly embedded

1-manifold. There is a dual graph of $\tilde{C}$ in $\tilde{\Sigma}$ whose vertices and edges correspond to components of $\tilde{\Sigma} - \tilde{C}$ and $\tilde{C}$ respectively; an edge e is incident to a vertex v if and only if the component of $\tilde{C}$ corresponding to e is contained in the closure of the component of $\tilde{\Sigma} - \tilde{C}$ corresponding to v. Since $\tilde{\Sigma}$ is 1-connected, it is not hard to show that the dual graph is 1-connected as well. The action of $\Gamma = \pi_1(\Sigma)$ on $\tilde{\Sigma}$ induces an action on this dual graph, and hence on a $\mathbf{Z}$-tree. We call this action the *dual action* to the curve system C.

It is easy to show that the dual action is minimal if and only if C is reduced in the sense that it is non-empty and has no homotopically trivial components. Under the dual action, the stabilizer of a vertex is the stabilizer of the corresponding component of $\tilde{\Sigma} - \tilde{C}$, which up to conjugacy is the image of the fundamental group of a component of $\Sigma - C$ in $\pi_1(\Sigma)$. Likewise, the stabilizer of an edge is (up to conjugacy) the image of the fundamental group of a component of C in $\pi_1(\Sigma)$. In particular, the stabilizer of every edge is an infinite cyclic group. It follows that the action dual to a reduced curve system is small.

2.3.3. An elementary argument, given in [MSh1] and essentially due to Stallings, shows that every small minimal action of a surface group on a $\mathbf{Z}$-tree can be obtained by the above construction.

2.3.4. As I said in 2.3.1, one can obtain interesting small actions of surface groups on $\mathbf{R}$-trees, including free actions, as limits of small actions on $\mathbf{Z}$-trees. This is the way these actions were described in [Sh]. However, in order to get a deeper understanding of these actions it is necessary to give a more direct description of them. They can in fact be described by generalizing the construction given in 2.3.2. This was done in [MSh3], and from a somewhat different point of view in Section 5 of [GiS1]. Here I will try to give a self-contained account of the approach of [GiS1], which involves Thurston's theory of measured foliations (see [FLP]). First it will be helpful to look at reduced curve systems a bit more closely.

2.3.5. A reduced curve system C can be described by combinatorial data in terms of a triangulation of Σ. In fact, given C one can construct a triangulation of Σ which is adapted to C in the sense that C intersects each 2-simplex σ in the way shown in Figure 2.3.5.1.

In somewhat fewer than a thousand words, this means that C meets each edge of σ, and that we can label the edges of σ as τ , τ' and τ'' in such a way that every component of $C \cap \sigma$ is an arc joining a point of τ to a point of τ' or τ''. In particular, C then intersects each 1-simplex τ in a set of finite cardinality $x_\tau > 0$. The positive integers x_τ clearly have the property that

(i) for any 2-simplex σ of Σ we may label the edges of σ as τ , τ' and τ'' in such a way that $x_\tau = x_{\tau'} + x_{\tau''}$.

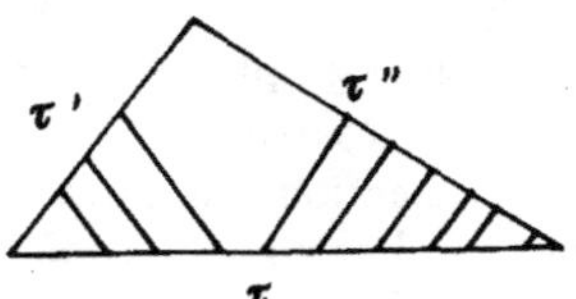

FIGURE 2.3.5.1

A family (x_τ) of positive integers indexed by the 1-simplices of a given triangulation will be called an integer *length system* for Σ if it satisfies condition (i). We call τ the *long edge* of σ relative to the given length system. For each vertex v of Σ, the *order* o_v of v with respect to the length system is defined to be the number of 2-simplices incident to v whose long edge is not incident to v. I'll say that a length system is *non-degenerate* if

(ii) $o_v > 2$ for every vertex v of Σ.

2.3.6. Given a reduced curve system, it is possible to choose a triangulation which is adapted to C and such that the corresponding length system is non-degenerate. Conversely, it is not hard to show that every non-degenerate integer length system is defined by some reduced curve system. So one can think of the small actions on $\mathbf{Z}$-trees described in 2.3.2 as being associated with $\mathbf{Z}$-valued length systems. This suggests defining small actions on $\mathbf{R}$-trees by means of *real-valued length systems.*

The definition of a real-valued length system is the same as that of an integer-valued length function except that the x_τ are positive *real* numbers. The definition of the long edge of a 2-simplex and the order of a vertex go through without change, and so does the definition of non-degeneracy. Now let me explain how a non-degenerate real-valued length system gives rise to an action of $\pi_1(\Sigma)$ on an $\mathbf{R}$-tree.

2.3.7. To construct the tree we note that for each 2-simplex σ there is an affine map $f_\sigma: \sigma \to \mathbf{R}$ which maps each edge τ of σ onto an interval of length x_τ. Such a map f_σ is unique up to composition with isometries of $\mathbf{R}$. We define the *length* $\lambda(\alpha)$ of an affine path $\alpha\colon [0,1] \to \sigma$ to be the length of the interval $f_\sigma(\alpha([0,1]))$. A piecewise-linear path α in Σ can be written as a composition $\alpha_1 * \cdots * \alpha_n$ of affine paths in 2-simplices, and it has a well-defined length $\lambda(\alpha) = \lambda(\alpha) + \cdots + \lambda(\alpha)$.

The length of a path $\tilde{\alpha}$ in the universal cover $\tilde{\Sigma}$ is defined to be the length of $p \circ \tilde{\alpha}$.

It can be shown that any two points x and y of $\tilde{\Sigma}$ are joined by a path of minimal length. The existence of such a minimal path is not at all obvious. It depends strongly on condition (ii) of 2.3.5; it also depends on the compactness of the surface Σ, which guarantees the existence of a group of simplicial homeomorphisms of $\tilde{\Sigma}$ which has compact quotient and preserves the induced length system of $\tilde{\Sigma}$.

If we denote the minimal length of a path from x to y by dist (x, y), it follows that *dist* is a pseudo-metric on $\tilde{\Sigma}$. This pseudo-metric gives rise to a metric space T by the standard construction: we say that two points $x, y \in \tilde{\Sigma}$ are *equivalent* if $\text{dist}(x, y) = 0$; the points of T are equivalence classes and the distance function on T is induced by dist.

It may be shown that T is an **R**-tree. The action of $\pi_1(\Sigma)$ on Σ by deck-transformations induces an action (by isometries) on T. The stabilizer of every non-degenerate segment is cyclic. When the given length system is integer-valued, this action is the completion of the action described in 2.3.2 in terms of the corresponding reduced curve system.

2.3.8. The above description of the tree associated to a real-valued length system is a paraphrase of the description given in [GiS1] in terms of Thurston's theory. His theory is needed to prove the assertions that I made in 2.3.7, including the existence of a minimal path between two points in $\tilde{\Sigma}$. Geometrically, the points of T are *leaves* of a *foliation with singularities* of $\tilde{\Sigma}$. Each leaf is locally Euclidean of dimension 1 except at the vertices. The foliation of $\tilde{\Sigma}$ induces a foliation of Σ.

A neighborhood of a vertex v in a leaf in Σ is a cone on o_v points. The assignment of a length to each piecewise-linear path may be interpreted as a *transverse measure* on the foliation. See Figure 2.3.8.1 for the local picture in the case $o_v = 3$.

The vertices v such that $o_v \neq 2$ are *singularities* of the foliation. Condition (ii) of 2.3.5 says that the singularities are all *non-degenerate*, i.e. that the situations shown in Figure 2.3.8.2 never occur up to homeomorphism. These degenerate situations correspond to the cases $o_v = 0$ and $o_v = 1$.

The action of $\pi_1(\Sigma)$ on T is often said to be *dual* to the measured foliation determined by the given length system.

2.3.9. Let me use the geometric picture to illustrate the comment that I made in 2.3.7, that the restrictions on the induced length system of $\tilde{\Sigma}$, including invariance under a group of simplicial homeomorphisms of $\tilde{\Sigma}$ which has compact quotient, are needed to prove the existence of a minimal path between two points of $\tilde{\Sigma}$. Let $f\colon (-1, 1) \to \mathbf{R}$ be a piecewise-linear continuous even function which

FIGURE 2.3.8.1

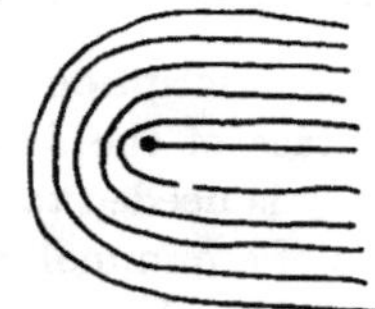

FIGURE 2.3.8.2

tends to $+\infty$ as $x \to \pm 1$. There is a foliation of $\mathbf{R}^2$ with no singularities, whose leaves are the lines $x = c$ for $|c| \geq 1$ and the curves $y = f(x) + b$ for $b \in \mathbf{R}$: see figure 2.3.8.3.

This foliation has a unique transverse measure for which the length of every vertical line segment in the y-axis, and of every horizontal line segment in the complement of $(-1, 1) \times \mathbf{R}$, is equal to its Euclidean length. It is possible to realize the resulting measured foliation by a length function on a suitable triangulation.

But now look at the points $P = (0, -1)$ and $Q = (0, -1)$. For any ϵ with $0 < \epsilon < 1$ we can join P and Q by an arc consisting of the line segments $[-1, -1 + \epsilon] \times \{0\}$ and $[1 - \epsilon, 1] \times \{0\}$ and an arc in the leaf. As $\epsilon \to 0$ the length of this arc (with respect to the measure) tends to 0. On the other hand, P and Q lie in different leaves, so there is no path of length 0, and hence no minimal

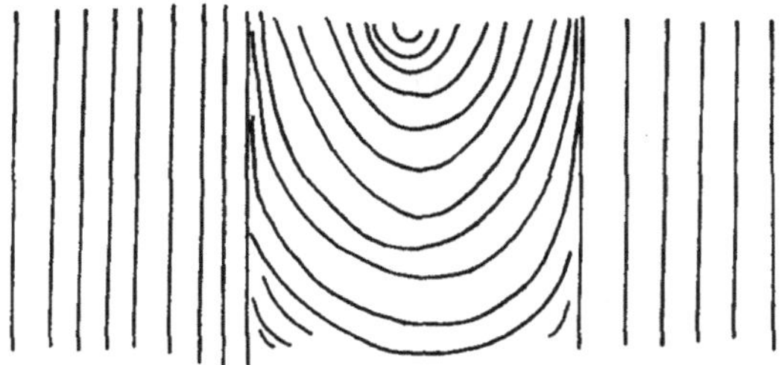

FIGURE 2.3.8.3

path, joining them. What's more, there is no reasonable way of associating a tree with this measured foliation.

2.3.10. If Σ is a compact orientable surface of positive genus, it may be shown that there always exsits a non-degenerate length system for some triangulation of Σ, and that for a generic choice of such a length system, the corresponding action of $\pi_1(\Sigma)$ on an **R**-tree is free. Furthermore, this construction, applied with a bit more care, gives free actions of the fundamental groups of all non-orientable surfaces except those of Euler characteristic 1, 0 and -1. These facts were established in [MSh3], although from a different point of view.

2.3.11. Let (x_τ) be a non-degenerate length system on Σ, and denote by $[l] \in \mathcal{SPL}(\pi_1(\Sigma))$ the projectivized length function defined by the action of $\pi_1(\Sigma)$ on the associated **R**-tree. It is a matter of linear algebra to show that we can approximate the numbers x_τ arbitrarily well by positive rational numbers x'_τ which define a length system in the same triangulation. After multiplying the x'_τ by a suitable integer we obtain an integer length system which defines an action on a **Z**-tree and a corresponding projectivized integer-valued length function $[l'] \in \mathcal{SPL}(\pi_1(\Sigma))$. If we choose the approximation x'_τ with a bit of care, it may be shown[3] that $[l']$ approaches $[l]$ as x'_τ approaches x_τ. This shows that the small action defined by the length system (x_τ) is, as I promised, a limit of actions of the type defined in 2.3.2.

2.3.12. The actions of surface groups on **R**-trees that are defined by non-degenerate length systems—or, if you prefer, are dual to measured foliations—are of course generally not polyhedral; nevertheless, these actions have a striking

[3] This is not at all obvious. It is equivalent to a result due to Thurston. You may find a proof of a similar result in a somewhat more general context in [GiSSk].)

finiteness property. Let us define a ***partition*** of a topological arc S to be a finite set $\mathcal{S}$ of sub-arcs of S such that S is the union of the segments in $\mathcal{S}$, and any two segments in $\mathcal{S}$ meet in at most a single point. If an action of a surface group Γ on a tree T is dual to a measured foliation, then for any segment $S \subset T$ there exist a partition $\mathcal{S}$ of S and an indexed family $(\gamma^s)_{s \in \mathcal{S}}$ such that the set $\{\gamma^s(s) \colon s \in \mathcal{S}\}$ is again a partition of S.

Note that each γ^s maps $s \subset S$ isometrically onto $\gamma^s \subset S$. So if we identify S isometrically with an interval in $\mathbf{R}$, each of the maps $\gamma^s|s$ is the restriction to s of a translation or a reflection in $\mathbf{R}$. If S' denotes the complement in S of the set of all endpoints of intervals in $\mathcal{S}$, we can define a measure-preserving map $\phi \colon S' \to S$ by setting $\phi|s = \gamma^s|s$ for each $s \in \mathcal{S}$. This map ϕ, is made up of a finite number of isometries. More precisely, its domain is the complement of a finite set of points in S, each component of the domain is mapped isometrically onto an interval in S, and the closure of these image intervals form a partition of S. Such a map is called an ***interval exchange*** on the topological arc S.

2.3.13. The property of a dual action described in 2.3.12 is a consequence of the geometric theory of measured foliations. If the given segment S has a partition into subsegments for which the assertion is true, then it is true for S as well. This observation allows one to assume that there is an arc $\tilde{A} \subset \tilde{\Sigma}$ whose interior avoids the singularities of the foliation and is transverse to the leaves, and such that the quotient map $\tilde{\sigma} \to T$ maps $\tilde{A}$ homeomorphically onto S. Likewise one can assume that the covering projection maps $\tilde{A}$ homeomorphically onto an arc A in Σ. So we have a natural homeomorphism between A and S.

Let us fix a transverse orientation for A in Σ. For a generic point $x \in A$, there is a unique arc α_x such that (i) α_x is contained in a leaf of the foliation and contains no singular points, (ii) $x \in \partial\alpha_x = \alpha_x \cap S$, and (iii) a small neighborhood of x in α_x lies on the positive side of A (with respect to the transverse orientation). This follows from the compactness of Σ and the existence of the transverse measure. Let $\phi(x)$ denote the endpoint $\neq x$ of α_x. It is not hard to show that the ***Poincaré first return map*** ϕ is an interval exchange map. By identifying S with A via the natural homeomorphism we can regard ϕ as an interval exchange on S. Furthermore, for each point $x \in \operatorname{dom} \phi$, the arc α_x may be thought of a as a path from x to $\phi(x)$, and therefore determines an element of $\Gamma = \pi_1(\Sigma, S)$. This element depends only on the component of dom ϕ containing x, so we can denote it γ_s. With these definitions of ϕ and γ_s it is straightforward to verify the assertion of 2.3.12.

2.4. Classification of small actions of surface groups. Richard Skora has recently proved that if Σ is a closed surface and if $\pi_1(\Sigma)$ acts minimally on an $\mathbf{R}$-tree in such a way that the stabilizer of every non-degenerate segment is cyclic, then the given action is associated to a non-degenerate length system by the construction that I described in 2.3.7. In more invariant terms this says that every action of $\pi_1(\Sigma)$ on an $\mathbf{R}$-tree with cyclic segment stabilizers is dual to a measured foliation on Σ. This result, which is the natural generalization of the elementary fact mentioned in 2.3.3 above, gives an affirmative answer[4] to [Sh, Question C].

I would like to give a sketch of Skora's proof, but using a point of view slightly different from Skora's and suggested by Culler and Vogtmann's paper [CuV2].

2.4.1. The starting point for Skora's proof is an idea that was first introduced by Stallings for $\mathbf{Z}$-trees, was applied for general $\mathbf{R}$-trees in [MSh2], and was further developed in [MO]. Suppose that we are given an arbitrary action of the fundamental group of a surface Σ on an $\mathbf{R}$-tree T. Let us fix any triangulation of Σ, and let the universal cover $\tilde{\Sigma}$ be given the triangulation induced by the first barycentric subdivision. Then $\Gamma = \pi_1(\Sigma)$ acts on $\tilde{\Sigma}$ by deck transformations, and it is not hard to construct a Γ-equivariant map $\tilde{f}\colon \tilde{\Sigma} \to T$. Furthermore, $\tilde{f}$ can be constructed so as to map each 1-simplex of Σ onto a segment in T, and so that each 2-simplex σ of Σ has an edge τ such that $\tilde{f}(\tau) = \tilde{f}(\sigma)$. The other two edges τ' and τ'' of σ are mapped onto subsegments whose union is $\tilde{f}(\tau)$. In particular we have

$$\operatorname{length}(\tilde{f}(\tau)) = \operatorname{length}(\tilde{f}(\tau')) + \operatorname{length}(\tilde{f}(\tau'')).$$

Hence we can define a measured foliation with (possibly degenerate) singularities by assigning to each 1-simplex τ of Σ the number $x_\tau = \operatorname{length}(\tilde{f}(\tau))$. The leaves of this foliation are the connected components of the pre-images under $\tilde{f}$ of points of T. The length of a path α in $\tilde{\Sigma}$ is the length of the path $\tilde{f} \circ \alpha$ in T. This measured foliation is $\pi_1(\Sigma)$-invariant on account of the equivariance of f, and hence induces a measured foliation on Σ.

One of the results proved by Morgan and Otal in [MO] (although stated in different language) is that if the action of $\pi_1(M)$ on T is non-trivial, then by modifying the map $\tilde{f}$ one can arrange that the foliation of Σ that it defines has non-degenerate singularities. When this has been done, the measured foliation has a dual tree T_0, and $\tilde{f}$ induces a Γ-invariant morphism $\phi\colon T_0 \to T$. (Morphisms of trees were defined in 1.7.3.)

2.4.2. None of this involves the hypothesis that the action of Γ on T is minimal, or—more important—that it is small. When these hypotheses hold, Skora shows

[4] In [Sh], I discussed the natural small actions of surface groups in terms of measured *laminations*. Here I am discussing them in terms of measured *foliations*. These are two alternative ways of thinking about a class of objects introduced by Thurston. The equivalence between the two points of view is not obvious, however. Measured laminations were used in [MSh3], and measured foliations in [GiS1].

that ϕ is an isomorphism and hence that the given action of Γ on T is dual to a measured foliation. Surjectivity follows from the minimality of the action. So one needs only to prove that if the action of Γ on T is small then ϕ is injective, which by 1.7.3 amounts to proving that it doesn't fold. The proof is slightly simpler under the stronger hypothesis that Γ acts freely on T, and it is this case that I will discuss.

Suppose that ϕ does fold, so that there are two segments $S_1, S_2 \subset T_0$ which meet in a common endpoint and are mapped isometrically by ϕ onto the same segment $S \subset T$. According to 2.3.12 there exist for $i = 1,2$ a partition $\mathcal{S}_i$ of S_i and an indexed family $(\gamma^s)_{s \in \mathcal{S}_i}$ such that the set $\{\gamma^s(s) \colon s \in \mathcal{S}_i\}$ is again a partition of S_i. As in 2.3.12 we get an interval exchange map E_i on each S_i. Conjugating E_1 and E_2 by the isometry $\phi_i = \phi | S_i \colon S_i \to S$, we get two interval exchanges E'_1 and E_2 on S. Now let x be a generic point of S. For each $n \geq 0$ and each $I = (i_1, \ldots, i_n) \in \{0,1\}^n$, we can define $x_I = E'_{i_n} \circ \cdots \circ E'_{i_1}(x)$. From the definition of the E'_i and the equivariance of ϕ you can see that for each $I \in \{0,1\}^n$ there is an element γ_I of Γ such that $x_I = \Gamma_I(x)$. Furthermore, γ_I is a word of length at most n in the elements $\gamma^s, s \in \mathcal{S}_1 \cup \mathcal{S}_2$.

Using the geometry of measured foliations one can arrange, after possibly replacing the group Γ by a finite-index subgroup, and if necessary replacing S by a shorter segment, that the function $I \to \gamma_I$ from $\{0,1\}^n$ to Γ is one-to-one. In this case, the cardinality of the set $\mathcal{G}_n = \{\gamma_I \colon I \in \{0,1\}^n$ is an exponentially growing function of n.

On the other hand, since an interval exchange is made up of finitely many translations and reflections, the cardinality of the set $X_n = \{x_I \colon I \in \{0,1\}^n$ has (at most) polynomial growth as a function of n. Hence when n is large enough, there are two elements $I, I' \in \{0,1\}^n$ such that $\gamma_I \neq \gamma_{I'}$ but $x_I = x_{I'}$, i.e. $\gamma_I(x) = \gamma_{I'}(x)$. This contradicts the hypothesis that the action is free.

2.5. Conjectures and Questions. We saw in 1.2.9 that any free product of subgroups of $\mathbf{R}$ acts freely on an $\mathbf{R}$-tree. A similar construction shows that the class of groups that act freely on $\mathbf{R}$-trees is closed under the formation of free products. In particular a free product of surface groups and free abelian groups admits a free action on an $\mathbf{R}$-tree provided that no factor has the form $\pi_1(\Sigma)$ where Σ is non-orientable and has Euler characteristic > -1.

2.5.1. In [Sh, Question B] I asked whether the following statement was true.

Every finitely generated group which acts freely on an $\mathbf{R}$*-tree is a free product of free abelian groups and surface groups.*

As there is now a great deal of evidence for the truth of this statement, I think it deserves to be called a conjecture. I will discuss some of the evidence in Section 5.

2.5.2. Since it is relatively elementary to show that an action of a surface group

$\pi_1(\Sigma)$ is defined by a measured foliation on Σ if and only if it is a limit of small simplicial actions, we can interpret Skora's theorem (2.4) as saying that $\mathcal{SPL}(\pi_1(\Sigma))$ has a dense subset consisting of projectivized integer-valued length functions. More generally, for any finitely generated group Γ, we can ask:

Does $\mathcal{SPL}(\Gamma)$ have a dense subset consisting of projectivized integer-valued length functions?

2.5.3. A weaker form of Question 2.5.2 was asked as Question D of [Sh]:

If the finitely generated group Γ admits a non-trivial small action on an $\mathbf{R}$-tree, does it admit a non-trivial small action on a $\mathbf{Z}$-tree?

By the Bass-Serre theory this is equivalent to asking whether Γ can be exhibited as either a free product of two proper subgroups with a small amalgamated subgroup, or an HNN extension with a small associated subgroup.

2.5.4. As I explained in [Sh] and will explain in more detail in Section 4 below, Question 2.5.3 is especially important for applications. It is also closely related to the conjecture about groups that act freely. In fact, Morgan and Skora [MSk] have shown that if Γ is a finitely presented group which is indecomposable (i.e. is not a non-trivial free product), and if Γ admits both a small action on an $\mathbf{Z}$-tree and a free action on an $\mathbf{R}$-tree, then Γ is either a free abelian group or a surface group. It follows easily from this result that if Question 2.5.3 has an affirmative answer for finitely presented groups, then Conjecture 2.5.1 is true with the additional hypothesis of finite presentation.

2.5.5. There is considerable evidence that Question 2.5.2 has an affirmative answer under a slightly stronger hypothesis. I will state this here as a conjecture:

Let Γ be a finitely presented group whose small subgroups are all finitely generated. Then $SPL(\Gamma)$ has a dense subset consisting of projectivized integer-valued length functions.

2.5.6. I will discuss the evidence for Conjectures 2.5.1 and 2.5.5 in Section 5.

Actually, the evidence points to a stronger conjecture which implies both the above conjectures, and which I find very natural and appealing in its own right. As the statement requires a bit of background I will defer it to Section 5.

2.5.7. Question 2.5.2 has an appealing analogue for arbitrary (not necessarily small) non-trivial actions:

Let Γ be a finitely generated group. Does $\mathcal{PL}(\Gamma)$ have a dense subset consisting of projectivized integer-valued length functions?

2.5.8. An affirmative answer would imply an affirmative answer to the following question, which was asked in [Sh] as Question A: *If Γ admits a non-trivial action on an $\mathbf{R}$-tree, does it admit a non-trivial action on a $\mathbf{Z}$-tree?* By the Bass-Serre theory this is equivalent to asking whether Γ admits a non-trivial splitting (1.3.3).

The evidence for an affirmative answer to Questions 2.5.7 and 2.5.8 is rather weak so far. I shall discuss it in Section 5.

2.6. Non-Archimedean trees. The questions that I have been discussing in this section for **R**-trees, and to which I will return in Section 5, take on a rather different flavor if we consider actions on Λ-trees where Λ is not a subgroup of **R**. Such actions, besides being a natural object of study in their own right, arise in connection with Tits buildings for general valued fields (Section 3) and degeneration of hyperbolic structures (see Section 4, especially 4.1.7, 4.2.3 and 4.4.1).

2.6.1. An arbitrary ordered abelian group Λ can be analyzed in terms of its convex subgroups. A subgroup Λ_0 of Λ is said to be *convex*, or *isolated*, if for any $x, y, z \in \Lambda$ such that $x \leq y \leq z$ and $x, z \in \Lambda_0$, we have $y \in \Lambda_0$. The group Λ is order-isomorphic to a subgroup of **R** if and only if it has no non-trivial proper convex subgroups. The simplest example of a group Λ which does have a non-trivial proper convex subgroup is the group $\Lambda = \mathbf{Z} \times \mathbf{Z}$ with the lexicographical order; here the unique non-trivial proper convex subgroup is $\{0\} \times \mathbf{Z}$.

In the general case, the convex subgroups of Λ form a linearly ordered set under inclusion. In most casea that arise in applications, there are only finitely many convex subgroups. The number of non-trivial convex subgroups (including Λ itself) is the *order-rank* of Λ. If Λ has order-rank n and its convex subgroups are $\{0\} = \Lambda_0 \leq \Lambda_1 \leq \cdots \leq \Lambda_n = \Lambda$, the quotient groups Λ_i/Λ_{i-1} have induced orders, and are order-isomorphic to subgroups of **R**.

2.6.2. Bass has shown in [Bas3] that it is largely possible to reduce the study of group actions on Λ-trees to the study of actions on Λ_i/Λ_{i-1}-trees. In particular, when the groups Λ_i/Λ_{i-1} are all cyclic, as in the case $\Lambda = \mathbf{Z}^n$ with the lexicographical order, these actions can be analyzed via the Bass-Serre theory. The results are quite different from the corresponding results for **R**-trees. For example, any free product with amalgamation of the form $F_1 *_C F_2$, where the F_i are free and C is an infinite cyclic group which is identified inder the amalgamation with a maximal cyclic subgroup of each F_i, admits a free action without inversions on a $\mathbf{Z} \times \mathbf{Z}$-tree, where $\mathbf{Z} \times \mathbf{Z}$ has the lexicographical order. By contrast, it was shown by Morgan in [M2], and is included in the results of Morgan and Skora discussed in 2.5.4, that a group of this type cannot act freely and without inversions on a **Z**-tree unless it is a free product of cyclic groups and surface groups, as predicted by Conjecture 2.5.1.

SECTION 3. THE BRUHAT-TITS TREE AND ITS APPLICATIONS

3.1. Valuations and Serre's construction. Many of the applications of dendrology involve the Bruhat-Tits building for PSL_2 of a valued field. In this section I am going to describe the Bruhat-Tits construction from a point of view that is essentially due to Serre, who presented it in his book [Se] for the case $\Lambda = \mathbf{Z}$. The general case was presented in my paper [MSh1] with John Morgan. In this section and in 4.1–4.2, I will talk about some applications of this construction to the study of matrix groups, group representations, 3-manifold topology and hyperbolic geometry.

3.1.1. Let me start with a standard definition from commutative algebra. Let K be a field. A *Krull valuation* of K is a homomorphism v of the multiplicative group $K^* = K - \{0\}$ onto an ordered abelian group Λ, called the *value group*, such that for any $x, y \in K^*$ with $x + y \neq 0$ we have

$$v(x+y) \geq \min(v(x), v(y)).$$

To understand this definition, one should bear two basic examples in mind. In both examples we have $\Lambda = \mathbf{Z}$.

First, let $K = \mathbf{Q}$. Any prime p defines a Krull valuation of $\mathbf{Q}$. For $0 \neq a \in \mathbf{Z}$ we define $v(a)$ to be the exponent of p in the prime factorization of a. Then v is a map from $\mathbf{Z} - \{0\}$ to the non-negative integers, and it has a well-defined extension to a map $v\colon \mathbf{Q}^* \to \mathbf{Z}$ given by $v(\frac{a}{b}) = v(a)v(b)$ for $a, b \in \mathbf{Z}$ and $b \neq 0$.

For the second example we take K to be the function field of a complex algebraic curve C. Any smooth point z of C defines a Krull valuation $v = v_z$ of K as follows. Let $f \in K$ be given. If f has a zero of order $n > 0$ at z we set $v(f) = n$; iff has a pole of order $n > 0$ at z we set $v(f) = -n$; and otherwise we set $v(f) = 0$.

If v is a Krull valuation of K then

$$\mathcal{O}_v = \{0\} \cup \{x \in K | v(x) > 0\}$$

is a sub-ring of K, the *valuation ring* defined by v. This ring has a unique maximal ideal, namely

$$\mathcal{M}_v = \{0\} \cup \{x \in K | v(x) > 0\}.$$

3.1.2. A Krull valuation v of K with value group Λ determines an action of the group $\mathrm{GL}_2(K)$ on a Λ-tree, the *Bruhat-Tits building* for $\mathrm{GL}_2(K)$. To define this tree we consider the set of all ($\mathcal{O}_v$-)*lattices* in the 2-dimensional vector space $V = K^2$. A lattice is by definition a finitely generated $\mathcal{O}_v$-submodule of V which spans V as a K-vector space. Any lattice is isomorphic to $\mathcal{O}_v$. We shall call two lattices L and L' *equivalent* if $L' = \theta\lambda L$ for some $\theta \in K^*$. We let T denote the set of all equivalence classes oflattices, and write $[L]$ for the equivalence class of a lattice L.

Given any two lattices L_1 and L_2 one can show that L_2 is equivalent to a lattice L_2' such that $L_2' \subset L_1$, and such that L_1/L_2' is isomorphic (as an $\mathcal{O}_v$-module) to $\mathcal{O}_v/\alpha\mathcal{O}_v$ for some non-zero element α of $\mathcal{O}_v$. One can also show that $v(\alpha)$ is uniquely determined by the equivalence classes of L_1 and L_2; we write $v(\alpha) = \text{dist}([L_1], [L_2])$. Finally, one can show that with this definition of dist: $T \times T \to \Lambda$, the set T becomes a Λ-tree. (You will find proofs of these facts in [Se] for the case $\Lambda = \mathbf{Z}$ and in [MSh1] for the general case.)

3.1.3. In the case $\Lambda = \mathbf{Z}$, there is a simple description of the link of a vertex s of T, i.e. the metric sphere of radius 1 about s. Let us write $s = [L_0]$, where L_0 is a lattice. Let k denote the *residue field* $\mathcal{O}/\mathcal{M} = \mathcal{O}_v/\mathcal{M}_v$. Since $\Lambda = \mathbf{Z}$, any element π of $\mathcal{O} = \mathcal{O}_v$ with $v(\pi) = 1$ generates $\mathcal{M}_v$. Hence the link of v consists of all classes represented by lattices $L \subset L_0$ such that L_O/L is isomorphic as an $\mathcal{O}$-module to $\mathcal{O}/\mathcal{M}$. Such lattices L are in bijective correspondence, via the natural map $\mathcal{O}^2 \to k^2$, with 1-dimensional subspaces of the k-vector space k^2. Thus the link of any vertex has the structure of a projective line over the field k. In particular, if k is a finite field of order q then T is a $(q+1)$-regular tree. (A graph is *k-regular*, where k is a positive integer, if every vertex has valence k.)

3.1.4. The natural (linear) action of $\text{GL}_2(K)$ on V induces an action (by isometries) of $\text{PGL}_2(K)$ on T. The restriction of this action to $PSL_2(K)$ an action without inversions.

It is easy to describe the stabilizers of points of T under the action of $\text{PSL}_2(K)$. First one observes that the action of $\text{PGL}_2(K)$ on lattices is transitive; hence the stabilizer in $\text{PSL}_2(K)$ of any point of T is conjugate in $\text{PGL}_2(K)$ to the stabilizer of the equivalence class of the standard lattice $L_0 = \mathcal{O}_v^2 \subset K^2 = V$. Next one checks that a unimodular matrix can fix $[L_0]$ only if it fixes L_0. But the stabilizer of L_0 is by definition the subgroup $\text{PSL}_2(\mathcal{O}_v)$ of $\text{PSL}_2(K)$. So the point-stabilizers in $\text{PSL}_2(K)$ are just the conjugates of $\text{PSL}_2(\mathcal{O}_v)$ by elements of $\text{PGL}_2(K)$.

3.2. Ihara's theorem. In [Se], Serre gives a number of applications of this construction. One celebrated application is a proof of a theorem due to Ihara about groups of matrices over p-adic numbers.

One nice way to define p-adic numbers is to use the concept of a Krull valuation. If p is any prime one can make $\mathbf{Q}$ into a metric space by using the valuation $v = v_p$ of $\mathbf{Q}$ that I defined above: the standard way is to set $\text{dist}(x, y) = p^{v(x-y)}$.

The set $\mathbf{Q}_p$ of p-adic numbers is by definition the completion of this metric space. The field operations of $\mathbf{Q}$ have unique continuous extensions to $\mathbf{Q}_p$, making $\mathbf{Q}_p$ a field; and $v = v_p$ extends uniquely to a valuation of $\mathbf{Q}_p$ with value group $\mathbf{Z}$. I shall denote this extension by v_p as well. It is easy to show that $\mathbf{Q}_p$ is locally compact in the topology defined by its complete metric. Furthermore, the valuation ring $\mathbf{Z}_p$ is compact.

It follows that $\text{PSL}_2(\mathbf{Q}_p)$ has in a natural way the structure of a locally compact topological group, and that $\text{PSL}_2(\mathbf{Z}_p)$ is a compact subgroup. Ihara's theorem asserts that *every torsion-free, discrete subgroup of $PSL_2(\mathbf{Q}_p)$ is free.*

Here is Serre's proof of Ihara's theorem. By the construction that I described in 3.1, the valuation v_p determines a natural action without inversions of $\mathrm{PSL}_2(\mathbf{Q}_p)$ on a $\mathbf{Z}$-tree $T_p = T_{v_p}$. The stabilizer in $\mathrm{PSL}_2(\mathbf{Q}_p)$ of any point (vertex) of T_p is a conjugate of $\mathrm{PSL}_2(\mathbf{Z}_p)$ by an element of $\mathrm{PGL}_2(\mathbf{Q}_p)$, and is therefore compact. Hence if $\Gamma \leq \mathrm{PSL}_2(\mathbf{Q}_p)$ is discrete then the stabilizer in Γ of any point of T_p is both compact and discrete, and is therefore finite. If Γ is torsion-free it follows that the point-stabilizers in Γ are trivial; that is, the action of Γ is free. But as we saw in 1.1.2 and 1.1.4, a group which acts freely and without inversions on a $\mathbf{Z}$-tree is free.

3.3. Ramanujan graphs. Lubotzky, Phillips and Sarnak [LPS] have discovered an amazing and beautiful application of the Bruhat-Tits tree to graph theory. We have seen that the group $\mathrm{PGL}_2(\mathbf{Q}_p)$ has a natural action on a $\mathbf{Z}$-tree T_p. For any subgroup Γ of $\mathrm{PSL}_2(\mathbf{Q}_p)$ we can restrict the action to Γ and form the quotient T_p/Γ, which is a graph. This construction is especially interesting when Γ is a *lattice*, i.e. when Γ is a discrete subgroup of the locally compact group $\mathrm{PGL}_2(\mathbf{Q}_p)$ and the volume of the quotient space $\mathrm{PGL}_2(\mathbf{Q}_p)1/\Gamma$–defined in terms of the Haar measure on $\mathrm{PGL}_2(\mathbf{Q}_p)$–is finite. In this case T_p/Γ is a finite graph. As in 3.2, if Γ happens to be torsion-free then it acts freely and without inversions on T_p. In this case the graph T_p/Γ is $(p+1)$-regular, since T_p is a $(p+1)$-regular tree by 3.1.3.

This construction turns out to give examples of finite graphs that are extremely interesting from the perspective of combinatorics and computer science.

3.3.1. The most natural lattices in $\mathrm{PSL}_2(\mathbf{Q}_p)$ are the arithmetic ones, which are constructed from quaternion algebras. For any two positive integers u and v, we may construct a 4-dimensional associative $\mathbf{Q}$-algebra $D = D_{uv}$ with a linear basis $\{1, i, j, k\}$ satisfying the relations

$$i^2 = -u, j^2 = -v, ij = -ji = k.$$

If u and v are fixed, then for all but finitely many primes p, the $\mathbf{Q}_p$-algebra $D \otimes \mathbf{Q}_p$ is isomorphic to the matrix algebra $\mathcal{M}_2(\mathbf{Q}_p)$. In this case $D = D_{uv}$ is said to be *unramified* at p. Note that when D is unramified, its group of units $D^* = D^*_{uv}$ is isomorphic to $\mathrm{GL}_2(\mathbf{Q}_p)$.

For any prime p there is a sub-algebra $D(\mathbf{Z}[\frac{1}{p}])$ of D consisting of all elements whose coefficients in the basis $\{1, i, j, k\}$ belong to the ring $\mathbf{Z}[\frac{1}{p}]$. The group of units $D^*(\mathbf{Z}[\frac{1}{p}])$ is a subgroup of $D^*(u, v)$. Hence when $D(u, v)$ is unramified at p we can identify $D^*(\mathbf{Z}[\frac{1}{p}])$ with a subgroup of $\mathrm{GL}_2(\mathbf{Q}_p)$. The image Γ of $D^*(\mathbf{Z}[\frac{1}{p}])$ in $\mathrm{PGL}_2(\mathbf{Q}_p)$ is a lattice.

3.3.2. This lattice Γ does not define an interesting graph, because it acts transitively on T_p. However, for every positive integer N prime to p there is a *congruence subgroup* $\Gamma(N)$ of Γ modulo N. To define $\Gamma(N)$ we note that the unique ring homomorphism $\mathbf{Z}[\frac{1}{p}] \to \mathbf{Z}/n\mathbf{Z}$ induces a homomorphism h from $D(\mathbf{Z}[\frac{1}{p}])$ to $D(\mathbf{Z}/n\mathbf{Z})$, where $D(\mathbf{Z}/n\mathbf{Z}) = D_{uv}(\mathbf{Z}/n\mathbf{Z})$ is the (finite) 4-dimensional algebra defined over $\mathbf{Z}/n\mathbf{Z}$) in the same way that D_{uv} is defined over Q. The kernel of $h|D^*(\mathbf{Z}[\frac{1}{p}])$ is a finite-index subgroup of $D^*(\mathbf{Z}[\frac{1}{p}])$. We define $\Gamma(N)$ to be the image of the latter subgroup in Γ. Since $\Gamma(N)$ has finite index in Γ, it is again a lattice. Furthermore, if p is not congruent to a square modulo p (as is the case for 50% of all integers N) then $\Gamma(N)$ is torsion-free, and so $T_p/\Gamma(N)$ is a finite $(p+1)$-regular graph.

3.3.3. These regular graphs have many remarkable properties. The most important can be expressed in terms of adjacency matrices. If $\mathcal{G}$ is a finite graph with vertex set V, its *adjacency matrix* is defined to be $A = (a_{vw})_{v,w\in V}$, where $a_{v,w}$ is equal to 1 if v and w are joined by an edge, and is equal to 0 otherwise. One can encode subtle information about a finite graph $\mathcal{G}$ in the eigenvalues of its adjacency matrix A, or equivalently in those of the matrix $\Delta = (\#V)I - A$, which can be interpreted as the matrix of a combinatorial Laplacian operator for $\mathcal{G}$. In particular, let λ_1 denote the smallest positive eigenvalue of Δ, and for any set $X \subset V$ define the *coboundary* δX of X to be the set of all edges having exactly one endpoint in X. It was proved by Tanner [T] and Alon and Milman [AloM] that for any subset X of V with $\#X \le \frac{1}{2}\#V$ we have

$$\#\delta X \ge \frac{\lambda_1}{2}\#X.$$

This is expressed by saying that the *Cheeger constant* of $\mathcal{G}$ is at least $\frac{\lambda_1}{2}$. Roughly speaking, if the Cheeger constant is large then there are a lot of edges joining any subgraph to its complement. As you may imagine, this makes graphs with large Cheeger numbers, known as *expanders*[5] ,important as components in communication networks. So it is of practical importance to have graphs for which λ_1 is large.

For simplicity, let's fix an integer $k > 1$ and look at k-regular graphs. It is a theorem due to Alon and Boppana that for any $\epsilon > 0$ there are only finitely many k-regular graphs with $\lambda_1 > k - \sqrt{2k-1} + \epsilon$. So in an asymptotic sense, the best estimate one can hope for is $\lambda_1 \ge k - \sqrt{2k-1}$.

3.3.4. A k-regular graph $\mathcal{G}$ is called a *Ramanujan graph* if every eigenvalue of its adjacency matrix is either $= \pm k$ or $\le \sqrt{2k-1}$. For a Ramanujan graph we automatically have $\lambda_1 \ge k - \sqrt{2k-1}$. Lubotzky, Phillips and Sarnak proved that the $(p+1)$-regular graphs $T_p/\Gamma(N)$ described in 3.3.2 are all Ramanujan graphs.

[5]For a more precise definition of an expander, see [L].

In particular this implies that for $k = p + 1$, where p is a prime, there are infinitely many k-regular Ramanujan graphs. Morgenstern [Mst] has generalized this to the case $k = q+1$ where q is a prime power. Apparently nothing is known for other values of k.

For q a prime power, these methods allow one to describe infinite families of $(q + 1)$-regular Ramanujan graphs quite explicitly, so that in principle they can be used for constructing networks.

3.3.5. The proof that the graphs $T_p/\Gamma(N)$ are Ramanujan graphs uses deep results in number theory and representation theory, including the Ramanujan Conjecture[6] (proved by Deligne using his proof of the Riemann-Weil conjecture) and the Jacquet-Langlands correspondence. Well, you get the idea. I had better refer you to Lubotzky's book [L] for details. The whole thing is an impressive demonstration of the power of pure mathematics.

3.4. Character varieties, ideal points and trees. I want to describe another way of applying the Bruhat-Tits construction. It involves some additional machinery that Marc Culler and I developed in our paper [CuS1]. Let me begin with some elementary background.

3.4.1. The group $\mathrm{SL}_2(\mathbf{C})$ has in a natural way the structure of a complex affine algebraic variety: it is the solution set of the equation $xw - yz = 0$ in the 4-dimensional affine space $\mathcal{M}$ of complex 2×2- matrices $\begin{pmatrix} x & y \\ z & w \end{pmatrix}$.

The group $\mathrm{PSL}_2(\mathbf{C})$ has likewise the structure of a complex affine variety. This can be seen, for example, by mapping the variety $\mathrm{SL}_2(\mathbf{C})$ into the affine space of linear transformations of the vector space $\mathcal{M}$ via the map $A \mapsto C_A$, where $C_A \colon \mathcal{M} \to \mathcal{M}$ is defined by $C_A(M) = AMA^{-1}$. For any $A, B \in \mathrm{SL}_2(\mathbf{C})$ we have $C_A = C_B$ if and only if $A = \pm B$. Thus the image of the map $A \mapsto C_A$ is identified with $\mathrm{PSL}_2(\mathbf{C})$. But the map $A \mapsto C_A$ is proper and is defined by polynomials in the affine coordinates; this implies that its image is again an affine variety.

3.4.2. For any finitely generated group Γ, consider the set $\mathrm{Hom}(\Gamma, \mathrm{PSL}_2(\mathbf{C}))$ of all representations of Γ in $\mathrm{PSL}_2(\mathbf{C})$. If we fix a set S of generators of Γ then a representation $\rho \in \mathrm{Hom}\ (\Gamma, \mathrm{PSL}_2(\mathbf{C}))$ is determined by the images of the elements of S under ρ. So we can identify $\mathrm{Hom}\ (\Gamma, \mathrm{PSL}_2(\mathbf{C}))$ with a subset of the algebraic variety $\mathrm{PSL}_2(\mathbf{C})^S$. It is easy to see that $\mathrm{Hom}\ (\Gamma, \mathrm{PSL}_2(\mathbf{C}))$ is in fact an algebraic subset of $\mathrm{PSL}_2(\mathbf{C})^S$: to each defining relation in Γ there corresponds a matrix equation, which translates into a system of algebraic equations in the affine coordinates. Up to isomorphism the structure of an algebraic set on $\mathrm{Hom}\ (\Gamma, \mathrm{PSL}_2(\mathbf{C}))$ is independent of the choice of generating set S.

[6] Yes, that's where they got the term.

3.4.3. Now let C be a curve in Hom $(\Gamma, \mathrm{PSL}_2(\mathbf{C}))$, i.e. an irreducible affine algebraic subset of complex dimension 1. Let K be the function field of C: concretely we can think of K as consisting of rational maps from C to $\mathbf{C}$, i.e. functions $C \to \mathbf{C}$ of the form $\frac{f}{g}$, where f and g are restrictions of polynomial functions on the ambient affine space, and g does not vanish identically on C. Next consider the group $\mathrm{PSL}_2(K)$. We may interpret an element of this group as a rational mapping from C to $\mathrm{PSL}_2(\mathbf{C})$. With each element $\gamma \in \Gamma$ we can associate the rational mapping $P(\gamma)\colon C \to \mathrm{PSL}_2(\mathbf{C})$ defined as follows: an arbitrary point of C is a representation $\rho\colon \Gamma \to \mathrm{PSL}_2(\mathbf{C})$. We define $P(\gamma)$ to map each $\rho \in C$ to $\rho(\gamma)$. If for each $\gamma \in \Gamma$ we now identify $P(\gamma)$ with an element of $\mathrm{PSL}_2(K)$, we obtain the *tautological representation* $P\colon \Gamma \to \mathrm{PSL}_2(K)$.

Any affine curve C has (up to isomorphism) a unique projective completion $\hat{C}$ such that all the (finitely many) points of $\hat{C} - C$ are smooth. I'll call these points *ideal points* of C. One of the basic ideas of [CuS1] is that if Γ is a finitely generated group and $C$is a curve in Hom $(\Gamma, \mathrm{PSL}_2(\mathbf{C}))$, ideal points of C tend to determine interesting actions of Γ on $\mathbf{Z}$-trees. It is not hard to see why this should be so. The inclusion map from C to $\hat{C}$ induces an isomorphism of function fields; that is, any rational function on C has a unique extension to $\hat{C}$ which is meromorphic at z. So we can identify the function field of $\hat{C}$ with K. Any point z of $\hat{C} - C$ then defines a valuation $v = v_z$ of K. This valuation in turn defines a $\mathbf{Z}$-tree $T = T_z$, and by pulling back the natural action of $\mathrm{PSL}_2(K)$ by the homomorphism P we obtain an action of Γ on T_z.

3.4.4. This action contains important information. For example, for any element γ of Γ we have a function $I_\gamma \in K$ defined by $I_\gamma = (\mathrm{trace}\ \rho(\gamma))^2$ for any $\rho \in$ Hom $(\Gamma, \mathrm{PSL}_2(\mathbf{C}))$. (For an element of $\mathrm{PSL}_2(\mathbf{C})$ the square of the trace is well-defined although the trace itself is not.) The restriction of I_γ to C extends uniquely to a function $\hat{I}_\gamma\colon \hat{C} \to \mathbf{C}$ This function $\hat{I}_\gamma$ is finite-valued at a given ideal point z if and only if γ fixes some point of T_z. This is not hard to prove: if you chase through the definitions you will find that $\hat{I}_\gamma$ is finite-valued at z if and only if $(\mathrm{trace} P(\gamma))^2 \in \mathcal{O}_v$, which by an elementary argument is equivalent to saying that $P(\gamma)$ is in a conjugate (within $\mathrm{GL}_2(K)$) of $\mathrm{PSL}_2(\mathcal{O}_v)$. But the conjugates of $\mathrm{PSL}_2(\mathcal{O}_v)$ are just the stabilizers of points of T.

A similar analysis shows that if $\hat{I}_\gamma$ has a pole at z then the order of the pole is $-2l(\gamma)$, where $l\colon \Gamma \to \mathbf{Z}$ is the length function defined by the action of Γ on T_z.

3.4.5. What this makes clear is that the action of Γ on the tree T_z reflects the behavior of the functions I_γ near the ideal point z. Now the important thing about the functions I_γ is that they are invariant under the natural action by conjugation of $\mathrm{GL}_2(\mathbf{C})$ on Hom $(\Gamma, \mathrm{PSL}_2(\mathbf{C}))$. So it is natural to expect that the I_γ can be defined on some sort of "quotient" of Hom $(\Gamma, \mathrm{PSL}_2(\mathbf{C}))$ under the action of $PGL_2(\mathbf{C})$, and that the trees T_z are best understood in terms of this quotient. This is quite true, as I shall explain.

3.4.6. Some care is required in order to define the quotient of Hom $(\Gamma, PSL_2(\mathbf{C}))$ under the action of $PGL_2(\mathbf{C})$. This is already apparent in the case of an infinite cyclic group $\Gamma = \langle t \rangle$. For every $z \in \mathbf{C}$ we have a representation ρ_z which maps t to $\begin{pmatrix} 1 & z \\ 0 & 1 \end{pmatrix}$. The points $\rho_z \in$ Hom $(\Gamma, PSL_2(\mathbf{C}))$, $z \in \mathbf{C}^*$ constitute a single orbit under the action of $PGL_2(\mathbf{C})$. But this orbit is not closed in Hom $(\Gamma, PSL_2(\mathbf{C}))$ since its closure contains the trivial representation ρ_0. So the quotient in the category of topological spaces is not even a Hausdorff space.

The correct approach is to form the quotient in the category of algebraic varieties. This can be done using the machinery of geometric invariant theory, or it can be done from the very elementary point of view described in [CuS1]. In any event, the upshot is that to every finitely generated group Γ we can canonically associate an affine algebraic set $X(\Gamma)$ and a morphism[7] τ: Hom $(\Gamma, PSL_2(\mathbf{C})) \to X(\Gamma)$ such that two points ρ_1 and ρ_2 have the same image under τ if and only if $I_\gamma(\rho_1) = I_\gamma(\rho_2)$ for every $\gamma \in \Gamma$. In particular, τ maps every $PGL_2(\mathbf{C})$-orbit to a point. What is more, τ is surjective, and every fiber of τ either is exactly a $PGL_2(\mathbf{C})$-orbit or consists entirely of representations which are reducible, i.e. are conjugate to representations of Γ by upper triangular matrices. (Thus if we stay away from the reducible representations, which are relatively degenerate examples, $X(\Gamma)$ behaves like a quotient in the naïve sense.) For any $\gamma \in \Gamma$ the function I_γ induces a function on $X(\Gamma)$ which I will also denote I_γ. The coordinate ring of $X(\Gamma)$ is generated by functions of the form I_γ: what this means in concrete terms is that we can take $X(\Gamma)$ to live in an affine space in such a way that the coordinate functions are of the form I_γ.

I like to think of $X(\Gamma)$ in terms of group characters. A representation $\rho\colon \Gamma \to PSL_2(\mathbf{C})$ determines a function $\chi\colon \Gamma \to PSL_2(\mathbf{C})$, its *character*, given by $\chi(\gamma) = (\text{trace } \rho(\gamma))^2$. The condition for two points ρ_1 and ρ_2 to have the same image under τ , that $I_\gamma(\rho_1) = I_\gamma(\rho_2)$ for every $\gamma \in \Gamma$, can be paraphrased by saying that ρ_1 and ρ_2 have the same character. So we can identify the points of $X(\Gamma)$ with characters of representations in $PSL_2(\mathbf{C})$, and τ is then the map that takes every representation to its character. Furthermore, for any $\gamma \in \Gamma$, the function $I_\gamma\colon X(\Gamma) \to \mathbf{C}$ is then simply the evaluation map $\chi \mapsto \chi(\gamma)$.

3.4.7. Once one has defined the space of characters $X(\Gamma)$ it is not hard to adapt the construction of 3.4.3 to curves in $X(\Gamma)$. To each ideal point x of an arbitrary curve $C \subset X(\Gamma)$, one we can associate an action of Γ on a $\mathbf{Z}$-tree T_x. Again, for any $\gamma \in \Gamma$, the extension $\hat{I}_\gamma$ of $I_\gamma|C$ to $\hat{C}$ is finite-valued at x if and only if

[7] A map between affine algebraic sets is a *morphism* if it is defined by polynomials in the affine coordinates.

γ fixes some point of T_x; and if $\hat{I}_\gamma$ has a pole at x then the order of the pole is $-2l(\gamma)$, where $l\colon \Gamma \to \mathbf{Z}$ is the length function defined by the action of Γ on T_x.

The construction of the action is a slight variant of the one described in 3.4.3. Since the natural map Hom $(\Gamma, \mathrm{PSL}_2(\mathbf{C})) \to X(\Gamma)$ is surjective, it maps some irreducible subvariety W of Hom $(\Gamma, \mathrm{PSL}_2(\mathbf{C}))$ onto a dense subset of C, and there is an induced monomorphism from the function field F of v to the function field K of W. Let us identify F with a subfield of K. The valuation v of F defined by the ideal point z can be extended–after possibly enlarging the value group by finite index, so that it is still isomorphic to $\mathbf{Z}$–to a valuation w of K. As in 3.4.3 we have a tautological representation $P\colon \Gamma \to \mathrm{PSL}_2(K)$. The valuation w defines a $\mathbf{Z}$-tree T, and $\mathrm{PSL}_2(K)$ has a natural action on T. Pulling back this action via P we get an action of Γ on T.

One clear advantage of working with a curve $C \subset X(\Gamma)$, as opposed to a curve in Hom $(\Gamma, \mathrm{PSL}_2(\mathbf{C}))$, is that the action of Γ on T_z is non-trivial for every ideal point z of C . Indeed, if the action were trivial, the function I_γ would be finite-valued at z for every $\gamma \in \Gamma$. In particular the coordinate functions would be finite-valued at z; this is impossible since $z \notin C$.

3.5. A finiteness theorem. As a first application of the theory that I have been discussing in 3.4, let me give a proof of a result that was first proved from a somewhat different point of view by Hyman Bass in [Bas1,2]. *Let Γ be a finitely generated group which admits no splitting in the sense of 1.3.3: that is, Γ is not an amalgamated free product of two proper subgroups, and admits no homomorphism onto $\mathbf{Z}$. Then up to conjugacy there are only finitely many irreducible representations of Γ in $PSL_2(\mathbf{C})$.*

To prove this we observe that the character space $X(\Gamma)$ is 0-dimensional. Indeed, if the dimension of $X(\Gamma)$ were > 0 then $X(\Gamma)$ would contain a curve C. Any ideal point of C would define a non-trivial action of Γ on a $\mathbf{Z}$-tree, which by the Bass-Serre theory would give a non-trivial splitting of Γ and a contradiction to the hypothesis. Thus $X(\Gamma)$ is a finite set. Since an irreducible representation is determined up to conjugacy by its character, the assertion follows.

3.6. The 3-manifold connection. The theory that I described in 3.4 is particularly well-adapted to the study of a 3-manifold M through its fundamental group (and was introduced in [CuS1] for this purpose). This is because, on the one hand, representations of $\pi_1(M)$ in $\mathrm{PSL}_2(\mathbf{C})$ are related to geometric structures on M, while on the other hand, actions of $\pi_1(M)$ on trees are related to the topology of M.

3.6.1. The work of W. Thurston shows that hyperbolic geometry plays a central role in 3-manifold topology. A *hyperbolic metric* (or hyperbolic structure) on an n-manifold M is a complete Riemannian metric of constant curvature -1. A hyperbolic metric on M gives an identification of the universal cover of M with hyperbolic n-space H^n, which up to isometry is the unique simply connected hyperbolic n-manifold. In the following discussion I will be thinking of H^n concretely as the (open) upper half-space $\mathbf{R}^{n-1} \times (0, \infty) \subset \mathbf{R}^n$ with the metric $x_n^{-2} \sum_{i=1}^n dx_i^2$.

Hyperbolic n-space is highly symmetric. Its group of isometries $\mathrm{O}(n,1)$ acts transitively on points, and the stabilizer of any point acts transitively on orthogonal frames in the tangent space. Any isometry $\gamma \in \mathrm{SO}(n,1)$ extends to a homeomorphism $\bar{\gamma}$ of $\bar{H}^n$, the one-point compactification of the closed upper half-space $\mathbf{R}^{n-1} \times [0, \infty)$, which is topologically a closed ball. The boundary $S^{(}n-1)_\infty$ of $\bar{H}^n$ is the one-point compactification of $\mathbf{R}^{n-1}$.

We can identify S^1_∞ with the real projective line, and S^2_∞ with the complex projective line or Riemann sphere. If $n = 2$ or 3 and if γ belongs to the orientation-preserving subgroup $\mathrm{SO}(n,1)$ of $\mathrm{O}(n,1)$, then the restriction of $\bar{\gamma}$ to S^{n-1}_∞ is a homography (linear fractional transformation) $z \mapsto \frac{az+b}{cz+d}$, where a, b, c, d are real if $n = 2$ and complex if $n = 3$, and $ad - bc = 1$. Thus $\mathrm{SO}(2,1)$ and $\mathrm{SO}(3,1)$ are respectively isomorphic to the groups $\mathrm{PSL}_2(\mathbf{R})$ and $PSLC$ of real and complex homographies.

3.6.2. If M^n is hyperbolic and orientable then $\pi_1(M)$ acts on the universal cover H^n by deck transformations which belong to $\mathrm{SO}(n,1)$, and thus with the hyperbolic metric on M there is associated a representation of $\pi_1(M)$ in $\mathrm{PSL}_2(\mathbf{C})$. This representation is discrete, in the sense that it is an isomorphism of $\pi_1(M)$ onto a discrete subgroup of $\mathrm{SO}(n,1)$.

These observations allow one to define a natural bijective correspondence between conjugacy classes of representations of a group Γ in $\mathrm{SO}(n,1)$ and n-dimensional *homotopy-hyperbolic structures* on an aspherical space $K = K(\Gamma, 1)$ with fundamental group Γ. A homotopy-hyperbolic structure is defined by a pair (M, ϕ) where M is a hyperbolic n-manifold and $\phi : K \to M$ is a homotopy equivalence. Two such pairs (M, ϕ) and (M', ϕ') define the same homotopy-hyperbolic structure if there is a homotopy equivalence $h\colon M \to M'$ such that $h \circ \phi$ is homotopic to ϕ'.

3.6.3. Let us consider an orientable 3-manifold M with a hyperbolic metric of finite volume. Then M contains a compact 3-manifold-with-boundary M_0, its *compact core*, such that every component of $M - \operatorname{int} M_0$ is isometric to the quotient of the *horoball* $\Omega = \mathbf{R}^2 \times [1, \infty) \subset H^3$ by a rank-2 free abelian discrete subgroup of $\mathrm{PSL}_2(\mathbf{C})$ consisting of elements of the form $\pm \begin{pmatrix} 1 & z \\ 0 & 1 \end{pmatrix}$. (The group consisting of all such elements clearly leaves Ω invariant.) These components are called *cusps*, or maybe I should say neighborhoods of cusps. Each of them is homemorphic to $T^2 \times [0, \infty)$. It follows that M is homeomorphic to the interior of M_0; in particular, $\pi_1(M)$ and $\pi_1(M_0)$ are identified in a natural way.

3.6.4. Mostow's rigidity theorem asserts that if a manifold M of dimension > 2 admits a hyperbolic structure of finite volume, then the homotopy-hyperbolic structure on M is unique; that is, the discrete representation of $\pi_1(M)$ in $\mathrm{PSL}_2(\mathbf{C})$ is unique up to conjugation by an element of $\mathrm{PSL}_2(\mathbf{C})$. However, if $n = 3$ and M is non-compact, i.e. has at least one cusp, then $\pi_1(M)$ always has useful *non-discrete* representations in $\mathrm{PSL}_2(\mathbf{C})$. In fact, Thurston has shown that if M has m cusps then the character χ_0 of the discrete representation of $\pi_1(M)$ is a smooth point of $X(\Gamma)$ and that the irreducible component of $X(\Gamma)$ containing χ_0 has dimension m. In particular, if M has finite volume but is not compact then $X(\Gamma)$ always contains curves, and the construction that I described in 3.4 can be applied to any ideal point of any of these curves.

3.6.5. As I said above, the second reason why the theory described in 3.4 is especially well-adapted to the study of a 3-manifold M is that actions of $\pi_1(M)$ on trees are related to the topology of M. This connection involves the same ideas of Stallings's that I invoked in 2.4.1. (In this 3-dimensional context these ideas were also discussed in [Sh].)

For simplicity let us suppose that M is compact—possibly with boundary—and orientable. For example, M may be the compact core of an orientable hyperbolic 3-manifold of finite volume. If $\Gamma = \pi_1(M)$ acts on a tree T, there is a Γ- equivariant map $\tilde{f}$ from the universal cover $\tilde{M}$ to T. In the case of a simplicial action on a simplicial tree T, we can choose $\tilde{f}$ so that the inverse image under $\tilde{f}$ of the set of midpoints of edges of T is a 2-manifold $\tilde{\Sigma}$. Again, $\tilde{\Sigma}$ may have a boundary, but it is properly embedded in M, in the sense that $\partial\Sigma = \Sigma \cap \partial M$. The equivariance of Σ implies that Σ is invariant under Γ. So we get a properly embedded compact 2-manifold $\Sigma = \tilde{\Sigma}/\Gamma \subset M$. If Γ acts without inversions we can take Σ to be orientable. Using fundamental results due to Papakyriakopoulos, one can also arrange that

(i) the fundamental group of each component of Σ maps injectively to $\pi_1(M)$.

Furthermore, we can choose $\tilde{f}$ so that Σ is non-degenerate in the sense that

(ii) no component of Σ is the boundary of a 3-ball, and
(iii) no component Σ_i of Σ is the frontier of a subset A of M such that $\pi_1(\Sigma_i)$ maps onto $\pi_1(A)$.

A properly embedded, compact, orientable 2-manifold $\Sigma \subset M$ that satisfies conditions (i)–(iii) is said to be *incompressible*. (At least that's my terminology. Some other authors use the term in a slightly weaker sense.)

If the given action of $\Gamma = \pi_1(M)$ on T is non-trivial, then any incompressible surface Σ obtained from the above construction is non-empty. This is because if Σ and hence $\tilde{\Sigma}$ were empty, $\tilde{f}$ would map $\tilde{M}$ to the star of some vertex s in the first barycentric subdivision of T; the vertex s would then be fixed by the entire group Γ.

3.7. Separating suraces in knot manifolds. The interaction among hyperbolic structures on int M_0, representations of $\pi_1(M)$ in $\mathrm{PSL}_2(\mathbf{C})$, actions of $\pi_1(M)$ on trees, and incompressible surfaces in M has had a number of applications in 3-manifold theory. An amusing early application was made in [CuS2] to the problem of finding interesting incompressible surfaces in knot complements.

3.7.1. If k is a non-trivial knot in the 3-sphere then there is always a connected orientable surface $F \subset S^3$, called a *Seifert surface*, whose boundary is k. We can take F to meet a tubular neighborhood N of k in an annulus. Now $M = S^3 - \mathrm{int}N$ is a compact 3-manifold bounded by a torus. The surface $F \cap M$ is properly embedded in M. Its boundary is a *longitude*, i.e. a simple closed curve in ∂M whose homology class represents a generator of $\ker H_1(\partial M, \mathbf{Z}) \to H_1(M, \mathbf{Z})$. It follows from the work of Papakyriakopoulos that one can always choose F so that $F \cap M$ is incompressible. For example this is always the case if we take F to have minimal genus among all Seifert surfaces.

So if M is the complement of a non-trivial knot in S^3, the obvious connected incompressible surfaces in M are those whose boundary consists of a single longitude; these are essentially Seifert surfaces. The question arises whether M always contains incompressible surfaces other than the obvious ones.

3.7.2. If Σ is an *arbitrary* incompressible surface with non-empty boundary in M, then $\partial\Sigma$ consist of a certain number of homotopically non-trivial simple closed curves in ∂M. As ∂M is a torus, the components of $\partial\Sigma$ must all represent the same homology class $\delta \in H_1(\partial M)$. Since δ is represented by a simple closed curve, it is an indivisible (or unimodular) element of $\pi_1(\partial M)$, and it is defined only up to sign since we have not specified any orientations. An indivisible element of $H_1(\partial M)$, defined up to sign, is often called a *slope*. A slope δ determined by a bounded incompressible surface in the way that I have described is called a *boundary slope*.

In [CuS2] it was shown that for every knot in S^3 (or in any rational homology sphere) there exists a boundary slope which is distinct from the (slope of a) longitude. In other words, there is always a closed incompressible surface whose boundary components are not longitudes. It is easy to show that such a surface always separates M into two pieces, in contrast to the Seifert surfaces which are always non-separating.

3.7.3. One consequence is a result originally conjectured by L. Neuwirth:

Every knot group is a non-trivial free product with amalgamation in which the amalgamated subgroup is free.

3.7.4. To prove the main result of [CuS2] one uses Thurston's results to reduce to the case in which int M admits a hyperbolic structure. In this case, it follows from 3.6.4 that the component of $X(\pi_1(M))$ containing the character of the discrete faithful representation is a curve C.

Let us fix a base point in ∂M. Since $\pi_1(\partial M)$ is abelian, each element $\alpha \in H_1(\partial M)$ corresponds to a unique element in $\pi_1(\partial M)$, whose image in $\pi_1(M)$ I will denote by $e(\alpha)$.

Each ideal point x of C determines a non-trivial action of $\pi_1(M)$ on a tree, which can be used to construct a non-empty incompressible 2-manifold Σ_x. One wishes to show that for some ideal point x, the boundary of Σ_x is non-empty and the boundary slope that it defines is not a longitude. To show this one considers the function $I_{e(\lambda)}$ on C, where λ denotes the longitudinal slope. By chasing through the definitions one discovers that if $\partial\Sigma_x = \emptyset$, or if the boundary slope defined by $\partial\Sigma_x$ is a longitude, then $I_{e(\lambda)}$ is finite-valued–i.e. does not have a pole–at x. Now one of the basic properties of the curve C provided by 3.6.4 is that for any non-zero element α of $H_1(\partial M)$, the function $I_{e(\alpha)}$ is non-constant. In particular, $I_{e(\lambda)}$ is non-constant; since it is finite-valued on C it must have a pole at some ideal point of C. This proves the Neuwirth conjecture.

3.8. Surgery on knots. A more ambitious application of the techniques described in 3.4 and 3.6 occurs in the proof of the Cyclic Surgery Theorem, which was proved by Culler, Gordon Luecke and myself [CuGLS]. Before explaining the statement I must give a bit of background.

3.8.1. A *solid torus* is a 3-manifold N homeomorphic to $D^2 \times S^1$. The simple closed curves $\{0\} \times S^1 \subset \operatorname{int} N$ and $(\partial D^2) \times S^1 \subset \partial N$ are well-defined up to isotopy and are called the *core* and *meridian* of N. If M is a 3-manifold whose boundary is a 2-torus, the operation known as *Dehn filling* consists of attaching a solid torus N to M by some homeomorphism between ∂M and ∂N. The topological type of the resulting closed manifold is determined if one specifies the slope α (see 3.7.2) of the simple closed curve in N which is attached to the meridian of N. I'll write M_α for the manifold obtained by the Dehn filling. The group $\pi_1(M_\alpha)$ is isomorphic to $\pi_1(M)/\langle e(\alpha)\rangle$, where $\langle\ \rangle$ denotes normal closure, and $e(\alpha)$ is defined as in 3.7.4.

Let M be a 3-manifold whose boundary is a torus, and α, β two indivisible elements of $H_1(\partial M)$. The 3-manifolds M_α and M_β are said to be related to each other by a *Dehn surgery.* Thus Dehn surgery is the operation of removing a solid torus from the interior of a manifold and sewing it back in a different way.

It is a classical result that any closed orientable 3-manifold can be obtained from the 3-sphere by a finite sequence of Dehn surgeries. A good deal of attention has been focused on the manifolds obtained from S^3 by a single Dehn surgery. Thus one considers Dehn fillings of manifolds $M = S^3 - N$, where $N \subset S^3$ is a solid torus. We can think of N as a tubular neighborhood of its core, which is a knot K. There is a natural basis of $H_1(\partial M)$ consisting of the meridian λ of N

and the longitude μ of K (see 3.7.1). (One must specify some orientations for the signs of λ and μ to be well-defined.)

It is convenient to parametrize slopes in ∂M by elements of $\mathbf{Q} \cup \{\infty\}$, by letting $\frac{a}{b}$ correspond to the slope $a\mu + b\lambda$.

One often writes $K(\frac{a}{b}) = M_{a\mu+b\lambda}$. The number $\frac{a}{b}$ is called a surgery coefficient. Note that $K(\infty) = M(\mu) = S^3$: this is the *trivial* surgery.

3.8.2. Any surgery on the trivial knot gives a lens space, i.e the quotient of S^3 by a cyclic group acting freely by isometries (in the round metric). A lens space clearly has a cyclic fundamental group. The cyclic surgery theorem deals with the question of which non-trivial surgeries on non-trivial knots can give manifolds with cyclic fundamental group.

3.8.3. Elementary examples are provided by torus knots. A *torus knot* is a knot in S^3 which can be isotoped into a standard torus. If we think of S^3 as the unit sphere $|z|^2 + |w|^2 = 2$ in $\mathbf{C}^2$, the standard torus Q is defined by $|z| = |w| = 1$. Torus knots arise in algebraic geometry in connection with singularities of algebraic curves: a plane curve of the form $z^p w^q = 1$, where p and q are relatively prime integers, has a singularity at the origin, and its intersection with a small Euclidean sphere $S \subset \mathbf{C}^2$ is a torus knot in S. For any torus knot K, there are infinitely many Dehn surgeries on K that give lens spaces.

Torus knots often play an exceptional role in knot theory because their complements are *Seifert fibered spaces*, i.e. they admit C^∞ foliations by 1-spheres. Seifert fibered spaces form a manageable and pleasant, but rather degenerate, class of 3-manifolds.

Rolfsen showed that certain *iterated torus knots*— which, like torus knots, arise as links of singularities of plane algebraic curves—also admit non-trivial Dehn surgeries that give lens spaces. This may regarded as a partial generalization of what happens for torus knots; however, for an iterated torus knot which is not a torus knot there is at most one surgery (as opposed to infinitely many) that can give a lens space.

3.8.4. A remarkable example was given by Fintushel and Stern. They showed that the surgeries on the so-called (-2,3,7)-pretzel knot with surgery coefficients 18 and 19 both yield lens spaces. As we shall see, the Cyclic Surgery Theorem sheds light on this example.

3.8.5. The general version of the theorem is best stated in terms of Dehn filling. A 3-manifold M is said to be *irreducible* if every smooth 2-sphere in M bounds a ball. A classical theorem due to Alexander implies that every knot complement in S^3 is irreducible.

Let M be an irreducible, compact, orientable 3-manifold whose boundary is a torus. Suppose that M is not a Seifert fibered space. Let α and α' be two *slopes in $H_1(\partial M)$. Suppose that $\pi_1(M_\alpha)$ and $\pi_1(M_{\alpha'})$ are both cyclic. Then the homological intersection number of α and α' (which is defined up to sign) has absolute value at most 1.*

3.8.6. This has the following formal consequences regarding non-trivial surgery on a knot K in S^3.

(3.8.6.1) *Suppose that K is not a torus knot. For any $r \in \mathbf{Q}$, if $\pi_1(K(r))$ is cyclic then r is an integer. There are at most two integers r for which $\pi_1(K(r))$ is cyclic, and if there are two they must be consecutive integers (as in the Fintushel-Stern example). Only for $r = 1$ or $r = -1$—and not both—can $K(r)$ possibly be simply connected.*

Combining this last fact with a result due to Bleiler and Scharlemann, one can show:

(3.8.6.2) *If K is a non-trivial knot which is invariant under a non-trivial periodic homeomorphism of S^3, then there is no $r \in \mathbf{Q}$ for which $K(r)$ is simply connected.*

3.8.7. I will briefly sketch the proof of the Cyclic Surgery Theorem in the case where the manifold M contains no closed incompressible surfaces. In this case Thurston's results imply that int M has a hyperbolic metric, and as in 3.6.4 and 3.7.4 we have a curve $C \subset X(\pi_1(M))$.

As I mentioned in 3.7.4, $I_{e(\alpha)}$ is non-constant whenever $1 \neq \alpha \in \pi_1(M)$; thus the degree of $I_{e(\alpha)}$ is a positive[8] integer in this case. We can interpret deg $I_{e(\alpha)}$ as the number of poles of $I_{e(\alpha)}$. Since $I_{e(\alpha)}$ has poles only at ideal points, we have deg $I_{e(\alpha)} = \sum_z P_{z,\alpha}$, where z ranges over the ideal points of C and and $P_{z,\alpha}$ denotes the order of the pole of $I_{e(\alpha)}$ at z, or 0 if $I_{e(\alpha)}$ does not have a pole at z.

By 3.4 and 3.6, an ideal point z determines an action of $\Gamma = \pi_1(M)$ on a tree, with which one can associate an incompressible surface $\Sigma_z \subset M$. Since we are assuming that M contains no closed incompressible surfaces, Σ_z must have non-empty boundary, and it therefore determines a boundary slope δ_z. One can show that there is a homomorphism $l_z \colon H_1(\partial M) \to \mathbf{Z}$, whose kernel is generated by β_z, such that $P_{z,\alpha} = |l_z(\alpha)|$ for every $\alpha \in H_1(\partial M)$. We can define a norm on the 2-dimensional vector space $V = H_1(\partial M; \mathbf{R})$ by $\|a\| = \sum_z l_z(a)$, where z ranges over the ideal points of C. We have deg $I_{e(\alpha)} = \|\alpha\|$ for every $\alpha \in H_1(\partial M)$. The unit ball of this norm is a compact convex polygon in V which is balanced (i.e. symmetric about 0) and whose vertices lie on lines spanned by boundary slopes.

Let's set $m = \min_{0 \neq \alpha \in L} \|\alpha\|$. Then the ball B of radius m with respect to our norm is again a convex balanced polygon whose vertices lies on lines spanned by boundary slopes. By definition, int B contains no non-zero points of the

[8] in ze English sense

lattice $L = H_1(\partial M)$. This implies, by a well-known elementary argument due to Minkowski, that the area of B is at most 4. (Here I am measuring area in V in such a way that V/L has area 1. If we identify V with $\mathbf{R}^2$ in such a way that $L = \mathbf{Z}^2$, we are looking at ordinary area on $\mathbf{R}^2$.)

The key step in the proof of the Cyclic Surgery Theorem, in the case we are considering, is to show that

(3.8.7.1) for any slope α such that $\pi_1(M_\alpha)$ is cyclic, we have $\|\alpha\| = m$, so that $\alpha \in \partial B$; and furthermore α is not a vertex of B.

Once (3.8.7.1) has been established the theorem follows easily. For if α and α' are two distinct slopes such that $\pi_1(M_\alpha)$ and $\pi_1(M_\alpha)$ are cyclic, then the four points $\pm\alpha, \pm\alpha'$ are the vertices of a parallelogram Π whose area is $2I$, where I denotes the absolute value of the homological intersection number of α and α'. It follows from (3.8.7.1) that $\Pi \subset B$ and hence that $I \leq \frac{1}{2}\text{Area } B \leq 2$, and that equality holds only if $B = \Pi$. But in the latter case α and α' would be vertices of B, contradicting the second assertion of (3.8.7.1). So we must have $I \leq 1$.

The first assertion of (3.8.7.1) is equivalent to saying that if α is a slope such that $\pi_1(M_\alpha)$ is cyclic, then for any non-zero element β of $H_1(\partial M; \mathbf{Z})$ we have $\deg I_{c(\alpha)} \leq \deg I_{c(\beta)}$. This is proved by showing that at every point of $\tilde{C}$ (or more accurately of its de-singularization) where I_α takes the value 4, the function I_β also takes the value 4, and with at least the same multiplicity. This is in turn proved in two cases, depending on whether the given point lies in C or an ideal point. First let's consider the case of a point $x \in C$.

It may be shown that every point of C is the character of some representation $\rho\colon M \to \mathrm{PSL}_2(\mathbf{C})$ with a non-cyclic image. For such a ρ we must have $\rho(\alpha) \neq 1$; for otherwise ρ would factor through a representation of $\pi_1(\Sigma_\alpha) = \pi_1(M)/\langle e(\alpha)\rangle$ with a non-cyclic image, and this is impossible since $\pi_1(\Sigma_\alpha)$ is cyclic. Hence for every point $x \in C$ for which $I_{e(\alpha)} = 4$, there is a representation ρ with character x such that $\rho(e(\alpha)) = \pm\begin{pmatrix} 1 & 1 \\ 0 & 1 \end{pmatrix}$. Now since $\pi_1(\partial M)$ is abelian, any element $e(\beta)$ of $\pi_1(\partial M)$ is represented by an element that commutes with $\pm\begin{pmatrix} 1 & 1 \\ 0 & 1 \end{pmatrix}$ and hence has the form $\pm\begin{pmatrix} 1 & b \\ 0 & 1 \end{pmatrix}$. In particular $I_{c(\beta)}(x) = 4$. By refining this argument—and doing a fair amount of hard technical work—one can show that $I_{c(\beta)}$ takes the value 4 at x with at least the same multiplicity as $I_{c(\beta)}$; and one can make the argument work on points of the de- singularization of $\tilde{C}$ that correspond to points of C.

The case of an ideal point is quite different. In this case one actually shows that the assertion is vacuously true; that is, one shows that if $\pi_1(M_\alpha)$ is cyclic, then $I_{c(\alpha)}$ cannot take the value 4, or any finite value for that matter, at an ideal point. This is one point at which the theory of 3.4 and 3.6 is crucial. If $I_{c(\alpha)}$ is finite-valued at an ideal point z then we have $\alpha = \delta_z$, so that α is a boundary slope. But one can show that $\pi_1(M_\alpha)$ cannot be cyclic when α is a boundary slope.

This is done by considering an incompressible surface Σ_0 which has boundary slope α and which has the smallest possible number of boundary components among all such surfaces. In particular Σ_0 is connected. Let's write $M_\alpha = M \cup N$, where N is a solid torus. Since Σ_0 has boundary slope α, its boundary components bound disjoint disks in N. The union of Σ_0 with these disks is a closed surface $\hat{\Sigma}_0 \subset M_\alpha$. If Σ_0 has positive genus, one proves that $\hat{\Sigma}_0$ is incompressible in M_α. In this case, $\pi_1(M_\alpha)$ contains the fundamental group of a closed orientable surface of positive genus, and is therefore non-cyclic. In the case where $\hat{\Sigma}_0$ is a sphere, one shows that this sphere decomposes M_α as a connected sum of two non- trivial lens spaces. So in this case $\pi_1(M_\alpha)$ is a free product of two non-trivial cyclic groups, and is therefore not cyclic.

The second assertion of (3.8.7.1) is now easy. I pointed out in defining B that the vertices of B lie on lines spanned by boundary slopes. So if $\alpha \in L \cap \partial B$ were a vertex of B then α would itself be a boundary slope. But we just saw that this implies that $\pi_1(M_\alpha)$ is non-cyclic.

Section 4. Higher-dimensional varieties of characters and degenerations of hyperbolic structures

4.1. Compactifying character varieties. In 3.4 I explained how an ideal point z of a curve in the space $X(\Gamma)$ of $\mathrm{PSL}_2(\mathbf{C})$-characters of a finitely generated group Γ defines a non-trivial action of Γ on a $\mathbf{Z}$-tree. I pointed out that for any $\gamma \in \Gamma$ the function $\hat{I}_\gamma$ is finite-valued at z if and only if γ fixes some point of T_z, and that if $\hat{I}_\gamma$ has a pole at z then the order of the pole is $-l(\gamma)$, where $l\colon \Gamma \to \mathbf{Z}$ is the length function defined by the action of Γ on T_z.

These properties have a convenient restatement in terms of the formalism of projectivized length functions introduced in 1.7.1. Let $[l_z] \in \mathcal{PL}(\Gamma) \subset \mathcal{P}^\Gamma$ be the projectivized length function defined by the action associated to z. (By 1.7.2, the integer-valued length function defined by the action coincides with the real-valued length function defined by its real completion.) On the other hand, let us define a continuous map $\Theta : X(\Gamma) \to \mathcal{P}^\Gamma$ by defining $\Theta(x)$ to be the image in $\mathcal{P}^\Gamma$ of $(|\mathrm{Re}\ \mathrm{arccosh}\ (\frac{1}{2}(I_c(x))^2 - 1)|)_{c \in \mathcal{C}(\Gamma)} \in [0, \infty)^{\mathcal{C}(\Gamma)}$. Here the curve C is understood to have the topology induced by the usual topology of $\mathbf{C}$. The multi-valued function arccosh has a well-defined real part up to sign. Geometrically[9] , if x is the character of a representation in $\mathrm{PSL}_2(\mathbf{C}) = \mathrm{SO}(3,1)$,

[9] In place of $|\mathrm{Re}\ \mathrm{arccosh}\ (\frac{1}{2}(I_c(x))^2 - 1)|$ one could use any globally defined expression which goes to infinity like a constant multiple of $\log I_c(x)$ as $I_c \to \infty$. In [MSh1], Morgan and I used

and $\gamma \in \Gamma$ is an element representing the conjugacy class c, the expression $|\mathrm{Re}\ \mathrm{arccosh}\ (\frac{1}{2}(I_c(x))^2 - 1)|$ gives the hyperbolic translation length of $\rho(\gamma)$ in H^3. In terms of the map θ, the properties of the action recalled above translate into the following fact: for any sequence (x_i) of points in C which converges to z in $\hat{C}$, the sequence $(\Theta(x_i))$ converges to $[l_z]$ in $\mathcal{P}^\Gamma$. So we can extend Θ to a continuous map from $\hat{C}$ to $\mathcal{P}^\Gamma$ by mapping each ideal point z to $[l_z]$.

4.1.1. This picture was generalized by Morgan and myself in [MSh1]. Instead of considering sequences of points on a fixed curve, we considered arbitrary sequences (x_i) tending to infinity in the locally compact space $X(\Gamma)$. It is not hard to show that any such sequence has a subsequence whose image under Θ converges to some point $[l] \in \mathcal{P}^\Gamma$. Morgan and I showed that the limit point $[l]$ always belongs to $\mathcal{PL}(\Gamma)$.

4.1.2. In contrast to the case where the sequence lies on a curve, the projectivized length function $[l]$ is not integer-valued in general. Instead, it takes values in a subgroup Λ of $\mathbf{R}$ whose $\mathbf{Q}$-rank $\dim_{\mathbf{Q}} \mathbf{Q}\Lambda$ is at most the dimension n of V. I will express this briefly by saying that the action has rank at most n.

An action on an $\mathbf{R}$-tree will be said to have rank $\leq n$ if its length function has rank $\leq n$. It follows from 1.7.2 that an action has rank $\leq n$ if and only if it is the real completion of a non-trivial action on a Λ-tree for some subgroup Λ of $\mathbf{R}$ whose $\mathbf{Q}$-rank is at most n.

4.1.3. Using 4.1.1 one can define a natural compactification $\hat{X} = \hat{X}(\Gamma)$ of $X = X(\Gamma)$, such that $\hat{X} - X$ is identified with a subset of $\mathcal{PL}(\Gamma)$. As a set, $\hat{X}$ is the disjoint union of X with a set $\mathcal{B} \subset \mathcal{PL}(\Gamma)$. The set $\mathcal{B}$ consists of all points which are limits in $\mathcal{P}^\Gamma$ of convergent sequences of the form $(\Theta(x_i))$ where $(x_i) \to \infty$ in X. The subsets X and $\mathcal{B}$ of $\hat{X}$ have respectively the complex topology and the topology inherited from $\mathcal{P}^\Gamma$. A sequence (x_i) in X converges in $\hat{X}$ to a point $[l] \in \mathcal{B}$ if and only if $\Theta(x_i) \to [l]$ in $\mathcal{P}^\Gamma$. These conditions characterize a compact topology on $\hat{X}$ in which X is an open dense subset.

4.1.4. Actually one can define a compactification of this type not only for the $\mathrm{PSL}_2(\mathbf{C})$ character space of a group, but for any complex affine algebraic set. (This is carried out in detail in [MSh1]. The construction depends on the choice of a countable set of generators for the coordinate ring as a $\mathbf{C}$-algebra; for the case of a character space $X(\Gamma)$ one takes the set to consist of all the functions $I_c, c \in \mathcal{C}(\Gamma)$.) In the case where the given algebraic set is a curve C, the compactification defined in this way is canonically isomorphic to the projective completion $\hat{C}$ described in 3.4.3.

$\log(|I_c(x)| + 2$. I have used the more complicated expression above because it is more directly related to the geometric applications that I'll be talking about later.

However, the point to be emphasized is that when the given algebraic set is a character space of a group, the *ideal points*, i.e. the points of $\mathcal{B} = \hat{X} - X$, correspond to non- trivial actions of the group on **R**-trees. This is what generalizes the theory of [CuS1], and–like the theory of [CuS1]–is useful for applications.

4.1.5. Before I talk about applications of this generalized theory, it will be best to say a few words about the proofs of the assertions of 4.1.1, and to discuss some further generalizations.

The proof of the main assertion of 4.1.1, while technically much more involved than the proof of the special case discussed in 3.4, is philosophically very similar. Suppose that (x_i) is a sequence tending to infinity in X, and that $\Theta(x_i)$ converges in $\mathcal{P}^{\Gamma}$. After passing to a subsequence we can assume that the x_i all lie in the same irreducible component V of X. The coordinate ring $\hat{\mathbf{Q}}[V]$, where $\hat{\mathbf{Q}} \subset \mathbf{C}$ denotes the algebraic closure of $\mathbf{Q}$, is an integral domain which contains all the functions $I_c, c \in \mathcal{C}(\Gamma)$. Its field of fractions is the function field $\hat{\mathbf{Q}}(V)$. After approximating the x_i by nearby generic points (without changing their limit) and again passing to a subsequence, we can achieve a nice situation in which for every function $f \in \hat{\mathbf{Q}}(V)$, the sequence $(f(x_i))$ has a limit in the extended complex line $\mathbf{C} \cup \{\infty\}$. In this situation there is a Krull valuation v of $\hat{\mathbf{Q}}(V)$ whose valuation ring $\mathcal{O}_v$ consists of all functions $f \in \hat{\mathbf{Q}}(V)$ such that $\lim_{i \to \infty}(f(x_i)) \neq \infty$. This valuation will of course play the role of the valuation associated to the ideal point in the case discussed in 3.4. The assumption that $x_i \to \infty$ in V implies that for some $c \in \mathcal{C}(\Gamma)$ we have $I_c \notin \mathcal{O}_v$. We have $v(\hat{\mathbf{Q}}^*) = 0$.

Since the natural map Hom $(\Gamma, \mathrm{PSL}_2(\mathbf{C})) \to X(\Gamma)$ is surjective, it maps some irreducible component W of $HGPSLC$ onto a dense subset of V, and there is an induced monomorphism from $\hat{\mathbf{Q}}(V)$ to the function field K of W over $\hat{\mathbf{Q}}$. Let us identify $\hat{\mathbf{Q}}(X)$ with a subfield of K. The valuation v can be extended–after possibly enlarging the value group by finite index–to a valuation w of K. As in 3.4.3 we have a tautological representation $P\colon \Gamma \to \mathrm{PSL}_2(K)$. The valuation w defines a Λ-tree T, where Λ is the value group of w, and $\mathrm{PSL}_2(K)$ has a natural action on T. Pulling back this action via P we get an action of Γ on T. One can check that this pulled-back action is non-trivial by checking that any element $\gamma \in \Gamma$ such that $I_\gamma \notin \mathcal{O}_v$ acts without a fixed point on T.

If we are lucky, Λ will be order-isomorphic to a subgroup of $\mathbf{R}$. In this case the action of Γ on T extends to an action on the real completion $\mathbf{R}T$. The identification of Λ with a subgroup of $\mathbf{R}$ is unique modulo a multiplicative constant, and hence the projectivized length function $l \in \mathcal{PL}(\Gamma)$ defined by the action of Γ on $\mathbf{R}T$ is uniquely determined by the valuation w. In this case one can check that the given sequence (x_i) has limit $[l]$.

In general Λ need not be order-isomorphic to a subgroup of $\mathbf{R}$, so one has to work harder. What turns out to be true in general is that there are convex subgroups (see 2.6.1) $\Lambda_0 \subset \Lambda_1$ of Λ such that Λ_1/Λ_0 is order-isomorphic to a subgroup of $\mathbf{R}$, and such that we have $w_-(I_c) = -\min(0, w_-(I_c) \in \Lambda_1$ for every $c \in \mathcal{C}(\Gamma)$, but $w_-(I_c) \notin \Lambda_0$ for some $c \in \mathcal{C}(\Gamma)$. The abelian group Λ/Λ_0 inherits an order from Λ, and by composing w with the projection $\Lambda \to \Lambda/\Lambda_0$ we get a valuation $\bar{w}\colon F^* \to \Lambda/\Lambda_0$. This gives an action of Γ on a Λ/Λ_0-tree, which may be shown to contain a Γ-invariant Λ_1/Λ_0-tree T. Since Λ_1/Λ_0 is order-isomorphic to a subgroup of $\mathbf{R}$, we can complete T to an $\mathbf{R}$-tree and proceed as before.

4.1.6. It is not hard to see from this construction why the points of $\mathcal{B} = \hat{X} - X$ are defined by actions of rank at most $n = \dim X$, as asserted in 4.1.2. In fact, for any component V of X, the transcendence degree of $\hat{\mathbf{Q}}(V)$ over $\hat{\mathbf{Q}}$ is $\dim V \leq n$, and from this it is a matter of elementary commutative algebra to deduce that any valuation of $\hat{\mathbf{Q}}(V)$ which is trivial on V has a value group Λ whose $\mathbf{Q}$-rank $\dim_{\mathbf{Q}}(\Lambda \otimes \mathbf{Q})$ is at most n. Enlarging a group by finite index and passing to a subgroup or quotient group do not increase the $\mathbf{Q}$-rank.

4.1.7. Notice that although the statement given in 4.1.1 involves only $\mathbf{R}$-trees, it is natural to prove it using Λ-trees for more general Λ. Indeed, the Λ-trees that arise in the proof contain important information. They describe the relative growth rates of the hyperbolic translation lengths of elements of Γ, in a sense that I will make precise in a moment.

If Λ is an ordered abelian group of $\mathbf{Q}$-rank $\leq n$, then the order-rank r of Λ (see 2.6.1) is also at most n. Let $\{0\} = \Lambda_0 \leq \ldots \Lambda_r = \Lambda$ be the convex subgroups of Λ. Each of the quotient groups Λ_k/Λ_{k-1} has order- rank 1 and hence admits an order-preserving embedding in $\mathbf{R}$. For any non-negative element λ of Λ, let us define the *height* of λ to be the least index k such that $\lambda \in \Lambda_k$. If λ and λ' are positive elements with the same height k, we define their *quotient* λ/λ' to be the real number $J(\lambda)/J(\lambda')$, where $J\colon \Lambda_k \to \mathbf{R}$ is the composition of the quotient map $\Lambda_k \to \Lambda_k/\Lambda_{k-1}$ with an embedding of Λ_k/Λ_{k-1} in $\mathbf{R}$. Since the embedding is unique up to a multiplicative constant, the quotient is well-defined. Let us set $\lambda/\lambda' = 0$ if height $\lambda <$ height λ', and $\lambda/\lambda' = \infty$ if height $\lambda >$ height λ'.

Now let (x_i) be a sequence tending to infinity in $X(\Gamma)$. After passing to a subsequence we can assume that for any two conjugacy classes c, c' in Γ, the sequence of quotients of hyperbolic translation lengths

$$\frac{|\text{Re arccosh }(\frac{1}{2}(I_c(x))^2 - 1)|}{|\text{Re arccosh }(\frac{1}{2}(I_{c'}(x))^2 - 1)|}$$

has a limit in $[0, \infty]$. The construction described in 4.1.5 gives a Λ-valued length function l defined by an action of Γ on a Λ-tree, where Λ is an ordered abelian group of finite $\mathbf{Q}$- rank, such that for any two elements $\gamma, \gamma' \in \Gamma$, the limit of the above sequence is equal to $l(c)/l(c')$. (in the quotient notation introduced above).

4.1.8. In [M1], Morgan generalized the theory described in 4.1.1–4.1.7. For any finitely generated group Γ and any $n \geq 2$ one can define a variety $X_n(\Gamma)$ of characters of representations of Γ in the isometry group $\mathrm{SO}(n,1)$ of hyperbolic n-space. Since by 3.6.1 the groups $\mathrm{SO}(2,1)$ and $\mathrm{SO}(3,1)$ are respectively isomorphic to $\mathrm{PSL}_2(\mathbf{R})$ and $\mathrm{PSL}_2(\mathbf{C})$, we can identify $X_3(\Gamma)$ with $X(\Gamma)$, and $X_2(\Gamma)$ with the subset $X_{\mathbf{R}}(\Gamma)$ of $X(\Gamma)$ consisting of all characters of representations in $PSLR \subset \mathrm{PSL}_2(\mathbf{C})$. In [M1] it is shown that for any $n \geq 2$ the space has a natural compactification $\hat{X}_n(\Gamma)$ by projectivized length functions, which specializes for $n = 3$ to the compactification $\hat{X}(\Gamma)$ described in 4.1.3, and for $n = 2$ to the closure of $X_{\mathbf{R}}(\Gamma)$ in $\hat{X}(\Gamma)$.

4.2. Degeneration of hyperbolic structures. From the geometric point of view there is a particularly interesting subset of $X_n(\Gamma)$, namely the set $\mathcal{D}_n(\Gamma)$ of characters of discrete representations in the sense of 3.6.2. It follows from the discussion in 3.6.2 that the points of this set are in bijective correspondence with homotopy-hyperbolic structures on the space $K = K(\Gamma, 1)$. As I mentioned in 3.6.4, the Mostow rigidity theorem implies that when Γ is isomorphic to the fundamental group of a finite-volume hyperbolic n-manifold, $\mathcal{D}_n(\Gamma)$ is a single point.

4.2.1. The closure of $\mathcal{D}_n(\Gamma)$ in $X_n(\Gamma)$ is a compactification $\hat{\mathcal{D}}_n(\Gamma)$ of $\mathcal{D}_n(\Gamma)$. It was shown in [MSh1] for $n = 2, 3$, and in [M1] for all $n \geq 2$ that the points of $\hat{\mathcal{D}}_n(\Gamma) - \mathcal{D}_n(\Gamma)$ are small projectivized length functions; that is, we have $\hat{\mathcal{D}}_n(\Gamma) - \mathcal{D}_n(\Gamma) \subset \mathcal{SPL}(\Gamma)$.

Furthermore, the map $\Theta|\mathcal{D}_n(\Gamma)\colon \mathcal{D}_n(\Gamma) \to \mathcal{P}^{\Gamma}$ has direct geometric meaning in terms of the hyperbolic manifolds M_x. For any $x \in \mathcal{D}_n(\Gamma)$, the homogeneous coordinate of $\Theta(x)$ corresponding to a conjugacy class c in Γ is the length of the closed geodesic in M_x corresponding to c. The small length functions in $\hat{\mathcal{D}}_n(\Gamma) - \mathcal{D}_n(\Gamma)$ contain information about the growth of lengths of closed geodesics as a hyperbolic structure degenerates.

4.2.2. A famous example occurs when Γ is the fundamental group of a closed orientable surface Σ of genus $g > 2$, and $n = 2$. In this case $\mathcal{D}_2(\Gamma)$ is the set of 2-dimensional homotopy-hyperbolic structures on Σ, known to analysts as Teichmüller space and denoted $\mathcal{T}_g$. The points of $\hat{\mathcal{T}}_g - \mathcal{T}_g$ are small projectivized length functions on $\pi_1(\Sigma)$, which by Skora's theorem are all defined by measured foliations on Σ. Thus $\hat{\mathcal{T}}_g$ is a natural compactification of $\mathcal{T}_g$ in which the ideal points are parametrized by measured foliations.

4.2.3. Just as the points of $\hat{\mathcal{D}}_n(\Gamma) - \mathcal{D}_n(\Gamma)$, where Γ is a finitely generated group, are defined by small actions on $\mathbf{R}$-trees, so the actions on Λ-trees associated as in 4.1.7 with sequences of characters of discrete representations are small actions. These actions contain finer information about how hyperbolic structures degenerate.

The small actions on Λ-trees that are defined by sequences in Teichmüller space have been studied by Morgan and Otal. They are associated to generalized measured foliations[10] in which the transverse measure takes values in Λ. So these generalize measured foliations contain interesting asymptotic information about Teichmüller space.

4.2.4. The compactification described in 4.2.2 was first discovered by Thurston from a quite different point of view, and bears his name. Thurston used the space $\hat{\mathcal{T}}_g$ to study outer automorphisms of surface groups. Every outer automorphism α of $\pi_1(\Sigma)$ is known to be induced by a self-homeomorphism of Σ. There is an analysis of the action of α on $\hat{\mathcal{T}}_g$ which is similar to the analysis of the action of a real Möbius transformation on the compactified upper half-plane.

In particular, one shows that α always has a fixed point in $\hat{\mathcal{T}}_g$. In the case where the fixed point is in $\mathcal{T}_g$ the automorphism has finite order and can be completely understood. On the other hand, a fixed point in $\hat{\mathcal{T}}_g - \mathcal{T}_g$ is a projectivized length function which is defined by a measured foliation and is invariant under α. Using such invariant foliations one can describe the action of α.

The most interesting case is the one in which α has *two* fixed points in $\hat{\mathcal{T}}_g - \mathcal{T}_g$. In this case there are two mutually transverse, projectively invariant measured foliations whose projectivized length functions are α-invariant. This information can even be realized geometrically in the strongest imaginable way: α is induced by a homeomorphism η which leaves each of the two transverse foliations invariant and pulls backs each transverse measure to a constant multiple of itself. Such a homeomorphism η is called a ***pseudo-Anosov map***. (Such maps were first studied by Thurston, who I believe named them in honor of Professor Ludwig von pseudo-Anosov, played by Sid Caesar.) The behavior of a pseudo-Anosov map with respect to the associated pari of transverse foliations leads to a rich theory of its dynamic behavior, which has become an exciting area of research.

4.2.5. There is an analogy between the compactification $\hat{Y}_n$ of outer space that I discussed in 2.1.1 and the Thurston compactification of Teichmüller space. The explanation for this analogy will become clear in 4.3. One goal of Culler and Vogtmann's program for studying $\mathrm{Out}(F_n)$ is to obtain a analysis for the action of an element of $\mathrm{Out}(F_n)$ on $\hat{Y}_n$ similar to Thurston's analysis for elements of the outer automorphism group of a surface group. The results of Bestvina and Handel that I discussed in 2.2.6 provide a step in this direction.

4.2.6. Since in general we have $\hat{\mathcal{D}}(\Gamma) - \mathcal{D}(\Gamma) \subset \mathcal{SPL}(\Gamma)$, the classification of

[10] Actually their results are stated in terms of laminations, not foliations. See the footnote to 2.4.

small actions of an arbitrary finitely generated group Γ is a central question in the study of degenerations of hyperbolic structures, particularly in the case where Γ admits discrete faithful representations on $SO(n,1)$ for some n.

We saw in 2.3 that the fundamental group of a closed orientable surface Σ_g has a wealth of non-trivial small actions on $\mathbf{R}$-trees. And we just saw that the length functions defined by these actions all appear in the Thurston boundary $\hat{\mathcal{T}}_g - \mathcal{T}_g = \hat{X}_2(\pi_1(\Sigma) - X_2(\pi_1(\Sigma)$. The opposite extreme occurs for a group Γ which admits *no* small non-trivial action on an $\mathbf{R}$-tree. For such a group Γ, and for any $n \geq 2$, we have $\hat{\mathcal{D}}_n(\Gamma) - \mathcal{D}_n(\Gamma) \subset \mathcal{SPL}(\Gamma) = \emptyset$; hence in this case the set $\mathcal{D}_n(\Gamma)$ is *compact.*

It follows that if Conjecture 2.5.5 is true, then for any finitely presented group Γ which admits no non-trivial splitting over a small subgroup, and for any integer $n \geq 2$, the space $\mathcal{D}_n(\Gamma)$ is compact. The condition in 2.5.5 that every small subgroup of Γ is finitely generated. is a harmless restriction here, because it is automatically satisfied whenever $\mathcal{D}_n(\Gamma) \neq \emptyset$, or more generally whenever Γ is isomorphic to a discrete subgroup of a Lie group. Of course if Question 2.5.3 had an affirmative answer in general, one could replace the hypothesis that Γ is finitely presented by the more natural and satisfactory hypothesis that it is finitely generated. This was the original motivation for Question 2.5.3.

4.2.7. It is worth pointing out that, since by 4.1.2 and 4.1.6 the set $\hat{X}(\Gamma) - X(\Gamma)$ consists of projectivized length functions of finite rank, for the application discussed in 4.2.6 it would be enough to prove Conjecture 2.5.5 for the finite-rank case; that is, to show that if Γ satisfies the hypotheses of 2.5.5 then every finite-rank point of $\mathcal{SPL}(\Gamma)$ is a limit of points of $\mathcal{SPL}(\Gamma)$ defined by small integer-valued length functions. I shall return to this case of Conjecture 2.5.5 in 5.1 and 5.2.

4.2.8. The program outlined in 4.2.6 has been largely carried out in the case $n = 3$. Let Γ be a finitely generated group such that $X_3(\Gamma) \neq \emptyset$, i.e. Γ is isomorphic to a discrete subgroup of $SO(3,1) = PSL_2(\mathbf{C})$. For simplicity suppose that the group Γ is torsion-free, so that Γ is isomorphic to the fundamental group of a hyperbolic 3-manifold M. Again for simplicity, suppose that M is orientable.

In [MSh1], Morgan and I showed that Question 2.5.3 has an affirmative answer whenever Γ is a finitely generated group which arises as the fundamental group of an orientable 3-manifold M (possibly with boundary). Furthermore, in this case the result has topological meaning in terms of M.

Consider an arbitrary orientable 3-manifold M with finitely generated fundamental group. To simplify the language I'll assume that M is irreducible (3.8.5); this is automatically true if M is hyperbolic. According to a theorem first proved in complete generality by Scott [Sc], M has a *compact core*[11], i.e. there is a

[11] This is a more general notion than the one I referred to in 3.6.3.

compact irreducible 3-manifold-with-boundary $M_0 \subset M$ such that the inclusion homomorphism $\pi_1(M_0) \to \pi_1(M)$ is an isomorphism. (In particular this implies that $\Gamma = \pi_1(M)$ is finitely presented, a result proved independently by Scott and myself.) Using the techniques of Stallings's that I referred to in 3.6.5, one can show that Γ admits a non-trivial small action on a $\mathbf{Z}$-tree if and only if M_0 contains a connected incompressible surface with a small fundamental group, or what is the same thing, with a non-negative Euler characteristic. What Morgan and I showed in [MSh2] is that if $\pi_1(M)$ admits a small non-trivial action on an $\mathbf{R}$-tree then M_0 contains such a surface. As I explained in [Sh], the proof involves extending Stallings's techniques to $\mathbf{R}$-trees, using codimension-1 *measured laminations* in place of surfaces, and applying a polynomial-vs.-exponential growth argument (similar to the one I described in 2.4.2) to approximate a lamination whose leaves have small fundamental groups by a surface of non-negative Euler characteristic.

When M is hyperbolic, the only possible connected incompressible surfaces of non-negative Euler characteristic in M_0 are disks and annuli. So the upshot, as far as hyperbolic geometry is concerned, is that if M is an orientable hyperbolic 3-manifold whose compact core contains no incompressible annuli or disks, then the space $D_3(\pi_1(M))$ is compact. By 3.6.2, this conclusion may be re-interpreted as saying that the space of all 3-dimensional homotopy-hyperbolic structures on M is compact. This result was first proved from an entirely different point of view by Thurston.

Note that although the project carried out in [MSh1] involved answering Question 2.5.3 for the case of a 3-manifold group, it did not lead to a proof of Conjecture 2.5.5 in this case. We showed that if $\pi_1(M)$ admits a non-trivial small action on an $\mathbf{R}$-tree then it admits a non-trivial small action on a $\mathbf{Z}$-tree; but we did not show that the given action is a limit of small simplicial actions. On the other hand, Morgan and Otal have good partial results on Conjecture 2.5.5 for 3-manifold groups.

4.2.9. Both the results of Thurston's that I have mentioned above—his classification of outer automorphisms of surface groups and his criterion for the compactness of the space of homotopy-hyperbolic structures on a 3-manifold—played central roles in his celebrated work on the existence of hyperbolic structures on 3-manifolds, which I discussed in 3.6. As I have explained, important components of both these theorems can be recovered through the study of group actions on $\mathbf{R}$-trees. This was my original excuse for getting interested in $\mathbf{R}$-trees.

4.3. The metric space approach; outer automorphism groups. There is a different approach to the theory that I described in 4.1 and 4.2.1. In place of algebro-geometric valuations and Tits buildings, this approach uses metric space geometry. The possibility of such an approach was suggested by Gromov and by Thurston. It has been carried out by Bestvina [Be] and, using a somewhat different point of view, by Paulin [Pau2]. Paulin's approach, which I will be describing, provides generalizations that apply not only to the study of hyperbolic manifolds, but to the theory of hyperbolic groups in the sense of Cannon and Gromov, which is discussed extensively elsewhere in this volume.

4.3.1. A representation of a group Γ in $SO(n,1)$ can be interpreted as an action of Γ by isometries on H^n. So in the compactification $\hat{X}_n$, the points of X_n are described by actions on H^n, whereas the points of $\hat{X}_n - X^n$ are described by actions of Γ on $\mathbf{R}$-trees. The common feature is that they are actions of Γ on metric spaces. Paulin's approach to the compactification is based on some general considerations involving Γ-metric spaces, where Γ is a given group; here by a Γ-metric space I mean a (real) metric space equipped with an action of Γ by isometries. Let's restrict attention to Γ-metric spaces having, say, at most the cardinality of the continuum, so that all equivariant isometry classes of Γ-metric spaces form a set, which I'll denote $\mathcal{U} = \mathcal{U}(\Gamma)$. Paulin begins by defining a topology on $\mathcal{U}$. The definition of this topology was suggested by F. Bonahon and is based on ideas due to Gromov and Thurston.

Let Y be a Γ-metric space. Given a finite set $K \subset Y$, a finite set $P \subset \Gamma$ and a positive number ϵ, we define a set $V(K,P,\epsilon) \subset \mathcal{U}$ as follows. A Γ-metric space Y belongs to $V(K,P,\epsilon)$ if and only if there is a map $f\colon K \to Y)$ such that

(i) For any two points $x_1, x_2 \in K$ we have $|\text{dist}\,(f(x_1), f(x_2)) - \text{dist}\,(x_1, x_2)| < \epsilon$, and

(ii) For any point $x \in K$, and any element $\gamma \in P$ such that $\gamma(x) \in K$, we have $\text{dist}\,(f(x), \gamma(f(x))) < \epsilon$.

If we let K and P vary over all finite subsets of Y and Γ, and ϵ over all positive numbers, the sets $V(K,P,\epsilon)$ satisfy the axioms for a basis of neighborhoods of Y in $\mathcal{U}$. In this way one defines a topology on $\mathcal{U}$.

In general this topology is pretty nasty: it is not even Hausdorff. But certain interesting subsets of $\mathcal{U}$ inherit nice subset topologies.

4.3.2. As a first example, consider the set $\mathcal{T}_0 = \mathcal{T}_0(\Gamma) \subset \mathcal{U}(\Gamma)$ consisting of all (equivariant isometry classes of) minimal, non-abelian actions of Γ on $\mathbf{R}$-trees. I have already implicitly described a topology on this set: in 1.5.4 I stated the result of Culler-Morgan and Alperin-Bass giving a bijective correspondence between this set and the set $\mathcal{L}_0(\Gamma) \subset \mathcal{L}(\Gamma)$ consisting of all non- abelian length functions. Of course $\mathcal{L}_0(\Gamma)$ inherits a topology from the product topology of $(0,\infty)^{\mathcal{C}(\Gamma)}$. Paulin shows in [Pau1] that the pull-back of this topology to $\mathcal{T}_0$ coincides with the subspace topology on $\mathcal{T}_0 \subset \mathcal{U}$. Paulin uses this to give an alternative proof of Culler and Morgan's result (see 2.1.2) that when Γ is not small, the space $\mathcal{SPL}(\Gamma)$, which is a subset of the image of $L_0(\Gamma)$ in $\mathcal{PL}(\Gamma)$, is compact.

One can also consider a slightly larger set than $\mathcal{T}_0(\Gamma)$, namely the set $\mathcal{T}(\Gamma) \subset$

$\mathcal{U}(\Gamma)$ consisting of all minimal actions of Γ on **R**-trees which are *semi-simple* in the sense that they are not exceptional abelian actions (1.6). It follows from 1.5.4 that there is a natural bijection between $\mathcal{T}(\Gamma)$ and $\mathcal{L}(\Gamma)$. I believe that the methods of [Pau1] also allow one to show that the topology induced on $\mathcal{T}(\Gamma)$ by this bijection coincides with the subspace topology on $\mathcal{T} \subset \mathcal{U}$, and to give a proof from this point of view of Culler and Morgan's result (see 1.7.1) that $\mathcal{PL}(\Gamma)$ is compact.

4.3.3. These compactness arguments are based on principles of which the applicability extends far beyond the case of **R**-trees. From the point of view of Gromov's theory, **R**-trees are 0-hyperbolic spaces. In [Pau2,3], a general compactness criterion is established for subsets of $\mathcal{U}(\Gamma)$ consisting of spaces that are hyperbolic in Gromov's sense. It is most conveniently stated (and proved) as a *sequential* compactness criterion, and this seems to cover all interesting applications.

Let (Y_i) be a sequence of Γ-metric spaces. Let $(\delta_i)_{i\geq 0}$ be a convergent sequence of non-negative numbers, and set $\delta = \lim \delta_i$. Suppose that Y_i is δ_i-hyperbolic for each i. Let $y_i \in Y_i$ be a base point for each i. Suppose that for every finite set $P \subset \Gamma$ and every $\epsilon > 0$ there exists an integer $N > 0$, such that for each $i \geq 0$ the closed convex hull of the set $P \cdot y_i \subset Y_i$ can be covered by at most N balls of radius ϵ. Then there is a subsequence of the Y_i that converges in $\mathcal{U}(\Gamma)$, and the limit is 50δ-hyperbolic.

(Here the closed convex hull of a subset $S \subset Y_i$ is defined to be the smallest closed convex subset of Y_i containing S. To say that a subset is convex means that every geodesic whose endpoints lie in the subset is itself contained in the subset.)

4.3.4. Using the compactness criterion 4.3.3, it is possible to recover the compactification of $X_n(\Gamma)$ that I discussed in 4.1. Let Γ be a finitely generated group with generators $u_1, \ldots, u_m$. Let (x_i) be an unbounded sequence of points in $X_n(\Gamma)$ for some $n \geq 2$. Each x_i is determined by a representation $\rho_i\colon \Gamma \to \mathrm{SO}(n,1)$. For each i and each $z \in H^n$ set $A_i(z) = \max_{1\leq j\leq m} \operatorname{dist}(z, \rho_i(u_j)(z))$. One can show that for each i there is a point $y_i \in H^n$ where the function A_i takes a smallest value λ_i. Since (x_i) is unbounded, we can assume after passing to a subsequence that $\lambda_i \to \infty$. Now let X_i denote the metric space whose underlying set is H^n, with the distance function obtained by multiplying hyperbolic distance by λ_i^{-1}. Then X_i is λ_i^{-1}-hyperbolic for each i. For each i the representation ρ_i gives X_i the structure of a Γ-metric space. Using the base points y_i, and making strong use of hyperbolicity, it is possible to verify the hypotheses of the compactness criterion of 4.3.3. This means that after passing to a subsequence we can arrange that the X_i converge in $\mathcal{U}(\Gamma)$ to a Γ-metric space T which is 0-hyperbolic, i.e. is an **R**-tree. The action of Γ on T defines a length function l

which is easily seen to be non-zero, so that we have a point $l \in \mathcal{PL}(\Gamma)$. One can then show that $(\Theta(x_i))$ converges to $[l]$ in $\mathcal{P}^\Gamma$.

One can also show from this metric space picture that if the x_i belong to $\mathcal{D}_n(\Gamma)$ then the action defining $[l]$ has small segment stabilizers (see 4.2.1). So the main properties of the compactifications $\hat{X}_n$ and $\hat{\mathcal{D}}_n$ can be established from this alternative point of view.

4.3.5. Because the metric-space approach to the compactification of the character variety sidesteps the use of valuations, it can be applied in situations where no algebraic variety is present. This was done by Paulin in his work [Pau3] on the outer automorphism group of a Gromov-hyperbolic group. He proved that if Γ is Gromov-hyperbolic, and if Γ admits no non-trivial small action on an **R**-tree, then $\mathrm{Out}(\Gamma)$ is finite. It follows that if Conjecture 2.5.5 is true, then any finitely presented Gromov-hyperbolic group which does not split over a small subgroup has a finite outer automorphism group. (The hypothesis from 2.5.5 that the small subgroups of Γ are finitely generated is automatically satisfied by a hyperbolic group. Indeed, Gromov showed that every small subgroup of a hyperbolic group is cyclic-by-finite.)

Paulin's proof of the above finiteness theorem is very similar to his approach to the compactification of $X_n(\Gamma)$ that I outlined in 4.3.4. If $\mathrm{Out}(\Gamma)$ is infinite, it contains a sequence (α_i) of distinct elements. Let Y denote the Cayley graph of Γ with respect to generators $u_1, \ldots, u_m$, and for each i let $\rho_i : \Gamma \to \mathrm{Out}(\Gamma)$ be defined by $\rho_i(\gamma)(y) = \alpha_i(\gamma) \cdot y$. As in 4.3.4, for each i one can find a point $y_i \in Y$ where the function $A_i(z) = \max_{1 \le j \le m} \mathrm{dist}\,(z, \rho_i(u_j)(z))$ takes a smallest value λ_i; here dist denotes the word metric on Y. Using the fact that the α_i are all distinct one can show that Since (x_i) is unbounded, we can assume after passing to a subsequence that $\lambda_i \to \infty$. Now let Y_i denote the metric space whose underlying set is Y, with the distance function obtained by multiplying the word metric on Y by λ_i^{-1}. For each i, the representation ρ_i gives X_i the structure of a Γ-metric space. Using hyperbolicity one checks that the conditions of the compactness criterion 4.3.3 hold, so that some subsequence of the (Y_i) converges to an **R**-tree with an action of Γ. Again this action can be shown to be non-trivial and to have small segment stabilizers.

4.3.6. It is clear from the above argument that if Γ is a hyperbolic group such that $\mathrm{Out}(\Gamma)$ is infinite, then the study of $\mathrm{Out}(\Gamma)$ is closely related to the study of actions of Γ on **R**-trees. This explains the role of **R**-trees in the study of outer automorphisms of a free group (see 2.1) and of a surface group (see 4.2.4). Furthermore, the analogy between the Culler-Vogtmann compactification of outer space and the Thurston compactification of Teichmüller space is made clear by 4.3.3–4.3.5: in both cases actions on **R**-trees arise as limits of actions of a group on δ-hyperbolic spaces as $\delta \to 0$.

4.3.8. Other intriguing approaches to the theory of compactifying character varieties described in 4.1 and 4.2.1 have been worked out by Basarab [Bab] from the point of view of model theory, and by Chiswell [Ch] using non-standard methods. These approaches use a logical perspective to clarify or simplify the approaches that I have described above. Still another very elegant approach, due to Brumfiel and based on the theory of ordered fields, combines some features of the valuation approach of 4.1 and the metric space approach of 4.3, and to some extent clarifies the relationship between them.

But I think a mystery remains. The theory of Bruhat-Tits buildings for algebraic groups over valued fields and the theory of Gromovian convergence of metric spaces both have wide applicability. When two theories have a common special case they often have a common generalization. I wonder if there is some general picture that includes both the Bruhat-Tits tree over an arbitrary valued field (such as the p-adic numbers), and the theory described in 4.3, as special cases. My feeling is that this might be conceptually useful, and could perhaps even lead somewhere.

4.4. Further thoughts. The material presented in this section leads to lots of interesting research questions. For one thing, of course, it provides additional motivation for the questions and conjectures of 2.5, which I will be discussing in Section 5. Here I would like to mention a few other natural directions for further research.

4.4.1. I explained in 4.2.8 how Morgan and I proved Thurston's compactness theorem in [MSh2] using $\mathbf{R}$-trees. In his work on the existence of hyperbolic structures, Thurston used a generalization of this compactness theorem which applies to manifolds that do contain incompressible annuli. He gave an ingenious argument which reduced it to the compactness theorem stated in 4.2.8, or more precisely to a relative version of the latter theorem which is proved by the same method (and was proved in [MSh2] using $\mathbf{R}$-trees). However, for reasons that I shall explain, it is very natural to try to proved the generalization directly using trees.

The generalized compactness theorem is stated in terms of the characteristic submanifold theory of [Jo] and [JaS]. The latter theory provides a picture of the incompressible annuli in a compact 3-manifold M. When M is the compact core of a hyperbolic 3-manifold, M contains a canonical submanifold Σ, each component of which is either an interval bundle meeting ∂M in the associated 0-sphere bundle or a solid torus meeting ∂M in a family of disjoint annuli. The components of the frontier of Σ are incompressible annuli, and every incompressible annulus in M is isotopic to one contained in Σ.

The generalized compactness theorem asserts that if A is any component of $M - \Sigma$, the restriction map $X_3(\pi_1(M)) \to X_3(\pi_1(A))$ maps $\mathcal{D}_3(\pi_1(M))$ to a set with compact closure.

In order to prove this, one has to show that if (x_i) is a sequence ending to infinity in $\mathcal{D}_3(\pi_1(M))$, then (x_i) has a subsequence whose image in $X_3(\pi_1(A))$ converges. By the construction of 4.1.7 we can associate with (x_i) a small action of $\mathcal{D}_3(\pi_1(M))$ on a Λ-tree, where Λ is some ordered abelian group of finite **Q**-rank. To prove the theorem it suffices to show that the restricted action of $\mathcal{D}_3(\pi_1(M))$ is trivial.

What is tantalizing is that Morgan and I proved in [MSh1] that for any small action of $\pi_1(M)$ acts on an **R**-tree, and any component A of $M-\Sigma$, the restricted action of $\pi_1(A)$ does have a fixed point. This is proved by a refinement of the argument that I mentioned in 4.2.8 and discussed in [Sh]. But our proof does not work for Λ-trees when Λ has order-rank greater than 1. In fact the general statement about actions seems to become false in this case. Nevertheless, it seems that there ought to be some way to adapt this approach to give a direct proof of the generalized compactness theorem in terms of trees.

4.4.2. The Cyclic Surgery Theorem, which I discussed in 3.8, gives strong information about how surgery on a knot in S^3 can yield a 3-manifold with a cyclic fundamental group. As I mentioned in 3.8.1, every closed orientable 3-manifold can be obtained from S^3 by a sequence of Dehn surgeries. Each surgery involves a solid torus which is the tubular neighborhood of some knot. An equivalent point of view is to think of all the surgeries as being done simultaneously by removing a finite union of disjoint solid tori from S^3 and sewing them back differently. Thus any closed orientable manifold can be obtained by Dehn surgery on a *link*, i.e. a finite disjoint union of knots in S^3.

If one could formulate and prove an analogue of the Cyclic Surgery Theorem for links, it might be useful in connection with the difficult problem of classifying orientable 3-manifolds with cyclic fundamental group. (This problem includes the Poincaré Conjecture, which asserts that any closed simply-connected 3-manifold is homeomorphic to S^3.)

Recall from 3.8.7 that the proof of the Cyclic Surgery Theorem, in the crucial case of a hyperbolic knot, involves looking at the curve in the $\mathrm{PSL}_2(\mathbf{C})$-character variety of the knot group containing the character of the discrete representation; and that one associates actions of the knot group on **Z**-trees, and hence incompressible surfaces in the knot manifold, with the ideal points of the curve, via the theory described in 3.4 and 3.6. The actions that arise in this way are not small, since the curve contains only one point which is the character of a discrete representation. Correspondingly, the incompressible surfaces that come up do not have small fundamental groups.

If we replace the hyperbolic knot by a hyperbolic link with n components, the irreducible component of the character variety containing the character of the faithful representation becomes n-dimensional. We have already seen that the theory described in 4.3 is the natural generalization to higher-dimensional varieties of characters of the theory described in 3.4, and that it involves considering **R**-trees in place of **Z**-trees. In [MSh2], the connection between trees and surfaces described in 3.6 was largely generalized to **R**-trees. In place of incompressible surfaces one uses incompressible measured laminations. While the immediate goal of [MSh2] was to apply the machinery to the small actions that arise as limits of discrete representations, the machinery is in principle of wider applicability. So one possible approach to generalizing the Cyclic Surgery Theorem to links would be to use this machinery.

I don't know whether this can be done. I mention it as an example of what a potentially rich subject dendrology seems to me to be.

SECTION 5. FREE AND SMALL ACTIONS ON R-TREES

In 2.5.1 and 2.5.5 I stated two conjectures about actions of groups on **R**-trees. In 4.2.6 and 4.3.5 I illustrated the implications of Conjecture 2.5.5 for the geometry of hyperbolic manifolds and the study of outer automorphism groups. In this section I will summarize some of the existing evidence for Conjectures 2.5.1 and 2.5.5. This evidence, as I shall argue, actually suggests a stronger conjecture, which would imply both of the Conjectures 2.5.1 and 2.5.5. This stronger conjecture has only recently occurred to me, and its present form is somewhat tentative; I will state it in 5.5.6.

5.1. Some known results. As I explained in 4.2.8, Morgan and I showed in [MSh2] that Question 2.5.3 has an affirmative answer when the finitely generated group Γ is the fundamental group of an orientable 3-manifold. We also proved Conjecture 2.5.1 for the case of a 3-manifold group; alternatively, this can be deduced via the results of [MSk] from the affirmative answer to Question 2.5.3 in the 3-manifold group case, using the fact that a finitely generated 3-manifold group is finitely presented (see 4.2.8). This special case of Conjecture 2.5.1 is far from vacuous, because any free product of surface groups and free abelian groups of rank ≤ 3 is in fact the fundamental group of an orientable 3-manifold M. Indeed, we can construct M as a connected sum of finitely many 3- manifolds, each of which is either an interval bundle over a surface or a product of circles and arcs.

The results of Morgan and Skora in [MSk], which I discussed in 2.5.4, not only relate Conjectures 2.5.1 and Question 2.5.3, but also give direct evidence for Conjecture 2.5.1, since they show that it is true for any group that splits over a small subgroup.

Another source of evidence for the Conjectures 2.5.1 and 2.5.5 is provided by the results in my joint papers [GiS1] with Gillet and [GiSSk] with Gillet and Skora. These papers deal with actions on Λ-trees, where Λ is an arbitrary subgroup of **R** whose **Q**-rank is at most 2. The **R**-completion of such an action is an action of Γ on an $\mathbf{R}$-tree; this completed action has rank ≤ 2 in the sense of 4.1.2.

It follows from the results of [GiS1] and [GiSSk] that Conjectures 2.5.1 and 2.5.5 are true if one restricts attention to actions of rank at most 2. More precisely, if a finitely generated group Γ admits a free action of rank ≤ 2 on an $\mathbf{R}$-tree, then Γ is a free product of surface groups and infinite cyclic groups. (If the $\mathbf{Q}$-rank is 1 then Γ is actually free.) If Γ is a finitely presented group whose small subgroups are all finitely generated, then any projectivized length function on Γ defined by a small action of rank 2 is the limit of a sequence of projectivized $\mathbf{Z}$-valued length functions. We saw in 4.2.7 that the finite-rank case of Conjecture 2.5.5 is particularly important for applications, so it seems encouraging that the case of rank ≤ 2 is true.

5.2. Strong convergence, standard actions and the ascending chain condition. The proofs of Conjectures of 2.5.1 and 2.5.5 in the rank-2 case use the main result of [GiS1], which is a structure theorem for a large class of actions on Λ- trees, where $\Lambda \leq \mathbf{R}$ has $\mathbf{Q}$-rank ≤ 2. Some recent evidence suggests that a similar structure theorem may hold with no restriction on the rank of the action. The ultimate goal of this section is to formulate the appropriate conjecture. I will begin by explaining the statement of the structure theorem[12] that is proved in [GiS1].

5.2.1. The theorem applies to actions of a group Γ on a Λ-tree (where $\Lambda \leq \mathbf{R}$ has $\mathbf{Q}$-rank ≤ 2) that satisfy the following *ascending chain condition.*

If $\sigma_1, \sigma_2 \ldots$ is a monotone decreasing sequence of segments in T with a common midpoint, and if Γ_i denotes the stabilizer of σ_i in Γ, then for all sufficiently large i we have $\Gamma_i = \Gamma_{i+1}$.

5.2.2. The form of the theorem is that if $\Lambda \leq \mathbf{R}$ has $\mathbf{Q}$-rank ≤ 2, then any action on a Λ-tree is a limit in a strong sense—much stronger, in the case of an action of a finitely presented group, than the sense of 2.2.7—of actions of a standard type. These standard actions constitute a common generalization of two types of actions that I have discussed in previous sections: (i) the polyhedral actions discussed in 1.2.10, and (ii) the actions described in 2.3.7 which are associated to length systems on triangulated surfaces (or equivalently are dual to measured foliations on surfaces).

Note that actions of both types (i) and (ii) are constructed from 1-connected simplicial complexes in which a length is assigned to every 1-simplex. In case (i) the complex is a simplicial tree. Thus the link of every vertex is 0-dimensional. In case (ii) the complex is the universal covering $\tilde{\Sigma}$ of the given surface Σ; thus

[12]In [Sh] I stated some of the consequences of the structure theorem. However, the statement of the structure theorem itself was not in final form when [Sh] was written.

Σ is itself a triangulated surface, so that the link of every vertex is a 1- sphere. In general the actions that are to be taken as standard are defined in terms of length systems on 1-connected complexes of a type that in [GiS1] are called singular surfaces.

5.2.3. A *singular surface* is, by definition, a simplicial complex $\tilde{\Sigma}$ of dimension 1 or 2 in which the connected components of the link of every vertex are points and combinatorial 1-manifolds. These 1-manifolds may be homemorphic to either S^1 or $\mathbf{R}$. If the link of every vertex of $\tilde{\Sigma}$ is a connected combinatorial 1-manifold, $\tilde{\Sigma}$ is called a ***surface with points at infinity***. In this case, the vertices whose links are non-compact, i.e. are homeomorphic to $\mathbf{R}$, are called ***points at infinity***. If $\tilde{\Sigma}$ is a surface with points at infinity, the complement of the set of points at infinity in $\tilde{\Sigma}$ is a topological 2-manifold.

5.2.4. Any discrete subgroup Γ of $\mathrm{PSL}_2(\mathbf{R})$ leads to a natural example of a simply-connected surface $\tilde{\Sigma}$ with points at infinity. As a topological space, we define $\tilde{\Sigma}$ to be the union of H^2 with the set of all fixed points in S^1_∞ of parabolic elements of Γ. There always exists a Γ-invariant triangulation of $\tilde{\Sigma}$ in which every 1-simplex is an arc contained in the closure in $\bar{H}^2 = H^2 \cup S^1_\infty$ of a hyperbolic geodesic. With respect to this triangulation, $\tilde{\Sigma}$ is a surface with points at infinity, and its points at infinity are precisely the parabolic fixed points of elements of Γ. If Γ is a lattice, i.e. if $\mathrm{PSL}_2(\mathbf{R})/\Gamma$ has finite volume, then $\tilde{\Sigma}$ is finite modulo Γ, i.e. there are only finitely many Γ-orbits of simplices.

5.2.5. If X is an arbitrary 1-connected triangulated space, the ***branches*** of X are defined to be the connected components of the complement of the 0-skeleton in $\tilde{\Sigma}$. If $\tilde{\Sigma}$ is a 1-connected singular surface, each branch of $\tilde{\Sigma}$ is either a closed 1-simplex or a surface with points at infinity. Thus we think of 1-connected singular surfaces as complexes obtained by gluing together 1-dimensional branches which are arcs and 2-dimensional branches which are surfaces with points at infinity, in some simply-connected pattern. If we use only 1-dimensional branches, we get a simplicial tree.

5.2.6. Just as in the case of a surface, we define a *length system* on a singular surface $\tilde{\Sigma}$ to be a family (x_τ) of positive real numbers indexed by the 1-simplices of Σ, with the property that for each 2-simplex σ of Σ, we can label the edges of σ as τ, τ' and τ'' in such a way that $x_\tau = x_{\tau'} + x_{\tau''}$. Again we call τ the *long edge* of σ. Note that if $\tilde{\Sigma}$ is a tree, there is no restriction on the assignment of lengths to 1-simplices; thus a polyhedral tree can be thought of as a simplicial tree with a length system.

As in the non-singular case, any piecewise-linear path in $\tilde{\Sigma}$ has a well-defined length with respect to any given length system on $\tilde{\Sigma}$.

To generalize the non-degeneracy condition (ii) of 2.3.5 to length systems on singular surfaces, a smidgen of care is required. In the case where $\tilde{\Sigma}$ is a surface with points at infinity, the definition proceeds much as in 2.3.5: we define the order o_v of a vertex to be the number of 2-simplices incident to v whose long edges are not incident to v. Since v may be a point at infinity, the cardinal o_v may be finite or infinite. We define the given length system on Σ to be ***non- degenerate*** if $o_v \geq 2$ for every vertex v of Σ. Now let Σ be an arbitrary 1-connected[13] singular surface. Any length system on $\tilde{\Sigma}$ restricts to a length system on each branch of $\tilde{\Sigma}$. A length system on $\tilde{\Sigma}$ is said to be ***non-degenerate*** if its restriction to every 2-dimensional branch is non-degenerate.

5.2.7. As I illustrated in 2.3.9 for the non-singular case, a non-degenerate length system λ on a singular surface $\tilde{\Sigma}$ need not define a tree. As in the non-singular case, a we need an extra condition, namely that there is a group of simplicial homeomorphisms Γ of Σ such that λ is invariant under Γ and $\tilde{\Sigma}$ is finite modulo Γ. If this condition holds then, as in the non-singular case, any two points of $\tilde{\Sigma}$ are joined by a path of minimal length; again, this defines a pseudo-distance on $\tilde{\Sigma}$, and the associated metric space is an $\mathbf{R}$-tree T. Furthermore, the action of Γ on $\tilde{\Sigma}$ induces an action on T. For the present exposition I will define an action of a group Γ on an $\mathbf{R}$-tree to be ***standard*** if it is defined in this way.

5.2.8. It is natural to define a corresponding class of standard actions on Λ-trees, where Λ is any subgroup of $\mathbf{R}$. Let Γ act on a singular surface $\tilde{\Sigma}$ in such a way that $\tilde{\Sigma}$ is finite modulo Γ. Let $\tilde{\Sigma}$ be equipped with a length system which is Λ-valued in the sense that $x_\tau \in \Lambda$ for every 1-simplex τ of $\tilde{\Sigma}$. The action of Γ on $\tilde{\Sigma}$ defines a standard action on an $\mathbf{R}$-tree T, and by the definition of T there is a natural Γ-invariant map $\chi\colon \tilde{\Sigma} \to T$. Let $X \subset T$ denote the image under χ of the 0-skeleton of Σ. It may be shown that the distance between any two points x and y of X is an element of Λ, so that the set

$$[x, y]_\Lambda = \{z \in T | \text{dist}\ (x, z) \in \Lambda\}$$

is isometric to an interval in Λ. It may also be shown that $T_0 = \bigcup_{x,y \in X} [x, y]_\Lambda$ is a Λ-tree; it is clearly invariant under the action of Γ. So if an action of Γ on $\tilde{\Sigma}$, and a length system, satisfy the above conditions, they determine an action of Γ on a Λ-tree T_0. I'll say that an action of a group on a Λ-tree is ***standard*** if it can be constructed in this way. The completion of a standard action on a Λ-tree is a standard action on an $\mathbf{R}$-tree.

5.2.9. As I have said, the gist of the main theorem of [GiS1] is that every action satisfying the hypotheses is a limit, in a strong sense, of standard actions. To

[13]The definitions can be extended to the non-simply- connected case with some extra work, but the statement of the structure theorem involves only the simply-connected case.

define the appropriate notion of limit we define a *category of group actions on* Λ-*trees* where Λ is any ordered abelian group. The category of Λ-trees was defined in 1.7.3. Let's think of a group action on a Λ-tree as a triple $\mathcal{T} = (T, \Gamma, \rho)$, where T is a Λ-tree, Γ is a group and ρ is a homomorphism from Γ to the group of automorphisms of T. A *morphism* from an action (T, Γ, ρ) to an action (T', Γ', ρ') is a pair $\phi = (f, h)$, where $f\colon T \to T'$ is a morphism of Λ-trees and h is a group homomorphism, such that $\rho'(h(\gamma))(f(x)) = f(\rho(\gamma)(x))$ for every $x \in T$ and every $\gamma \in \Gamma$.

There is also a notion of standard morphism between standard actions. Suppose that $\mathcal{T}_1$ and $\mathcal{T}_2$ are standard actions. Then $\mathcal{T}_i = (T_i, \Gamma_i, \rho_i)$ is defined by a 1-connected singular surface $\tilde{\Sigma}_i$, a non-degenerate λ-valued length system on $\tilde{\Sigma}_i$, and an action of Γ_i on $\tilde{\Sigma}_i$, such that $\tilde{\Sigma}$ is finite modulo Γ. Let χ_i denote the natural map from the 0-skeleton Y_i of $\tilde{\Sigma}_i$ to T_i. A morphism $phi = (f, h)\colon \mathcal{T}_1 \to \mathcal{T}_2$ is said to be *standard* if there is a continuous map $F\colon \tilde{\Sigma}_1 \to \tilde{\Sigma}_2$ such that (i) $F(\gamma \cdot x) = h(\gamma) \cdot F(x)$ for every point $x \in \Sigma$ and every $\gamma \in \Gamma$, (ii) $F(Y_1) \subset Y_2$ and (iii) $\chi_2 \circ (F|Y_1) = f \circ \chi_1$.

5.2.10. Now suppose that $(\mathcal{T}_i; \phi_{ij})$ is a direct system in the category of actions on Λ-trees. This means, first, that we have a family of actions $\mathcal{T}_i$ indexed by some filtered ordered set I. (For most real-life applications we can take I to be the natural numbers.) Second, whenever $i < j$ we have a morphism $\phi_{ij}\colon \mathcal{T}_i \to \mathcal{T}_j$, and we have $\phi_{jk} \circ \phi_{ij} = \phi_{jk}$ whenever $i < j < k$. For each i let's write $\mathcal{T}_i = (T_i, \Gamma_i, \rho_i)_{i \in I}$. The system $(\mathcal{T}_i; \phi_{ij})$ is said to *converge strongly* if for every $i \in I$ and for every segment $S \subset T_i$ there is an index $j \geq i$ such that the set $f_{ij}(S)$ is mapped isometrically into T_k by f_{jk} for every $k \geq j$. (Thus the segment S may be crumpled up (see 1.7.3) to a certain stage, but beyond some stage the crumpling stops.) In particular, for any $i \in I$ and any two points $x, y \in T_i$, the distance between $f_{ij}(x)$ and $f_{ij}(y)$ is independent of j for all sufficiently large $j \geq i$.

If the direct system $(\mathcal{T}_i; \phi_{ij})$ converges strongly, then there exist a tree T, and morphisms of Λ-trees $f_i\colon T_i \to T$ for all $i \in I$, such that (i) $f_j \circ f_{ij} = f_j$ whenever $i < j$, (ii) $T = \bigcup_{i \in I} f_i(T_i)$, and (iii) for any $i \in I$ and any two points $x, y \in T_i$ we have dist $(f_{ij}(x), f_{ij}(y)) = \mathrm{dist}(f_i(x), f_i(y))$ for all sufficiently large $j \geq i$. The tree T and the maps f_i are unique up to isometry making all imaginable diagrams commute. Furthermore, if Γ is the direct limit group of the system Γ_i, and h_i denotes the natural homomorphism from Γ_i to Γ, there is a unique $\rho\colon \Gamma \to$ Aut (T) such that for each i the pair $\phi_i = (f_i, h_i)$ is a morphism from $\mathcal{T}_i$ to the action $\mathcal{T} = (T, \Gamma, \rho)$. The action $\mathcal{T}$ is called the *limit* of the strongly convergent system $(\mathcal{T}_i, \phi_{ij})$.

5.2.11. The limit of a strongly convergent direct system is in particular a direct limit in the category of actions. However, a direct system may well have a direct limit without converging strongly. An example is given by the sequence $(\mathcal{T}_i)_{i\geq 1}$, where $\mathcal{T}_i$ is the trivial action of the trivial group on a tree T_i; topologically, T_i is a cone on a 3-element set $\{x_i, y_i, z_i\}$, with cone point t_i, and the edges joining t_i to x_i, y_i and z_i have lengths $1 - \frac{1}{i}$, $\frac{1}{i}$ and $\frac{1}{i}$ respectively. See Figure 5.2.11.1.

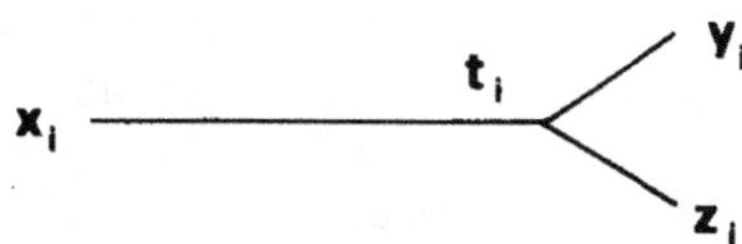

FIGURE 5.2.11.1

For $i < j$ there is a unique morphism from T_i to T_j mapping x_i, y_i and z_i to x, y and z respectively. This defines a morphism $\phi_{ij} \colon \mathcal{T}_i \to \mathcal{T}_j$. The direct system $(\mathcal{T}_i; \phi_{ij})$ is not strongly convergent since the distance from y_i to z_i approaches but never equals 0. However, the direct limit exists and is isometric to the unit interval.

5.2.12. Here is the main result of [GiS1]:

Let Λ be a subgroup of $\mathbf{R}$ and let $\mathcal{T}$ be an action on a Λ-tree. Suppose that either Λ has $\mathbf{Q}$-rank 1, or that Λ has $\mathbf{Q}$-rank 2 and the action $\mathcal{T}$ satisfies the ascending chain condition 5.2.1. Then $\mathcal{T}$ is the limit of a strongly convergent direct system $(\mathcal{T}_i, \phi_{ij})$, where the $\mathcal{T}_i$ are standard actions and the ϕ_{ij} are standard morphisms.

If the given action is on a countable tree we can take the strongly convergent direct system to be indexed by the natural numbers. (This is the case that is explicitly done in [GiS1], but the proof in general is essentially the same.)

5.2.13. Let me now sketch the arguments used in [GiS1] and [GiSSk] to deduce from the above structure theorem that Conjectures 2.5.1 and 2.5.5 are true in the case where the given action has rank at most 2. To deduce Conjecture 2.5.1 in this case, we observe that the given group admits a free action $\mathcal{T} = (T, \Gamma, \rho)$ on a Λ- tree, where $\Lambda \leq \mathbf{R}$ has $\mathbf{Q}$-rank at most 2. A free action automatically satisfies the ascending chain condition. The theorem therefore exhibits $\mathcal{T}$ as the limit of a strongly convergent system $(\mathcal{T}_i)$, where the standard actions $\mathcal{T}_i = (T_i, \Gamma_i, \rho_i)$

must themselves be free. It then follows from the definition of a standard action that eahc group Γ_i admits a free action, with compact quotient, on a 1-connected singular surface. By convering space theory, Γ_i is the fundamental group of a compact singular surface, and is therefore a free product of infinite cyclic groups and surface groups. The group Γ is the direct limit of the Γ_i. It may be shown that a direct limit of groups, each of which is a free product of infinite cyclic groups and surface groups, is itself a free product of infinite cyclic groups and surface groups.

Note that this argument appears to establish a stronger result than Conjecture 2.5.1 in the case of $\mathbf{Q}$-rank ≤ 2, because in place of the free abelian factors predicted by 2.5.1 we have infinite cyclic factors. This is an illusion, however, because if Λ has $\mathbf{Q}$-rank at most n then it is elementary to show that any free abelian group which admits a free action of rank $\leq n$ must have rank at most n. Thus when $n = 2$ one expects to have free abelian factors of ranks 1 and 2. As a free abelian group of rank 2 is the fundamental group of a 2-torus, it can appear among the surface group factors.

To establish Conjecture 2.5.5 in the case of rank ≤ 2 one must show that if $\mathcal{T} = (T, \Gamma, \rho)$ is a small action on a Λ-tree, where $\Lambda \leq \mathbf{R}$ has $\mathbf{Q}$-rank at most 2 and Γ is a finitely presented group with the property that all its small subgroups are finitely generated, then the action is a limit, in the sense of 2.2.7, of small actions of Γ on $\mathbf{Z}$-trees. The hypothesis that the action $\mathcal{T}$ is small, together with the restriction on the small subgroups of Γ, guarantees that $\mathcal{T}$ satisfies the ascending chain condition. Using the countability of Γ we can reduce to the case where T is countable, so that the theorem exhibits $\mathcal{T}$ as the limit of a strongly convergent direct system $(\mathcal{T}_i, \phi_{ij})$ of small standard actions indexed by the positive integers. Let us write $\mathcal{T}_i = (T_i, \Gamma_i, \rho_i)$. Using the hypothesis that Γ is finitely presented, we can modify the system $(\mathcal{T}_i, \phi_{ij})$ so as to arrange that the Γ_i are equal to Γ and the group homomorphisms involved in the morphisms ϕ_{ij} are all the identity. It is then easy to show that T is the limit of the actions $\mathcal{T}_i$ in the sense of 2.2.7. This reduces the proof to the case where the given action $\mathcal{T}$ is standard. In this case the proof uses the ideas sketched in 2.3.11: one exhibits the length system defining the action $\mathcal{T}$ as a limit of constant multiples of integer-valued length systems. Each of these defines a small action on a $\mathbf{Z}$-tree, and by doing the approximation with some care one can prove that $\mathcal{T}$ is the limit of these actions.

5.2.14. In [Sh] I discussed a direct proof of Conjecture 2.5.1 for rank-2 actions, extracted from a preliminary version of [GiS1]. The proof of the structure theorem is a refinement of the latter argument. We are given an action $\mathcal{T} = (T, \Gamma, \rho)$, where T is a Λ-tree for some $\Lambda \subset \mathbf{R}$ of $\mathbf{Q}$-rank at most 2. In the case where Λ is free abelian of rank 2 one uses the contractible complex $K = K(T)$, defined in [Sh, Section 6], on which the automorphism group of the Λ-tree T acts.

Let us consider any connected subcomplex Y of K which is Γ-invariant and finite modulo Γ. One can use the ideas explained in [Sh, Section 6] to show that Y can be equivariantly deformed into a Γ-invariant singular surface Σ, also finite modulo Γ. Now by the definition of K, every 1-simplex of K corresponds to a segment of T; by associating to each 1-simplex of Σ the length of the corresponding segment in T one can define a Λ-valued length system on Σ. One can choose Σ so that this length system is non- degenerate. In general Σ need not be simply connected, but its universal cover $\tilde{\Sigma}$ inherits a Λ-valued length system. The action of Γ on Σ induces an action on $\tilde{\Sigma}$ of some group $\tilde{\Gamma}$ which is an extension of Γ by $\pi_1(\Sigma)$: that is, $\tilde{\Gamma}$ maps homomorphically onto Γ with kernel $\pi_1(\Sigma)$. Furthermore, $\tilde{\Sigma}$ is finite modulo $\tilde{\Gamma}$ and has an induced $\tilde{\Gamma}$-invariant, non-degenerate length system. By 5.2.7 and 5.2.8, the 1-connected singular surface $\tilde{\Sigma}$, with this length system and this action, defines a standard action of $\tilde{\Gamma}$ on a Λ-tree.

Let us write K as a monotone union[14] of subcomplexes Y_i that are finite modulo Γ. With each Y_i we can associate a standard action $\mathcal{T}_i$ by the above construction. This construction is not quite canonical, but by choosing the $\mathcal{T}_i$ with a bit of care one can arrange that there are natural morphisms that make the $\mathcal{T}_i$ into a direct system, and that this system converges strongly to $\mathcal{T}$.

The case where Λ has rank 2 is not free abelian requires a further refinement. In this case we need to write Λ as a monotone union of subgroups L_j that are free abelian of rank 2; with each of these subgroups we can associate a complex $K_j = K(T, L_j)$. These complexes are not contractible or even connected, but their homotopy-theoretic direct limit is contractible. One can then apply the above construction for each j; by doing this with some care one obtains a doubly indexed direct system of actions which converge strongly to $\mathcal{T}$.

5.3 Automorphisms of free groups revisited. Since the notion of strong convergence was introduced in [GiS1], it has gradually become clear that it arises naturally in other settings than that of rank-2 actions. I know of two different instances of this.

To explain the first instance, let me return to the example of the Bestvina-Handel theory that I discussed in 2.2.2–2.2.5, and re-examine it from a slightly different point of view. Recall from 2.2.2 that we thought of the Cayley graph T_0 of F_3 with respect to the generators x, y and z as a $\mathbf{Z}$-tree, and we considered the length function l_0 associated with the natural action of F on T_0. We associated a matrix A to the positive automorphism α, and considered the positive eigenvalue λ of A corresponding to its unique eigenvector v_0 in the first octant. We saw that the sequence $(\lambda^{-n} l_0 \circ \alpha^n)_{n \geq 0}$ converges to the length function of a non-polyhedral

---

[14] If T is uncountable then K has uncountably many vertices. In this case we need a transfinite monotone union. Only the countable case was done explicitly in [GiS1].

free action l_∞.

A slight variant of this approach is to regard the Cayley graph as a polyhedral tree T'_0 by assigning to each edge a length equal to one of the coordinates ξ_0, η_0 or ζ_0 of v_0, according to whether the given edge is labeled with the generator x, y or z. The action of F_3 on T'_0 defines a length function l'_0. The sequence $(\lambda^{-n} l'_0 \circ \alpha^n)_{n \geq 0}$ again converges to l_∞, but in a tamer way than $(\lambda^{-n} l_0 \circ \alpha^n)$: the arguments of 2.2.2–2.2.5 show that we have $l'0(W) = \lambda^{-n} l'_0 \circ \alpha^n(W) = l_\infty(W)$ for every cyclically legal word W and every $n \geq 0$, and that for an arbitrary $\gamma \in F$ we have $\lambda^{-n} l'_0 \circ \alpha^n(W) = l_\infty(W)$ for all sufficiently large n.

We can say more. For each $n \geq 0$ let T'_n denote the polyhedral tree obtained from T'_0 by multiplying the length of every edge by λ^{-n}. As a set, each T_i is identified with the Cayley graph of F_3, and it contains a copy X_i of the 0-skeleton of the Cayley graph, whose points are indexed by the elements of F_3. Let us write $X_i = \{x_i^\gamma : \gamma \in F_3\}$. For each $i \geq 1$ we can define a map $g_i \colon X_i \to X_{i+1}$ by $g_i(x_\gamma) = x_{i(\gamma)}$. It follows from our choice of lengths of edges in the polyhedral trees T_i that g_i extends to a morphism $f_i \colon T_i \to T_{i+1}$ which maps each edge homeomorphically onto a simplicial arc in T_i. We can now define a direct system $(\mathcal{T}_i)$ of actions indexed by the natural numbers: we have $\mathcal{T}_i = (T_i, F_3, \rho_i)$, where ρ_i is the natural action of F_3 on its Cayley graph, interpreted as an action on T_i; and the morphism from f_i to f_j when $j > i$ is $f_{j-1} \circ \cdots \circ f_i$. The arguments of 2.2.2–2.2.5 are easily adapted to show that this direct system is strongly convergent and that its limit is l_∞.

This is very striking, because the example of 2.2.2 gives the simplest example that I know of a non-polyhedral free action of a free group on an **R**-tree, and it now turns out that this action is the limit of a strongly convergent direct system of polyhedral actions.

5.4. Contracting outer space and related spaces. The other situation in which strong convergence has arisen naturally is related to the contractibility of outer space, which I mentioned in 2.1.1. Several years after contractibility was proved by Culler-Vogtmann and Gersten, another proof was announced by Michael Steiner. Besides proving contractibility for Y_n he appears to have proved it for many other naturally defined subsets of $\mathcal{PL}(F_n)$, including $\hat{Y}_n$, $\mathcal{SPL}(F_n)$, and $\mathcal{PL}(F_n)$ itself. The method is to provide a geometric contraction of $\mathcal{PL}(F_n)$ which is so geometrically natural that it induces contractions of all these subsets.

I would like to describe briefly an elegant version of this theory which has been given by Richard Skora in [Sk2].

5.4.1. A key ingredient in this version of the theory is a construction for factoring any morphism between **R**-trees through a family of intermediate trees. Suppose that T and T' are **R**-trees and that $\phi \colon T \to T'$ is a surjective morphism. For each $t \in [0, \infty)$ we define a pseudo-distance $D_t \colon T \times T \to \mathbf{R}$ as follows. Let x and y be any two distinct points of T. Let us fix a homeomorphism $\alpha = \alpha_{xy}\colon [0,1] \to [x,y]$. Let $\mathcal{B}$ denote the set of all paths $\beta\colon [0,1] \to T'$ which are piecewise geodesics, i.e. are morphisms from the $\mathbf{R}$-tree $[0,1]$ to T', and which have initial point $\beta(0) = \phi(x)$ and terminal point $\beta(1) = \phi(y)$. Any $\beta \in \mathcal{B}$ is a rectifiable path in the metric space T and thus has a well-defined finite length. We have $\phi \circ \alpha \in \mathcal{B}$, and length $\phi \circ \alpha = \mathrm{dist}_T(x,y)$. Now let $\mathcal{B}_t$ denote the set of all paths $\beta \in \mathcal{B}$ which are uniform t-approximations to $\phi \circ \alpha$, i.e. which satisfy $\mathrm{dist}_{T'}(\beta(u), \phi \circ \alpha(u)) < t$ for every $t \in [0,1]$. We define $D_t(x,y)$ to be the infimum of the lengths of all paths in $\mathcal{B}_t$. (This infimum can be shown to be realized as a minimum.) It is clear that $\mathcal{B}_0 = \{\phi \circ \alpha\}$ and hence that $D_0(x,y) = \mathrm{dist}_T(x,y)$. It is also clear that $\mathcal{B}_\infty = \mathcal{B}$ and hence that $D_\infty(x,y) = \mathrm{dist}_{T'}(\phi(x), \phi(y))$.

For each t, let T_t denote the metric space determined by the pseudo-distance D_t. Then T_0 and T_∞ are canonically identified with T and T'. Note also that whenever $0 \le s \le t \le \infty$ we have $D_t \le D_s$. Hence there is a natural distance-decreasing map $\phi_{st}\colon T_s \to T_t$. We have $\phi_{tu} \circ \phi_{st} = \phi_{su}$ whenever $s \le t \le u$, and $\phi = \phi_{0\infty}$.

In [Sk2] it is proved that the T_s are all $\mathbf{R}$-trees and that the ϕ_{st} are all morphisms. If T and T' are equipped with (isometric) actions of a group Γ and if ϕ is Γ-equivariant, then by naturality we have an action of Γ on each T_s, and the ϕst are also Γ-equivariant.

5.4.2. In order to apply the above construction to prove contractibility results one needs a way of choosing a canonical base point in a given $\mathbf{R}$-tree equipped with a non-abelian action of a group Γ with a given finite generating set S. In [Sk2] this is done by a construction similar to one that I mentioned in 4.3.4. Given an action of Γ on T and a point $x \in T$ we set $A(x) = \max_{\gamma \in S} \mathrm{dist}\,(x, \gamma \cdot x)$. It can be shown that the function $A(x)$ always takes a minimum value l_0 on T. The set X of all points $x \in T$ for which $A(x) = l_0$ is a subtree. If the action is non-abelian then X is finite. The barycenter (1.2.7) of X is a canonical base point of T.

5.4.3. Now suppose that we are given a point $[l] \in \mathcal{PL}(F_n)$. We can represent $[l]$ by a non-trivial minimal action of F_n on an $\mathbf{R}$-tree T, and if $[l]$ is abelian we can take the action to be non-exceptional, i.e. we can take $T = \mathbf{R}$. If the action is non-abelian, then using the standard generators of F_n we can define a base point $x_0 \in T$ by the above construction. If the action is abelian and non-exceptional we take x_0 to be an arbitrary point of T. Now let T_0 denote the Cayley graph of F_n with respect to the standard generators. Each edge e of T_0 is labeled with a generator u_e from the standard generating set, and we can give T_0 the structure of a polyhedral $\mathbf{R}$-tree by assigning to e the length $\mathrm{dist}_T(x_0, u_e(x_0))$. Now T_0 has a natural base point—the vertex labeled by the identity element—and a natural action of F_n defined by left multiplication. It follows from the definitions of the lengths of edges in T_0 that there is a unique F_n-equivariant morphism $\phi\colon T_0 \to T$. Applying the construction of 5.4.1 with this choice of ϕ, we get a tree T_t with an F_n-action for each $t \in [0,\infty)$, and a morphism $\phi_{st}\colon T_s \to T_t$ whenever $s \le t$. The action of F_n on T_t defines a point $[l_t] \in \mathcal{PL}(F_n)$. Note that $[l_\infty] = [l]$ and that $[l_0]$ is defined by the natural polyhedral action of F_n on T_0.

We can now define a map $H\colon \mathcal{PL}(F_n)[0,1] \to \mathcal{PL}(F_n)[0,1]$ by $H([l],t) = [l_{t^{-1}-1}]$. In [Sk2] it is shown that H is continuous. By construction we have $H([l],0) = [l]$ and $H([l],1) \in \Delta$, where $\Delta \subset \mathcal{PL}(F_n)$ consists of all projectivized length functions obtained from the standard action of F_n on its Cayley graph by assigning positive lengths to the (standard) generators. It is also straightforward to check that $H([l],1) = [l]$ for each $[l] \in \Delta$, so that H is a deformation retraction of $\mathcal{PL}(F_n)$ to Δ. But by the definition of Δ there is a natural bijection between Δ and the quotient of the positive cone in $\mathbf{R}^n$ by homotheties. This bijection is a homeomorphism, so Δ is contractible. The contractibility of $\mathcal{PL}(F_n)$ follows. It is also not hard to show that the sets Y_n, $\hat{Y}_n$, $\mathcal{SPL}(F_n)$, and the set of all projectivized length functions defined by free actions of F_n are invariant under H; that is, if W denotes any of these sets we have $H(W \times [0,1]) \subset W$. It follows that these sets are all contractible.

5.4.4. Now consider any small minimal action of F_n on an $\mathbf{R}$-tree T. Such an action cannot be abelian. Hence the construction of 5.4.3 gives a family $(\mathcal{T}_t)_{t\geq 0}$ of actions of F_n on $\mathbf{R}$-trees, and morphisms $\phi_{st}\colon \mathcal{T}_s \to \mathcal{T}_t$ for $s \le t$. As I pointed out in 5.4.1, we have $\phi_{tu} \circ \phi_{st} = \phi_{su}$ whenever $s \le t \le u$; so the $\mathcal{T}_t$ for $t < \infty$ and the ϕ_{st} for $s \le t \le \infty$ constitute a direct system. It is easy to see from the construction of 5.4.1 that this direct system converges strongly and that its limit is the given action of F_n on T. If you prefer to think countably, you may think of the given action as the limit of the strongly convergent system $(\mathcal{T}_n)$, where n ranges over the natural numbers.

If the given action is a limit of free polyhedral actions, i.e. if its projectivized length function $[l_\infty]$ is in $\hat{Y}_n$, then for every t the projectivized length function $[l_t]$ defined by $\mathcal{T}_t$ lies in $\hat{Y}_n$. We have $[l_0] \in Y_n$. By the continuity of H, there is a smallest t, say $t = t_0$, for which $[l_t] \notin Y_n$. Thus the small action $\mathcal{T}_{t_0}$ is not polyhedral but it is the limit of a strongly convergent system of free polyhedral actions.

5.4.5. The construction in 5.3, which looked very different from the one just described in 5.4.4, also exhibited certain actions corresponding to points of $\hat{Y}_n - Y_n$ as limits of strongly converging systems of free polyhedral actions. The construction that I have just given certainly suggests that there should be a wealth of points in Y_n that correspond to actions which are limits of this type. I don't think every action corresponding to a point of $\hat{Y}_n - Y_n$ is such a limit. Some points of $\hat{Y}_n - Y_n$ are defined by actions dual to measured foliations, and such actions do not appear to be limits of strongly convergent systems of polyhedral actions. However, it may well be the case that for every action corresponding to a point of $\hat{Y}_n$, or more generally for every small minimal action of F_n, the system $(T_t)_{0\leq t<\infty}$ consists of actions that are standard in the sense of 5.2.7.

5.4.6. This leads to the conjecture that every small action of a free group is the limit of a strongly convergent direct system of actions which are standard in the sense of 5.2.7.

5.4.7. Some of Jiang's results in [Ji] are relevant to Conjecture 5.4.6. Jiang gives an intricate and ingenious proof that if a finitely generated free group acts freely and minimally on an $\mathbf{R}$-tree T, then there are only finitely many orbits of branch points under the action. Here a point $x \in T$ is called a ***branch point*** if there are two segments $S, S' \subset T$ whose interiors contain x but such that x is not an interior point of $S \cap S'$. Jiang also proves an estimate for the number of orbits of branch points; this estimate is best possible and can realized by a standard action. Skora has pointed out that if Conjecture 5.4.6 were true, one could deduce Jiang's finiteness result, and his estimate, as corollaries. So one may regard these results of Jiang's as additional evidence that Conjecture 5.4.6 is true.

5.5. A tentative conjecture. There is an obvious parallel between Conjecture 5.4.6 and the main theorem of [GiS1], which I stated in 5.2.12. This suggests that there may be a general structure theorem for group actions on $\mathbf{R}$-trees which satisfy the ascending chain condition. However, a simple-minded example shows that some care is needed formulating the appropriate statement. If a free action of a finitely generated group Γ on an $\mathbf{R}$-tree is the limit of a strongly convergent direct system of actions that are standard in the sense of 5.2.7, then the arguments that I sketched in 5.2.13 show that Γ is a free product of surface groups and infinite cyclic groups. But any free abelian group Γ is isomorphic to a subgroup of $\mathbf{R}$ and therefore acts freely on the $\mathbf{R}$-tree $T = \mathbf{R}$ by translations; and if Γ has rank > 2 then it is not a free product of surface groups and cyclic groups.

So we need to broaden the definition of "standard action." One may hope to prove that every group action on an $\mathbf{R}$-tree which satisfies the ascending chain condition is the limit of a strongly convergent direct system of actions which are standard in a suitably generalized sense. I would like to propose a tentative definition, which I have worked out with the help of Marc Culler, Henri Gillet and Richard Skora.

5.5.1. Let X be a triangulated space. Let μ be a function that assigns a non-negative real number $\mu(\alpha)$ to every polyhedral path $\alpha\colon [0,1] \to X$. I'll call μ a ***path measure*** on X if (i) it is invariant under re-parametrization, i.e. $\mu(\alpha \circ h) =$

$\mu(\alpha)$ for any polyhedral path μ and any homeomorphism $h\colon [0,1] \to [0,1]$; and (ii) μ is additive under composition; that is, if α_1 and α_2 are polyhedral paths with $\alpha_1(1) = \alpha_2(0)$, their composition $\alpha_1 * \alpha_2$ satisfies $\mu(\alpha_1 * \alpha_2) = \mu(\alpha_1) + \mu(\alpha_2)$.

If Y is a subcomplex of X then any path measure on Y restricts to a path measure on Y.

5.5.2. If the triangulated space X consists of single closed edge, then for any positive real number λ there is a natural path measure on X: the length of a polyhedral path α is the ordinary length in $\mathbf{R}$ of the path $h \circ \alpha$, where h is an affine homeomorphism of X onto an interval of length λ in $\mathbf{R}$.

5.5.3. Let $\tilde{\Sigma}$ be a surface with points at infinity (5.2.3). Any non-degenerate length system on $\tilde{\Sigma}$ determines a path measure on $\tilde{\Sigma}$: for any polyhedral path α we define $\mu(\alpha)$ to be the length of α with respect to the given length system.

5.5.4. If X is a triangulated space and J is a piecewise-linear homeomorphism from X to a product $Y \times \mathbf{R}$, where Y is another triangulated space, then J defines a length function on X: for any path α we define $\mu(\alpha)$ to be the length in $\mathbf{R}$ of the path $pJ(\alpha)$, where $p\colon Y \times \mathbf{R} \to \mathbf{R}$ denotes projection to the second factor.

5.5.5. Let X be a 1-connected triangulated space. I will say that a path measure μ on X is *standard* if every branch (5.2.5) Y of X is either a closed edge or a surface with points at infinity, or is homeomorphic to a product; and the restriction of μ to Y is defined by one of the three corresponding constructions 5.5.2–5.5.4.

If every branch is of one of the first two types then X is a singular surface, and in this case a standard path measure is essentially the same thing as a non-degenerate length system.

5.5.6. Let X be a triangulated space with a standard path measure, and let Γ be a group of automorphisms of X such that X is finite modulo Γ. For any polyhedral path α, let's call $\mu(\alpha)$ the *length* of α. It should be easy to prove, generalizing 5.2.7, that any two points of X are joined by a path of minimal length, and that if we make X into a pseudo-metric space by defining the distance between two points to be the length of the minimal path between them, then the corresponding metric space is an $\mathbf{R}$-tree. The group γ then has an induced action on this tree. My tentative proposal is to define an action to be *standard* if it arises in this way.

As I have said, with this tentative definition comes a tentative conjecture, that every action satisfying the ascending chain condition is the limit of a strongly convergent direct system of standard actions. When I say that this is tentative, I mean that I would not be surprised if it were necessary to revise the definition of standard action slightly in order to make it true.

This tentative conjecture would imply Conjectures 2.5.1 and 2.5.2, via the same arguments that are used in [GiS1] and [GiSSk] (and sketched in 5.2.13 above) to deduce the rank-2 case of these conjectures from the main theorem of [GiS1].

5.6. Unsmall actions. Let me briefly discuss the question posed in 2.5.7, whether $\mathcal{PL}(\Gamma)$, where Γ is a finitely generated group, has a dense subset consisting of projectivized integer-valued length functions. What I believe to be true, at least, is that every action satisfying the ascending chain condition is a limit (in the sense of 2.2.7) of simplicial actions. In fact, in the rank-2 case this is proved in [GiSSk] by an argument almost identical to the one that I sketched in 5.2.13 for the small case. (In the rank-1 case, for example for a **Q**-tree, one does not need to assume the ascending chain condition.) This argument would go through with no restriction on the rank if the tentative conjecture of 5.5.6 were known.

For actions not satisfying the ascending chain condition there seems to be rather little known about this question. Morgan and I proved in [MSh2] that Question 2.5.8 has an affirmative answer if Γ is a finitely generated fundamental group of a 3-manifold: that is, if such a group Γ admits a non-trivial action on an **R**-tree, then it admits a non-trivial action on a **Z**-tree. However, we did not show that the given action is a limit of simplicial actions.

In general there seems to be rather little known in general about actions on **R**-trees not satisfying the ascending chain condition. In [GiS2], Gillet and I did prove the following result:

Let Γ be a group which acts without inversions on a Λ-tree, where Λ is a subgroup of **R**. *Then the cohomological dimension of Γ is at most $1 + r + d$, where r is the* **Q**-*rank of Λ and d is the supremum of the cohomological dimensions of the stabilizers in Γ of points of T.*

References

[AloM] N. Alon and V. D. Milman, λ_1 *isoperimetric inequalities for graphs and superconcentrators*, J. Comb. Th. B **38** (1985), 78–88.

[AlpB] R. Alperin and H. Bass, *Length functions of group actions on trees*, Combinatorial Group Theory and Topology (S. M. Gersten and J. R. Stallings, eds.), Ann. of Math. Studies 111, Princeton Univ. Press, 1987, pp. 265–378.

[AlpM] R. Alperin and K. Moss, *Complete trees for groups with a real-valued length function*, J. London Math. Soc. **31 (2)** (1985), 55–68.

[Bab] S. Basarab, *Morgan-Shalen compactification of affine algebraic varieties over local fields*, Proceedings of the 7th Easter Conference on Model Theory (B. Dahn and H. Wolter, eds.), Sekt. Math. der Humboldt-Univ. Berlin, 1989, pp. 4–18.

[Bas1] H. Bass, *Groups of integral representation type*, Pacific J. Math. **86** (1980), 15–52.

[Bas2] ———, *Finitely generated subgroups of* GL_2, The Smith Conjecture (H. Bass and J. W. Morgan, eds.), Academic Press, 1984.

[Bas3] ———, *Group actions on non-archimedean trees*, Proceedings of the Workshop on Arboreal Group Theory, September 1988, MSRI Publications, Springer-Verlag (to appear).

[Be] M. Bestvina, *Degenerations of hyperbolic space*, Duke Math. J. **56** (1988), 143–161.

[BeH] M. Bestvina and M. Handel, *Train tracks for automorphisms of the free group* (to appear).

[BiNS] R. Bieri, W. D. Neumann and R. Strebel, *A geometric invariant of discrete groups*, Invent. Math. **90** (1987), 451–477.

[Bro] K. S. Brown, *Trees, valuations and the Bieri-Neumann-Strebel invariant*, Invent. Math. **90** (1987), 479–504.

[Ch] I. M. Chiswell, *Non-standard analysis and the Morgan-Shalen compactification*, Pre-print.

[CuGLS] M. Culler, C. McA. Gordon, J. Luecke and P. B. Shalen, *Dehn surgery on knots*, Ann. of Math. **125** (1987), 237–300.

[CuM] M. Culler and J. W. Morgan, *Group actions on* $\mathbf{R}$*-trees*, Proc. London Math. Soc. **55 (3)** (1987), 571–604.

[CuS1] M. Culler and P. B. Shalen, *Varieties of group representations and splittings of 3-manifolds*, Ann. of Math. **117** (1983), 109–146.

[CuS2] M. Culler and P. B. Shalen, *Bounded, separating, incompressible surfaces in knot manifolds*, Invent. Math. **75** (1984), 537–545.

[CuV1] M. Culler and K. Vogtmann, *Moduli of graphs and automorphisms of free groups*, Invent. Math. **84** (1986), 91–119.

[CuV2] M. Culler and K. Vogtmann, *The boundary of outer space in rank two*, Proceedings of the Workshop on Arboreal Group Theory, September 1988, MSRI Publications, Springer-Verlag (to appear).

[FLP] A. Fathi, F. Laudenbach, V. Poenaru, *Travaux de Thurston sur les surfaces*, Séminaire Orsay, Astérisque, Soc. Math. de France, Paris, 1979, pp. 66–67.

[GiS1] H. Gillet and P. B. Shalen, *Dendrology of groups in low* $\mathbf{Q}$*-ranks*, J. Diff. Geom **32** (1990), 605–712.

[GiS2] ———, *Cohomological dimension of groups acting on* $\mathbf{R}$*-trees*, Proceedings of the Workshop on Arboreal Group Theory, September 1988, MSRI Publications, Springer-Verlag (to appear).

[GiSSk] H. Gillet, P. B. Shalen and R. K. Skora, *Simplicial approximation and low-rank trees*, Pre-print.

[Go] E. S. Golod, *On nil-algebras and finitely approximable p-groups*, Izv. Akad. Nauk. SSR Ser. Mat. **28** (1964), 273–276 (Russian); English transl in Amer. Math. Soc. Transl. Ser. 2 **48** (1965), 103–106.

[GuS] N. Gupta and S. Sidki, *On the Burnside problem for periodic groups*, Math. Z. **182** (1983), 385–388.

[H] N. Harrison, *Real length functions in groups*, Trans. Amer. Math. Soc **174** (1972), 77–106.

[JaS] W. H. Jaco and P. B. Shalen, *Seifert fibered spaces in 3-manifolds*, Memoirs Amer. Math. Soc **21 (220)** (1979).

[Ji] R. Jiang, *Branch points and free actions on* $\mathbf{R}$*-trees*, Proceedings of the Workshop on Arboreal Group Theory, September 1988, MSRI Publications, Springer-Verlag (to appear).

[Jo] K. Johannson, *Homotopy equivalences of 3-manifolds with boundaries*, Lecture Notes in Mathematics 761, Springer-Verlag, 1979.

[LPS] A. Lubotzky, R. Phillips and P. Sarnak, *Ramanujan graphs*, Combinatorica **8** (1988), 261–277.

[L] A. Lubotzky, *Discrete groups, expanding graphs and invariant measures*, Pre-print.

[M1] J. W. Morgan, *Group actions on trees and the compactifications of the space of classes of* $SO(n,1)$*-representations*, Topology **25** (1986), 1–34.

[M2] ———, *Group actions on trees and hyperbolic geometry*, to appear in CBMS Lecture Notes series.

[MO] J. W. Morgan and J.-P. Otal, *Non-Archimedean measured laminations and degenerations of surfaces*, Pre-print.

[MSh1] J. W. Morgan and P. B. Shalen, *Valuations, trees and degeneration of hyperbolic structures I*, Ann. of Math. **120** (1984), 401–476.

[MSh2] ———, *Degenerations of hyperbolic structures III*, Ann. of Math. **127** (1988), 457–519.

[MSh3] ———, *Actions of surface groups on* **R**-*trees*, Topology, (to appear).

[MSk] J. W. Morgan and R. K. Skora, *Groups acting freely on* **R**-*trees*, Pre-print.

[Mst] M. Morgenstern book Ramanujan diagrams and explicit construction of bounded concentrators, Thesis, Hebrew University, Jerusalem.

[Par] W. Parry, *Axioms for translation length functions*, Proceedings of the Workshop on Arboreal Group Theory, September 1988, MSRI Publications, Springer-Verlag (to appear).

[Pau1] F. Paulin, *Gromov topology for* **R**-*trees*, Topology and its Applications, (to appear).

[Pau2] ———, *Topologies de Gromov équivariantes, structures hyperboliques et arbres réels*, Invent. Math **94** (1988), 53–80.

[Pau3] F. Paulin, *Outer automorphisms of hyperbolic groups and small actions on* **R**-*trees*, Pre-print, ENS de Lyon.

[Sc] G. P. Scott, *Compact submanifolds of 3-manifolds*, J. London Math. Soc. **7** (1973), 246-250.

[ScW] G. P. Scott and C. T. C. Wall, *Topological methods in group theory*, Homological Group Theory (C. T. C. Wall, ed.), London Math. Soc. Lecture Notes 36, Cambridge Univ. Press, 1979.

[Se] J.-P. Serre, *Trees*, Springer-Verlag, New York, 1980.

[Sh] P. B. Shalen, *Dendrology of groups: an introduction*, Essays in Group Theory (S. Gersten, ed.), MSRI Pub. 8, Springer-Verlag, 1987.

[Sk1] R. K. Skora, *Splittings of surfaces*, Bull. Amer. Math. Soc. **23** (1990), 85–90.

[Sk2] ———, *Deformations of length functions in groups*, Pre-print.

[SmV] J. Smillie and K. Vogtmann, *A generating function for the Euler characteristic of* $\mathrm{Out}(F_n)$, J. Pure Appl. Alg. **44** (1987), 329–348.

[T] R. M. Tanner, *Explicit concentrators from generalized* N-*gons*, SIAM J. Alg. Disc. Meth. **5** (1984), 278–285.

[V] K. Vogtmann, *Local structure of some* $\mathrm{Out}(F_n)$-*complexes*, Proc. Edinburgh Math. Soc., (to appear).

COCYCLES BORNÉS ET ACTIONS DE GROUPES SUR LES ARBRES RÉELS

JEAN BARGE
Institut Fourier, Laboratoire de Mathématiques, BP 74
38402 St Martin d'Hères Cedex, France

et

ETIENNE GHYS
Ecole Normale Supérieure de Lyon, 46, Allée d'Italie
69364 Lyon Cedex, France

1. Introduction.

Les arbres réels (voir définition ci-dessous) sont apparus, pour la première fois, dans le travail de J. Morgan et P. Shalen [5] comme une nouvelle interprétation des points à l'infini de l'espace de Teichmüller des surfaces de Riemann de genre supérieur ou égal à 2. En particulier, ces arbres étaient munis d'une action isométrique du groupe fondamental d'une telle surface et cette action s'avérait "fréquemment" être une action libre. Ce fait surprenant (il est bien connu qu'un groupe agissant librement sur un arbre simplicial est en fait un groupe libre) a conduit à se demander quels groupes, disons de type fini, peuvent agir librement sur un arbre réel. C'est clair pour $\mathbf{Z}^n$ qui se réalise comme groupe de translations de la droite affine et, par une construction standard, pour les groupes qui sont produits libres de groupes abéliens libres et de groupes de surfaces orientables. Dans [6], P. Shalen demande si ce sont les seuls. Dans la note qui suit, nous associons à toute action d'un groupe Γ sur un arbre réel T, une large famille de 2-cocycles bornés. Si, de plus, l'action de Γ sur T est libre, les classes de cohomologie bornée définies par les 2-cocycles précédents ne sont pas toutes nulles à moins que Γ ne soit abélien sans torsion. En particulier, un groupe de type fini, non isomorphe à $\mathbf{Z}^n$, qui agit librement sur un arbre réel, possède nécessairement un second groupe de cohomologie bornée non trivial.

Un arbre réel T est un espace métrique vérifiant les deux conditions suivantes :

1) Pour toute paire de points distincts A et B de T, il existe un unique plongement γ, du segment réel $[0, d(A, B)]$ dans T, qui soit une isométrie sur son image et qui joigne A et B (*i.e.* $\gamma(0) = A$ et $\gamma(d) = B$). Notons AB l'image de γ.

2) Soient A, B, C trois points de T. Si $AB \cap BC = \{B\}$, alors $AC = AB \cup BC$.

Un arbre réel peut être globalement et même localement très compliqué. Néanmoins, l'enveloppe convexe d'un nombre fini de points de l'arbre possède la même structure combinatoire que dans un arbre simplicial. Par exemple, les triangles sont comme indiqués (figure 1), les quadrilatères (figure 2) *etc.*

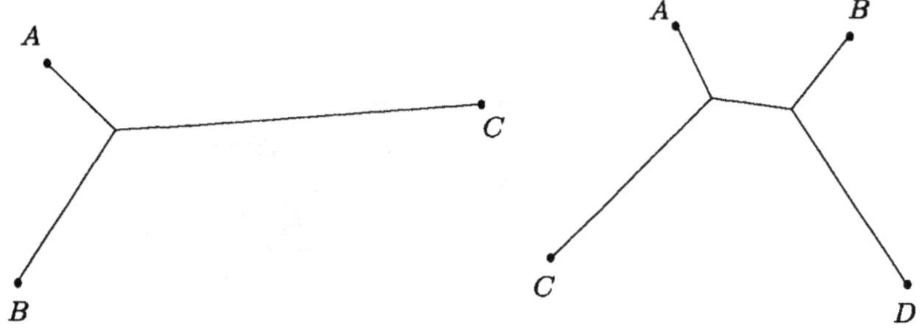

Figure 1 Figure 2

2. Classes de cohomologie bornée.

Soit donc Γ un groupe opérant par isométries sur un arbre réel T.

Nous noterons $\vec{u}$ un segment non dégénéré, *i.e.* non réduit à un point, et orienté, de l'arbre réel T. Nous allons nous servir de l'unité $\vec{u}$ pour "mesurer" la longueur des segments de T, d'une façon inspirée de la construction de R. Brooks [3] pour les groupes libres.

Si A et B sont deux points de T, on note $\Phi_{\vec{u}}(A, B)$ le plus grand entier (positif ou nul) n pour lequel il existe n éléments $\gamma_1, \gamma_2, \ldots, \gamma_n$ du groupe Γ vérifiant les propriétés suivantes :

1) les segments $\gamma_i\ \vec{u}$ sont d'intérieurs disjoints deux à deux.

2) chacun des segments $\gamma_i\ \vec{u}$ est contenu dans AB.

3) $\gamma_i\ \vec{u}$ est orienté de A vers B.

On note alors $\text{mes}_{\vec{u}}(A, B)$ l'antisymétrisée de la fonction $\Phi_{\vec{u}}(A, B)$, c'est-à-dire la fonction définie par :

$$\text{mes}_{\vec{u}}(A, B) = \Phi_{\vec{u}}(A, B) - \Phi_{\vec{u}}(B, A) .$$

Cette fonction est presque additive. Précisément, si A, B, C sont trois points d'un même segment géodésique de T, avec B entre A et C, on a :

$$|\mathrm{mes}_{\vec{u}}(A,C) - \mathrm{mes}_{\vec{u}}(A,B) - \mathrm{mes}_{\vec{u}}(B,C)| \leqslant 1 \; .$$

Il en résulte que si maintenant A, B, C sont trois points quelconques de T, le "périmètre" du triangle ABC est presque nul (voir figure 1)

$$|\mathrm{mes}_{\vec{u}}(A,B) + \mathrm{mes}_{\vec{u}}(B,C) + \mathrm{mes}_{\vec{u}}(C,A)| \leqslant 3 \; .$$

Choisissons un point base, noté $*$, dans T et posons, pour γ_1, γ_2 dans Γ :

$$c_{\vec{u},*}(\gamma_2\gamma_1) = \mathrm{mes}_{\vec{u}}(*, \gamma_1, \gamma_2(*)) - \mathrm{mes}_{\vec{u}}(*, \gamma_1(*)) - \mathrm{mes}_{\vec{u}}(*, \gamma_2(*))$$

c'est-à-dire encore le périmètre du triangle de sommets $*, \gamma_1(*), \gamma_1\gamma_2(*)$.

On peut alors énoncer :

PROPOSITION.

i) *La fonction* $c_{\vec{u},*} : \Gamma \times \Gamma \longrightarrow \mathbb{Z}$ *définit un 2-cocycle (non homogène) borné sur* Γ.

ii) *La classe de cohomologie définie par* $c_{\vec{u},*}$ *est nulle.*

iii) *La classe de cohomologie bornée définie par* $c_{\vec{u},*}$ *ne dépend pas du point base* $*$ *dans* T.

Preuve. — i) et ii). La cochaîne $c_{\vec{u},*}$ est le cobord de la 1-cochaîne $f_* : \Gamma \longrightarrow \mathbb{Z}$ définie par :

$$f_*(\gamma) = \mathrm{mes}_{\vec{u}}(*, \gamma(*)) \; .$$

Cette cochaîne $c_{\vec{u},*}$ est donc un cocycle dont la valeur absolue est bornée par 3. Notons que f_* n'est pas a priori bornée de sorte que la classe de cohomologie bornée de $c_{\vec{u},*}$ peut être non nulle (comme nous nous en assurerons d'ailleurs plus loin).

iii). Soit $*'$ un autre point de T. La différence $f_* - f_{*'}$ est bornée. En effet $f_*(\gamma) - f_{*'}(\gamma)$ n'est autre que le périmètre du quadrilatère de sommets $(*, \gamma(*), \gamma(*'), *')$ (figure 2).

3. Actions libres.

Nous supposerons de plus que le groupe Γ agit librement sur l'arbre réel T.

Rappelons que toute isométrie γ de Γ, sans point fixe, possède un axe $\mathcal{A}_\gamma$, c'est-à-dire un sous-arbre invariant isométrique à $\mathbb{R}$ et sur lequel γ agit par translation. Cet axe est l'ensemble des points x de T tels que $d(x, \gamma(x)) \leqslant d(t, \gamma(t))$ pour tout t de T.

Soit γ_0 un élément non trivial du groupe Γ. Soit $\vec{u}$ le segment $A\gamma_0(A)$ où A est un point de l'axe $\mathcal{A}_{\gamma_0}$ de γ_0. On note alors $[c_{\gamma_0}]$ la classe de cohomologie bornée de $c_{\vec{u},*}$ (dont on s'assure facilement qu'elle ne dépend pas du choix du point A sur l'axe de γ_0).

LEMME. — *Si la classe $[c_{\gamma_0}]$ est nulle, il existe un homomorphisme de Γ dans* **R** *qui prend la valeur 1 sur γ_0.*

Preuve. — Il y a d'abord l'observation générale suivante [2]. Si c est une classe de cohomologie bornée, nulle en cohomologie ordinaire, alors c est la classe d'une 2-cochaîne bornée qui s'écrit df où $f : \Gamma \to \mathbf{R}$ est une application qui vérifie, pour une constante K :

$$|f(\gamma_1\gamma_2) - f(\gamma_1) - f(\gamma_2)| \leqslant K$$

pour tous γ_1, γ_2 de Γ.

L'inégalité ci-dessus montre que pour tout γ de Γ la suite $\frac{1}{n}f(\gamma^n)$ possède une limite, notée $\bar{f}(\gamma)$ lorsque n tend vers l'infini. De plus, l'application $f - \bar{f}$ est bornée sur Γ ce qui montre que c est aussi la classe bornée de $d\bar{f}$. On a alors équivalence entre les propriétés suivantes :

i) $c = 0$.

ii) $f = b + \Psi$ où b est bornée et Ψ est un morphisme.

iii) $\bar{f}$ est un morphisme.

Le lemme résulte de l'assertion suivante :

$$\lim_{n\to\infty} \frac{1}{n} \operatorname{mes}_{\overrightarrow{A\gamma_0(A)}}(*, \gamma_0^n(*)) = 1 .$$

Comme observé précédemment, la limite du premier membre ne dépend pas du choix du point base $*$. Choisissons donc $* = A$. Il est clair que :

$$\Phi_{\overrightarrow{A\gamma_0(A)}}(A, \gamma_0^n(A)) = n$$

et montrons que $\Phi_{\overrightarrow{A\gamma_0(A)}}(\gamma_0^n(A), A) = 0$.

En effet, il existerait sinon un élément γ de Γ tel que $\overrightarrow{\gamma_0(A), \gamma\gamma_0(A)}$ soit contenu dans l'axe $\mathcal{A}_{\gamma_0}$ de γ_0 avec une orientation opposée. Quitte à modifier γ par une puissance convenable de γ_0, nous obtenons la figure suivante :

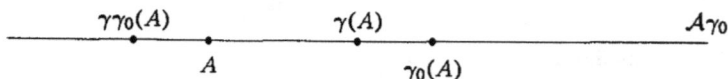

sur laquelle il est clair que le milieu du segment $A\gamma(A)$ est fixe par γ. Ceci est contraire au fait que nous supposons l'action libre.

Nous obtenons alors la proposition :

PROPOSITION. — *Soit Γ un groupe qui opère librement par isométries sur un arbre réel T. Alors, ou bien Γ est abélien sans torsion ou bien le second groupe de cohomologie bornée de Γ ne s'injecte pas dans le second groupe de cohomologie ordinaire (à coefficients réels).*

Pour donner un énoncé plus concret, nous utiliserons un résultat de [1], [4]. Soit Γ' le premier groupe dérivé de Γ et $\gamma \in \Gamma'$. On note $|\gamma|$ le nombre minimum de commutateurs de Γ dont le produit est égal à γ. On définit alors $||\gamma||$ par :

$$||\gamma|| = \lim_{n\to\infty} \frac{1}{n}|\gamma^n| \,.$$

Le résultat dont il s'agit est le suivant : le second groupe de cohomologie bornée de Γ s'injecte dans la cohomologie usuelle si et seulement si cette "longueur stable des commutateurs" est nulle pour tous les éléments de Γ'.

PROPOSITION. — *Soit Γ un groupe non abélien tel que la longueur stable des commutateurs soit nulle sur Γ'. Alors Γ n'opère pas librement sur un arbre réel.*

Bibliographie

[1] BAVARD C. — *Longueur stable des commutateurs*, Prépublication E.N.S. Lyon, 1989.

[2] BESSON G. — *Séminaire de cohomologie bornée*, E.N.S. Lyon, fév., 1988.

[3] BROOKS P. — *Some remarks on bounded cohomology in Riemannian surfaces and related topics*, Ann. of Math. Stud., **91** (1981), 53–65.

[4] MATSUMOTO S., MORITA S. — *Bounded cohomology of certain groups of homeomorphisms*, Proc. A.M.S., **94** (1985), 539–544.

[5] MORGAN J., SHALEN P. — *Valuations, trees and degenerations of hyperbolic structures, I*, Ann. of Math., **120** (1984), 401–476.

[6] SHALEN P. — *Dendrology of groups; an introduction in "Essays in group theory"*, S..M. Gersten editor, Pub. MSRI, **8** (1987), 265–319.

A QUICK INTRODUCTION TO BURNSIDE'S PROBLEM

VLAD SERGIESCU
Institut Fourier, Laboratoire de Mathématiques, BP 74
38402 St Martin d'Hères Cedex, France

ABSTRACT

The main purpose of this exposition is to describe an interesting and fairly elementary geometric construction due to Gupta and Sidki in connection with a classical problem in group theory. This is related in several ways to the topics discussed during the workshop. It also provides motivation for further research on open problems.

1. Introduction.

Let us start by stating the :

1.1. *General Burnside Problem* (1902) ([2]). — Let G be a finitely generated group such that each element is of finite order. Is G a finite group?

Here is some evidence for a positive answer :

i) G is an abelian group. In this case the structure theorem implies that G is a direct sum of finite groups, so it is finite itself.

ii) G is a group of *exponent 2*, *i.e.* $g^2 = e$ for $g \in G$. It is well-known that such a group is abelian so G is finite.

iii) G is a group of *exponent 3*. This case, already considered by Burnside is more interesting. Note first that if $g, h \in G$ then $[g, hgh^{-1}] = e$. Indeed one has :

$$[g, hgh^{-1}] = g^{-1}hg^{-1}h^{-1}ghgh^{-1} = g^2hg^2h^2ghgh^2$$
$$= g^2hg^2h(hghgh)h = g^2hg^2hg^2h = e$$

as $(hg)^3 = e$ and so $hghgh = g^{-1} = g^2$ and $(g^2h)^3 = e$.

A consequence of the preceding identity is that $[g, [h, g]] = e$.

If G is generated by two elements g_1, g_2, consider the exact sequence

$$1 \to G' \to G \to G/G' \to 1 \ .$$

The group G' is generated as a *normal* subgroup by the commutator $[g_1, g_2]$. Since g_1 and g_2 commute with $[g_1, g_2]$ it follows that G' is a finite cyclic group. As G/G' is a finite abelian group we conclude that G is finite.

When G has $n \geqslant 3$ generators $g_1, \ldots, g_n$, one uses an induction argument applied to the sequence

$$1 \to \langle g_1 \rangle \to G \to G/\langle g_1 \rangle \to 1 ,$$

where $\langle g_1 \rangle$ is the normal subgroup generated by g_1. We leave the (easy) details to the reader.

Despite this and other evidence, the answer to the General Burnside Problem is negative as was shown by Golod (1965).

1.2. THEOREM ([3]). — *For each prime $p \geqslant 2$ there is a finitely generated p-group which is infinite.*

Golod's construction came as a by product of some joint work with Shafarevich in number theory. Another line of such groups was started by Aliochin (1972) [1]; further related developments are due to Suschanski [16], Grigortchuk [4], Merzliakov [12] and Gupta-Sidki [7].

Aliochin's group was initially described as generated by two finite automata (see also [10]) of order p and p^2. It should be emphasized that this is a different kind of groups as the automatic groups discussed during the workshop.

All above groups have a faithful action on some regular tree. Moreover, a theorem of L. Kaloujnin [9], states that any finitely generated residually finite p-group can be embedded in the automorphism group of a regular tree and in fact in a subgroup isomorphic to the inverse limit of the Sylow subgroups of the symetric groups of order p^k (see [9], [10]).

Gupta-Sidki's construction is arguably the simplest negative answer to Burnside's problem. In [7] they proved the :

1.3. THEOREM ([7]). — *For each prime $p \geqslant 3$, there exists an infinite p-group with two generators of order p.*

When $p = 2$, they also show there exists a 2-group with one generator of order 2 and one of order 4 [8].

Finally, Gromov's work on hyperbolic groups exibits lots of negative examples for the General Burnside Problem. For instance, the fundamental group of any negatively curved manifold has a quotient which is an infinite torsion group.

2. The main construction.

In this section we describe the Gupta-Sidki group which corresponds to the case $p = 3$. For an arbitrary prime p the definition and the proofs are similar but less digest.

2.1. *Notations* . — Denote by T the infinite ternary tree with root $*$ (fig .1). It's vertices are labelled using a triadic expansion. When u is a vertex of T, let T_u be the subtree of T with root u which is isomorphic to T.

If a group G acts on T, we let G_u be a "copy" of G acting on the subtree T_u. Note that G_u is not the stabilizer of u. For $g \in G$, let g_u be the corresponding element in G_u.

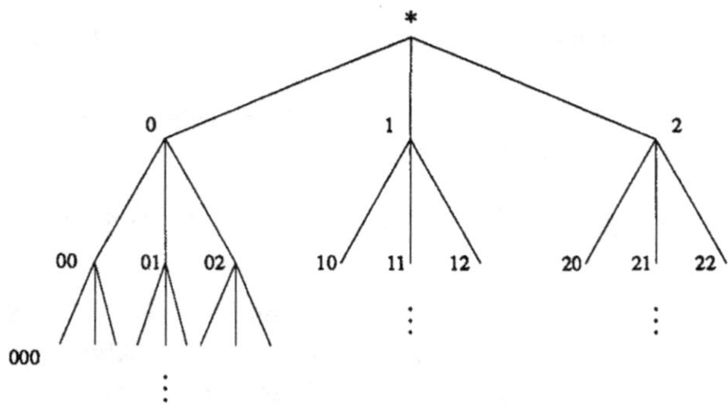

fig. 1

2.2. *Definition of the group G.* — We shall first define two automorphisms t and a of T.

The automorphism t will just cyclically permute the vertices 0, 1 and 2 and then also permute the corresponding subtrees T_0, T_1, T_2 in a straitforward way. If one thinks about T as being tetraedric-like in the 3-space, t amounts to a rotation around the central axis. One has the following formula for t :

$$t(i_1, \ldots, i_k) = (i_1 + 1 \bmod 3, i_2, \ldots, i_k) .$$

The automorphism a has a more involved definition, which uses an infinite repetition device.

The vertices 0, 1 and 2 are fixed. It acts on T_0 by t_0 and on T_1 by t_1^2. Now we repeat this process on T_2, *i.e.* a acts on T_{20} by t_{20}, on T_{21} by t_{21}^2, on T_{220} by t_{220}, on T_{221} by t_{221}^2, *etc.* Note that a fixes the vertex $222 \cdots 2$.

To give a formula for a, let $(i_1, \ldots, i_k)$ a vertex of T and n such that $i_1 = i_2 = \cdots = i_{n-1} = 2$, $i_n \neq 2$. Then one has :

$$a(i_1, \ldots, i_k) = (i_1, \ldots, i_n, \overline{i_{n+1}}, i_{n+2}, \ldots, i_k)$$

where $\overline{i}_{n+1} = i_{n+1} + 1 \bmod 3$ if $i_n = 0$ and $\overline{i}_{n+1} = i_{n+1} + 2 \bmod 3$ if $i_n = 1$.

It is almost obvious that $t^3 = a^3 = e$.

2.2.1. Definition. — *We call G the group of automorphisms of the ternary tree T generated by t and a.*

2.3. *Remark.* — At this point one can already wonder about the reasons of this construction; in particular why we did not just considered the binary analogue of G?

A partial answer is that working with automorphisms of the ternary tree appears most naturally in the light of Kaloujnin's result quoted in section 1. Another stronger motivation will come after proposition 3.2 bellow.

Consider now the binary analogue $\mathcal{G}$ of G generated by automorphisms $\mathcal{T}$ and $\mathcal{A}$. It is true that $\mathcal{G}$ is infinite (see the proof of prop. 3.2). Thus the element $\mathcal{AT}$ is not a torsion element : if $(\mathcal{AT})^n = e$ one has

$$\mathcal{G} = \{\mathcal{A}, \mathcal{T}, \mathcal{AT}, \mathcal{ATA}, \ldots, (\mathcal{AT})^{n-1}\mathcal{A}, \mathcal{TAT}, \ldots, (\mathcal{TA})^{n-1}\mathcal{T}\}$$

showing that $\mathcal{G}$ is finite.

It is nevertheless instructive to keep in mind this example while going through the proofs.

3. Proofs.

This section contains the proof of the following

3.1. THEOREM. — *The ternary group G is torsion and infinite.*

We shall first show that G is infinite. Let us note that the action of the generators t and a on the set of vertices $\{0, 1, 2\}$ shows that G cyclically permutes these vertices. Let $\varphi : G \to \mathbb{Z}_3$ the corresponding morphism.

3.2. LEMMA. — *The kernel $H = \ker\varphi$ is generated by a, $b = tat^{-1}$ and $c = t^2at^{-2}$ (see fig.2).*

Proof. — Let $g = a^{i_1}t^{j_1}a^{i_2}t^{j_2}\ldots a^{i_n}t^{j_n}$ in G. Writing

$$g = a^{i_1}t^{j_1}a^{i_2}t^{-j_1} \,.\, t^{j_1+j_2}a^{i_3}t^{-j_1-j_2}t^{j_1+j_2+j_3}a^{i_4}t^{-j_1-j_2-j_3}\ldots a^{i_n}t^{-j_1-\cdots-j_{n-1}}t^{j_1+\cdots+j_n}$$

one sees that $g = \omega(a, b, c)t^{j_1+\cdots j_n}$ where $\omega(a, b, c) \in H$ and so $\ker\varphi$ is generated by a, b and c. ■

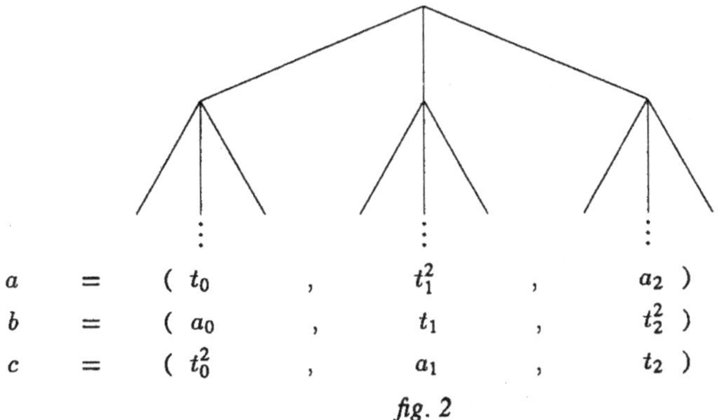

fig. 2

For $h \in H$ we shall write $h = ({}_0h, {}_1h, {}_2h)$ where ${}_ih$ is the restriction of h to T_i . We now prove the

3.3. PROPOSITION. — *G is an infinite group.*

Proof. — Let $\psi : H \to G_0$ be the map $\psi(h) = {}_0h$. As $\psi(a) = t_0$, $\psi(b) = a_0$, $\psi(c) = t_0^2$ and G_0 is isomorphic to G, it follows that ψ is surjective. Thus H is a subgroup of finite index which surjects on G. This implies that G is infinite. ∎

3.4. *Remark.* — The argument used above to show that G is an infinite group motivates a posteriori the definition of G. Once we choose t as the simplest automorphism of order 3, the definition of a is rather natural in order to get the map ψ *surjective*.

Before going to prove G is a torsion group, we will introduce an appropriate notion of length.

For an element $h \in H$, let $||h||$ be the usual length relative to the system of generators a, b, c. As $a^2 = a^{-1}$, $b^2 = b^{-1}$, $c^2 = c^{-1}$, this is the same as the syllable length of h, *i.e.* the minimum number of syllables a^i, b^j, c^k needed to write h.

For an arbitrary element $g \in G$, we write $g = ht^\varepsilon$ and set $|g| = ||h||$ if $\varepsilon = 0$ and $|g| = ||h|| + 1$, if $\varepsilon \neq 0$ mod 3. For example, $|a^2| = ||a^{-1}|| = 1$, $|at| = ||a|| + 1 = 2$. Finally one has a corresponding notion of length in a group G_u, relative to the generators a_u, b_u, c_u, t_u.

We now prove the

3.5. PROPOSITION. — *G is a 3-group.*

Proof (see also [7],[8]). — We use induction on the length. If $|g| = 1$, then $g = a^\varepsilon, b^\varepsilon, c^\varepsilon, \varepsilon = \pm 1$, if $g \in H$ and $g = t^\varepsilon$ if $g \notin H$. In both cases $g^3 = e$.

We now suppose that if $|g| \leqslant n$ then $g^{3^n} = e$ and let g an element of length $|g| = n + 1 \geqslant 2$. We distinguish two cases :

i) $g \in H$, *i.e.* $g = \omega(a, b, c)$.

ii) $g \notin H$, *i.e.* $g = \omega(a, b, c)t^\varepsilon$, $\varepsilon = \pm 1$.

Case i). — Let us write $g = ({}_0g, {}_1g, {}_2g)$ and prove that each component is of order 3^{n+1}. We do this for ${}_0g$. Let us write ${}_0g = \omega(t_0, a_0, t_0^2)$. As an element of G_0, write ${}_0g = \widetilde{\omega}(a_0, b_0, c_0)t_0^\varepsilon$. Note that the way one creates syllables in $\widetilde{\omega}$ from $\omega(t_0, a_0, t_0^2)$ shows they correspond to syllables in the letter b in ω. As $|g| \geqslant 2$, at least one syllable in a or c occurs in g. Thus $||\widetilde{\omega}||_0 < ||\omega|| = n + 1$.

One concludes that either $|{}_0g|_0 < n+1$ and then ${}_0g^{3^n} = e$ and a fortiori ${}_0g^{3^{n+1}} = e$, or $|{}_0g|_0 = n + 1$ and we are in the case *ii)*; this last situation occurs for example for $g = ab : g_0 = t_0a_0 = t_0a_0t_0^{-1}t_0 = b_0t_0$ and so $|g_0|_0 = 2 = |g|$.

Case ii). — To explain the strategy here consider $g = at$, $|g| = 2$. Then

$$g^3 = atatat = atat^{-1}t^2at^{-2} = abc\ .$$

As $g^3 \in H$, one can write $g^3 = ({}_0g, {}_1g, {}_2g)$ and then

$${}_0g = t_0a_0t_0^{-1} = b_0\ ,\ {}_1g = t_1^2t_1a_1 = a_1\ ,\ {}_2g = a_2t_2t_2^2 = a_2\ .$$

Note that $|{}_0g| = |{}_1g| = |{}_2g| = 1 < |g|$ and as ${}_0g, {}_1g, {}_2g$ have order 3, g will have order 9.

In the general case let $g = \omega(a,b,c)t^\varepsilon$, $\varepsilon = \pm 1$. To keep notation easier we take $\varepsilon = 1$. Then

$$g^3 = \omega t\omega t\omega t = \omega t\omega t^{-1}t^2\omega t^{-2} = \omega(a,b,c)\omega(b,c,a)\omega(c,a,b)$$

as t conjugates a to b, b to c and c to a. If we set $g^3 = ({}_0g, {}_1g, {}_2g)$, then

$${}_0g = \omega(t_0, a_0, t_0^2)\cdot\omega(a_0, t_0^2, t_0)\cdot\omega(t_0^2, t_0, a_0)\ .$$

Writing ${}_0g = \widetilde{\omega}(a_0, b_0, c_0)t_0^\eta$ we claim that $|\widetilde{\omega}|_0 \leqslant |\omega|$ and $\eta = 0 \bmod 3$. It then follows that $|\widetilde{\omega}|_0 \leqslant |\omega| < |g| = n+1$, ${}_0g = \widetilde{\omega}$ is of order 3^n by the induction assumption and, as the same is true for ${}_1g$ and ${}_2g$, one has $g^{3^{n+1}} = e$.

To prove the above claim, we note that when forming $\widetilde{\omega}$ from ${}_0g$ as in 3.2, syllables in $\widetilde{\omega}$ correspond to blocks in the letter a_0 in ${}_0g$ which appear once for each syllable in ω and thus $|\widetilde{\omega}| \leqslant |\omega|$. As t_0 and t_0^2 appear once in ${}_0g$ for each syllable in ω it also follows that the total degree in t_0 is 0 mod 3. The proof is completed. ■

We end this section with the following

3.6. PROPOSITION. — *G is a group of infinite exponent : for each n, there exists an element of order* 3^n.

Proof (see [7] for more details). — Let $\alpha(t,a) \in G$ an element or order 3^n. To find an element of order 3^{n+1}, observe that if $(\alpha_0, \mathrm{id}, \mathrm{id})$ defines an element of H, then $g = (\alpha_0, \mathrm{id}, \mathrm{id})t$ verifies $g^3 = (\alpha_0, \alpha_1, \alpha_2)$. Thus g^3 is of order 3^n and g of order 3^{n+1}.

In fact the above argument will still work if one can find elements $d', d'' \in H$ s.t. $\big(\alpha(d',d'')_0, \mathrm{id}, \mathrm{id}\big) \in H$ and $\alpha(d',d'') = \big(\alpha(t_0,a_0), *, *\big)$. We will chose $d' = [a,t]$ and $d'' = \big[[a,t],t\big]$. Direct inspection then shows that $\alpha(d',d'') = \big(\alpha(t_0,a_0), *, *\big)$. To show $\big(\alpha(d',d'')_0, \mathrm{id}, \mathrm{id}\big) \in H$ we note that $\alpha(d',d'') \in G'$. But $G'_0 \times \mathrm{id} \times \mathrm{id}$ is exactly the normal subgroup of H generated by $\gamma = ([a_0,t_0], \mathrm{id}, \mathrm{id})$: this is easy once one sees that $\gamma = [cb, ca^{-1}]$, so γ is in H. ■

4. Final remarks.

4.1. — The group G has it's first homology $H_1(G;\mathbb{Z}) \simeq \mathbb{Z}_3 \oplus \mathbb{Z}_3$ (this was pointed to me by Christophe Bavard). To see this, note that as $G/H \simeq \mathbb{Z}_3$ by 3.2, it is sufficient to show that $a \notin G'$. The commutator G' is generated as a *normal* subgroup by $[a,t] = (t_0, t_1a_1, a_2^{-2}t_2)$. When restricted to the second level of the ternary tree T, the action of $[a,t]$ equals that of (t_0,t_1,t_2) and commutes with the generators t and a. Thus G' is generated by $[a,t]$ as a *subgroup* when restricted to the second level and so does not contain the generator a.

Question : compute $H_2(G;\mathbb{Z})$.

4.2. — One can prove that G is not finitely presented (this would follow for instance if $H_2(G)$ is not finitely generated). In fact the General Burnside Problem is still *open* for finitely presented groups.

4.3. — As shown in 3.6, the group G is not of finite exponent. Whether a finitely generated group of finite exponent is finite is known as the *Burnside Problem.* The negative answer to this question is due to Adjan and Novikov [13]. More geometric, shorter examples are due to Ol'sanskii [14] who also constructed an infinite finitely generated group, *all* of whose proper subgroups are finite of order p for a large prime p [15], thus answering Schmidt's problem.

4.4. — In [6], Gromov gives a method to build various counterexamples to the General Burnside Problem. He also suggests that while hyperbolic techniques appear to be too simple to cover Ol'sanskii's results "further development of geometric language will take care of these questions".

4.5. — The group G of Gupta-Sidki is close to an example due to Grigortchuk who also lead to the solution of two other problems of geometric nature : Milnor's problem on the existence of groups with subexponential, non-polynomial growth and Day's problem on the existence of an amenable group which can not be obtained from finite groups and abelian groups by "elementary" operations [5].

4.6. — We finally mention the *Restricted Burnside Problem* : is the number of finite groups with exponent m and d generators finite? This was affirmatively answered by Kostrikin [11] for a prime m and recently by Zelmanov in the general case.

Bibliography

(References [8], [10] and [11] provide overall information on Burnside's problem)

[1] ALIOCHIN S.V. — *Finite automata and the Burnside problem for periodic groups*, Mat. Zametki, **11** (1972), 319–328 [Math. Notes, 11, 199–203].

[2] BURNSIDE W. — *On an unsettled question in the theory of discontinuous groups*, Quart. J. Pure Appl. Math., **33** (1902), 230–238.

[3] GOLOD A.S. — *On nil-algebras and finitely approximable p-groups*, Izv. Akad. Nauk SSSR, Ser. Mat., **28** (1964), 273–276.

[4] GRIGORTCHUK R.I. — *On the Burnside problem for periodic groups*, Funkcional Anal. i Prilozen, **14** (1980), 53–54 [Functional Anal. Appl., 14, 41–43].

[5] GRIGORTCHUK R.I. — *Degrees of growth of finitely generated groups and the theory of invariant means*, Izv. Akad. Nauk SSSR, Ser. Mat., **48** (1984), n° 5 [Math. USSR Izvestiya 25, (1985), n° 2, 259–300.

[6] GROMOV M. — *Hyperbolic groups*, in Essays in Group Theory, S.M. Gersten editor, MSRI Pub., **8** (1987), 75–264.

[7] GUPTA N., SIDKI S. — *On the Burnside problem for periodic groups*, Math. Z, **182** (1983), 385–388.

[8] GUPTA N. — *On groups in which every element has finite order*, Am. Math. Monthly, **96** (1989), 297–308.

[9] KALOUJNIN L. — *Sur le groupe P_∞ des tableaux infinis*, C. R. Acad. Sci. Sér. I Math., **224** (1947), 1097–1099.

[10] KARGAPOLOV M., MERZLIAKOV I. — *Eléments de la théorie des groupes*, Ed. Mir., 1985, Moscou.

[11] KOSTRIKIN A.I. — *Around Burnside*, Nauka, 1986, Moscou.

[12] MERZLIAKOV I. — *Infinite finitely generated periodic groups*, Dokl. Akad. Nauk. SSSR, **268** (1983), 803–805.

[13] NOVIKOV P.S., ADJAN S.I. — *Infinite periodic groups I, II, III*, Dokl. Akad. Nauk. SSSR, 245, Ser. Mat., **32** (1968), 212–244, 251–254, 709–734.

[14] OL'SANŠKII A.J. — *On the Novikov-Adjan theorem*, Math. SSSR Sbornik, **46** (1982), 203–206.

[15] OL'SANSKII A.J. — *Groups of bounded period with subgroups of prime order*, Algebra i Logika, **21** (1982), 553–618.

[16] SUSCHANSKI V.I. — *Periodic p-groups of permutations and the unrestricted Burnside problem*, Dokl. Akad. Nauk. SSSR, **247** (1979), 557-561.

[17] VAUGHAN-LEE M.R. — *The restricted Burnside problem*, Bull. London Math. Soc., **17** (1985), 113–133.

V. MISCELLANEOUS

PROJECTIVE ASPECTS OF THE HIGMAN–THOMPSON GROUP

PETER GREENBERG

1. INTRODUCTION

This paper is about some of the subgroups of a certain group of homeomorphisms of the circle, the Higman-Thompson group G. This group has a long history ([2], [6]) in logic, algebra and dynamical systems. In algebra, it is closely related to the first example of a finitely presented infinite simple group; in logic, it was studied in relation to the word problem; it's dynamical aspects were studied by Ghys and Sergiescu ([6]). Recently, a remarkable connection of G with the classical braid group was found ([9]).

The group G is usually defined as the automorphism group of a certain algebro-combinatorial object, or as a group of piecewise linear homeomorphisms. Apparently, W. Thurston first observed that there is a "projective" definition.

In the sequel, $S^1 = \mathbb{R} \cup \{\infty\}$; $\mathrm{PSL}_2\,\mathbb{Z}$ acts on S^1 by linear fractional transformations, and $\widehat{\mathbb{Q}} = \mathbb{Q} \cup \{\infty\}$ is the orbit of 0. The circle inherits it's counterclockwise orientation from the left-to-right orientation of the line. By an *interval* (a, b), $a, b \in S^1$ we mean the set of $x \in S^1$ so that a, x, b are in counterclockwise order.

1.1. Definition. G is the group of homeomorphisms $g : S^1 \longrightarrow S^1$ such that there exist $r_1, \ldots, r_k \in \widehat{\mathbb{Q}}$, in counterclockwise order, and $A_i \in \mathrm{PSL}_2\,\mathbb{Z}$, $i = 1, \ldots, k$ so that $g|_{(r_i, r_{i+1})} \equiv A_i$, $g|_{(r_k, r_1)} \equiv A_k$. The r_i are called the *breakpoints* of g.

F is the subgroup of G consisting of elements fixing ∞, and F_C is the subgroup of F whose elements are the identity in some neighborhood of ∞.

This "projective" definition of G suggests the following family of subgroups.

1.2. Definition. Let Γ be a subgroup of $\mathrm{PSL}_2\,\mathbb{Z}$ which is a lattice in $\mathrm{PSL}_2\,\mathbb{R}$, and which acts freely on the hyperbolic plane. Then we denote by G_Γ (resp. $F_\Gamma, F_{C\Gamma}$) the group of elements of G which, on the intervals between the breakpoints, agree with some element of Γ.

We denote by $\mathbb{H}$ the hyperbolic plane. As Γ is a lattice, acting freely on $\mathbb{H}$, the quotient $\mathbb{H}/\Gamma$ is a Riemann surface of finite area, with genus g_Γ and number of cusps ν_Γ; the set $\widehat{\mathbb{Q}}$ breaks up into ν_Γ orbits under the action by Γ, and we will occasionally use "cusp" to denote an orbit.

We define in this paper a tree $\widetilde{X}_\Gamma$ on which G_Γ acts "at infinity", and study G_Γ with it's aid. In particular, G_Γ depends, up to isomorphism, only on g_Γ and ν_Γ. Perhaps the main application is corollary 3.2, which states that the G_Γ and F_Γ are not finitely generated, if $g_\Gamma > 0$, contrary to the case of G and F, which are finitely presented ([5]). I do not know whether G_Γ and F_Γ are finitely generated when $g_\Gamma = 0$.

The ideal of an action "at infinity" is central for this paper. A precise definition, in it's context, is 2.4. Here, we briefly explain in what sense G acts "at infinity" on the hyperbolic plane $\mathbb{H}$.

Recall that the circle S^1 can be viewed as the set of points at infinity of $\mathbb{H}$, and that the action of $\mathrm{PSL}_2\,\mathbb{R}$ on S^1 is the continuous extension of the action of the group of isometries of $\mathbb{H}$. Let $g \in G$, with breakpoints $r_1, r_2, \ldots, r_k$. Note that g has a natural extension, not to all of $\mathbb{H}$, but to the union of the convex closures of the (r_i, r_{i+1}) (fig. 1.3) in $\mathbb{H}$. The complement of this union indeed extends to infinity, but it has finite area. Thus, one might say that G acts "at infinity" on $\mathbb{H}$.

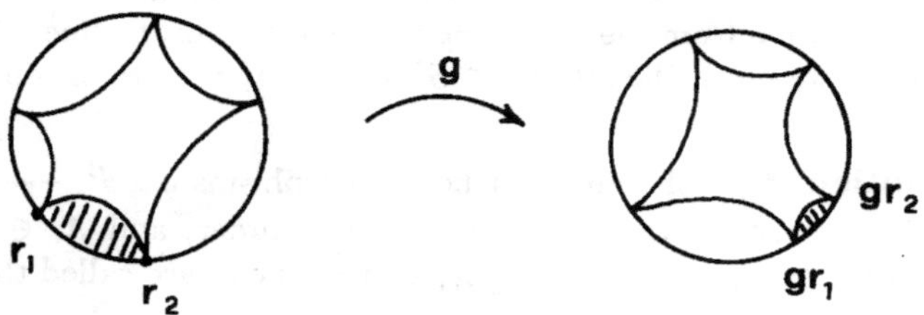

Figure 1.3

The organization of the paper is as follows. In section 2, we define the tree $\widetilde{X}_\Gamma$. In section 4, we study the geometry on S^1 implied by the action of G_Γ, and verify a "flexibility condition". This condition is what is needed for the application of certain homological techniques due to Mather, Thurston, McDuff and Segal (nicely exposited in [6]), whose consequences include the corollary 3.2 cited above. In section 5, we conjecture a presentation for the F_Γ, following ideas of Brown and Stein ([14]).

A final remark: the techniques of this paper most probably extend to general lattices Γ in $\mathrm{PSL}_2\,\mathbb{R}$. One must replace $\widehat{\mathbb{Q}}$ by the cusps of Γ, and the presence of torsion demands the concept of orbifold (as in the case for $\mathrm{PSL}_2\,\mathbb{Z}$ itself).

2. Graph and Riemann surface

In this section we associate, to a lattice $\Gamma \subset \mathrm{PSL}_2\,\mathbb{Z}$, a graph $\widetilde{X}_\Gamma \subseteq \mathbb{H}$ whose "automorphisms at infinity" are the elements of G_Γ. As motivation, let us recall the construction of a well-known graph associated to $\mathrm{PSL}_2\,\mathbb{Z}$.

In the upper half plane, let $L = \{z : \mathrm{Im}\, z = 1\}$ and consider (fig. 2.1) the set of circles which are the images of L under $\mathrm{PSL}_2\,\mathbb{Z}$. These circles alternate with triangular

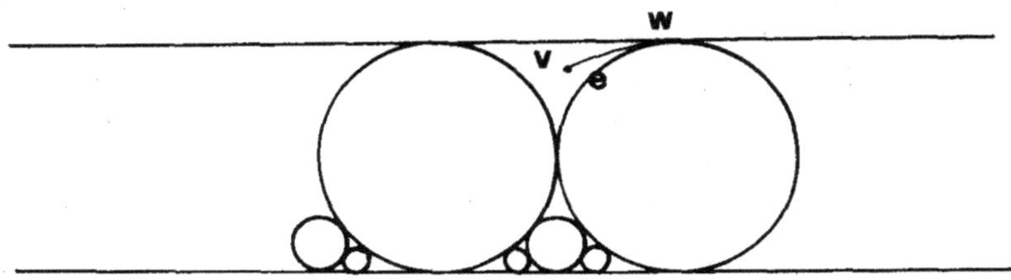

Figure 2.1

regions to fill $\mathbb{H}$. Choose one such region, let v be it's center of mass, and draw a geodesic segment (labelled e in 2.1) to one of the three corners of the region; let w denote the other endpoint of e. In $\mathrm{PSL}_2\,\mathbb{Z}$ there is an element of order 3 fixing v, and sending w to the other two corners of the triangle. Applying all of $\mathrm{PSL}_2\,\mathbb{Z}$ to these segments, we obtain a graph $\widetilde{X}$ in $\mathbb{H}$, whose edges are labelled e and whose vertices, v or w (fig. 2.2).

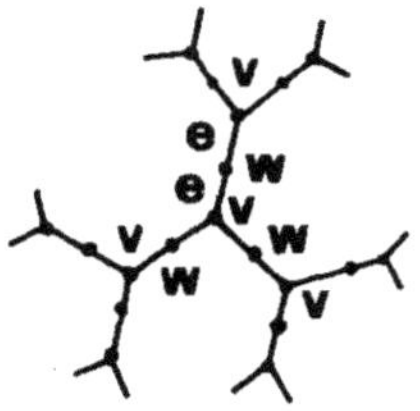

Figure 2.2

2.3. Facts. $\widetilde{X}$ is a labelled, cyclically oriented (that is, the edges emanating from a given vertex are cyclically ordered) locally finite tree. The index of each vertex is at least 2, and

- (*i*) the group of automorphisms of $\widetilde{X}$ (as a labelled, cyclically oriented tree) is $\mathrm{PSL}_2\,\mathbb{Z}$;
- (*ii*) every component C of $\mathbb{H} - \widetilde{X}$ has a single limit point $r_C \in \widehat{\mathbb{Q}}$, and every $r \in \widehat{\mathbb{Q}}$ occurs as the limit point of some component. The stabilizer of r_C in $\mathrm{PSL}_2\,\mathbb{Z}$ is infinite cyclic, and acts by translation on the subgraph ∂C of $\widetilde{X}$.

Let $g \in G$, with breakpoints $r_1, \dots, r_k \in \widehat{\mathbb{Q}}$, and let R denote the union of the convex closures of the intervals (r_i, r_{i+1}). Recall (fig. 1.3) that g extends to a function $g : R \to \mathbb{H}$. Let X_1 be the largest subgraph of $\widetilde{X}$ contained in R. By fact (*ii*), X_1 is a cofinite subgraph of $\widetilde{X}$, that is, X_1 contains all but a finite number of vertices and edges from $\widetilde{X}$. Note that the infinite components of a cofinite subgraph of a cyclically oriented tree have a natural cyclic ordering.

2.4. Definition. An *automorphism at infinity* of a labelled, cyclically oriented locally finite tree T is an equivalence class of isomorphisms $g : T_1 \to T_2$ between cofinite subgraphs of T, preserving the cyclic order of the components; we set $g : T_1 \to T_2$ equivalent to it's restriction to a cofinite subgraph of T_1.

The set of automorphisms at infinity of a tree T form a group. It follows from 2.3 that:

2.3 Proposition. *G is the group of automorphisms at infinity of $\widetilde{X}$.*

We now consider the subgroups G_Γ of G. Bowditch and Epstein ([1]) define a generalization of the graph $\widetilde{X}$.

2.6 Proposition ([1]). *On the surface $\mathbb{H}/\Gamma$ there is a finite graph X_Γ whose edges are geodesic segments with distinct endpoints. Every vertex has index at least 3. Further, X_Γ is a deformation retract of $\mathbb{H}/\Gamma$.*

We label X_Γ by assigning to each edge and to each vertex a different label. Now let $\widetilde{X}_\Gamma$ be the universal cover of X_Γ, with it's edges and vertices labelled according to the labels for X_Γ. Then facts 2.3 hold for $\widetilde{X}_\Gamma$, with $\mathrm{PSL}_2\,\mathbb{Z}$ replaced by Γ. Consequently:

2.7. Proposition. *G_Γ is the group of automorphisms at infinity of $\widetilde{X}_\Gamma$.*

2.8. Corollary. *Up to isomorphism, G_Γ depends only on g_Γ and ν_Γ.*

Here are some examples.

2.9. Examples.

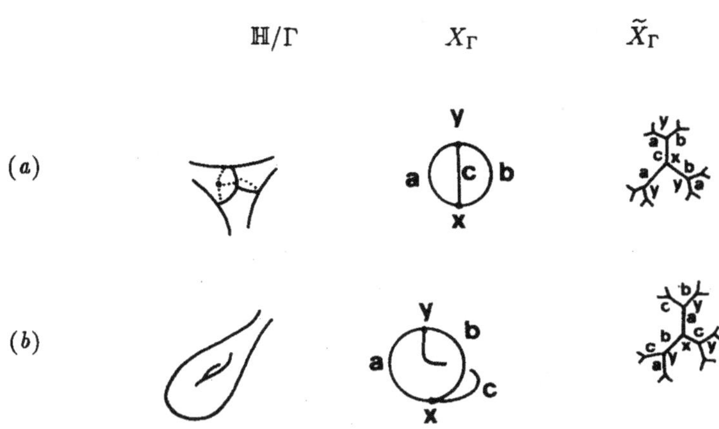

Note that the X_Γ in (a) and (b) are not isomorphic, as labelled, cyclically oriented graphs. Note also how the periodic labelling on the subgraph ∂C of a component of $\mathbb{H} - \widetilde{X}_\Gamma$ identifies the cusp to which it's limit point corresponds.

To relate the tree $\widetilde{X}$ to the construction of X_Γ and $\widetilde{X}_\Gamma$, we must think of $\mathbb{H}/\operatorname{PSL}_2\mathbb{Z}$ as an orbifold. We continue with 2.9 to include this example:

3. Results in homology

In the next section we will discuss the geometry imposed on S^1 by the group G_Γ, with the aim of proving a local flexibility condition (4.9). This condition, of interest in it's own right, is moreover what is necessary in order to invoke theorems of McDuff and Segal ([6], [7], [12]) for calculations of the homology of G_Γ, F_Γ and $F_{C\Gamma}$. We record the results of these calculations here, in part to justify the section which follows.

3.1. Proposition. *There is a space BP_Γ (described below) with a map m :*

$S^1 \to BP_\Gamma$, and maps

$$(i) \qquad K(F_{C\Gamma}, 1) \longrightarrow \Omega\widetilde{BP}_\Gamma$$

$$(ii) \qquad K(G_\Gamma, 1) \longrightarrow (BP_\Gamma^{S^1})_m \times_{S^1} ES^1$$

which induce isomorphisms in integral homology.

The space BP_Γ is weakly homotopy equivalent to a wedge $Q_\Gamma \vee S^3 \vee \cdots \vee S^3$ (ν_Γ copies of S^3) where Q_Γ is the total space of the circle bundle with Euler number $2-2g_\Gamma-\nu_\Gamma$, over a compact surface of genus g_Γ, with ν_Γ 3-disks removed. The map m is the inclusion of a fiber.

In (i), Ω denotes "loops". The space $(BP_\Gamma^{S^1})_m$ is the component of m in the space of maps from S^1 to BP_Γ. The circle acts on this space by reparametrization, and $(B\Gamma_\Gamma^{S^1})_m \times_{S^1} ES^1$ is the homotopy quotient.

For example, if $g_\Gamma = 0$ then Q_Γ is weakly homotopy equivalent to a lens space $S^3/(\mathbb{Z}/\nu_\Gamma - 2)$, with ν_Γ 3-disks removed. However, if $g_\Gamma > 0$, then BP_Γ has infinite fundamental group, as well as non zero second homology and third homotopy groups. Consequently, the first homology groups of $\Omega\widetilde{BP}_\Gamma$ and $(BP_\Gamma^{S^1})_m \times_{S^1} ES^1$ are infinitely generated, and we obtain:

3.2. Corollary. *G_Γ, F_Γ and $F_{C\Gamma}$ are not finitely generated if $g_\Gamma > 0$.*

This should be compared with the fact that G and F are finitely presented ([5]). I do not know what happens when $g_\Gamma = 0$.

Now let us return to geometry.

4. Intervals in G_Γ-geometry

The group G_Γ acts on the circle S^1. We propose to study the local geometry on S^1 suggested by this action. As it happens, it is simpler to work with the universal cover $\widetilde{S}^1$, and with the central extension $\widetilde{\Gamma}$.

4.1. Definition. Let $p : \widetilde{S}^1 \to S^1$ be the projection of the universal cover. Let $\widetilde{\Gamma}$ be the group of homeomorphisms γ of $\widetilde{S}^1$ which cover some element $g \in \Gamma$; that is, $p\gamma(x) = gp(x)$ for all $x \in \widetilde{S}^1$. Denote by T the "translation one unit to the right" which generates the covering group of p, which is the kernel of the projection $\widetilde{\Gamma} \to \Gamma$, and the center of $\widetilde{\Gamma}$.

Abusing language, we denote by $\mathbb{Q}$ the inverse image $p^{-1}(\widehat{\mathbb{Q}}) \subset \widetilde{S}^1$ of the rational points of the circle. Then $\mathbb{Q}$ breaks up into ν_Γ orbits under the action of $\widetilde{\Gamma}$; again abusing language, we'll call each orbit a cusp. The basic geometric objects we consider are intervals $[a, b]$, $a, b \in \mathbb{Q}$; passage to the universal cover permits us to consider intervals which wind several times around the circle. From now on, intervals will be assumed to have rational endpoints. The group G_Γ suggests an equivalence relation on intervals.

4.2. Definition. We say the intervals $[a,b]$ and $[c,d]$ are ***equivalent***, $[a,b] \sim [c,d]$ if there exist $x_i, y_i \in \mathbb{Q}$, $a = x_0 < x_1 < \cdots < x_n = b$, $c = y_0 < \cdots < y_n = d$, and $g_i \in \widetilde{\Gamma}$ so that $g_i x_i = y_i$, $g_i x_{i+1} = y_{i+1}$, $0 \le i \le n-1$.

The basic question of G_Γ-geometry is: when are two intervals equivalent? Of course, it is necessary that corresponding pairs of endpoints belong to the same cusp. We introduce a category C_Γ to shed light on equivalence.

4.3. Definition. The objects of the category C_Γ are $x_1, \ldots, x_{\nu_\Gamma}$; there is an object for each cusp. The morphisms from x_i to x_j are equivalence classes of intervals $[a,b]$ with a in cusp i, b in cusp j. We allow the degenerate interval $[a,a]$, which is the identity morphism. Composition is by concatenation; $[a,b] \circ [b',c] = [a,c']$, where $c' = g(c)$, for some $g \in \widetilde{\Gamma}$ such that $g(b') = b$.

If Γ has only one cusp, C_Γ is a monoid. Even in the general case, one can discuss generators and relations for the morphisms of C_Γ, which, as we will now see, can be read off from the graph $\widetilde{X}_\Gamma$. We begin by describing generators.

Let e be an edge in $\widetilde{X}_\Gamma$, and v one of the vertices of e. Thinking of e as directed away from v, e belongs to the boundary of a "left" and a "right" component of $\mathbb{H} - \widetilde{X}_\Gamma$ (see fig. 4.4) called L and R. Let $\text{Int}(e,v) = [r_L, r_R] \subseteq S^1$. Now let a be the label of e, x be the label of v, and let $I(a,x)$ be the morphism of C_Γ corresponding to any lift of $\text{Int}(e,v)$ to $\widetilde{S}^1$. It is not hard to see that the $I(a,x)$, whose definition is independent of the choice of e representing a, generate the morphisms of C_Γ. The vertices of X_Γ will provide relations.

Let x be a vertex in X_Γ, and let $a_0, a_1, \ldots, a_k$ be the edges which share x, listed in counterclockwise order. Let $x_1, \ldots, x_k$ be the other vertices of a_i, $i = 1, \ldots, k$. Then we have the *subdivision relation*

$$I(a_1, x_1) \cdots I(a_k, x_k) = I(a_0, x). \tag{4.4}$$

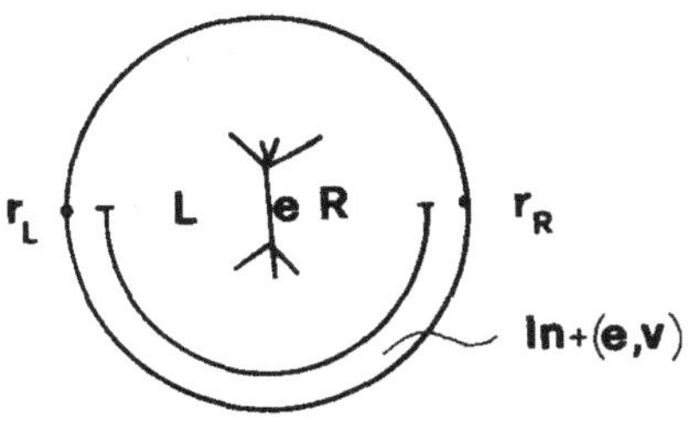

Figure 4.4

It is not hard to see that the $I(a,x)$, with the subdivision relations, give a presentation for C_Γ.

There is another interesting relation, however. Let $x_i \in \mathbf{Q}$ be an element of the i-th cusp. Let $E_i = [x_i, Tx_i]$ (recall that T generates the covering group of $\widetilde{S}^1$ over S^1). Let a be an edge in X_Γ, with vertices x, y. Suppose that $I(a,x)$ is a morphism from x_i to x_j. Then $I(a,x)\cdot I(a,y) = E_i$. The E_i will be important in the sequel.

The generators and relations (4.4) can be used in the simplest example, to understand equivalence.

4.5. Example. Recall (fig. 2.2) the tree $\widetilde{X}$ constructed for $\mathrm{PSL}_2\,\mathbf{Z}$. Evidently, $C_{\mathrm{PSL}_2\mathbf{Z}}$ is generated by $I(e,v)$ and $I(e,w)$. The subdivision relations (4.4) give $I(e,v) = I(e,w)$, and $I(e,v) = I(e,w)I(e,w)$. Hence $C_{\mathrm{PSL}_2\mathbf{Z}} = \{1, I\}$, where $I = I(e,v)$, and $I^2 = 1$; any two intervals are equivalent in G-geometry. (I thank G. Mess and R. Kulkarni for remarking that this can be seen directly).

For more complicated examples, the question of equivalence is difficult, so we "stabilize" to obtain an easier question.

4.6. Definition. Let $[a,b]$, $[c,d]$ be intervals in $\widetilde{S}^1$. Then $[a,b]$ and $[c,d]$ are *stably equivalent*, $[a,b] \sim_S [c,d]$ if there are intervals $[x_i,y_i]$ and $[z_i,w_i]$, $0 \le i \le k$, so that $x_0 = a$, $z_0 = c$, $b = y_k$, $d = w_k$, $x_{i+1} < y_i$, $z_{i+1} < w_i$, and $[x_i,y_i] \sim [z_i,w_i]$, $[y_{i+1},x_i] \sim [w_{i+1},z_i]$. We allow $y_0 = a$, $w_0 = c$. The number k is called the number of zig-zags.

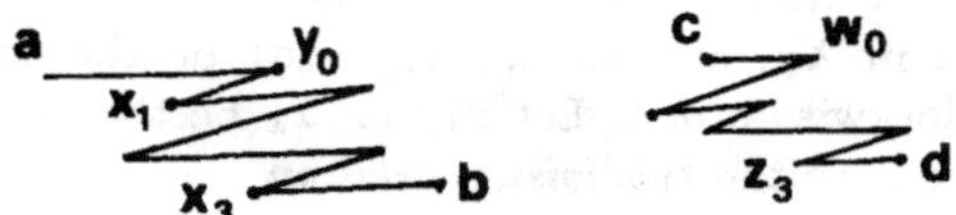

Number of zig-zags is 3
Figure 4.7

The following proposition shows that the question of stable equivalence is straightforward. Let $y_i \in \mathbf{Q}$ be an element of the i-th cusp, and let $P_i \subseteq \widetilde{\Gamma}$ be the stabilizer of y_i.

4.8. Proposition. *Let $[a,b]$, $[c,d]$ be intervals. Then $[a,b] \sim_S [c,d]$ if and only if there are $g,h \in \widetilde{\Gamma}$, with $ga = c$, $hb = d$, and such that $g \equiv h$ in $\widetilde{\Gamma}/\overline{(P_1,\ldots,P_{\nu_\Gamma})}$, the quotient of $\widetilde{\Gamma}$ by the normal closure of the stabilizers of the rational points.*

Proof. Left to the reader; recall that if $gx = g'x$, $x \in \mathbf{Q}$, $g,g' \in \widetilde{\Gamma}$ then $g \equiv g'$ modulo the stabilizer of x.

Finally, we can state the flexibility condition, whose verification justifies the results of section 3. Note first that equivalence implies stable equivalence of intervals. The converse is not a *a priori* true.

4.9. Definition. The G_Γ-geometry is *flexible* if, for any intervals I, J, $I \sim_S J$ implies $I \sim J$.

Flexibility is similar to the question of whether scissors congruence is implied by stable scissors congruence (see Sah [11], especially the theorem of Zylev).

4.10. Theorem. *The G_Γ-geometry in flexible.*

The proof is rather general in scope, using the E_i defined before 4.5 to prove that the category C_Γ has cancellation. This should be compared with the geometrical proofs of "germ-connectedness" in [7], [8]. Two lemmas will be used.

4.11. Lemma. *Let $a, b, c \in \mathbb{Q}$, with $a < c < b$. Then*

(a) contraction. *There are $a', b' \in \mathbb{Q}$, $a < b' < c < a' < b$, such that* $[a', b] \sim [a, b] \sim [a, b']$.

(b) expansion. *There are $c', c'' \in \mathbb{Q}$, $c' < a$, $c'' > b$, such that* $[c, b] \sim [c', b]$, $[a, c] \sim [a, c'']$.

4.12. Lemma. *If $\alpha : x_i \to x_j$ is a morphism in C_Γ, not the identity, then α can be factored $\alpha = \gamma E_j$, $\alpha = E_i\gamma'$, with E_i (resp. E_j) the morphism $[y_i, Ty_i]$, where $y_i \in \mathbb{Q}$ is in the i-th cusp, and T generates the covering group of $\widetilde{S}^1$ over S^1.*

Proofs of lemmas. Assuming 4.11, E_i may be contracted as much as necessary to prove 4.12. We prove the existence of a' in 4.11; the other cases in (a) and (b) follow similarly.

We may assume that $[a, b]$ is the lift to S^1 of some $\mathrm{Int}(e, v)$, as the general case then follows by induction. We abuse notation, letting $[a, b] = \mathrm{Int}(e, v)$ (see fig. 4.13).

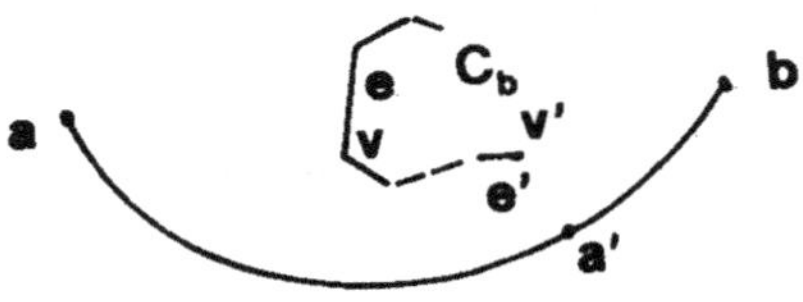

Figure 4.13

Let Y_b be the graph which is the boundary of $\overline{C}_b$, where C_b is that component of $\mathbb{H} - \widetilde{X}_\Gamma$ whose limit point is b. The generator T_b of the stabilizer of b in Γ acts on Y_b by translation. Consequently, the labels on Y_b are periodic, and we may find a pair (e', v') as close to b as we like, with the same labels as (e, v). Thus applying T_b sufficiently often to $[a, b]$, we can obtain a $[a', b]$ with a' as close to b as we want. □

We now proceed with the proof of theorem 4.10. Suppose then that $[a, b] \sim_S [c, d]$, with x_i, y_i, w_i, z_i, $0 \leq i \leq k$ as in definition 4.6 (fig. 4.14). We will prove that $[a, b] \sim [c, d]$ by showing that the number of zig-zags k can always be reduced.

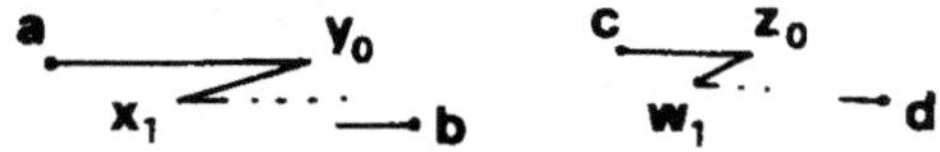

Figure 4.14

To begin, by 4.11.(b) we can assume, as in fig. 4.14, that $a < x_1 < y_0$, $c < w_1 < z_0$. Then $[a, x_1] \sim_S [c, w_1]$, and to reduce the number of zig-zags it suffices that $[a, x_1] \sim [c, w_1]$.

It is interesting to write, in the category C_Γ, what it is that we must prove. Let $[a, x_1] = \alpha$, $[x_1, y_0] = \beta$, and $[a, y_0] = \gamma = \alpha\beta$. Similarly, let $[c, z_1] = \alpha'$, $[z_1, w_0] = \beta'$, $[c, w_0] = \gamma' = \alpha'\beta'$. By assumption, $[x_1, y_0] \sim [w_1, z_0]$, whence $\beta = \beta'$ in C_Γ and we must prove:

4.15. Proposition (cancellation). *$\alpha\beta = \alpha'\beta$ implies $\alpha = \alpha'$, and $\delta\alpha = \delta\alpha'$ implies $\alpha = \alpha'$, for $\alpha, \beta, \alpha', \delta, \delta'$ nonidentity morphisms in C_Γ.*

Proof. We prove the first assertion, the proof of the second being similar. By induction, it suffices to take $\beta = I(e, v)$ a generator. Now (fig. 4.16) $\alpha\beta = \alpha'\beta$ implies that there is a morphism ω, such that $\alpha' = \alpha\omega$ and $\omega\beta = \beta$. We wish to show that $\alpha = \alpha'$.

Writing (by 4.12) $\alpha = \gamma E_i$, whence $\alpha' = \gamma E_i\omega$, it suffices to prove that $E_i\omega = E_i$. Write $E_i = I(e, v)I(e, v')$ (v' is the label of the "other" vertex of an edge in X_Γ whose label is e). Then $E_i\omega = \omega E_i = \omega I(e, v)I(e, v') = \omega\beta I(e, v') = \beta I(e, v') = E_i$. □

5. Conjectured presentation

By 3.2, F_Γ is not finitely generated if $g_\Gamma > 0$. Borrowing ideas of K. Brown,

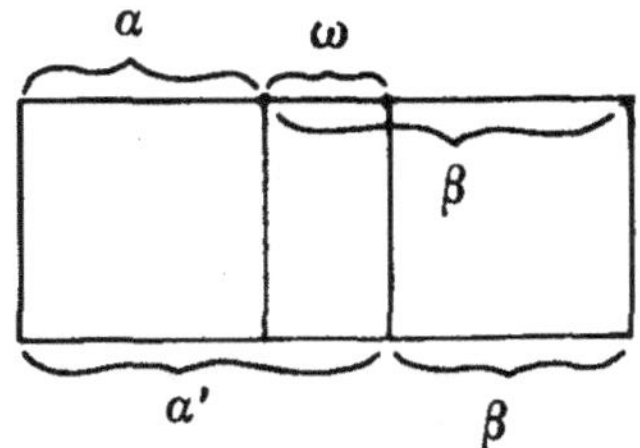

Figure 4.16

R. Geoghegan and M. Stein we briefly describe a conjectural infinite presentation for F_Γ.

We require some notation: a *subdivision* of S^1 is a finite subset $S = \{x_i\}$ of $\widetilde{\mathbb{Q}} - \{\infty\}$, $1 \le i \le k$, so that $\infty = x_0, \ldots, x_{k+1} = \infty$ are in counterclockwise order, and the union of the $[x_i, x_{i+1}]$ cover S^1 exactly once, and such that $[x_i, x_{i+1}]$ is an $\mathrm{Int}(e_i, v_i)$, $0 \le i \le k$, for some edge and vertex e_i, v_i in $\widetilde{X}_\Gamma$. Letting a_i, b_i be the labels of e_i, v_i, the *word associated* to S is $w(S) = I(a_0, b_0)\cdots I(a_k, b_k)$. If $w = I(a_0, b_0)\cdots I(a_k, b_k)$ is any word, we write $a_i(w) = a_i$, $b_i(w) = b_i$. An *admissible* word w is a word $w = w(S)$, for some subdivision S.

The point is that an element of F_Γ is precisely an ordered pair of subdivisions, with the same associated word.

For each admissible word w, choose a subdivision $S(w)$ so that $w(S(w)) = w$, and let S^1_w be a copy of S^1. For any word $w = I(a_0, b_0)\cdots I(a_k, b_k)$, let $e_\ell w$ be the word obtained by applying the subdivision relation (4.4) to $I(a_\ell, b_\ell)$. Note the canonical homeomorphism $h^w_\ell : S^1_w \longrightarrow S^1_{e_\ell w}$.

5.1. Conjecture. *F_Γ has presentation with generators h^w_ℓ, and relations*

$$h_i^{e_j w} h_j^w = h_{j+L(b_i(w))-1}^{e_i w} h_i^w, \quad i < j$$

where w runs over all admissible words $I(a_0, b_0)\cdots I(a_k, b_k)$, and $0 \le \ell \le k$, and $L(b)$ denotes the number of edges emanating from the vertex b in X_Γ.

Bibliography

1. Bowditch and Epstein, *Natural triangulations associated to a surface*, Topology **21** (1987), 91–117.
2. Brown K.S, *Finiteness properties of groups*, J. Pure and Applied Alg. **44** (1987), 45–75.
3. Brown K.S, *Presentations for groups acting on simply connected complexes*, J. Pure and Applied Alg. **32** (1984), 1–10.
4. Brown K.S, *Cohomology of groups*, Springer GTM, Berlin, 1982.
5. Brown K.S. and Geoghegan R, *An infinite-dimensional torsion-free FP_∞ group*, Invent. Math. **77** (1984), 367–381.

6. Ghys E. and Sergiescu V, *Sur un groupe remarquable de difféomorphismes du cercle*, Comm. Math. Helv. **62** (1987), 185–239.
7. Greenberg P, *Pseudogroups from group actions*, Amer. J. Math. **109** (1987), 893–906.
8. Greenberg P, *Pseudogroups of C^1, piecewise projective homeomorphisms*, Pacific Math. J. **129** (1987), 67–75.
9. Greenberg P. and Sergiescu V, *An acyclic extension of the braid group*, Comm. Math. Helv (to appear).
10. Higman G, *Finitely presented infinite simple groups*, Notes on Pure Math. 8, Australian National University, Canberra, 1974.
11. Sah C.H, *Hilbert's 3rd problem, scissors congruence*, Pitman Pub., London, 1979.
12. Segal G, *Classifying spaces related to foliations*, Topology **17** (1978), 367–382.
13. Shimura G, *Introduction to the arithmetic theory of automorphic functions*, Princeton University Press, 1971.
14. Stein M, *Groups of piecewise-linear homeomorphisms*, Cornell preprint.

Hyperbolic Geometry and the Subgroups of the Modular Group

*Ravi S. Kulkarni**

Abstract

Special types of fundamental domains for the subgroups of finite index in the modular group are constructed. These have the property that the side–pairing transformations are independent in the sense of Rademacher. A parallel arithmetic procedure using generalized Farey sequences is devised. The main applications are explicit constructions of fundamental domains for certain congruence subgroups.

Contents

§1 Statement of the Problem. This work is essentially an application of hyperbolic geometry to a problem in number theory. Usually "geometry" and "number theory" are quite different modes of thought. Very rarely as in the case of the classical modular group they come close together and it is very interesting to see how one is related to the other.

* *Partially supported by an NSF grant, and a PSC-CUNY award.*

(1.1) Recall that the classical (inhomogeneous) modular group is $\Gamma \approx PSL_2(\mathbf{Z})$. An element in Γ is sometimes denoted by its matrix form $A = \pm\begin{pmatrix} a & b \\ c & d \end{pmatrix}$. Here a, b, c, d are integers satisfying $ad-bc = 1$. The $\pm$ indicates that A and $-$A define the same element in the group. Some of its subgroups of arithmetic interest are the *congruence subgroups* such as

$$\Gamma(N) = \{\begin{pmatrix} a & b \\ c & d \end{pmatrix} | a \equiv d \equiv 1\ (N),\ \ b \equiv c \equiv 0\ (N)\},$$

$$\Gamma_0(N) = \{\begin{pmatrix} a & b \\ c & d \end{pmatrix} | c \equiv 0\ (N)\},\ \ \Gamma^0(N) = \{\begin{pmatrix} a & b \\ c & d \end{pmatrix} | b \equiv 0\ (N)\}$$

$$\Gamma^1(N) = \{\begin{pmatrix} a & b \\ c & d \end{pmatrix} | a \equiv d \equiv 1\ (N),\ \ c \equiv 0\ (N)\}.$$

The modular group acts on the upper half plane $\mathbf{H} = \{z \in \mathbf{C} | Im\, z > 0\}$ via

$$z \mapsto \frac{az+b}{cz+d}. \tag{1.1.1}$$

We shall denote an element of Γ either by a transformation of the type (1.1.1) or by its matrix whenever it is convenient.

As is wellknown

$$\mathcal{D} = \{z \in \mathbf{H} | -\frac{1}{2} \leq Re\, z \leq \frac{1}{2}, |z| \geq 1\} \tag{1.1.2}$$

is a fundamental domain for this action in a very precise sense that its translates by Γ tessallate $\mathbf{H}$ and two distinct tiles have disjoint interiors. $\mathcal{D}$ is a hyperbolic triangle with vertices at $\rho = exp(\frac{\pi i}{3})$, ρ^2 and ∞. This tessellation is called the *modular tessellation*. The indicated side–pairing of $\mathcal{D}$ shows that the quotient $\mathbf{H}/\Gamma$ is a kind of "triangular pilow cover" — namely, if

$$\mathcal{D}^* = \{z \in \mathbf{H} | 0 \leq Re\, z \leq \frac{1}{2}, |z| \geq 1\} \tag{1.1.3}$$

which is the hyperbolic triangle with vertices at $i = \sqrt{-1}$, ρ, and ∞, then $\mathbf{H}/\Gamma$ is obtained by taking two copies of $\mathcal{D}^*$ and sowing them along the edges.

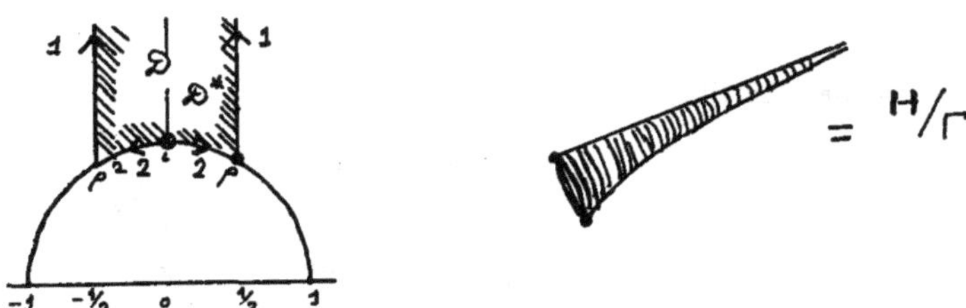

$\mathcal{D}^*$ itself is a fundamental domain for the extended modular group Γ^* which is the group generated by reflections in the edges of $\mathcal{D}^*$. It contains Γ as a subgroup of index 2. In fact Γ is the full subgroup of orientation–preserving transformations in Γ^* in its action on $\mathbf{H}$. It is useful to note that Γ^* may also be identified with the group $\mathrm{PSL}_2^*(\mathrm{Z})$ of 2×2 integer matrices with determinant 1 or -1, modulo its center $< -\mathrm{I} >$ where I denotes the identity matrix. Under this identification an element $A = \begin{pmatrix} a & b \\ c & d \end{pmatrix}$, with determinant 1, i.e. an element of Γ, acts as in (1.1.1), whereas with determinant -1 it acts by

$$z \mapsto \frac{a\bar{z}+b}{c\bar{z}+d}, \quad a,b,c,d \in \mathbf{R}. \tag{1.1.4}$$

(1.2) A natural question in this context is to find "good" fundamental domains for the congruence subgroups. This question is of substantial arithmetic interest. Constructions of fundamental domains for $\Gamma_0(p)$ or $\Gamma^0(p)$ where p is a prime go back at least to Fricke, cf. [Fr] ch. 3, p. 349, cf. also [S], p. 88, [Z]. In the standard texts and treatises one also finds some ad hoc constructions for $\Gamma_0(\mathrm{N})$ where N is not necessarily a prime for low values of N, and the same for $\Gamma(\mathrm{N})$. However no general method of explicit constructions of these fundamental domains depending on the arithmetic nature of N appears to be known.

It is of course known and easy to see that for any subgroup Φ of finite index in Γ a fundamental domain can be constructed which is a union of the tiles in the modular tessellation. A domain of this type must be a union of the translates of $\mathcal{D}$ by a set of coset representatives of Φ and any such union would serve as (a possibly disconnected) fundamental domain. With more care it may be shown that one may choose the coset representatives so that the tiles form a convex hyperbolic polygon. Indeed $\mathbf{H}/\Phi$ is a surface which also admits a tessellation by $(\Gamma : \Phi)$ copies of $\mathcal{D}$.

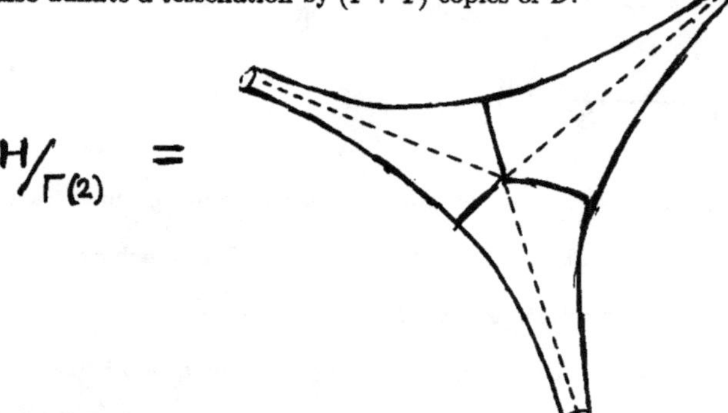

By appropriately cutting along its edges we cut $\mathbf{H}/\Phi$ into a space which is isometric to a convex hyperbolic polygon and then appropriately developing it in $\mathbf{H}$ one obtains such a fundamental domain. For a more algebraic discussion on fundamental domains see [Ran], ch. 2.

(1.3) In a quite different direction Rademacher posed the following question, cf. [R]. It is wellknown that as an abstract group Γ is a free product of $\mathbf{Z}_2$ and $\mathbf{Z}_3$. In fact $\Gamma \approx \langle A \rangle * \langle B \rangle$, where

$$A = \begin{pmatrix} 0 & 1 \\ -1 & 0 \end{pmatrix}, \quad B = \begin{pmatrix} 0 & 1 \\ -1 & -1 \end{pmatrix}. \tag{1.1.5}$$

It follows from a theorem of Kurosh that a subgroup of Γ is a free product of copies of $\mathbf{Z}_2's$ and $\mathbf{Z}_3's$ and $\mathbf{Z}'s$. If the subgroup is of finite index then the number of factors is also finite. This suggests a notion. Let Φ be a subgroup of finite index. Then a system of generators x_i, $i \in I$ is said to be *independent* if Φ is an internal free product of the cyclic subgroups $< x_i >$, $i \in I$. Rademacher asked for a construction of independent systems of generators for $\Gamma_0(N)$. He gave a construction for $\Gamma_0(p)$. He applies a very general method known as the Reidemeister–Schreier process which generates a system of generators for a subgroup of a group with a given presentation in terms of the coset representatives of the subgroup. In general however there are many redundant relations among the generators of the subgroup obtained in this way. So when applied to the modular group there is a further substantial work to obtain an independent system of generators, and moreover for number-theoretic purposes one would need these generators in a matrix form. For further work in this direction cf. [C], [F]. In these works there is no attempt to relate this problem to constructions of fundamental domains.

(1.4) Now there is another general method of finding generators and relations for a group once one knows its fundamental domain in some action. The method goes under the name *Poincare's theorem on fundamental polygons.* Poincare in fact stated it for Fuchsian groups, cf. [Pon]. For an exposition see [B], which is sufficient for our purposes. For higher dimensional generalizations see [M] and a forthcoming book by Ratcliff, cf. [Rat]. For a much more general result see [Mac]. For a Fuchsian group once a hyperbolic polygon which is a fundamental domain for the group is constructed such that its translates by the group form a locally finite tessellation of **H** then its side-pairing transformations form a system of generators for the group. Moreover the relations can be read from the total angles at the vertex cycles generated by the side–pairing. However this procedure when applied to subgroups of the modular group in general does not lead to an *independent* system of generators. In fact it is easy to see that the side-pairing transformations for the Fricke's fundamental domains for $\Gamma^0(p)$ or $\Gamma_0(p)$ referred to earlier do *not* form an independent system of generators.

(1.5) A motivation for our approach to Rademacher's problem was to construct fundamental domains for $\Gamma(N)$ and $\Gamma_0(N)$ so that the side-pairing transformations form an independent set of generators. In the process however a certain method is developed which is partly geometric and partly arithmetic and which applies to all subgroups of finite index in the modular group. The geometric part is based on the Poincare's theorem mentioned above and it is elaborated in §3. Observe however that a purely geometric constuction can hardly be expected to solve our original problem, namely to construct fundamental domains for the congruence subgroups which are good enough to provide an independent set of generators – the *arithmetic* nature of N has to come in somewhere. The arithmetic aspects which enter here are some of the most elementary and intriguing parts of number theory which historically appear to have led to the consideration of the modular group in the first place, and which underlie some sophisticated number theory in which the modular group enters. We collect these facts in §2.

A more detailed account of this work will appear in $[K]_2$. The following more leisurely account working out several motivating examples was presented at the meeting at ICTP in Trieste, Italy on Hyperbolic Groups in March '90. As just a free product of $\mathbf{Z}_2$ and $\mathbf{Z}_3$ the modular group is perhaps the simplest complicated hyperbolic group (in the sense of Gromov, cf. [G]). It is quite amazing how this group contains the essence in embryonic form of so much arithmetic and geometry.

§2 Some Elementary Number–theoretic Facts. We now explain three classical facts from elementary number theory which are closely connected with the modular group and which will be of relevance to us in our constructions.

(2.1) Farey Sequences Let n be a natural number. Classically the n-th Farey sequence $\mathcal{F}_n$ is the finite sequence of rationals in [0, 1] which in their reduced form have denominators at most n and which are arranged in an increasing order, cf. [HW], ch. 3. Thus the first five Farey sequences are

$$\frac{0}{1}, \frac{1}{1}.$$

$$\frac{0}{1}, \frac{1}{2}, \frac{1}{1}.$$

$$\frac{0}{1}, \frac{1}{3}, \frac{1}{2}, \frac{2}{3} \frac{1}{1}.$$

$$\frac{0}{1}, \frac{1}{4}, \frac{1}{3}, \frac{1}{2}, \frac{2}{3}, \frac{3}{4}, \frac{1}{1}.$$

$$\frac{0}{1}, \frac{1}{5}, \frac{1}{4}, \frac{1}{3}, \frac{2}{5}, \frac{1}{2}, \frac{3}{5}, \frac{2}{3}, \frac{3}{4}, \frac{4}{5}, \frac{1}{1}.$$

A remarkable property of these sequences is that if $x_i = \frac{a_i}{b_i}$ and $x_{i+1} = \frac{a_{i+1}}{b_{i+1}}$ are two consecutive terms of $\mathcal{F}_n$ then they satisfy a "modular relation": $a_{i+1}b_i - a_ib_{i+1} = 1$. Also $\mathcal{F}_n$ can be built inductively. Given $\mathcal{F}_n$ the next sequence $\mathcal{F}_{n+1}$ consists of $\mathcal{F}_n$ and the terms inserted in the following way: between two consecutive terms $\frac{a_i}{b_i}$ and $\frac{a_{i+1}}{b_{i+1}}$ in $\mathcal{F}_n$ satisfying $b_i + b_{i+1} = n + 1$ insert the term $\frac{a_i+a_{i+1}}{b_i+b_{i+1}}$. These facts have simple geometric analogues in terms of the modular tessellation as we shall see.

(2.2) Continued Fractions For a rational number x let

$$x = a_0 + \cfrac{1}{a_1 + \cfrac{1}{a_2 + \cfrac{1}{a_3 + \cfrac{1}{\dots + \cfrac{1}{a_k}}}}} \tag{2.2.1}$$

be its continued fraction expansion. Here as usual $a_0 = [x]$, $a_1 = [\frac{1}{x-a_0}]$ etc. We shall abbreviate (2.2.1) to $x = [a_0; a_1, a_2, \dots, a_k]$. If x is not an integer the a_i's, $i \geq 1$, are positive, and $a_k \geq 2$. We define *the depth of* x by

$$a_1 + a_2 + \dots + a_k, \tag{2.2.2}$$

and denote it by $\Delta(x)$. If x is an integer then $\Delta(x) = 0$. Clearly $\Delta(x)$ depends only on the congruence class of x mod 1. Now suppose for simplicity that x lies in (0, 1). Let

$$y_i = [0; a_1, a_2, ..., a_i], \ \ 1 \leq i \leq k$$

be the convergents of x. So $y_k = x$. It is convenient to put

$$p_{-1} = 1, \ q_{-1} = 0, \ p_0 = 0, \ q_0 = 1,$$

(2.2.3) $$p_i = a_i p_{i-1} + p_{i-2}, \ q_i = a_i q_{i-1} + q_{i-2}, \ 1 \leq i \leq k.$$

$$y_{-1} = \infty, \ y_0 = 0.$$

Then as is wellknown $y_i = \frac{p_i}{q_i}$ (reduced fractions), where we regard 0 as $\frac{0}{1}$ and ∞ as $\frac{1}{0}$. Moreover we have a "modular relation":

(2.2.4) $$p_i q_{i-1} - q_i p_{i-1} = (-1)^{i-1},$$

And also

(2.2.5) $$p_i q_{i-2} - q_i p_{i-2} = (-1)^i a_i,$$

and

(2.2.6) $$0 < y_2 < y_4 < ... < x < ... < y_3 < y_1.$$

The a_i's, y_i's, and $\Delta(x)$, and the above equations and inequalities also have an interpretation in terms of the modular tessellation as we shall see.

(2.3) Integral Binary Quadratic Forms Consider the equation $Q(x, y) = ax^2 + 2bxy + cy^2 = n$ where a, b, c, n are integers. The problem: *given* a, b, c *which* n *are* *representable as $Q(x, y)$?* is a basic problem in number theory. It has an obvious built-in symmetry: if x, y are subjected to a "modular substitution": $x = ux' + vy'$, $y = sx' + ty'$ where u, v, s, t are integers satisfying $ut - vs = 1$ then $Q(x, y) = Q'(x', y')$ where Q' is the transform of Q by the modular substitution. The deeper aspects of this problem involve the class numbers of quadratic number fields on the one hand, and also by a theorem of Latimer and Macduffee the conjugacy class structure in the extended modular group on the other, cf. [N], p. 53. The two special cases which are of importance in our constructions of fundamental domains for the congruence subgroups are $Q(x, y) = x^2 + y^2$ and $Q(x, y) = x^2 + xy + y^2$. The representability of p by these two forms where p is a prime depends on whether p splits or not in the fields Q(i), $i = \sqrt{-1}$, and Q(ρ), $\rho = exp(\frac{\pi i}{3})$ respectively. The fact that the rings of integers in these fields are principal ideal domains is equivalent to the fact that there are unique conjugacy classes of subgroups of order 2 and 3 respectively in the modular group.

§3 The Extended Modular Tessellation.

(3.1) Recall that $\mathcal{D}^*$ is the hyperbolic triangle with vertices at $i = \sqrt{-1}$, ρ, and ∞, cf. (1.1.3). It is a fundamental domain for the extended modular group Γ^*. The translates of $\mathcal{D}^*$ by Γ^* form *the extended modular tessellation* $\mathcal{T}^*$ of **H**.

It will be convenient to adopt the following terminology:

The elements in the Γ^*–orbit of i, ρ, ∞ will be called the *even vertices,* the *odd vertices,* and the *cusps* of $\mathcal{T}^*$ respectively. Notice that the cusps are precisely the rational numbers counting ∞. The elements in the Γ^*–orbit of the edge joining i to ∞, resp. the edge joining ρ to ∞ will be called the *even edges* resp. the *odd edges* of $\mathcal{T}^*$. Each of these edges has infinite hyperbolic length. Each of the edges in the Γ^*–orbit of the edge joining i to ρ has finite hyperbolic length. These edges will be called the *f-edges* of $\mathcal{T}^*$.

The hyperbolic line joining 0 to ∞ consists of two even edges. Its Γ^*—translates will be called the *even lines*. Also the hyperbolic line joining -1 to 1 (or ∞ to $\frac{1}{2}$) consists of two odd edges and two f—edges. Its Γ^*—translates will be called the *odd lines*.

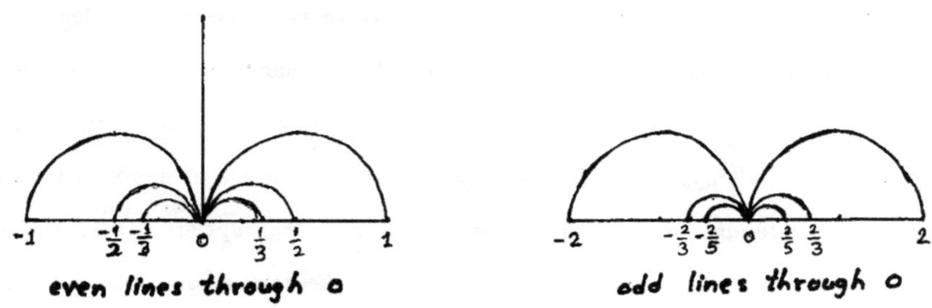

Notice that the f—edges form a cubic (i. e. trivalent) tree in which each edge is further divided into two edges by introducing an extra vertex of valence 2 at its midpoint.

The modular tessellation $\mathcal{T}$ whose tiles are the Γ—translates of $\mathcal{D}$, cf. (1.1.2), is the one usually drawn in the texts on the modular group. It uses only the odd lines. On the other hand the even lines also provide another interesting tessellation of $\mathbf{H}$. All of its tiles are ideal triangles, i. e. hyperbolic triangles all of whose angles are zero. It will be denoted by $\mathcal{I}$.

(3.2) There is a simple *arithmetic* procedure to draw the extended modular tessellation. Although the co-ordinates of the odd vertices of $\mathcal{T}^*$ involve quadratic irrationalities this procedure involves only rational numbers. Moreover any finite portion of the tessellation can be drawn using only ruler and compass!

Observe that Γ is transitive on the set of rationals counting ∞. An obvious invariant of its action on the pairs $\{\frac{a}{b}, \frac{c}{d}\}$ of rationals in reduced forms is $|ad - bc|$. In fact any such pair is equivalent under Γ to a unique pair of the form $\{\infty, \frac{c'}{d'}\}$ where $d' = |ad - bc|$ and $0 \leq c' < d'$. It follows that for pairs $\{\frac{a}{b}, \frac{c}{d}\}$ with $|ad - bc| = 1$ or 2 the quantity $|ad - bc|$ is a complete invariant. From this remark it immediately follows that *the even, resp odd, lines are precisely the hyperbolic lines whose endpoints are a pair of rational numbers* $\{\frac{a}{b}, \frac{c}{d}\}$ *satisfying* $|ad - bc| = 1$ *resp.* 2. This gives a rational procedure for drawing $\mathcal{T}^*$.

(3.3) The Farey sequences are closely connected with the tessellation $\mathcal{I}$. Consider a hyperbolic polygon P of finite area which is bounded by even lines and which contains 0 and ∞ as two of its vertices. (The last condition is only a partial normalization.) The vertices of P all lie in $\mathbf{R} \cup \{\infty\}$ which may be identified with a circle and so these vertices have a natural cyclic order. Since P has finite area it has only finitely many vertices. In fact recall that the area of an ideal triangle is π. So the area of P is a multiple of π. If P contains k tiles of $\mathcal{I}$ then its area is $k\pi$, and it has $k+2$ vertices. So the vertices of P is a finite sequence of cyclically ordered rationals of the form

$$\{\infty,\ x_0,\ x_1,\ \ldots\ x_n\ \infty\} \tag{3.3.1}$$

where

i) x_0 and x_n are integers, and some $x_i = 0$,

ii) $x_i = \frac{a_i}{b_i}$ are rational numbers in their reduced forms and ordered according to their magnitudes, such that $|a_i b_{i+1} - b_i a_{i+1}| = 1, i = 1,\ 2,\ \ldots,\ n-1$.

A hyperbolic polygon of the above type will be called ***an ideal polygon*** and a sequence of the type (3.3.1) satisfying ***i***) and ***ii***) will be called ***a generalized Farey sequence*** or gFS for short. Thus gFS's are in 1-1 correspondence with ideal polygons. By adjoining ∞ on both sides we can consider $\mathcal{F}_n$ as a gFS. The operation of inserting a vertex while passing from $\mathcal{F}_n$ to $\mathcal{F}_{n+1}$ amounts to adjoining externally a tile of $\mathcal{I}$ to an ideal polygon corresponding to $\mathcal{F}_n$.

(3.4) Now let x be a rational number lying in $(0, 1)$. Certainly x belongs to some gFS; e. g. if $x = \frac{a}{b}$ in reduced form then x belongs to the gFS corresponding to $\mathcal{F}_b$. Equivalently

x is a vertex of some ideal polygon P. Notice that two distinct even lines never intersect. So if P_1 and P_2 are two ideal polygons each containing x as a vertex then their intersection is also an ideal polygon containing x as a vertex. Since the area of an ideal polygon is always a multiple of π it follows that *there is a unique minimal ideal polygon which contains x as a vertex.* We denote this polygon by $P_0(x)$.

Theorem *Let x be a rational number lying in (0, 1). It is possible to read the continued fraction of x from the vertices of* $P_0(x)$ *and conversely the continued fraction of x determines* $P_0(x)$. *This polygon is contained in the strip bounded by the vertical lines* $x = 0$ *and* $x = 1$. *In the notation of (2.2) the number of tiles of $\mathcal{I}$ in* $P_0(x)$ *is* $\Delta(x)$. *All the convergents y_i of x are among the vertices of* $P_0(x)$. *For $i \geq 1$ there are precisely $a_i - 1$ vertices of* $P_0(x)$ *between y_{i-2} and y_i.*

For a formal proof of this theorem see $[K]_2$, §8. (The last assertion follows from (3.5) below.) Here we work out an example which will hopefully give the reader a fair feeling about this result. Let $x = \frac{13}{42} = [0; 3, 4, 3]$. Here $\Delta(x) = 10$. So $P_0(x)$ is a union of 10 tiles of $\mathcal{I}$, and it has 12 vertices including 0, 1, and ∞, and the convergents $y_1 = \frac{1}{3}$, $y_2 = \frac{4}{13}$, and $y_3 = x = \frac{13}{42}$. The interval $(y_3, y_1) = (\frac{13}{42}, \frac{1}{3})$ contains two extra vertices $\frac{9}{29}$ and $\frac{5}{16}$. The interval $(y_0, y_2) = (0, \frac{4}{13})$ contains three extra vertices $\frac{1}{4}$, $\frac{2}{7}$ and $\frac{3}{10}$. Finally the interval $(y_1, y_{-1}) = (\frac{1}{3}, \infty)$ contains two extra vertices $\frac{1}{2}$ and 1. (A precise procedure for inserting extra terms is worked out in $[K]_2$, §8.)

For the sake of clear visualization the following picture illustrates only the combinatorial pattern and is not drawn to the scale.

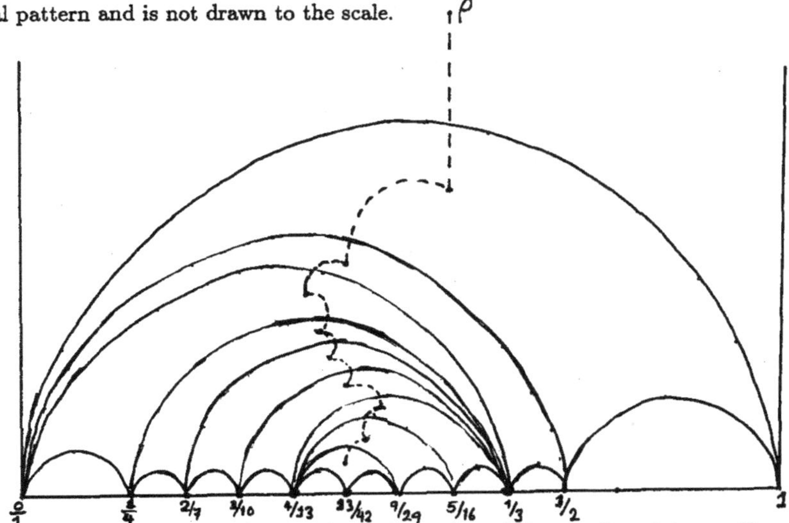

A certain path of f–edges is shown by a dotted line in the picture. One should imagine a diver who is diving from the spring-board at ρ into the "infinitely deep" sea whose bottom is covered by the pearls of real numbers. Each tile of $\mathcal{I}$ is a layer of uniform density in which a diver can have a (fairly) clear perception. If each f–edge is assigned a length $\frac{1}{2}$ then the diver will have to go the distance $\Delta(x) - 1$ before a rational number x becomes clearly visible! This should explain the terminology why $\Delta(x)$ is called the *the depth of x*. If one applies the same philosophy to an arbitrary real number in $(0, 1)$ then one gets into the area of diophantine approximations which was explained beautifully by Artin, cf. [A], in terms of the geodesic flow in the hyperbolic geometry of the surface $\mathbf{H}/\Gamma$. Going along a uniquely determined infinite branch of the cubic tree of the f–edges provides a combinatorial substitute for a geodesic, and its finite branches contain information about the rationals!!

(3.5) Cusp–width and the relation (2.2.5). Let P be an ideal polygon. Under the canonical projection $\pi : \mathbf{H} \mapsto \mathbf{H}/\Gamma$ P is mapped onto $\mathbf{H}/\Gamma$. Let x_o be a vertex of P. Now a neighborhood of x_o is wrapped around a neighborhood of the "puncture" on $\mathbf{H}/\Gamma$ an integral number of times by π. This integer is obviously just the integer which is the number of tiles of $\mathcal{I}$ which are incident with x_o. This number is called *the cuspwidth of P at* x_o. A natural question is how to compute this cuspwidth in terms of the corresponding gFS.

Proposition *Let* $x_{i-1} = \frac{a_{i-1}}{b_{i-1}}$, $x_i = \frac{a_i}{b_i}$, $x_{i+1} = \frac{a_{i+1}}{b_{i+1}}$ *be three consecutive vertices of an ideal polygon P. Then the cusp-width of P at* x_i *is* $|a_{i+1}b_{i-1} - a_{i-1}b_{i+1}|$.

Proof In view of the remarks on the invariants of Γ in its action on the pairs of rational numbers in (3.2) it is easy to see that by translating P by an appropriate element of Γ we may assume that $x_i = \infty$. Then the gFS-property of the vertices implies that x_{i-1} and x_{i+1} must be integers, so $b_{i-1} = b_{i+1} = 1$. Moreover the cusp-width at ∞ is simply $|x_{i+1} - x_{i-1}| = |a_{i+1} - a_{i-1}|$. Hence the assertion. q.e.d.

Notice that the relation (2.2.5) in the continued fraction of a rational number x involves a factor $(-1)^i$. This factor is arising from the fact that the even convergents increase and underapproximate x whereas the odd convergents decrease and overapproximate x. (With another convention about continued fraction this factor could be eliminated.) In either case the relation (2.2.5) is essentially a certain cuspwidth and it is this fact which is at the basis of the last assertion in the above theorem.

§4 Special Polygons and Farey Symbols. (4.1) We now formalize an abstract notion which is useful for constructing fundamental domains for subgroups of finite index in Γ. For this purpose in order to account for the 3-torsion in a subgroup we need to consider some special triangles. Notice that the hyperbolic triangle with vertices 0, ∞ and ρ is bounded by an even line and two odd edges making an internal angle $\frac{2\pi}{3}$. The

Γ–translates of this triangle will be called *special triangles.*

(4.2) Definition *A special polygon* is a convex hyperbolic polygon P which is a union of an ideal polygon P_0 and a certain number of special triangles which are adjoined externally to P_0 together with a *side–pairing* satisfying certain rules which are explained below.

Notice first of all that P is bounded by even lines and pairs of odd edges making an internal angle $\frac{2\pi}{3}$, and as a subset of **H** it has a canonical orientation. So its boundary ∂P also has a canonical orientation so that a bug travelling along ∂P always finds P on his or her left. After a side–pairing is specified in each pair two sides may be identified by a unique element of Γ always in an orientation–reversing manner. Here "side" has a technical meaning which is made precise below. Sofar ∂P consists of even and odd *edges* which are equipped with a canonical orientation.

A *side-pairing* is an involution on edges so that no edge is carried into itself and the following *side–pairing rules* hold.

S_1) An odd edge e is always paired to the odd edge f which makes an internal angle $\frac{2\pi}{3}$ with e. Both e and f are considered as *sides* of P, and are called its *odd sides.*

S_2) Let e, f be two even edges in ∂P forming an even line. Then either *i*) e is paired to f, in which case both e and f are considered as *sides* of P, and are called its *even sides,* or else *ii*) e, f form a *free side* of P and this free side is paired with another such free side of P.

The points of intersection of the adjacent sides including those on ∂H are called *the vertices* of P. A point of intersection of the adjacent even, resp. odd, sides is *an even resp. odd* vertex, wheras a vertex lying in ∂H is *a free* vertex. We finally assume that

S_3) 0 and ∞ are two of the vertices of P.

This concludes the definition of a *special polygon*. We emphasize that a special polygon does not contain any f-edge in its boundary.

(4.3) The general shape of a special polygon looks as in the following picture. The side–pairing is shown by side–labels.

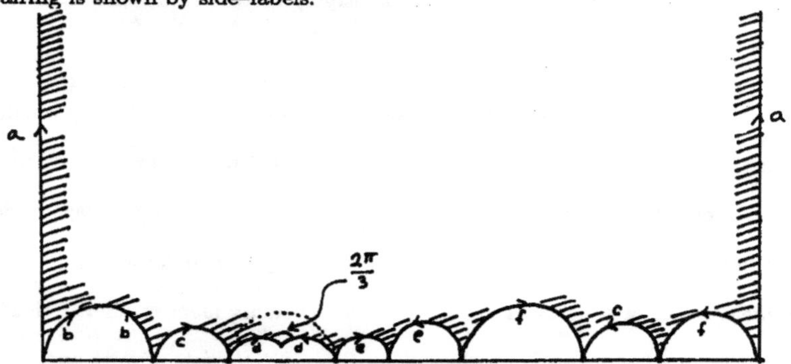

(4.4) The free vertices of a special polygon P form a gFS. We now adorn this gFS with an extra structure which encodes the side-pairing information. Suppose the gFS is given as in (3.3.1). We regard $\infty = x_{-1} = x_{n+1}$. If the complete hyperbolic geodesic joining x_i to x_{i+1}, $i = -1, 0, 1, ...n+1$ consists of two sides of P which are paired then we indicate this information by

(4.4.1) $$\underset{\circ}{\underbrace{x_i \quad x_{i+1}}}.$$

In this case we shall call $\{x_i \quad x_{i+1}\}$ *an even interval* of the gFS . If x_i and x_{i+1} are the endpoints of two odd edges which are two sides of P and which are paired then we indicate this information by

(4.4.2) $$\underset{\bullet}{\underbrace{x_i \quad x_{i+1}}}.$$

In this case we shall call $\{x_i \quad x_{i+1}\}$ *an odd interval* of the gFS. If x_i and x_{i+1} are the endpoints of a free side e of P and $x_{i'}$ and $x_{i'+1}$ are the endpoints of the free side of P paired to e then we indicate this information by

$$(4.4.3) \qquad \underbrace{x_i \quad x_{i+1}}_{a} \qquad\qquad \underbrace{x_{i'} \quad x_{i'+1}}_{a}.$$

Here a is a numerical symbol. If the $a's$ occur at all they will be numbered from 1 to some positive integer r, it being understood that different pairs of associated free sides carry different numerical symbols. Of course the specific numerical values for the labels have no significance. We shall call each of $\{x_i \quad x_{i+1}\}$ and $\{x_{i'} \quad x_{i'+1}\}$ *a free interval* of the g.F.S..

Definition (Without any reference to P,) a g.F.S. (3.3.1) adorned with an extra structure on each consecutive pair of $x_i's$ of the type (4.4.1) - (4.4.3), will be called *a Farey symbol.*

Thus a typical Farey symbol may look like

$$\{\underbrace{\infty \quad x_0}_{\circ} \underbrace{\quad x_1}_{1} \underbrace{\quad x_2}_{\bullet} \underbrace{\quad x_3}_{\circ} \underbrace{\quad x_4}_{1} \underbrace{\quad x_5}_{2} \underbrace{\quad x_6}_{\bullet} \underbrace{\quad \infty}_{2}\}$$

It is obvious that special polygons are in a natural 1–1 correspondence with the Farey symbols.

(4.5) We note a simple formula relating the length of a Farey symbol and the area of the corresponding special polygon. If there are k terms in a Farey symbol with b odd intervals then the corresponding special polygon P has k free vertices and b odd vertices. So it consists of $k-2$ ideal triangles and b special triangles. Hence its area is $\frac{1}{3}\pi(3(k-2)+b)$. After we relate a special polygon to a subgroup of finite index in Γ, cf. (5.3) below, then it will follow that the index of this subgroup will be $3(k-2)+b$.

§5 Admissible Fundamental Domains. **(5.1)** It is easy to see that the stabilizer of the hyperbolic line joining 0 to ∞ in Γ is the subgroup $< z \mapsto -\frac{1}{z} >$ which is of order 2. The same is true for the stabilizer of the hyperbolic line joining 1 to -1. From this it easily follows that the stabilizer of an even or odd edge or *oriented* even line in Γ is the identity subgroup.

Now let P be a special polygon. From the above remarks it follows that *if e, f are two sides which are paired then there is a unique element in Γ which carries e into f in an orientation–reversing manner.* The elements of Γ obtained this way will be called *the side-pairing transformations* of P. Also the subgroup of Γ generated by the side-pairing transformations will be denoted by Φ_P.

(5.2) Let Φ be a subgroup of finite index in Γ. Throughout we shall consider only those fundamental domains for Φ which are convex hyperbolic polygons consisting of the tiles of the extended modular tessellation $\mathcal{T}^*$. It is wellknown that for such a fundamental domain there are elements of Φ which identify its sides and these side–pairing transformations generate Φ. Such a fundamental domain will be called *admissible* if its side–pairing transformations form an independent set of generators in the sense of Rademacher as defined in (1.3).

(5.3) The main geometric part of our method is contained in the following theorem.

Theorem *Let P be a special polygon, and Φ_P the associated subgroup of Γ as defined in (5.1). Then P is an admissible fundamental domain for Φ_P. Conversely every subgroup Φ of finite index in Γ admits an admissible fundamental domain which is a special polygon P so that $\Phi = \Phi_P$.*

Proof (*A sketch*) The first part is a typical application of Poincare's theorem. (In fact the standard fundamental domain and the corresponding presentation for the modular group may have motivated this theorem in the first place.) In the first place the usual considerations involving simple connectivity show that the Φ_P–translates of P tessellate **H** and the tiles are in 1–1 correspondence with the elements of Φ_P. This means that P is a fundamental domain for Φ_P. Now Poincare's theorem provides a presentation for Φ_P, cf. [B] or [M]. The side–pairing transformations are the generators for Φ_P. As for the relations one has to see how the vertices are identified when we identify sides according to the side–pairing. This divides the vertices into disjoint *vertex–cycles*. The *total angle* at a vertex cycle is the sum of the interior angles at the vertices in the cycle. This angle must be of the form $\frac{2\pi}{l}$ where l is an integer, and it provides a relation of the type w^l where w is a word in the generators which occur in the succession of side–identifications which create the vertex–cycle in the first place. If $l = \infty$ i. e. the total angle is zero then there is no nontrivial relation. For a special polygon the vertex–cycle of an even or odd vertex consists of the vertex itself and provides the relations of the form w^2 resp. w^3 where w is a single generator. On the other hand a cycle of free vertices of P has total angle zero so these vertex–cycles do not lead to a nontrivial relation at all. In other words the side–pairing transformations are independent, and so P is an admissible fundamental domain for Φ_P.

For the converse part we start with $\mathbf{H}/\Phi$. This surface comes equipped with a tessellation induced from $\mathcal{T}^*$, and so we can talk of even and odd vertices, edges ... etc. One notices that the subgraph of f–edges is actually a deformation retract of the surface. Choose a maximal tree, say T, in this subgraph. One shows that cutting along certain even or odd edges incident with the terminal vertices of T one can cut the surface open into a space which is abstractly isometric to an ideal polygon, say P. (Notice that one *never* cuts along the f–edges in this process.) Now appropriately isometrically developing P in $\mathbf{H}$ and defining the side–pairing one obtains a special polygon which is an admissible fundamental domain for Φ. See $[K]_2$, §3 for details. q.e.d.

(5.4) We shall illustrate the above theorem in the case of some subgroups of small index. The reader may find it of interest to compare the fundamental domains so obtained with the ones which appear in the standard texts on this subject. Notice in the first place however that special polygons or Farey symbols provide an efficient method of constructing subgroups of finite index in Γ together with an independent system of generators already given in a matrix form.

The standard fundamental domain $\mathcal{D}$ for Γ is *not* admissible. An admissible one is given by

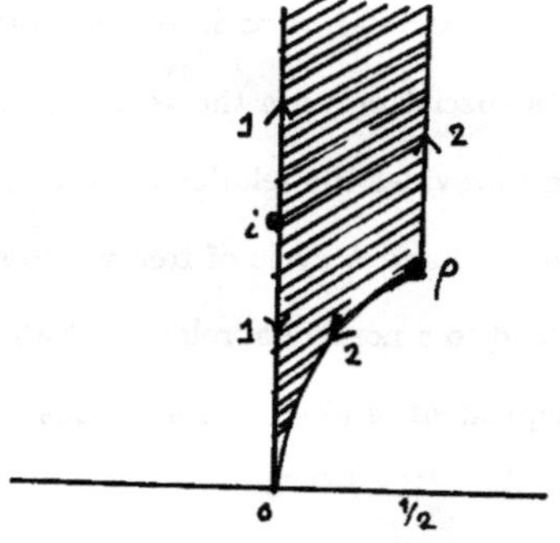

The corresponding Farey symbol is $\{\underbrace{\infty, 0}_{\circ}, \underbrace{\,\infty}_{\bullet}\}$.

There is a unique subgroup of index 2 in Γ. An admissible fundamental domain for this subgroup is

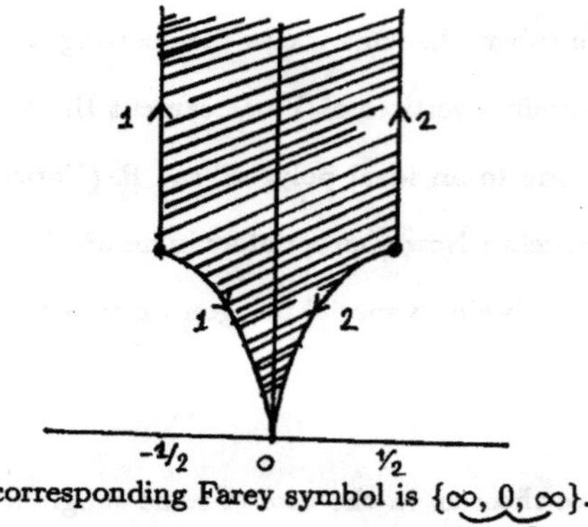

The corresponding Farey symbol is $\{\underbrace{\infty, 0}_{\bullet}, \underbrace{\,\infty}_{\bullet}\}$.

There is a unique normal subgroup of index 3 in Γ. An admissible fundamental domain for this subgroup is

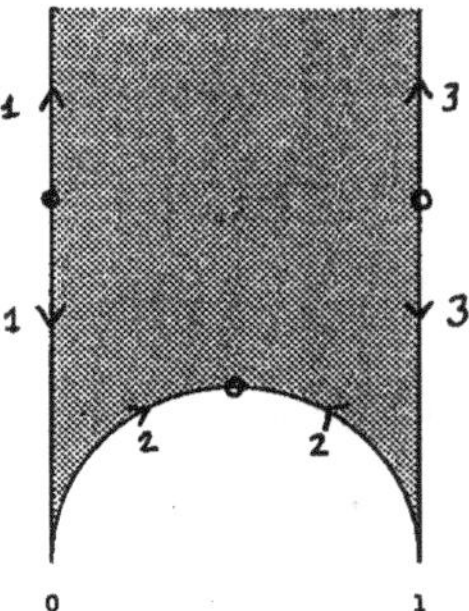

The corresponding Farey symbol is $\{\infty, 0, 1, \infty\}$. Its independent generators are

$$< z \mapsto -\frac{1}{z},\ z \mapsto \frac{z-1}{2z-1},\ z \mapsto \frac{z-2}{z-1} > .$$

There are three non–normal subgroups of index 3 in Γ. These are $\Gamma^0(2)$, $\Gamma_0(2)$ and the socalled Jacobi's θ–subgroup Γ_θ. They form a single conjugacy class. Admissible fundamental domains for these subgroups are

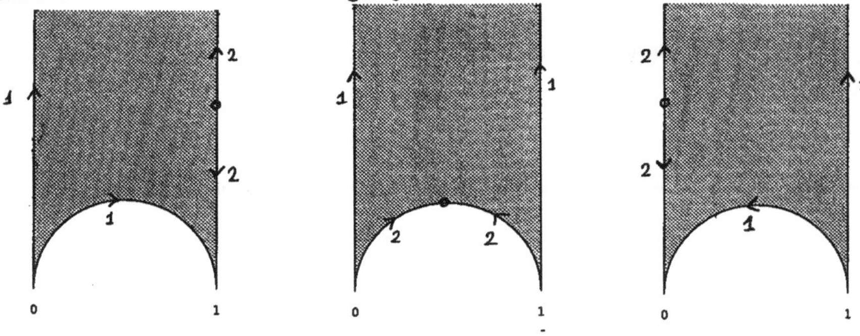

The corresponding Farey symbols are

$$\{\underbrace{\infty, 0}_{1}, \underbrace{\,1}_{1}, \underbrace{\,\infty}_{0}\}, \qquad \{\underbrace{\infty, 0}_{1}, \underbrace{\,1}_{0}, \underbrace{\,\infty}_{1}\}, \qquad \{\underbrace{\infty, 0}_{0}, \underbrace{\,1}_{1}, \underbrace{\,\infty}_{1}\}.$$

The corresponding independent systems of generators are

$$\begin{aligned}
\Gamma_\theta &= < z \mapsto -\frac{1}{z},\ z \mapsto \frac{2z-1}{z} >, \\
\Gamma^0(2) &= < z \mapsto \frac{-z}{z-1},\ z \mapsto \frac{z-2}{z-1} >, \\
\Gamma_0(2) &= < z \mapsto z+1,\ z \mapsto \frac{z-1}{2z-1} > .
\end{aligned}$$

The subgroup $\Gamma(2)$, i. e. the principal congruence subgroup of level 2, has index 6 and has no torsion. In particular its fundamental domain must have the area 2π. So a special polygon for it must be a union of two tiles of $\mathcal{I}$. A simple trial and error shows that the special polygon

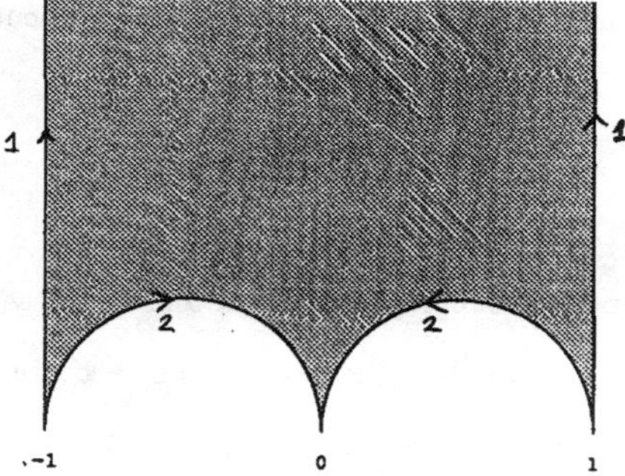

has area 2π and the side–pairing transformations are

$$z \mapsto 2z, \qquad z \mapsto \frac{z}{-2z+1}.$$

Since these lie in $\Gamma(2)$ we conclude that the above special polygon is an admissible fundamental domain for $\Gamma(2)$. The form of the domain clearly shows that $\Gamma(2)$ has 3 cusps and genus 0, i. e. $\mathbf{H}/\Gamma(2)$ is a sphere with three punctures. The corresponding Farey symbol here is $\{\underbrace{\infty, -1}_{1}\underbrace{, 0}_{2}\underbrace{, 1}_{2}\underbrace{, \infty}_{1}\}$.

As a final example consider the commutator subgroup Γ' of Γ. From simple algebraic considerations we know that this subgroup also has index 6 and has no torsion. So a special polygon for it must be again a union of two tiles of $\mathcal{I}$. We can easily compute some elements of Γ'. For example if A and B are as in (1.1.5) then

$$\mathrm{C} = (\mathrm{A},\mathrm{B}) = \begin{pmatrix} 2 & 1 \\ 1 & 1 \end{pmatrix},$$

$$\mathrm{D} = (\mathrm{A},\mathrm{B}^2) = \begin{pmatrix} 1 & 1 \\ 1 & 2 \end{pmatrix}.$$

Since C carries 0 to 1 and ∞ to 2 we might try the following special polygon

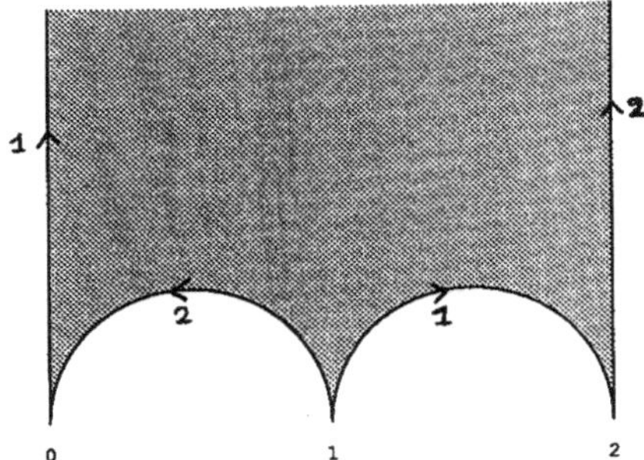

This special polygon has the side–pairing transformations

$$\mathrm{C}: z \mapsto \frac{2z+1}{z+1}, \quad \mathrm{E}: z \mapsto \frac{3z-1}{z}.$$

One easily checks that $\mathrm{E} = \mathrm{CD}^{-1}$. So these side–pairing transformations lie in Γ'.

Conclusion: this is the special polygon for Γ'. The form of the domain clearly shows the handle! So Γ' has one cusp and genus 1, i. e. $\mathbf{H}/\Gamma'$ is a torus with one puncture. The corresponding Farey symbol here is $\{\underbrace{\infty, 0}_{1}, \underbrace{1}_{2}, \underbrace{2}_{1}, \underbrace{\infty}_{2}\}$.

(5.5) We make some remarks on the computations of the side–pairing transformations in a special polygon P. For simplicity let us call a transformation *free, even, or odd* if it pairs free, even, or odd sides respectively. The transformation $z \mapsto \frac{az+b}{cz+d}$ carries ∞ to $\frac{a}{c}$ and 0 to $\frac{b}{d}$. So if $x = \frac{b}{d}$, $y = \frac{a}{c}$, $u = \frac{q}{s}$, $v = \frac{p}{r}$ satisfy $ad - bc = 1$ and $ps - qr = 1$ then the unique element in Γ which carries x into u and y into v is QM^{-1} where

$$Q = \begin{pmatrix} p & q \\ r & s \end{pmatrix}, \quad and \ M = \begin{pmatrix} a & b \\ c & d \end{pmatrix}. \tag{5.5.1}$$

From this remark it is easy to determine the free and even side–pairing transformations. For example an even generator which identifies the two even edges contained in the even line with endpoints $\frac{b}{d}$, $\frac{a}{c}$ satisfying $ad - bc = 1$ is

$$\begin{pmatrix} ac - bd & -a^2 - b^2 \\ c^2 + d^2 & -ac + bd \end{pmatrix},$$

It is slightly tricky to determine the odd generators. Notice that an odd generator pairs two odd edges making an internal angle $\frac{2\pi}{3}$ at an odd vertex. The two free vertices on these odd edges are endpoints of an even line. Let these endpoints be $\frac{b}{d}$, $\frac{a}{c}$. We know that the transformation A : $z \mapsto \frac{az+b}{cz+d}$ carries ∞ to $\frac{a}{c}$ and 0 to $\frac{b}{d}$. Now 0, ∞, ρ form a model special triangle in the sense of (4.1). Here ρ is a point of intersection of the odd lines joining ∞ to $\frac{1}{2}$ and 0 to 2. Since A carries $\frac{1}{2}$ to $\frac{a+2b}{c+2d}$ and 2 to $\frac{2a+b}{2c+d}$ it follows that it maps ρ onto the point of intersection of the odd lines joining $\frac{a}{c}$, $\frac{a+2b}{c+2d}$ and $\frac{b}{d}$, $\frac{2a+b}{2c+d}$. This point of intersection is an odd vertex of P. Thus the odd transformation fixing this odd vertex and pairing the incident odd edges is the transformation which carries the odd line with endpoints $\frac{b}{d}$, $\frac{2a+b}{2c+d}$ into the one with endpoints $\frac{a}{c}$, $\frac{a+2b}{c+2d}$ in fact carrying $\frac{b}{d}$ into $\frac{a}{c}$ and $\frac{2a+b}{2c+d}$ into $\frac{a+2b}{c+2d}$. Now A carries ∞, $\frac{1}{2}$ to $\frac{a}{c}$, $\frac{a+2b}{c+2d}$ resp., whereas B: $z \mapsto \frac{bz-a-b}{dz-c-d}$ carries ∞, $\frac{1}{2}$ to $\frac{b}{d}$, $\frac{2a+b}{2c+d}$ resp. (Note that *det* B = 1 so B is in Γ.) So our required odd generator is AB^{-1}. As a matrix it is given by

$$\begin{pmatrix} -ac - ad - bd & a^2 + ab + b^2 \\ -c^2 - cd - d^2 & ac + bc + bd \end{pmatrix}.$$

§6 A Graph–theoretic Method. (6.1) There is a neat graph–theoretic method to understand the totality of subgroups of finite index in Γ. It is based on the cubic tree of f-edges. Let P be a special polygon and let T denote the graph of all the f-edges contained in P. Since T is a connected subgraph of the cubic tree it is itself a finite tree. Also T intersects ∂P only in its terminal vertices. In fact T has exactly one terminal vertex on each even line in ∂P and also each odd vertex in ∂P is also a terminal vertex of T. The orientation of P induces a *cyclic order* on the trivalent vertices of T all of which lie in the interior of P. Finally the side–pairing of P induces a natural *involution* on the terminal vertices of T. These properties are abstracted in a notion of a *tree diagram* developed below. On the other hand by identifying the terminal vertices which are paired by the involution one gets a certain graph. Its properties are abstracted in a notion of a *bipartite cuboid graph.*

(6.2) *A tree diagram* is a finite tree T such that *i*) all the internal vertices are of valence 3, *ii*) there is a prescribed cyclic order on the edges incident at each trivalent vertex, *iii*) the terminal vertices are partitioned into two possibly empty subsets R and B where the vertices in R (resp. B) are called red (resp. blue) vertices, *iv*) there is an involution σ on R.

A bipartite cuboid graph is a finite graph whose vertex set is divided into two disjoint subsets V_0 and V_1 such that *i*) every vertex in V_0 has valence 1 or 2, *ii*) every vertex in V_1 has valence 1 or 3, *iii*) there is a prescribed cyclic order on the edges incident at each trivalent vertex in V_1, *iv*) every edge joins a vertex in V_0 with a vertex in V_1.

The equivalence of two tree diagrams or two bipartite cuboid graphs is defined in an obvious way.

(6.3) Obviously a special polygon or equivalently a Farey symbol gives rise to a tree diagram or a bipartite cuboid graph as explained in (6.1).

Starting from a tree diagram one obtains a bipartite cuboid graph by identifying the red vertices which are related by the involution and introducing "dummy" vertices of valence 2 on the edges joining two trivalent vertices or joining a trivalent vertex to a blue vertex. Conversely starting from a bipartite cuboid graph we can cut it along a minimal set of vertices of valence 2 so as to obtain a tree which can be turned into a tree diagram in an obvious way.

Finally let us start with a tree diagram T. Consider its trivalent vertices and blue vertices as odd vertices. As above introduce "dummy" vertices of valence 2 on the edges joining two trivalent vertices or joining a trivalent vertex to a blue vertex and consider these bivalent vertices as well as the red vertices as even vertices. Now imbed this tree into the tree of f-edges so that each edge joining an even vertex with an odd vertex is mapped onto an f-edge taking care that *i*) even resp. odd vertices of T go into even resp. odd vertices of $\mathcal{T}^*$, *ii*) the f-edge joining i to ρ is in the image and *iii*) the cyclic order at the trivalent vertices in the image matches with the induced orientation. It is clear that such an embedding is unique up to ambient isotopy and in fact if we equip T with a metric in which each edge joining an even vertex with an odd vertex has the same length as an f-edge then this embedding is unique once the image of one edge is fixed. By appropriately introducing even lines and odd edges at the terminal vertices of this embedding we obtain a special polygon.

(6.4) It is fairly clear from the above discussion that these four classes of objects – special polygons, Farey symbols, tree diagrams, and bipartite cuboid graphs with appropriate notions of equivalence – are in finite-to-one correspondence with each other. Moreover – and this is the main point –

Theorem *The conjugacy classes of subgroups of finite index in* Γ *are in 1–1 correspondence with the equivalence classes of bipartite cuboid graphs.*

We shall not discuss the proof here, cf. $[K]_2$. An earlier version of this result in a quite different language appears as a special case of a more general result on non–cocompact finitely genetrated fuchsian groups, cf. $[K]_1$. (As abstract groups these latter groups are just finite free products of finitely many cyclic groups.) In a completely different context this result has come up in the computation of the volumes of the moduli spaces of Riemann surfaces by R. Penner, cf. [P].

(6.5) We draw some tree diagrams corresponding to the examples discussed in (5.4). The tree diagram for Γ is

o———●

For the unique subgroup of index 2 it is

●———●

There are two tree diagrams for subgroups of index 3, one corresponding to the normal subgroup and the other corresponding to the three non–normal ones which form a single conjugacy class.

resp.

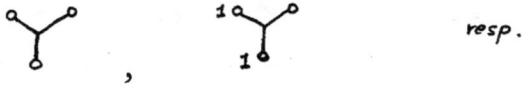

Finally the tree diagrams corresponding to $\Gamma(2)$ and Γ' are

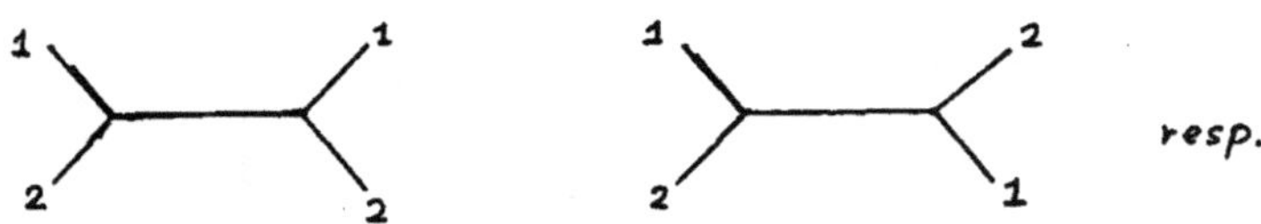

§7 Fundamental Domains for $\Gamma_0(N)$, $\Gamma(N)$, and $\Gamma^1(N)$. (7.1) First of all the index d of $\Gamma_0(N)$ in Γ is given by

$$d = N \prod_{p|N} (1 + \frac{1}{p}).$$

where p runs over all primes dividing N. Let $\mathbf{X}_N$ denote $\mathbf{H}/\Gamma_0(N)$. The genus (g), the number of cusps (t), and the number of branch points of order 2 and 3 (a and b) of $\mathbf{X}_N$ as functions of N are wellknown, cf. [S], ch. 4. Let $r = 2g + t - 1$. Then $\Gamma_0(N)$ is a free product of a copies of $\mathbf{Z}_2$, b copies of $\mathbf{Z}_3$, and r copies of $\mathbf{Z}$. These quantities are related by the Riemann–Hurwitz formula

$$d = 3a + 4b + 6r - 6 = 3a + 4b + 12g + 6t - 12.$$

As may be seen from the account in [S] the values of a and b can be read from the prime factorization of N, and the congruences of the primes in this factorization *mod* 4 and 3. As is wellknown these congruences contain the information about the solvability of the equations $x^2 + y^2 \equiv 0 \ (mod\, \mathrm{N})$ and $x^2 + xy + y^2 \equiv 0 \ (mod\, \mathrm{N})$ respectively in coprime integers. The actual solutions of these equations enter in the construction of the admissible fundamental domains for $\Gamma_0(N)$.

(7.2) For constructing the admissible fundamental domains for $\Gamma_0(N)$ we shall see that we actually do not need the values of g and t. Once we know d, a, and b (and hence r by the Riemann–Hurwitz formula) the following procedure starts.

Let

$$n = a + b + 2r - 2. \tag{7.2.1}$$

A gFS of the form

$$\{\infty,\ 0 = x_0,\ x_1,\ \dots,\ x_n = 1,\ \infty\}, \tag{7.2.2}$$

where $x_i = \frac{a_i}{b_i}$ (reduced fractions with a_i, b_i nonnegative) is said to be *semibalanced for* N *if*

i) there are a *values* i, $0 \leq i \leq n-1$, *s.t.*

$$b_i^2 + b_{i+1}^2 \equiv 0\ (N),$$

ii) there are b *values* i, $0 \leq i \leq n-1$, *s.t.*

$$b_i^2 + b_i b_{i+1} + b_{i+1}^2 \equiv 0\ (N),$$

iii) the remaining $2r-2$ *values of* i, $0 \leq i \leq n-1$, *are paired* $i \leftrightarrow i^*$ *s.t.*

$$b_i b_{i^*} + b_{i+1} b_{i^*+1} \equiv 0\ (N).$$

The significance of this notion is the following result.

(7.3) Theorem *Let* $N \geq 2$ *and* n *as in (13.1.1). Then there exists a gFS in the form (7.2.2) which is semi-balanced for* N. *Moreover there exists a canonical structure of a Farey symbol on this gFS such that the corresponding special polygon is an admissible fundamental polygon for* $\Gamma_0(N)$.

The idea of the proof is as follows. The gFS is made into a Farey symbol so that there are a even intervals, b odd intervals, and the remaining intervals are paired in the way suggested in the definition of semi-balance for N . (The interval $\{\infty, 0\}$ is paired with $\{\infty, 1\}$. This ensures that ∞ is a cusp of X_N of width 1. This is as it should be since $z \mapsto z + 1$ belongs to $\Gamma_0(N)$.) The conditions of semibalance for N ensure that the side–pairing transformations do lie in $\Gamma_0(N)$. This implies that the subgroup defined by this special polygon indeed is a subgroup of $\Gamma_0(N)$. Finally the value of n is so arranged that this subgroup has the same index as $\Gamma_0(N)$, cf. the remark in (4.5). So the subgroup defined by this special polygon is $\Gamma_0(N)$.

(7.4) Let us illustrate this result for N = 3, 4, 5.

The case $\Gamma_0(3)$: The index d of $\Gamma_0(3)$ is 4. The existence of $z \mapsto z + 1$ in all $\Gamma_0(N)$ which pairs $\{\infty, 0\}$ with $\{\infty, 1\}$ shows that we have always $r \geq 1$. The Riemann–Hurwitz formula now already implies that $a = 0$, $b = 1$ and $r = 1$. So $n = 1$, i. e. the Farey symbol is just $\{\underbrace{\infty, 0}_{1}, \underbrace{0, 1}_{\bullet}, \underbrace{1, \infty}_{1}\}$. The corresponding special polygon and the tree diagram are

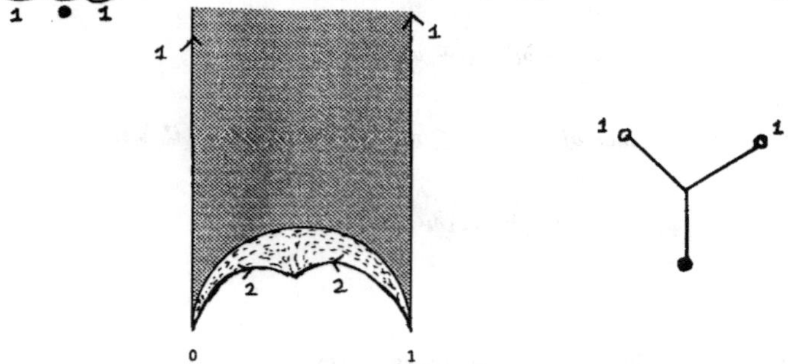

The transformation pairing the odd sides is, cf. (5.5), $\begin{pmatrix} -1 & 1 \\ -1 & 0 \end{pmatrix}$. So $\Gamma_0(3)$ is a free product of $< \begin{pmatrix} 1 & 1 \\ 0 & 1 \end{pmatrix} >$ and $< \begin{pmatrix} -1 & 1 \\ -1 & 0 \end{pmatrix} >$.

The case $\Gamma_0(4)$: Here $d = 6$. One can compute by a procedure given in [S] that $a = b = 0$. (This also shows up in the fact that there is no solution of $x^2 + y^2 \equiv 0 \ (mod\, 4)$

or $x^2 + xy + y^2 \equiv 0 \ (mod\, 4)$ in coprime integers x and y.) So by the Riemann–Hurwitz formula we have $r = 2$. So $n = 2$. The only possibility for a gFS is $\{\infty,\ 0,\ \frac{1}{2},\ 1,\ \infty\}$. Since we know that $a = b = 0$ and that $\{\infty, 0\}$ is paired with $\{\infty, 1\}$ we see that the corresponding Farey symbol must be $\{\underbrace{\infty,\ 0}_{1},\ \underbrace{}_{2}\!\!\frac{1}{2},\ \underbrace{}_{2}1,\ \underbrace{}_{1}\infty\}$. The corresponding special polygon and the tree diagram are

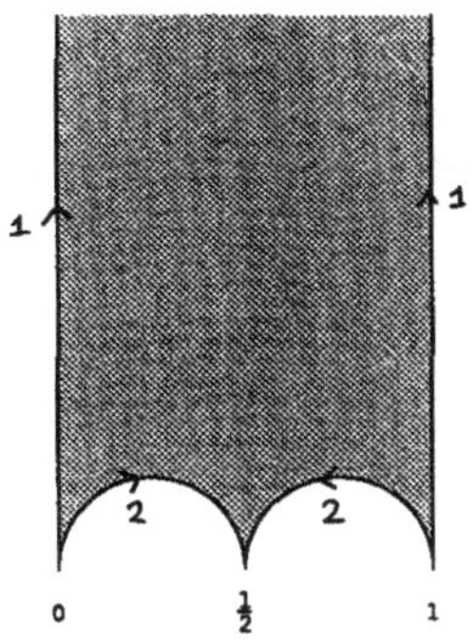

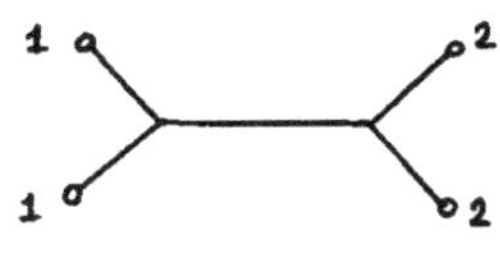

Again by using (5.5) we see that $\Gamma_0(4)$ is a free group on two generators which may be taken to be $\begin{pmatrix} 1 & 1 \\ 0 & 1 \end{pmatrix}$ and $\begin{pmatrix} 1 & -1 \\ 4 & -3 \end{pmatrix}$.

The case $\Gamma_0(5)$: The index of $\Gamma_0(5)$ is 6. Since $5 \equiv 1 \ mod\ 4$ but $\equiv -1 \ mod\ 3$ one concludes $a = 2,\ b = 0$. (Note also that $1^2 + 2^2 = 5$.) It follows by the Riemann–Hurwitz formula that $r = 1$. So $n = 2$. This determines the Farey symbol uniquely as

$\{\underbrace{\infty,\ 0}_{1},\ \underbrace{}_{\circ}\!\!\frac{1}{2},\ \underbrace{}_{\circ}1,\ \underbrace{}_{1}\infty\}$.

The corresponding special polygon and the tree diagram are

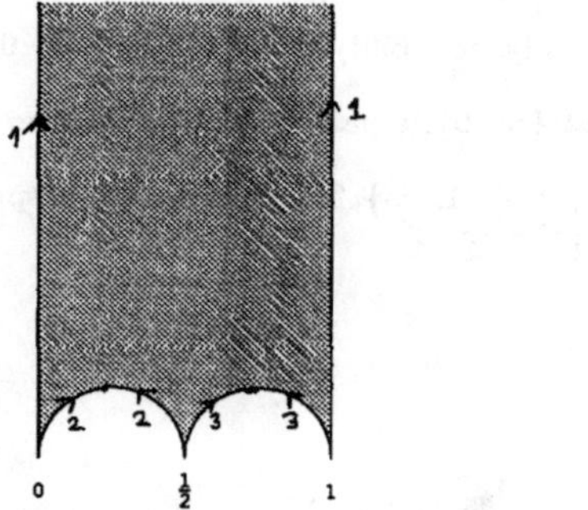

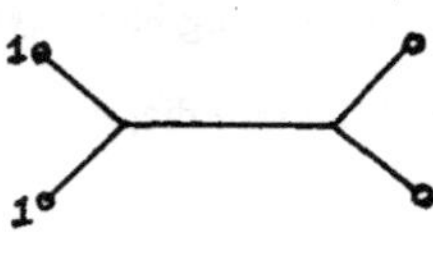

We also see that $\Gamma_0(5)$ is a free product of three factors $<\begin{pmatrix}1 & 1\\ 0 & 1\end{pmatrix}>$, $<\begin{pmatrix}2 & -1\\ 5 & -2\end{pmatrix}>$, and $<\begin{pmatrix}3 & -2\\ 5 & -3\end{pmatrix}>$.

In general of course we may not be lucky to have apriori arguments leading to unique choices. There is some trial and error involved. It seems possible to write a computer program for generating Farey symbols. By hand and without much effort the author computed the Farey symbols for all N≤ 25 and all primes $p \leq 100$. It would be of interest to divise better number-theoretic algorithms reducing the trial and error in the above procedure.

(7.5) The cases of $\Gamma(\mathrm{N})$ and $\Gamma^1(\mathrm{N})$ go together. A substitute for a semibalanced gFS in these cases is the following notion. The gFS (7.2.2) is *balanced for N* if for each $i = 0, 1, \ldots, n-1$, there exists an $i^* = 0, 1, \ldots, n-1$, such that either

$$b_i \equiv b_{i^*+1}\ (\mathrm{N}), \text{ and } b_{i+1} \equiv -b_i^*\ (\mathrm{N}),$$

or

$$b_i \equiv -b_{i^*+1}\ (\mathrm{N}), \text{ and } b_{i+1} \equiv b_i^*\ (\mathrm{N}).$$

Here n is even and (i, i^*) are $\frac{n}{2}$ disjoint pairs containing all vertices except $x_n = 1$. We equip the structure of a Farey symbol on the g.F.S. given by (7.2.2) by pairing the intervals

$\{\infty,\ 0\}$ with $\{1,\ \infty\}$ and $\{x_i,\ x_{i+1}\}$, $0 \le i \le n-1$ with the corresponding $\{x_{i^*},\ x_{i^*+1}\}$. We shall also call this Farey symbol *balanced for* N.

Theorem *Let $N \ge 4$ and*

$$n = \frac{N^2}{6}\prod_{p|N}(1-\frac{1}{p^2}). \tag{7.5.1}$$

Then there exists a gFS in the form (7.2.2) which is balanced for N and the corresponding Farey symbol is a Farey symbol for $\Gamma^1(N)$. Moreover consider the gFS

$$y_{ij} = x_i + j,\ 0 \le i \le n,\ \ 0 \le j \le N-1. \tag{7.5.2}$$

Then this gFS can be equipped with the structure of a Farey symbol for $\Gamma(N)$.

For a proof see §12 of $[K]_2$. The argument is similar to but more elaborate than the one for $\Gamma_0(N)$. The cases N = 2 and 3 require a different treatment due to the fact that in these cases $\Gamma^1(N)$ contains torsion. We illustrate the result for the case N = 4. In $[K]_2$, appendix 2 the Farey symbols for $\Gamma^1(N)$ are worked out for N$\le$ 12. See also appendix 5 there and the next section for a different more efficient approach.

The case $\Gamma^1(4)$ *and* $\Gamma(4)$. We have $\Gamma_0(4) = \Gamma^1(4)$ so we already know a gFS for it from (7.4). The above result says that from the union of this gFS and its three successive translates by $z \mapsto z+1$ we can construct a Farey symbol for $\Gamma(4)$. One easily sees that this Farey symbol is $\{\underbrace{\infty,\ 0}_{1},\ \underbrace{\ ,\tfrac{1}{2}}_{2},\ \underbrace{\ ,1}_{3},\ \underbrace{\ ,\tfrac{3}{2}}_{3},\ \underbrace{\ ,2}_{4},\ \underbrace{\ ,\tfrac{5}{2}}_{4},\ \underbrace{\ ,3}_{5},\ \underbrace{\ ,\tfrac{7}{2}}_{5},\ \underbrace{\ ,4}_{2},\ \underbrace{\ ,\infty}_{1}\}$.

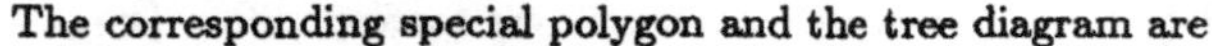

The corresponding special polygon and the tree diagram are

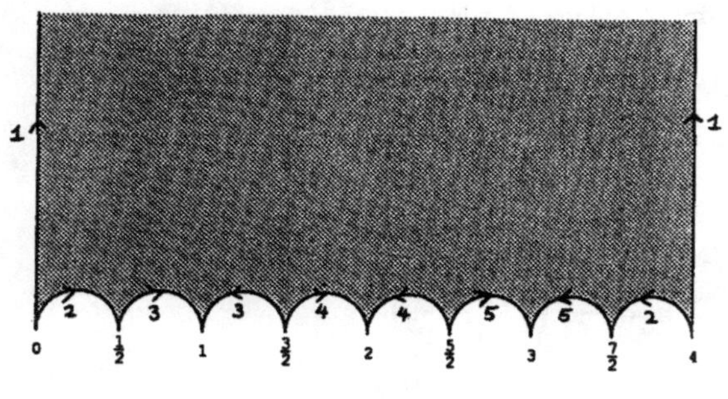

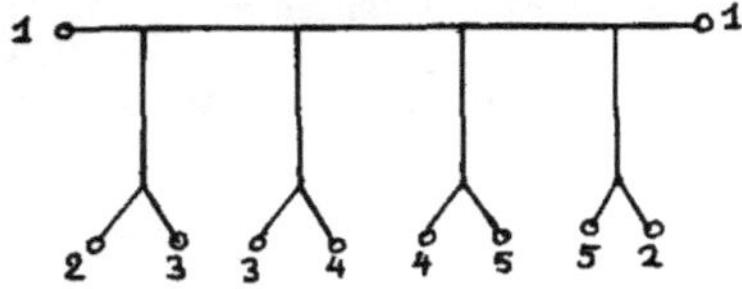

Finally $\Gamma(4)$ is a free group on five generators which may be taken to be

$$\begin{pmatrix} 1 & 1 \\ 0 & 1 \end{pmatrix}, \begin{pmatrix} 1 & -4 \\ 4 & -15 \end{pmatrix}, \begin{pmatrix} 3 & -4 \\ 4 & -5 \end{pmatrix},$$

$$\begin{pmatrix} 7 & -16 \\ 4 & -9 \end{pmatrix}, \begin{pmatrix} 11 & -36 \\ 4 & -13 \end{pmatrix}.$$

§8 Fundamental Domains for Subgroups of Γ^*. (8.1) Let Φ^* be a subgroup of Γ^* and $\Phi = \Gamma \cap \Phi^*$. Then $(\Phi^* : \Phi) \leq 2$ and if it equals 2 then $\mathbf{H}/\Phi$ admits an antoholomorphic involution θ such that $< \theta > \backslash \mathbf{H}/\Phi = \mathbf{H}/\Phi^*$. A natural question is to study the canonical projection $\mathbf{H}/\Phi \to \mathbf{H}/\Phi^*$ and extend the previous theory to finding "good" fundamental domains for Φ^*. This is of some importance even if one is primarily interested in the subgroups of Γ —for, the normalizer of a subgroup of Γ in Γ^* may be strictly larger than that in Γ and may contain some interesting information. This is true in particular for $\Gamma_0(N)$, $\Gamma(N)$, and $\Gamma^1(N)$ all of which are normalized by $z \mapsto -\bar{z}$. We shall use the following notation.

$$\Gamma^*(N) = <\Gamma(N),\ z \mapsto -\bar{z}>, \quad \Gamma_0^*(N) = <\Gamma_0(N),\ z \mapsto -\bar{z}>,$$

$$\Gamma^{1*}(N) = <\Gamma^1(N),\ z \mapsto -\bar{z}> .$$

(8.2) Let Φ^*, Φ and θ be as in (8.1). An antiholomorphic map of a Riemann surface is called *a reflection* if it has fixed points. It is easy to see that if θ is a reflection on $\mathbf{H}/\Phi$ then Φ^* itself contains reflections. Now an easy analysis shows that there are precisely two conjugacy classes of reflections in Γ^*, namely *even reflections* which fix an even line and which are conjugate to $z \mapsto -\bar{z}$, and *odd reflections* which fix an odd line and which are conjugate to $z \mapsto 1-\bar{z}$. On the other hand in a subgroup of Γ^* in general there are several conjugacy classes of reflections which are roughly speaking classified by the boundary components of $\mathbf{H}/\Phi^*$. A detailed analysis shows that there are precisely six possible types for the neighborhoods of boundary components of $\mathbf{H}/\Phi^*$.

These are depicted in the following pictures.

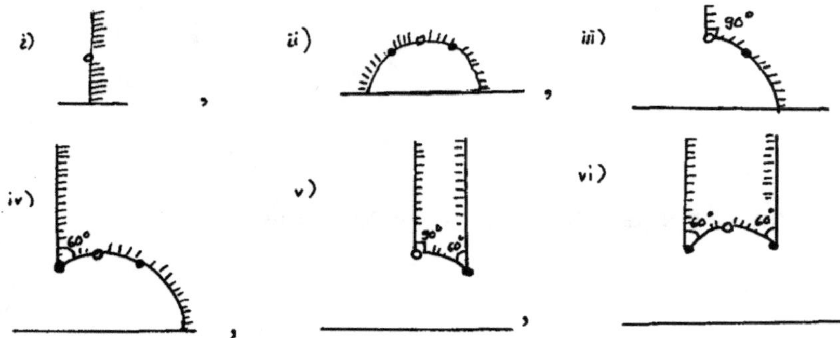

The fifth type occurs only for Γ^*. The sixth type occurs only for a particular subgroup Φ_0^* which has index 2 in Γ^* and which is the group generated by reflections in the edges of the hyperbolic triangle with vertices ∞, ρ, and ρ^2. The other four types are the general types. The first and second ones are the neighborhoods of an even and odd line respectively. The third is a neighborhood of an even edge making an angle $\frac{\pi}{2}$ with a geodesic segment consisting of an f-edge and an odd edge, whereas the fourth is a neighborhood of an odd edge making an angle $\frac{\pi}{3}$ with a geodesic segment consisting of two f-edges and an odd edge.

(8.3) Let Φ^*, Φ and θ be as in (8.1), and suppose that θ is a reflection. We also suppose that Φ^* is different from Γ^* or Φ_0^* which was discussed in (8.2). An interesting question here is *whether it is possible to lift θ to a reflection of a suitable fundamental domain for Φ*. (This question first arose when we noticed experimentally that it was possible to construct a special polygon with a reflection symmetry for $\Gamma_0(p)$ where p is a prime but not necessarily so for $\Gamma_0(N)$ where N is not a prime. The first such case is N = 8.) The above analysis makes it fairly clear that *if Φ^* is such that $\mathbf{H}/\Phi^*$ has two boundary components which are odd lines then it will not admit a special polygon with a reflection symmetry*. This then suggests that we should modify the notion of a special polygon. An appropriate modification is a following notion.

Let us call an *ideal *-polygon* a convex hyperbolic polygon of finite area which is bounded by even as well as odd lines. Moreover in addition to the special triangles considered in (4.1) also consider the *special *-triangles* of the following two types. The first –*even*– type is a Γ^*–translate of a hyperbolic triangle with vertices ∞, i, and 1. The second –*odd*– type is a Γ^*–translate of a hyperbolic triangle with vertices ∞, ρ^2, and 1.

*A special *-polygon* is convex hyperbolic polygon P^* which is a union of an ideal *-polygon $P^*_\circ$ and a finite number of special triangles and special *-triangles which are adjoined externally to $P_\circ$ all along even lines. Moreover there is a *side–pairing* satisfying the following rules in addition to the ones for a special polygon which are described below.

An element γ in Γ^* is said to *pair* e and f if $\gamma(e) = f$ or $\gamma(f) = e$ and $\gamma(int\,P) \cap int\,P = \phi$. Note that if γ is in Γ then the pairing is orientation–reversing, and if γ is a conjugation then the pairing is orientation–preserving. Notice also that given two even (resp. odd) lines e and f in ∂P there is a unique element in Γ which pairs them in an orientation–reversing way, and a unique conjugation in Γ^* which pairs them in an orientation–preserving way,

The side–pairing for a special *-polygon for even or odd lines also involves a specification whether it preserves or reverses the orientation. (Notice that now it is possible to pair an even or odd line to itself of course necessarily by a reflection.) Moreover a geodesic segment on the boundaries of a special *-triangle is paired with itself. (So the corresponding side–pairing transformation for such a segment is necessarily a reflection.)

(8.4) If P^* is a special $*$-polygon then by $\Phi^*_{P^*}$ we denote the subgroup of Γ^* generated by the side–pairing transformations. Notice however that we cannot expect these side–pairing transformations to form an independent system of generators in the sense of Rademacher. For the transformations arising from the boundary of an even special $*$-triangle form a $\mathbf{Z}_2 \times \mathbf{Z}_2$, whereas those arising from the boundary of an odd special $*$-triangle form a S_3, i. e. the symmetric group on three letters. So a reasonable substitute for Rademacher's notion of independent system of generators appears to be the following.

A system of generators $< x_i >$, $i \in I$ for a group is said to be *quasi–independent* if the only relations are of the following types

$$(8.4.1) \qquad a)\ x_i^2,\ x_j^3,\quad b)\ x_k^2,\ x_l^2,\ (x_k x_l)^2,\quad c)\ x_u^2,\ x_v^2,\ (x_u x_v)^3,$$

where all subscripts $i, j, \ldots$ are distinct. Obviously a group admits a quasi–independent system of generators iff it is isomorphic to a free product of groups isomorphic to $\mathbf{Z}_2$, $\mathbf{Z}_3$, $\mathbf{Z}$, $\mathbf{Z}_2^2$ and S_3.

(8.5) The following then is an analogue of (5.3) for the subgroups of Γ^* which is proved in an analogous manner.

Theorem *Let P^* be a special $*$-polygon, and $\Phi^*_{P^*}$ the subgroup of Γ^* generated by the side–pairing transformations. Then $\Phi^*_{P^*}$ is subgroup of finite index in Γ^*, P^* is a fundamental domain for $\Phi^*_{P^*}$, and the side-pairing transformations form a quasi-independent system of generators for $\Phi^*_{P^*}$. Conversely if Φ^* is a subgroup of finite index in Γ^* but $\neq \Gamma^*$ or Φ^*_0, cf. (8.2), then it admits a fundamental domain which is a special $*$-polygon. There are only finitely many choices of such fundamental domains. In particular Φ^* is an internal free product of subgroups isomorphic to $\mathbf{Z}_2$, $\mathbf{Z}_3$, $\mathbf{Z}$, $\mathbf{Z}_2^2$ and S_3.*

(8.6) Corresponding to the notion of a special *-polygon there are notions of a *-gFS and a *-Farey symbol which are more or less obvious modifications of the notions of a gFS and a Farey symbol. Namely in the notation of (3.3) in a *-gFS we allow for the possibility that $|a_ib_{i+1} - b_ia_{i+1}| = 1$, or 2. Moreover its further decoration to turn it into a *-Farey symbol reflect the modified side–pairing rules in (8.3). For more details on this, on the form of the side–pairing transformations which modifies (5.5), and on the notions of semibalance or balance of a *-Farey symbol for a natural number N see §14 of $[K]_2$.

(8.7) Now we can state the answer to the question raised in (8.3) for our groups of interest.

Theorem $\Gamma_0(p)$ *when p is a prime admits a Farey symbol semibalanced for p which admits a reflection symmetry around $\frac{1}{2}$. In particular it admits a special polygon with a reflection symmetry as a fundamental domain. In any case for* $\Gamma_0(\mathrm{N})$ *as well as for* $\Gamma^1(\mathrm{N})$ *where N is not necessarly a prime there is a *-Farey symbol which admits a reflection symmetry around $\frac{1}{2}$.*

The advantage of this result over (7.3) or (7.5) is that it reduces the work of constructing fundamental domains for $\Gamma_0(\mathrm{N})$, $\Gamma^1(\mathrm{N})$ and $\Gamma(\mathrm{N})$ by 50%! There are some fine number–theoretic as well as geometric points in the proof of this result. When p is a prime $\mathbf{H}/\Gamma_0(p)$ has only two cusps and the fixed point set of the antiholomorphic involution on this surface induced by $z \mapsto -\bar{z}$ has only two components one of which is an even line and the other an odd line. So $\mathbf{H}/\Gamma_0^*(p)$ has two boundary components only one of which is an odd line. So we can cut the latter surface open into a polygon which is almost a special polygon except for one odd line as a boundary component. Hence gluing two copies of this polygon along the odd-line-components we obtain a space which is isometric to a special polygon and appropriately developing it in $\mathbf{H}$ we obtain a special polygon with a reflection symmetry as a fundamental domain for $\Gamma_0(p)$. This may not be possible for $\Gamma_0(\mathrm{N})$ when N is not a prime, and for $\Gamma^1(\mathrm{N})$ even when N happens to be a prime. Nevertheless one proves that for $\mathrm{N} \geq 4$ neither $\Gamma_0^*(\mathrm{N})$ nor $\Gamma^{1*}(\mathrm{N})$ contain a pair of distinct reflections which fix an even vertex or an odd vertex. This ensures that $\mathbf{H}/\Gamma_0^*(\mathrm{N})$ has no boundary components which contain "corners" with angles $\frac{\pi}{2}$ or $\frac{\pi}{3}$. So it is still possible to cut it open into a polygon and glue two copies of this polygon in such a way that one still obtains at least a special *-polygon. The cases $\mathrm{N} = 2, 3$ are not covered by this argument but they can be treated easily directly.

Here are two examples:

The case $\Gamma_0(8)$: A Farey symbol for $\Gamma_0(8)$ is

$$\{\underbrace{\infty, 0, \frac{1}{4}, \frac{1}{3}, \frac{1}{2}, 1, \infty}_{1\quad 2\quad 2\quad 3\quad 3\quad 1}\}.$$

It is easy to check that there is no Farey symbol in this case in which the denominators are symmetric around $\frac{1}{2}$. On the other hand there is a *-Farey symbol which has such symmetry. Namely it is

$$\{\underbrace{\infty, 0, \frac{1}{4}, \frac{1}{2}, \frac{3}{4}, 1, \infty}_{1\quad 2\quad 3\quad 3\quad 2\quad 1}\}.$$

The case $\Gamma^1(5)$: A Farey symbol for $\Gamma^1(5)$ is

$$\{\underbrace{\infty, 0, \frac{1}{3}, \frac{2}{5}, \frac{1}{2}, 1, \infty}_{1\quad 2\quad 3\quad 3\quad 2\quad 1}\}.$$

Again it is easily checked that there is no Farey symbol in this case in which the denominators are symmetric around $\frac{1}{2}$. On the other hand a *-Farey symbol with this symmetry is

$$\{\underbrace{\infty, 0, \frac{2}{5}, \frac{1}{2}, \frac{3}{5}, 1, \infty}_{1\quad 2\quad 3\quad 3\quad 2\quad 1}\}.$$

References

[A] Artin E. Ein Mechanisches System mit Quasi-ergodischen Bahnen, *Collected Papers, Addison Wesley, Reading, Mass. (1965), 499-501.*

[B] Beardon A.F., The Geometry of Discrete Groups, *Grad. Texts in Math., Springer Verlag, vol. 91(1983).*

[C] Chuman Y., Generators and Relations of $\Gamma_0(N)$, *Jour. Math. Kyoto Univ. 13-2(1973), 381-390.*

[F] Frasch H., Die Erzeugenden der Hauptkongruenz gruppen für primzahlstufen, *Math. Ann. 108(1932), 229-252.*

[Fri] Fricke R., Die elliptischen Funktionen und ihre Anwendungen, part II, *Teubner(1922), reprinted (1972).*

[G] Gromov M., Hyperbolic Groups, *in Group Theory, ed. by S. Gersten and J. Stallings, MSRI Publications(1988).*

[Gun] Gunning R.C., Lectures on Modular Forms, *Ann. of Math. Studies No.48, Princeton University Press(1960).*

[HW] Hardy G.H., and Wright E.M., An Introduction to The Theory of Numbers, *Oxford University Press(1960), Fourth Edition.*

$[K]_1$ Kulkarni R. S., An extension of a theorem of Kurosh and applications to fuchsian groups, *Mich. Math. Jour., 30 (1983), 259-272.*

$[K]_2$ Kulkarni R. S., An Arithmetic–Geometric Method in the Study of the Subgroups of the Modular Group, *to appear in the Amer. Jour. of Math.*

[Mac] Macbeath A. M., Groups of Homeomorphisms of a Simply Connectzeed Space, *Ann. of Math. 79(1964, 473–388.*

[M] Maskit B., On Poincare's Theorem for Fundamental Polygons, *Advances in Math., vol 7,(1971), 219-230.*

[N] Newman M., Integral Matrices, *Pure and Applied Mathematics monographs, vol. 45, Academic Press(1972).*

[P] Penner R., *Moduli Space of a Punctured Surface and Perturbative Series, Bull. A. M. S. 15(1986), 73–77.*

[Pon] Poincare H., Memoire sur les groups Fuchsiens, *Acta Math. 1(1882), 1-62.*

[R] Rademacher H., Über die Erzeugenden der Kongruenzuntergruppen der Modulgruppe, *Abh. Hamburg 7(1929), 134-148.*

[Ran] Rankin R. A., Modular Forms and Functions, *Cambridge University Press(1977).*

[Rat] Ratcliff J., *a forthcoming book.*

[S] Schoeneberg B., Elliptic Modular Functions, *Die Grundlehren der Math. Wissen., vol. 203, Springer Verlag (1974).*

[Z] Zagier D., Modular parametrizations of Elliptic Curves, *Canad. Math. Bull. 28(1985), 372-384.*

IDENTITIES AMONG RELATIONS OF GROUP PRESENTATIONS

Stephen J Pride
Department of Mathematics, University of Glasgow,
University Gardens, Glasgow, G12 8QW, U.K.

Introduction

My aim in this article is to give an account of the theory of (identity) sequences over group presentations.

In §1, I will discuss the algebraic theory. In §2 I will give an elementary and complete exposition of the theory of *pictures* (these are geometric objects which are very useful for studying sequences). In §3 I will discuss examples of so-called combinatorially aspherical presentations.

There is some inevitable overlap between the material in this paper and the material in [12] (and also [21]). However, I hope that my treatment is sufficiently different to be of interest. Moreover, many new results have been obtained since [12] was written.

It is a great pleasure to thank W. Bogley and J. Howie for their help in the preparation of this paper. Bogley made detailed and very useful comments on an initial draft, and supplied me with many additional references. Howie brought several additional references to my attention. He also outlined a rigorous proof of Theorem 2.3, and showed me the sort of example described in the exercise at the end of §2.3.

Notation

The following notation and terminology is to remain fixed throughout.

$\mathcal{P} = <\mathbf{x};\mathbf{r}>$ will be a group presentation. Here $\mathbf{x}$ is a set ("generators") and $\mathbf{r}$ is a set of cyclically reduced words on $\mathbf{x}\cup\mathbf{x}^{-1}$ ("relators"). We assume that if $R\in\mathbf{r}$ then no cyclic permutatition of $R^{\pm 1}$ except for R itself belongs to $\mathbf{r}$.

We let $\mathbf{w}$ denote the set of *all* words on $\mathbf{x}\cup\mathbf{x}^{-1}$ (reduced or not). Two words in $\mathbf{w}$ are *freely equal* if one can be obtained from the other by a finite number of insertions and deletions of inverse pairs $x^{\varepsilon}x^{-\varepsilon}$ ($x\in\mathbf{x}$, $\varepsilon=\pm 1$). The free equivalence class of $W\in\mathbf{w}$ will be denoted by $[W]$. The *free group* F on $\mathbf{x}$ then consists of the free equivalence classes, where the multiplication is defined by $[W][V]=[WV]$. We let N denote the normal closure of $\{[R]:r\in\mathbf{r}\}$ in F, and we let $G=F/N$. (G is the *group defined by the presentation* $\mathcal{P}$.)

If $\mathbf{s}$ is a subset of $\mathbf{r}$ then we let $\mathbf{s}^{\mathbf{w}}$ denote the set of all words of the form $WS^{\varepsilon}W^{-1}$ ($W\in\mathbf{w}$, $S\in\mathbf{s}$, $\varepsilon=\pm1$).

We will use the term *graph* in the sense of Serre [57]. Thus, a graph consists of two sets $\mathbf{v}$ (vertices), $\mathbf{e}$ (edges) and three functions, the initial and terminal functions $\iota,\tau:\mathbf{e}\to\mathbf{v}$, and the inversion function $^{-1}:\mathbf{e}\to\mathbf{e}$, satisfying the following rules: $\iota(e^{-1})=\tau(e)$, $(e^{-1})^{-1}=e$, $e^{-1}\neq e$ for all $e\in\mathbf{e}$. An *orientation* $\mathbf{e}^{+}$ of $\mathbf{e}$ is a choice of one element from each edge pair e, e^{-1}. A graph is *simple* if for any two (not necessarily distinct) vertices x,y there is at most one edge e with $\iota(e)=x$, $\tau(e)=y$.

The following symbols are introduced in the text, and are listed here for the reader's convenience.

$\Pi\sigma$, the product of the terms of the sequences σ, §1.1.
$<\sigma>$, the Peiffer equivalence class of σ, §1.1.
(Σ,F,∂), the crossed module associated with $\mathcal{P}$, §1.1.
π_2, the module of equivalence classes of identity sequences, §1.1.
$\{e_R:R\in\mathbf{r}\}$, the standard generators of the free crossed module (Σ,F,∂), §1.2.
P_1, P_2, the free $\mathbb{Z}G$–modules on $\{t_x:x\in\mathbf{x}\}$, $\{t_R:R\in\mathbf{r}\}$, §1.3.
j, the standard homomorphism from $\pi_2(\mathcal{P}_0)$ to $\pi_2(\mathcal{P})$ ($\mathcal{P}_0$ a subpresentation), §1.4.
$\mathring{R}$,$p(R)$, the root, period of the relator R, §1.5.
ζ_R, trivial sequence, §1.5.
T, the module generated by the equivalence classes of trivial sequences, §1.5.
$\exp_R(\ )$, $\exp_{(R,V)}(\ )$, $\mathrm{wexp}_{(R,V)}(\ )$, exponent sum functions, §1.6.
$\mathbb{P}$, a picture, §2.1.
$W(\Delta)$, $W(\gamma)$, $W(\mathbb{P})$, the label on a disc, path, picture, §2.1.
$\sigma(\underline{\gamma})$, the sequence associated with a spray $\underline{\gamma}$, §2.1.
$J(\mathbf{X})$, the submodule of π_2 associated with a collection $\mathbf{X}$ of spherical pictures, §2.3.

1. The algebra of identity sequences

1.1. *Sequences*

We will be interested in finite sequences σ of elements of $\mathbf{r}^{\mathbf{w}}$.

Let $\sigma=(c_1,\ldots,c_m)$, where $c_i\in\mathbf{r}^{\mathbf{w}}$ ($i=1,\ldots,m$). We define $\Pi\sigma$ to be the product $c_1c_2\ldots c_m$. We say that σ is an *identity sequence* if $\Pi\sigma$ is freely equal to 1. We define the *inverse* σ^{-1} of σ to be $(c_m^{-1},\ldots,c_1^{-1})$, and for $W\in w$ we define the *conjugate* $W\sigma W^{-1}$ of σ by W to be $(Wc_1W^{-1},\ldots,Wc_mW^{-1})$.

We define operations on sequences as follows. Let σ be as above where $c_i=W_iR_i^{\varepsilon_i}W_i^{-1}$ ($W_i\in\mathbf{w}$, $R_i\in\mathbf{r}$, $\varepsilon_i=\pm1$, $i=1,\ldots,m$).

(SUB) *Substitution* Replace each W_i by a word freely equal to it.

(DEL) *Deletion* Delete two consecutive terms if one is identically equal to the inverse of the other.

(INS) *Insertion* The opposite of deletion.

(EX) *Exchange* Replace two consecutive terms c_i, c_{i+1} by either

c_{i+1}, $c_{i+1}^{-1}c_ic_{i+1}$ or by $c_ic_{i+1}c_i^{-1}$, c_i.

Two sequences σ, σ' will be said to be (*Peiffer*) *equivalent* (denoted $\sigma \sim \sigma'$) if one can be obtained from the other by a finite number of applications of the operations (SUB), (DEL), (INS), (EX). The equivalence class containing σ will be denoted by $<\sigma>$.

We can define a binary operation + on the set Σ of all equivalence classes by the rule

$$<\sigma_1> + <\sigma_2> = <\sigma_1\sigma_2>.$$

(Here $\sigma_1\sigma_2$ is the juxtaposition of the two sequences σ_1,σ_2.) Under this operation Σ is a group. The identity (zero element) is the equivalence class of the empty sequence, and the inverse (negative) $-<\sigma>$ of $<\sigma>$ is $<\sigma^{-1}>$.

We note that although we have used additive notation, the group Σ is *not* commutative.

We let π_2 denote the subgroup of Σ consisting of all elements $<\sigma>$ where σ is an identity sequence. (Occasionally, when we want to emphasise the presentation $\mathscr{P}$, we will write $\pi_2(\mathscr{P})$.)

It is clear that F acts on Σ by the rule

$$[W]\cdot<\sigma> = <W\sigma W^{-1}> \qquad ([W]\in F,\ <\sigma>\in\Sigma).$$

It is also clear that the mapping

$$\partial:\Sigma\to F \qquad <\sigma>\mapsto[\Pi\sigma] \qquad (<\sigma>\in\Sigma)$$

is a group homomorphism.

The following two basis results hold.

(a) $\partial([W]\cdot<\sigma>)=[W]\partial(<\sigma>)[W]^{-1}$ for all $[W]\in F$, $<\sigma>\in\Sigma$.

(b) $<\sigma_1>+<\sigma_2>=\partial(<\sigma_1>)\cdot<\sigma_2>+<\sigma_1>$ for all $<\sigma_1>$, $<\sigma_2>\in\Sigma$.

To prove the second of these, let $\sigma_1=(c_1,\ldots,c_m)$, $\sigma_2=(d_1,\ldots,d_n)$. Starting with $\sigma_1\sigma_2$ and applying a succession of exchange operations, we obtain the sequence

$$(c_1,\ldots,c_{m-1},\ c_md_1c_m^{-1},\ldots,c_md_nc_m^{-1},c_m).$$

Then applying a further succession of exchange operations we obtain

$$(c_1,\ldots,c_{m-2},c_{m-1}c_md_1c_m^{-1}c_{m-1}^{-1},\ldots,c_{m-1}c_md_nc_m^{-1}c_{m-1}^{-1},c_{m-1},\ c_m).$$

Continuing this way we finally end up with $((\Pi\sigma_1)\sigma_2(\Pi\sigma_1)^{-1})\sigma_1$.

The triple (Σ,F,∂) is an example of a *crossed module*.

1.2 Crossed modules

A *crossed module* is a triple (A,X,∂) where A is a group (written additively), X is a group (written multiplicatively), X acts (on the left) on A, ∂ is a homomorphism from A to X. We require that the following two axioms hold:

(CM1) $\partial(x\cdot a)=x\partial(a)x^{-1}$ for all $x\in X$, $a\in A$

(CM2) $a+b=\partial(a)\cdot b+a$ for all $a,b\in A$.

Crossed modules were introduced by J.H.C. Whitehead in [61].

A *mapping* from one crossed module (A_1,X_1,∂_1) to another (A_2,X_2,∂_2) is a pair (φ,ψ), where $\varphi:A_1\to A_2$, $\psi:X_1\to X_2$ are group homomorphisms, $\varphi(x\cdot a)=\psi(x)\cdot\varphi(a)$ for all $x\in X_1$, $a\in A_1$, and where the diagram

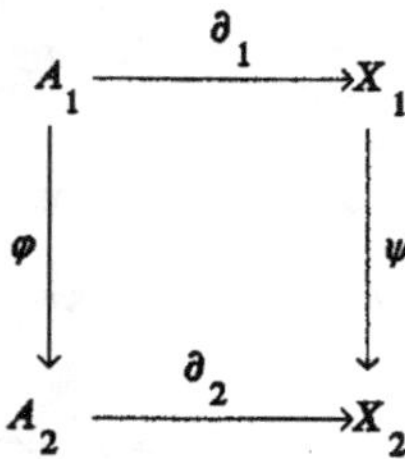

commutes.

A crossed module (A,X,∂) is said to be *free on* E (where E is a subset of A) if given any crossed module (B,Y,∂'), any function $\phi:E\to B$, and any group homomorphism $\psi:X\to Y$ such that the diagram

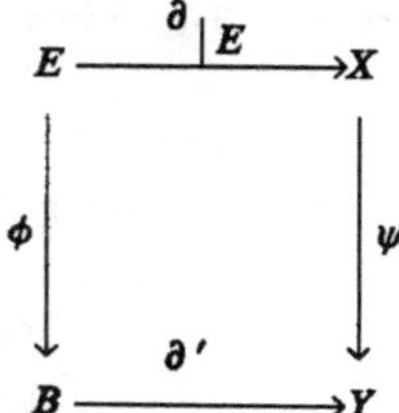

commutes, there is a unique mapping $(\varphi,\psi):(A,X,\partial)\to(B,Y,\partial')$ of crossed modules such that $\varphi|_E=\phi$.

The following result will not play any role in the sequel, but is included here for interest. (The result is due to Whitehead, and is a special case of a much more general result – see [9], [26], [55], [61].)

Theorem 1.1. *(Σ,F,∂) is free on $E=\{e_R:R\in\mathrm{r}\}$, where $e_R=\langle(R)\rangle$.*

The proof of this theorem is fairly straightforward, and is left to the reader. The only tricky part is to show that the required extension of a given function $E\to B$ is well–defined.

Let (A,X,∂) be a crossed module, and let $K=\mathrm{Ker}\partial$, $L=\mathrm{Im}\partial$. By (CM1), L is normal in X. By (CM2) (take $a\in K$, $b\in A$) K is contained in the centre $Z(A)$ of A. Also by (CM2) (take $a\in A$, $b\in Z(A)$), L acts trivially on $Z(A)$. It follows that we may consider the group $\bar{X}=X/L$, and then K is a left $\bar{X}$–module, with $\bar{X}$–action given by $xL\cdot a=x\cdot a$.

Let A' be the derived subgroup of A. Then A' is X–invariant, so we get an induced action of X on the quotient A/A'. Now L acts trivially on A/A' (for if $y\in L$ then $y=\partial(u)$ for some $u\in A$, so for any $a\in A$ we have $y\cdot a=\partial(u)\cdot a=u+a-u\equiv a$ mod A'). Thus A/A' can be given a left $\bar{X}$–module structure via the action of X.

The homomorphism $\partial:A\to L$ gives an induced homomorphism $\partial^*:A/A'\to L/L'$. Now L/L' is a left $\bar{X}$–module, with $\bar{X}$–action given by

$$xL\cdot yL'=xyx^{-1}L' \quad (x\in X,\ y\in L),$$

and under this action ∂^* is an $\tilde{X}$–homomorphism.

Let υ denote the composition $K \xrightarrow{\text{inclusion}} A \longrightarrow A/A'$.

Lemma 1.1. (i) *The sequence*

$$K \xrightarrow{\upsilon} A/A' \xrightarrow{\partial^*} L/L' \longrightarrow 0 \tag{1.1}$$

is exact.

(ii) *If* $A \xrightarrow{\partial} L$ *has a section (that is, if there is a group homomorphism* $\beta : L \longrightarrow A$ *such that* $\partial\beta = id_L$*) then* $K \cap A' = 0$*, and so the sequence* (1.1) *is short exact.*

(For a proof of (ii), see [12, p.160].)

The sequence (1.1) is natural in the following sense. Let

$$(\varphi,\psi) : (A,X,\partial) \longrightarrow (A_1,X_1,\partial_1)$$

be a mapping of crossed modules. Let $K_1 = \operatorname{Ker}\partial_1$, $L_1 = \operatorname{Im}\partial_1$. Then $\varphi(K) \subseteq K_1$ and $\psi(L) \subseteq L_1$. Also, we have induced homomorphisms

$$A/A' \xrightarrow{\hat{\varphi}} A_1/A_1' \qquad a + A' \mapsto \varphi(a) + A_1'$$

$$L/L' \xrightarrow{\hat{\psi}} L_1/L_1' \qquad uL' \mapsto \psi(u)L_1'.$$

Then the diagram

$$\begin{array}{ccccc} K & \xrightarrow{\upsilon} & A/A' & \xrightarrow{\partial^*} & L/L' \\ \downarrow{\scriptstyle\varphi} & & \downarrow{\scriptstyle\hat{\varphi}} & & \downarrow{\scriptstyle\hat{\psi}} \\ K_1 & \xrightarrow{\upsilon_1} & A_1/A_1' & \xrightarrow{\partial_1'} & L_1/L_1' \end{array}$$

commutes.

For further results on crossed modules, see [5], [10], [11], [15].

1.3. Some exact sequences concerning the module π_2

We return to our presentation $\mathcal{P} = \langle \mathbf{x};\mathbf{r} \rangle$, and the corresponding crossed module (Σ,F,∂). Recall that the group G defined by the presentation is F/N, where $N = \operatorname{Im}\partial$.

Since N is free, (being a subgroup of the free group F), the map $\partial : \Sigma \longrightarrow N$ has a section. Thus, by Lemma 1.1 (ii), we have the following result (first proved by Papakyriakopoulos [49, p.208]).

Lemma 1.2. $\pi_2 \cap \Sigma' = 0$.

The G–module Σ/Σ' turns out to be isomorphic to the free module $P_2 = \bigoplus_{R \in \mathbf{r}} \mathbb{Z}Gt_R$ (under the correspondence $e_R + \Sigma' \leftrightarrow t_R$ $(R \in \mathbf{r})$), and the short exact sequence (1.1) then becomes

$$0\longrightarrow\pi_2\longrightarrow P_2\longrightarrow N/N'\longrightarrow 0. \tag{1.2}$$

(See [12, p.162] for these matters.) The map $\pi_2\longrightarrow P_2$ is given by

$$<\sigma>\mapsto \mathrm{eval}(\sigma) \quad (<\sigma>\in\pi_2),$$

where, if $\sigma=(W_1R_1^{\varepsilon_1}W_1^{-1},\dots,W_nR_n^{\varepsilon_n}W_n^{-1})$, then

$$\mathrm{eval}(\sigma)=\sum_1^n \varepsilon_i[W_i]Nt_R.$$

We remark that the G–module N/N' is called the *relation module* of the presentation $\mathscr{P}$.

There are two other short exact sequences related to (1.2).

Let $P_0=\mathbb{Z}G$, and regard $\mathbb{Z}$ as a left $\mathbb{Z}G$–module with trivial G–action. There is the *augmentation map* $P_0\longrightarrow\mathbb{Z}$ which takes each element of G to 1. The kernel of this map is the *augmentation ideal*, denoted IG. Thus we have a short exact sequence

$$0\longrightarrow IG\longrightarrow P_0\longrightarrow\mathbb{Z}\longrightarrow 0$$

Now let $P_1=\bigoplus_{x\in\mathbf{x}}\mathbb{Z}Gt_x$. Then there is a short exact sequence

$$0\longrightarrow N/N'\longrightarrow P_1\longrightarrow IG\longrightarrow 0 \tag{1.3}$$

(see [, p.199]). The map $P_1\longrightarrow IG$ is given by

$$t_x\mapsto 1-[x]N \quad (x\in\mathbf{x}),$$

and the map $N/N'\longrightarrow P_1$ is given by

$$[W]N'\longrightarrow\sum_{x\in\mathbf{x}}\rho\left(\frac{\partial[W]}{\partial x}\right)t_x \quad ([W]\in N).$$

(Here $\frac{\partial}{\partial x}:\mathbb{Z}F\longrightarrow\mathbb{Z}F$ is Fox derivation [46, §II.3], and $\rho:\mathbb{Z}F\longrightarrow\mathbb{Z}G$ is induced by the natural epimorphism $F\longrightarrow G$.)

The above sequences are very useful for homological calculations. Note that if we put the three sequences together we get an exact sequence

$$0\longrightarrow\pi_2\longrightarrow P_2\xrightarrow{\partial_2}P_1\xrightarrow{\partial_1}P_0\longrightarrow\mathbb{Z}\longrightarrow 0.$$

We deduce in particular that $cdG\leq 3$ if and only if π_2 is projective. Also, if $\pi_2=0$, then $cdG\leq 2$. (Presentations for which $\pi_2=0$ are called *aspherical* and will be discussed in more detail later.)

Theorem 1.2. *Let B, C be right, left $\mathbb{Z}G$–modules, respectively. There are exact sequences*

$$0\longrightarrow H_3(G,B)\longrightarrow B\otimes_{\mathbb{Z}G}\pi_2\longrightarrow Ker(1\otimes\partial_2)\longrightarrow H_2(G,B)\longrightarrow 0 \tag{1.4}$$

$$0\longrightarrow H^2(G,C)\longrightarrow coker(Hom_{\mathbb{Z}G}(\partial_2,1))\longrightarrow Hom_{\mathbb{Z}G}(\pi_2,C)\longrightarrow H^3(G,C)\longrightarrow 0 \tag{1.5}$$

Proof. The short exact sequences (1.2), (1.3) give rise to long exact Tor sequences

$$\dots\longrightarrow\mathrm{Tor}_1^{\mathbb{Z}G}(B,P_2)\longrightarrow\mathrm{Tor}_1^{\mathbb{Z}G}(B,N/N')\longrightarrow B\otimes_{\mathbb{Z}G}\pi_2\longrightarrow B\otimes_{\mathbb{Z}G}P_2\xrightarrow{\vartheta}B\otimes_{\mathbb{Z}G}N/N'\longrightarrow 0,$$

$$\ldots\longrightarrow \mathrm{Tor}_1^{\mathbb{Z}G}(B,P_1)\longrightarrow \mathrm{Tor}_1^{\mathbb{Z}G}(B,IG)\longrightarrow B\otimes_{\mathbb{Z}G}N/N'\xrightarrow{\phi} B\otimes_{\mathbb{Z}G}P_1\longrightarrow B\otimes_{\mathbb{Z}G}IG\longrightarrow 0.$$

Since P_1 and P_2 are free, $\mathrm{Tor}_1^{\mathbb{Z}G}(B,P_2)=\mathrm{Tor}_1^{\mathbb{Z}G}(B,P_1)=0$. Also, by dimension shifting, $\mathrm{Tor}_1^{\mathbb{Z}G}(B,N/N')=H_3(G,B)$, $\mathrm{Tor}_1^{\mathbb{Z}G}(B,IG)=H_2(G,B)$ (see [29, p.199 and p.189]). Thus we obtain from the first sequence above the short exact sequence

$$0\longrightarrow H_3(G,B)\longrightarrow B\otimes_{\mathbb{Z}G}\pi_2\longrightarrow \ker\vartheta\longrightarrow 0,$$

and we deduce from the second that $H_2(G,B)\cong\mathrm{Ker}\phi$. Now since ϑ is surjective there is a short exact sequence

$$0\longrightarrow\ker\vartheta\longrightarrow\ker\phi\vartheta\longrightarrow\ker\phi\longrightarrow 0.$$

Using the fact that $\phi\vartheta=1\otimes\partial_2$ we obtain (1.4).

The proof of (1.5) is similar.

The above theorem appears to be well–known. The sequence (1.4), with B the trivial G–module $\mathbb{Z}$, seems to be due to Hopf [62].

1.4. Subpresentations

Let $\mathcal{P}_0=\langle \mathbf{x}_0;\mathbf{r}_0\rangle$ be a subpresentation of $\mathcal{P}$, and let $(\Sigma_0,F_0,\partial_0)$ be the associated crossed module. There is an obvious mapping of crossed modules

$$(\varphi,\psi):(\Sigma_0,F_0,\partial_0)\longrightarrow(\Sigma,F,\partial)$$

where

$$\varphi(\langle\sigma\rangle_0)=\langle\sigma\rangle \qquad (\langle\sigma\rangle_0\in\Sigma_0)$$
$$\psi([W]_0)=[W] \qquad ([W]_0\in F_0).$$

Restricting φ gives a homomorphism

$$j:\pi_2(\mathcal{P}_0)\longrightarrow\pi_2(\mathcal{P}).$$

As one might expect, j is not in general injective (see the exercise at the end of §2.3 below).

Whitehead problem. Suppose $\mathcal{P}$ is aspherical (i.e. $\pi_2(\mathcal{P})=0$). Is $j:\pi_2(\mathcal{P}_0)\longrightarrow\pi_2(\mathcal{P})$ injective for every subpresentation $\mathcal{P}_0$ of $\mathcal{P}$? In other words, is every subpresentation of an aspherical presentation aspherical?

This problem has received quite a lot of attention. See, for example, [8], [32], [39] and the references cited there.

We now discuss a theorem of Gutierréz and Ratcliffe [27, Theorem 1]. (See also [5], [10] for generalizations.)

Consider the presentation $\mathcal{P}=\langle \mathbf{x};\mathbf{r}\rangle$ and suppose that $\mathbf{r}$ is expressed as a union $\mathbf{r}=\mathbf{r}_1\cup\mathbf{r}_2$. For $\lambda=1,2$, let $\mathcal{P}_\lambda=\langle\mathbf{x};\mathbf{r}_\lambda\rangle$ and let $j_\lambda:\pi_2(\mathcal{P}_\lambda)\longrightarrow\pi_2(\mathcal{P})$ be the natural homomorphism as discussed above. Note that $\mathrm{Im}j_\lambda$ is a submodule of $\pi_2(\mathcal{P})$.

Let $N_\lambda(\lambda=1,2)$ be the normal closure of $\{[R]:R\in\mathbf{r}_\lambda\}$ in the free group F. Now F acts on $\dfrac{N_1\cap N_2}{[N_1,N_2]}$ via conjugation:

$$[W]\cdot[U][N_1,N_2]=[WUW^{-1}][N_1,N_2] \quad ([W]\in F,\ [U]\in N_1\cap N_2).$$

It is easy to show that $N(=N_1N_2)$ acts trivially, and so we get an induced action of $G=F/N$ on $\frac{N_1\cap N_2}{[N_1,N_2]}$. We can define a G–homomorphism

$$\eta:\pi_2(\mathscr{P})\longrightarrow\frac{N_1\cap N_2}{[N_1,N_2]}$$

by the following rule. Let $<\sigma>\in\pi_2(\mathscr{P})$, and let V be the product (taken in order) of the elements of σ which belong to $\mathbf{r}_1^{\mathbf{W}}$. Then $\eta(<\sigma>)=[V][N_1,N_2]$. It is not hard to show that η is well–defined.

Theorem 1.3. *Let $\xi:Imj_1\oplus Imj_2\longrightarrow\pi_2(P)$ be induced by the inclusions Imj_1, $Imj_2\longrightarrow\pi_2(\mathscr{P})$. Then the sequence*

$$Imj_1\oplus Imj_2\xrightarrow{\xi}\pi_2(\mathscr{P})\xrightarrow{\eta}\frac{N_1\cap N_2}{[N_1,N_2]}\longrightarrow 0$$

is exact. If $\mathbf{r}_1$ and $\mathbf{r}_2$ are disjoint then ξ is injective, and so the sequence is short exact.

Proof. Clearly $\eta\xi=0$.

Surjectivity of η. Let $[W]\in N_1\cap N_2$. Then $[W]=[\Pi\sigma_1]$, where σ_1 is a sequence of elements of $\mathbf{r}_1^{\mathbf{W}}$, and $[W]=[\Pi\sigma_2]$, where σ_2 is a sequence of elements of $\mathbf{r}_2^{\mathbf{W}}$. Hence $\sigma_1\sigma_2^{-1}$ is an identity sequence, and clearly $\eta(<\sigma_1\sigma_2^{-1}>)=[W][N_1,N_2]$.

$Ker\eta\subseteq Im\xi$. Let $\eta(<\sigma>)\in[N_1,N_2]$. Applying exchange operations if necessary, we may assume that $\sigma=\sigma_1\sigma_2$, where σ_1 consists of elements of $\mathbf{r}_1^{\mathbf{W}}$ and σ_2 consists of elements of $\mathbf{r}_2^{\mathbf{W}}$. Then $\eta(<\sigma>)=[\Pi\sigma_1]$. Hence $\Pi\sigma_1$ is freely equal to a product

$$[\Pi\omega_1,\Pi\kappa_1]^{\varepsilon_1}\ \dots\ [\Pi\omega_n,\Pi\kappa_n]^{\varepsilon_n}$$

where the ω_i are sequences of elements of $\mathbf{r}_1^{\mathbf{W}}$, the κ_i are sequences of elements of $\mathbf{r}_2^{\mathbf{W}}$, and the ε_i are ±1. Hence

$$<\sigma_1>-\sum_{i=1}^{n}\varepsilon_i(<\omega_i>+<\kappa_i>-<\omega_i>-<\kappa_i>)$$

belongs to $\pi_2(P)$. In fact, since $<\kappa_i>-<\omega_i>-<\kappa_i>=-\partial(<\kappa_i>)\cdot<\omega_i>$ (i=1,2,...,n), the above element belongs to $\mathrm{Im}j_1$. Similarly

$$\left[\sum_{i=1}^{n}\varepsilon_i(<\omega_i>+<\kappa_i>-<\omega_i>-<\kappa_i>)\right]+<\sigma_2>$$

belongs to $\mathrm{Im}j_2$. Hence $<\sigma_1\sigma_2>\in\mathrm{Im}\xi$.

Finally, suppose that $\mathbf{r}_1$ and $\mathbf{r}_2$ are disjoint. Then by (1.1) we have an embedding $\pi_2(\mathscr{P})\longrightarrow P_2^1\oplus P_2^2$, where $P_2^\lambda=\bigoplus_{R\in\mathbf{r}_\lambda}\mathbb{Z}Gt_R$ ($\lambda=1,2$). Under this embedding $\mathrm{Im}j_\lambda$ is mapped into P_2^λ ($\lambda=1,2$), and so $\mathrm{Im}j_1$, $\mathrm{Im}j_2$ generate their direct sum in $\pi_2(\mathscr{P})$.

1.5. Trivial sequences and weak equivalence

Let $R \in \mathbf{r}$. Then we can write $R = \mathring{R}^{p(R)}$ where $\mathring{R}$ is not a proper power and $p(R)$ is a positive integer ($\mathring{R}$ is the *root* of R, and $p(R)$ is the *period*).

The identity sequences

$$\zeta_R = (R, \mathring{R}R^{-1}\mathring{R}^{-1}) \quad (R \in \mathbf{r})$$

will be called the *trivial sequences*, and the elements

$$<\zeta_R>$$

of π_2 will be called the *trivial elements*. The submodule of π_2 generated by the trivial elements will be denoted by T.

We introduce new operations on sequences.

(W/DEL) *Weak deletion* Delete two adjacent terms if their product is freely equal to 1.

(W/INS) *Weak insertion* The opposite of weak deletion.

We will say that two sequences are *weakly equivalent* if and only if one can be obtained from the other by a finite number of operations (SUB), (EX), (W/DEL), (W/INS).

It is convenient to consider certain special weak deletions and insertions.

(W/DEL*) Delete two adjacent terms of the form $WR^{\varepsilon}W^{-1}$, $W\mathring{R}^{k}R^{-\varepsilon}\mathring{R}^{-k}W^{-1}$ ($W \in \mathbf{w}$, $R \in \mathbf{r}$, $\varepsilon = \pm 1$, k some integer).

(W/INS*) The opposite of (W/DEL*).

Lemma 1.3. *Two sequences are weakly equivalent if and only if one can be transformed to the other by a finite number of operations (SUB), (EX), (W/DEL*), (W/INS*).*

Proof. It suffices to show that an operation (W/DEL) can be expressed in terms of an operation (SUB) and an operation (W/DEL*).

Suppose a sequence σ has two adjacent terms $V_1R_1^{\varepsilon_1}V_1^{-1}$, $V_2R_2^{\varepsilon_2}V_2^{-1}$ whose product is freely equal to 1, and suppose that σ' is obtained by removing these two terms. Now in F we have

$$[V_1^{-1}V_2][R_2]^{\varepsilon_2}[V_1^{-1}V_2]^{-1} = [R_1]^{-\varepsilon_1}.$$

Using standard results about free groups we deduce that $R_1 = R_2$, $\varepsilon_1 = -\varepsilon_2$ and $[V_1^{-1}V_2] = [\mathring{R}_1]^k$ for some k. Thus σ' can be obtained from σ by first applying a substitution operation to replace V_2 by $V_1\mathring{R}_1^k$, and then applying an operation (W/DEL*).

Theorem 1.4. *σ and σ' are weakly equivalent if and only if $<\sigma> - <\sigma'> \in T$.*

Proof. Suppose $<\sigma> - <\sigma'> \in T$. Then there exist $W_1,\ldots,W_n \in \mathbf{w}$, $\varepsilon_1,\ldots,\varepsilon_n \in \{-1,1\}$, and $R_1,\ldots,R_n \in \mathbf{r}$ such that σ' is equivalent to

$$\sigma'' = \sigma(W_1\zeta_{R_1}^{\varepsilon_1}W_1^{-1})\ldots(W_n\zeta_{R_n}^{\varepsilon_n}W_n^{-1}).$$

Now a succession of n weak deletions will convert σ'' to σ, so σ and σ' are weakly equivalent, as required.

To prove the converse, it is enough to show that if σ' is obtained from σ by an operation (W/DEL*) then $<\sigma> - <\sigma'> \in T$. Thus, suppose that $\sigma=\sigma_1\sigma_2\sigma_3$ and $\sigma'=\sigma_1\sigma_3$, where $\sigma_2=(WR^{\varepsilon}W^{-1}, W\mathring{R}^kR^{-\varepsilon}\mathring{R}^{-k}W^{-1})$ ($W\in \mathbf{w}, R\in \mathbf{r}, \varepsilon=\pm 1$, k some integer). It is not hard to show that $<\sigma_2> \in T$. Hence, since elements of π_2 are central, we have

$$<\sigma> = <\sigma_1> + <\sigma_2> + <\sigma_3> = <\sigma'> + <\sigma_2>.$$

1.6. Exponent sums

Let σ be a sequence.

For $R\in \mathbf{r}$, consider all terms of σ which belong to $\{R\}^{\mathbf{W}}$. Each such term has the form $WR^{\varepsilon}W^{-1}$ for some $W\in\mathbf{w}$, and some $\varepsilon=\pm 1$. We define the *exponent sum of R in* σ, denoted $\exp_R(\sigma)$, to be the sum of the ε's over all such terms.

Also, for a pair (R,V) with $R\in\mathbf{r}$, $V\in\mathbf{w}$, consider all terms of σ which have the form $WR^{\varepsilon}W^{-1}$ for some $W\in\mathbf{w}$ *with* $[W]\equiv[V]$ *mod*N, and for some $\varepsilon=\pm 1$. We define the *exponent sum of* (R,V) *in* σ, denoted $\exp_{(R,V)}(\sigma)$ to be the sum of the ε's over all such terms. The *weak exponent sum of* (R,V) *in* σ, denoted $\mathrm{wexp}_{(R,V)}(\sigma)$, is defined similarly, except that we consider all terms of the form $WR^{\varepsilon}W^{-1}$ where $[W]\equiv[V][\mathring{R}]^k$ mod$N$ for some k.

Remarks. (a) $\mathrm{wexp}_{(R,V)}(\sigma)=\sum_{k=1}^{m}\exp_{(R,V\mathring{R}^k)}(\sigma)$, where m is the order of $[\mathring{R}]N$ in G.

(b) $\exp_R(\sigma)=\sum_{V\in\mathbf{w}}\exp_{(R,V)}(\sigma)$.

(c) If $\sigma\sim\sigma'$ then $\exp_{(R,V)}(\sigma)=\exp_{(R,V)}(\sigma')$ (and similarly for $\exp_R(\ )$ and $\mathrm{wexp}_{(R,V)}(\ )$ by (a) and (b) above).

Theorem 1.5. (i) $<\sigma> \in\Sigma'$ *if and only if* $\exp_{(R,V)}(\sigma)=0$ *for all pairs* (R,V).

(ii) $<\sigma>\in\Sigma'+T$ *if and only if* $\mathrm{wexp}_{(R,V)}(\sigma)=0$ *for all pairs* (R,V).

Proof. (i) If $<\sigma>\in\Sigma'$ then there are sequences κ_i, ω_i $(i=1,\dots,n)$ such that σ is equivalent to

$$\sigma'=\kappa_1\omega_1\kappa_1^{-1}\omega_1^{-1}\dots\kappa_n\omega_n\kappa_n^{-1}\omega_n^{-1}.$$

Clearly $\exp_{(R,V)}(\sigma')=0$ for all pairs (R,V).

Conversely, suppose $\exp_{(R,V)}(\sigma)=0$ for all pairs (R,V). Let $US^{\varepsilon}U^{-1}$ be the first term of σ. Then there must be a term $U'S^{-\varepsilon}U'^{-1}$ where $[U']\equiv[U]$ mod N. By using a succession of exchange operations, if necessary, we can assume that this is the *second* term of σ. These exchange operations will replace U' by another word U'', but the new word will still define an element of F congruent to $[U]$ mod N. Then by a substitution operation we can replace U'' by WU, where $[W]\in N$. Since $[W]\in N$, $[W]=\partial(<\omega>)$ for some sequence ω. We can now write

$$<\sigma> = <US^{\varepsilon}U^{-1}> - \partial(<\omega>)\cdot <US^{\varepsilon}U^{-1}> + <\sigma_0>,$$

where σ_0 has two less terms than σ, and where $\exp_{(R,V)}(\sigma_0)=0$ for all pairs (R,V). Since

$$\begin{aligned} &<US^{\varepsilon}U^{-1}> - \partial(<\omega>)\cdot <US^{\varepsilon}U^{-1}> \\ &= <US^{\varepsilon}U^{-1}> + <\omega> - <US^{\varepsilon}U^{-1}> - <\omega> \\ &\in \Sigma', \end{aligned}$$

we have $<\sigma> \equiv <\sigma_0>$ mod Σ'. By induction, we may assume that $<\sigma_0> \in \Sigma'$.

(ii) is proved similarly.

Definitions. (i) $\mathcal{P}$ is called *aspherical* (respectively, *combinatorially aspherical* (CA)) if $\exp_{(R,V)}(\sigma)$ (respectively, $\text{wexp}_{(R,V)}(\sigma)$) is 0 for all pairs (R,V) and all identity sequences σ.

(ii) $\mathcal{P}$ is called *Cockroft* if $\exp_R(\sigma)=0$ for all $R\in\mathbf{r}$ and all identity sequences σ.

Obviously

$$\text{aspherical} \Rightarrow \text{CA} \Rightarrow \text{Cockroft}.$$

Theorem 1.6. (i) *$\mathcal{P}$ is aspherical if and only if $\pi_2=0$.*

(ii) *$\mathcal{P}$ is CA if and only if $\pi_2=T$.*

Proof. This follows from Theorem 1.5 and Lemma 1.2.

Remark. An identity sequence σ with $\exp_{(R,V)}(\sigma)=0$ for all pairs (R,V) is said by Papakyriakopoulos [49] to have the *identity property*. The terminology is changed slightly in [12]. In that paper "identity property" is used for identity sequences σ with $\text{wexp}_{(R,V)}(\sigma)=0$ for all pairs (R,V). If $\exp_{(R,V)}(\sigma)=0$ for all pairs (R,V) then the sequence is said in [12] to have the *primary* identity property.

Theorem 1.6(i) is due to Papakyriakopoulos [49], and Theorem 1.6(ii) seems first to have been proved in [12].

Examples of aspherical and CA presentations will be given in Section 3 below. The Cockroft property does not appear to have received much attention (however, see [7], [8], [23]).

We now give some results concerning Cockroft and (combinatorially) aspherical presentations.

Theorem 1.7. *Suppose $\mathcal{P}$ is Cockroft.*

(i) *For each $R\in\mathbf{r}$, the order of $[\mathring{R}]N$ is precisely $p(R)$.*

(ii) *Let B, C be right, left $\mathbb{Z}G$-modules respectively, with* trivial *G-action*:

(a) $H_3(G,B)=B\otimes_{\mathbb{Z}G}\pi_2$, $H^3(G,C)=\text{Hom}_{\mathbb{Z}G}(\pi_2,C)$.

(b) $H_2(G,B)=\ker(1\otimes\partial_2)$, $H^2(G,C)=\text{coker}(\text{Hom}_{\mathbb{Z}G}(\partial_2,1))$.

Proof. (i) Let ℓ be the order of $[\mathring{R}]N$ (note that $\ell|p(R)$). Then there is a sequence σ such that $[R]^{\ell}=\partial(<\sigma>)$. Thus the sequence ω consisting of R^{-1} followed by $\frac{p(R)}{\ell}$ copies of σ is an identity sequence. So

$$0=\exp_R(\omega)=-1+\frac{p(R)}{\ell}\exp_R(\sigma),$$

which implies that $p(R)\,|\,\ell$.

(ii) We make use of Theorem 1.2. Observe that the Cockroft property implies (in fact, is equivalent to) the assertion that the image of π_2 under the embedding

$$\pi_2 \longrightarrow P_2 \quad <\sigma> \mapsto \mathrm{eval}(\sigma)$$

lies in $IG \cdot P_2$. From this one easily deduces that the maps

$$B \otimes_{\mathbb{Z}G} \pi_2 \longrightarrow \ker(1 \otimes \partial_2)$$
$$\mathrm{coker}(\mathrm{Hom}_{\mathbb{Z}G}(\partial_2, 1)) \longrightarrow \mathrm{Hom}_{\mathbb{Z}G}(\pi_2, C)$$

are both zero.

Theorem 1.7(i) is essentially due to Huebschmann [38] (see Proposition 1).

Theorem 1.8. *$\mathcal{P}$ is aspherical if and only if $\mathcal{P}$ is CA and no element of* **r** *is a proper power.*
Proof. If $\mathcal{P}$ is aspherical then $cdG \leq 2$, so G is torsion–free. Thus, since $\mathcal{P}$ is Cockroft, no element of **r** can be a proper power, by Theorem 1.7(i).

The above result is obtained in [13] (Proposition 1.3).

Theorem 1.9. *Suppose $\mathcal{P}$ is CA. For $R \in \mathbf{r}$, let G_R denote the cyclic subgroup of G generated by $[\mathring{R}]N$. Let B, C be right, left $\mathbb{Z}G$-modules, respectively.*
(i) *For all $n \geq 3$,*

$$H_n(G,B) = \bigoplus_{R \in \mathbf{r}} H_n(G_R, B),$$

$$H^n(G,C) = \prod_{R \in \mathbf{r}} H^n(G_R, C).$$

(ii) *The conjugates of the groups $G_R (R \in \mathbf{r})$ are the maximal finite subgroups of G; distinct maximal finite subgroups intersect trivially.*

Part (i) is essentially due to Lyndon [44] (see also [38]), while part (ii) follows from (i) and a theorem due to Serre (quoted in [38]).

2. The geometry of identity sequences

Sequences can be studied very successfully using geometric objects called *pictures*. These geometric objects are a very useful tool in combinatorial group theory, and can be used in a variety of different ways. See [5], [6], [12], [14], [17], [21], [25], [34], [35], [36], [53], [54], [56], [58].

2.1. Pictures

A *picture* $\mathbb{P}$ over $\mathcal{P}$ consists of the following:

(a) A disc D^2 with basepoint 0 on ∂D^2.

(b) Disjoint discs $\Delta_1,\ldots,\Delta_n$ in the interior of D^2. The disc Δ_λ $(\lambda=1,\ldots,n)$ has $p(\lambda)$ basepoints $0_{\lambda 1},\ldots,0_{\lambda p(\lambda)}$ on $\partial\Delta_\lambda$, encountered in the order $0_{\lambda 1},0_{\lambda 2},\ldots,0_{\lambda p(\lambda)}$ if we start at $0_{\lambda 1}$ and travel once around $\partial\Delta_\lambda$ in the clockwise direction.

(c) A finite number of disjoint arcs $\alpha_1,\ldots,\alpha_m$. Each arc lies in the closure of $D^2-\bigcup_{\lambda=1}^{n}\Delta_\lambda$ and is either a simple closed curve having trivial intersection with $\partial D^2\cup\partial\Delta_1\cup\ldots\cup\partial\Delta_n$, or is a simple non–closed curve which joins two points of $\partial D^2\cup\partial\Delta_1\cup\ldots\cup\partial\Delta_n$, neither point being a basepoint. Each arc has a normal orientation, indicated by a short arrow meeting the arc transversely, and is labelled by an element of $\mathbf{x}\cup\mathbf{x}^{-1}$.

(d) If we travel around $\partial\Delta_\lambda$ in the clockwise direction from $0_{\lambda r}$ to $0_{\lambda,r+1}$ and read off the labels on the arcs encountered (with the understanding that if we cross an arc, labelled y say, in the direction of its normal orientation then we read y, whereas if we cross the arc against the orientation we read y^{-1}) then we obtain a word S_λ (*independent of* r) where S_λ is not a proper power and $S_\lambda^{p(\lambda)}$ belongs to $\mathbf{r}\cup\mathbf{r}^{-1}$. We call $S_\lambda^{p(\lambda)}$ *the label on* Δ_λ, and denote it by $W(\Delta_\lambda)$. The *label on* $\mathbb{P}$ (denoted $W(\mathbb{P})$) is the word we read off by travelling around ∂D^2 once in the clockwise direction starting at 0.

The *components of* $\mathbb{P}$ are the connected components of $\bigcup_{\lambda=1}^{n}\Delta_\lambda\cup\bigcup_{\mu=1}^{m}\alpha_\mu$.

We remark that when we refer to the discs of $\mathbb{P}$ we mean the discs $\Delta_1,\Delta_2,\ldots,\Delta_n$, but not the ambient disc D^2. We define $\partial\mathbb{P}$ to be ∂D^2.

We say that $\mathbb{P}$ is *spherical* if no arc meets $\partial\mathbb{P}$.

Remarks. (a) We do not distinguish between pictures which are isotopic.

(b) Most authors only allow one basepoint on each disc of a picture. The reader is referred to §2.4 below for a discussion of this.

Example 1. $\mathcal{P}=\langle a,b;a^3,b^2,(ab)^2\rangle$

(The broken lines in this example represent a *spray*, which will be defined shortly.)

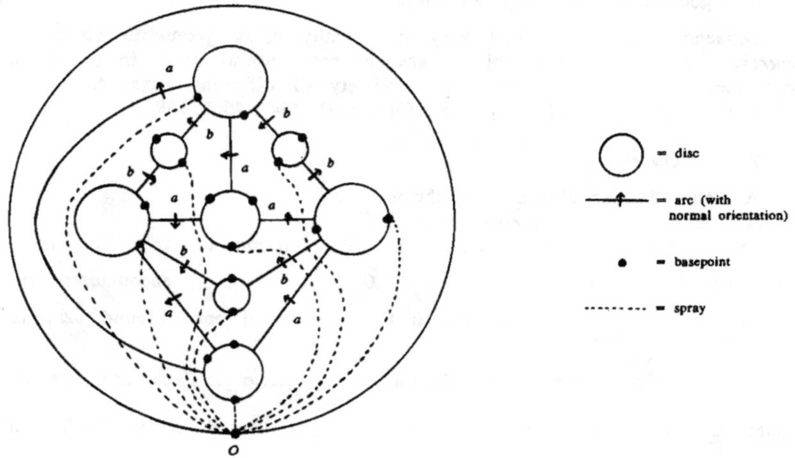

(This, and other examples, are described in [5, Proposition III.2.1].)

Example 2. $\mathscr{P} = \langle a,b,c;[a,b],(bc)^2,[c,a],c^2 \rangle$

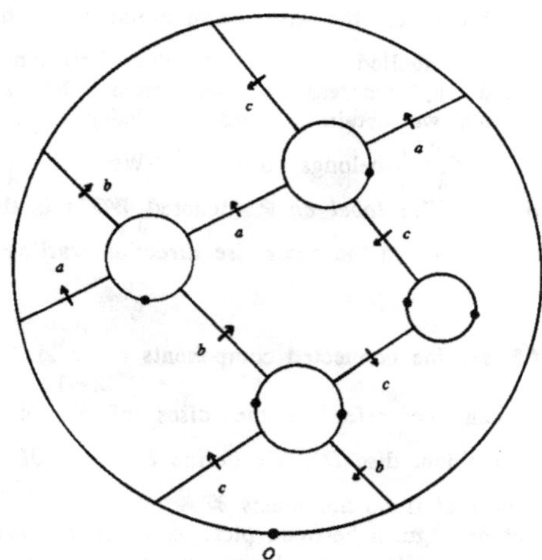

The first example above is spherical.

The above examples have only one component. In general a picture will have several components. Note that a component may consist of a single arc (which is then either a simple closed curve, or is a simple non–closed curve starting and ending on the boundary of the picture).

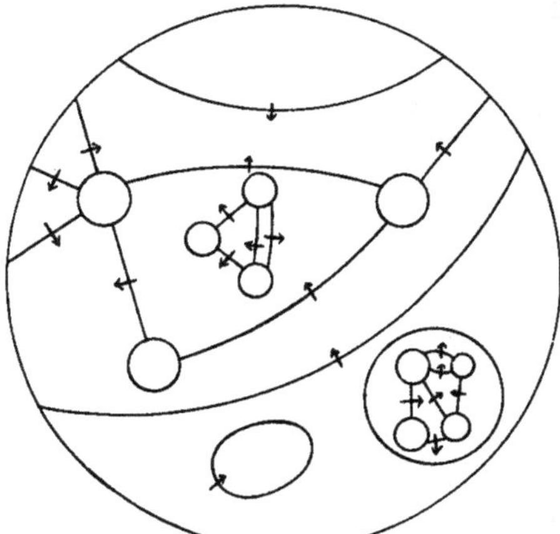

We define an *atomic picture* to be one with a single disc Δ, and which has the property that any arc α not meeting Δ is such that Δ and the basepoint 0 of the picture lie on opposite sides of α.

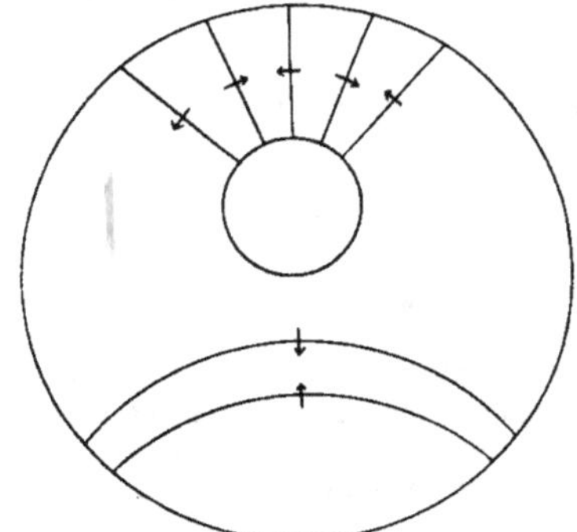

The *mirror image* of a picture $\mathbb{P}$ will be denoted by $-\mathbb{P}$. We can form the *sum* $\mathbb{P}+\mathbb{P}'$ of two pictures $\mathbb{P}$, $\mathbb{P}'$ in the obvious way.

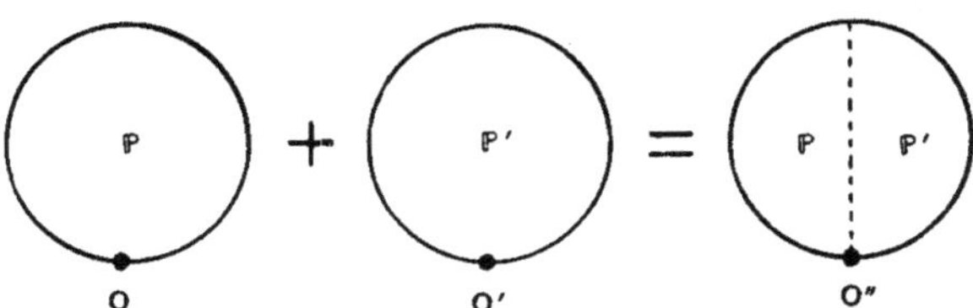

We will write $\mathbb{P}-\mathbb{P}'$ for $\mathbb{P}+(-\mathbb{P}')$.

Let $\mathbb{P}$ be a picture as defined above. A *transverse path* in $\mathbb{P}$ is a path in the closure of $D^2-\bigcup_{\lambda=1}^{n} \Delta_\lambda$ which intersects the arcs of $\mathbb{P}$ only finitely many times (moreover, if the path intersects an arc then it crosses it, and doesn't just touch it). Since we will only ever consider transverse paths, we will from now on drop the use of the adjective "transverse".

If we travel along a path γ from its initial point to its terminal point we will cross various arcs, and we can read off the labels on these arcs, giving a word $W(\gamma)$, the *label on* γ.

A *spray* for $\mathbb{P}$ is a sequence $\underline{\gamma}=(\gamma_1,\gamma_2,\ldots,\gamma_n)$ of simple paths satisfying the following: for $\lambda=1,2,\ldots,n$, γ_λ starts at 0 and ends at a basepoint of $\Delta_{\vartheta(\lambda)}$, where ϑ is a permutation of $\{1,2,\ldots,n\}$ (depending on $\underline{\gamma}$); for $1\le\lambda<\mu\le n$, γ_λ and γ_μ intersect only at 0; travelling around 0 clockwise in $\mathbb{P}$ we encounter the paths in the order $\gamma_1,\gamma_2,\ldots,\gamma_n$. The *sequence* $\sigma(\underline{\gamma})$ *associated with* $\underline{\gamma}$ is

$$(W(\gamma_1)W(\Delta_{\vartheta(1)})W(\gamma_1)^{-1},\ldots,W(\gamma_n)W(\Delta_{\vartheta(n)})W(\gamma_n)^{-1}).$$

Example 1 (ctd).

Denote $a^3,b^2,(ab)^2$ by R,S,V respectively. Then the sequence associated with the spray illustrated is

$$(a^{-1}Va,a^{-2}Va^2,a^{-2}b^{-1}a^{-1}S^{-1}aba^2,\ a^{-2}S^{-1}a^2,R^{-1},abR^{-1}b^{-1}a^{-1},\ abaS^{-1}a^{-1}b^{-1}a^{-1},V)$$

A picture will be said to *represent* a sequence σ if there is a spray for the picture whose associated sequence is σ. Note that if $\mathbb{P}$ represents σ then $-\mathbb{P}$ represents σ^{-1}. Note also that if $\mathbb{P}_1,\mathbb{P}_2$ represent σ_1,σ_2 respectively then $\mathbb{P}_1+\mathbb{P}_2$ represents $\sigma_1\sigma_2$.

We now present some elementary but basic facts concerning pictures.

Theorem 2.1. (i) *Every sequence can be represented by a picture.*

(ii) *Every identity sequence can be represented by a spherical picture.*

Proof. (i) It is easy to see that a sequence of length 1 can be represented by an atomic picture. Thus an arbitrary sequence can be represented by a sum of atomic pictures.

(ii) Let σ be an identity sequence and let $\mathbb{P}_0$ be any picture representing σ. If $\mathbb{P}_0$ is not spherical then reading around $\partial\mathbb{P}_0$ from the basepoint we will encounter two consecutive arcs with opposite labels. We then modify $\mathbb{P}_0$ to obtain $\mathbb{P}_1$ as follows.

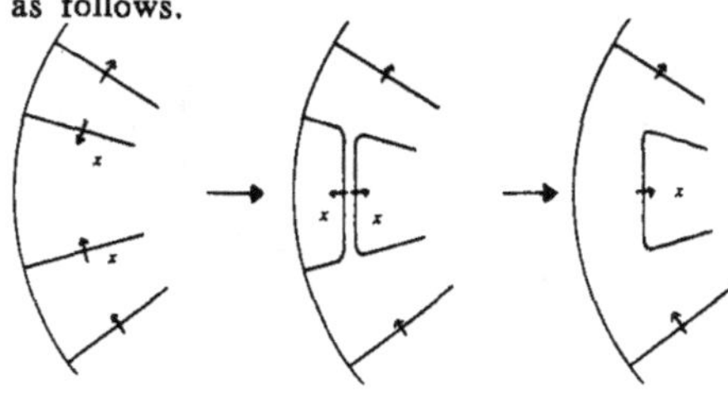

Then $\mathbb{P}_1$ still represents σ. Repeating the above procedure for $\mathbb{P}_1$, and so on, we will eventually obtain a spherical picture.

Remark. The operations used in the proof of (ii) above are examples of certain standard operations on pictures, to be discussed in detail in §2.2.

Theorem 2.2. *If* $\mathbb{P}$ *is a picture and if* $\underline{\gamma}$ is a spray for $\mathbb{P}$, then $\partial(<\sigma(\underline{\gamma})>)=[W(\mathbb{P})]$.

Proof. Cut along the paths of $\underline{\gamma}$ and cut out the discs of $\mathbb{P}$ to obtain a new picture $\mathbb{P}'$ with $W(\mathbb{P}')=W(\mathbb{P})(\Pi\sigma(\underline{\gamma}))^{-1}$.

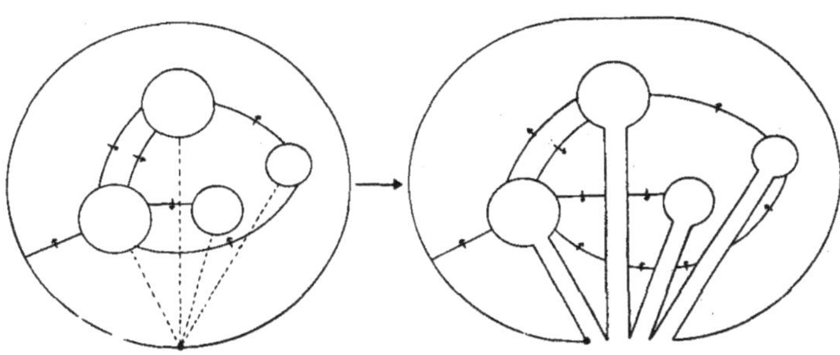

Now $\mathbb{P}'$ has no discs. It is easy to show (by induction on the number of arcs) that the label on a picture with no discs is freely equal to 1.

Corollary. *If* $\mathbb{P}$ *is a spherical picture and if* $\underline{\gamma}$ *is a spray for* $\mathbb{P}$, *then* $\sigma(\underline{\gamma})$ *is an identity sequence.*

Theorem 2.3. *Let* $\mathbb{P}$ *be a picture and let* γ *be a closed path in* $\mathbb{P}$. *Then* $[W(\gamma)]\in N$.

The proof of this is roughly as follows.

First note that if γ is a *simple* closed path then the part of $\mathbb{P}$ enclosed by γ can be regarded as a picture in its own right, and so $[W(\gamma)]\in \mathrm{Im}\partial=N$, by Theorem 2.2.

In the general case, there is a path γ' arbitrarily close to γ such that γ' intersects the arcs of $\mathbb{P}$ in precisely the same places as γ, and moreover, γ' has only finitely many (transverse) self-intersections. Any such self-intersection splits γ' into two closed paths each with fewer self-intersections than γ'. Now use induction.

Theorem 2.4. *Let* $\underline{\gamma}$, $\underline{\gamma}'$ *be two sprays for a picture* $\mathbb{P}$. *Then* $\sigma(\underline{\gamma})$ *and* $\sigma(\underline{\gamma}')$ *are weakly equivalent.*

Proof. We claim that

$$\mathrm{wexp}_{(R,W)}(\sigma(\underline{\gamma}))=\mathrm{wexp}_{(R,W)}(\sigma(\underline{\gamma}'))$$

for all pairs (R,W). To see this, observe that for each term γ of $\underline{\gamma}$ there is a corresponding term γ' of $\underline{\gamma}'$, where γ, γ' join the basepoint of $\mathbb{P}$ to the same disc Δ of $\mathbb{P}$. If S^{ε} ($S\in\mathbf{r}$, $\varepsilon=\pm1$) is the label on Δ then it suffices to show that $[W(\gamma')]\equiv[W(\gamma)S^{k}]$ mod N for some k. But this is clear,for we may join the end point of γ to the endpoint of γ' by a suitable path β with $W(\beta)=S^{k}$ for some k. Then apply Theorem 2.3 to the closed path $\gamma\beta\gamma'^{-1}$.

It now follows from Theorem 1.5 that $<\sigma(\underline{\gamma})\sigma(\underline{\gamma}')^{-1}>\in S'+T$. But by Theorem 2.2, $\sigma(\underline{\gamma})\sigma(\underline{\gamma}')^{-1}$ is an identity sequence. Thus

$$<\sigma(\underline{\gamma})\sigma(\underline{\gamma}')^{-1}>\in(S'+T)\cap\pi_2.$$

But $(S'+T)\cap\pi_2=T$ by Lemma 1.2, and so $\sigma(\underline{\gamma})$ and $\sigma(\underline{\gamma}')$ are weakly equivalent by Theorem 1.4.

2.2. *Operations on pictures (1)*

Certain basic operations ("deformations") can be applied to a picture $\mathbb{P}$ as follows.

(A) Deletion of a closed arc which encircles no discs or arcs of $\mathbb{P}$ (such a closed arc is called a *floating circle*).

$(A)^{-1}$ Insertion of a floating circle.

(B) Deletion of an arc α of $\mathbb{P}$ which starts and ends on $\partial\mathbb{P}$ and which is such that all other arcs and discs of $\mathbb{P}$ lie on the same side of α as the basepoint (such an arc α is called a *floating semicircle*).

$(B)^{-1}$ Insertion of a floating semicircle.

A *cancelling pair* is a spherical picture with exactly two discs.

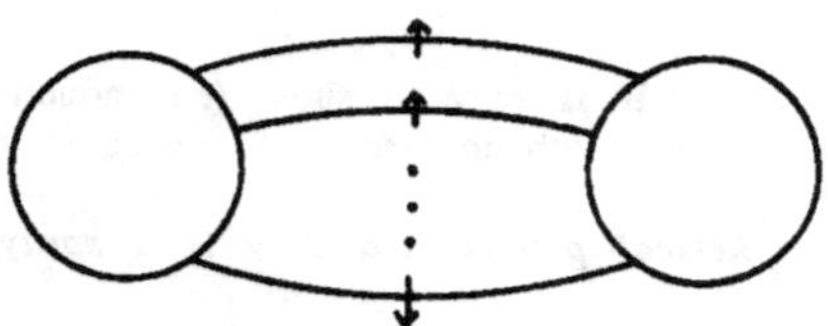

(C) (Deletion of a cancelling pair). If there is a simple closed path β in $\mathbb{P}$ such that the part of $\mathbb{P}$ encircled by β is a cancelling pair, then remove that part of $\mathbb{P}$ encircled by β.

$(C)^{-1}$ (Insertion of a cancelling pair) The opposite of (C).

(D) (Bridge move)

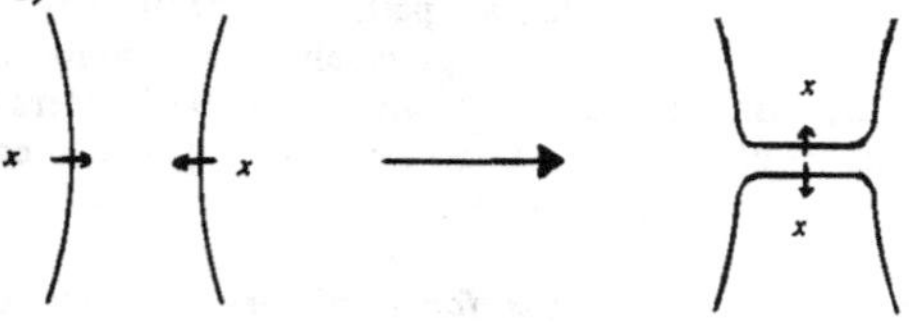

Two pictures will be said to be *equivalent* if either: (a) the pictures are both spherical and one can be transformed to the other by a finite number of operations (A), $(A)^{-1}$, (C), $(C)^{-1}$, (D); or (b) the pictures are not both spherical and one can be transformed to the other by a finite number of operations (A), $(A)^{-1}$, (B), $(B)^{-1}$, (C), $(C)^{-1}$, (D).

Our aim is to prove the following basic result.

Theorem 2.5. *Let σ, σ' be sequences represented by pictures $\mathbb{P}$, $\mathbb{P}'$ respectively. Then σ and σ' are weakly equivalent if and only if $\mathbb{P}$ and $\mathbb{P}'$ are equivalent.*

The "if" part of this theorem can be obtained from the following lemma by induction on the number of elementary operations used to transform $\mathbb{P}$ to $\mathbb{P}'$.

Lemma 2.1. *Suppose $\mathbb{P}_1$ is obtained from $\mathbb{P}_0$ by an elementary operation. Let σ_0, σ_1 be sequences associated with sprays for $\mathbb{P}_0$, $\mathbb{P}_1$ respectively. Then σ_0, σ_1 are weakly equivalent.*

Proof. Since we are working up to weak equivalence, it suffices, by Theorem 2.4, to show that there are sprays $\underline{\gamma}_0$, $\underline{\gamma}_1$ for $\mathbb{P}_0$, $\mathbb{P}_1$ respectively such that $\sigma(\underline{\gamma}_0)$ and $\sigma(\underline{\gamma}_1)$ are weakly equivalent.

If $\mathbb{P}_1$ is obtained from $\mathbb{P}_0$ by a bridge move, or by insertion or deletion of a floating circle or floating semicircle, then we may choose $\underline{\gamma}_0$, $\underline{\gamma}_1$ such that $\sigma(\underline{\gamma}_0)=\sigma(\underline{\gamma}_1)$. If $\mathbb{P}_1$ is obtained by inserting or deleting a cancelling pair then we may choose $\underline{\gamma}_0$, $\underline{\gamma}_1$ so that one of $\sigma(\underline{\gamma}_0)$, $\sigma(\underline{\gamma}_1)$ is obtained from the other by a reduction.

We now prove the "only if" part of the theorem.

Suppose that σ and σ' are weakly equivalent. We will show shortly how to construct a picture $\mathbb{P}''$ which is equivalent to $\mathbb{P}$ and represents σ'. It will then suffice to show that $\mathbb{P}''$ and $\mathbb{P}'$ are equivalent. This will require the following result.

Lemma 2.2. *If $\mathbb{P}_0$, $\mathbb{P}_1$ are pictures representing the same sequence then $\mathbb{P}_0 - \mathbb{P}_1$ is equivalent to the empty picture.*

Granted this lemma, the fact that $\mathbb{P}'$ and $\mathbb{P}''$ are equivalent can be deduced as follows. Since $-\mathbb{P}''+\mathbb{P}''$ is equivalent to the empty picture, $\mathbb{P}'-\mathbb{P}''+\mathbb{P}''$ is equivalent to $\mathbb{P}'$. But since $\mathbb{P}'-\mathbb{P}''$ is equivalent to the empty picture, $\mathbb{P}'-\mathbb{P}''+\mathbb{P}''$ is equivalent to $\mathbb{P}''$.

Proof of Lemma 2.2. Since $\mathbb{P}_0$ and $\mathbb{P}_1$ have sprays whose associated sequences are identical, it follows that $\mathbb{P}_0 - \mathbb{P}_1$ has a spray $(\gamma_1^{(0)},\gamma_2^{(0)},\ldots,\gamma_n^{(0)},\gamma_n^{(1)},\ldots,\gamma_2^{(1)},\gamma_1^{(1)})$ such that the associated sequence has the form $(c_1,c_2,\ldots,c_n,c_n^{-1},\ldots,c_2^{-1},c_1^{-1})$. Suppose $\gamma_\lambda^{(0)},\gamma_\lambda^{(1)}$ join the basepoint of $\mathbb{P}_0 - \mathbb{P}_1$ to discs $\Delta_\lambda^{(0)},\Delta_\lambda^{(1)}$ respectively $(\lambda=1,\ldots,n)$. Then we can draw disjoint simple closed paths $\beta_1, \beta_2, \ldots, \beta_n$ in $\mathbb{P}_0 - \mathbb{P}_1$, where for $\lambda = 1, \ldots, n$, β_λ lies within a suitably small neighbourhood of $\gamma_\lambda^{(0)} \cup \gamma_\lambda^{(1)} \cup \Delta_\lambda^{(0)} \cup \Delta_\lambda^{(1)}$, and where the part of $\mathbb{P}_0 - \mathbb{P}_1$ inside β_λ contains the two discs $\Delta_\lambda^{(0)} \cup \Delta_\lambda^{(1)}$ and has the form $\mathbb{A}_\lambda - \mathbb{A}_\lambda$, $\mathbb{A}_\lambda$ atomic. Now each $\mathbb{A}_\lambda - \mathbb{A}_\lambda$ can be transformed, using bridge moves and the elimination of a cancelling pair, to a picture with no discs. It follows that $\mathbb{P}_0 - \mathbb{P}_1$ is equivalent to a picture with no discs. Such a picture is easily proved to be equivalent to the empty picture by induction on the number of arcs.

It remains to construct $\mathbb{P}''$. It suffices, using Lemma 1.3 and induction, to deal with the case when σ' is obtained from σ by an operation (SUB), (EX), (W/INS*), (W/DEL*).

Let $\underline{\gamma} = (\gamma_1, \gamma_2, \ldots, \gamma_n)$ be a spray for $\mathbb{P}$ whose associated sequence is σ.

A substitution operation on σ amounts to changing the labels on the paths in $\underline{\gamma}$ by a succession of free insertions and free deletions. These can be realised as follows.

Free insertion

Free deletion

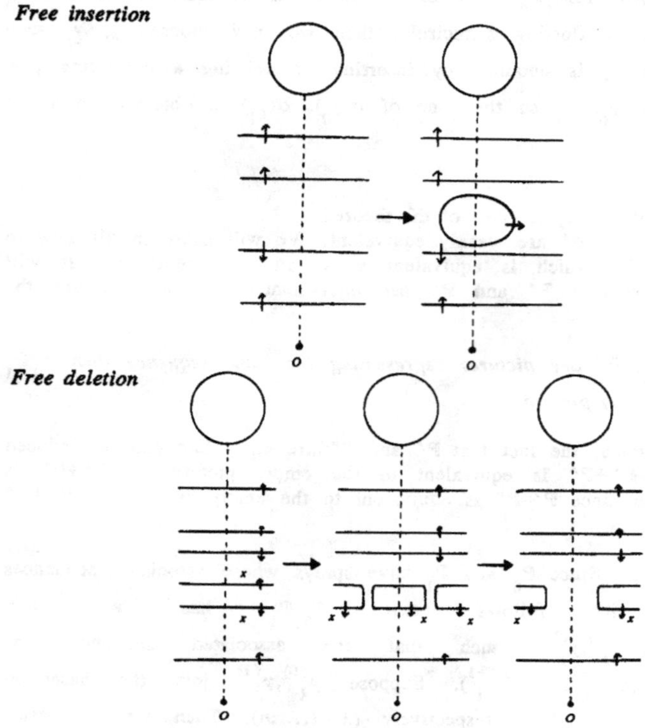

Now consider an exchange operation. Suppose that σ' is obtained from σ by replacing two consecutive terms c_λ, $c_{\lambda+1}$ by $c_\lambda c_{\lambda+1} c_\lambda^{-1}, c_\lambda$ (the other case, where c_λ, $c_{\lambda+1}$ is replaced by $c_{\lambda+1}, c_{\lambda+1}^{-1} c_\lambda c_{\lambda+1}$ is similar). We leave $\mathbb{P}$ unchanged, but we replace $\underline{\gamma}$ by another spray whose terms are the same as those of $\underline{\gamma}$ except for the λth and $(\lambda+1)$st terms, which are replaced by γ_λ^*, $\lambda_{\lambda+1}^*$as depicted below.

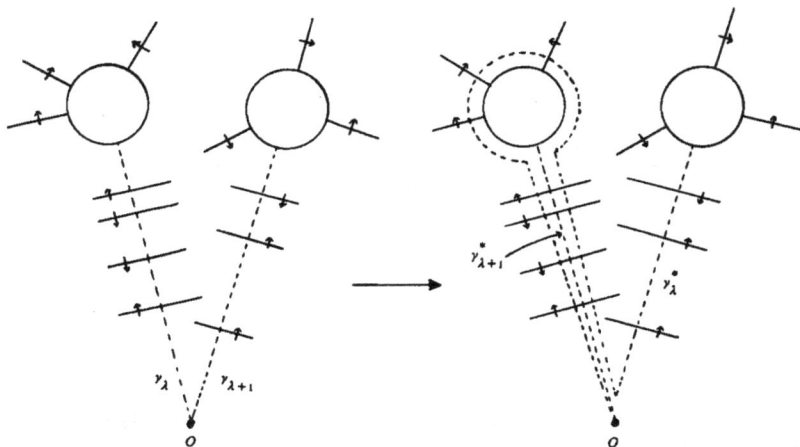

Next, suppose that σ has two consecutive terms c_λ, $c_{\lambda+1}$ where $c_\lambda = WR^\varepsilon W^{-1}$, $c_{\lambda+1} = W\mathring{R}^k R^{-\varepsilon}\mathring{R}^{-k}W^{-1}$ ($W \in \mathbf{w}$, $R \in \mathbf{r}$, $\varepsilon = \pm 1$, k some integer), and suppose that σ' is obtained by removing these terms. By performing a succession of bridge moves on $\mathbb{P}$, and replacing $\gamma_{\lambda+1}$ by a new path $\gamma_{\lambda+1}^*$ (attached to the same disc as $\gamma_{\lambda+1}$, but possibly at a different basepoint) we can reduce to the case $k=0$:

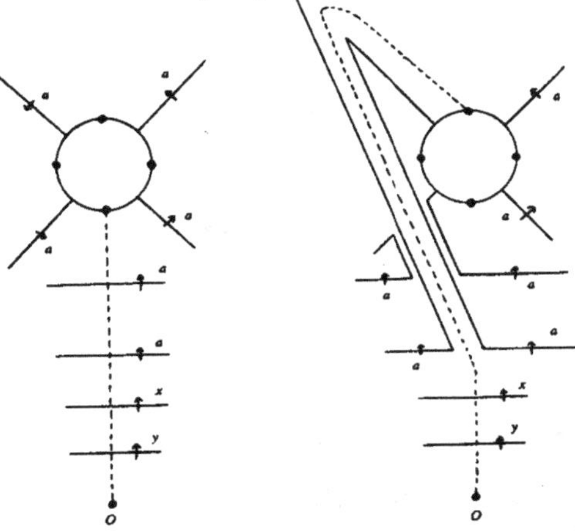

Then there is a simple closed path β lying in a small neighbourhood of γ_λ, $\gamma^*_{\lambda+1}$ and the discs to which they are attached, such that the part of $\mathbb{P}$ enclosed by β has the form $\mathbb{A}$–$\mathbb{A}$, $\mathbb{A}$ atomic. We can perform a succession of bridge moves inside β, followed by the deletion of a cancelling pair to obtain our required picture $\mathbb{P}''$.

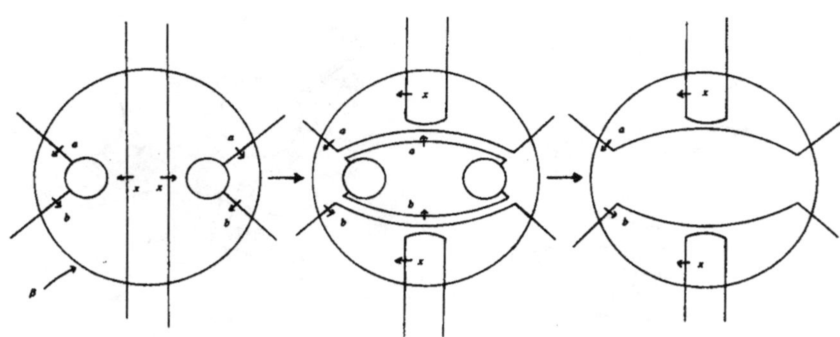

Finally, suppose that σ' is obtained from σ by a weak insertion. We illustrate what to do in this case by a simple example. Suppose that σ' is obtained by inserting the pair $xya^4y^{-1}x^{-1}$, $xya^{-2}a^{-4}a^2y^{-1}x^{-1}$ $(x,y,a \in \mathbf{x},\ a^4 \in \mathbf{r})$ between the μth and $(\mu+1)^{\text{st}}$ terms of σ. Then we obtain a picture representing σ' by inserting floating circles and a cancelling pair into $\mathbb{P}$ as follows.

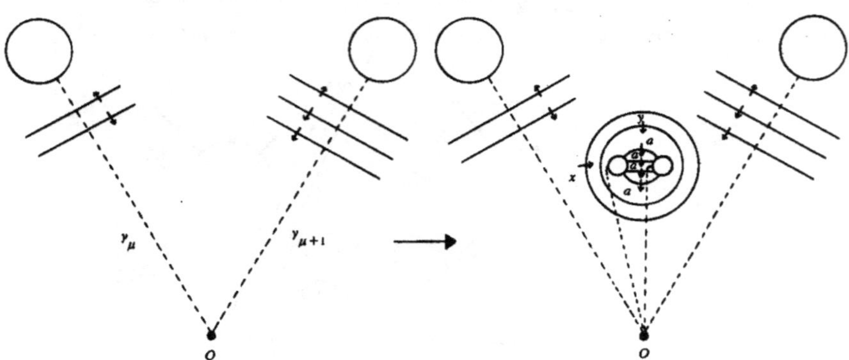

This completes the proof of Theorem 2.5.

2.3. *Operations on pictures* (2)

We can "relativize" the material of §2.2.

Consider a collection $\mathbf{X}=\{\mathbb{P}_\lambda:\lambda\in\Lambda\}$ of spherical pictures. For each λ, let σ_λ be an identity sequence arising from some spray for $\mathbb{P}_\lambda$. Let $J(\mathbf{X})$ be the submodule of π_2 generated by the elements $<\sigma_\lambda>$ $(\lambda\in\Lambda)$.

We introduce two further operations on pictures as follows.

(E) (Deletion of an **X**–picture) If there is a simple closed path β in a picture such that the part of the picture enclosed by β is a copy of $\mathbb{P}_\lambda$ on $-\mathbb{P}_\lambda$ for some λ, then delete that part of the picture enclosed by β.

$(E)^{-1}$ (Insertion of an **X**–picture) The opposite of (E).

Two pictures will be said to be *equivalent* (*rel***X**) if either: (a) the pictures are both spherical and one can be transformed to the other by a finite number of operations $(A)^{\pm1}$, $(C)^{\pm1}$, (D), $(E)^{\pm1}$; or (b) the pictures are not both spherical and one can be transformed to the other by a finite number of operations $(A)^{\pm1}$, $(B)^{\pm1}$, $(C)^{\pm1}$, (D), $(E)^{\pm1}$.

Theorem 2.6. *Let σ, σ' be sequences represented by pictures $\mathbb{P}$, $\mathbb{P}'$ respectively. Then $<\sigma'>-<\sigma>\in J(\mathbf{X})+T$ if and only if $\mathbb{P}$ and $\mathbb{P}'$ are equivalent (rel***X***).*

To prove the "if" part, we only need to generalize Lemma 2.1 to cover the operations (E), $(E)^{-1}$. But if $\mathbb{P}_1$ is obtained from $\mathbb{P}_0$ by insertion or deletion of an **X**–picture, then we may choose sprays $\underline{\gamma}_0$, $\underline{\gamma}_1$ for $\mathbb{P}_0$, $\mathbb{P}_1$ respectively such that

$$<\sigma(\underline{\gamma}_0)>-<\sigma(\underline{\gamma}_1)>=\varepsilon[W]N\cdot<\sigma_\lambda>$$

for some $\lambda\in\Lambda$, $\varepsilon\in\{-1,1\}$, $W\in\mathbf{w}$.

To prove the "only if" part, suppose $<\sigma'>-<\sigma>\in J(\mathbf{X})+T$. Then (taking account of Theorem 1.4), we have that σ' is weakly equivalent to a sequence

$$\sigma''=\sigma(W_1\sigma_{\lambda_1}^{\varepsilon_1}W_1^{-1})\dots(W_n\sigma_{\lambda_n}^{\varepsilon_n}W_n^{-1}),$$

where $\lambda_1,\dots,\lambda_n\in\Lambda$, $W_1,\dots,W_n\in\mathbf{w}$, $\varepsilon_1,\dots,\varepsilon_n\in\{-1,1\}$. By inserting floating circles and copies of the pictures $\mathbb{P}_{\lambda_1},\dots,\mathbb{P}_{\lambda_n}$ or their mirror images (according to the signs of $\varepsilon_1,\dots,\varepsilon_n$) we can obtain from $\mathbb{P}$ a picture $\mathbb{P}''$ representing σ''. Note that $\mathbb{P}''$ is equivalent (rel**X**) to $\mathbb{P}$. By Theorem 2.5, $\mathbb{P}'$ is equivalent to $\mathbb{P}''$. Hence $\mathbb{P}'$ is equivalent (rel**X**) to $\mathbb{P}$, as required.

Corollary 1. *Let* **X** *be a collection of spherical pictures. Then $\pi_2=J(\mathbf{X})+T$ if and only if every spherical picture is equivalent (rel***X***) to the empty picture.*

As a special case of Corollary 1 (take **X** empty) we have the following result (which could also be obtained directly from Theorem 2.5).

Corollary 2. *$\mathcal{P}$ is CA if and only if every spherical picture over $\mathcal{P}$ is equivalent to the empty picture.*

These corollaries are of considerable importance. Given a group presentation $\mathcal{P}$ we look for a collection $\mathbf{X}=\{\mathbb{P}_\lambda:\lambda\in\Lambda\}$ of "obvious" spherical pictures over $\mathcal{P}$. We then try to show geometrically that every spherical picture is equivalent (rel**X**) to the empty picture. If this is the case then π_2 is generated (as a module) by the trivial elements, together with the elements $<\sigma_\lambda>$ ($\lambda \in \Lambda$) (where σ_λ is the sequence arising from some spray for $\mathbb{P}_\lambda$). If we cannot find any "obvious" spherical pictures over $\mathcal{P}$, then we may suspect that $\mathcal{P}$ is CA, and we try to verify this.

Example 3. Let $\mathcal{X}$ be a simple graph with vertex set **x** and edge set **e**, and let $\mathbf{e}^+$ be an orientation of **e**. Let ϕ be a function from $\mathbf{e}^+$ to {2,3,4,...}, and let

$$\mathcal{P}(\mathcal{X},\phi) = <\mathbf{x};\ x^2\ (x \in \mathbf{x}),\ (\iota(e)\tau(e))^{\phi(e)}\quad (e \in \mathbf{e}^+)>.$$

Then $\mathcal{P}(\mathcal{X},\phi)$ is called a *Coxeter presentation* (and the group it defines is called a *Coxeter group*). For $e \in \mathbf{e}^+$, with $\iota(e)=x$, $\tau(e)=y$ say, we have a spherical picture

$\mathbb{P}_e$:

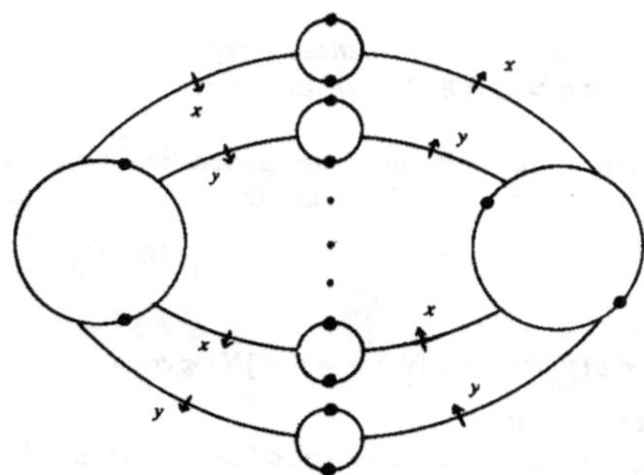

It is shown in [ ] (see Theorem 1) that if $\mathbf{X} = \{\mathbb{P}_e : e \in \mathbf{e}^+\}$ then $\pi_2 = J(\mathbf{X}) + T$ provided the following condition holds: for any three edges e_1, e_2, $e_3 \in \mathbf{e}^+$ which form a triangle,

$$\frac{1}{\phi(e_1)} + \frac{1}{\phi(e_2)} + \frac{1}{\phi(e_3)} \le 1.$$

For other examples, see [5], [52], [53], and §3 below.

Exercise Show that if $\mathbb{P}$ is the picture in Example 1, then $\pi_2 = J(\mathbb{P}) + T$.

Exercise Let $\mathcal{P} = <a;a^2,a>$, $\mathcal{P}_0 = <a;a^2>$. Show:

(a) $\mathcal{P}_0$ is CA

(b) $\pi_2(\mathcal{P}) = J(\mathbb{P})$, where $\mathbb{P}$ is the picture

(c) the natural map $j:\pi_2(\mathscr{P}_0)\to\pi_2(\mathscr{P})$ (see §1.5) is 0.

2.4. Remark on basepoints

In our definition of picture the number of basepoints on a disc is equal to the period of the label on that disc. However, most authors only allow one basepoint on each disc.

Let us use the term "*–picture" for pictures where each disc has only one basepoint. There is the obvious notion of a spray for *–pictures. Then we have:

Theorem 2.4* *If γ, γ' are two sprays for a *-picture, then $\sigma(\underline{\gamma})$, $\sigma(\underline{\gamma}')$ are (Peiffer) equivalent.*

In considering operations on *–pictures one has to be careful about the meaning of a "cancelling pair". A cancelling pair (sometimes called a folding pair) is a spherical *–picture with exactly two discs *such that the basepoints of the discs both lie in the same region.* Thus

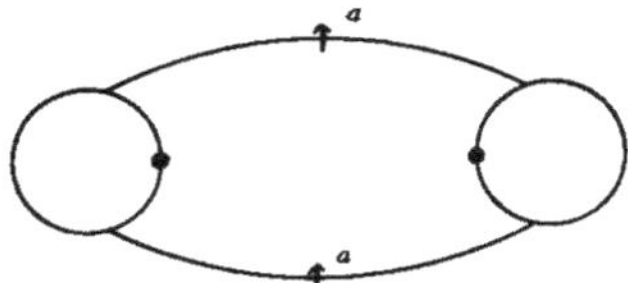

is a cancelling pair, whereas

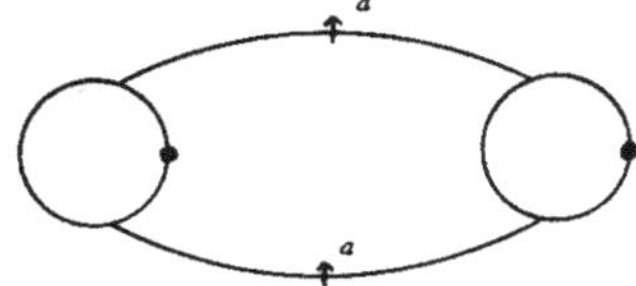

is not. We then have the notion of equivalence of *–pictures, and the analogue of Theorem 2.5 is

Theorem 2.5* *Let σ,σ' be sequences represented by *-pictures $\mathbb{P}$, $\mathbb{P}'$ respectively. Then σ and σ' are equivalent if and only if $\mathbb{P}$ and $\mathbb{P}'$ are equivalent.*

The advantage of *–pictures is that they reflect exactly the basic notion of (Peiffer) equivalence. The disadvantage is that one always has to be very careful about cancelling pairs.

3. Asphericity

The aim of this section is to give examples of aspherical and CA presentations. We remark that there are other notions of asphericity in combinatorial group theory. However, these will not be discussed here. The interested reader is referred to [13].

Call a picture *reduced* if it does not contain two discs Δ, Δ' joined by an arc α, such that the word obtained by reading around $\partial\Delta$ once in the clockwise direction starting at $\alpha\cap\Delta$ is the same as that obtained by reading around $\partial\Delta'$ once in the anticlockwise direction starting at $\alpha\cap\Delta'$:

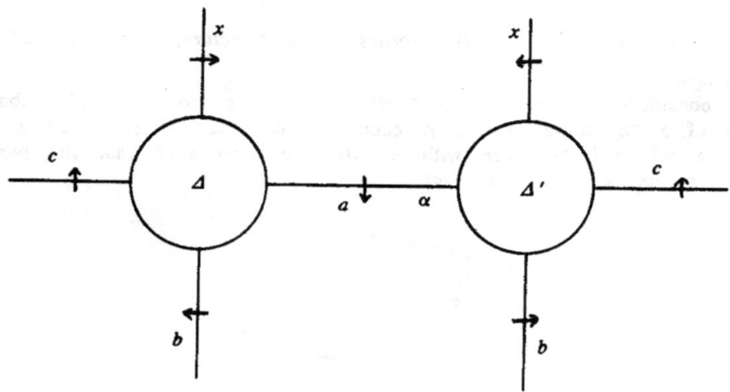

It is easy to see that if a picture is not reduced then by applying bridge moves we can create a cancelling pair, which can then be deleted, giving a picture with two fewer discs. Using this remark, and Theorem 2.6, Corollary 2, we deduce that $\mathcal{P}$ is CA if the following condition is satisfied:

(*) *no spherical picture over $\mathcal{P}$ having at least one disc is reduced.*

Remark. Presentations in which each relator has period 1, and for which (*) holds are called *diagrammatically reducible* (DR) (see [2], [24]).

Some examples where (*) holds include:

(a) *One-relator (and more generally "staggered") presentations.* See Proposition III.9.7 of [46]. (Note that that proposition is phrased in terms of diagrams on spheres. The dual of such a diagram is essentially a connected spherical picture – however, the arcs are now labelled by words rather than generators. This duality between diagrams and pictures is relevant to some of the other examples mentioned below.)

The combinatorial asphericity of one–relator presentations was originally proved by Lyndon in his pioneering paper [44].

(b) *Presentations satisfying small cancellation conditions* $C(p)$, $T(q)$ $\left(\frac{1}{p}+\frac{1}{q}=\frac{1}{2}\right)$. See Lyndon [45] (one essentially uses Corollary 2.5 of [45].)

(c) *Presentations satisfying the small cancellation condition* $W(6)$. See

Juhász [41].

(d) *Some presentations considered by Kanevskii.* Let Λ be a set of ordered triples of distinct elements of a set $\mathbf{x}$, where if two triples have more than one element in common then they coincide. Associated with this set–up we have a presentation

$$< \mathbf{x};\ x^2\ (x \in \mathbf{x}),\ (xyz)^2\ ((x,y,z) \in \Lambda) > \tag{3.1}$$

Such presentations were considered by Kanevskii in [42] (see also [43], [51]). Kanevskii was interested in the torsion in the group defined by such a presentation. Now it turns out that for the presentation (3.1), (*) holds, and so Kanevskii's result (and in fact a more general result) follows from Theorem 1.9(ii).

The fact that (3.1) satisfies (*) can be proved as follows. Suppose there were a connected reduced spherical picture $\mathbb{P}$ over the presentation with at least one disc. Contract the discs of $\mathbb{P}$ to points (vertices) to obtain a tesselation of the 2–sphere. The contraction of discs labelled $x^{\pm 2}$ for some $x \in \mathbf{x}$ gives vertices of valence 2 which can be ignored. Then in the resulting tesselation all vertices have valence 6, and (since $\mathbb{P}$ was reduced) each region has at least 3 sides. This contradicts Euler's formula.

The above can be generalized. Instead of (3.1) we can consider a presentation

$$< \mathbf{x};\ x^2\ (x \in \mathbf{x}),\ (xyz)^{n(x,y,z)}\ ((x,y,z) \in \Lambda) >,$$

where $n(x,y,z) \geq 2$.

There is a useful "weight test" which gives sufficient conditions for (*) to hold. Let $\mathcal{P}^{st}$ be the *star-graph* (also known as the *star-complex*, or *coinitial graph* of $\mathcal{P}$) defined as follows. The vertex set is $\mathbf{x} \cup \mathbf{x}^{-1}$ and the edge set is the set $\mathbf{r}^*$ of all cyclic permutations of elements of $\mathbf{r} \cup \mathbf{r}^{-1}$. The initial, terminal and inversion functions are specified for $R \in \mathbf{r}^*$ by:

$\iota(R)$ = first symbol of R

$\tau(R)$ = inverse of last symbol of R

inverse edge of $R = R^{-1}$.

Now it turns out that $\mathcal{P}$ satisfies (*) if we can find a real–valued function ("weight function") ϑ on the edge set $\mathbf{r}^*$ of $\mathcal{P}^{st}$ such that the following three conditions are satisfied:

(a) $\vartheta(R^{-1}) = \vartheta(R)$ for all $R \in \mathbf{r}^*$.

(b) For every cyclically reduced closed path in $\mathcal{P}^{st}$, the sum of the ϑ values of the edges making up the path is at least 2.

(c) For each $R \in \mathbf{r}$, with $R = x_1 x_2 \ldots x_n$, say ($x_i \in \mathbf{x} \cup \mathbf{x}^{-1}$, $i = 1,\ldots,n$),

$$\sum_{i=1}^{n} (1 - \vartheta(x_i \ldots x_n x_1 \ldots x_{i-1})) \geq 2.$$

See [24], [50]. The weight test had its origins in a "colouring test" due to Sieradski [59]. A more refined test ("Zyklentest") has recently been given in [37].

For other uses of the star–graph see [18], [19], [20], [22], [28], [30], [31], [51].

Exercise. Show that $< a,b,c,x,y;\ abx^2,\ bay^2,\ (bc^{-1})^p,\ (ca)^2 >$ $(p,q \geq 2)$ satisfies (*).

A presentation of a classical knot obtained by discarding one of the relators from the Wirtinger presentation is aspherical. (This follows from work of Papakyriakopoulos [48] – see [12, p.183] for a discussion). Such presentations are examples of ***LOT-presentations***, defined as follows.

Let $\mathcal{T}$ be a tree, with vertex set **x**, and edge set **e**, and let $\mathbf{e}^+$ be an orientation of **e**. Let $\phi:\mathbf{e}^+\to\mathbf{x}$ be some function. Associated with the pair $(\mathcal{T},\phi)$ we have the LOT–presentation

$$<\mathbf{x};\ \iota(e)\phi(e)\tau(e)^{-1}\phi(e)^{-1}\quad(e\in\mathbf{e}^+)>.$$

Open problem. Is every LOT–presentation aspherical?

This question is a test case for the Whitehead problem mentioned in §1.4, for LOT–presentations are subpresentations of aspherical presentations. To see this, let $\mathcal{P}=<\mathbf{x};\mathbf{r}>$ be an LOT–presentation. In the group $G=F/N$ defined by $\mathcal{P}$ all elements of the generating set $\{[x]N:x\in\mathbf{x}\}$ are conjugate. Thus if x_0 is some fixed element of **x** then the group defined by the presentation $\mathcal{P}_0=<\mathbf{x};\mathbf{r},x_0>$ is trivial. Moreover, $\mathcal{P}_0$ has the same number of generators as relators. Thus $\mathcal{P}_0$ is aspherical (see [13, Lemma 1.6]).

For partial results on the asphericity of LOT–presentations, see [2], [33], [37]. The paper [4] is also relevant.

Ol'shanskii [47] has shown that certain free Burnside groups have CA presentations (see also [1], [40]). Further examples of CA presentations can be found in [3], [16]. Necessary and sufficient conditions are given in [13] (see Theorems 3.1, 4.3) for presentations of *HNN* extensions and amalgamated products to be CA. In [52], Theorems 2, 3, sufficient conditions are given for the combinatorial asphericity of presentations in which each relator involves at most two (types of) generators.

Finally, we mention briefly the work on *relative* presentations in [6]. A relative presentation is a triple $<H,\mathbf{x};\mathbf{r}>$, where H is a group, **x** is a set, and **r** is a set of words on $H\cup\mathbf{x}\cup\mathbf{x}^{-1}$. One can define the concept of a picture over a relative presentation, and one can then consider relative presentations for which an analogue of (*) holds. Such relative presentations are called *aspherical* in [6]. Group–theoretic and cohomological consequences of the asphericity condition are discussed, and numerous examples are given.

We also mention the papers [34], [35], [36], which are in a similar vein as [6].

References

1. I.S. Ashmanov and A. Yu Ol'shanskii, On abelian and central extensions of aspherical groups, *Soviet Math. (Iz. VUZ)* **29** (1985), 65-82.
2. W.A. Bogley, Retractive maps and local collapsibility, Ph.D. Thesis, University of Oregon, 1987.
3. W.A. Bogley, An identity theorem for multi-relator groups, *Math. Proc. Camb. Phil. Soc.* (to appear).
4. W.A. Bogley, On the intersection of efficient normal factors of a finitely generated free group, preprint, 1990.
5. W.A. Bogley and M.A. Gutiérrez, Mayer Vietoris sequences in homotopy of 2-complexes and in homology of groups, *J. Pure Appl. Alg.* (to appear).
6. W.A. Bogley and S.J. Pride, Aspherical relative presentations, *Proc. Edin. Math. Soc.* (to appear).
7. J. Brandenberg and M. Dyer, On J.H.C. Whitehead's aspherical question I, *Comment. Math. Helvetici* **56** (1981), 431-446.
8. J. Brandenberg, M. Dyer and R. Strebel, On J.H.C. Whitehead's aspherical question II, in: *Low Dimensional Topo l ogy* (S.L. Lomonaco, ed.), *Contemporary Mathematics* **20** (1983), 65-78.
9. R. Brown, On the second relative homotopy group of an adjunctive space: an exposition of a theorem of J.H.C. Whitehead, *J. London Math. Soc.* (2) **22** (1980), 146-152.
10. R. Brown, Coproducts of crossed P-modules: applications to second homotopy groups and to the homology of groups, *Topology* **23** (1984), 337-3435.
11. R. Brown, Some problems in non-abelian homotopical and homological algebra, in: *Homotopy Theory and Related Topics* (M. Mimura, ed.) *Springer Lecture Notes in Mathematics* **1419** (1990), 105-129.
12. R. Brown and J. Huebschmann, Identies among relations, in: *Low-Dimensional Topology* (R. Brown and T.L. Thickstun, eds.), *LMS Lecture Note Series* **48** (1982), 153-202.
13. I.M. Chiswell, D.J. Collins and J. Huebschmann, Aspherical group presentations, *Math. Z.* **178** (1981), 1-36.
14. D.J. Collins and J. Huebschmann, Spherical diagrams and identities among relations, *Math. Ann.* **261** (1982), 155-183.
15. M. Dyer, Subcomplexes of two complexes and projective crossed modules, in: *Combinatorial Group Theory and Topology* (S. Gersten and J. Stallings, eds.), *Annals of Maths. Studies* **111** (1987), Princeton University Press, 255-264.
16. E. Dyer and A.J. Vasquez, Some small aspherical spaces, *J. Austral. Math. Soc.* **16** (1973), 332-352.
17. A.J. Duncan and J. Howie, The genus problem for one-relator products of locally inducible groups, preprint, Heriot-Watt University, Edinburgh, 1989.
18. M. Edjvet and J. Howie, Star-graphs, projective planes and free subgroups in small cancellation groups, *Proc. London Math. Soc.* (3) **57** (1988), 301-328.
19. M.S. El-Mosalamy, Free subgroups of small cancellation groups, *Israel J. Math.* **56** (1986), 345-348.
20. M.S. El-Mosalamy, *Applications of star-complexes in group theory*, Ph.D. Thesis, University of Glasgow, 1987.
21. R.A. Fenn, Techniques of Geometric Topology, *London Mathematical Society Lecture Note Series* **57** (1983), CUP.
22. E.J. Fennessey and S.J. Pride, Equivalences of two-complexes, with

applications to NEC-groups, *Math. Proc. Camb. Phil. Soc.* **106** (1989), 215-228.

23. S.M. Gersten, Amalgamations and the Kervaire problem, *Bulletin Amer. Math. Soc.* (New Series), **17** (1987), 105-108.
24. S.M. Gersten, Reducible diagrams and equations over groups, in: *Essays in Group Theory* (S.M. Gersten, ed.), *MSRI Publications* **8** (1987), 15-73.
25. F. Gonzalez Acuña and H. Short, Knot surgery and primeness, *Math. Proc. Camb. Phil. Soc.* **99** (1986), 89-102.
26. M.A. Gutiérrez and P.S. Hirschhorn, Free simplicial groups and the second relative homotopy group of an adjunction space, *J. Pure Appl. Alg.* **39** (1986), 119-123.
27. M.A. Gutiérrez and J.G. Ratcliffe, On the second homotopy group, *Quart. J. Math. Oxford* (2) **32** (1981), 45-55.
28. P. Hill, S.J. Pride and A.D. Vella, On the $T(q)$-conditions of small cancellation theory, *Israel J. Math.* **52** (1985), 293-304.
29. P.J. Hilton and U. Stammbach, *A Course in Homological Algebra*, *Graduate Texts in Mathematics* **4**, Springer-Verlag, 1971.
30. A.H.M. Hoare, A. Karrass and D. Solitar, Subgroups of finite index in Fuchsian groups, *Math. Z.* **120** (1971), 289-298.
31. A.H.M. Hoare, A. Karrass and D. Solitar, Subgroups of infinite index in Fuchsian groups, *Math. Z.* **125** (1972), 59-69.
32. J. Howie, Some remarks on a problem of J.H.C. Whitehead, *Topology* **22** (1983), 475-485.
33. J. Howie, On the asphericity of ribbon disc complements, *Trans. Amer. Math. Soc.* **289** (1985), 281-302.
34. J. Howie, The quotient of a free product of groups by a single high-powered relator. I. Pictures. Fifth and higher powers, *Proc. London Math. Soc.* (3) **59** (1989), 507-540.
35. J. Howie, The quotient of a free product of groups by a single high-powered relator. II. Fourth powers, *Proc. London Math. Soc.* (3) **61** (1990), 33-62.
36. J. Howie and R.M. Thomas, On the asphericity of presentations for the groups $(2,3,p;q)$ and a conjecture of Coxeter, preprint, University of Leicester, 1990.
37. G. Huck and S. Rosebrock, Hyperbolische Tests auf diagrammitische Reduzierbarkeit für standard 2-Komplexe, preprint, J.W.Goethe Universität, Frankfurt/M, 1990.
38. J. Huebschmann, Cohomology theory of aspherical groups and of small cancellation groups, *J. Pure Appl. Alg.* **14** (1979), 137-143.
39. J. Huebschmann, Aspherical 2-complexes and an unsettled problem of J.H.C. Whitehead, *Math. Ann.* **258** (1981), 17-37.
40. S.V. Ivanov and A. Yu Ol'shanskii, Some applications of graded diagrams in the combinatorial group theory, to appear in *Proceedings Groups - St Andrews*, 1989.
41. A. Juhász, Small cancellation theory with a unified small cancellation condition I, *J. London Math. Soc.* **40** (1989), 57-80.
42. D.S. Kanevskii, The structure of groups connected with automorphisms of cubic surfaces, *Math. USSR Sbornik* **32** (1977), 252-264.
43. D.S. Kanevskii, On cubic planes and groups connected with cubic varieties, *J. Algebra* **80** (1983), 559-565.
44. R.C. Lyndon, Cohomology theory of groups with a single defining relation, *Ann. of Math.* **52** (1950), 650-665.
45. R.C. Lyndon, On Dehn's algorithm, *Math. Ann.* **166** (1966), 208-228.

46. R.C. Lyndon and P.E. Schupp, *Combinatorial Group Theory*, Springer-Verlag, 1977.
47. A. Yu Ol'shanskii, On a theorem of Novikov-Adian, *Mat. Sb.(N.S.)* **118** (**160**) (1982), 203-235.
48. C.D. Papakyriakopoulos, On Dehn's lemma and the asphericity of knots, *Ann. of Math.* **66** 1-26.
49. C.D. Papakyriakopoulos, Attaching 2-dimensional cells to a complex, *Annals of Math.* **78** (1963), 205-222.
50. S.J. Pride, Star-complexes and the dependence problems for hyperbolic complexes, *Glasgow Math. J.* **30** (1988), 155-170.
51. S.J. Pride, Involutary presentations, with applications to Coxeter groups, NEC groups and groups of Kanevskii, *J. Algebra* **120** (1989), 200-223.
52. S.J. Pride, The (co)homology of groups given by presentations in which each defining relator involves at most two types of generators, *J. Austral. Math. Soc.* (to appear).
53. S.J. Pride and R. Stöhr, Relation modules of groups with presentations in which each relator involves exactly two types of generators, *J. London Math. Soc.* (2) **38** (1988), 99-111.
54 S.J. Pride and R. Stöhr, The (co)homology of aspherical Coxeter groups, *J. London Math. Soc.* (to appear).
55. J.G. Ratcliffe, Free and projective crossed modules, *J. London Math. Soc.* (2) **22** (1980), 66-74.
56. C.P. Rourke, Presentations of the trivial group, in: *Topology of Low-Dimensional Manifolds* (R. Fenn, ed.) *Springer Lecture Notes in Mathematics* **722** (1979), 134-143.
57. J.-P. Serre, *Trees*, Springer-Verlag, 1980.
58. H.B. Short, *Topological methods in group theory: the adjunction problem*, Ph.D. Thesis, University of Warwick, 1984.
59. A. Sieradski, A colouring test for asphericity, *Quart. J. Math. Oxford* (2) **34** (1983), 97-106.
60. J.H.C. Whitehead, Combinatorial homotopy I, *Bull. Amer. Math. Soc.* **55** (1949), 213-245.
61. J.H.C. Whitehead, Combinatorial homotopy II, *Bull. Amer. Math. Soc.* **55** (1949), 453-496.
62. H. Hopf, Beitrage zur Homotopietheorie, *Comm. Math. Helv.* **17** (1945), 307-326.

CLOSURES OF TOTALLY GEODESIC IMMERSIONS IN MANIFOLDS OF CONSTANT NEGATIVE CURVATURE

NIMISH A. SHAH
School of Mathematics, Tata Institute of Fundamental Research
Bombay 400 005, INDIA

Abstract

Using techniques of Lie groups and ergodic theory, it can be shown that in a compact manifold of constant negative curvature, the closure of a totally geodesic, complete (immersed) submanifold of dimension atleast 2 is a totally geodesic immersed submanifold. The main purpose of this article is to illustrate some important ideas involved in this method, by giving a proof for the simplest case of a codimension-1 totally geodesic immersed submanifold.

1 Introduction

In this article we prove the following theorem :

Theorem A *Let M be a compact, connected, oriented riemannian manifold with constant negative curvature and dimension $n \geq 3$. Let D be a complete, oriented riemannian manifold, whose connected components are $(n-1)$-dimensional and simply connected and let $\phi : D \to M$ be a totally geodesic immersion. Then $\phi(D)$ is either compact or dense in M.*

Let $\mathcal{F}(D)$ be the oriented orthonormal $(n-1)$-frame bundle over D, $\mathcal{F}(M)$ be the oriented orthonormal n-frame bundle over M and $\phi_* : \mathcal{F}(D) \to \mathcal{F}(M)$ be the immersion induced from ϕ. Then $\phi_*(\mathcal{F}(D))$ is either compact or dense in $\mathcal{F}(M)$.

A riemannian immersion $\phi : D \to M$ is called *totally geodesic* if $\phi \circ \gamma$ is a geodesic in M for every geodesic γ in D.

We shall prove this theorem using Lie groups, discrete subgroups and ergodic transformations on homogeneous spaces. As we shall see in §2, the Theorem A can be reformulated in the group theoretic setup as follows :

Theorem B *Let $G = SO_0(1,n)$, $\Gamma \subset G$ be a discrete subgroup such that $\Gamma \backslash G$ is compact and let $H = SO_0(1,n-1)$, where $n \geq 3$. Then every H-invariant subset of $\Gamma \backslash G$ is either dense or it is a union of finitely many closed H-orbits.*

Certain techniques for studying the closures of orbits have been developed in [M2], [DM1], [DM2] and [DM3]. We shall give an elementry proof of Theorem B closely following the line of arguments in these references.

In the last section we shall discuss some related results of a more general nature.

2 Group theoretic interpretation

For convenience we recall some known facts about hyperbolic spaces and their groups of isometries; (see also [Fl, Preliminaries]).

2.1 *The hyperbolic n-space and its isometry group*

Let $SO_0(1,n)$ denote the connected component of the group of linear transformations of $\mathbb{R}^{n+1}$ preserving the bilinear form

$$\langle \mathbf{x}, \mathbf{y} \rangle = x_0 y_0 - \sum_{i=1}^{n} x_i y_i.$$

$SO_0(1,n)$ acts on $\mathbb{R}^{n+1}$ in the standard way and its orbit through the point $\mathbf{f}_0 = {}^t(1,0,\ldots,0) \in \mathbb{R}^{n+1}$ is a sheet of the hyperboloid

$$\Sigma^n = \{\mathbf{x} \in \mathbb{R}^{n+1} : \langle \mathbf{x}, \mathbf{x} \rangle = 1 \text{ and } x_0 > 0\}.$$

The bilinear form $-\langle \cdot, \cdot \rangle$ restricted to the tangent bundle $\mathrm{T}(\Sigma^n) \subset \Sigma^n \times \mathbb{R}^{n+1}$ is positive definite. With this riemannian structure, Σ^n has the constant sectional curvature -1 and $SO_0(1,n)$ is the group of its oriented isometries.

It is a well-known fact that all equi-dimensional, simply connected, complete riemannian manifolds of a fixed constant sectional curvature are isometric. Hence we call any n-dimensional, simply connected, complete riemannian manifold with constant curvature -1 (for example, $(\Sigma^n, -\langle \cdot, \cdot \rangle)$); *the Hyperbolic n-space* and denote it by $\mathbb{H}^n$.

2.2 *Identifications*

The stabilizer of $\mathbf{f}_0$ in $SO_0(1,n)$ consists of matrices of the form

$$\begin{pmatrix} 1 & \mathbf{0}_{1\times n} \\ \mathbf{0}_{n\times 1} & k \end{pmatrix}, \quad k \in SO(n),$$

where $\mathbf{0}_{i\times j}$ is an $i \times j$ matrix with all entries zero. We obtain the identification,

$$SO_0(1,n)/SO(n) \sim \mathbb{H}^n \tag{1}$$

given by $gSO(n) \sim g\mathbf{f}_0$, for all $g \in SO_0(1,n)$.

Notations. Let M be an oriented riemannian manifold of dimension n. The oriented orthonormal n-frame bundle over M is denoted by $\mathcal{F}(M)$ and the orhtonormal k-frame bundle over M is denoted by $\mathcal{F}^k(M)$, where $1 \leq k \leq n$.

Remark 2.1 Let E be an oriented n-dimensional euclidean vector space. Given any orthonormal $(n-1)$-frame $[\mathbf{v}_1, \ldots, \mathbf{v}_{n-1}]$ in E, there exists unique $\mathbf{v}_n \in E$ such that $[\mathbf{v}_1, \ldots, \mathbf{v}_n]$ is an oriented orthonormal n-frame in E. This shows that for M as above, there is a canonical isomorphism, $\mathcal{F}^{n-1}(M) \simeq \mathcal{F}(M)$.

For $1 \leq i \leq n$, let $\mathbf{f}_i = {}^t(0, \ldots, 1, \ldots, 0) \in \mathbb{R}^{n+1}$, with 1 in the $(i+1)^{\text{th}}$ co-ordinate and 0 in all the others; here ${}^tX$ denotes the transpose of a matrix X. The tangent space to Σ^n at $\mathbf{f}_0$, denoted by $\mathrm{T}_{\mathbf{f}_0}(\Sigma^n)$, is spanned by $\{\mathbf{f}_1, \ldots, \mathbf{f}_n\}$. Now $SO(n)$ acts simply transitively on the set of all oriented orthonormal n-frames in $\mathrm{T}_{\mathbf{f}_0}(\Sigma^n)$. Hence $SO_0(1,n)$ acts simply transitively on $\mathcal{F}(\Sigma^n)$ and we have the identification,

$$SO_0(1,n) \sim \mathcal{F}(\Sigma^n) = \mathcal{F}(\mathbb{H}^n) \tag{2}$$

given by $g \sim [g\mathbf{f}_1, \ldots, g\mathbf{f}_n]_{g\mathbf{f}_0} \subset \mathrm{T}_{g\mathbf{f}_0}(\Sigma^n)$, for all $g \in SO_0(1,n)$.

2.3 Totally geodesic submanifolds of $\mathbb{H}^n$

We want to describe all totally geodesic immersions in to $\mathbb{H}^n$. Observe that if L is a riemannian manifold and σ is an isometry of L, then each connected component of the σ-fixed set in L is a totally geodesic submanifold of L.

For $1 \leq k \leq n-1$, consider the standard inclusions

$$\Sigma^k \hookrightarrow \Sigma^n \quad \text{and} \quad SO_0(1,k) \hookrightarrow SO_0(1,n).$$

Using the above remark it is easy to verify that Σ^k is a totally geodsic submanifold of Σ^n.

Let Ψ be a simply connected, complete riemannian manifold of dimension k and $\phi : \Psi \to \mathbb{H}^n$ be a totally geodesic immersion. Then there exits an isometry $g \in SO_0(1,n)$ such that $\phi(\Psi) = g \cdot \Sigma^k$. In view of the identification 1, we have

$$g \cdot SO(1,k)\,(SO(n)) \sim \phi(\Psi) \subset \mathbb{H}^n. \tag{3}$$

Suppose Ψ as above has dimension $(n-1)$. The derivative $\mathrm{D}\phi : \mathrm{T}(\Psi) \to \mathrm{T}(\Sigma^n)$ induces the immersion $\phi_* : \mathcal{F}(\Psi) \to \mathcal{F}^{n-1}(\Sigma^n)$. Now there exits an isometry $g \in SO_0(1,n)$ such that $\phi(\Psi) = g\Sigma^{n-1}$ and $\phi_*(\mathcal{F}(\Psi)) = g\mathcal{F}(\Sigma^{n-1}) \hookrightarrow \mathcal{F}^{n-1}(\Sigma^n)$. In view of the identifications 1 and 2 and Remark 2.1, we have

$$\begin{aligned} \phi(\Psi) &\sim g \cdot SO_0(1,n-1)/SO(n-1) \hookrightarrow SO_0(1,n)/SO(n), \\ \phi_*(\mathcal{F}(\Psi)) &\sim g \cdot SO_0(1,n-1) \hookrightarrow SO_0(1,n). \end{aligned} \tag{4}$$

2.4 *Totally geodesic immersions in manifolds of constant negative curvature*

Let M be a connected, oriented, n-dimensional, complete riemannian manifold with constant sectional curvature -1. Then the universal covering space of M is isometric to $\mathbb{H}^n$. Now there exits a discrete group Γ consisting of oriented isometries acting properly discontinuously on $\mathbb{H}^n$ such that M is isometric to $\Gamma\backslash\mathbb{H}^n$. Since $\Gamma \subset SO_0(1,n)$, by identifications 1 and 2,

$$\begin{aligned} \Gamma\backslash SO_0(1,n)/SO(n) &\sim \Gamma\backslash\mathbb{H}^n \sim M, \\ \Gamma\backslash SO_0(1,n) &\sim \Gamma\backslash\mathcal{F}(\mathbb{H}^n) \sim \mathcal{F}(M). \end{aligned} \tag{5}$$

Let Ψ be a simply connected, complete, $(n-1)$-dimensional riemannian manifold and $\phi : \Psi \to M$ be a totally geodesic immersion. The derivative $\mathrm{D}\phi : \mathrm{T}(\Psi) \to \mathrm{T}(M)$ induces the immersion $\phi_* : \mathcal{F}(\Psi) \hookrightarrow \mathcal{F}(M)$, where $\mathcal{F}(M)$ is identified with $\mathcal{F}^{n-1}(M)$ by Remark 2.1.

Let $p : \mathbb{H}^n \to M$ be a locally isometric covering. Since Ψ is simply connected, there exits a totally geodesic immersion $\tilde{\phi} : \Psi \to \mathbb{H}^n$ such that $\phi = p \circ \tilde{\phi}$. Hence due to identifications 4 and 5, there exits an isometry $g \in SO_0(1,n)$ such that

$$\begin{aligned} \phi(\Psi) &\sim \Gamma g SO_0(1,n-1)SO(n) \subset \Gamma\backslash SO_0(1,n)/SO(n) \sim M, \\ \phi_*(\mathcal{F}(\Psi)) &\sim \Gamma g SO_0(1,n-1) \subset \Gamma\backslash SO_0(1,n) \sim \mathcal{F}(M). \end{aligned} \tag{6}$$

Using this dual language, Theorem A can be easily derived from Theorem B. The next four sections are devoted to giving a proof of Theorem B. Some notations and preliminaries are set up in §3. The main results needed to prove Theorem B are given in §4 and §5. And the proof of the theorem is completed in §6.

3 Some important subgroups of $SO_0(1,n)$

Let $B = \begin{pmatrix} 1 & 0 \\ 0 & -\mathrm{Id}_{n\times n} \end{pmatrix}$. Then for all $\mathbf{v},\mathbf{w} \in \mathbb{R}^{n+1}$, $\langle \mathbf{v},\mathbf{w}\rangle = {}^t\mathbf{v}B\mathbf{w}$. Hence $G = SO_0(1,n)$ is the connected component of the identity of the group

$$\left\{g \in GL(n+1,\mathbb{R}) : {}^t gBg = B\right\}$$

and its Lie algebra

$$\mathcal{G} = \left\{X \in \underline{gl}(n+1,\mathbb{R}) : {}^t XB + BX = 0\right\}.$$

There is a right Adjoint action Ad of G on $\mathcal{G}$ given by

$$X \cdot \mathrm{Ad}\, g = g^{-1} X g \quad (X \in \mathcal{G},\ g \in G).$$

Let $D = SO_0(1,1) \subset G$ and $\mathcal{D} \subset \mathcal{G}$ be the associated Lie subalgebra. Let $\alpha = \begin{pmatrix} 0 & 1 \\ 1 & 0 \end{pmatrix}$. Then $\exp t\alpha = \begin{pmatrix} \cosh t & \sinh t \\ \sinh t & \cosh t \end{pmatrix}$, for all $t \in \mathbb{R}$. Now

$$\begin{aligned} \mathcal{D} &= \left\{ d(t) \stackrel{\text{def}}{=} \begin{pmatrix} t\alpha & 0_{2\times n-1} \\ 0_{n-1\times 2} & 0_{n-1\times n-1} \end{pmatrix} : t \in \mathbb{R} \right\}, \\ D &= \left\{ d_t \stackrel{\text{def}}{=} \begin{pmatrix} \exp t\alpha & 0_{2\times n-1} \\ 0_{n-1\times 2} & \mathrm{Id}_{n-1\times n-1} \end{pmatrix} : t \in \mathbb{R} \right\}. \end{aligned}$$

With repect to the right Adjoint action of D, the Lie algebra $\mathcal{G}$ decomposes into the direct sum of simultaneous eigenspaces as $\mathcal{G} = \mathcal{G}^+ \oplus \mathcal{G}^0 \oplus \mathcal{G}^-$, where

$$\begin{aligned} \mathcal{G}^+ &= \left\{ n^+(\mathbf{v}) \stackrel{\text{def}}{=} \begin{pmatrix} 0 & 0 & {}^t\mathbf{v} \\ 0 & 0 & -{}^t\mathbf{v} \\ \mathbf{v} & \mathbf{v} & 0_{n-1\times n-1} \end{pmatrix} : \mathbf{v} = \begin{pmatrix} x_1 \\ \vdots \\ x_{n-1} \end{pmatrix} \in \mathbb{R}^{n-1} \right\}, \\ \mathcal{G}^- &= \left\{ n^-(\mathbf{v}) \stackrel{\text{def}}{=} {}^t n^+(\mathbf{v}) : \mathbf{v} \in \mathbb{R}^{n-1} \right\}, \\ \mathcal{G}^0 &= \mathcal{D} \oplus \mathcal{M}, \\ \mathcal{M} &= \left\{ m(A) \stackrel{\text{def}}{=} \begin{pmatrix} 0_{2\times 2} & 0_{2\times n-1} \\ 0_{n-1\times 2} & A_{n-1\times n-1} \end{pmatrix} :\ A + {}^tA = 0 \right\}. \end{aligned}$$

For all $\mathbf{v}, \mathbf{w} \in \mathbb{R}^{n-1}$, $t \in \mathbb{R}$ and $(n-1) \times (n-1)$ skew symmetric matrices A, we have following commutation relations:

$$\begin{aligned} n^+(\mathbf{v}) \cdot \mathrm{Ad}\, d_t &= n^+(e^t \mathbf{v}), \\ n^-(\mathbf{v}) \cdot \mathrm{Ad}\, d_t &= n^-(e^{-t} \mathbf{v}), \\ m(A) \cdot \mathrm{Ad}\, d_t &= m(A) \end{aligned} \tag{7}$$

$$\begin{aligned} \left[n^+(\mathbf{w}), n^+(\mathbf{v})\right] &= 0, \\ \left[m(A), n^+(\mathbf{v})\right] &= n^+(A \cdot \mathbf{v}), \\ \left[n^-(\mathbf{w}), n^+(\mathbf{v})\right] &= 2a({}^t\mathbf{w} \cdot \mathbf{v}) + 2m(\mathbf{w} \cdot {}^t\mathbf{v} - \mathbf{v} \cdot {}^t\mathbf{w}). \end{aligned} \tag{8}$$

Now $\mathcal{G}^+$, $\mathcal{G}^-$ and $\mathcal{M}$ are Lie subalgebrs of $\mathcal{G}$. Let N^+, N^- and M be the connected Lie subgroups associated to $\mathcal{G}^+$, $\mathcal{G}^-$ and $\mathcal{M}$ respectively.

Remark 3.1 The maps $\exp : \mathcal{G}^{\pm} \to N^{\pm}$ are group isomorphisms, hence $N^{\pm}$ are vector groups. Let $u = \exp(n^{+}(\mathbf{v})) \in N^{+}$. Then by Eq. 7, $d_t^{-1} u d_t \to 1$ as $t \to -\infty$. Similary if $v \in N^{-}$ then $d_t^{-1} v d_t \to 1$ as $t \to +\infty$.

The group M is isomorphic to $SO(n-1)$ and the group DM is the centralizer of D in G.

Remark 3.2 Due to Eq. 8, the Lie subalgebras $\mathcal{G}^{+}$ and $\mathcal{G}^{-}$ generate the Lie algebra $\mathcal{G}$. Hence the subgroup generated by N^{+} and N^{-} is dense in G.

4 Ergodic properties of actions on homogeneous spaces

4.1 Ergodic transformations

Definition 4.1 Let X be a topological space and μ be a Borel measure on X. A measure preserving transformation T of (X, μ) is called ***ergodic*** if the following holds: for any measurable set $E \subset X$ if $\mu(T(E) \,\Delta\, E) = 0$ then either $\mu(E) = 0$ or $\mu(X \setminus E) = 0$, where $A \,\Delta\, B \stackrel{\text{def}}{=} A \cup B \setminus A \cap B$.

The following property of ergodic transformations makes the concept of ergodicity very useful for applications.

Lemma 4.1 *Let X be a second countable topological space and μ be a Borel measure on X such that $\mu(E) > 0$ for any non-empty open subset E of X. Let T be an ergodic transformation on (X, μ). Then for μ-almost all $x \in X$, the set $\{T^n x\}_{n \in \mathbf{N}}$ is dense in X.*

Proof. For a nonempty open subset E of X, define

$$X(E) = \bigcup_{n=0}^{\infty} T^{-n}(E).$$

Then $T(X(E)) \supset X(E)$. Now T preserves the measure μ, hence

$$\mu(T(X(E)) \,\Delta\, X(E)) = 0.$$

Since $\mu(E) > 0$, by the ergodicity of T-action $\mu(X(E)) = 1$.

Let $\mathcal{B}$ be a countable open base of X. Let

$$Y = \bigcap_{E \in \mathcal{B} \setminus \emptyset} X(E).$$

Then for all $y \in Y$ the set $\{T^n y\}_{y \in \mathbf{N}}$ is dense in X and $\mu(Y) = 1$. □

Remark 4.1 Let $X = \Gamma \backslash G$. Since Γ is discrete and X is compact, there exists a probability measure μ on X which is invariant under the right action of G on X (see [R, Chap. I]).

Lemma 4.2 (Mautner, cf. [M1]) *The right action of $d = d_1 \in D$ on $X = \Gamma\backslash G$ is an ergodic transformation on (X, μ).*

Proof. Since μ is finite and G-invariant, there is a continuous unitary representation ρ of G on the Hilbert Space $\mathcal{H} = \mathcal{L}^2(X, \mu)$, defined such that for all $\xi \in \mathcal{H}$, $g \in G$ and μ-almost all $x \in X$,

$$[\xi \cdot \rho(g)]\,(x) = \xi(xg).$$

Suppose E is a measurable subset of X such that $\mu(E \cdot d \,\Delta\, E) = 0$. Let χ_E denote the charecteristic function of E. Then $\xi = \chi_E \in \mathcal{H}$ and for all $k \in \mathbb{Z}$,

$$\xi \cdot \rho(d^k) = \chi_{(E\cdot d^{-k})} = \chi_E = \xi.$$

Let $u \in N^+$. Since ρ is unitary, for all $k \in \mathbb{Z}$,

$$\langle \xi \cdot \rho(u),\, \xi\rangle = \langle \xi \cdot \rho(d^k)\rho(u),\, \xi \cdot \rho(d^k)\rangle = \langle \xi \cdot \rho(d^k u d^{-k}),\, \xi\rangle.$$

By Remark 3.1, $d^k u d^{-k} \to 1$ as $k \to +\infty$. Hence by continuity of ρ,

$$\langle \xi \cdot \rho(u),\, \xi\rangle = \langle \xi, \xi\rangle.$$

Thus $\xi \cdot \rho(u) = \xi$ for all $u \in N^+$. Similarly, we can show that $\xi \cdot \rho(w) = \xi$ for all $w \in N^-$. Now by Remark 3.2, $\xi \cdot \rho(g) = \xi$ for all $g \in G$. Thus $\chi_E = \xi$ is constant almost every where on X. Hence $\mu(E) = 1$ or 0. This shows that d acts ergodically on (X, μ). □

Lemma 4.2 and Lemma 4.1 imply that almost all orbits of D are dense in $\Gamma\backslash G$. For our purpose we will need its following consequence regarding individual orbits (see [D2, Preliminaries] for a general statement and references).

Lemma 4.3 *Every orbit of the subgroup N^+D acting on $X = \Gamma\backslash G$ is dense.*

Proof. Let $x, y \in X$. Since $d = d_1 \in D$ acts ergodically on (X, μ), by Lemma 4.1, there exist sequences $x_i \to x$, $x_i \in X$ and $n_i \to \infty$, $n_i \in \mathbb{N}$ such that $x_i d^{n_i} \to y$ as $i \to \infty$. Let the sequence $g_i \to 1$, $g_i \in G$ be such that $x_i = xg_i$. Since $\mathcal{G} = \mathcal{G}^+ \oplus \mathcal{G}^0 \oplus \mathcal{G}^-$, for all large $i \in \mathbb{N}$ there exist $w_i \in N^-$, $v_i \in N^+$ and $z_i \in DM$, such that $g_i = v_i z_i w_i$ and $w_i, v_i, z_i \to 1$ as $i \to \infty$.

Now for all large $i \in \mathbb{N}$,

$$x_i d^{n_i} = x v_i d^{n_i} (d^{-n_i} z_i d^{n_i})(d^{-n_i} w_i d^{n_i}).$$

By Remark 3.1, as $i \to \infty$,

$$d^{-n_i} z_i d^{n_i} = z_i \to 1 \quad \text{and} \quad d^{-n_i} w_i d^{n_i} \to 1.$$

Therefore $x v_i d^{n_i} \to y$ as $i \to \infty$. Since x, y are arbitrary, this shows that for all $x \in X$ the orbit xN^+D is dense in X. □

4.2 *Minimal closed invariant sets*

It was shown by G.A. Margulis in [M2] that minimal closed invariant sets of the action of unipotent subgroups can be used very effectively for studying orbit closures in homogeneous spaces of Lie groups.

Definition 4.2 Let F be a semi-group acting on a topological space X by continuous transformations. If a closed subset Z of X is invariant under the action of F and no proper closed subset of Z is invariant under the F-action then Z is called *minimal closed F-invariant*. Thus, if Z is a minimal closed F-invariant set then every orbit of F in Z is dense.

Remark 4.2 Any compact F-invariant subset of X contains a minimal closed F-invariant subset. To see this, use Zorn's lemma along with the fact that the intersection of any totally ordered (with respect to set inclusion) family of compact sets is nonempty. This remark will be used in §6.

5 Orbits of unipotent groups under linear actions

Let $H = SO_0(1, n-1) \subset G$. Now $D = SO_0(1,1) \subset H$. Put $N_1 = N^+ \cap H$ and $M_1 = M \cap H$.

Let Y be a closed H-invariant subset of $\Gamma\backslash G$. Now H contains the subgroup $N_1 D$ and by Lemma 4.3 we know that every orbit of the subgroup N^+D is dense in $\Gamma\backslash G$. Let N_2 be a one-parameter subgroup of N^+ such that $N^+ = N_2 N_1$. In §6 we show that under certain 'local' condition, Y contains an orbit of N_2. This will imply that $Y = \Gamma\backslash G$. The next proposition is a crucial step for obtaining, under that condition, a N_2-invariant subset in Y. It will be convenient to introduce some notations to state and prove the proposition.

Let $\mathcal{H}$ be the Lie algebra corresponding to H. Let $\mathcal{H}^+ = \mathcal{G}^+ \cap \mathcal{H}$, $\mathcal{M}_1 = \mathcal{M} \cap \mathcal{H}$ and $\mathcal{H}^- = \mathcal{G}^- \cap \mathcal{H}$. Then $\mathcal{H} = \mathcal{H}^+ \oplus \mathcal{D} \oplus \mathcal{M}_1 \oplus \mathcal{H}^-$. Also $\mathcal{H}^+$ and $\mathcal{M}_1$ are the Lie subalgebras corresponding to N_1 and M_1 respectively.

Let $\mathcal{P}$ be the ortho-complement of $\mathcal{H}$ in $\mathcal{G}$ with respect to the symmetric bilinear form $Q : \mathcal{G} \times \mathcal{G} \to \mathbb{R}$, defined by $Q(X,Y) = \operatorname{tr}(XY)$. Now Q is non-degenerate on $\mathcal{G}$ as well as on $\mathcal{H}$ and it is invariant under the right Adjoint action of G on $\mathcal{G}$. Therefore $\mathcal{G} = \mathcal{P} \oplus \mathcal{H}$ and $\mathcal{P}$ is invariant under the Adjoint action restricted to H. Let $\mathcal{P}^+ = \mathcal{P} \cap \mathcal{G}^+$, $\mathcal{P}^0 = \mathcal{P} \cap \mathcal{G}^0$ and $\mathcal{P}^- = \mathcal{P} \cap \mathcal{G}^-$. Then

$$\mathcal{P} = \mathcal{P}^+ \oplus \mathcal{P}^0 \oplus \mathcal{P}^-.$$

Let N_2 be the connected Lie subgroup corresponding to the Lie subalgebra $\mathcal{P}^+$. Now $\mathcal{G}^+ = \mathcal{P}^+ \oplus \mathcal{H}^+$ and $N^+ = N_2 N_1$.

Let $[\mathbf{e}_1, \ldots, \mathbf{e}_{n-1}]$ denote the standard ordered basis of $\mathbb{R}^{n-1}$. Then the set $\{n^+({}^t\mathbf{e}_k) : 1 \le k \le n-2\}$ is a basis of $\mathcal{H}^+$, $p^+ \stackrel{\text{def}}{=} n^+({}^t\mathbf{e}_{n-1})$ is a basis of $\mathcal{P}^+$, $p^- \stackrel{\text{def}}{=} n^-({}^t\mathbf{e}_{n-1})$ is a basis of $\mathcal{P}^-$ and the set

$$\left\{p_k^0 \stackrel{\text{def}}{=} m(X_k - {}^tX_k) : X_k = \begin{pmatrix} 0_{n-2\times n-1} \\ \mathbf{e}_k \end{pmatrix},\ 1 \le k \le n-2\right\}$$

is a basis of $\mathcal{P}^0$.

Proposition 5.1 (Margulis, cf. [DM1, Lemma 2.2]) *Let $\{q_i\}_{i\in\mathbb{N}} \subset \mathcal{P}\setminus\mathcal{P}^+$ be a sequence such that $q_i \to 0$ as $i \to \infty$. Then there exist a one-parameter subgroup $\{u_t\}_{t\in\mathbb{R}} \subset N_1$, a sequence $t_i \to \infty$ and a non-constant polynomial φ such that if $\{q_i\}_{i\in\mathbb{N}}$ is replaced by a suitable subsequence then for every $s \in \mathbb{R}$, as $i \to \infty$,*

$$q_i \cdot \mathrm{Ad}(u_{st_i}) \to \varphi(s)p^+.$$

Proof. For each $i \in \mathbb{N}$, let $\theta_i \in \mathbb{R}$, $\{\sigma_{k,i} : 1 \le k \le n-2\} \subset \mathbb{R}$ and $\delta_i \in \mathbb{R}$ be such that

$$q_i = \theta_i\, p^- + \sum_{k=1}^{n-2} \sigma_{k,i}\, p_k^0 + \delta_i\, p^+.$$

Now as $i \to \infty$: $\theta_i \to 0$, $\delta_i \to 0$ and $\sigma_{k,i} \to 0$ for all $1 \le k \le n-2$. Since $\{q_i\}_{i\in\mathbb{N}} \cap \mathcal{P}^+ = \emptyset$, there exists $k \in \{1,\ldots,n-2\}$ such that replacing $\{q_i\}_{i\in\mathbb{N}}$ by a subsequence, we get $\theta_i \ne 0$ or $\sigma_{k,i} \ne 0$ for all $i \in \mathbb{N}$. Consider the one-parameter subgroup

$$\left\{u_t \stackrel{\text{def}}{=} \exp n^+(t\cdot {}^t\mathbf{e}_k) : t \in \mathbb{R}\right\} \subset N_1.$$

Then by Eq. 8,

$$\begin{aligned} q_i \cdot \mathrm{Ad}\, u_t &= q_i + t\cdot\left[q_i, n^+({}^t\mathbf{e}_k)\right] + (t^2/2)\cdot\left[[q_i, n^+({}^t\mathbf{e}_k)], n^+({}^t\mathbf{e}_k)\right] \\ &= q_i + (\theta_i\, t)\cdot p_k^0 + (\sigma_{k,i}\, t + \theta_i\, t^2/2)\cdot p^+. \end{aligned}$$

For each $i \in \mathbb{N}$, let $t_i > 0$ be such that

$$\max\{|\sigma_{k,i}|\, t_i,\ |\theta_i|\, t_i^2\} = 1.$$

Replacing $\{q_i\}_{i\in\mathbb{N}}$ by a subsequence, there exist $\lambda_1, \lambda_2 \in \mathbb{R}$ such that as $i \to \infty$, $\theta_i\, t_i^2 \to \lambda_1$ and $\sigma_{k,i}\, t_i \to \lambda_2$. Note that $\max\{|\lambda_1|, |\lambda_2|\} = 1$. Since $\theta_i\, t_i^2 \to \lambda_1$ and $t_i \to \infty$, we have $\theta_i\, t_i \to 0$ as $i \to \infty$.

Let φ be a polynomial defined by $\varphi(s) = \lambda_1 s + \lambda_2\, s^2$, $s \in \mathbb{R}$. Then φ is non-constant and for every $s \in \mathbb{R}$, as $i \to \infty$,

$$q_i \cdot \mathrm{Ad}\, u_{st_i} \to \varphi(s)p^+.$$

□

6 Proof of Theorem B

Let $X = \Gamma\backslash G$ and Y be the closure of the given H-invariant subset in X. Then Y is H-invariant. We want to show that either $Y = X$ or Y is a union of finitely many closed H-orbits.

Let Y_1 be a minimal closed H-invariant subset of Y and Z be a minimal closed N_1-invariant subset of Y_1. The existance of these sets follows from Remark 4.2.

Since $\mathcal{G} = \mathcal{P} \oplus \mathcal{H}$, there exist a neighbourhood Ψ of 0 in $\mathcal{G}$ and a neighbourhood Ω of 1 in G such that the map $(q, \mathbf{y}) \mapsto \exp q \cdot \exp \mathbf{y}$, $(q \in \mathcal{P} \cap \Psi, \mathbf{y} \in \mathcal{H} \cap \Psi)$ is a diffeomorphism onto Ω.

Fix $z \in Z$ for rest of the proof. Let $g \in \Omega$ be such that $zg \in Y$. Write $g = (\exp q)h$ for some $q \in \mathcal{P} \cap \Psi$ and $h \in H$. Since Y is H-invariant, $z \exp q \in Y$. Define

$$\mathcal{Q} = \{q \in \mathcal{P} \cap \Psi : z \exp q \in Y\}.$$

If we choose Ψ small enough then one of the following possibilities occurs :

I. $0 \in \overline{\mathcal{Q} \setminus \mathcal{P}^+}$.

II. $0 \in \overline{\mathcal{Q} \setminus \{0\}}$ and $\mathcal{Q} \subset \mathcal{P}^+$.

III. $\mathcal{Q} = \{0\}$.

We shall prove that **a**) if Case I occurs then Y_1 is dense in $\Gamma\backslash G$,**b**) if Case III occurs then Y_1 is a closed H-orbit and it is a connected component of Y and **c**) the occurrence of Case II leads to a contradiction. This shows that either $Y = X$ or every connected component of Y is a closed H-orbit. Note that since Y is compact, it has only finitely many connected components. This will prove Theorem B.

Case I : (cf. [DM3, Prop. 8, Case a)])

In this case there exists a sequence $\{q_i\}_{i\in\mathbb{N}} \subset \mathcal{P} \setminus \mathcal{P}^+$ such that $q_i \to 0$, as $i \to \infty$ and $z \exp q_i \in Y$, for all $i \in \mathbb{N}$.

Step 1 Replacing $\{q_i\}_{i\in\mathbb{N}}$ *by a suitable subsequence, there exist a one-parameter subgroup* $\{u_t\}_{t\in\mathbb{R}} \subset N_1$*, a sequence* $t_i \to \infty$ *and a non-constant polynomial* φ *such that for every* $s \in \mathbb{R}$*, as* $i \to \infty$,

$$q_i \cdot \operatorname{Ad} u_{st_i} \to \varphi(s)p^+.$$

This is just a restatement of Proposition 5.1.

Step 2 (cf. [M2, Lemma 1]) *For every* $s \in \mathbb{R}$, $Z \exp(\varphi(s)p^+) \subset Y$.

Proof. Fix $s \in \mathbb{R}$. Put $y_i = z \exp q_i \in Y$ and $z_i = z u_{st_i} \in Z$, for all $i \in \mathbb{N}$. Since Z is compact, by passing to subsequences, we may assume that as $i \to \infty$, $z_i \to z'$ for some $z' \in Z$. Now by Step 1, as $i \to \infty$,

$$y_i u_{st_i} = z_i \exp(u_{-st_i} q_i u_{st_i}) \to z' \exp(\varphi(s) p^+)$$

Put $v = \exp(\varphi(s)p^+) \in N_2$. Since $y_i u_{st_i} \in Y$ and Y is closed, $z'v \in Y$.

Since Z is minimal closed N_1-invariant, $z'N_1$ is dense in Z. Now N_2 normalizes N_1, therefore

$$Zv = \overline{z'N_1}v \subset \overline{Yv^{-1}N_1 v} = \overline{YN_1} = Y.$$

Step 3 *Y contains an orbit of N_2.*

Proof. Note that $\varphi(0) = 0$. Hence we can choose $F = \{\exp(tp^+) : t \geq 0\}$ or $\{\exp(tp^+) : t \leq 0\}$, so that

$$F \subset \left\{\exp(\varphi(s)p^+) : s \in \mathbb{R}\right\}.$$

Now by Step 2, $ZF \subset Y$.

Since Y is compact, by Remark 4.2, $\overline{ZF}$ contains a minimal closed F-invariant subset Z_1. If $v \in N_2$ then there exists $w \in F$ such that $wv \in F$ and hence

$$Z_1 = Z_1(wv) = Z_1 v$$

Thus Z_1 is N_2 invariant. This proves Step 3.

Since Y is N_1D invariant, by Step 3, Y contains an orbit of $N_2(N_1D) = N^+D$. Now by Lemma 4.3, $Y = X = \Gamma\backslash G$, as we wanted to show in this case.

Case III : (cf. [DM3, Prop. 8, Case b)])

In this case zH contains a neighbourhood of z in Y and hence it is an open subset of Y. Now $Y_1 \setminus zH$ is a closed H-invariant subset of Y_1. Since Y_1 is closed minimal H-invariant, $Y_1 = zH$. Thus Y_1 is a closed orbit of H and is a connected component of Y. This is what wanted to show in this case.

Case II : (cf. [DM3, Prop. 8, Case c)])

In this case there exists neighbourhood Ω of 1 in G such that

$$Y \cap z\Omega \subset z(N_2 H \cap \Omega) \tag{9}$$

and there exists a sequence $\{v_i\}_{i\in\mathbb{N}} \subset N_2 \setminus \{1\}$ such that $v_i \to 1$, as $i \to \infty$ and $zv_i \in Y$ for all $i \in \mathbb{N}$. Since Y is compact and Γ is discrete, we can choose Ω small enough so that $\Omega\Omega^{-1} \cap G_y = \{1\}$ for all $y \in Y$, where G_y denotes the stabilizer of y in G.

Step 1 (cf. [M2, Lemma 4]) *Given any compact set $K \subset N_1$, there exists $u \in N_1 \setminus K$ such that $zu \in z\Omega$.*

Proof. Since Z is compact and N_1 is non-compact, there exists a sequence $\{u_i\}_{i\in\mathbb{N}} \subset N_1$ such that as $i \to \infty$, $zu_i \to z' \in Z$ and $u_i \to \infty$. Since Z is minimal closed N_1-invariant, $z'N_1$ is dense in Z. Let $u' \in N_1$ be such that $z'u' \in z\Omega$. Hence for all large enough $i \in \mathbb{N}$, $(zu_i)u' \in z\Omega$ but $u_iu' \notin K$. This proves Step 1.

Step 2 $\quad zN_1 \cap z\Omega \subset z(N^+DM_1 \cap \Omega)$.

Proof. Let $u \in N_1$ be such that $zu \in Z\Omega$. Then by Eq. 9 there exists $g \in N_2H \cap \Omega$ such that $zu = zg$.

Since $v_i \to 1$, there is $i_0 \in \mathbb{N}$ such that $gv_{i_0} \in \Omega$. Since $zv_{i_0} \in Y$, we have

$$zv_{i_0}u = zuv_{i_0} = zgv_{i_0} \in Y \cap z\Omega.$$

Therefore by Eq. 9 there exist $v \in N_2$ and $h \in H$ such that $vh \in \Omega$ and $zgv_{i_0} = zvh$. Hence $gv_{i_0} = vh$, because $\Omega\Omega^{-1} \cap G_z = \{1\}$.

Now according to the notations in §5,

$$p^+ \cdot \mathrm{Ad}(gv_{i_0}) = p^+ \cdot \mathrm{Ad}(vh) = p^+ \cdot \mathrm{Ad}\, h \in \mathcal{P}.$$

Put $q = p^+ \cdot \mathrm{Ad}\, g$. Since $q \in \mathcal{P}$ and $v_{i_0} \in N_2 \setminus \{1\}$, from Eq. 8 it follows that, $q \cdot \mathrm{Ad}\, v_{i_0} \in \mathcal{P}$ only if $q \in \mathcal{P}^+$. Writing $g = v'h'$ for suitable $v' \in N_2$ and $h' \in H$, we get $q = p^+ \cdot \mathrm{Ad}\, h'$. Since $\mathcal{P}^+$ is the fixed point space of $\mathrm{Ad}(N_1)$ in $\mathcal{P}$ and N_1DM_1 is the normalizer of N_1 in H, we have $q \in \mathcal{P}^+$ only if $h' \in N_1DM_1$. Hence $g \in N_2N_1DM_1 = N^+DM_1$. This completes the proof of Step 2.

By Steps 1 and 2, there exist $u \in N_1 \setminus \overline{\Omega}$ and $g \in (N^+DM_1 \cap \Omega)$ such that $zu = zg$. Then $\delta = gu^{-1} \in G_z \setminus \{1\}$. By Remark 3.1, there exists $t_0 > 0$ such that $d_{t_0}ud_{t_0}^{-1} \in \Omega$ and for all $t > 0$, $d_t(N^+DM_1 \cap \Omega)d_t^{-1} \subset \Omega$. Now $d_{t_0}G_zd_{t_0}^{-1} = G_{zd_{t_0}^{-1}}$. Hence

$$d_{t_0}\delta d_{t_0}^{-1} = (d_{t_0}gd_{t_0}^{-1})(d_{t_0}u^{-1}d_{t_0}^{-1}) \in (G_{zd_{t_0}^{-1}} \cap \Omega\Omega^{-1}) \setminus \{1\}.$$

This contradicts the choice of Ω, for $zd_{t_0}^{-1} \in Y$. Hence Case II does not occur.

This completes the proof of Theorem B. □

Remark 6.1 Theorem B is still valid if we assume that $\Gamma\backslash G$ admits a finite G-invariant measure, even though it need not be compact. In order to extend our proof in this case, we will need to show that any closed N_1-invariant subset of $\Gamma\backslash G$ contains a minimal compact N_1-invariant subset. A result due to S.G. Dani and G.A. Margulis achieves precisely this (see [DM1, Corollary 1.5] and [M3]). Now with the help of the proof of Proposition 8 in [DM3], the reader may be able to verify, without much difficulty, the Theorem B under the above assumption.

7 General Results

Using the ideas from [M2], [DM1] and [DM2] and using the method of the proof of the Main theorem in [Sh], the following result can be proved :

Theorem C *Let $G = SO_0(1,n)$ and Γ be a discrete subgroup of G such that $\Gamma\backslash G$ admits a finite G-invariant measure (i.e. Γ is a lattice in G). Let $H = SO_0(1,k)$ for some $2 \leq k \leq n$ and Y be a closed H-invariant subset of $\Gamma\backslash G$. Then Y has finitely many connected components; each of them is of the form $xLCC'$, where C' is a compact subset of $C_G(H)$, the centralizer of H in G, and $L = SO_0(1,m)$, $k \leq m \leq n$, C a compact subgroup of $C_G(L)$ and $x \in X$ are such that $\overline{xH} = xLC$.*

In particular, if Y is the closure of a single orbit of H then $Y = y(g^{-1}LCg)$, where $y \in Y$, $g \in C_G(H)$ and L and C are as above.

This theorem has the following geometric consequence.

Theorem D *Let M be a complete, connected riemannian manifold with constant negative curvature and finite riemannian volume. Let D be a complete riemmanian manifold, whose connected components are simply connected and of dimension $k \geq 2$ and let $\phi : D \to M$ be a totally geodesic immersion. Then there exists a complete riemannian manifold L and a totally geodesic immersion $\psi : L \to M$ such that the following holds :*

1. *L has finitely many components, possibly of different dimensions, and each one of them has finite riemannian volume.*
2. *$\psi(L)$ is the closure of $\phi(D)$ in M.*
3. *Let $\tilde{\phi} : D \to L$ be a riemannian immersion such that $\phi = \psi \circ \tilde{\phi}$. Let $\tilde{\phi}_* : \mathcal{F}^k(D) \to \mathcal{F}^k(L)$ be the immersion of the orthonormal k-frame bundles, which is induced from the derivative of $\tilde{\phi}$. Then $\tilde{\phi}_*(\mathcal{F}^k(D))$ is dense in $\mathcal{F}^k(L)$.*

Note that the closure of a geodesic in M need not be the image of a closed immersion. To give an example of such a geodesic, let $p : \widetilde{M} \to M$ be the universal cover of M and let $\tilde{\gamma}_-$ and $\tilde{\gamma}_+$ be two distinct geodesics in $\widetilde{M}$ such that $\gamma_\pm = p \circ \tilde{\gamma}_\pm$ are closed compact geodesics in M. Since $\widetilde{M}$ is isomorphic to the Hyperbolic n-space, there exists a geodesic $\tilde{\gamma}$ in $\widetilde{M}$ such that $\tilde{\gamma}(-\infty) = \tilde{\gamma}_-(-\infty)$ and $\tilde{\gamma}(+\infty) = \tilde{\gamma}_+(+\infty)$. Then the geodesic $\gamma = p \circ \tilde{\gamma}$ of M winds around γ_+ in one direction and γ_- in the opposite direction. Clearly, $\gamma_- \cup \gamma \cup \gamma_+$ is the closure of γ in M but it is not the image of a closed immersion into M.

In the group theoretic setup, the geodesics in M correspond to orbits of the subgroup $SO_0(1,1)$ in $\Gamma\backslash G$. Note that $SO_0(1,1)$ does not contain any unipotent element other than identity, while our proof of Theorem B depends crucially on the behaviour of the orbits of nontrivial unipotent one-parameter subgroups contained in H. In fact, Theorem C proves a particular case of the following very general conjecture due to M.S. Raghunathan.

Conjecture 1 (Raghunathan) *Let G be a Lie group, Γ be a lattice in G and H be a subgroup generated by unipotent elements of G contained in it. Then for every $x \in \Gamma\backslash G$, there exists a closed subgroup L of G containing H such that, $\overline{xH} = xL$ and xL supports a finite L-invariant measure.*

An element $u \in G$ is called *unipotent* if the map $\mathrm{Ad}\, u$ is a unipotent automorphism of the Lie algebra of G. We note that if a connected subgroup H of G is semisimple and has no connected nontrivial compact normal subgroup, then H is generated by unipotent elements of G contained in it.

We refer the reader to the survey articles by S.G. Dani [D1] and G.A. Margulis [M1,M4] for various developements related to Raghunathan's conjecture. Recently, this conjecture has been proved by Marina Ratner. She first classiffied all finite ergodic invariant measures of H on $\Gamma\backslash G$ (see [Ra1]) and then proved the following stronger theorem, which implies Raghunathan's conjecture.

Theorem E (Ratner [Ra2], see also [DS,Sh]) *Let G be a Lie group, Γ be a lattice in G and $\{u_t : t \in \mathbb{R}\}$ be a unipotent one-parameter subgroup of G. Then for every $x \in \Gamma\backslash G$ there exists a closed subgroup L such that xL is closed, xL admits an L-invariant probability measure σ and the $\{u_t\}_{t\in\mathbb{R}}$-orbit through x is uniformly distributed with respect to σ; that is, for all bounded continuous functions f on $\Gamma\backslash G$,*

$$\lim_{T\to\infty} \frac{1}{T}\int_0^T f(xu_t)\, dt = \int_{xL} f\, d\sigma.$$

Acknowledgements : I wish to thank S.G. Dani and Gopal Prasad for a number of stimulating conversations and their remarks regarding Raghunathan's Conjecture. Thanks are due to Etienne Ghys, who suggested to me the geometric implications of Raghunathan's conjecture. I express my thanks to A. Haefliger for providing me an opportunity to give a talk in the 'Workshop on Group theory from a Geometrical View Point'. I also thank the ICTP for its hospitality and support to participate in this workshop.

References

[DS] Dani, S.G., Smillie, J.: Uniform distribution of horocycle orbits for Fuchsian Groups. *Duke Math. J.* **51**, 185-194 (1984)

[D1] Dani, S.G.: Dynamics of flows on homogeneous spaces: A survey. Proceedings of Coloquio de Systemas Dinamicos (Guanajuato, 1983), *Aportacione Mat.* 1, Soc. Mat. Mexicana, Maxico City, pp. 1-30, 1985

[D2] ———: Orbits of Horosphirical Flows. *Duke Math. J.* **53**, 177-188 (1986)

[DM1] Dani, S.G., Margulis, G.A.: Values of quadratic forms at primitive integral points. *Invent. Math.* **98**, 405-424 (1989)

[DM2] ———: Orbit closures of generic unipotent flows on homogeneous spaces of $SL(3,\mathbb{R})$. *Math. Ann.* **286**, 101-128 (1990)

[DM3] ———: Values of quadratic forms at integral points: an elementary approach. *L'Enseignement Math.* **36**, 143-174 (1990)

[Fl] Flaminio, L.: An extension of Ratner's rigidity theorem to n-dimensional hyperbolic space. *Ergod. Th. & Dynam. Sys.* **7**, 73-92 (1987)

[M1] Margulis, G.A.: Lie groups and ergodic theory. In: Avramov, L.L. (ed.) *Algebra - Some Current Trends.* Proceedings Varna 1986. (Lect. Notes Math., vol 1352, pp.130-146) Berlin Heidelberg New York: Springer 1988

[M2] ———: Discrete subgroups and ergodic theory. In: Aubert, K.E., Bombieri, E., Goldfield, D. (eds.) *Number theory, trace formulas and discrete groups,* Symposium in honor of Atale Selberg, Oslo, 1987. New York London: Academic Press 1989

[M3] ———: Compactness of minimal closed invariant sets of actions of unipotent groups. *Geometriae Dedicata,* Special volume in honor of Jacques Tits, (to be published in 1991)

[M4] ———: Dynamical and ergodic properties of subgroup actions on homogeneous spaces with applications to number theory. A Planary address in ICM, Kyoto, 1990.

[R] Raghunathan, M.S.: *Discrete subgroups of Lie groups.* Berlin Heidelberg New York: Springer 1972

[Ra1] Ratner, M.: Invariant measures for unipotent translations on homogeneous spaces. *Proc. Natl. Acad. Sci. USA* **87**, 4309-4311 (1990)

[Ra2] ———: Raghunathan's topological conjecture and distributions of unipotent flows. Preprint.

[Sh] Shah, N. A.: Uniformly distributed orbits of certain flows on homogeneous spaces. To appear in *Math. Annalen.*

LIST OF CONTRIBUTORS

H. Abels (Universität Bielefeld, Bielefeld, Germany)
J. Barge (Institut Fourier, St. Martin d'Hères, France)
B. Bowditch (I.H.E.S., Bures sur Yvette, France)
M. R. Bridson (Cornell University, Ithaca, USA)
K. S. Brown (Cornell University, Ithaca, USA)
P. de la Harpe (Université de Genève, Switzerland)
T. Delzant (Université Louis Pasteur, Strasbourg, France)
E. Ghys (Ecole Normale Supérieure de Lyon, France)
P. A. Greenberg (C.I.E.A.- I.P.N., Mexico City, Mexico)
A. Haefliger (Université de Genève, Switzerland)
T. Januskiewicz (University of Wroclaw, Poland)
R. S. Kulkarni (Graduate Center, City University of New York, U.S.A.)
L. Paris (University of Wisconsin, U.S.A.)
F. Paulin (Ecole Normale Supérieure de Lyon, France)
S. J. Pride (University of Glasgow, U.K.)
V. Sergiescu (Institut Fourier, Saint-Martin d'Hères, France)
N. A. Shah (TIFR, Bombay, India)
P. B. Shalen (University of Illinois at Chicago, U.S.A.)
H. Short (Graduate Center, City University of New York, U.S.A.)
J. R. Stallings (University of California, Berkeley, U.S.A.)

www.ingramcontent.com/pod-product-compliance
Lightning Source LLC
Chambersburg PA
CBHW070709220326
41598CB00026B/3675

* 9 7 8 9 8 1 0 2 1 4 3 0 2 *